Lecture Notes in Artificial Intelligence 9729

Subseries of Lecture Notes in Computer Science

More information about this series at http://www.springer.com/series/1244

Petra Perner (Ed.)

Machine Learning and Data Mining in Pattern Recognition

12th International Conference, MLDM 2016
New York, NY, USA, July 16–21, 2016
Proceedings

Editor
Petra Perner
Institute of Computer Vision
and applied Computer Sciences, IBaI
Leipzig, Saxony
Germany

ISSN 0302-9743 ISSN 1611-3349 (electronic)
Lecture Notes in Artificial Intelligence
ISBN 978-3-319-41919-0 ISBN 978-3-319-41920-6 (eBook)
DOI 10.1007/978-3-319-41920-6

Library of Congress Control Number: 2016943411

LNCS Sublibrary: SL7 – Artificial Intelligence

Printed on acid-free paper

This Springer imprint is published by Springer Nature
The registered company is Springer International Publishing AG Switzerland

Preface

The 12th event of the International Conference on Machine Learning and Data Mining MLDM 2016 was held in New York (www.mldm.de) running under the umbrella of the World Congress "The Frontiers in Intelligent Data and Signal Analysis, DSA2016" (www.worldcongressdsa.com).

For this edition the Program Committee received 169 submissions. After the peer-review process, we accepted 56 high-quality papers for oral presentation. The topics range from theoretical topics for classification, clustering, association rule, and pattern mining to specific data-mining methods for the different multimedia data types such as image mining, text mining, video mining, and Web mining. Extended versions of selected papers will appear in the international journal *Transactions on Machine Learning and Data Mining* (www.ibai-publishing.org/journal/mldm).

A tutorial on "Data Mining," a tutorial on "Case-Based Reasoning," a tutorial on "Intelligent Image Interpretation and Computer Vision in Medicine, Biotechnology, Chemistry, and Food Industry," and a tutorial on "Standardization in Immunofluorescence" were held before the conference.

We were pleased to give out the best paper award for MLDM for the fifth time this year. Three announcements are mentioned at www.mldm.de. The final decision was made by the Best Paper Award Committee based on the presentation by the authors and the discussion with the auditorium. The ceremony took place during the banquet. This prize is sponsored by ibai solutions (www.ibai-solutions.de), one of the leading companies in data mining for marketing, Web mining, and e-commerce.

We would like to thank all reviewers for their highly professional work and their effort in reviewing the papers. We would also like to thank the members of Institute of Applied Computer Sciences, Leipzig, Germany (www.ibai-institut.de), who handled the conference as secretariat. We appreciate the help and understanding of the editorial staff at Springer, and in particular Alfred Hofmann, who supported the publication of these proceedings in the LNAI series.

Last, but not least, we wish to thank all the speakers and participants who contributed to the success of the conference. We hope to see you in 2017 in New York at the next World Congress (www.worldcongressdsa.com) on "The Frontiers in Intelligent Data and Signal Analysis, DSA2017," which combines under its roof the following three events: International Conferences on Machine Learning and Data Mining (MLDM), the Industrial Conference on Data Mining (ICDM), and the International Conference on Mass Data Analysis of Signals and Images in Medicine, Biotechnology, Chemistry and Food Industry (MDA).

July 2016 Petra Perner

Organization

Program Chair

Petra Perner	IBaI Leipzig, Germany

Program Committee

Sergey V. Ablameyko	Belarus State University, Belarus
Patrick Bouthemy	Inria-VISTA, France
Michelangelo Ceci	University of Bari, Italy
Xiaoqing Ding	Tsinghua University, China
Christoph F. Eick	University of Houston, USA
Ana Fred	Technical University of Lisbon, Portugal
Giorgio Giacinto	University of Cagliari, Italy
Makato Haraguchi	Hokkaido University of Sapporo, Japan
Dimitrios A. Karras	Chalkis Institute of Technology, Greece
Adam Krzyzak	Concordia University, Montreal, Canada
Thang V. Pham	Intelligent Systems Lab Amsterdam (ISLA), The Netherlands
Linda Shapiro	University of Washington, USA
Tamas Sziranyi	MTA-SZTAKI, Hungary
Alexander Ulanov	HP Labs, Russian Federation
Patrick Wang	Northeastern University, USA

Additional Reviewers

Zejin Jason Ding	Hewlett Packard Enterprise
Long Ma	Georgia State University
Lumin Zhang	Facebook

Contents

Evolving a Low Price Recovery Strategy for Distressed Securities

Robert E. Marmelstein(✉), Alexander L. Hunt, and Christoper Eroh

Computer Science Department, East Stroudsburg University,
East Stroudsburg, PA 18301, USA
rmarmelstein@esu.edu, {ahunt,ceroh1}@live.esu.edu

Abstract. This paper investigates methods to evolve an automated agent that executes a niche trading stock strategy. Unlike trading strategies that seek to exploit broad market trends, we choose a very specific strategy on the assumption that it will be easier to learn, require less input data to do so, and more straightforward to evaluate the agents performance. In this case, we select a Low Price Recovery Strategy (LPRS), which involves picking stocks that have a high likelihood of quickly recovering after a steep, one day decline in share price. A series of intelligent agents are evolved through the use of a Genetic Programming approach. The inputs to our algorithms included traditional stock performance metrics, sentiment indicators available from online sources, and associated classification rules. The essential aspects of the research discussed include: identification of opportunities, feature selection and extraction, design of various genetic programs for evolving the agent, and testing approaches for the agents. We demonstrate that the evolved agent yields results outperform a randomized version of the LPRS and the benchmark Standard & Poor's 500 (S&P500) stock market index.

1 Introduction

The last two decades have seen substantial work done in development of algorithms to perform rapid, automated trading of securities. Given the need to adapt to dynamically changing market conditions and the tremendous amount of data available to make trading decisions, the machine learning community has been in the forefront of these efforts. For many trading firms, the use of neural networks has proved extremely popular [6][8]. Such "black box" techniques excel in performance and the underlying design is much easier to keep secret for the purposes of competitive advantage. However, from a knowledge acquisition perspective, it is preferable to learn rules which can explain why a given decision was made. Once promising rules are identified, they can form the basis for more sophisticated, intelligent agents which use those underlying rules to make trading decisions. Evolutionary Algorithm techniques, such as Genetic Algorithms (GA) and Genetic Programming (GP), are a proven way to combine existing rules or discover new ones in order to construct such trading agents.

P. Perner (Ed.): MLDM 2016, LNAI 9729, pp. 1–14, 2016.
DOI: 10.1007/978-3-319-41920-6_1

1.1 Problem Description

This research evolves automated agents to trade using a Low Price Recovery Strategy. The main idea of the LPRS is to identify and purchase selected stocks which suffer a steep decline in share price, but have a high likelihood of rebounding within a short period of time. The candidate stocks selected for this strategy possess three characteristics. First, the stock share price must decline by at least 10% in value on a single trading day. While a lower percent decline would yield more candidates, we chose 10% as our decline threshold in order to ensure a resonable profit if a rebound occurs within a sixty (60) day time period. The trading day on which the decline occurs is known as the baseline day.

A second critera is that the company must have a pre-decline market capitalization value of at least $10B dollars. We incorporated this rule because we wanted to evaluate the impact of online sentiment on the likelihood of recovery. While using small capitalization stocks would have increased the size of our candidate data set, these tend to have little social media chatter relating to the decline event. Measurable levels of chatter are more likely with larger capitalization companies than smaller ones. Third, the stock share type must be common stock only.

Once candidates have been identified, they are then classified as either recovered or non-recovered. A "successful" recovery occurs when a stock gains back 75% of the value it lost within 60 days after the baseline day. Note that a stock just has to breach this threshold once to be successful. Thus, a stock that rebounds and then declines further within the 60 day period is still labeled as recovered. Any candidate stock that does not meet this condition is classified as non-recovered. We chose the 60 day period because it is long enought to give the stock adequate time to recover and short enough to reinvest the proceeds up to six (6) times in a given year. Figure 1 shows an example recovery situation for the Best Buy Corporation (stock symbol: BBY).

In our LPRS, the agent must decide to purchase the stock (or not) the day after the baseline day; recommended purchases are executed immediately. This snap decision is based on the available information at that time. If a stock is purchased, its target price (current price + 75% of the absolute decline amount) is computed. The stock is then held in the portfolio until the target price is met or until the 60 day hold period has passed. If the former condition occurs, the stock holding is considered to be sold the first time the target price is reached. If the latter condition occurs, the holding is sold for the stock's closing price on the last day of the holding period. Note that a non-recovered stock may result in either a capital loss or gain. If a gain, it will be less than the gain if the stock had met our recovery criteria.

2 Approach

We employ the Genetic Programming paradigm to evolve several LPRS agent variations. GP can be viewed as an extension of the GA, a biologically inspired

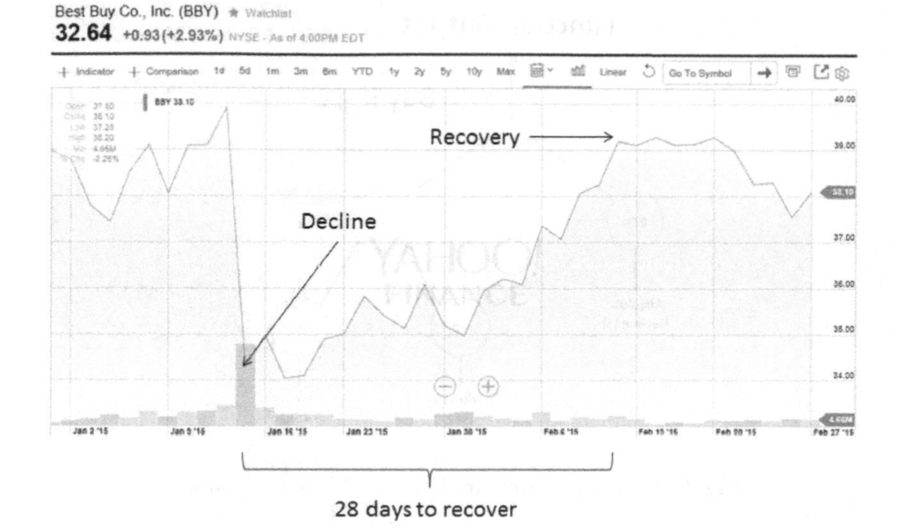

Fig. 1. Low Price Recovery Example (BBY)

model for testing and selecting the best choice among a set of results, each represented by a string. However, GP goes a step farther–it generates a program to solve a problem. The utility of the generated program is tested using a predefined fitness function. Two steps are employed over multiple generations to evolve a successful program. The first is selection of the fittest set of programs, using a competitive approach (such as a tournament). The second step is breeding the selected program, using crossover and mutation techniques, to create a new population of even fitter solutions for the next generation.

A crucial aspect of any GP is designing the fitness function to measure the degree to which a program is achieving the desired goal. In this experiment, we use different fitness function to evolve multiple variations of the LPRS agent. The GP chromosome in our application encodes trading rules as a function in the form of a tree. Figure 2 shows an example of such a tree structure. Each chromosome is evaluated and the value returned by the function (root node) is converted into a Boolean value. In our case, if the function returns a positive value, it is interpreted as a buy signal (true); when a negative value is returned, no purchase is made. Note the tree structure is composed of a combination of primitive functions (e.g., +, -, sin, exp, etc.), variables and variables. The complete set of these primative functions are described in section 2.3.

We chose GP as our evolutionary computation paradigm because it has more flexibility in exploring the algorithmic search space to creating novel trading strategies than a GA. Extensive tutorials on both GAs and GP can be found at [2] [7] [11] [12] [18].

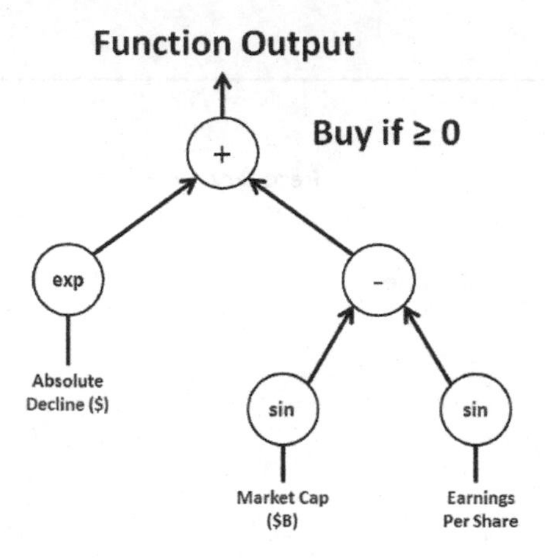

Fig. 2. Function represented as a GP tree structure

2.1 LPRS Agent Fitness Functions

Our GP utilized two types of fitness functions to evolve the LPRS agent. The first function type is based on a Confusion Matrix (CM) [10], which measures the agents ability to correctly classify each candidate stock as recovered or non-recovered. In machine learning, a CM contains the true positives, false positives and false negatives for every class in the data set. In our approach, the classification results over the training set are put in a confusion matrix (X). Each diagonal value of the matrix (x_{ii}) represents the true positives the ith class–that is, the percentage of class i that was correctly classfied. The fitness value is computed by multiplying the diagonal values of the matrix together (see Equation 1). Thus, the fitness metric reflects the agents ability to correctly discriminate between classes, even if it comes at the expense of overall classification accuracy.

$$Fitness = \prod_{i=1}^{n} x_{ii} \tag{1}$$

The second type of fitness function returns the overall profit (P) ratio earned by the agent. For this case, a simulation is elaborated with the LPRS agent's portfolio having an initial cash balance and no stocks on a predetermined start date. Beginning with this date, trade decisions are made in temporal order, as candidates are identified. Stock holdings are sold according to the rules enumerated in section 1.1. Capital gains on trades are added to the available cash reserve in the account; likewise, capital losses are subtracted. Using this approach, purchases are only carried out if the sufficient free cash exists in the portfolio to cover the transaction. The fitness result is returned based on the percent increase in portfolio value within 60 days after the last security purchase is made. The overall fitness of the LPRS is the ratio of the starting and ending portfolio balances (see Equation 2).

$$Fitness = \left(\frac{PortfolioValue(\$) - Ending}{PortfolioValue(\$) - Starting} \right) - 1 \quad (2)$$

Unlike the CM fitness, P fitness is highly dependent on the time-sensitive aspects of trading, both in terms of order and spacing of stock purchases. For example, if there is not sufficient cash available to buy a stock, it is not purchased regardless of what the buy signal is. Thus, performance is dependent on the kinds of resource and timing constraints that real world traders must contend with.

2.2 Types of LPRS Agents

Each LPRS agent evaluates a candidate stock using some set of features. The set of features utilized is dependent on the agent type. For any given candidate, the output of the LPRS agent is a buy signal. A positive buy signal indicates the stock should be purchased; no purchase is made for a negative signal. In cases where a profit (P) fitness function is used, the stock is not necessarily purchased even if the LPRS agent emits a positive buy signal. For this experiment, the following types of LPRS agents were used:

- Random. This agent selects candidate stocks for purchase at random; each candidate has a 50% probability that the agent will generate a positive buy signal. This agent-type is used only with the P fitness function. Note that this agent type is solely intended as a baseline for performance comparison with other agent types.
- Metric-based. This agent is evolved from a combination of fundamental stock metrics, features derived from social media sources, and low level primitive operators.
- Rule-based. For this type of agent, each metric in the dataset is evaluated using a Classification and Regression Tree (CART) [13] technique on its ability to discriminate between the two classes. The most promising rules, described in section 2.3, are input as features to evolve the GP.
- Majority-rule. For this type of agent, we select the nine (9) most promising agents for a majority-rules scheme. That is, if a majority of the selected LPRS agent output a positive buy signal, then the buy-signal of the conglomerate is positive; otherwise, it is negative.

2.3 Genetic Program Characteristics

In this section, we describe the inputs to the Genetic Programming framework. For our experiment, we chose the AForge.Net framework [9], which provides a wide variety of machine learning C# libraries, including GA and GP. The primitive functions supported by the frameworks default GP chromosome are: addition, subtraction, multiplication, division, sine, cosine, exponent, natural logarithm, and square root. These primitives operate on a set of features, which appear as leaves on the tree (see Figure 2). The choice of features is very important to the success of a GP; in some cases, the features themselves may be implemented as functions. The feature set employed for the LPRS agents are described in the subsections that follow.

Candidate Stocks. For this research, we surveyed the biggest decliners on the New York Stock Exchange (NYSE) for the period from Jan 1, 2011 through April 30, 2015. From the total list of decliners, we identified 207 transactions that met the criteria described in Section 1.1. These cases were almost evenly split between the two classes, with 101 (48.8%) in the recovery class and 106 (51.2%) in the non-recovery class. We partitioned this set of candidates into two subsets for training and testing the agents, respectively.

Stock Metrics. For each candidate transaction, the following metrics were available as features for GP training:

- Decline date
- Closing share price on decline date
- Absolute price decline (on decline date)
- Percentage price decline
- Recovery or Sell Date
- Final Share Price
- Trade volume in millions of trades (on decline date)
- Market Capitalization of company in billions of $ (pre decline)
- Earnings per Share (prior quarter)
- Current Ratio [16]
- Dividend Per Share (quarterly)
- Price-Earnings (PE) Ratio [16]

Several other features, derived from online sources, were used as well (below). The purpose of these features was to get a snapshot of the prevailing sentiment about each security on the day of its decline.

- Google Trend Index. The weighted stock symbol value from Google Trends on day of decline. This is the trend value on the decline day divided by the average Google trend value for the previous three (3) months.
- Message Post Index. The normalized number of message board posts on Yahoo Finance about the stock on the day of the decline. This value is normalized with respect to the mean percentage increase in posts noted for all candidate stocks on the day of decline.
- Analyst Count. The total number of financial analysts following the company that issued the security.
- Analyst Sentiment. The cumulate change in sentiment of analysts over the past 120 days. Each update to analyst ratings on the stock changed the sentiment value, on a scale between -1 and $+1$ depending on the specific ratings change.

Buying Rules. The following are decision rules that were generated using the CART decision tree algorithm. These rules were chosen based on their ability to discriminate each candidate into the correct class. Each of the below four (4) rules were used as input to the GP to evolve the rule-based LPRS agents.

A. Buy signal true if: absolute price decline is greater than $4.01; false otherwise.
B. Buy signal true if: PE ratio is less than 9.3; false otherwise.
C. Buy signal true if: volume is greater than 40.26M OR less than 0.7707M shares; false otherwise.
D. Buy signal true if: the percent price decline is between 10.1% and 11.5%; false otherwise.

Additionally, for the rule-based LPRS agent, the following metrics were also input to the GP: percentage price decline, market capitalization, volume, sentiment count, Message Post Index and Google Trend Index.

3 Related Work

Over the past two decades, both GAs and GPs have been applied to a wide range of financial trading problems[5]. In many cases, the resulting evolved solutions have been shown to outperform both human traders and benchmark trading indices (e.g., S&P500) for specific problems. This approach differed from ours in that the pool of candidate stocks was much smaller[1] used GPs to evolve stock trading rules for the S&P500 index as a whole from 1929 through 1995. After transaction costs were factored in, it was found the rules did not earn consistent excess returns over a simple buy-and-hold strategy in the out-of-sample test periods. The rules were able to identify periods to be invested in the index when daily returns were positive and volatility is low and out when the reverse was true. Becker & Seshadri [3] were later able to improve upon their work and actually outperform a buy-and-hold strategy for the S&P500 index. Potvin, Soriano & Vallée [15] employed a GP approach to trade a small pool of fourteen (14)l Canadian stocks, with each stock selected from a distinct commercial sector. Like [1], their evolved rules were unable to outperform a buy-and-hold strategy. In contrast to these researchers, our approach did not use either indices or a specific pool of stock. Instead, we utilized the LPRS criteria to continuously indentify specific buying opportunities from the entire NYSE. The only real decision our evolved agent makes is the buy decision; unlike these other approaches, our sell decision is automatic. Thus our approach does not result in excessive churning of stock purchases. Further, when our P fitness function is utilized, the cash balance of the portfolio serves as a limiting factor for potential stock purchases.

A number of researchers have also experimented with using public sentiment as a basis for predicting the performance of markets and individual stocks. Bollen, Mao, & Zeng [4] showed that the general public mood, as derived from Twitter feeds, was correlated to the Dow Jones Industrial Average (DJIA) over time. Nuij et al. [14] incorporated mined specific categories of news events and incorporated them into a GP to evolve trading rules for the FTSE350 and

S&P500 indices. They found that augmenting news information with more traditional technical financial indicators generated higher returns than if the news events had not been included. While our approach likewise utilizes sentiment to make buying decisions, we attempt to measure the prevailing sentiment for individual stocks (vs. global stock indices). This drove our decision to ignore stocks which had a low market capitalization since there appeared to be insufficient sentiment information available for these stocks. Vu, Chang, Ha & Collier [17] had considerable success in using Twitter sentiment to predict the daily up/down changes of widely held technology stocks (AMZN, APPL, GOOG, and MSFT). A key advantge of this approach is that the large size of the companies helps ensure a critical mass of sentiment was available. Even so, large companies tend to have fairly stable stock prices; as such, they are typically not LPRS candidates.

4 Experimentation and Analysis of Results

In this section, we describe the setup of the experiment, including all test cases. We then present and analyze the results of each test case. The primary objective of our analysis is to compare approaches for evolving the most effective LPRS agent. As part of our analysis, we also compare the performance of our evolved agents against the performance of the S&P500 index and a random purchase decision for LPRS candidates.

4.1 Experiment Setup

We experimented with a number of different strategies for evolving each type of LPRS agent. In general, these strategies varied by:

- Agent type
- Fitness function type
- Buying strategy (buy all or selective)

For some cases, the fitness function type was alternated, such that one was used for training and the other for testing. The motivation here was to see how effective an agent trained on one fitness function would be on the other. Table 1 provides a summary of test cases.

The breakdown of transactions by class was discussed in section 2.3. For this experiment, the training period was from January 1, 2011 through 31 May 2014. The test period was from 1 June 2014 through 30 April 2015. The training and test data sets had 165 and 42 transactions, respectively. The parameters for the evolved GP were as follows:

- Population size = 100
- Maximum generations = 100
- Crossover Rate = 75%
- Mutation Rate = 10%
- Selection method is Roulette wheel

The results reported for each test case are based on ten (10) agents evolved using the GP.

Table 1. Summary of Test Cases

Exp ID	Metrics Type: Stock	
	Train	Test
SM-CM-All	Fitness: CM; BuyMode: All	Fitness: CM; BuyMode: All
SM-CM-Sel	Fitness: CM; BuyMode: All	Fitness: CM; BuyMode: Selective using GP-Output
SM-CM-Vote	Fitness: CM; BuyMode: All	Fitness: CM; BuyMode: Selective using GP-Vote
SM-P-Sel	Fitness: Profit ; BuyMode: Selective using GP-Output	Fitness: Profit ; BuyMode: Selective using GP-Output
SM-P-Vote	Fitness: Profit ; BuyMode: Selective using GP-Output	Fitness: Profit ; BuyMode: Selective using GP-Vote
Exp ID	Metrics Type: Rule-Based	
	Train	Test
RB-CM-All	Fitness: CM; BuyMode: All	Fitness: CM; BuyMode: All
RB-CM-Sel	Fitness: CM; BuyMode: All	Fitness: CM; BuyMode: Selective using GP-Output
RB-CM-Vote	Fitness: CM; BuyMode: All	Fitness: CM; BuyMode: Selective using GP-Vote
RB-P-Sel	Fitness: Profit ; BuyMode: Selective using GP-Output	Fitness: Profit ; BuyMode: Selective using GP-Output
RB-P-Vote	Fitness: Profit ; BuyMode: Selective using GP-Output	Fitness: Profit ; BuyMode: Selective using GP-Vote

4.2 Experiment Results

For the CM-ALL experiment, we evolved a series of ten GPs using the CM fitness function and the training data set. In each case, the final, evolved GP was then selected as the LPRS agent and run against the test data set. This was done for both the Stock Metric (SM) and Rule-based (RB) feature sets. The averaged results of these runs are shown in Table 2. The REC and NonRec columns in the table indicate the percentage of items in each class that were correctly classified by the GP. Though not shown in Table 2, the fitness for each test case may be computed using Equation 1.

Table 2. Confusion Matrix Training and Test Results

Test Case	Statistic	Training		Test	
		REC %	NonREC %	REC %	NonREC %
SM-CM-All	Mean	67.6%	62.9%	57.9%	49.6%
	STDev	6.6%	6.5%	13.0%	16.0%
RB-CM-All	Mean	70.8%	67.9%	70.0%	48.1%
	STDev	6.3%	3.8%	4.5%	7.8%

When we use the CM fitness function, we are most concerned with evolving an LPRS agent that can distinguish a recovery situation from a non-recovery one. While such an LPRS agent is not explicitly trained to maximize profit, it is reasonable to expect it to be profitable. Table 3 shows the extent to which this may be true. Here we run the LPRS agents evolved for the SM-CM-ALL and RB-CM-ALL cases to determine the increase in portfolio value each would reap.

The P fitness function, given in Equation 2, is reported against both the training and test datasets (labeled in Table 3 as Test-1 and Test-2, respectively). Note that there is no need to provide the data for the REC and NoREC columns in Table 3, because the results would be identical to those in Table 2.

Table 3. CM Training and Profit Test Results

Test Case	Statistic	Test-1 (on Training data)			Test-2 (Test Data)		
		Fitness	REC %	NonREC %	Fitness	REC %	NonREC %
SM-CM-Sel	Mean	0.624	N/A		0.164	N/A	
	STDev	0.078			0.051		
RB-CM-Sel	Mean	0.607	N/A		0.182	N/A	
	STDev	0.094			0.124		

Next, we selected the top 9 out of 10 LPRS agents from the previous experiment based on the Table 2 results. We then combined these to create a composite LPRS agent, where the REC/NoREC determination is made based on what the majority of agents decide. The results of this test case are shown in Table 4. As is the case for Table 3, the fitness is computed using the P fitness function and we also track the CM classification accuracy for each composite LPRS. Since we employ majority rule to make this decision on the data, this experiment consisted of a single run for each test case. For this reason, there was no need to report the standard deviation statistic.

For the last two experiments, we focused on LPRS agents evolved to maximize profit during training. The results for the SM and RB feature set are given in Table 5. Even though profit is the guiding factor here, the REC/NoREC results are included to give a sense of how critical class discrimination is to profit.

Table 4. Best of Nine (9) Votes - CM Training and Profit Test Results

Test Case	Statistic	Test-1 (on Training data)			Test-2 (Test Data)		
		Fitness	REC %	NonREC %	Fitness	REC %	NonREC %
SM-CM-Vote	Mean	0.7355	76.7%	65.1%	0.2289	64.3%	63.0%
	STDev	N/A					
RB-CM-Vote	Mean	0.4724	67.8%	69.7%	0.3519	67.9%	51.9%
	STDev	N/A					

Table 5. Profit Training and Test Results

Test Case	Statistic	Training			Test		
		Fitness	REC %	NonREC %	Fitness	REC %	NonREC %
SM-P-SEL	Mean	0.816	90.4%	12.9%	0.188	85.4%	25.5%
	STDev	0.035	3.5%	5.2%	0.035	5.9%	6.8%
RB-P-SEL	Mean	0.832	91.2%	21.5%	0.181	88.1%	17.0%
	STDev	0.034	5.5%	10.6%	0.030	8.7%	13.3%
Random	Mean	0.315	N/A		0.137	N/A	
	STDev	0.129			0.028		
S&P 500	Mean	0.530	N/A		0.086	N/A	
	STDev	N/A					

Table 6. Best of Nine (9) Votes - Profit Training and Test Results

Test Case	Statistic	Test-1 (on Training data)			Test-2 (Test Data)		
		Fitness	REC %	NonREC %	Fitness	REC %	NonREC %
SM-P-Vote	Mean	0.8174	91.3%	10.6%	0.1725	85.7%	29.2%
	STDev	N/A					
RB-P-Vote	Mean	0.902	98.8%	17.7%	0.2098	95.2%	4.8%
	STDev	N/A					

Finally, Table 6 shows the performance of the majority rule approach for LPRS agents evolved for the profit fitness function. Similar to Table 4, we selected the top nine LPRS agents from those evolved for the Table 5 test case. Thus, the results in Table 6 reflect a single, composite LPRS agent.

4.3 Analysis of Results

While Tables 2 through 6 contain the raw experiment results, Figure 3 compares the performance results for each approach side by side. For this figure, we converted the raw profit fitness performance into an annualized Return on Investment (ROI) percentage. Taking duration out of the training/test performance results enables us to better compare how the LPRS results generalize to new data.

In the CM-ALL experiment (Table 2), we were only concerned with class discrimination performance. Given that the classes were nearly equally represented, the agents were able to marginally improve the apriori class probability ($\approx 50\%$), with a training classification accuracy SM-CM-ALL and RB-CM-ALL of 65% and 69% respectively. For the test data, these percentages decreased to 54% and 59%, respectively. For these cases, the discrimination performance did not generalize well and the classification accuracy on the test data was barely better than guessing.

For the P-SEL cases, the evolved agents also did poorly in terms of training classification, with accuracy rates of SM-P-SEL and RB-P-SEL of 50% and 55%, respectively. However, unlike the CM-ALL approach, these rates generalized well with the test data. Here, the bias was heavily skewed in favor of classifying cases as REC. For this case, fitness was measure in terms of profit. In terms of the increase in overall portfolio value, this approach achieved better fitness than both Random and the S&P500. This indicates that the LPRS strategy is more profitable if the agent is biased toward stock recovery.

The best overall results were achieved with P-Vote approach. As Figure 3 shows, these LPRS agents did well in terms of profit on the training data and as well, or better, on the test data (generalized well). For these two cases, the RB-P-Vote did slightly better than the SM-P-Vote agent. Based on the classification accuracy, it appears that both these agents are biased toward classifying stocks as REC. This characteristic supports the counter-intuitive assertion that accurate class discrimination is not essential to good portfolio performance.

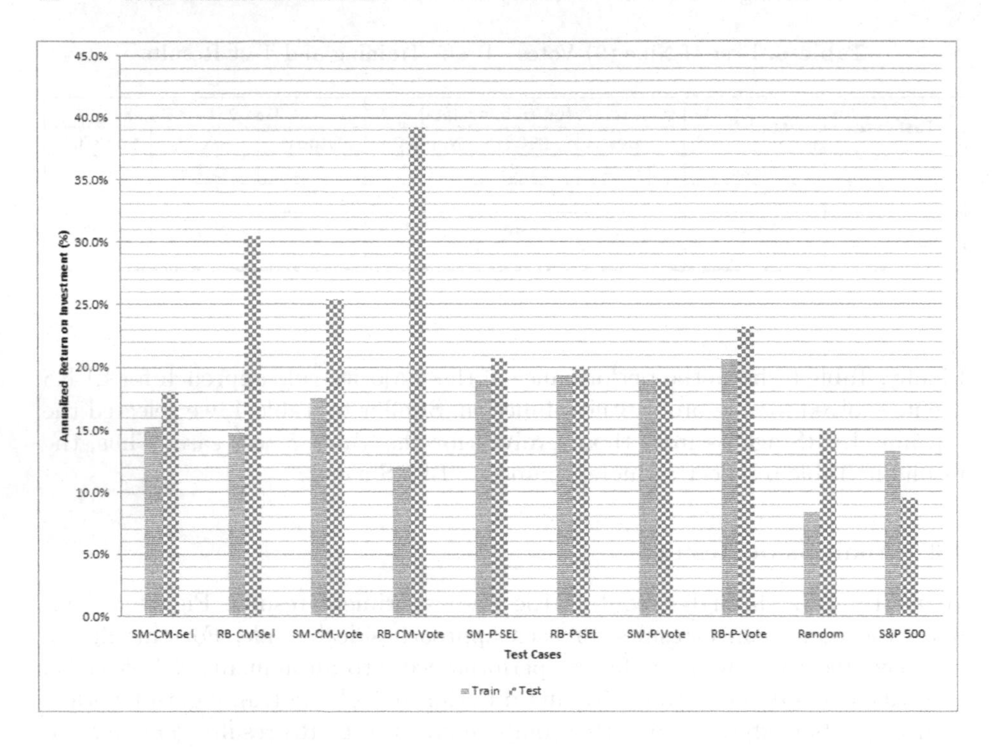

Fig. 3. Comparison View of Annualized ROI Results

For the most part, the Random approach was outperformed by all the LPRS agents. That being said, the SM-CM-Sel results (Table 3) did not convincingly outperform Random on the test data, especially when taking the large fitness standard deviations into account. Likewise, the results for SM-P-SEL cases (Table 5) were only marginally better than the Random results. The performance of the S&P500 index was competitive with the LPRS agents on the training data. However, with regard to the test data, all the LPRS approaches (including the Random approach) outperformed this benchmark index by substantial margins.

These results also yielded a number of surprises. For the RB-CM-SEL case, Buying Rule A (from section 2.3) was very prominent in the evolved solutions, appearing repeatedly. In fact, only two distinct solutions were involved. This phenomenon indicates that Buying Rule A is a very powerful discrimination rule. Another surprise was that the evolved LPRS agents consistently performed better on test cases than on the training cases.

5 Summary

In this paper, we investigated a number of different approaches for evolving agents to conduct a Low Price Recovery Strategy for stock trading. We targeted the LPRS because our observation had been that stocks experiencing major declines will often bounce back in a relatively short period of time. This phenomenon has been shown to be fairly common in our research, with nearly 50% of the candidate stocks recovering. Given this, our evolved LPRS agents were all able to beat a strategy of buying random candidates. Additionally, their performance also bested the S&P500 index, which is considered an important benchmark when evaluating the performance of mutual stock funds. In general, the best LPRS performance was observed when the agents were trained to maximize the return of the portfolio. From a profit perspective, this approach proved superior to simply training agents to discriminate between the REC and NoREC classes. Within those subgroups, the majority rules strategy yielded the best results, with the agent evolved for the RB-P-Vote case having the best overall performance, both in terms of fitness and generalization of results (from training to test). It was also found that the absolute decline in stock price (Buying Rule A) was an especially important factor in choosing stocks that would eventually recover.

This research has demonstrated the utility of evolving agents for specific trading strategies. The fact that these agents can outperform benchmarks like the S&P500 index is especially promising. Our future work in this area will focus on variations on this technique to determine if performance can be improved still further. In particular, instead of a binary (yes/no) trading decision, we want to use the output of the evolved function as a basis for prioritizing buying opportunities. Additionally, we will continue to investigate popular sentiment metrics which potentially have short-term predictive value for investments.

References

1. Allen, F., Karjalainen, R.: Using genetic algorithms to find technical trading rules. Journal of Financial Economics **51**(2), 245–271 (1999)
2. Banzhaf, W., Nordin, P., Keller, R.E., Francone, F.D.: Genetic programming: an introduction, vol. 1. Morgan Kaufmann San Francisco (1998)
3. Becker, L.A., Seshadri, M.: Gp-evolved technical trading rules can outperform buy and hold (2003)
4. Bollen, J., Mao, H., Zeng, X.: Twitter mood predicts the stock market. Journal of Computational Science **2**(1), 1–8 (2011)
5. Chen, S.-H.: Genetic algorithms and genetic programming in computational finance. Springer Science & Business Media (2012)
6. Enke, D., Thawornwong, S.: The use of data mining and neural networks for forecasting stock market returns. Expert Systems with Applications **29**(4), 927–940 (2005)
7. Golberg, D.E.: Genetic algorithms in search, optimization, and machine learning, 1st edn., vol. 1989 (1989)
8. Kimoto, T., Asakawa, K., Yoda, M., Takeoka, M.: Stock market prediction system with modular neural networks. In: 1990 IJCNN International Joint Conference on Neural Networks, pp. 1–6. IEEE (1990)
9. Kirillov, A.: Aforge.net framework (2013)

10. Kohavi, R., Provost, F.: Glossary of terms. Machine Learning **30**(2–3), 271–274 (1998)
11. Koza, J.R.: Genetic programming: on the programming of computers by means of natural selection, vol. 1. MIT press (1992)
12. Koza, J.R.: Survey of genetic algorithms and genetic programming. In: Wescon Conference Record, pp. 589–594. Western Periodicals Company (1995)
13. Lewis, R.J.: An introduction to classification and regression tree (cart) analysis. In: Annual Meeting of the Society for Academic Emergency Medicine in San Francisco, California, pp. 1–14 (2000)
14. Nuij, W., Milea, V., Hogenboom, F., Frasincar, F., Kaymak, U.: An automated framework for incorporating news into stock trading strategies. IEEE Transactions on Knowledge and Data Engineering **26**(4), 823–835 (2014)
15. Potvin, J.-Y., Soriano, P., Vallée, M.: Generating trading rules on the stock markets with genetic programming. Computers & Operations Research **31**(7), 1033–1047 (2004)
16. Rosenberg, C.N.: Stock Market Primer. Grand Central Publishing (1991)
17. Vu, T.-T., Chang, S., Ha, Q.T., Collier, N.: An experiment in integrating sentiment features for tech stock prediction in twitter (2012)
18. Whitley, D.: A genetic algorithm tutorial. Statistics and Computing **4**(2), 65–85 (1994)

A Spectral Clustering Based Outlier Detection Technique

Yuan Wang[1], Xiaochun Wang[1(✉)], and Xia Li Wang[2]

[1] School of Software Engineering, Xi'an Jiaotong University, Xi'an 710049, China
wy734174981@gmail.com, xiaocchunwang@mail.xjtu.edu.cn
[2] School of Information Engineering, Changan Univeristy, Xi'an 710061, China
xlwang@chd.edu.cn

Abstract. Outlier detection shows its increasingly high practical value in many application areas such as intrusion detection, fraud detection, discovery of criminal activities in electronic commerce and so on. Many techniques have been developed for outlier detection, including distribution-based outlier detection algorithm, depth-based outlier detection algorithm, distance-based outlier detection algorithm, density-based outlier detection algorithm and clustering-based outlier detection. Spectral clustering receives much attention as a competitive clustering algorithms emerging in recent years. However, it is not very well scalable to modern large datasets. To partially circumvent this drawback, in this paper, we propose a new outlier detection method inspired by spectral clustering. Our algorithm combines the concept of *kNN* and spectral clustering techniques to obtain the abnormal data as outliers by using the information of eigenvalues and eigenvectors statistically in the feature space. We compare the performance of our methods with distance-based outlier detection methods and density-based outlier detection methods. Experimental results show the effectiveness of our algorithm for identifying outliers.

Keywords: Outlier detection · Distance-based outlier detection · Density-based outlier detection · Spectral clustering · Eigenvalues

1 Introduction

With the rapid development of information technology, people have been able to easily access and store large amounts of information from the real world. However, how to find important and useful information from these massive and high-dimensional data has become an urgent problem. Therefore, data mining and database technologies come into being consequently.

Outlier detection is an important data mining technique, which focuses on discovering the small portion of data objects in a data set that, being inconsistent with the conventional pattern of the majority of the data set, however, may imply important information. Hawkins proposed a relatively widely accepted definition: "Outlier is an observation that deviates so much from other observations as to arouse suspicion that it was generated by a different mechanism." [1] Currently, outlier detection has important applications in areas like intrusion detection [2,3], fraud detection [4], etc..

P. Perner (Ed.): MLDM 2016, LNAI 9729, pp. 15–27, 2016.
DOI: 10.1007/978-3-319-41920-6_2

In recent years, there have been a broad range of definitions for outliers and researchers have developed many types of algorithms for outlier detection, mainly including distribution-based outlier detection algorithm, depth-based outlier detection algorithm, distance-based outlier detection algorithm, density-based outlier detection algorithm and clustering-based outlier detection [5,6]. However, none of them has been proved to be completely applicable to every situation. Methods based on the statistical distribution are firstly presented using the classic statistical approaches with the assumption that the database corresponds to a given distributional model. However, the model is usually not known a priori for modern sophisticated large databases. As a result, they have a limited number of applications. Being an improvement, depth-based outlier detection methods assign a depth value to each data object and regard the data objects in the shallow layers to be more likely to be outliers than those in the deep layers. Unfortunately, these methods suffer high computational complexity for more than a few dimensions. Distance-based methods, also known as adjacency-based methods, believe that data objects are outliers if they are far away from the majority of data points and address more globally-oriented outliers in the database [7]. Density-based methods usually assign each data object a measure of outlier degree as the classic LOF algorithm did and then regard those data objects which possess the largest outlier degrees as outliers [8]. In comparison to distance-based methods, these methods address more locally-oriented outliers. Finally, clustering-based methods obtain outliers as a by-product and regard the outliers as the data items that reside in the smallest clusters [9].

Recently, spectral clustering has been widely applied to pattern recognition and data mining because it can obtain global optimal solution on sample spaces of arbitrary shape [10,11,12]. Meanwhile, since it is only related to the number of data points, not the dimensionality, spectral clustering can help solve the curse of dimensionality suffered by distance-based methods and density-based methods on high-dimensional data space. In this paper, we propose a spectral clustering based outlier mining approach, which, when compared with distance-based methods and density-based methods on some standard test datasets, manifests its effectiveness and efficiency.

The rest of the paper is organized as follows. In section 2, we present some preliminaries of spectral clustering. In section 3, the proposed spectral clustering based outlier mining approach is introduced. In section 4, we present the results of our experiments conducted to evaluate the performance of our algorithm. Finally, the conclusions are given in section 5.

2 Preliminaries

2.1 Spectral Clustering

Clustering is the process to divide points in a data set into a number of categories of clusters so that the similarity between two points within the same cluster and the dissimilarity between two points belonging to respectively two different clusters is as high as possible. Cluster analysis plays an important part in data mining. Spectral clustering receives much attention as a competitive clustering algorithms emerging in recent years, which is mainly applied to image segmentation. Basically, possessing numerous advantages, spectral clustering treats points in a dataset as the vertices of a weighted undirected graph and the similarity between two vertexes as the weights of edges in the graph, and converts the problem of clustering into an optimal partitioning problem of a graph. The goal is to find a graph partitioning methodology to make the weights of edges connecting two different sub-graphs as large as possible and the weights of edges within a sub-graph as small as possible. The most common graph division criteria proposed include Mini Cut [13], Normalized Cut [14,15] and Ratio Cut [16]. Mini Cut produces a good result on image segmentation but is prone to skew-segmentation, while Normalized Cut and Ratio Cut take both the minimum sum of the weights of the cutting edges and the balance of division into account.

In this paper, we utilize the *k*-way Normalized Cut partitioning approach, for a given data set, which contains *n* points, x_1, x_2, …, x_n, the goal of clustering is to divide this *n* points into *k* clusters so that the similarities between two points belonging to a same cluster are maximized and those between two points belonging to different clusters are minimized. Treating data point x_i in the dataset as a vertex v_i in a graph and the similarity between x_i and x_j, W_{ij}, as the weight of the edge connecting x_i and x_j, and we obtain a weighted undirected graph, $G = (V, E)$ with V the set of all the vertexes and E the set of all the edges in the graph G. Suppose that the multi-way spectral clustering partitions the graph G into k sub-graphs, A_1, A_2, …, A_k, then the target function of Normalized Cut (*Ncut*) that makes the optimal partitioning can be described as:

$$Ncut(A_1, A_2, \cdots, A_k) = \sum_{i=1}^{k} \frac{cut(A_i, \overline{A}_i)}{vol(A_i)} \tag{1}$$

where $cut(A, B) = \sum_{i \in A, j \in B} W_{ij}$, $\overline{A}_i = \{v_p \mid v_p \in V, v_p \notin A_i\}$, $vol(A_i) = \sum_{p \in A_i, q \in V} W_{pq}$.

After mathematic transformations, the solution for the above *Ncut* problem can be finally converted to a problem of finding eigenvalues (and eigenvectors) of a Laplacian matrix and the smallest series of eigenvalues corresponds to the optimal partitioning of the graph. Thus the discrete clustering problem becomes to find the eigenvectors on a contiguous data space.

In the general frame of the spectral clustering, the following three Laplacian matrixes are commonly used,

$$L = D - W \tag{2}$$

$$L_{sys} = D^{-\frac{1}{2}} L D^{-\frac{1}{2}} = I - D^{-\frac{1}{2}} W D^{-\frac{1}{2}} \tag{3}$$

$$L_{rw} = D^{-1} L = I - D^{-1} W \tag{4}$$

where W is the similarity matrix, D is the corresponding diagonal matrix, $D_{ij}=\sum^{N}_{j=1}W_{ij}$, and I is the identify matrix. In this paper, we adopt the second form of Laplacian matrix, and notice that $L_{sys}=I-D^{-1/2}WD^{-1/2}$, hence the sum of eigenvalues of $D^{-1/2}WD^{-1/2}$ and the corresponding eigenvalues of L_{sys} are one and the corresponding eigenvectors are equal. Therefore, finding the eigenvectors of $D^{-1/2}WD^{-1/2}$ which correspond to the k largest eigenvalues is equivalent to finding the eigenvectors of L_{sys} which corresponds to the k smallest eigenvalues. With these ideas in mind, the general frame of spectral clustering works as following,

1. Given a set of data points, $x_1, x_2, \ldots, x_n$, calculate the similarity between each two points according to a certain kind of similarity definition (e.g., the Gaussian kernel function) and construct the similarity matrix $\boldsymbol{W}$;
2. Compute the corresponding Laplacian matrix L_{sys}, and find its k eigenvectors corresponding to the k smallest eigenvalues and use these k eigenvectors as columns to construct a feature matrix in a k-dimensional space, $H \in R^{n \times k}$;
3. Treat each row of feature matrix, H, as a data point in k-dimensional space and perform K-means clustering on these points. Each original d-dimensional data point is assigned the same cluster number as the k-dimensional feature vector of the corresponding row in the feature matrix H.

2.2 An Enhancement for Spectral Clustering

From the general framework of spectral clustering given in the previous subsection, it can be seen a similarity definition should first be provided to establish the similarity matrix. The most commonly used one is the Gaussian kernel function, which has the following form,

$$W_{ij} = e^{-\frac{\|x_i - x_j\|^2}{2\sigma^2}} \tag{5}$$

where $\|x_i - x_j\|$ denotes the Euclidean distance between two points x_i and x_j, and σ is the width parameter, controlling the radial function scope of the Gaussian kernel function. Obviously, the similarity is closer to 1 if two points are very close; otherwise, the similarity is closer to 0 if two points are far away from each other.

With respect to parameter σ, in the original NJW algorithm, it was proposed to perform spectral clustering respectively using several pre-set values and choose the one that yields the best clustering results. To improve the running time performance, some researchers suggested to determine σ by empirical formulas. However, all these practices require a combination of the domain knowledge and are not applied to all the situations. Therefore, researchers start to focus on how to automatically determine this parameter according to the data set itself. Under this line, Zelnik-Manor and Perona proposed the "Self-Tuning" algorithm [17]. Based on "Local Scale" idea, the algorithm constructs a self-adaptive parameter $\sigma_i = \|x_i - x_p\|$ (where x_p is the k-th nearest neighbor of x_i, usually $k = 7$ is used) for each data point using its own neighborhood information. The similarity between two points is then replaced by the following form,

$$W_{ij} = e^{-\frac{\|x_i - x_j\|^2}{\sigma_i \sigma_j}} \quad (6)$$

Empirical evidence shows that this latter approach can address those data set with a multi-scale nature where a universal σ can not do, and isolate accurately the dense clusters embedded in a sparse background.

2.3 Outlier Detection

There are three parts of the unsupervised outlier detection literature that are related to our study: distance-based outlier detection, density-based outlier detection and clustering-based outlier detection.

Knorr and Ng proposed distance-based outlier detection methods as a good way to detect outliers residing in relatively sparse regions. Beginning with this work, various versions of distance-based outlier definition have been developed. For an example, given two integers, *n* and *k*, outliers are the data items whose distance to their *k*-th nearest neighbor is among top *n* largest ones [18], referred to as the DB-max method in the following. For another example, given two integers, *n* and *k*, outliers are the data items whose average distance to their *k* nearest neighbors is among top *n* largest ones [19,20] , referred to as the DB method in the following.

Distance-based outlier detection techniques work well for detecting global outliers in simply-structured data sets that contain one or more clusters with similar density. However, for many real world data sets which have complex structures in the sense that different portions of a database can exhibit very different characteristics, they might not be able to find all interesting outliers. To deal with this situation, Breunig et al. pioneered the density-based outlier detection research by assigning to each object a degree of being an outlier, called the Local Outlier Factor (LOF), for judging the outlyingness of every object in the data set based on ratios between the local density around an object and the local density around its neighboring objects [8]. The LOF method works by first calculating the LOF for each object in the data set. Next, all the objects are ranked according to their LOF values. Finally, objects with top-*n* largest LOF values are marked as outliers.

A problem associated with distance-based as well as density based outlier detection algorithms is their strong sensitiveness to the setting of some parameters. The situation could be worse for the detection of outliers in high-dimensional feature space since data points cannot be visualized there. This is where clustering algorithms can be of some help. Being a very important data mining tool, the main concern of clustering algorithms is to find clusters by optimizing some criterion, such as minimizing the intra-cluster distance and maximizing the inter-cluster distance. As a by-product, data items in small groups can often be regarded as outliers (noise) that should be removed to make clustering more reliable.

3 The Proposed Spectral Clustering Based Outlier Mining Algorithm

In the above section, outlier definitions and spectral clustering are presented. In view of the curse of dimensionality problem suffered by distance-based and density-based algorithms on high-dimensional data space, while spectral clustering shields this problem by computing similarity between two data objects using Euclidean distance but suffer high computation cost and high memory storage cost. For a given dataset with N number of objects, the size of the similarity matrix used by spectral clustering is $N \times N$. When N is very large, the size of the similarity matrix will become too large to fit into the main memory.

3.1 A Simple Idea

It is generally believed outliers comprise a small portion of the whole dataset and reside in small clusters in sparse region and behave differently relative to the majority of the normal data. To take the dimension and the number of data objects in a database both into consideration for clustering based outlier detection, we propose a pre-process to find for each data point its k-nearest neighbors (*kNN*) and then perform a spectral clustering process on these k+1 data point. By this way, it is more easily to examine the abnormal behavior of a small number of data points by taking advantage of a small neighborhood of each data point and its *kNN* while keeping the tempo- and spatio-computational cost as low as possible. The result of this pre-processing step is N new small datasets consisting of each data point and its k nearest neighbors. Next, we perform spectral clustering on every new dataset, which results in N new sets of k-dimensional feature data in the eigen-space. For these k-dimensional feature data in the eigen-space, their smallest eigenvalue is zero for all the data. Our clustering based outlier detection algorithm is based on the observation that the values of the second smallest eigenvalues associated with outliers have lowest frequency of occurrence. However, the opposite is not true. That is, the values of the second smallest eigenvalues associated with some inliers have very low frequency of occurrence as well. From a statistical point of view, we select those points (for example 15%~20% of the total number) as outlier candidates whose corresponding second smallest eigenvalues appear least frequently. From our experience, there are some inliers among the outlier candidates. To remove inliers, we then compute the distance of each outlier candidate to its k-th nearest neighbor as its outlier index to rank the outlier candidates (i.e., the distance-based outlier score for the DB-max method). Finally, top n ranked outlier candidates with the largest outlier indices will be returned as outliers in the database.

To illustrate this observation, a synthetic 2-dimensional dataset is plotted in Fig. 1. It consists of 73 data points and, in our pre-process stage, produces 73 size-reduced mini datasets. After performing spectral clustering process on these 73 new groups of data, we plot the second smallest eigenvalues for each new data set in Fig. 2. From the plot, it can be clearly seen that six points, labeled by 68, 69, 70, 71, 72, 73, have their second smallest eigenvalues occurring with lowest frequency (i.e., one time, while other values appear more than one time) among all such eigenvalues as plotted in Fig. 2. Then back to Fig. 1, we can see they correspond to six outstanding outliers.

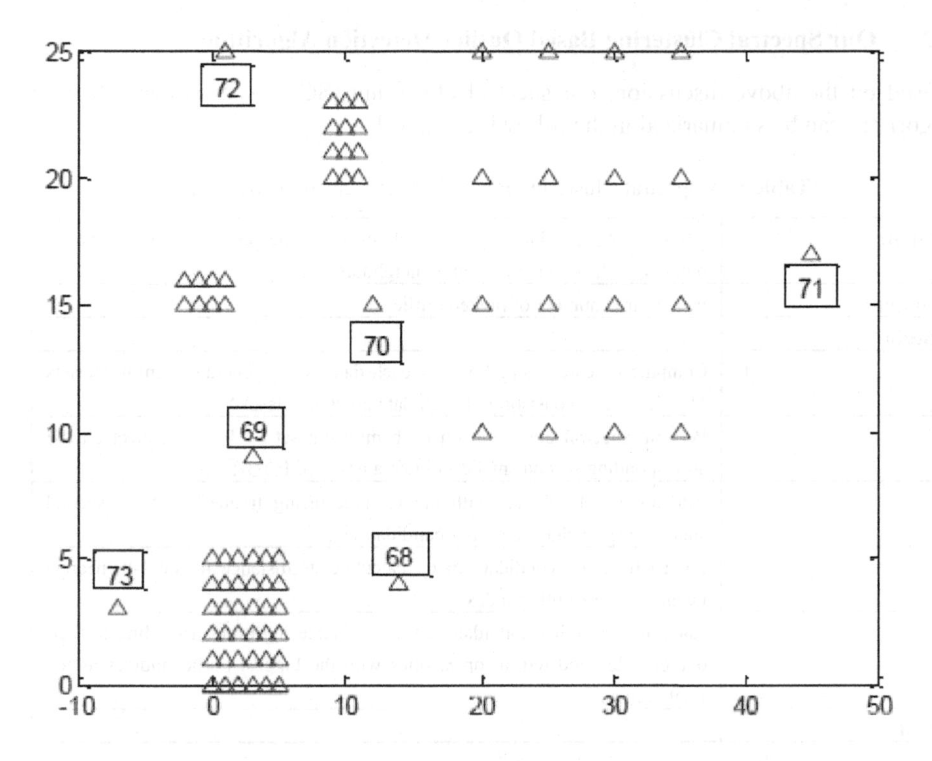

Fig. 1. Synthetic dataset for illustration of our working idea

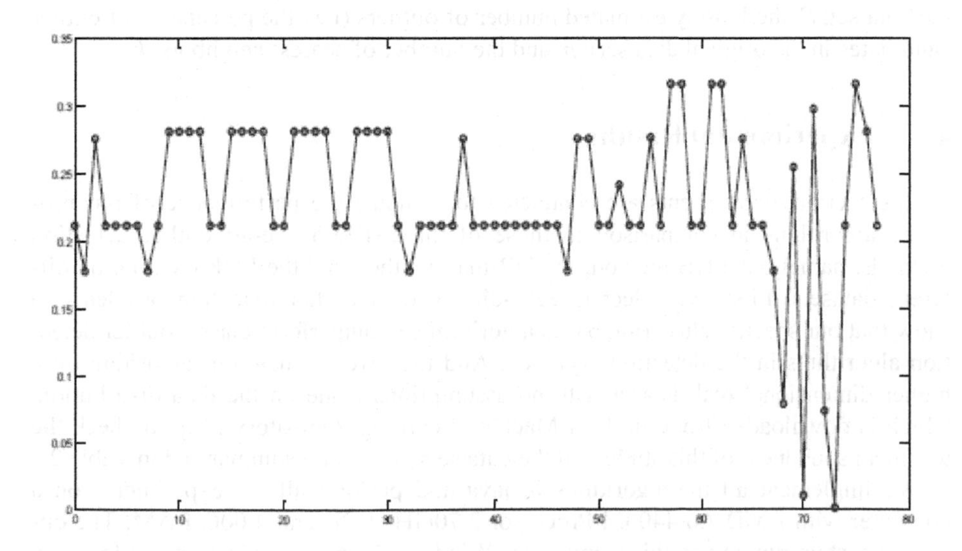

Fig. 2. The distribution of the second smallest eigenvalues of synthetic dataset

3.2 Our Spectral Clustering Based Outlier Detection Algorithm

Based on the above discussion, our spectral clustering (SC)-based outlier detection algorithm can be summarized in the following Table 1.

Table 1. A Spectral Clustering Based Outlier Detection Algorithm

Input:	*S*: a set of *N* data objects; *p*: a loosely estimated percentage of the number of outliers; *k*: the number of nearest neighbors
Output:	*m*: a desired number of ranked outliers;
Begin:	
1:	Compute *k* nearest neighbors for each data point; generate *N* mini datasets $\{S_1, S_2, \ldots, S_N\}$ consisting of each data point and its *kNN*;
2:	Perform spectral clustering on each mini dataset S_i, $1 \leq i \leq N$; collect all the corresponding second smallest eigenvalues λ_{i2}, $1 \leq i \leq N$;
3:	Add top *p* data objects with the least occurring frequency of the second smallest eigenvalues to outlier candidates *C*;
4:	For each outlier candidate in *C*, calculate its distance to the *k*-th nearest neighbor as the outlier index;
5:	Rank all the outlier candidates in a non-increasing order according to their outlier index and return top *m* ones with the biggest outlier indices as the final outliers.
End	

To summarize, the numerical parameters the algorithm needs from the user include the data set, *S*, the loosely estimated number of outliers (i.e., the percentage of outlier candidates in the original data set), *p*, and the number of nearest neighbors, *k*.

4 Experimental Results

In this section, experiments are conducted to evaluate the performance of our proposed algorithms in comparison to those of three-state-of-the-art outlier detection methods, namely, the DB method, the DB-max method and the LOF method, on different datasets. First, we select three 2-dimensional outlier detection problems to show that our spectral clustering based algorithm can outperform classic outlier detection algorithms in the detection accuracy. And then we evaluate our algorithm on a higher dimensional real dataset with no assumptions made on the data distribution, which is downloaded from the UCI Machine Learning Repository [21], to check the technical soundness of this study. All the data sets are briefly summarized in Table 2.

We implement all the algorithms in java and perform all the experiments on a computer with AMD A6-4400M Processor 2.70GHz CPU and 4.00G RAM. The operating system running on this computer is Windows 7. In our evaluation, we focus on the outlier detection accuracy rate of these four outlier detection algorithms on different data sets. The results show that, overall, our spectral clustering based outlier detection algorithm is superior over other state-of-the-art outlier detection algorithms.

Table 2. Descriptions of all datasets

Data Name	Data Size	Dimension	# of outliers
syn_ Data1	89	2	11
syn_ Data2	78	2	6
lymphography	148	18	6

4.1 Performance of Our Algorithm on Synthetic Data

In this subset ion, we use two synthetic datasets to show that the proposed spectral clustering based outlier detection method can effectively identify local and global outliers in various scenarios.

All two synthetic datasets, syn_Data1 and syn_Data2 are shown in Fig. 3. The first synthetic dataset, syn_ Data1, consists of 89 instances, including one large uniformly distributed cluster surrounded by eleven planted easy-identified outliers denoted by yellow circles. This is a global outlier detection task. The best experimental results are obtained with the parameter *k*'s being determined by error and trial to be 5, 5, 5, 14 for DB, DB-max, LOF and our method, respectively and depicted in Fig. 4.

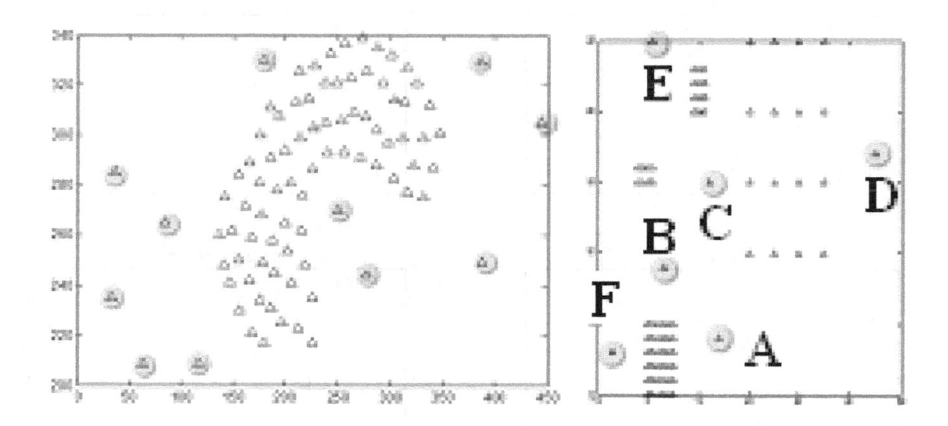

Fig. 3. Synthetic Datasets (left) syn_Data1, (right) syn_Data2

From the figure, it can be seen that, LOF method and our method can correctly detect all the outliers (denoted by red circles) while DB and DB-max methods both miss one.

syn_Data2 contains 78 instances, including five planted global outliers (A,D,E), two local outliers (B, C), and four clusters of different densities consisting of 36, 8, 12 and 16 uniformly distributed instances. To demonstrate the effectiveness of our approach in finding both global and local outliers, the same set of experiments is conducted and the best experimental results are obtained with the parameter *k*'s being determined by error and trial to be 5, 5, 5, 5 for DB, DB-max, LOF and our method, respectively and depicted in Fig. 5. For this case, DB and DB-max both miss the two local outliers.

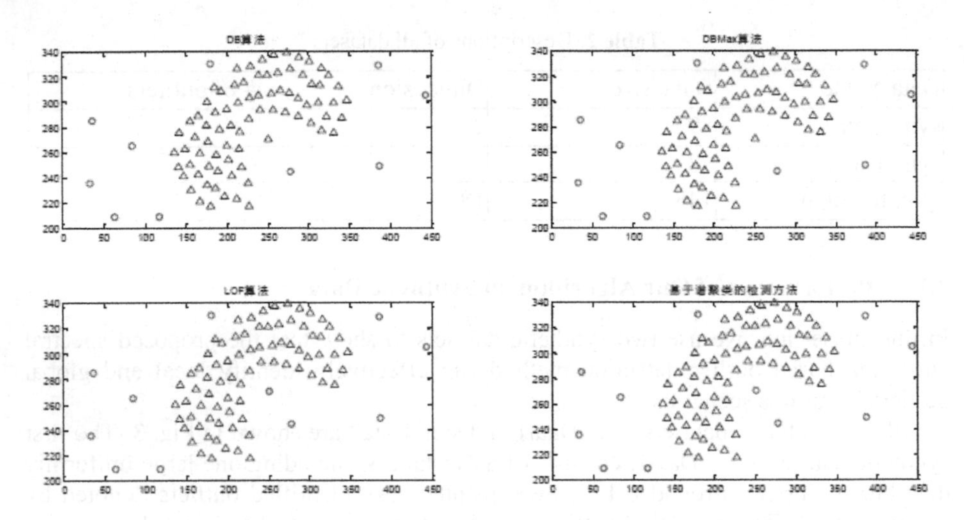

Fig. 4. Experimental results for syn_Data1 (upper left) DB method, (upper right) DB-max, (lower left) LOF method, (lower right) our method

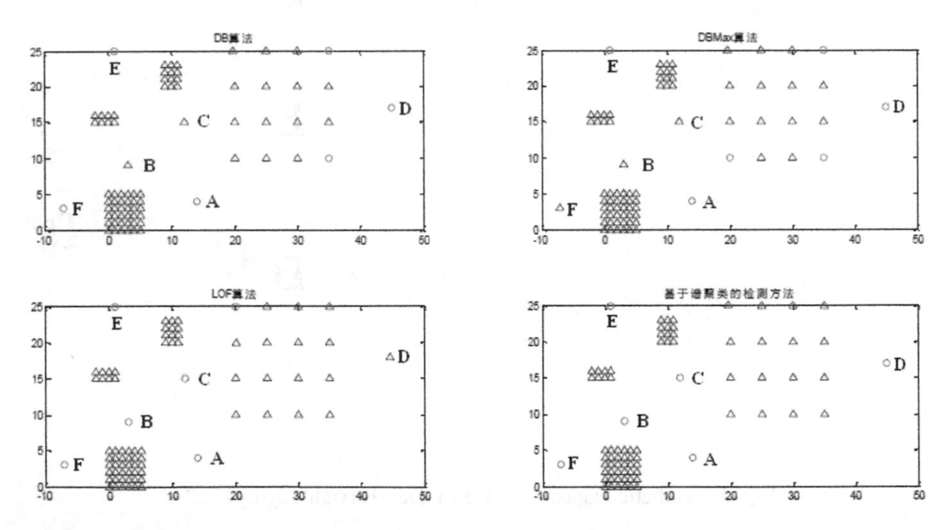

Fig. 5. Experimental results for syn_Data2 (upper left) DB method, (upper right) DB-max, (lower left) LOF method, (lower right) our method

LOF misses one global outlier. Our spectral clustering based outlier detection algorithm identifies all six outliers correctly.

4.2 Performance of Our Algorithm on Real Data

As pointed out by Aggarwal and Yu, one way to test how well the outlier detection algorithm works is to run the method on the dataset and test the percentage of points which belongs to the rare classes [22]. To evaluate the effectiveness and accuracy of our proposed method on real data, we compare the algorithms by their performance on detecting rare classes in a real dataset, namely, lymphography, which is downloaded from UCI [21]. The lymphography dataset has 148 instances with 18 attributes and contains a total of 4 classes. Classes 2 and 3 have the largest number of instances (81 and 61, respectively). The remaining two classes have totally 6 instances (2 and 4, respectively) and are regarded as outliers (i.e., rare classes) for they are small in size. To quantitatively measure the performance of an outlier detection scheme, a metric called recall is employed here. Assuming that a dataset $D=D_o \cup D_n$ where D_o denotes the set of all outliers and D_n denotes the set of all normal data. Given any integer $m \geqslant 1$, if O_m denotes the set of outliers among objects in the top m positions returned by an outlier detection scheme, recall is defined as

$$recall = \frac{|O_m|}{|D_o|} \quad (7)$$

We report the corresponding detecting results of four methods in terms of recall in Table 3 with the parameter k's being determined by error and trial to be 5, 5, 5, 8 for DB, DB-max, LOF and our method, respectively, and m=6. From the experimental results, it can be seen that our method performs the best.

Table 3. Experimental results for lymphography data

Dataset	DB	DB-max	LOF	Our Method
lymphography	0.33	0.50	0.50	0.67

4.3 Discussion

For an undirected, weighted graph with weight matrix, the multiplicity n of the eigenvalue 0 of the matrix equals the number of n connected components in the graph in the ideal case. The matrix has as many eigenvalues 0 as there are connected components, and the corresponding eigenvectors are the indicator vectors of the connected components. If the graph only consists of one connected component, eigenvalue 0 has multiplicity 1 and the first eigenvector is the constant vector. Therefore, if each point and its k nearest neighbors all belong to the same cluster in the ideal situation, the smallest eigenvalues for all the point's k+1 nearest neighbor set are zero and the rest eigenvalues are very similar if not of exactly the same values, resulting in their high frequency of appearance. However, in real situations where outliers may exist, the second smallest eigenvalues in the set of eigenvalues of the spectral clustering for each point's k+1 nearest neighbor set should behave (in terms of frequency of appearance) quite differently for outliers than for those of the majority of normal data points. To summarize, a low frequency of the second eigenvalue in the set of eigenvalues of the spectral clustering for each point's k+1 nearest neighbor set is used as an indication that this point is an outlier.

Distance-based and density-based outlier detection methods are good outlier detectors. However, they are very sensitive to parameter k and a small change in k can lead to changes in the scores and, correspondingly, the ranking. From Section 3, we know that our method is based on a low frequency of the second eigenvalue in the set of eigenvalues of the spectral clustering for each point's k+1 nearest neighbor set. k is a very critical parameter of our proposed method. For a suitable value to be chosen for k, our goal is to promote eigenvalues other than the second smallest one to appear as frequently as possible. In other words, a desired k should be large enough for each point's k+1 nearest neighbor set to include normal points so much that these normal points dominate the neighborhood. From the experiments, we see that, for syn_Data1 and the real data, where outlying groups have outliers more than one data point, the optimal k's for our method are much larger than those for the other three methods, while for syn_Data2, where there is only one data point for each outlying group, the optimal k's for our method is the same as those for the other three methods.

5 Conclusions

In this paper, we have proposed an effective spectral clustering based outlier detection method that can detect both global and local outliers. Traditional spectral clustering-based outlier detection algorithms have a quadratic running time complexity with data sizes and are very time- and space-consuming for modern large datasets. To partially circumvent this problem, we apply the spectral clustering process upon N mini-datasets, each consisting a data point and its k-nearest neighbors. To identify potential outliers, our algorithm first locates those data points in the eigen space whose second smallest eigenvalue appears least frequently as outlier candidates that deviate from the main patterns. Candidate outliers are then ranked based on the notion of distance-based outlier scores that are assigned to each data point. To demonstrate the utility of our proposed outlier detection mechanism, we have performed a detailed comparison of its performance with state-of-the-art distance-based and density-based outlier detection methods. Experimental results show the ability of our algorithm to rank the best candidates for being an outlier with high recall. Our study also manifests that, in reality, it is usually difficult to detect all the outliers that fit user's intuitions. Thus, it is more meaningful to incorporate our proposed outlier detection method as a component into current outlier detection framework. Finally, the size of the nearest neighborhood parameter, k, is very critical to the behavior of the proposed method. Though a qualitative analysis for how to choose it is given in the above, a more quantitative determination for it should be the focus of our continuing effort in the future.

Acknowledgment. The authors would like to thank the Chinese National Science Foundation for its valuable support of this work under award 61473220 and all the anonymous reviewers for their valuable comments.

References

1. Hawkins, D.M.: Identification of Outliers, Monographs on Applied Probability and Statistics. Chapman and Hall, London (1980)

2. Eskin, E., Arnold, A., Prerau, M., Portnoy, L., Stolfo, S.: A geometric framework for unsupervised anomaly detection: detecting intrusions in unlabeled data. In: Data Mining for Security Applications (2002)
3. Lane, T., Brodley, C.E.: Temporal sequence learning and data reduction for anomaly detection. ACM Transactions on Information and System Security **2**(3), 295–331 (1999)
4. Sheng, B., Li, Q., Mao, W., Jin, W.: Outlier detection in sensor networks. In: Proceedings of ACM International Symposium on Mobile Ad Hoc Networking and Computing, pp. 219–228 (2007)
5. Hodge, V.J., Austin, J.: A Survey of Outlier Detection Methodologies. Artificial Intelligence Review **22**, 85–126 (2004)
6. Chandola, V., Banerjee, A., Kumar, V.: Anomaly Detection: A Survey. ACM Computing Surveys **41**(3), Article 15 (2009)
7. Knorr, E.M., Ng, R.T.: Algorithms for mining distance-based outliers in large datasets. In: Proceedings of the 24th VLDB Conference, New York, USA, pp. 392–403 (1998)
8. Breuning, M.M., Kriegel, H.P., Ng, R.T., Sander, J.: LOF: identifying density-based local outliers. In: Proceedings of the 2000 ACM SIGMOD International Conference on Management of Data, pp. 93–104 (2000)
9. Jiang, M.F., Tseng, S.S., Su, C.M.: Two-Phase Clustering Process for Outliers Detection. Pattern Recognition Letters **22**, 691–700 (2001)
10. Malik, J., Belongie, S., Leung, T., et al.: Contour and texture analysis for image segmentation. International Journal of Computer Vision **43**(1), 7–27 (2001)
11. Bach, F.R., Jordan, M.I.: Blind one-microphone speech separation: a spectral learning approach. In: Proceedings of NIPS 2004, Vancouver, BC, pp. 65–72 (2004)
12. Weiss, Y.: Segmentation using eigenvectors: a unified view. In: International Conference on Computer Vision, Corfu, pp. 975–982 (1999)
13. Ding, C., He, X., Zha, H., et al.: A min-max cut algorithm for graph partitioning and data clustering. In: Proceedings of International Conference on Data Mining, California, pp. 107–114 (2001)
14. Shi, J., Malik, J.: Normalized Cuts and Image Segmentation. IEEE Transactions on Pattern Analysis and Machine Intelligence **22**(8), 888–905 (2000)
15. Stoer, M., Wagner, F.: A simple min-cut algorithm. Journal of the ACM **44**(4), 585–591 (1997)
16. Hagen, L., Kahng, A.: New spectral methods for ratio cut partitioning and clustering. IEEE Trans. Computer-Aided Design **11**(9), 1074–1085 (1992)
17. Zelnik-Manor, L., Perona, P.: Self-tuning spectral clustering. In: Proceedings of NIPS 2004, Vancouver, BC, pp. 1601–1608 (2004)
18. Knorr, E.M., Ng, R.T., Tucakov, V.: Distance-based outliers: algorithms and applications. VLDB Journal: Very Large Databases **8**(3–4), 237–253 (2000)
19. Ramaswamy, S., Rastogi, R., Shim, K.: Efficient algorithms for mining outliers from large data sets. In: Proceedings of the ACM SIGMOD Conference, pp. 427–438 (2000)
20. Angiulli, F., Pizzuti, C.: Fast outlier detection in high dimensional spaces. In: Elomaa, T., Mannila, H., Toivonen, H. (eds.) PKDD 2002. LNCS (LNAI), vol. 2431, pp. 15–26. Springer, Heidelberg (2002)
21. UCI: The UCI KDD Archive. University of California, Irvine, CA. http://kdd.ics.uci.edu/
22. Aggarwal, C., Yu, P.: Outlier detection for high-dimensional data. In: Proceedings of SIGMOD 2001, Santa Barbara, CA, USA, pp. 37–46 (2001)

A Learning Framework to Improve Unsupervised Gene Network Inference

Turki Turki[1,2(✉)], William Bassett[2], and Jason T.L. Wang[2(✉)]

[1] Computer Science Department, King Abdulaziz University, P.O. Box 80221, Jeddah 21589, Saudi Arabia
tturki@kau.edu.sa

[2] New Jersey Institute of Technology Bioinformatics Program and Department of Computer Science, University Heights, Newark, NJ 07102, USA
{ttt2,wab2,wangj}@njit.edu

Abstract. Network inference through link prediction is an important data mining problem that finds many applications in computational social science and biomedicine. For example, by predicting links, i.e., regulatory relationships, between genes to infer gene regulatory networks (GRNs), computational biologists gain a better understanding of the functional elements and regulatory circuits in cells. Unsupervised methods have been widely used to infer GRNs; however, these methods often create missing and spurious links. In this paper, we propose a learning framework to improve the unsupervised methods. Given a network constructed by an unsupervised method, the proposed framework employs a graph sparsification technique for network sampling and principal component analysis for feature selection to obtain better quality training data, which guides three classifiers to predict and clean the links of the given network. The three classifiers include neural networks, random forests and support vector machines. Experimental results on several datasets demonstrate the good performance of the proposed learning framework and the classifiers used in the framework.

Keywords: Feature selection · Graph mining · Network analysis · Applications in biology and medicine

1 Introduction

Network inference through link prediction is a major research topic in computational social science [9,12,18] and biomedicine [1,6,8]. For example, computational biologists develop different methods for reconstructing gene regulatory networks (GRNs) using high throughput genomics data. Maetschke *et al.* [21] categorized the existing GRN reconstruction algorithms into three groups: unsupervised, supervised and semi-supervised. While supervised algorithms are capable of achieving the highest accuracy among all the network inference methods, these algorithms require a large number of positive and negative

P. Perner (Ed.): MLDM 2016, LNAI 9729, pp. 28–42, 2016.
DOI: 10.1007/978-3-319-41920-6_3

training examples, which are difficult to obtain in many organisms [21,24,29]. Unsupervised algorithms infer networks based solely on gene expression profiles and do not need any training example; however, the accuracy of the unsupervised algorithms is low [21]. In our previous work [24,29], we studied supervised and semi-supervised methods. Here we explore ways for improving the accuracy of unsupervised methods.

Specifically we propose a learning framework to clean the links of the GRNs inferred by unsupervised methods using time-series gene expression data. These methods include BANJO (Bayesian Network Inference with Java Objects) [36], TimeDelay-ARACNE (Algorithm for the Reconstruction of Accurate Cellular Networks) [37], tlCLR (Time-Lagged Context Likelihood of Relatedness) [10,20], DFG (Dynamic Factor Graphs) [16], BPDS (Boolean Polynomial Dynamical Systems) [30], MIDER [31], Jump3 [14], ScanBMA [35], and Inferelator [3]. BANJO models networks as a first-order Markov process; it searches through all possible networks, seeking the network with the best score. TimeDelay-ARACNE infers networks from time-series data using mutual information from information theory.

The tlCLR method also uses mutual information and depends on ordinary differential equations to model time-series data. DFG models experimental noise as a fitted Gaussian and then infers networks based on an assumed underlying, idealized gene expression pattern. Jump3 uses a non-parametric procedure based on decision trees to reconstruct GRNs. ScanBMA is a Bayesian inference method that incorporates external information to improve the accuracy of GRN inference. Inferelator uses ordinary differential equations that learn a dynamical model for each gene using time-series data. Recent extensions of Inferelator incorporate prior knowledge into the tool, and are resilient to noisy inputs.

The main drawback of the unsupervised methods is that they often create missing and spurious links [21]. The learning framework proposed here consists of several steps to clean the links. First, since an inferred network is sizable, we develop a graph sparsification technique to generate dual sample graphs, which represent significant portions of the network constructed by an unsupervised method. Graph sparsification is a technique for generating sample graphs from a large network, which speeds up the learning process from the network [2,17,23]. Next, we use principal component analysis (PCA) as a dimensionality reduction technique for feature learning. PCA is commonly used in the bioinformatics community to select important features from data [26,34]. Finally, we build three classifiers using the important features learned from the dual sample graphs and apply these classifiers to predicting and cleaning the links in the noisy network constructed by an unsupervised method. The three classifiers include neural networks, random forests and support vector machines. Like PCA, these three classifiers are commonly used in bioinformatics[32]. As a case study, we focus on Inferelator [3] in the paper and show how to use the proposed framework to predict and clean the links constructed by Inferelator, which is one of the most widely used unsupervised methods in the field.

The rest of this paper is organized as follows. Section 2 presents our learning framework, describing the techniques of graph sparsification, feature construction and principal component analysis, as well as the proposed link prediction and cleaning algorithm. Section 3 reports experimental results, comparing the three classifiers used in the framework. The results demonstrate the effectiveness of the framework, showing how it improves the accuracy of Inferelator. Section 4 concludes the paper.

2 The Learning Framework

2.1 Graph Sparsification

The input of the proposed learning framework is a weighted directed graph $G = (V, E)$ that represents the topological structure of (a subgraph of) the gene regulatory network (GRN) constructed by Inferelator based on a time series gene expression dataset. E is the set of edges or links, and V is the set of vertices or nodes in G, where each link represents a regulatory relationship and each node represents a gene. Each edge $e = (u, v) \in E$ is associated with a weight, denoted by $W(e)$, where $0 < W(e) \leq 1$.

Our graph sparsification method, named GeneProbe (reminiscent of LinkProbe [5] for social network analysis), takes as input the graph G and two genes of interest: an origin or regulator gene, and a destination or regulated gene. GeneProbe creates six sets of genes, described below, and produces as output an inference subgraph that contains all genes in the six sets and all edges in E that connect the genes in the six sets.

Two Sets of k-backbone Genes. These include one set of k-backbone hub genes and one set of k-backbone authority genes. The k-backbone hub genes include all genes whose weighted outgoing degree is greater than or equal to a user-specified positive real value $k_{hub} \in \mathbb{R}^+$. The weighted outgoing degree of a gene or node u is defined as the sum of edge weights for all outgoing edges of u. Likewise, the k-backbone authority genes include all genes whose weighted incoming degree is greater than or equal to a user-specified value $k_{authority} \in \mathbb{R}^+$. The weighted incoming degree of a node u is defined as the sum of edge weights for all incoming edges of u. (In the study presented here, $k_{hub} = 15$ and $k_{authority} = 10$.) Intuitively we select few "highly social" individuals who would represent "social hubs/authorities" for inference across geographical regions. The genes most likely to be regulators (with the largest weighted outgoing degrees) are selected as the "hubs" of the network G for inclusion in the inference subgraph. Furthermore, the genes most likely to be regulated genes (with the largest weighted incoming degrees) are selected as the "authorities" of the network G for inclusion in the inference subgraph.

Formally, let $W\text{-}out(u)$ ($W\text{-}in(u)$, respectively) denote the weighted outgoing (incoming, respectively) degree of node u. Then

$$W\text{-}out(u) = \sum_{e \in E\text{-}out(u)} W(e) \tag{1}$$

$$W\text{-}in(u) = \sum_{e \in E\text{-}in(u)} W(e) \tag{2}$$

where $W(e)$ is the weight of edge e, and $E\text{-}out(u)$ ($E\text{-}in(u)$, respectively) denotes the set of edges leaving (entering, respectively) u. GeneProbe retrieves all genes $u \in G$ where $W\text{-}out(u) \geq k_{hub}$ and $W\text{-}in(u) \geq k_{authority}$.

Two Sets of d-local Genes. These include one set of d-local genes for the origin and one set of d-local genes for the destination. The d-local genes are the genes adjacent to each of the two genes of interest, i.e., the origin and destination, with incident edge weights greater than or equal to a user-specified positive real value $d \in \mathbb{R}^+$. (In the study presented here, $d = 0.95$.) Intuitively, the d-local genes represent the genes most likely to be regulated by and most likely to regulate the two genes of interest.

Two Sets of Random Walk Metropolis Genes. These include one set of random walk metropolis genes for the origin and one set of random walk metropolis genes for the destination. The random walk metropolis (RWM) genes provide a stochastic path from the genes of interest back to a k-backbone gene (if possible). The RWM does not differentiate between k-backbone hub and k-backbone authority genes. All of the genes encountered along the RWM path are added to the inference subgraph. For the origin or regulator gene, the random walk is a walk along outgoing edges towards the k-backbone, whereas the random walk for the destination or regulated gene is a backtrack to the k-backbone along incoming edges. Each step along the random walk metropolis is selected based on a randomized chance until a k-backbone gene is reached (or a maximum number of tries is exceeded).

The randomized chance at each step along the random walk for the regulator gene (i.e., origin) can be characterized as follows. Given a current gene u, we select a random edge from the list of outgoing edges of gene u. Let w represent the gene at the end of the randomly selected outgoing edge. The random walk will proceed from gene u to gene w if a randomly selected number between 0 and 1 is less than or equal to the minimum of 1 and the weighted outgoing degree of w divided by the weighted outgoing degree of u. That is, w is accepted as the next state with the probability of less than or equal to an acceptance rate α_{out}. Otherwise, another random outgoing edge of gene u is selected and similar calculations are performed. This move can be formalized in Equation (3). $P(u \rightarrow w)$ is the probability that a random walk proceeds from u to w where

$$\alpha_{out} = P(u \rightarrow w) = min\left\{1, \frac{W\text{-}out(w)}{W\text{-}out(u)}\right\} \tag{3}$$

This process is repeated until a maximum number of tries is reached (or a k-backbone gene is reached). Note that given enough chances in a connected gene regulatory network, the random walks will always reach a k-backbone gene. It logically follows that a setting that includes few k-backbone genes will likely generate many RWM genes and vice versa.

The randomized chance at each step along the random walk for the regulated gene (i.e., destination) can be characterized as follows. Given a current gene v, we select a random edge from the list of incoming edges of gene v. Let w represent the gene at the end of the randomly selected incoming edge. The random walk will backtrack from gene v to gene w if a randomly selected number between 0 and 1 is less than or equal to the minimum of 1 and the weighted incoming degree of w divided by the weighted incoming degree of v. That is, w is accepted as the next state with the probability of less than or equal to an acceptance rate α_{in}. Otherwise, another random incoming edge of gene v is selected and similar calculations are performed. This move is formalized in Equation (4).

$$\alpha_{in} = P(w \leftarrow v) = min\left\{1, \frac{W\text{-}in(w)}{W\text{-}in(v)}\right\} \tag{4}$$

where $P(w \leftarrow v)$ is the probability that a random walk moves backward from v to w. This process is repeated until a maximum number of tries is reached (or a k-backbone gene is reached). Figure 1 illustrates an inference subgraph.

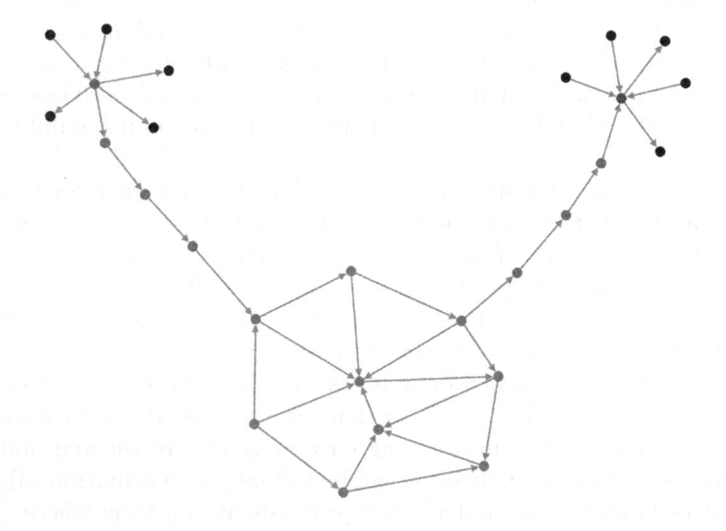

Fig. 1. Example of an inference subgraph containing an origin (green), a destination (purple), d-local genes (black), k-backbone genes (blue) and random walk metropolis genes (red).

2.2 Feature Selection

An inference subgraph may still have missing and spurious links. We select a more reliable sample from the inference subgraph where the weight of each edge in the sample is greater than or equal to 0.5. For each pair of genes u, v in the sample graph, we create a feature vector B by concatenating the gene expression profiles of u and v, as in [7,29]. That is,

$$B = [u^1, u^2, \dots, u^p, v^1, v^2, \dots, v^p] \tag{5}$$

where $u^1, u^2, \dots, u^p$ are the gene expression values of u, and $v^1, v^2, \dots, v^p$ are the gene expression values of v. Each gene expression value is a feature.

We employ principal component analysis (PCA) to reduce the dimensionality of the feature vectors of a sample graph [28]. Specifically, we combine the feature vectors into a $2p \times N$ matrix X where $2p$ is the total number of features and N is the number of links in the sample graph. Let the rank of the matrix X be r where the rank represents the maximum number of uncorrelated column vectors in X [33]. We represent X through singular value decomposition (SVD) as

$$X = U \cdot S \cdot V^T \tag{6}$$

Both U and V are orthogonal matrices. Each column of U is one of the eigenvectors of the covariance matrix $X \cdot X^T$ where X^T is the transpose of X. Each column of V is one of the eigenvectors of the matrix $X^T \cdot X$. The $r \times r$ matrix S contains eigenvalues of X on the diagonal line of S.

In our case, each column vector of the matrix X represents a (present or missing) link of the sample graph. That is, each link of the sample graph is a vector in the $2p$-dimensional Euclidean space. The dot product between two column vectors reflects the extent to which the two corresponding links share similar feature occurrences. Thus, we can use the dot product to get pairwise link distances. Let M contain pairwise link distances, i.e., M_{ij} is the dot product distance between link L_i and link L_j. Then M can be derived by:

$$M = X^T \cdot X \tag{7}$$

which can be generalized as:

$$\begin{aligned} M &= (U \cdot S \cdot V^T)^T \cdot (U \cdot S \cdot V^T) && (8)\\ &= (V \cdot S \cdot U^T) \cdot (U \cdot S \cdot V^T) && (9)\\ &= V \cdot S^2 \cdot V^T && (10)\\ &= (V \cdot S) \cdot (V \cdot S)^T && (11) \end{aligned}$$

The new representation of the matrix M shows that the pairwise link comparison matrix M can be obtained through the dot product of $(V \cdot S)$ and $(V \cdot S)^T$. That is, the ith row of the $N \times r$ matrix $(V \cdot S)$ is an r-dimensional vector representing the ith link in the sample graph. This result indicates that after performing projection transformation with respect to the matrix V, we can keep pairwise distances between link-vectors as in the original setting.

SVD reduces the dimensionality (from $2p$ to r where $r < min(2p, N)$) of a link-vector. However, the reduced r-dimensional link-vector might still contain redundant dimensions. In practice, the dimensionality of these link-vectors could be further reduced without losing characteristics of the link-vectors in the original $2p$-dimensional Euclidean space. Specifically, based on the optimal solution of the squared-error criterion of PCA [28], an r-dimensional vector could be projected onto a k-dimensional subspace, $k < r$, spanned by the eigenvectors corresponding to the k largest eigenvalues of the covariance matrix $X \cdot X^T$. As a result, we can obtain X_k, which is an approximation of the original matrix X, by keeping the k largest eigenvalues of the covariance matrix $X \cdot X^T$ and replacing the remaining eigenvalues with zeros. Then, Equation (6) can be rewritten as

$$X_k = U_k \cdot S_k \cdot V_k^T \tag{12}$$

Therefore, the $N \times k$ matrix $(V_k \cdot S_k)$ replaces the matrix $(V \cdot S)$ in Equation (11), where the ith row of the matrix $(V_k \cdot S_k)$ is a k-dimensional vector that represents the ith link in the sample graph. (In the study presented here, $k = 10$.)

2.3 The Link Prediction Algorithm

After explaining the concepts of graph sparsification and feature selection, we now describe how the proposed learning framework (i.e., link prediction algorithm) works. The main assumption here is that the network G constructed by Inferelator is not accurate, and there are many missing and spurious links in G. A missing link or edge e_m refers to a regulatory relationship that exists in the ground truth but is not inferred by Inferelator, and hence $e_m \notin G$. A spurious link e_s refers to a regulatory relationship that does not exist in the ground truth, but is inferred by Inferelator, and hence $e_s \in G$. The goal here is for our link prediction algorithm to detect these missing and spurious links, so as to clean them. To achieve this goal, the algorithm predicts whether there is a link between two nodes and uses the predicted outcome to replace the result obtained from Inferelator if the predicted outcome differs from Inferelator's result.

Let $G = (V, E)$ be the gene regulatory network (GRN) constructed by Inferelator based on a time series gene expression dataset. Our algorithm first creates two subgraphs $G_+ = (V_+, E_+)$ and $G_- = (V_-, E_-)$ where V_+ (V_-, respectively) contains incident nodes of the edges in E_+ (E_-, respectively), the weight of each edge in E_+ (E_-, respectively) is greater than or equal to (less than, respectively) the median of the weights of the edges in E, $E_+ \cap E_- = \emptyset$ and $E_+ \cup E_- = E$. Thus, the edges in G_+ are likely to be positive instances and the edges in G_- are likely to be negative instances. Note, however, that in practice these two subgraphs G_+ and G_- have low quality data, i.e., they contain many missing and spurious links.

Our link-prediction algorithm consists of five steps.

Step 1: Suppose the algorithm aims to predict whether there is a link from node/gene u to node/gene v. There are two cases to consider. In case 1, the gene pair (u, v) is in G_+. Then the algorithm creates an inference subgraph $I_+ \subseteq G_+$ by invoking GeneProbe and using G_+, the origin u and the destination v as input. In addition, the algorithm randomly selects a pair of genes x, y in G_-, and creates an inference subgraph $I_- \subseteq G_-$ by invoking GeneProbe and using G_-, the origin x and the destination y as input. In case 2, the gene pair (u, v) is in G_-. Then the algorithm creates an inference subgraph $I_- \subseteq G_-$ by invoking GeneProbe and using G_-, the origin u and the destination v as input. In addition, the algorithm randomly selects a pair of genes x, y in G_+, and creates an inference subgraph $I_+ \subseteq G_+$ by invoking GeneProbe and using G_+, the origin x and the destination y as input. Without loss of generality, we assume that case 1 holds and will use case 1 to describe the following steps. Thus, the algorithm creates dual graph sparsifications I_+ and I_- in step 1.

Step 2: Create a sample graph $I_+^{'} \subseteq I_+$ where $I_+^{'}$ does not contain the testing gene pair (u, v), and the weight of each edge in $I_+^{'}$ is greater than or equal to 0.5. We consider the edges in $I_+^{'}$ to have higher quality and are more likely to be positive instances. Suppose there are K edges in $I_+^{'}$. We then randomly select K edges from I_- and use the randomly selected edges to form a sample graph $I_-^{'} \subseteq I_-$. In training the three classifiers including neural networks, random forests and support vector machines, we will use the edges in $I_+^{'}$ as positive training examples, and use the edges in $I_-^{'}$ as negative training examples. The dual sample graphs $I_+^{'}$ and $I_-^{'}$ together form the training dataset.

Step 3: Construct a feature vector for the testing gene pair (u, v) by concatenating the gene expression values of u and v, as shown in Equation (5). Also, construct a feature vector for each gene pair (p, q) in the training dataset by concatenating the gene expression values of p and q.

Step 4: Reduce the dimensionality of the feature vectors constructed in step 3 using principal component analysis (PCA), as described in Section 2.2.

Step 5: Use the training examples (reduced feature vectors obtained from step 4) to train three classifiers including neural networks, random forests and support vector machines. Use the trained models to predict whether there is a link from gene u to gene v.

3 Experiments and Results

We conducted a series of experiments to evaluate the performance of the proposed learning framework and the three classifiers used in the framework. Below, we describe the datasets and experimental methodology used in our study, and then present the experimental results.

3.1 Datasets

We adopted the five time-series gene expression datasets available in the DREAM4 100-gene *in silico* network inference challenge [10,22,25,27]. Each dataset contains 10 times series, where each time series has 21 time points, for 100 genes. Each gene has $(10 \times 21) = 210$ gene expression values. Each link consists of two genes, and hence is represented by a 420-dimensional feature vector; cf. Equation (5). Through principal component analysis, each reduced feature vector has only 10 dimensions.

Each time-series dataset is associated with a gold standard file, where the gold standard represents the ground truth of the network structure for the time-series data. Each link in the gold standard represents a true regulatory relationship between two genes. For a given time-series dataset, Inferelator [3] constructs a directed network, in which each link has a weight and represents an inferred regulatory relationship between two genes.

Table 1 presents details of the five networks, true and inferred, used in the experiments. The table shows the numbers of true present and missing links in each gold standard, and the numbers of inferred present and missing links in each network constructed by Inferelator. Each network contains 100 nodes or genes, which form 9,900 ordered gene pairs totally.

Table 1. Networks used in the experiments

	Net1	Net2	Net3	Net4	Net5
Directed	Yes	Yes	Yes	Yes	Yes
Nodes	100	100	100	100	100
True present links	176	249	195	211	193
True missing links	9,724	9,651	9,705	9,689	9,707
Inferred present links	6,232	6,066	6,186	5,930	6,180
Inferred missing links	3,668	3,834	3,714	3,970	3,720

For each network, we created four sets of testing data. Each testing dataset contains 50 randomly selected links from the gold standard. Among the 50 links, 25 are true present links and 25 are true missing links in the gold standard. The label (+1 or -1) of each selected link is known, where +1 represents a true present link and -1 represents a true missing link. These testing data were excluded from the training datasets used to train the classifiers studied in the paper. There were 20 testing datasets totally.

3.2 Experimental Methodology

We considered three classification algorithms, namely neural networks (NN), random forests (RF) and support vector machines (SVM). Software used in this work included: the neuralnet package in R [11], the random forest package in R [19], and the SVM program with the polynomial kernel of degree 2 in the LIBSVM package [4]. The principal component analysis (PCA) program was based on the prcomp function in R [13]. The graph sparsification method (GeneProbe) was implemented in C++. In addition, we used R to write some utility tools for performing the experiments.

The performance of each classification algorithm was evaluated as follows. We trained each classification algorithm as described in Section 2.3. For each link in a testing dataset, we used the trained model to predict its label. In evaluating our link prediction algorithm, we define a true positive to be a true present link that is predicted as a present link. A false positive is a true missing link that is predicted as a present link. A true negative is a true missing link that is predicted as a missing link. A false negative is a true present link that is predicted as a missing link. In evaluating Inferelator, we define a true positive to be a true present link that is an inferred present link. A false positive is a true missing link that is an inferred present link. A true negative is a true missing link that is an inferred missing link. A false negative is a true present link that is an inferred missing link. Let TP (FP, TN, FN, respectively) denote the number of true positives (false positives, true negatives, false negatives, respectively) for a testing dataset. We adopted the balanced error rate (BER) [29], defined as

$$BER = \frac{1}{2} \times \left(\frac{FN}{TP + FN} + \frac{FP}{FP + TN} \right) \tag{13}$$

We applied each classification algorithm to each testing dataset and recorded the BER for that testing dataset. The lower BER a classification algorithm has, the better performance that algorithm achieves. We define the *improvement rate*, denoted IR, on a testing dataset to be $(P^* - P) \times 100\%$ where P^* is the BER of Inferelator and P is the BER of a classification algorithm (NN, RF or SVM) for that dataset. Statistically significant performance differences were calculated using Wilcoxon signed rank tests [15]. As in [28], we considered p-values below 0.05 to be statistically significant.

3.3 Experimental Results

Table 2 shows the improvement rate (IR) each classification algorithm achieves on each of the 20 testing datasets. A positive (negative, respectively) IR for a classification algorithm indicates that the algorithm performs better (worse, respectively) than Inferelator. The larger the positive IR a classification algorithm has, the more improvement over Inferelator that algorithm achieves. For each testing dataset, the classification algorithm with the best performance, i.e., the largest positive IR, is shown in boldface.

It can be seen from Table 2 that SVM (support vector machines) outperforms Inferelator, NN (neural networks) and RF (random forests). SVM improves Inferelator on 16 testing datasets, and the improvement is statistically significant according to Wilcoxon signed rank tests ($p < 0.05$). NN improves Inferelator on 12 testing datasets; however, the improvement is not statistically significant according to Wilcoxon signed rank tests ($p > 0.05$).

Table 2. Improvement rates of three classification algorithms on twenty datasets

Dataset	NN	RF	SVM
Net1.test1	+11.1%	+0.90%	**+17.9%**
Net1.test2	**+17.0%**	+1.80%	+12.3%
Net1.test3	+7.10%	+4.90%	**+13.6%**
Net1.test4	+5.20%	+2.10%	**+9.20%**
Net2.test1	**+5.80%**	+1.80%	−1.60%
Net2.test2	**+14.8%**	−6.20%	+0.90%
Net2.test3	**+4.40%**	−12.1%	+1.20%
Net2.test4	−0.60%	+1.20%	**+5.20%**
Net3.test1	**+0.20%**	−9.40%	−1.10%
Net3.test2	+0.00%	**+8.00%**	+4.00%
Net3.test3	−8.00%	−6.00%	**+2.00%**
Net3.test4	−2.00%	−10.0%	**+8.00%**
Net4.test1	**+10.4%**	−4.80%	+3.00%
Net4.test2	−5.70%	−6.00%	**+4.50%**
Net4.test3	**+10.8%**	+2.50%	+5.20%
Net4.test4	−3.00%	**+8.20%**	+2.60%
Net5.test1	+1.20%	−5.00%	**+7.10%**
Net5.test2	−1.70%	−13.1%	**+2.90%**
Net5.test3	−7.00%	−13.3%	−3.50%
Net5.test4	−3.00%	−7.20%	−1.10%

We also carried out experiments to evaluate the performance of different SVM kernel functions, including the linear kernel (SVM_L), polynomial kernel of degree 2 (SVM_P2), polynomial kernel of degree 15 (SVM_P15), Gaussian kernel (SVM_G), and sigmoid kernel (SVM_S). Figure 2 shows the BER values, averaged over the 20 testing datasets, for the different kernel functions. It can be seen that SVM with the polynomial kernel of degree 2 (SVM_P2) used in this study performs the best.

Finally we conducted experiments to evaluate the effectiveness of the components of the proposed learning framework. There are two core components: graph sparsification (GeneProbe) and feature selection (PCA). Figure 3 compares the approach with graph sparsification (GS) only, the approach with feature selection (FS) only, and our proposed approach, which combines both graph sparsification (GS) and feature selection (FS). Each bar represents the average BER over the 20 testing datasets. The classifier used to generate the results was the SVM program with the polynomial kernel of degree 2. It can be seen from Figure 3 that the proposed approach combining GS and FS performs the best.

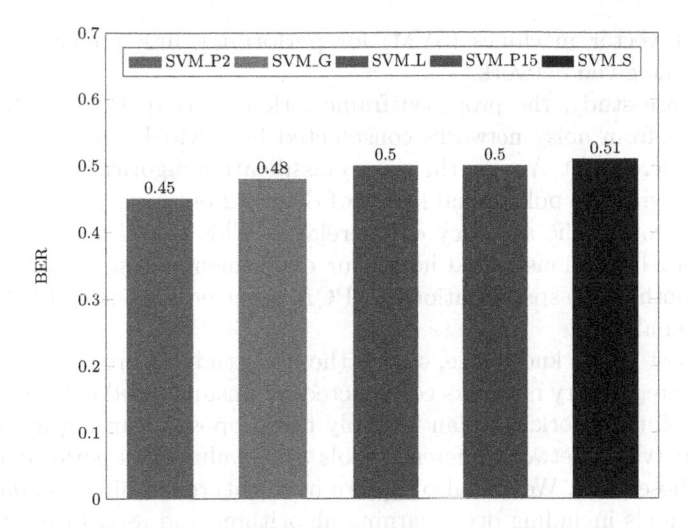

Fig. 2. Performance evaluation of different SVM kernel functions.

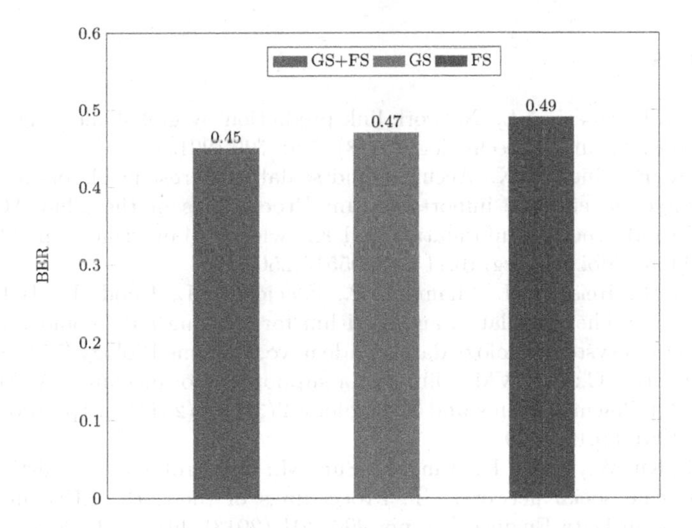

Fig. 3. Effectiveness of the components of the proposed learning framework.

4 Conclusion

Given gene regulatory networks constructed by unsupervised network inference methods, our goal is to predict and clean the links in the networks. To achieve this goal, we propose a learning framework, which employs (i) a graph sparsification technique (GeneProbe) for generating inference subgraphs from a given network, and (ii) principal component analysis (PCA) for selecting significant features from high-dimensional feature vectors. The selected feature values are then used to train three classifiers including neural networks (NN), random forests (RF) and support vector machines (SVM) for performing link prediction and link cleaning in the given network.

In our case study, the proposed framework is able to learn better quality training data from noisy networks constructed by a widely used network inference tool (Inferelator). Among the three classification algorithms studied in the paper, SVM with the polynomial kernel of degree 2 outperforms NN and RF in terms of improving the accuracy of Inferelator. This kernel is the best among all SVM kernel functions tested here. Our experimental results also show that combining both graph sparsification and PCA is better than using PCA or graph sparsification alone.

To the best of our knowledge, ours is the first study to predict and clean the links in gene regulatory networks constructed by unsupervised network inference methods. In future work, we plan to apply the proposed learning framework to other unsupervised network inference tools and evaluate its performance when used with those tools. We will also explore new feature learning and data classification methods including deep learning algorithms and assess their feasibility for our framework.

References

1. Barzel, B., Barabási, A.L.: Network link prediction by global silencing of indirect correlations. Nature Biotechnology **31**(8), 720–725 (2013)
2. Bogdanov, P., Singh, A.K.: Accurate and scalable nearest neighbors in large networks based on effective importance. In: Proceedings of the 22nd ACM International Conference on Information and Knowledge Management, pp. 1009–1018 (2013). http://doi.acm.org/10.1145/2505515.2505522
3. Bonneau, R., Reiss, D.J., Shannon, P., Facciotti, M., Hood, L., Baliga, N.S., Thorsson, V.: The Inferelator: an algorithm for learning parsimonious regulatory networks from systems-biology data sets de novo. Genome Biology **7**(5), R36 (2006)
4. Chang, C., Lin, C.: LIBSVM: a library for support vector machines. ACM Transactions on Intelligent Systems and Technology 2(3), 27 (2011). http://doi.acm.org/10.1145/1961189.1961199
5. Chen, H., Ku, W., Wang, H., Tang, L., Sun, M.: LinkProbe: probabilistic inference on large-scale social networks. In: Proceedings of the 29th IEEE International Conference on Data Engineering, pp. 290–301 (2013). http://dx.doi.org/10.1109/ICDE.2013.6544833
6. Clauset, A., Moore, C., Newman, M.E.: Hierarchical structure and the prediction of missing links in networks. Nature **453**(7191), 98–101 (2008)
7. De Smet, R., Marchal, K.: Advantages and limitations of current network inference methods. Nature **8**(10), 717–729 (2010)
8. Elloumi, M., Iliopoulos, C.S., Wang, J.T.L., Zomaya, A.Y.: Pattern Recognition in Computational Molecular Biology: Techniques and Approaches. Wiley (2015)
9. Getoor, L., Diehl, C.P.: Link mining: a survey. SIGKDD Explorations **7**(2), 3–12 (2005). http://doi.acm.org/10.1145/1117454.1117456
10. Greenfield, A., Madar, A., Ostrer, H., Bonneau, R.: DREAM4: combining genetic and dynamic information to identify biological networks and dynamical models. PLoS ONE **5**(10), e13397 (2010). http://dx.doi.org/10.1371%2Fjournal.pone.0013397

11. Günther, F., Fritsch, S.: Neuralnet: training of neural networks. Nature **2**(1), 30–38 (2010)
12. Hasan, M., Zaki, M.: A survey of link prediction in social networks. In: Aggarwal, C.C. (ed.) Social Network Data Analytics, pp. 243–275. Springer, US (2011). http://dx.doi.org/10.1007/978-1-4419-8462-3_9
13. Hothorn, T., Everitt, B.S.: A Handbook of Statistical Analyses Using R. CRC Press (2014)
14. Huynh-Thu, V.A., Sanguinetti, G.: Combining tree-based and dynamical systems for the inference of gene regulatory networks. Bioinformatics **31**(10), 1614–1622 (2015). http://dx.doi.org/10.1093/bioinformatics/btu863
15. Kanji, G.K.: 100 Statistical Tests. Sage (2006)
16. Krouk, G., Mirowski, P., LeCun, Y., Shasha, D., Coruzzi, G.: Predictive network modeling of the high-resolution dynamic plant transcriptome in response to nitrate. Genome Biology **11**(12), R123 (2010). http://dx.doi.org/10.1186/gb-2010-11-12-r123
17. Leskovec, J., Faloutsos, C.: Sampling from large graphs. In: Proceedings of the 12th ACM SIGKDD International Conference on Knowledge Discovery and Data Mining, NY, USA, pp. 631–636 (2006). http://doi.acm.org/10.1145/1150402.1150479
18. Leskovec, J., Huttenlocher, D., Kleinberg, J.: Predicting positive and negative links in online social networks. In: Proceedings of the 19th International Conference on World Wide Web, NY, USA, pp. 641–650 (2010). http://doi.acm.org/10.1145/1772690.1772756
19. Liaw, A., Wiener, M.: Classification and regression by randomForest. R News **2**(3), 18–22 (2002). http://CRAN.R-project.org/doc/Rnews/
20. Madar, A., Greenfield, A., Vanden-Eijnden, E., Bonneau, R.: DREAM3: network inference using dynamic context likelihood of relatedness and the Inferelator. PLoS ONE **5**(3), e9803 (2010). http://dx.doi.org/10.1371%2Fjournal.pone.0009803
21. Maetschke, S., Madhamshettiwar, P.B., Davis, M.J., Ragan, M.A.: Supervised, semi-supervised and unsupervised inference of gene regulatory networks. Briefings in Bioinformatics **15**(2), 195–211 (2014). http://dx.doi.org/10.1093/bib/bbt034
22. Marbach, D., Schaffter, T., Mattiussi, C., Floreano, D.: Generating realistic in silico gene networks for performance assessment of reverse engineering methods. Nature **16**(2), 229–239 (2009)
23. Mathioudakis, M., Bonchi, F., Castillo, C., Gionis, A., Ukkonen, A.: Sparsification of influence networks. In: Proceedings of the 17th ACM SIGKDD International Conference on Knowledge Discovery and Data Mining, NY, USA, pp. 529–537 (2011). http://doi.acm.org/10.1145/2020408.2020492
24. Patel, N., Wang, J.T.L.: Semi-supervised prediction of gene regulatory networks using machine learning algorithms. Journal of Biosciences 40(4), 731–740 (2015). http://dx.doi.org/10.1007/s12038-015-9558-9
25. Prill, R.J., Marbach, D., Saez-Rodriguez, J., Sorger, P.K., Alexopoulos, L.G., Xue, X., Clarke, N.D., Altan-Bonnet, G., Stolovitzky, G.: Towards a rigorous assessment of systems biology models: the DREAM3 challenges. PLoS ONE 5(2), e9202 (2010). http://dx.doi.org/10.1371%2Fjournal.pone.0009202
26. Ringnér, M.: What is principal component analysis? Nature **26**(3), 303–304 (2008)
27. Schaffter, T., Marbach, D., Floreano, D.: GeneNetWeaver: in silico benchmark generation and performance profiling of network inference methods. Bioinformatics 27(16), 2263–2270 (2011). http://dx.doi.org/10.1093/bioinformatics/btr373

28. Turki, T., Roshan, U.: Weighted maximum variance dimensionality reduction. In: Martínez-Trinidad, J.F., Carrasco-Ochoa, J.A., Olvera-Lopez, J.A., Salas-Rodríguez, J., Suen, C.Y. (eds.) MCPR 2014. LNCS, vol. 8495, pp. 11–20. Springer, Heidelberg (2014)
29. Turki, T., Wang, J.T.L.: A new approach to link prediction in gene regulatory networks. In: Jackowski, K., et al. (eds.) IDEAL 2015. LNCS, vol. 9375, pp. 404–415. Springer, Heidelberg (2015)
30. Vera-Licona, P., Jarrah, A.S., García-Puente, L.D., McGee, J., Laubenbacher, R.C.: An algebra-based method for inferring gene regulatory networks. BMC Systems Biology 8, 37 (2014). http://dx.doi.org/10.1186/1752-0509-8-37
31. Villaverde, A.F., Ross, J., Morn, F., Banga, J.R.: MIDER: network inference with mutual information distance and entropy reduction. PLoS ONE **9**(5), e96732 (2014). http://dx.doi.org/10.1371%2Fjournal.pone.0096732
32. Wang, J.T.L., Zaki, M.J., Toivonen, H.T.T., Shasha, D.: Data Mining in Bioinformatics. Springer (2005)
33. Wang, J.T.L., Liu, J., Wang, J.: XML clustering and retrieval through principal component analysis. International Journal on Artificial Intelligence Tools **14**(4), 683 (2005). http://dx.doi.org/10.1142/S0218213005002326
34. Yeung, K.Y., Ruzzo, W.L.: Principal component analysis for clustering gene expression data. Bioinformatics **17**(9), 763–774 (2001). http://dx.doi.org/10.1093/bioinformatics/17.9.763
35. Young, W., Raftery, A.E., Yeung, K.Y.: Fast Bayesian inference for gene regulatory networks using ScanBMA. BMC Systems Biology **8**, 47 (2014). http://dx.doi.org/10.1186/1752-0509-8-47
36. Yu, J., Smith, V.A., Wang, P.P., Hartemink, A.J., Jarvis, E.D.: Advances to Bayesian network inference for generating causal networks from observational biological data. Bioinformatics **20**(18), 3594–3603 (2004). http://bioinformatics.oxfordjournals.org/content/20/18/3594.abstract
37. Zoppoli, P., Morganella, S., Ceccarelli, M.: TimeDelay-ARACNE: reverse engineering of gene networks from time-course data by an information theoretic approach. BMC Bioinformatics **11**, 154 (2010). http://dx.doi.org/10.1186/1471-2105-11-154

A Closed Frequent Subgraph Mining Algorithm in Unique Edge Label Graphs

Nour El Islem Karabadji[1,2], Sabeur Aridhi[3(✉)], and Hassina Seridi[2]

[1] Preparatory School of Science and Technology, P.O. Box 218, 23000 Annaba, Algeria

[2] LabGED Laboratory, Badji Mokhtar University, P.O. Box 12, 23000 Annaba, Algeria

{karabadji,seridi}@labged.net

[3] Aalto University, School of Science, P.O. Box 12200, 00076 Espoo, FI, Finland

sabeur.aridhi@aalto.fi

Abstract. Problems such as closed frequent subset mining, itemset mining, and connected tree mining can be solved in a polynomial delay. However, the problem of mining closed frequent connected subgraphs is a problem that requires an exponential time. In this paper, we present ECE-CloseSG, an algorithm for finding closed frequent unique edge label subgraphs. ECE-CloseSG uses a search space pruning and applies the strong accessibility property that allows to ignore not interesting subgraphs. In this work, graph and subgraph isomorphism problems are reduced to set inclusion and set equivalence relations.

Keywords: Unique edge labels · Closed frequent subgraph · Graph/subgraph isomorphism · Set inclusion/equivalence

1 Introduction

The problem of interesting pattern mining is a main task in pattern mining and has several application domains such as social network analysis [6], bioinformatics [1,13] and Web mining [14]. It consists in finding patterns that satisfy a set of constraints. The frequency is one of the most used constraints [4,5,7,10]. A frequent pattern in a given collection/database $\mathcal{D}$ is a pattern that occurs at least in δ structures of the database where δ is a given support threshold. In general, the size of the set of frequent patterns is too large. This is due to the downward closure property (all generalizations or sub-patterns of a frequent pattern must be frequent). Thus, the enumeration and the analysis of such a big set of frequent patterns is a challenging problem. To overcome this issue, many research works have focused on special types of frequent subgraphs that allow to restore the set of all frequent subgraphs. These particular patterns are closed and maximal frequent subgraphs. In this context, many efficient algorithms for mining closed and maximal patterns have been proposed such as CloseGraph [18], SPIN [9], Margin [16] and ISG [15]. While mining closed and maximal patterns like itemsets,

P. Perner (Ed.): MLDM 2016, LNAI 9729, pp. 43–57, 2016.
DOI: 10.1007/978-3-319-41920-6_4

keys and trees needs a polynomial delay [17], [11], and [12], the situation gets more complicated when complex patterns such as graphs are considered. The task of mining graph patterns is called frequent subgraph mining (FSM) and includes two main phases: (1) generation of candidates and (2) frequency test. Frequency test has a variable cost according to pattern complexity. For example, itemsets use set inclusion relation to test frequency, whereas trees use subtree isomorphism. In contrast to these cases that need only a polynomial delay, an exponential delay is required for graph databases (using subgraph isomorphism, which is an *NP-complete* problem).

Recently, there has been an increased interest to identify a practically relevant tractable graph class. Well-behaved outer-planar and unique edge label graphs have been encoded using itemset codes that preserve subgraph isomorphism orders (i.e., comparability). These codes are Block and Bridge Preserving (BBP) [8] and Edges and Converses Edges triplets (ECE) [15] that allow solving the isomorphism problem in a polynomial time.

In this paper, we propose ECE-CloseSG (**E**dges and **C**onverses **E**dges **R**epresentation for **Close**d **S**ub**G**raphs), a novel algorithm that investigates the search space strong accessibility to reduce the number of generated subgraphs and to find closed frequent unique edge label connected subgraphs. Moreover, ECE-CloseSG uses a set-theoretical representation in the candidate generation step and ECE encoding in the frequency closeness computation step. The strong accessibility property allows to jump from a closed subgraph to its immediate closed successors, which allows a significant reduction on the visited candidates. We notice that the set-theoretical representation facilitates the candidate generation task, and the ECE encoding allows graph (respectively subgraph) isomorphism problem to be reduced to set equivalence (respectively set inclusion) where the problem can be solved in polynomial time in the worst case. We compared the performance of ECE-CloseSG to a naive algorithm that tries to visit all the frequent subgraphs and outputs only the closed ones.

This paper is organized as follows. The next section presents the used notations. In Section 3, the proposed approach for closed frequent subgraphs in unique edge label graphs is presented. In Section 4, comparative results on synthetic datasets are reported. In Section 5, a survey of closed and maximal frequent subgraph mining algorithms is presented.

2 Preliminary

In this section, we present some basic notations and definitions.

2.1 Notations

Definition 1. ***(Graph)*** *A graph $G = (V, E)$ is a collection of objects where V is the set of vertices, and $E \subseteq V \times V$ is the set of edges.*

We define a labelled graph by adding a set of labels Ψ and a function $\Gamma : V \cup E \rightarrow \Psi$ that assigns for each vertex and for each edge a label of Ψ.

Definition 2. ***(Labelled graph)*** *A graph $G = (V, E, \Psi, \Gamma)$ is a labelled graph where V is a set of vertices, $E \subseteq V \times V$ is a set of edges, Ψ is a set of labels and Γ is a labelling function for all edges and nodes.*

An undirected graph is a graph in which edges have no orientation. An edge $e = (u, v)$ has two end-points u and v. Two edges are adjacent if they share the same end points. The degree of a vertex v (denoted by $deg_G(v)$) is the number of its incident edges.

Definition 3. ***(Unique edge label graph)*** *A graph $G = (V, E, \Psi, \Gamma)$ is a unique edge label graph if and only if each edge label occurs at most once.*

Example 1. Figure 1 shows some examples of unique edge label graphs. The graph G_4 is not a unique edge label graph because edges (a, b) and (a, c) share the same label.

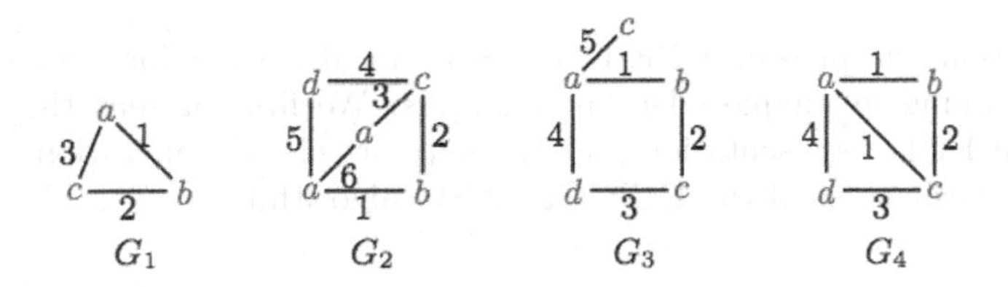

Fig. 1. Example of undirected labelled connected graphs.

Definition 4. ***(Isomorphism)*** *A graph isomorphism from $G_1 = (V_1, E_1)$ to $G_2 = (V_2, E_2)$ is a bijective function φ from E_1 to E_2 such that: $\forall\ u, v \in V_1 \iff (\varphi(u), \varphi(v)) \in E_2$.*

Definition 5. ***(Subgraph isomorphism)*** *$G_1 = (V_1, E_1)$ is a subgraph isomorphism into $G_2 = (V_2, E_2)$ if there is a bijection function φ from G_1 to G'_2 where G'_2 is a subgraph ($\subseteq_{sgi}$) of G_2.*

2.2 Closure Operator and Set System

Let $\mathcal{P}$ be a poset system (a set of subsets). A mapping $\sigma : \mathcal{P} \rightarrow \mathcal{P}$ is called a closure operator if it satisfies for all $X, Y \in \mathcal{P}$ that

- $X \subseteq \sigma(X)$(extensivity);
- $X \subseteq Y \rightarrow \sigma(X) \subseteq \sigma(Y)$ (monotonicity);
- $\sigma(X) = \sigma(\sigma(X))$(idempotence).

In data mining, a different closeness definition is used. The support-closed pattern of a dataset is defined as follows. Given a transactional database, a pattern is closed if there is no specific pattern that has the same support of the original pattern.

In the following, we give a definition of a set system and we present some of its properties.

Definition 6. *(Set system). A set system is an ordered pair* $(R, \mathcal{P})$*, where* R *is the ground finite set and* $\mathcal{P}$ *is a non-empty subset of the power set of* R*,* $\mathcal{P} \subseteq 2^R$*. A non-empty set system* $(R, \mathcal{P})$ *is:*

- **a closed system** if $R \in \mathcal{P}$, and $X, Y \in \mathcal{P}$ implies $X \cap Y \in \mathcal{P}$;
- **accessible** if for all $X \in \mathcal{P} \backslash \{\emptyset\}$ there is an $e \in X$ such that $X \backslash \{e\} \in \mathcal{P}$;
- **strongly accessible** if for every $X, Y \in \mathcal{P}$ satisfying $X \subset Y$, there is an $e \in Y \backslash X$ such that $X \cup \{e\} \in \mathcal{P}$;
- **independent** if $Y \in \mathcal{P}$ and $\forall X \subseteq Y \rightarrow X \in \mathcal{P}$;
- **a confluent** if $\forall I, X, Y \in R$ with $\emptyset \neq I \subseteq X$ and $I \subseteq Y$ it holds that $X \cup Y \in \mathcal{P}$.

3 Closed Frequent Subgraph Mining in Unique Edge Label Graphs

In this section, we present ECE-CloseSG, an algorithm for closed frequent subgraph mining in unique edge label graphs. We first present the subgraph system, the ECE representation and the support closure operation. Then, we present the basic steps of the ECE-CloseSG algorithm.

3.1 Connected Subgraph System

The main objective of the proposed approach is to restrict the search space. To this end, the strong accessibility of the subgraph system, and non-redundancy of the data $\mathcal{D}$ are required. A subgraph system considered in this work is the family of edge sets that induce connected subgraphs of $G_i \in \mathcal{D}$. A subgraph system $\mathcal{P}_i$ of a given graph $G_i(V_i, E_i) \in \mathcal{D}$ satisfies $\mathcal{P}_i \subseteq \mathcal{P}(E_i)$, where $\mathcal{P}(E_i)$ denotes the power set of the set of edges E_i. Here, we study the strong accessibility of a connected subgraph system $\mathcal{P}_i$. A subgraph system $\mathcal{P}_i$ is strongly accessible means that every $X \in \mathcal{P}_i$ can be reached from all $Y \subset X$ with $Y \in \mathcal{P}_i$ via extension with single edge inside $\mathcal{P}_i$.

Theorem 1. *A unique edge label subgraph system* $\mathcal{P}_i$ *is strongly accessible.*

Proof. Let X be a connected subgraph where $X \in \mathcal{P}_i$, if X is a graph that contains a cycle then we just drop an edge e from the cycle and the result $X \backslash e \in \mathcal{P}_i$. Otherwise, if X does not contain a cycle (a tree), we just drop one of its leaves. Given two graph $X_1, X_2 \in \mathcal{P}_i$ with $X_2 \subset X_1$. Assume that there is no edge $e \in X_1 \backslash X_2$ such as $X_2 \cup e \in \mathcal{P}_i$. So, X_2 and $X_1 \backslash X_2$ are two disconnected components in $G[X_1]$ which contradicting the choice of X_2.

This property improves the enumeration process because any closed frequent subgraph can be reached from anyone that has been already found by one edge augmentation [2].

Table 1. ECE 3-edge graph encoding

Graph	ECE Items	COMMONID	TYPEID
G_1	$(a, 1, a), (a, 2, a), (a, 3, a), (1, a, 2), (2, a, 3), (1, a, 3)$	(1,2,3)	((1,2,3),150)
G_2	$(a, 1, a), (a, 2, a), (a, 3, a), (1, a, 2), (2, a, 3), (1, a, 3)$	(1,2,3)	((1,2,3),200)
G_3	$(a, 1, a), (a, 2, a), (a, 3, a), (1, a, 2), (2, a, 3)$	(1,2,3)	((1,2,3),100)

3.2 Edges and Converse Edges (ECE) Representation

Graph representation has a significant influence on memory usage and execution time. In this work, ECE-CLOSESG uses two different graph representations, where the algorithm swings between: (1) a set-theoretical (vertices, and edges sets) to facilitate the augmentation and the connectivity tasks, and (2) an ECE codes to simplify the frequency tests.

The ECE encoding is a mapping of different parts of graphs to set of items. Every edge $e = (u, v)$ is represented by a 3-tuple $(et_u,\ et_e,\ et_v)$, where et_u, et_v are the labels of the two vertices connected by e; et_e is the label of e. Moreover, each 3-edge connected substructure is represented by two sign items (i.e., items that show the 3-edge connectivity and their structural form: a linear chain, a triangle, or a spike). Thus, to ensure unambiguous decoding of an ECE code to the right graph. Sign items can be presented by a first unique item that is assigned to each three connected edges noted the COMMONID item, and a second item noted TYPEID item to precise the type of the three edges code (i.e., triangle, spike and linear chain). Table 1 presents ECE codes of the three edge graphs (G_1, G_2, G_3) illustrated in Figure 2.

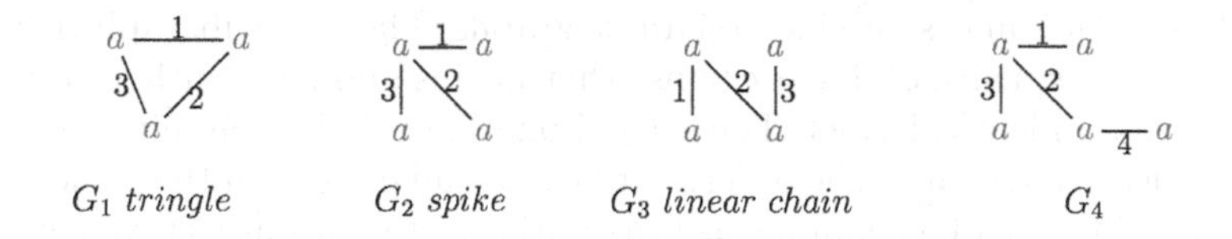

Fig. 2. Example of connected graphs.

Table 1 shows an identical ECE code of two different graphs G_1 and G_2 where the codes of G_1 and G_2 include the ECE code of G_3. However, this encoding allows preserving graphs' equivalence relation. Additionally, it is trivial to note that a graph ECE code is the union of the ECE codes of its 3-edge connected subgraphs.

As mentioned earlier, we propose to use ECE codes to apply subset relation that replaces the subgraph isomorphism test. This will reduce the frequency test complexity. Unfortunately, there is a failure case, and ECE encoding does not preserve incompatibilities and affects frequency test's accuracy. Therefore, two incomparable graphs can have two comparable ECE codes. This occurs in graphs having more than two linear chains' subgraphs. Figure 3 presents two connected incomparable graphs g_1(i.e., g), and g_2(i.g., g') and their 3-edge connected subgraphs (parts (a) $\mathcal{E}(g_1)$ and (b) $\mathcal{E}(g_2)$). The two graphs are not isomorphic, but g_2 contains all 3-edge subgraphs of g_1 (i.e., $\mathcal{E}(g_1) \subset \mathcal{E}(g_2)$). Thus, g_1 ECE code is included in g_2 ECE code. According to this ECE failure, g_1 is a subgraph of all graphs G_i that include g_2, but this is not the case when subgraph isomorphism is used. Assume that the database $\mathcal{D}$ contains two graphs G_1 and G_2. The two graphs contain g_1 and g_2 respectively. For a frequency threshold =2, g_1 is listed frequent, but it occurs only once at the first graph G_1.

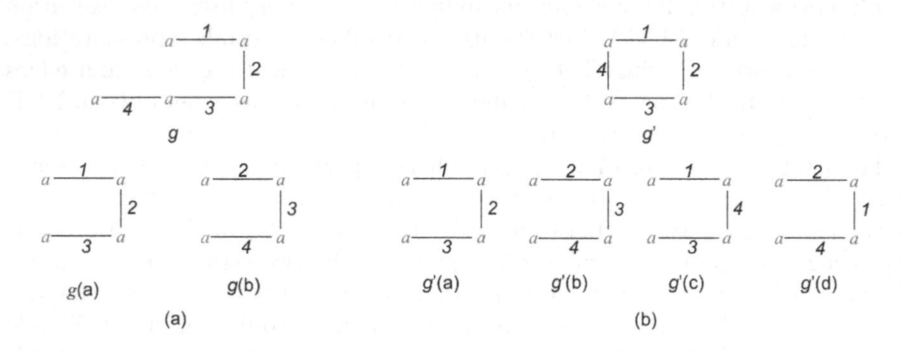

Fig. 3. Encoding failure example.

For two incomparable graphs g and g', we have an ECE encoding failure when subgraphs $\mathcal{E}(g')$ of g' contain all subgraphs $\mathcal{E}(g)$ of g. Specifically, there are two equal sized subgraphs $s \subseteq_{sgi} g$ and $s' \subseteq_{sgi} g'$ having the same edges and $\mathcal{E}(s) \subset \mathcal{E}(s')$, but s and s' are incomparable graphs. The first subgraph (i.e., s) is a simple path (sequence of linear chains) that begins and ends with two distinct vertices with an identical label. The second one (i.e., s') is a circuit that shares the same edges of s in the same order, but begins and ends with the same vertex. Closing the last edge of a chain toward the initial vertex produces two new linear chains' subgraphs. Therefore, to detect a failure with respect to two graphs g and g', we first seek couples of suspicious edges. Then, the suspected pairs of edges are tested. These edges are identified following characteristics: (1) two edges (e_1, and e_2) are not connected if $ECE(g)$ does not admit a converse edge that represents a connection between e_1 and e_2, and (2) in $ECE(g)$, there is no COMMONID item that includes the two edges. Additionally, any edge expansion requires adding: (a) one edge code item, (b) some edge converse codes (according to the degrees of the vertices), and (c) some corresponding sign items. For instance, the graph G_4 in Figure 2 is an augmentation from G_2 by adding the edge $e = (a, 4, a)$. The code of G_4 ($ECE(G_4)$) is constructed by appending $ECE(G_2$, the code of the edge e, one converse code and two sign items ($ECE(G_4)$=$ECE(G_2) \cup \{(a, 4, a), (2, a, 4), (1, 2, 4), ((1, 2, 4), 100), (2, 3, 4), ((2, 3, 4), 100)\}$).

Moreover, the support set of X with respect to $\mathcal{D}$, noted $\mathcal{D}[X]$, is a set of transactions containing X set, where the support-closure operator is defined by $\sigma(X) = \cap\mathcal{D}[X]$. According to this support-closure operator, the result $ECE(g)$ can be ambiguous, when it contains a 3-edge COMMONID item without the presence of its TYPEID item. For example, let $\mathcal{D}' = \{ECE(G_1), ECE(G_2), ECE(G_3)\}$, $\sigma((a,1,a)) = \{(a,1,a),(a,2,a),(a,3,a),(1,a,2),(2,a,3),(1,a,3),(1,2,3)\}$, the three item sets share the three edge triplets $(a,1,a)$, $(a,2,a)$, $(a,3,a)$, the converse edge triplets $(1,a,2)$, $(2,a,3)$, and the 3-edge COMMONID item $((1,2,3))$, but they do not share any of the TYPEID items. The $\cap\mathcal{D}[(a,1,a)]$ would form an ambiguous itemset which would form more than one possible graph. A postprocessing phase is required for finding the maximal frequent itemsets that form closed no conflicting itemsets. This phase begins by forming an initial set of components based on the couple of edges that belong to each converse edge triplet, and recursively extending these components until they cannot be extended. According to our example, the components are $\{(a,1,a),(a,2,a),(1,a,2)\}$ and $\{(a,2,a),(a,3,a),(2,a,3)\}$.

Proposition 1. *Let $\mathcal{D}$ and $\mathcal{D}'$ two graphs, $ECE(\mathcal{D})$ a set of sets, $G_i \in \mathcal{D}$ a connected graph, g a subgraph of G_i such that $g \in \mathcal{P}_i$. If $y = \sigma(ECE(g))$ is an ambiguous itemset then it contains at least two maximal non-ambiguous itemsets t_0 and t_1 such that $\mathcal{D}'[i] = \mathcal{D}'[t_0] = \mathcal{D}'[t_1]$.*

Proof. Clearly for an unambiguous itemset y there exist at least one 3-edge COMMONID item without its TYPEID item, and this COMMONID item refers to at least two converse edges (in the case of a linear chain). This implies to start the postprocessing by two components, which are recursively extended to generate two maximal non-ambiguous itemsets t_0, t_1 such that $\mathcal{D}'[y] = \mathcal{D}'[t_0] = \mathcal{D}'[t_1]$.

Assuming the proposition 1, and the fact that ECE encoding with respect to the failure detection test is a bijective function from $\mathcal{P}_i \rightarrow ECE(\mathcal{P}_i)$, we deduce that for any connected graph $g \in \mathcal{P}_i$, the set of closed connected graphs that contain g is $C_g = \{ECE^{-1}(\sigma(ECE(g)))\}$.

3.3 Support Closure Operation

While the strong accessibility guarantees the existence of a chain between each consecutive closed connected subgraphs pair, calculating closure allows to jump directly from an immediate successor of a closed one g to its closed subgraph successor g^* [2]. We notice that the itemsets closure of an itemset I is the intersection of the transactions containing I as a subset [17]. Unfortunately, for connected subgraphs, the result of a subgraph closure is not unique and can be a disconnected graph that does not even belong to the system. Therefore, we conclude that there is no closure operator in P_i, and we define the closure support operation within $\mathcal{P}_i$ with respect to $\mathcal{D}$ as follows:

Definition 7. *Let P_i be a subgraph system and $\mathcal{D}$ be a dataset. The support closure of $\mathcal{P}_i$ with respect to $\mathcal{D}$ is defined by:*

$$\sigma(i) = max\ \Sigma(i)$$

Algorithm 1. | ECE-CLOSESG

Input: a graph $G_j \in \mathcal{D}$, support closure operator σ.
Output: $\sigma(\mathcal{F}_j)$: closed frequent subgraphs of G_j.

1: $E_F \leftarrow$ All frequent 1-edge graphs in G_j
2: $i \leftarrow 0$
3: $C_i \leftarrow \emptyset$
4: $\sigma(\mathcal{F}_j) \leftarrow C_0$
5: $IN_i \leftarrow$ GET-IND-GENERATOR (C_i, E_F, G_j)
6: **while** $IN_i \neq \emptyset$ **do**
7: $\quad C_{i+1} \leftarrow$ GET-CLOSED (IN_i, E_F, G_j)
8: $\quad \sigma(\mathcal{F}_j) \leftarrow \sigma(\mathcal{F}_j) \cup C_{i+1}$
9: $\quad IN_{i+1} \leftarrow$ GET-IND-GENERATOR (C_{i+1}, E_F, G_j)
10: $\quad i \leftarrow i+1$
11: **end while**

return $\sigma(\mathcal{F}_j)$

where $\Sigma(i) = \{i' \in \mathcal{P}_i : i \subseteq i' \text{ et } \mathcal{D}[i] = \mathcal{D}[i']\}$

i.e., the set of subgraphs $\Sigma(i)$ *may have more than one maximal.*

From this definition, we clearly notice that the closure of a connected graph $g \in \mathcal{P}_i$ consists of all connected subgraphs of maximal size in $\mathcal{P}_i$ and having g as predecessors and similar occurrences in $\mathcal{D}$.

The set-theoretical representation simplifies the augmentation and connectivity test operations. However, the main motivation to use it is the transformation facility of graphs to ECE codes; that reduces the graph/subgraph isomorphism complexity to equivalence sets and inclusion sets tests. We recall that the closure of a graph g consists of a set of ECE codes. These ECE codes are the result of a post-processing of an ambiguous ECE code X calculated by intersecting graphs $G_i \in \mathcal{D}$ ECE codes that contain the graph inductive generator g code (i.e., $X = \cap\mathcal{D}[ECE(G_i)]$ $s.t$ $ECE(g) \subseteq ECE(G_i)$). The post-processing phase consists in replacing an ambiguous ECE code by the largest connected and not ambiguous itemsets I included in $ECE(g)$. Moreover, the code X may correspond to an unconnected subgraph $g' \notin G_i$. To avoid this latter problem, only the sub-code $I \subseteq X$ that contains $ECE(g)$ is considered. We note for the best case, where there is a unique closed subgraph Y (i.e., an unambiguous ECE code), calculating closure needs at least $|\mathcal{D}|$ inclusion tests and generation operations.

3.4 The ECE-CloseSG Algorithm

The main objective of our proposed subgraph mining algorithm is to restrict the number of visited subgraphs and to reduce the complexity of graph/subgraph isomorphism tests. Algorithms 1, 2 and 3 illustrate our algorithm.

Algorithm 2. | GET-IND-GENERATOR

Input: a graph $G_j \in \mathcal{D}$, a set of closed graphs C_i, and a set of frequent edges E_F.
Output: IN: a set of ECE codes of inductive graph generators.

```
1: IN ← ∅
2: for each c ∈ C_i do
3:     for each e ∈ E_F do
4:         if c ∪ e is connected then
5:             g' ←AUGMENTATION (c, e)
6:             ECE(g') ← ENCODE (g')
7:             if IS-FREQUENT (ECE(g'), D') then
8:                 IN ← IN ∪ ECE(g')
9:             end if
10:        end if
11:    end for
12: end for

    return IN
```

Starting from the closed set C_0 that contains only an empty subgraph (i.e., $\emptyset$) that still considered closed frequent when the data $\mathcal{D}$ is non-redundant [2]. Then, for each set of closed frequent subgraphs C_i, each closed $c \in C_i$ is extended to generate all possible inductive generator subgraphs (i.e., using GET-IND-GENERATOR function). Thus, a level IN_i of inductive generator subgraphs is generated from each level of closed frequent subgraphs C_i. Next, closure of each subgraph in IN_i is calculated to generate the next level of the closed frequent subgraphs C_{i+1} (i.e., using GET-CLOSED function). In order to avoid redundancy, all closed and inductive subgraphs already visited at previous levels are eliminated. Finally, these steps are repeated until no new generator inductive subgraph is generated (i.e., $IN_i = \emptyset$).

More specifically, ECE-CLOSESG algorithm consists of four steps:

1. the graph data $\mathcal{D}$ is encoded to a transactional database $\mathcal{D}'$ containing a list of itemsets.
2. starting at the closed set C_0 which contains only an empty subgraph $\emptyset$, all codes of inductive generator subgraphs are generated from C_0. Algorithm 2 illustrates the pseudo-code of the GET-IND-GENERATOR function that allows generating inductive generator subgraphs. This function receives as input a set of closed graphs C_i, the frequent edges E_F and the graph G_i, and generates all the codes of the inductive generator graphs IN. For each closed subgraph $g \in C_i$, all extensions that generate connected subgraphs are applied and encoded as an ECE code. The frequent ones are added to the set IN.
3. for each set of ECE codes IN_i, the closure is computed. Algorithm 3 shows the pseudo-code of computing the closure function (noted GET-CLOSED). This function receives as input a set of ECE codes, the frequent edges E_F and the graph G_j. It generates the set of closed connected graphs C_{i+1}. For each of the ECE codes ($ECE(X) \in IN_i$), the intersection of all codes in $\mathcal{D}'$ that

Algorithm 3. | GET-CLOSED

Input: a graph $G_j \in \mathcal{D}$, a set of ECE codes IN_i, and a set of frequent edges E_F.
Output: C_i: a set of closed connected subgraphs.

```
 1: for each ECE(X) ∈ IN_i do
 2:     ECE(Y) ← ∩D'[ECE(X)]
 3:     if ECE(Y) is an ambiguous code then
 4:         MaxECE ← GET-SUB-CODES (ECE(Y))
 5:         for each ECE(Z) ∈ MaxECE do
 6:             g'' ← RECONSTRUCT (ECE(Z), G_j)
 7:             C_i ← C_i ∪ g''
 8:         end for
 9:     else
10:         g'' ← RECONSTRUCT (ECE(Y), G_j)
11:         C_i ← C_i ∪ g''
12:     end if
13: end for

        return C_i
```

contain the code $ECE(X)$ with respect to a negative test of an encoding failure test. Further, for each code $ECE(X)$ a new code $ECE(Y)$ is generated ($ECE(Y) = \cap\mathcal{D}[ECE(X)]$). This code may correspond to an ambiguous and/or disconnected graph. Then, the ambiguity and the connectivity of each newly generated code $ECE(X)$ are checked. If it is an ambiguous code, $ECE(X)$ will be replaced by a set of unambiguous sub-codes $MaxECE$, and for each of these codes a graph g'' is reconstructed. If the code is unambiguous, then the corresponding graph g'' is generated. Finally, each graph is reconstructed and added to the set of closed subgraphs C_i.

4. each set of closed subgraphs C_i is added to the global closed set $\sigma(\mathcal{F}_j)$ and both step 2 and step 3 are repeated until no new inductive subgraph generator is generated. The global closed set $\sigma(\mathcal{F}_j)$ is returned.

4 Experimental Study

The proposed closed frequent subgraph mining algorithm is implemented in Java and tested on synthetic datasets, which are generated by GraphGen [3], a graph generator that generates graph data based on five parameters: T (the number of graphs), S the size (i.e., the number of edges) of each graph. The size is defined as a normal distribution with the input as the mean and five as the variance. ETE (respectively ETV) represents the number of distinct edges (respectively the number of the labels of the vertices), L represents the density of each graph. The latter is defined as the number of edges in the graph divided by the number of edges in a complete graph, i.e., $L = |E|/(|V|(|V|-1)/2)$. After data generating, a post-processing step is invoked, the edge labels of each graph are randomly modified such that graph satisfies the constraint of unique edge labels.

Table 2. Experimental results with frequency support=2% and ETV=20.

Dataset						Runtime (s)	
$\|\mathcal{D}\|$	S	ETE	L	$\|\mathcal{F}\|$	$\|\mathcal{C}\|$	ECE-CloseSG	NAIVE
	20	20	0.1	2867	1420	1.76	10.32
$T = 100$	20	20	0.4	5032	1660	4.86	20.59
	20	20	0.7	6058	1059	12.44	28.31
	20	20	0.1	4961	2004	5.43	62.07
$T = 200$	20	20	0.4	6342	2207	9.05	43.97
	20	20	0.7	7224	1897	15.08	63.38
	30	40	0.1	5690	3364	29.37	199.86
$T = 300$	30	40	0.4	7351	4144	60.87	338.92
	30	40	0.7	7731	3954	124.12	439.60
	30	40	0.1	5692	3590	77.75	1129.30
$T = 400$	30	40	0.4	8395	5585	183.29	1199.69
	30	40	0.7	8909	5149	440.35	1291.04
	30	40	0.1	6045	4243	153.05	3488.83
$T = 500$	30	40	0.4	9234	6721	487.61	2481.30
	30	40	0.7	9693	5982	944.77	3591.94

We run our experiments on a 2.9 GHz Intel i7 PC with 8 GB of RAM. Running times of ECE-CloseSG are analyzed for frequency support value of 2%. The set of graphs are generated by varying the parameters of the graph generator as follows: (1) the number of graphs T from 100 to 500; (2) S as 20 and 30 that overlap with a high probability the intervals [10,30] and [15,45] respectively; (3) Vertex and edge labels: ETV 20, ETE 20 and 40; (4) density ranges from 0.1 to 0.7. This variation of parameters aims to prove a picture of ECE-CloseSG behaviour over different graph datasets.

Table 2 lists the results of the proposed algorithm on synthetic datasets. It presents:

1. the number of frequent connected subgraphs generated by NAIVE, a naive algorithm that explores all the search space to find all frequent ones first, and then filters them to list only the closed ones.
2. the runtime and the number of closed frequent connected subgraphs listed by ECE-CloseSG.
3. the runtime of the NAIVE algorithm.

The presented results show a significant difference in the size of the set of frequent subgraphs $\mathcal{F}$ compared to the closed frequent ones $\mathcal{C}$. Moreover, Table 2 illustrates a total domination of the proposed algorithm compared to the NAIVE algorithm. ECE-CloseSG performs 10 times faster than the NAIVE algorithm. In addition, we observe that increasing the graph size and the densities allow to increase the runtime of ECE-CloseSG to list the closed subgraphs. Finally, ECE-CloseSG requires more important runtime with respect to the value of the density, which increases the number of triples in a graph, where the encoding and decoding, as well as inclusion and equivalence tests, will be more complex.

5 Related Works

Mining only closed and maximal frequent subgraphs is the common way to reduce the huge number of frequent subgraphs. A typical closed and maximal frequent subgraph mining task consists of two steps. The first step lies in finding all frequent subgraphs $\mathcal{F}$. The second step consists in filtering the frequent subgraphs in order to keep only the closed and maximal ones. This approach seems to be not efficient due to the huge number of subgraphs to be visited in order to find the maximal frequent ones. To avoid exploring all frequent subgraphs, existing closed and maximal frequent subgraph mining algorithms use pruning techniques to reduce the search space. In this section, we discuss most popular algorithms that were proposed to solve this problem.

CloseGraph [18] is a closed frequent subgraph mining algorithm that uses adjacency lists to store graphs and to test frequency, whereas an ordered sequence's edges code (called DFS lexicographic order) is used to generate candidates and to detect redundancy. CloseGraph adopts a pruning technique to restrict the search space. This pruning technique uses an equivalent occurrences property based on an early termination condition. This property allows to decide whether a descendant super-graph of a given graph is a closed one or not. The early termination condition means that the search process will be completely stopped for some descendant branches, which effectively reduces the search space.

In the case of maximal frequent subgraph mining, several algorithms such as SPIN [9], Margin [16] and ISG [15] have been proposed.

SPIN [9] is based on the fact that most of the frequent subgraphs are trees (acyclic graphs). In order to reduce the computation cost, SPIN mines all frequent trees from a graph database and then built groups of frequent subgraphs from the mined trees. Each group of frequent subgraphs consists in an equivalence class, where each class is composed of subgraphs that share the same canonical spanning tree. We mention that this grouping step is not efficient because we still need to list all frequent subgraphs to construct maximal and frequent ones. To avoid this problem, some optimization techniques (i.e., Bottom-Up Pruning, Tail Shrink, and External- Edge Pruning) could be integrated to speed up the mining process.

Margin [16] is a subgraph mining algorithm that mines only maximal frequent subgraphs. Margin is based on the fact that maximal frequent subgraphs lie in the middle of the search space, which implies a high computational cost. To overcome this problem, Margin restricts the search space by visiting only the set of promising subgraphs $\widehat{\mathcal{F}}$ that encloses the frequent and infrequent subgraphs lie on the border. For each graph $G_i \in \mathcal{D}$, Margin works as follows. First, it explores the search space (the subgraph system) in a depth-first way to find the representative subgraph R_i. Then, it applies a method called ExpandCut on the cut CR_i and R_i where CR_i is a super infrequent graph of R_i. ExpandCut finds the nearby cuts and recursively calls itself for each newly found cut, until no new cut can be found. Given a set of cuts $(C|P)$ in which ExpandCut is applied, the frequent subgraphs P for each cut is reported as promising frequent subgraphs $\widehat{\mathcal{F}}$, and then only maximal local frequent subgraphs $\mathcal{ML}$ are kept from the promising ones ($\widehat{\mathcal{F}}$). Finally, the maximal global frequent subgraphs $\mathcal{M}$ in the database $\mathcal{D}$ are listed by removing the set of graphs in $\mathcal{ML}$, which are proper subgraphs of other frequent graphs in $\mathcal{ML}$.

ISG [15] is an algorithm for mining maximal frequent subgraphs over a graph database with unique edge labels. In order to list these maximal frequent subgraphs, an itemset mining technique is used. The idea of ISG is to transform the problem of maximal graph mining to maximal itemset mining. First, the graph database $\mathcal{D}$ is transformed to a list of itemsets $\mathcal{D}'$, where each graph in $\mathcal{D}$ is encoded as a set of items. These items represent the edges and the converse edges of the graph and 3-edge substructure that are contained in the graph (i.e., triangle, spike, and linear chain). These 3-edge blocks are called secondary structures and are added to avoid the problem of edge triplets and converse edge triplets conversion to a graph. The edge triplets and converse edge triplets together do not guarantee that the given maximal frequent itemset can be converted into a graph. Each secondary structure is assigned to a unique item identifier in addition to the unique identifiers of the edge and the converse edge triplets that compose the 3-edge block. Second, the transaction database $\mathcal{D}'$ is used as input to a maximal itemset mining algorithm. The set of the maximal code itemsets $\mathcal{M}_I$ is enumerated. Generally, there is no guarantee that all these codes correspond to unique connected subgraphs (i.e., there is no bijection between $\mathcal{M}_I$ and a subgraph in $\mathcal{P}$). The main problem over $\mathcal{M}_I$ is that there are codes for which the conversion generates disconnected graphs. To avoid these problems, a post-processing step after the conversion of the codes that can not be converted unambiguously is required. This step consists in converting a disconnected code to a set of its connected graphs, and for the case of conflicting maximal frequent itemset, it is broken to form non-conflicting subsets. After this post-processing step, the set of generated subgraphs is filtered, and the subgraphs that are contained in other subgraphs are pruned. Finally, the set of maximal connected unique edge label subgraphs is mined. We mention that the ECE encoding does not preserve the incomparability of subgraphs in a particular case. Thus, we have noticed a failure that affects the frequency test, which explains the incomplete output. This phenomenon leads to produce false frequency results when two incomparable graphs encoding can be comparable.

Besides the fact that subgraph mining related tasks are complex, and graph/subgraph isomorphism test is a hard problem; current works do not show up a significant optimization in the size of visited subgraphs during the mining process.

6 Conclusion

In this paper, we have illustrated that the pattern search space system and the complexity of frequency test affect the enumeration process of closed and maximal frequent pattern mining algorithms. We proposed ECE-CloseSG, an algorithm that mines the unique edge label connected subgraphs based on an encoding of the input graphs into a set of ECE items. This encoding allowed us to reduce graph and subgraph isomorphism problems to set-equivalence and set-inclusion tests, respectively. We investigated the strong accessibility property in order to restrict the search space. The efficiency of the ECE-CloseSG algorithm depends on the position of the border between infrequent and frequent nodes of the search space. For a dense (respectively sparse) search space, the border lies in the higher (respectively lower) levels.

In future works, we plan to do a comparative study of our approach with existing closed frequent subgraph mining algorithms.

References

1. Aridhi, S., Sghaier, H., Zoghlami, M., Maddouri, M., Nguifo, E.M.: Prediction of ionizing radiation resistance in bacteria using a multiple instance learning model. Journal of Computational Biology **23**(1), 10–20 (2016)
2. Boley, M., Horváth, T., Poigné, A., Wrobel, S.: Listing closed sets of strongly accessible set systems with applications to data mining. Theoretical Computer Science **411**(3), 691–700 (2010)
3. Cheng, J., Ke, Y., Ng, W.: Graphgen: A graph synthetic generator (2006)
4. Chi, Y., Muntz, R.R., Nijssen, S., Kok, J.N.: Frequent subtree mining-an overview. Fundamenta Informaticae **66**(1–2), 161–198 (2005)
5. Ganter, B., Reuter, K.: Finding all closed sets: A general approach. Order **8**(3), 283–290 (1991)
6. Giatsidis, C., Thilikos, D., Vazirgiannis, M.: Evaluating cooperation in communities with the k-core structure. In: 2011 International Conference on Advances in Social Networks Analysis and Mining (ASONAM), pp. 87–93 (2011)
7. Gunopulos, D., Khardon, R., Mannila, H., Saluja, S., Toivonen, H., Sharma, R.S.: Discovering all most specific sentences. ACM Transactions on Database Systems (TODS) **28**(2), 140–174 (2003)
8. Horváth, T., Ramon, J., Wrobel, S.: Frequent subgraph mining in outerplanar graphs. Data Mining and Knowledge Discovery **21**(3), 472–508 (2010)
9. Huan, J., Wang, W., Prins, J., Yang, J.: Spin: mining maximal frequent subgraphs from graph databases. In: Proceedings of the Tenth ACM SIGKDD International Conference on Knowledge Discovery and Data Mining, pp. 581–586. ACM (2004)
10. Mannila, H., Toivonen, H.: Levelwise search and borders of theories in knowledge discovery. Data Mining and Knowledge Discovery **1**(3), 241–258 (1997)
11. Nourine, L., Petit, J.M.: Extending set-based dualization: Application to pattern mining. ECAI **242**, 630–635 (2012)
12. Nourine, L., Petit, J.M.: Dualization on partially ordered sets: Preliminary results. In: Choong, Y.W., Kotzinos, D., Spyratos, N., Tanaka, Y. (eds.) ISIP 2014. CCIS, vol. 497, pp. 23–34. Springer, Heidelberg (2016)
13. Saidi, R., Aridhi, S., Nguifo, E.M., Maddouri, M.: Feature extraction in protein sequences classification: A new stability measure. In: Proceedings of the ACM Conference on Bioinformatics, Computational Biology and Biomedicine, BCB 2012, pp. 683–689. ACM, New York (2012)
14. Srivastava, J., Cooley, R., Deshpande, M., Tan, P.N.: Web usage mining: Discovery and applications of usage patterns from web data. SIGKDD Explor. Newsl. **1**(2), 12–23 (2000)

15. Thomas, L., Valluri, S., Karlapalem, K.: Isg: Itemset based subgraph mining. Tech. rep., Technical Report, IIIT, Hyderabad, December 2009
16. Thomas, L.T., Valluri, S.R., Karlapalem, K.: Margin: Maximal frequent subgraph mining. ACM Transactions on Knowledge Discovery from Data (TKDD) **4**(3), 10 (2010)
17. Uno, T., Kiyomi, M., Arimura, H.: Lcm ver. 2: Efficient mining algorithms for frequent/closed/maximal itemsets. In: FIMI, vol. 126 (2004)
18. Yan, X., Han, J.: Closegraph: mining closed frequent graph patterns. In: Proceedings of the Ninth ACM SIGKDD International Conference on Knowledge Discovery and Data Mining, pp. 286–295. ACM (2003)

Using Glocal Event Alignment for Comparing Sequences of Significantly Different Lengths

Vinh-Trung Luu(✉), Mathis Ripken, Germain Forestier, Frédéric Fondement, and Pierre-Alain Muller

MIPS, Université de Haute Alsace, 12, rue des frères Lumière, 68093 Mulhouse Cedex, France
{trung.luu-vinh,mathis.ripken,germain.forestier,frederic.fondement, pierre-alain.muller}@uha.fr

Abstract. This work takes place in the context of conversion rate optimization by enhancing the user experience during navigation on e-commerce web sites. The requirement is to be able to segment visitors into meaningful clusters, which can then be targeted with specific call-to-actions, in order to increase the web site turnover. This paper presents an original approach, which equally combines global- and local-alignment techniques (Needleman-Wunsch and Smith-Waterman) in order to automatically segment visitors according to the sequence of visited pages. Experimental results on synthetic datasets show that our approach out-performs other typically used alignment metrics, such as hybrid approaches or Dynamic Time Warping.

Keywords: Web mining · Sequential pattern mining · Clustering

1 Introduction

Conversion rate optimization is considered as one of the most promising approaches for improving the turnover of e-commerce web sites. A lot of researches have already focused on understanding web browsing event-patterns, in order to improve the online content delivery. Clustering visitors into meaningful segments, associated to targeted call-to-actions and related item recommendation, is one of the techniques typically used for cross- and up-selling. As clustering aims at organizing similar items into the same group with no prior knowledge of item class, it is seen as an approach of unsupervised learning. In web usage mining context, the similarity of page visits and their order in a session is one of the relevant information to cluster. For example, cluster analysis helps to reach people who are interested in some specific kind of goods or services so that the owner can recommend to such groups other related things, or offer them some discounts. The clustering result can also be applied to advertising placement organization on web sites, based on page visiting frequency in each cluster.

In this paper, we present our work for computing the similarity between event-sequences of significantly different lengths. Our proposal is based on

P. Perner (Ed.): MLDM 2016, LNAI 9729, pp. 58–72, 2016.
DOI: 10.1007/978-3-319-41920-6_5

a new way of equally combining global- and local-alignment techniques (*i.e.* Needleman-Wunsch [21] and Smith-Waterman [29]). The originality of our measure is to take into account the length of longest sequence in the pair of compared sequences.

Thus, regardless of the difference in sequence lengths, the result provided by our metric is accurate and can be used to perform clustering. Experimental results show that our approach outperforms other typically used similarity measures, such as hybrid approaches or Dynamic Time Warping (DTW), in the context of event-sequences of different lengths. This paper is divided into five sections with the following structure: Section 2 explains the proposed method. Section 3 describes experimental results. The discussion of these results is in Section 4. Section 5 presents related work. Finally, Section 6 concludes the paper and gives some future research directions.

2 Proposed Method

Before discussing our sequence alignment approach, we introduce a few basic concepts: Given a finite set Σ whose elements are characters, called alphabet, any possible string of length $k > 0$ over Σ is a k-tuple built by characters from Σ. For example, if $\Sigma = \{A, B, C\}$, a set $S = \{s_1, s_2, \ldots, s_n\}$ of n finite strings over Σ can consist of $s_1 = AB, s_2 = ABC, \ldots, s_n = ACB$. In our model, each web session contains a series of page visits is assumed to be a sequence (*i.e.,* visits which are ordered). Hence, each sequence s_i is composed as a string from Σ, representing a session. A set of navigation sequences S, as mentioned, contains sessions from multiple visitors. To group sessions based on visit order of visitors, our method works as follows: S is processed to create clusters containing comparable sequences that are dissimilar to sequences in other clusters. For this purpose, our alignment-based similarity measure is proposed. An alignment over a set of sessions $S = \{s_1, s_2, \ldots, s_n\}$ can be described as another set $S_a = \{s_{1a}, s_{2a}, \ldots, s_{na}\}$ of equal length sessions which built by adding necessary gap "-"to s_i, for $1 \leq i \leq n$ [7]. Next, elements at the same *index* of session strings are compared and scored by a scoring scheme, so-called similarity definition.

Sequence similarity definition of specific context (or application-dependent) is essential to perform a relevant similarity evaluation. For instance, the correspondence of DNA sequences [26] is not identical to time-series [18], and both of them are different to web session similarity. Therefore, session similarity measure has to be adapted to web usage situation. At first, it has to deal with the variety of session lengths and thus traditional vector distances like Euclidean, Manhattan or even Hamming cannot be applied. Such distances require equal-length sequences like $s_1 = ABCD$, $s_2 = ABCD$. Secondly, as sessions are expected to be differentiated by page visit orders, the appropriate metric has to take into account this order. Thus, metrics such as Levenshtein and VLVD[25] are inappropriate as they consider the visit of page$A$ before B and vice versa to be the same. As a result, $s_1 = ABCD$ and $s_2 = BDCA$ are identical. These statistical approaches count the occurrence of element in each sequence to measure their similarity, regardless of element order. Additionally, the continuity of common pages between two browsing behaviors is a significant factor to evaluate their correspondence. Therefore, LCS or SAM[11] should not be used as it consider sessions started by page A and ended by B, that are common pages, like $s_1 = AB$ and $s_2 = ACDFB$ to be the same, regardless of how many unique pages between them. As the matter of fact, there is a meaningful difference in web visitor interest when one hits C,D and F between A and B and the other hits no pages between those two but they are not counted in this kind of metrics. In summary, an applicable approach in web usage mining context should be able to (1) process sequences with variable length, (2) take the order and succession of common pages into consideration. Such measures are suitable to compute the similarity between two sets of visits.

The Needleman-Wunsch (NW) method is a dynamic programming algorithm for sequence alignment which was developed by Needleman and Wunsch in 1970. Dynamic programming makes it possible to find the optimal alignment of sequences, is easy to implement and popular in computer science. When aligning elements of a sequence, matching and mismatching scoring scheme are given. A corresponding score matrix is then established to find the highest score of all possible alignments. In NW, this alignment is carried out from beginning to end of each sequence, it is called a *global alignment.* Global alignment is appropriate to work with sequences of similar length to find their best alignment. However, sequences may inherently not have the same length but might contains similar subsequences. Thus, a *local alignment* is relevant to detect them. To address this issue, the Smith-Waterman (SW) method, introduced by Smith and Waterman in 1981, performs the alignment by taking high comparable regions within sequences into account, regardless of the dissimilar parts and even the difference of sequence lengths. These two dynamic programming algorithms are commonly adopted for aligning protein or nucleotide sequences [14,19].

As NW and SW alignment methods are somehow opposite, each one has their own advantages and drawbacks. NW finds the optimal similarity of the entire sequence, while SW detects regions of likeliness between two sequences. As a result, a combination of both methods is better than using a single one, since the correspondence between sequences can be evaluated correctly (*i.e.* globally and locally). We pointed out the effectiveness of the NW and SW rules combination compared to DTW [24] and hybrid metric [6,8] in our previous work [16]. Following this finding, we propose a new similarity metric called *combination metric.* This metric is based on NW, SW and the size of the longest sequence:

$$S(s_i, s_j) = \left[\frac{NW(s_i, s_j)}{l}\right] + \left[\frac{SW(s_i, s_j)}{(2 * l)}\right] \quad (1)$$

with $NW(s_i, s_j)$ and $SW(s_i, s_j)$ respectively NW and SW scores between the two sequences s_i and s_j, l the length of longest sequence in the pair (*i.e.* $max(|s_i|, |s_j|)$), the NW scoring scheme of +1 for matching and -1 for non-matching pair of items in sequences, the SW scoring scheme of +2 for matching and -1 for non-matching inside matching, ignore non-matching outside.

Another way to combine NW and SW was proposed previously in *hybrid metric* scores similarity [6,8] between two sequences s_i and s_j, is defined as:

$$S(s_i, s_j) = (1 - p) * SW(s_i, s_j) + p * NW(s_i, s_j) \quad (2)$$

with the defined parameter $p = |s_i|/|s_j|$. From this definition (Eq. 2) , one can notice that hybrid metric does not equally take both NW and SW into account because of the difference in sequence lengths. As a consequence, the advantage of *combination metric* (Eq. 1) over the *hybrid metric* (Eq. 2) is that the similarity measure works better when sequence lengths are very different. In this case, hybrid metric only focuses on SW, while combination metric focuses on both NW and SW. Consequently, the more different in lengths the sequence pair are, the more the hybrid metric will focus on SW to measure their similarity. Therefore, the hybrid metric would consider two sequences of different classes to be in same class, if their difference in length and SW are important enough. On the other hand, two sequences of same class with comparable lengths may not be similar enough to be in the same cluster because their similarity score by hybrid metric is smaller than in the previous case. To illustrate this scenario, we created cluster dendrograms with a toy example. The Figure 1a shows the similarity evaluation of the hybrid metric in case of sequences with quite different length. As the two sequences of blue class are not similar enough to be merged in agglomerative hierarchical clustering, the resulting clustering is of poor quality. As illustrated in Figure 1b, this case does not happen using the combination metric as both NW and SW are considered equally, regardless of the length difference between the sequences. Consequently, two clusters green and blue of perfect accuracy are obtained when the dendrogram is cut.

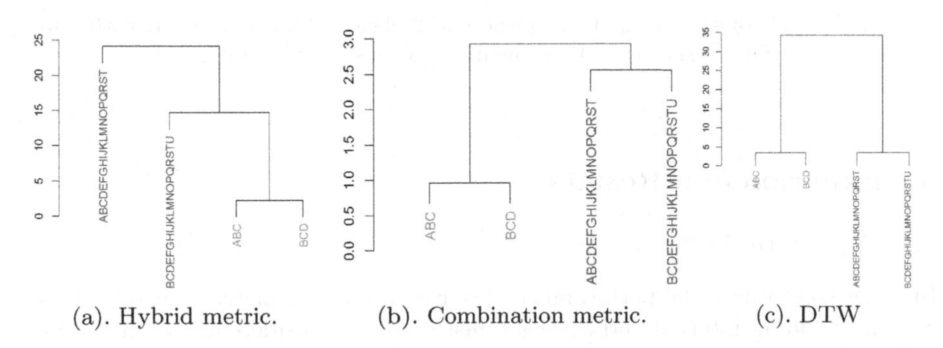

Fig. 1. Example of clustering of 4 sequences of 2 classes (blue and green) with quite different length for hybrid (a), combination (b) and DTW (c) metrics.

Dynamic Time Warping (DTW) is known to be effective to find good alignment between time-series. In the context of sequence pairs of quite different lengths and no duplicate as in Figure 1, DTW works mostly as good as combination method. As illustrated in Figure 1c, it makes blue and green sequences merged in agglomerative hierarchical clustering. However, DTW minimizes distance of one sequence to another by allowing flexible transformation so that time-series with similar shapes can be detected. This feature leads to a problem when identical consecutive elements in sequences are merged. Thus, the warping path of sequence pairs is vertical or horizontal. In other words, a single element from one sequence is aligned with many successive and duplicate elements in the other as they are all identical. This feature is a drawback in web usage mining as sessions containing duplicate web pages should not be "skipped" but mined for web visitor behavior. Our proposed metric considers them to be traversal pattern and takes this duplication into account while DTW regards them as only one page no matter how many visits are duplicated. Figure 2b and 2c respectively illustrate the similarity evaluation of the combination metric and DTW in case of sequences with duplicate elements. Furthermore, with sequence pairs which contain duplicate elements and quite different in length like in Figure 2, hybrid metric is less effective than the combination, as presented in Figure 2a.

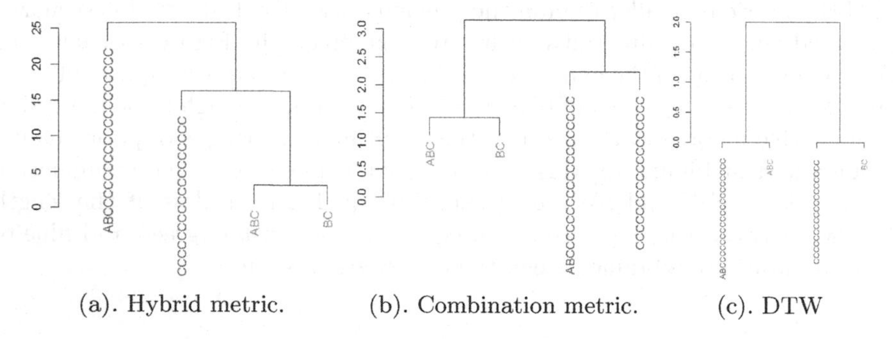

(a). Hybrid metric. (b). Combination metric. (c). DTW

Fig. 2. Example of clustering of 4 sequences of 2 classes (blue and green) with duplicated elements for hybrid (a) and combination (b) and DTW (c) metrics.

3 Experimental Results

3.1 Synthetic Data

In order to evaluate the performance of our combination metric, clustering validation including internal and external measures are considered. As some appropriate classified data is not currently available for external validation, we used internal validation on generated synthetic datasets. Nevertheless, Liu et al. [13] analyzed both kind of validations and revealed the relevance of internal validation measure in many aspects over some other external validation measures. Rendon et al. [27] also concluded that they can get better precision by validity internal indexes than external ones, on their datasets and scenarios. To evaluate the performance of the proposed metric and competitors (*i.e. hybrid metric* and *DTW*), we generated 10 synthetic datasets randomly. Each of them contains more than 500 sequences of sessions (about 520 in average) which are grouped into three defined classes with following features:

- Class 1: About 170 sequences of lengths [20 – 22], sharing a common sub-sequence, for instance: **ABC**DU3YU31DQ6Q4FO2JGHW, **ABC**DSI5OPHH9EDGLPFLST, **ABC**DAFF5UAK7GEX3XJIU. Generated sequences in other classes are related to this common sub sequence;
- Class 2: About 170 sequences of lengths [3 – 4], sharing a common sub-sequence (that is also sub-sequence of common sub-sequence in Class 1), for instance: **ZBC**, **ABC**5, **BC**0;
- Class 3: About 170 sequences of lengths [18 – 20], mostly containing identical and consecutive symbols (that appear in the common sub-sequence in Class 1), for instance: DDDDDDDDB**BC**CCCCCCC, DDDDDDDD**C**CAAAA-AAA, BBBBBBB**B**BADBBBBBBBBB;

Note that all the datasets used in the experiments are available to download here[1]. As shown in Section 2, sequences with common subsequence but different lengths such as Class 1 and 2 are likely to be misclassified by hybrid metric. Yet sequences with same consecutive symbols in Class 3 are likely to be confused with sequences of Class 2 using DTW. These classes are assumed to be representative of behaviors that can be witnessed when analyzing real web sessions.

Using these sequence datasets, similarity matrices for each metric were computed. In order to implement agglomerative hierarchical clustering [9], these matrices are then used as input with three well known hierarchical methods: *single-linkage* (sing.), *complete-linkage* (compl.) and *average-linkage* (avg.). This variety of hierarchical methods contributes to the effectiveness of the evaluation. Table 1 shows the means (μ) and standard deviations (σ) of clustering result precision for the three methods with the three hierarchical methods over the 10 datasets. Note that as hierarchical clustering is deterministic, running the experiments multiple times is not required. Thus, the means and standard deviations correspond to the execution on the 10 different datasets.

Table 1. Results for the three methods on the 10 datasets.

	Original datasets								
	Hybrid			DTW			Combination		
	sing.	compl.	avg.	sing.	compl.	avg.	sing.	compl.	avg.
μ	**100%**	**58.5%**	**100%**	**90.5%**	**85.5%**	**89.7%**	**100%**	**98.2%**	**100%**
σ	±0%	±8.5%	±0%	±9.7%	±1.4%	±9.8%	±0%	±5.7%	±-0%

The correlation of experimental results in Table 1 illustrated by dendrograms in Figure 3, 4 and 5 on sample set of the sequences (for sake of clarity) with leave

[1] https://www.dropbox.com/sh/b6wxv5opn1u3n6n/AAB8ObwvqBPbDsnXvB9xZ_yca

values as defined classes. By cutting dendrograms at the desired level, clusters are separated into frames and their quantity matching the number of defined classes. As shown in Figure 3, there is one cluster with unexpected accuracy containing sequences from both Class 1 and 3. The number of inaccurate clusters in Figure 4 is two as they correspond to Class 2, 3 and Class 1, 3 sequences. However, clusters in Figure 5 achieve a perfect accuracy. These figures also illustrate that combination metric works very well compared to hybrid and DTW.

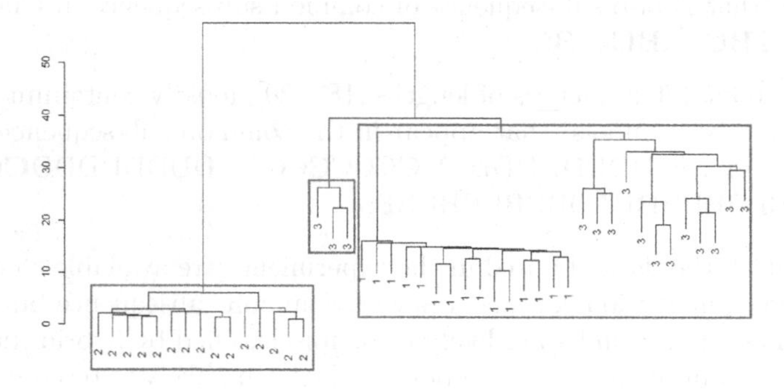

Fig. 3. Hierarchical clustering using hybrid metric on original dataset.

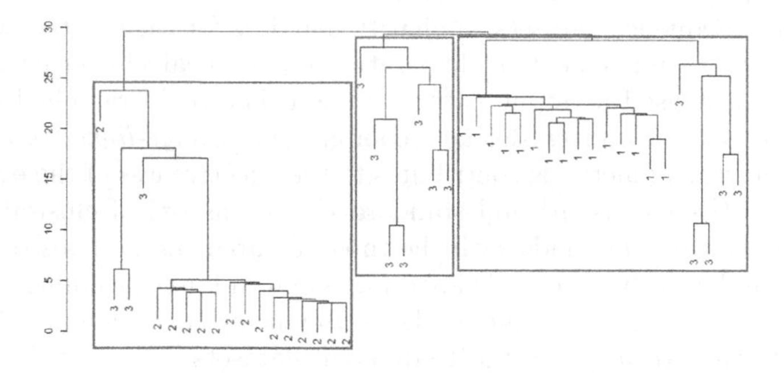

Fig. 4. Hierarchical clustering using DTW metric on original dataset.

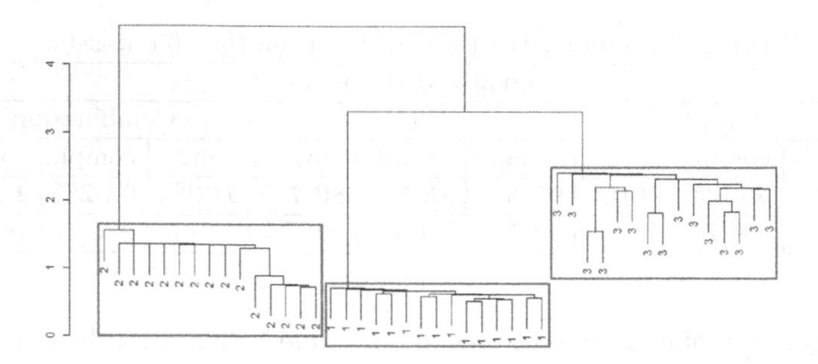

Fig. 5. Hierarchical clustering using combination metric on original dataset.

Following, we present additional results performed to consider two popular aspects of data in web usage mining: the noise and unbalanced density of classes (*i.e.* classes with important difference in number of elements).

Noise: About 15 sequences of lengths [3, 24] were randomly generated from alphabet and numbers, for instance: APE8V98MDTIH77I, H96YXT7N, M9AK-KAA, etc. were added to the original datasets with 3 classes. The accuracy of clustering results on these datasets is presented in Table 2.

Table 2. Results for the methods on the 10 datasets with noise.

	Datasets with noise								
	Hybrid			DTW			Combination		
	sing.	compl.	avg.	sing.	compl.	avg.	sing.	compl.	avg.
μ	**90.4%**	**84.3%**	**90%**	**65.8%**	**73.1%**	**86.7%**	**100%**	**89.2%**	**100%**
σ	±11.6%	±3%	±7.2%	±1%	±9%	±11%	±0%	±6%	±-0%

Similarly to the previous results, dendrograms on sample of sequences with leave values as defined classes are presented on Figure 6, 7 and 8. These figures illustrate the correlation of experimental results presented in Table 2. Dendrograms are cut at the desired level to make cluster quantity match the defined number of classes and separate them into red frames.

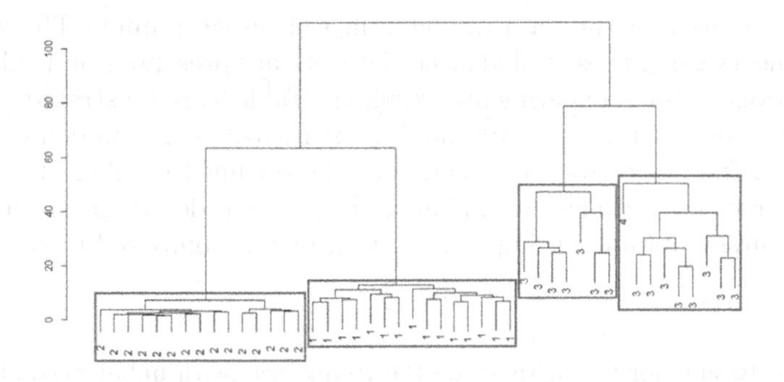

Fig. 6. Hierarchical clustering using hybrid metric on dataset with noise.

Unbalanced Density: The number of sequences of Class 1, 2 and 3 are respectively around 320, 170 and 10 in the first three datasets. In the next four datasets, number of sequences in Class 1, 2 and 3 are respectively around 170, 320 and 10. Lastly, in the remaining three datasets, sequence numbers are around 10, 170 and 320 for Class 1, 2 and 3. As the number of users having the same usage can be very different according to specific behaviors, this kind of datasets are assumed

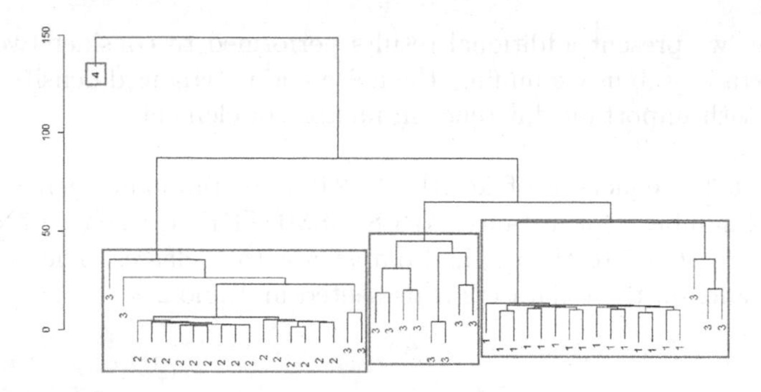

Fig. 7. Hierarchical clustering using DTW metric on dataset with noise.

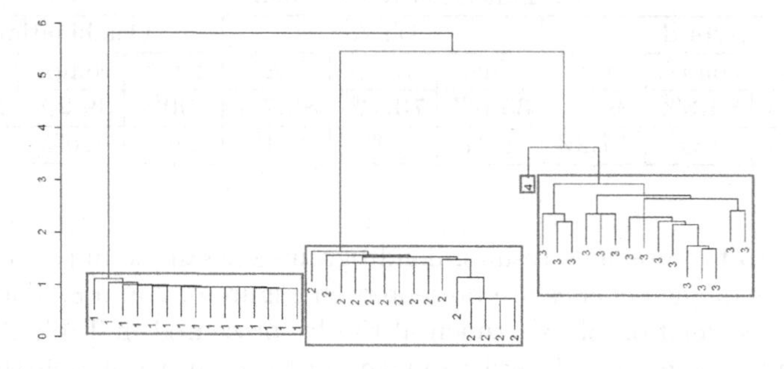

Fig. 8. Hierarchical clustering using combination metric on dataset with noise.

to be representative of the data available in web usage mining. The results of the experiments using these unbalanced datasets are presented in Table 3.

As mentioned above, experimental results in Table 3 are illustrated by sample data dendrograms in Figure 9, 10 and 11, with leave values as defined classes. Similarly, red frames separate elements into class defined number of clusters by cutting the tree at desired levels. Similar to the previous dendrograms, it presents the best accuracy obtained using combination metric compared to the others.

Table 3. Results for the methods on the 10 datasets with unbalanced classes.

	Unbalanced density datasets								
	Hybrid			DTW			Combination		
	sing.	compl.	avg.	sing.	compl.	avg.	sing.	compl.	avg.
μ	**81.6%**	**73.2%**	**84.8%**	**90.9%**	**91.1%**	**91.2%**	**100%**	**93.7%**	**91.1%**
σ	±19%	±18.9%	±24.5%	±16.5%	±12.7%	±12.8%	±0%	±10%	±-0%

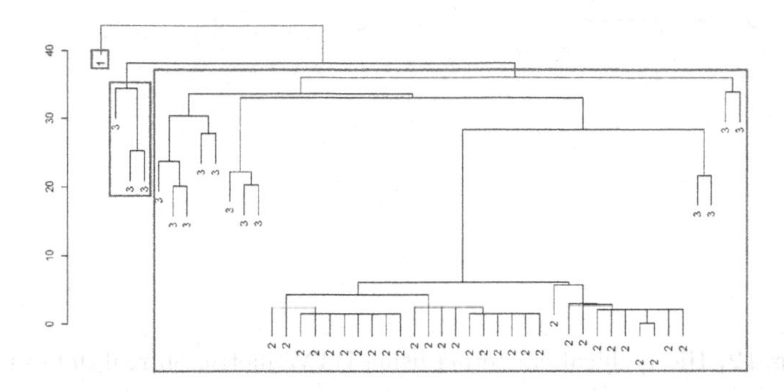

Fig. 9. Hierarchical clustering using hybrid metric on unbalanced dataset.

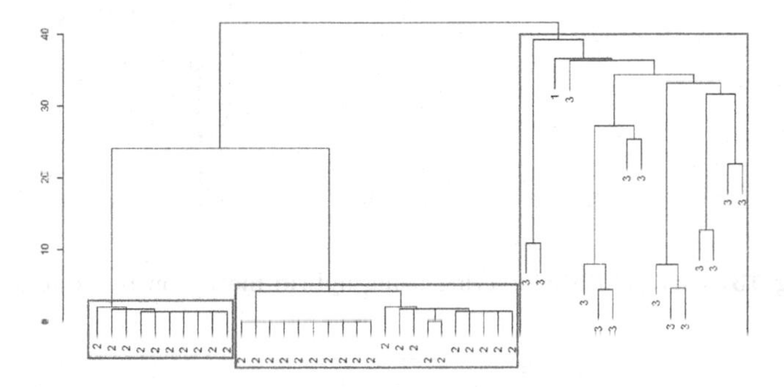

Fig. 10. Hierarchical clustering using DTW metric on unbalanced dataset.

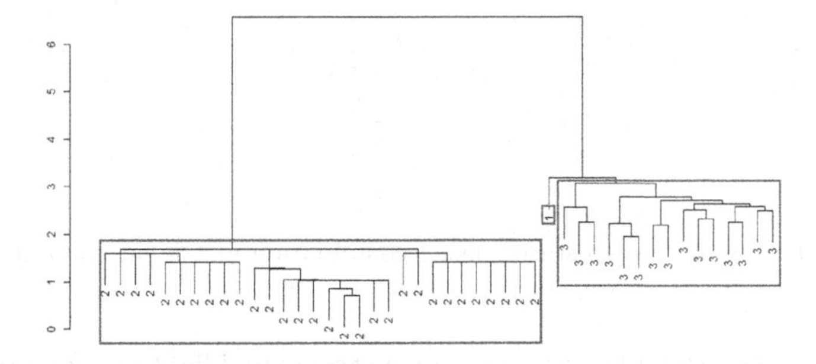

Fig. 11. Hierarchical clustering using combination metric on unbalanced dataset.

3.2 Real Data

The dataset used for our external validation was collected from a commercial website. The dataset was provided by the Beampulse company which commercializes a product written in Javascript and Java, which collects information about web visitors behaviours such as page visit order, activity time or duration of page visit. As shown in dendrograms in Figure 12, 13 and 14 that is a sample extracted from experimental result on 1500 individual sessions, each contains

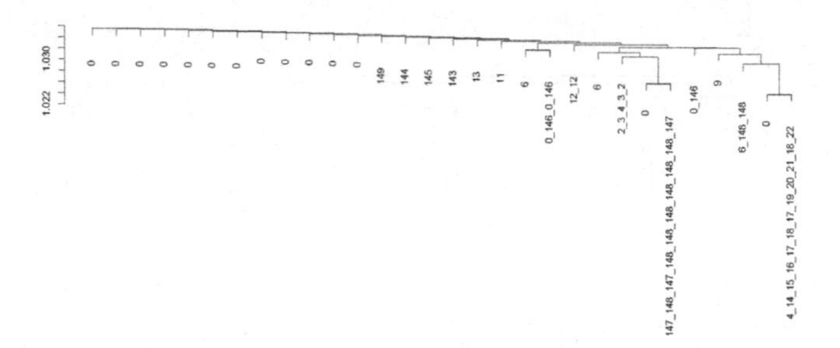

Fig. 12. Hierarchical clustering using DTW metric on real dataset

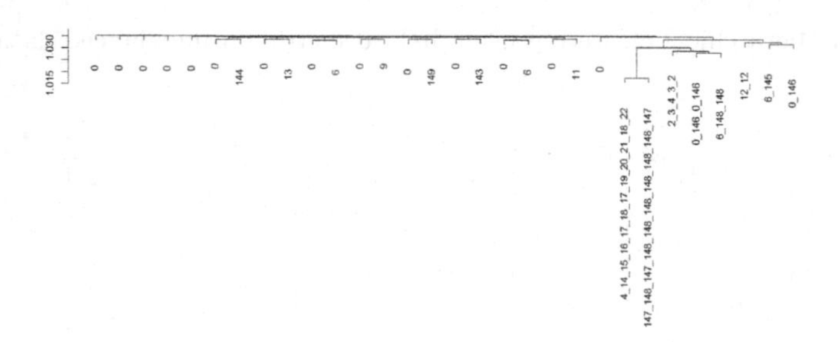

Fig. 13. Hierarchical clustering using hybrid metric on real dataset

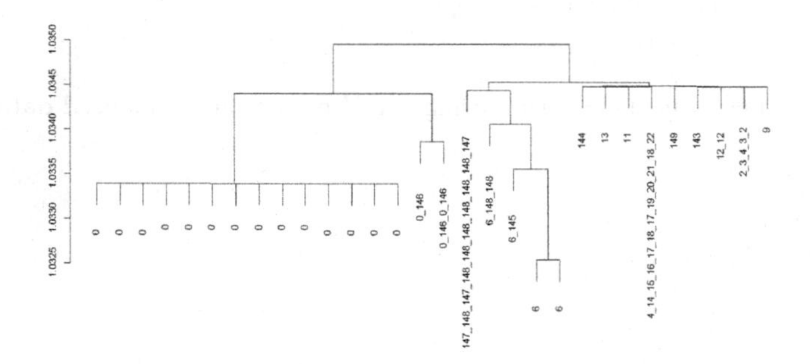

Fig. 14. Hierarchical clustering using combination metric on real dataset

numbered page(s), the advantages of our metric is highlighted compared to the other methods. Using our metric, sessions with similar pages, similar page order and similar length are likely to be grouped in the same cluster. However, such features are not obtained using hybrid and DTW metrics.

4 Discussion

As shown in Table 1, throughout the 10 normal datasets (*i.e.*, with neither noise nor unbalanced clusters density), similarity matrix produced by DTW outputs the lowest precision clustering using single- and average-linkage. However, DTW precision is higher than hybrid metric by complete linkage. Hybrid metric is as good as combination metric using single and average-linkage but is significantly worse using complete-linkage where combination metric is mostly stable.

As noise may impact the clustering algorithm performance, it is good to obtain datasets without noise before clustering. However, clustering algorithms should have the ability to deal with noise because of the difficulty to avoid its presence, especially in big datasets. Our metric maintains a perfect precision using average and single-linkage, and also provides the highest precision using complete-linkage in Table 2, where 10 datasets including noise are used as input. Meanwhile, hybrid handles noise more accurately than DTW in this context.

Similarly, clustering methods are challenged by various density datasets because of its importance in the clustering process. On unbalanced datasets, the good clustering precision obtained using our metric remain almost the same using single and average hierarchical methods. In contrast to DTW, hybrid metric is highly influenced by unbalanced density. Again, our metric reached the best accuracy using single-linkage and always works better than the other two methods using complete and average linkage (see Table 3).

5 Related Work

The goal of web usage mining is to identify hidden pattern from visitor browsing data. This involves clustering of different visits having similar navigational patterns. One of the most popular approaches to discover these clusters is session classification. Among various classification forms, many previous studies have focused on sequence alignment algorithms to evaluate the similarity of sessions. Mandal et al. [17] presented the calculation of distance between two sessions using Cosine measure but it requires sessions with exactly the same length, which is more suitable to vectors than web accesses. Furthermore, these approaches also ignore regions of local similarity of session pages. In [5], Chitraa et al. intended to find usage patterns using k-means algorithm application, yet clustering is to discover hidden pattern without specific input parameters.Similarly, defining k by the granularity that clusters should be in order to group web users proposed by Jianfeng et al. in [28] is inappropriate to browsing sessions.

Pairwise alignment is commonly used to compare sequences optimally. There exist sequence alignment algorithms, both global and local adopted through pairwise alignment such as NW [12] and SW [31]. These two algorithms take into account the similarity between sequences in different alignment [30] and have their own strengths and drawbacks [10]. Lu et al. [15] studied how to generate significant usage patterns using NW, however it ignores consecution that is essential to evaluate similarity of web session pairs. In contrary, a local alignment algorithm such as SW can only detect partial similarities [2].

Consequently, there have been previous works on integrating one into the other to take advantage of the combination. For example, Brudno et al. [4] proposed a system to align genomes with biological features glocally. Chordia et al. [6] described a hybrid metric, concerning a consolidation of global and local sequence similarity scoring. Correspondingly, the same metric was developed by

Dimopoulos et al. [8] to measure the similarity of two sequences. This formula computes the distance between sequences by taking global and local alignment and their weights into consideration. These weights are in inverse proportion to each other, depending on how different sequences are in length. Specifically, local alignment weight would be greater if sequence lengths are different. As local alignment scoring does not take the difference in sequence lengths into account, this computation may work in some specific situations but not in web accesses similarity because that difference reveals the dissimilarity in visitor browsing behavior. Similarly, Algiriyage et al. [1] used Levenshtein distance as a similarity metric for web session pairs but it does not identify the succession of common visits. On the other hand, with regard to sequential characteristic of web session, there has been approach such as [3] compares homogeneity of sequence pair by frequency of common items occurrence but does not take into account their order, that is key feature to differentiate sessions. DTW, that is a popular algorithm in comparing two sequences of events [20] or data points [23], is also used in symbolic sequence comparison. However, this kind of approach is not effective in web usage mining since DTW ignores duplicated elements. Consequently, it is not able to evaluate the dissimilarity in browsing behavior comparing a specific web page loaded many times with the same page loaded only one time.

Note that the lack of available benchmarks in the domain of web usage mining makes the comparison of the different existing methods difficult. In order to address this issue, we released with this paper all datasets (*i.e.* original, with noise and with unbalanced density) that were used for the experiments (see supplementary material). We hope that these datasets will be used in future research to compare new contributions.

6 Conclusion

In this paper, we have presented our contribution to event-sequence comparison, with a specific focus on sequences of significantly different lengths. This new way of combining global- and local-alignment techniques is based on the equal combination of both approaches. We experimentally evaluated this approach in context of clustering web site visitors, by analyzing the browsing patterns. Under those settings, it was observed that our sequence similarity metric outperformed other related techniques. In the close future, we plan to introduce mutations in the current datasets to better stress the combination technique in the presence of noise. We also want to compare the approach with other techniques such as PAM/k-medoids, ROCK or Ward. Furthermore, other distance measures such as Hamming or Levenshtein, which have been studied in a variety of sequence comparisons including spectra [22] should be considered in upcoming works.

Supplementary Materials

All the datasets used in the experiments are available here: http://germain-forestier.info/src/mldm2016/.

References

1. Algiriyage, N., Jayasena, S., Dias, G.: Web user profiling using hierarchical clustering with improved similarity measure. In: Moratuwa Engineering Research Conference (MERCon), pp. 295–300. IEEE (2015)
2. Aruk, T., Ustek, D., Kursun, O.: A comparative analysis of smith-waterman based partial alignment. In: 2012 IEEE Symposium on Computers and Communications (ISCC), pp. 000250–000252. IEEE (2012)
3. Bouguessa, M.: A practical approach for clustering transaction data. In: Perner, P. (ed.) MLDM 2011. LNCS, vol. 6871, pp. 265–279. Springer, Heidelberg (2011)
4. Brudno, M., Malde, S., Poliakov, A., Do, C.B., Couronne, O., Dubchak, I., Batzoglou, S.: Glocal alignment: finding rearrangements during alignment. Bioinformatics **19**(Suppl. 1), i54–i62 (2003)
5. Chitraa, V., Thanamni, A.S.: An enhanced clustering technique for web usage mining. International Journal of Engineering Research and Technology **1**. ESRSA Publications (2012)
6. Chordia, B.S., Adhiya, K.P.: Grouping web access sequences using sequence alignment method. Indian Journal of Computer Science and Engineering (IJCSE) **2**(3), 308–314 (2011)
7. Della Vedova, G.: Multiple Sequence Alignment and Phylogenetic Reconstruction: Theory and Methods in Biological Data Analysis. Ph.D. thesis, Citeseer (2000)
8. Dimopoulos, C., Makris, C., Panagis, Y., Theodoridis, E., Tsakalidis, A.: A web page usage prediction scheme using sequence indexing and clustering techniques. Data & Knowledge Engineering **69**(4), 371–382 (2010)
9. Duraiswamy, K., Mayil, V.V.: Similarity matrix based session clustering by sequence alignment using dynamic programming. Computer and Information Science **1**(3), 66 (2008)
10. Giegerich, R., Wheeler, D.: Pairwise sequence alignment. BioComputing Hypertext Coursebook **2** (1996)
11. Hay, B., Wets, G., Vanhoof, K.: Clustering navigation patterns on a website using a sequence alignment method. Intelligent Techniques for Web Personalization: IJCAI, 1–6 (2001)
12. Likic, V.: The needleman-wunsch algorithm for sequence alignment. Lecture given at the 7th Melbourne Bioinformatics Course, Bi021 Molecular Science and Biotechnology Institute, University of Melbourne (2008)
13. Liu, Y., Li, Z., Xiong, H., Gao, X., Wu, J.: Understanding of internal clustering validation measures. In: International Conference on Data Mining, pp. 911–916. IEEE (2010)
14. Liu, Y., Hong, Y., Lin, C.Y., Hung, C.L.: Accelerating smith-waterman alignment for protein database search using frequency distance filtration scheme based on cpu-gpu collaborative system. International Journal of Genomics **2015** (2015)
15. Lu, L., Dunham, M., Meng, Y.: Discovery of significant usage patterns from clusters of clickstream data. In: Proc. of WebKDD, pp. 21–24. Citeseer (2005)
16. Luu, V.-T., Forestier, G., Fondement, F., Muller, P.-A.: Web site audience segmentation using hybrid alignment techniques. In: Li, X.-L., Cao, T., Lim, E.-P., Zhou, Z.-H., Ho, T.-B., Cheung, D. (eds.) PAKDD 2015. LNCS, vol. 9441, pp. 29–40. Springer, Heidelberg (2015). doi:10.1007/978-3-319-25660-3_3
17. Mandal, O.P., Azad, H.K.: Web access prediction model using clustering and artificial neural network. International Journal of Engineering Research and Technology **3**. ESRSA Publications (2014)

18. Meesrikamolkul, W., Niennattrakul, V., Ratanamahatana, C.A.: Shape-based clustering for time series data. In: Tan, P.-N., Chawla, S., Ho, C.K., Bailey, J. (eds.) PAKDD 2012, Part I. LNCS, vol. 7301, pp. 530–541. Springer, Heidelberg (2012)
19. Muhamad, F.N., Ahmad, R., Asi, S.M., Murad, M.: Reducing the search space and time complexity of needleman-wunsch algorithm (global alignment) and smith-waterman algorithm (local alignment) for dna sequence alignment. Jurnal Teknologi **77**(20) (2015)
20. Nakamura, A., Kudo, M.: Packing alignment: alignment for sequences of various length events. In: Huang, J.Z., Cao, L., Srivastava, J. (eds.) PAKDD 2011, Part II. LNCS, vol. 6635, pp. 234–245. Springer, Heidelberg (2011)
21. Needleman, S.B., Wunsch, C.D.: A general method applicable to the search for similarities in the amino acid sequence of two proteins. Journal of Molecular Biology **48**(3), 443–453 (1970)
22. Perner, P.: A novel method for the interpretation of spectrometer signals based on delta-modulation and similarity determination. In: 2014 IEEE 28th International Conference on Advanced Information Networking and Applications (AINA), pp. 1154–1160. IEEE (2014)
23. Petitjean, F., Forestier, G., Webb, G., Nicholson, A.E., Chen, Y., Keogh, E., et al.: Dynamic time warping averaging of time series allows faster and more accurate classification. In: International Conference on Data Mining, pp. 470–479. IEEE (2014)
24. Petitjean, F., Gançarski, P.: Summarizing a set of time series by averaging: From steiner sequence to compact multiple alignment. Theoretical Computer Science **414**(1), 76–91 (2012)
25. Poornalatha, G., Raghavendra, P.S.: Web user session clustering using modified k-means algorithm. In: Lloret Mauri, J., Buford, J.F., Suzuki, J., Thampi, S.M., Abraham, A. (eds.) ACC 2011, Part II. CCIS, vol. 191, pp. 243–252. Springer, Heidelberg (2011)
26. Qi, Z., Redding, S., Lee, J.Y., Gibb, B., Kwon, Y., Niu, H., Gaines, W.A., Sung, P., Greene, E.C.: Dna sequence alignment by microhomology sampling during homologous recombination. Cell **160**(5), 856–869 (2015)
27. Rendón, E., Abundez, I., Arizmendi, A., Quiroz, E.: Internal versus external cluster validation indexes. International Journal of Computers and Communications **5**(1), 27–34 (2011)
28. Si, J., Li, Q., Qian, T., Deng, X.: Discovering K web user groups with specific aspect interests. In: Perner, P. (ed.) MLDM 2012. LNCS, vol. 7376, pp. 321–335. Springer, Heidelberg (2012)
29. Smith, T.F., Waterman, M.S.: Identification of common molecular subsequences. Journal of Molecular Biology **147**(1), 195–197 (1981)
30. Yan, R., Xu, D., Yang, J., Walker, S., Zhang, Y.: A comparative assessment and analysis of 20 representative sequence alignment methods for protein structure prediction. Scientific Reports **3** (2013)
31. Zahid, S.K., Hasan, L., Khan, A.A., Ullah, S.: A novel structure of the smith-waterman algorithm for efficient sequence alignment. In: International Conference on Digital Information, Networking, and Wireless Communications (DINWC), pp. 6–9. IEEE (2015)

Using Support Vector Machines for Intelligent Service Agents Decision Making

Babak Khosravifar and Mohamed Bouguessa(✉)

Department of Computer Science, University of Quebec at Montreal, Montreal, QC, Canada
bouguessa.mohamed@uqam.ca

Abstract. Analyzing online web services' behavior is a difficult task. The challenge is to effectively manage their cooperation and group work while as rational systems they always seek to maximize their overall utilities. Existing approaches either manage to enhance the quality of service provided by web services or group them together to corporate a stronger web service by gathering web services of the similar functionalities. Although these approaches have shown to be useful, they are not practical in the sense that enhancing the quality of service is a costly process that is not always the best option. In this paper, we present an efficient approach that applies support vector machine to equip web services with this machine learning algorithm to train the previously generated data and effectively make decisions to cooperate with one another. In our experiments, we applied three kernel functions to create the normal model and compared their overall performance together as well as the benchmark, that is, rational web services without learning abilities. The results show that the Gaussian kernel outperforms the other two learning models as well as the benchmark non-learning model by maintaining high true join recommendations rate while producing low false not-join recommendations.

Keywords: Web services · Cooperative actions · Community of services · SVM

1 Introduction

Online services have substantially grown in recent years and are now employed by various industries, including travel planning, search engines, and online computation. These services are provided by intelligent autonomous entities. Intelligent modelling has attracted a great deal of attention from researchers whose objective is to effectively manage these web services while maintaining high quality service from rational intelligent entities. To this end, the research question is how to engineer these web services to receive large amounts of inquiries while maintaining limited functionalities. Improving functionalities is not always practical because they are costly and can cause frequent idle stages. Therefore, to

P. Perner (Ed.): MLDM 2016, LNAI 9729, pp. 73–87, 2016.
DOI: 10.1007/978-3-319-41920-6_6

address the research question, researchers propose different frameworks to engineer the functionality and service quality of these web services while they are in operation.

Some work [6] has proposed using communities of web services with similar functionalities. The idea is to integrate the cooperated services and deliver high quality service as a result of cooperative task. This way, when a service request is filed to the community, the service is provided within a reasonable service time, cost, and quality, thus allowing the community to handle a large number of service requests while maintaining a good reputation. The main challenge is for the community of web services to effectively manage task allocation and cooperative tasks in order to maintain quality service.

The idea of community is not necessarily practical if there are limited resources to bind web services together to present an abstract web service with strong abilities. Adversely, the idea of gathering web services in the form of a community is a practical way to ensure high quality service without investing too much into enhancing the functionality of each individual web service. However, the challenge is how to find the appropriate web services to construct a strong community. In fact, for rational entities of autonomous web services, this is a twofold challenge. Rational web services look for an appropriate community of web services to join and improve their overall reputation, income, or returning customers. Meanwhile, the joined web services might decide to leave a community if they feel they will obtain a better reputation and income by acting alone or joining another community. From the community point of view, the community manager continuously seeks web services that will increase the overall income. Meanwhile, the manager entity might ask a web service to leave if its presence negatively influences the overall community reputation, income, or returning customers.

In this paper, we address the question of how to engineer the continuous joining of community members. The leaving concept is left for our future work and it needs to built up based on the joined policies and community framework. We use a machine learning technique to effectively use the previously collected data to make prudent decisions for both communities of web services and single web services regarding their cooperative functionalities. To do this, we use a two-class Support Vector Machine (SVM) [10] with (1) Gaussian; (2) polynomial; and (3) linear kernels. SVMs are generalized linear classifiers. They are known for simultaneously minimizing the empirical classification error while maximizing the geometric margin. The SVMs are appropriate for two-class tasks and are directly practical in the context of behavioural analyses of web services since data for both categories (successful and unsuccessful join) is available. To analyze web services' behavior, we use a community framework that hosts web services in communities and enables continuous joins to or leaves from communities. There are also single web services that do not find a proper community to join and prefer to act as individual web services. We apply all three classifiers to equipped web services and investigate the impact of using machine learning algorithms on the overall outcome of these rational entities. The benchmark for our comparison is the outcome of web services that are rational but not equipped with any machine learning algorithm. That means these web services make decisions based only on available data regarding the surrounding environment and act rationally, which means they pick the action that leads to the best outcome. Our findings show that the impact of applying machine learning classifier algorithms to such models is substantial. Web services by far enhance their overall outcomes while considering learning-based decisions.

The remainder of this paper is organized as follows: In Section 3, we present our model. An empirical evaluation of our proposal is given in Section 4. Section 5, concludes the paper.

2 Related Work

Network of web service agents is a typical example of multi-agent environments (see [13]) that run continuous business interactions like service exchange. In this context, the reputation management have been intensively stressed [8], [9], [17] aiming to facilitate and automate the good service selection. In [16], the authors have developed a framework aiming to allocate tasks to web services based on the trust policies expressed by the users. The framework allows the users to select a web service matching their needs and expectations. In [12], the authors proposed to compute the reputation of a web service according to the personal evaluation of the previous users. In general, the common characteristic of these methods is that the reputation of the web service is measured by a combination of data collected from users. To this end, the credibility of the user that provides this data should be taken into account. If the user tries to provide a fake rating, then its credibility will be decreased and the rating of this user will have less importance in the reputation of the web service. In [14], the authors have designed a multi-agent framework based on an ontology for Quality of Service (QoS). The ontology provides a basis that allows the providers to advertise their offerings, the users to express their preferences and the ratings of services to be gathered and shared. The users' ratings according to the different qualities are used to compute the reputation of the web service. In [8], service-level agreements are discussed in order to set the penalties over the lack of QoS for the web services.

In general, in all the mentioned models, web services are considered to act individually and not in collaboration with other web services. In such systems, the service selection process is very complicated due to their relatively high number in the network. Further example, web services can easily distract the system by leaving and joining the network when they have incentives to do so. For example, when their reputation is fall off for some reason, which is a rational incentive for such web services that manage to start as new once they have shown a low efficiency. Meanwhile, it is hard to manage the large amount of data in web services settings. Considering these inefficiencies, there are proposed frameworks [6] that introduce the concept of gathering web services together into communities. Then we could address the problems that would come through when web services individually act in the environment. In [3], the authors propose a dependable (i.e., intrusion-tolerant) infrastructure for cooperative web services coordination that is based on the tuple space coordination model. In [4] and [5], there are proposed frameworks to group web services to strengthen the combined group in terms of quality of service. In this paper, we consider group of web services cooperating together as communities of web services.

Communities are in general formed to get stronger and more publicized in the system, so they do not resign and register as new. In such methodology, users interconnect with the community as the service provider and there would be a web service assigned through the community. Regarding the aforementioned issues, there have been some proposals that try to gather web services and propose the concept of community-based multi-agent systems (CMSs) [6], [7], [11]. In [6], the authors propose a reputation-based architecture for CMSs and classify the involved metrics that affect the reputation of a community. They derive the involved metrics by processing some historical performance data recorded in a run-time logging system. The purpose is to be able to analyze the reputation in different points of views, such as users to CMSs, CMSs to web services, and web services to CMSs. The authors discuss the effect of different factors while diverse reputation directions are analyzed. However, they do not derive the overall reputation of a CMS from the proposed metrics. Failing to assess the general reputation for the community leads to failure in efficient service selection. Moreover, authors assume that the run-time logging mechanism (the logging file, which holds the feedback submitted by the service consumers) is an accurate source of information. In general, in open reputation-feedback mechanisms, always the feedback file is subject to be the target by selfish entities. To this end, the feedback mechanism should be supervised and its precise assessment should be guaranteed.

In [11], the authors have proposed a framework that explores the possibilities that the active communities act truthfully and provide their actual information upon request. In [7], a layered reputation assessment system is proposed mainly addressing the issue of anonymity. In this work, the focus is on the layered policies that are applied to measure the reputation of different types of web services, specially the new comers. Although, the proposed work is interesting in terms of anonymous reputation assessment, the layered structure does not optimally organize a community-based environment that gathers web services and also the computational expenses seem to be relatively high.

All the proposed frameworks share a common aspect, which is managing the quality of service provided by rational agent-based web services. Most of them address the problem by creating a trusted environment, where users can find the target web service that can handle the service request. But, this approach relies on the capabilities of the web service and does not handle the cases where the web services face overhead in service requests. The frameworks that use communities of web services also do not investigate the consequences of web services joining together and providing the service in the form of group task. Neglecting such important factors might end up in hosting web services without effectively managing them. The reputation of the community of web services would remain high bit with big investment on resources. In current approaches, the resource management is a major missing aspect. In this work, we address this missing aspect and apply a machine learning technique to substantially enhance the performance of web services either acting alone or as a member of a community.

3 The Proposed Model

3.1 Preliminaries

In this paper, we discuss topics that are already represented in this domain. However, for clarification and stating our assumptions (i.e., in the proposed framework, we assume that all involved entities are acting rationally. Therefore, given their status, their upcoming actions are deterministic), we briefly introduce them and denote their role in our proposed framework.

Web Services. Web services are rational entities that are programmable to act on the behalf of humans. They are intended to become part of people's daily life. Thus, the current frequency of requesting them for different types of online services will be very high. However, web services suffer from limited storage and processing power that can simply cause overhead in their internal processing. The consequences of such overhead may include lower quality of service, delay in delivering the service, or rejection of the requested service. All of these occur on a daily basis when we run weather forecasting, trip planning, car rental searching, and other online service related applications.

In the proposed framework, the programmable web services are aimed at providing services either directly to human end-users or to other web services to accomplish a part of the whole task. To simplify our discussions and to not deviate from the main objective of this paper, we assume a torrent of services is filed for the whole web service network that involves some communities of web services with a variety of cardinality and some single web services that act solely. In general, communities of web services provide higher availability, performance, reliability, and recovery for end-users, but they are costly. So, the service requester looks for the best quality at the lowest cost.

Communities of Web Service. The idea of a community of web services has emerged within the last decade. There are still many frameworks that do not integrate web services into communities. This is because the concept of a community, as an abstract web service designed to manage a variety of tasks in a high-quality manner, is a new entity. This objective is hard to achieve because professional task allocation has certain costs that some frameworks tend not to pay. They produce better results by representing web services as independent entities. The frameworks that gather similar functional web services propose protocols that handle task allocation and task distribution. The community representative is responsible for maintaining the reputation and performance of the community as a whole. In this paper, we only consider frameworks that involve communities. We consider them as abstract web services; however, there might be other web services that choose to act as individual web services. In further details, we highly emphasize on managing web service's cooperative tasks using a decision-making system that is trained on previously collected data.

3.2 Feature Engineering

To address the research problem, we collect data for training purposes. The steps for training the collected data are explained in Section 4. In this subsection, we introduce the nature of the collected data and the features obtained for use in our training model. The raw data collected from a group of real web services is presented in [1], [15], [18]. We categorize web services with similar functionalities to better train the learning model. Among all of the web service characteristics, we considered the following features to aggregate in our engineered feature vector:

Throughput for Community C (Th_C): This is the first internal feature that we collect, and it denotes the rate of processing a service request. This value varies in different web services and the higher it is, the stronger the web service is. For communities, the expected throughput value of collaborating web services (web services w in community C) can be estimated as described by Equation 1. In this equation, w is the index we use for web service and C is the index we use for communities. Therefore, $|C|$ is the cardinality of the community and if $|C| = 1$, the community only hosts a single web services and is technically an individual web service.

$$Th_C = \begin{cases} Th_w & |C| = 1 \\ \sum_{w \in C}(Th_w) & |C| > 1 \end{cases} \tag{1}$$

Availability for Community C (A_C): This is the second internal feature that we collect, and it denotes the percentage of time that a service is operating. This feature is a probability within the interval of $[0, 1]$ and the higher the value is, the higher the chance is that the web service is available to provide the requested service. For communities, the expected availability of collaborating web services (considering they follow a parallel system) can be estimated in Equation 2. In this equation, A_w denotes the availability for web service w.

$$A_C = \begin{cases} A_w & |C| = 1 \\ 1 - \prod_{w \in C}(1 - A_w) & |C| > 1 \end{cases} \tag{2}$$

Execution Time for Community C (Et_C): This is the last internal parameter that we collect, and it denotes the amount of the time a service requires to respond to various types of requests. This parameter is a positive integer and the higher it is, the longer it takes to respond to a requested service. For communities, the expected execution time of collaborating web services can be estimated as:

$$Et_C = \begin{cases} Et_w & |C| = 1 \\ max_{w \in C}(Et_w) & |C| > 1 \end{cases} \tag{3}$$

The internal features reflect how well a web service functions when a request is made by a service user. However, we need external features in order to compare the functionality scores of different web services. Using these external features, we create a complete feature vector that represents the characteristics of web services.

External Parameter 1 ($Ex1$): This is the first external parameter that we compute, and it denotes how similar the web service's execution time is to that of other web services. This parameter is calculated as a difference between the web service's response time (third internal parameter) and the maximum value obtained from the response time associated with other web services. This parameter is computed in the same way for the communities, as they represent an abstract web service. The communities that integrate consistent and coherent web services hold better value for this parameter. Therefore, the higher the external parameter is, the better service provide the web service is considered.

External Parameter 2 ($Ex2$): This is the second external parameter that we compute, and it denotes the web service's rate of performing tasks compared to that of other competing web services. It is the difference between the web service's throughput metric and the maximum value of throughput of the other web services. Therefore, like the first external parameter, the higher this external parameter is, the better service provide the web service is considered.

To better analyze the feature vectors, all parameters are normalized to fit the interval of $[0, 1]$. So in the rest of this paper, we assume all feature vectors hold values within the mentioned range.

3.3 Join Consequences

It is worth explaining what the consequences are of joining a single web service versus joining an already existing community of web services. The first perspective is from single web service's point of view, joining a community should be beneficial in the sense that the web service's overall outcome increases. This is the case in which the web service's three internal parameters improve by joining a community. In other words, the requested service is better handled when the web service belongs to a community, and the chance of rejecting the task or providing low quality service is substantially decreased. Therefore, the objective of the web service is to find a community that enhances the overall outcome of the web service based on what we compute in Equation 4. This does not always take place because if the web service does not appropriately match the community, then all requests for the web service are directed to the web service as a single independent node, as if it were acting alone. Meanwhile, the web service is obliged to pay the cost of being a member of the community. Thus, joining a community is tricky because a web service could end up paying the community fee while the overall outcome is not enhanced by the community. The communities and web services earn revenue by performing tasks. The total gain is a function of the quality of the rate of tasks being performed. The utility of a collaborative group of services U_C is the revenue of the community. Equation 4 is an estimation of the gain of the community, where β and γ are variables that are subject to set and be fixed for all entities. In this work, we manually set the weight parameters as a result of running different sets of experiments. However, the more efficient way of setting such problems is listed in our future work.

$$U_C = (\beta(A_C - Et_C) - \gamma(Ex1_C - Ex2_C)) \times Th_C \quad (4)$$

The second perspective is from the community point of view. The community representative's goal is to gather web services that perfectly match together, allowing the task allocation mechanism to simply distribute the tasks among the web services. The community reputation is a crucial parameter of the community and when this parameter fails, the member web services leave the community to either seek a better community or to function alone. Thus, the community representative always looks for web services that bring creditability to the community and that can handle group work with other web services. As with the previous perspective, this is tricky because the joining of an irrelevant web service with a community can distract the task allocation and reduce the functionality of the community. Having web services as single nodes within the community will reduce the quality of service, which directly affects the community reputation.

In this paper, we address the join challenge from both perspectives. We use a two-class SVM with different kernels to provide effective recommendations to the single web services seeking a community to join and to the communities looking for suitable web services. We show that web services who are equipped with the learning model act better than the ones who do not use recommendations. Both types of web services are rational, which means they seek to improve their overall outcome. But the learning model web services outperform the other type due to better control of the network using the past data. Given a set of training examples, each marked as belonging to one of the two categories, an SVM training algorithm builds a model that assigns new examples to one category or the other, making it a non-probabilistic binary linear classifier.

3.4 The Learning Model

In this section, we describe our training process and the way we build the two-class SVM classifier that is further used by equipped web services to decide whether or not to join a particular community. For this purpose, we first filtered the training real data and extracted our desired internal features and computed the two external features. The engineered data contains a number of web services acting alone as single nodes and a number of communities that gather some web services and act as abstract web services. We concentrate on a training process with a theoretical point of view, and in Section 4, we provide details of our case study and discuss statistics of the training process as well as details of our findings.

In our learning model we use three different kernels. These kernels are linear (Equation 5), polynomial (Equation 6, where d is the size of the feature vector), and Gaussian (Equation 7). Using the kernels, we create different classifiers. For example, using Gaussian kernel, the classifier function is computed in Equation 8. In this equation, the parameters α is manually set to 0.1 and like previously mentioned weight parameters is a future case study.

$$K(x_j, x_k) = x_j^T x_k \tag{5}$$

$$K(x_j, x_k) = 1 + (x_j^T x_k)^d, \quad d = |x_j| \tag{6}$$

$$K(x_j, x_k) = exp(\frac{-||x_j - x_k||^2}{2\sigma^2}) \tag{7}$$

$$f(x) = \sum_{i=1}^{n} \alpha_i y_i exp(\frac{-||x_j - x_k||^2}{2\sigma^2}) + b \tag{8}$$

Using different kernel functions, the SVM creates a trained model that web services refer to identify whether they have to join and cooperate with others or they obtain better utility if they act alone. The learning model generalizes the need to cooperate with others because the labeled data states whether similar web services have made right or wrong decisions in terms of participating in a cooperative work. In the following section, we expand more details about a real case study and investigate the impact of applying a machine learning algorithm on the overall performance of web services that are active in a competitive environment.

4 Experimental Case Study

The objectives of the conducted experiments are to check the effect of applying machine learning algorithms in web services' decision making on cooperating with communities of web services. In our experiments, we compare the performance of different kernel functions, that are linear, polynomial, and Gaussian. We take rationality as benchmark and compare web services that are equipped with learning algorithm with the ones that are rational and take actions to enhance their overall performance. The performance evaluation is conducted on a real-world dataset obtained from web services' online actions described in Section 4.1.

4.1 Dataset

Web service activities are collected from 4,532 web services in 64 different time slots. This dataset is provided by [18] and Table 1 gives statistics of the collected data. We group the web services together and measure their pre/post utilities before and after cooperation as a community of web services. Out of 4,532 web services, we chose 80 web services and gathered all combinations of them as grouped web services to form the communities. All combinations of size 2 out of 80 were 3,160 where we picked 20 of them to activate for the size 3 community and at the end, we picked 13 communities to extend their cardinality to 4. This way,

Table 1. Summary of collected dataset.

Community size	Training data points	Activated percentage
1	4532	80/4532
2	3160	20/3160
3	1540	13/1540
4	714	-

we created communities of different sizes and train their gained utilities as a result of joint cooperation.

Out of the training dataset, the created communities are of sizes 2, 3, and 4. Thus, we obtained 9,946 abstract web services (either single or grouped to form a community) that we extract actions in 64 different time slots. The consequitive time slots help up to appropriately track web services' progress while working alone or as a member of a community. Moreover, we have evidence that can make impressions whether the join action of a web service to a community is beneficial. We measured the change on utility of web services as a result of any action. Therefore, we labled good and bad decisions based on different situations and created the training input data with decision label. Based on Equation 4 the utility (U_C) of these communities are estimated and the gain matrix $|9946 \times 9946|$ of all possible ways of merging these 9,946 communities, is generated and each community has the ordered preference among other communities known in the set.

4.2 Experimental Protocol

In our experimental studies, we trained the learning model with 9,946 input data point, which include 9,946 made decisions with corresponding outcome labeled either as good or bad decisions. In our experiments, join is a twofold action and it takes place when both sides either initiate the join request or accept the join request. For example, the web service number 10 may decide to join the community 5, that hosts web services 6 and 72. The join takes place only when both sides accept such action, and the result is expanding community 5 to host web services 6, 10, 72 and single web service 10 is deactivated in the web service network. The join action could take place between two communities (abstract web services). For example, the community 5 that hosts web services 6, 10, 72 may initiate the join to the community 7 that hosts web services 12, 14, 17. Upon approval of the community 7, it hosts web services 6, 10, 12, 14, 17, 72 and the community 5 is deactivated.

We split 8,000 of the data points (each associated with an action) to use for training the learning model. The rest of the data is used for validation as well as some additional web services for testing. The learning model is created based on three different kernel functions (linear, polynomial, and Gaussian). We compare their overall performance in the form of Receiver Operating Characteristics

(ROC) curve. This curve plots false positive in x-axis and true positive in y-axis. That means, false joining decisions are counted for x-axis and true joining decisions are counted for y-axis. The outcome of the curve is to plot a point for each threshold, that is ranged from 0 to 1. We compare the three curves representing each kernel function and compare their area underneath the curve. Meanwhile, we run the same experiments for the web services that are not equipped with a learning model and only act based on their rational decision maker. These web services are used as benchmark because they hold rational attitudes, but they may not pick the best community for their long term actions. To this end, their overall performance may be lower than the ones that are equipped with a learning algorithm.

4.3 Results

Join Applications. The concept we discuss here is web services' overall tendency to join other web services (or existing communities) to cooperate and practice in group work. This is done when a web service knows of a web service or a community that its expected utility will increase if the joining action takes place. The web service could also get to know about the target web service or the community from a recommendation provided by the learning model. Therefore, the web services that are not using the learning model are limited to only the network of web services around them. Acting rationally, a web service always compares its current utility with the expected utility as a result of joining or not joining a new group and decides based on utility maximizing strategy. This strategy might prevent the joining application if the surrounding web services or communities are not good options for a join.

Figure 1 illustrates four graphs associated with each group: (1) not learning equipped, but rational web services; (2) rational and learning equipped using linear kernel function; (3) rational and learning equipped using polynomial kernel function; and (4) rational and learning equipped using Gaussian kernel function. The x-axis is the time within which the group of web services are active and the y-axis is the percentage of the web services that tend to join other web services or communities. Recall that the join event takes place when both sides (inviter and invitee) agree on the join.

It is clear from the graphs that the polynomial and Gaussian kernel models tend to join with higher rates. This means that these groups find better options and tend to expand their cooperative activities. In contrast, the non-learning model as well as linear kernel do not find good options to join and cooperate, thus, their tendency rate is relatively lower.

Figure 2(a) is the same group of web services with the same experiment, except that we measured the percentage of the web services that joined a web service or a community. It is clear that the join percentage substantially drops when the invitee group does not accept the join inquiry. Accepting a join invitation via the invitee depends on the join layer we set for the invitee. If this layer is 1 (which is this case), the join takes place when the invitee also believes that the inviter is the best option for cooperation. If the layer score is higher,

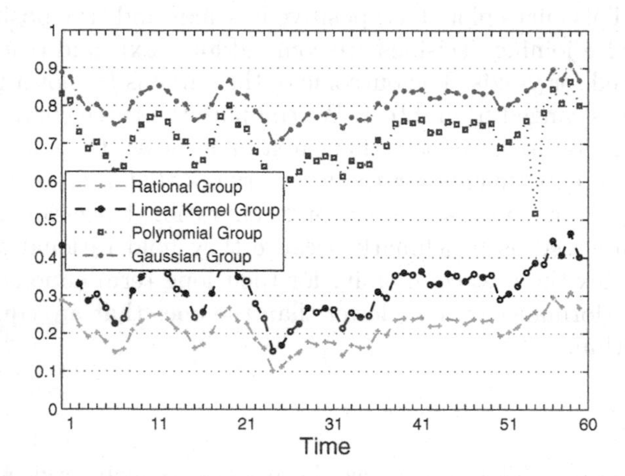

Fig. 1. Tendency to join other web services plotted for four different groups of web services.

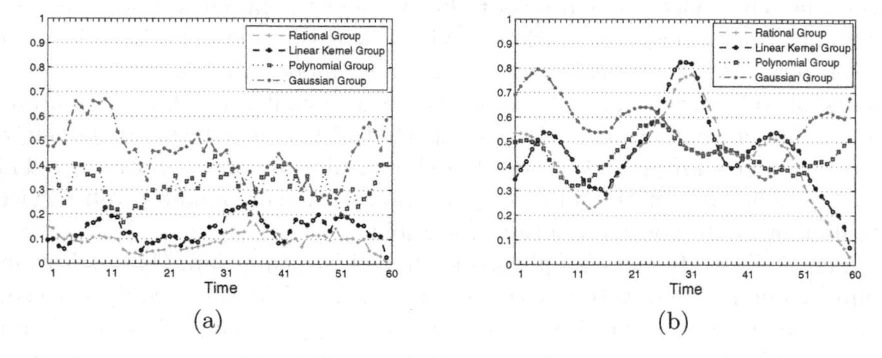

Fig. 2. (a): Web services' joined percentage for four different groups of web services. (b): Successful joined inquiry percentile counted for four different groups of web services.

the invitee accepts the invitation as long as the inviter is among the top group of best cooperative web services. Further details of these layers as well as their performance is illustrated in the form of ROC curves later in this Section.

The rational not learning web services are normally not exposed to the best cooperative choices of web services. Therefore, there is high chance that a join inquiry is not accepted. In contrast, the web services with learning algorithm can better analyze the join action and the percentage of rejecting the cooperation inquiry is lower. Figure 2(b) illustrates the successful joining inquiry trend for all four groups. It is clear from the graphs that learning model helps web services to effectively analyze the joining inquiry and appropriately find the best choices of web services or communities to join. But among the learning models, Gaussian

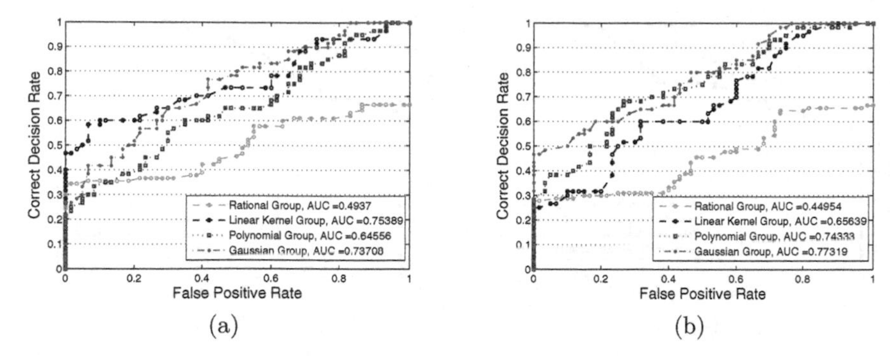

Fig. 3. (a): ROC curve sketched for layer score 1. In this layer, web services only accept the join invitation if the inviter is the best option in their list. (b): ROC curve sketched for layer score 2. In this layer, web services accept the join invitation if the inviter is the among the first two best options in their list.

kernel function seems to outperform other kernel functions in the sense that it provides better learning model representative of the input training data points.

We continue this direction in further details by analyzing how well the kernel functions operate in building the normal model and thus, help equipped web services to effectively analyze their choices of join/leave invitation by calling the kernel function $f(x)$ with their feature vectors as input. We analyze their performances in different layers within which denotes the extent to which web services (or communities) carefully respond to the invitations. For example, if the layer score is 1, it means that the invitee web service (or community) will only accept the invitation inquiry if the inviter web services is the best choice of the community. That means, in layer 1, the join event will take place only when both inviter and invitee seek the best opportunity in this joining action. This event is rare, but takes place when both inviter and invitee find each other either in their surrounding network or through a recommendation. Layer score 2 means that the invitee will accept both best and second best options of join and so on for other layers. In our experiments, we analyzed the layers scored from 1 to 4 that are respectively illustrated in Figures 3 and 4.

In Figure 3, the linear kernel performs well in providing correct recommendations. It means that, using this kernel function, web services get 60% of the recommendations useful while suffering only 8% from false positives regarding not join recommendations. The Gaussian kernel also provides good trend of recommendations and holds relatively close area underneath curve (also seen in Figure 4). But referring to Figure 2, the joining percentage of Gaussian kernel is higher than that of linear kernel. By increasing the layer score, the Gaussian kernel outperforms others by holding the highest area underneath curve. It shows that this kernel generates better classifier that the created decision boundary is a strong source of decision making for web services that tend to join other single or community of web services.

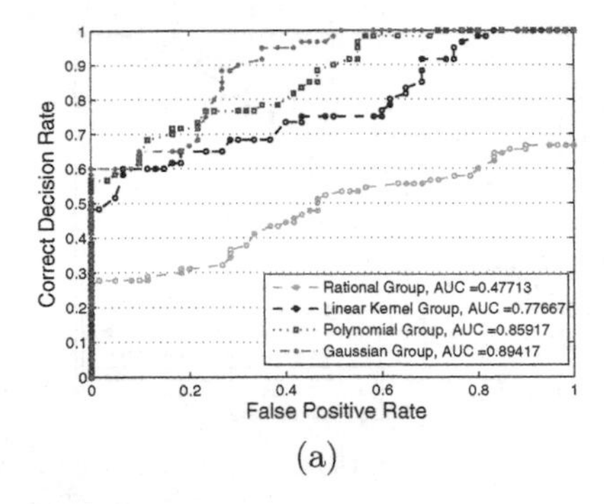

(a)

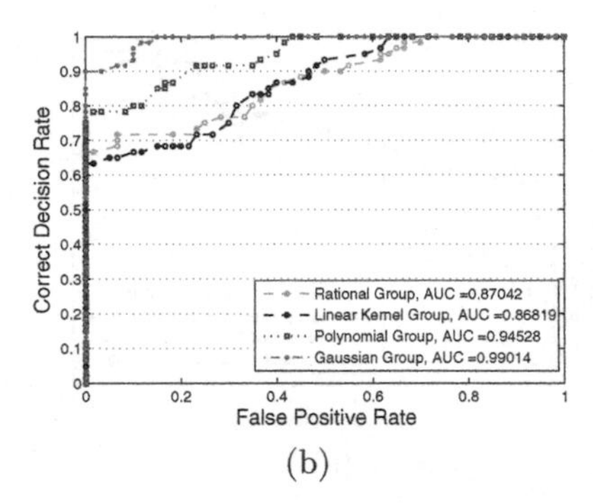

(b)

Fig. 4. (a): ROC curve sketched for layer score 3. In this layer, web services accept the join invitation if the inviter is the among the first three best options in their list. (b): ROC curve sketched for layer score 4. In this layer, web services accept the join invitation if the inviter is the among the first four best options in their list.

5 Conclusion

The contribution of this paper is applying SVM to facilitate the process of finding optimal cooperators in regards to enhancing web services' overall performance as service providers. In the deployed infrastructure, web services can operate alone as single service provider. But, they might reduce their quality of service over time and hence, they cannot keep high reputation. We considered communities of web services that gather web services of similar functionalities and discussed their management issues. In this paper, we applied SVM algorithm with three different kernel functions to improve the management process of communities gathering web services. We applied all the steps of the algorithm and explained it in our particular domain of interest. The discussion is accompanied by empirical results that highlight web services' and communities' performance in handling continuous tasks. The results show that Gaussian kernel function outperforms other two kernel functions as well as non-learning model.

Our plan for future work is to advance the learning process such that web services that risk to join can deploy different learning algorithms that enables them to measure their improvement chance. We also need to check the impact of different learning parameters (i.e., α, β, and γ) in the results we find and regulate their automated setting/update mechanism. We also need to improve the learning algorithm to adapt itself with the new chunk of input data. At the end, we need to expand the proposed framework to handle leave matters in addition of what is done in terms of join matters. Web services that are a member of a community might consider to leave to act alone or join another community and this is a decision that impacts web services' overall utility in competitive enterprise network.

References

1. Ardagna, D., Pernici, B.: Adaptive Service Composition in Flexible Processes. IEEE Transactions on Software Engeneering **33**(6), 369–384 (2007)
2. SVM as a Convex Optimization Problem. http://www.cs.cmu.edu/~epxing/Class/10701-08s/recitation/svm.pdf
3. Alchieri, E., Bessani, A., Fraga, J.: A Dependable Infrastructure for Cooperative Web Services Coordination. International Journal of Web Services Research **7**(2) (2010)
4. Zheng, Z., Ma, H., Lyu, M.R., King, I.: Collaborative Web Service QoS Prediction via Neighborhood Integrated Matrix Factorization. IEEE Transactions on Services Computing **6**(3), 289–299 (2012)
5. Olivaa, L., Ceccaronia, L.: REST Web Services in Collaborative Work Environments, pp. 419–427. IOS Press (2009)
6. Elnaffar, S., Maamar, Z., Yahyaoui, H., Bentahar, J., Thiran, Ph.: Reputation of communities of web services - preliminary investigation. In: Proc. of the 22nd IEEE Int. Conf. on Advanced Information Networking and App., AINA 2008, pp. 1603–1608 (2008)
7. Fourquet, E., Larson, K., Cowan, W.: A Reputation Mechanism for Layered Communities. ACM SIGecom Exchanges **6**(1), 11–22 (2006)
8. Jurca, R., Faltings, B.: Obtaining Reliable Feedback for Sanctioning Reputation Mechanisms. Journal of Artificial Intelligence Research **29**, 391–419 (2007)
9. Kalepu, S., Krishnaswamy, S., Loke, S.W.: A QoS metric for selecting web services and providers. In: Proc. 4th Int. Conf. on Web Information Systems Engineering Workshops, pp. 131–139 (2003)
10. Kang, D.-K., Fuller, D., Honavar, V.: Learning classifiers for misuse and anomaly detection using a bag of system calls representation. In: Systems, Man and Cybernetics (SMC) Information Assurance Workshop. Proceedings from the Sixth Annual IEEE, pp. 118–125 (2005)
11. Kastidou, G., Larson, K., Cohen, R.: Exchanging reputation information between communities: a payment function approach. In: Proceedings of IJCAI 2009, pp. 195–200 (2009)
12. Malik, Z., Bouguettaya, A.: Evaluating rater credibility for reputation assessment of web services. In: Benatallah, B., Casati, F., Georgakopoulos, D., Bartolini, C., Sadiq, W., Godart, C. (eds.) WISE 2007. LNCS, vol. 4831, pp. 38–49. Springer, Heidelberg (2007)
13. Maximilien, E.M.: Multiagent system for dynamic web services selection. In: The 1st Workshop on Service-Oriented Computing and Agent-based Eng., SOCABE 2005, pp. 25–29 (2005)
14. Maximilien, E.M., Singh, M.P.: Conceptual Model of Web Service Reputation. SIGMOD Record **31**(4), 36–41 (2002)
15. Menasc, E., Daniel, A.: QoS Issues in Web Services. IEEE Internet Computing **6**(6), 72–75 (2002)
16. Shakshuki, E., Zhonghai, L., Jing, G.: An Agent-based Approach to Security Service. International Journal of Network and Computer Applications **28**(3), 183–208 (2005)
17. Xiong, L., Liu, L.: PeerTrust: Supporting Reputation-based Trust for Peer-to-Peer Electronic Communities. IEEE Transactions on Knowledge and Data Engineering **16**(7), 843–857 (2004)
18. Zhang, Y., Zheng, Z., Lyu, M.R.: WSPred: a time-aware personalized QoS prediction framework for web services. In: IEEE 24th International Symposium on Software Reliability Engineering (ISSRE), pp. 210–219 (2013)

List Price Optimization Using Customized Decision Trees

Kiran Rama(✉), Shashank Shekhar, John Kiran, Raghava Rau,
Sam Pritchett, Anit Bhandari, and Parag Chitalia

VMware Inc., Palo Alto, USA
{rki,sshekhar,kjohn,drau,spritchett,anitb,pchitalia}@vmware.com

Abstract. There are many data mining solutions in the market which cater to solving pricing problems to various sectors in the business industry. The goal of such solutions is not only to give an optimum pricing but also maximize earnings of the customer. This paper illustrates the application of custom data mining algorithms to the problem of list price optimization in B2B (Business to Business). Decision trees used are typically binary and pick the right order based on impurity measures like Gini/entropy and mean squared error (for example in CART). In our study we take a novel approach of non-binary decision trees with order of splits being the choice of business and stopping criteria being the classical. We exploit proxies for list price changes as discount %age and Special Pricing Form (SPF) discounting. We calculate transaction thresholds, anchor discounts and elasticity determinants for each Stock Keeping Unit (SKU) segment to arrive at recommended list price which gets used by pricing unit.

Keywords: Non-binary decision tree · List price · SPF · B2B · Anchor discount · Classification · Regression · Entropy · Log-loss

1 Introduction

VMware (VMW) is a virtualization, end user computing and cloud company with annual revenues of over USD 6 BB (as of 2015) and a market cap of USD 25 BB [1]. VMware sells products in the Software Defined Data Center (vSphere, VSAN, NSX for computing, storage & network virtualization respectively), end user computing (Airwatch, Horizon, and Fusion/Workstation) and cloud. These are all sold to B2B customers.

The prices of products at VMW have been rarely changed over time. However Sales representatives could offer discounts via SPF flag which a special discretionary discount was typically given to large orders. The Pricing Business Unit was keen to figure out a way to get to optimal list prices at VMW.

1.1 Objectives

The Advanced Analytics & Data Sciences team came up with the following objectives with the Pricing Business Unit.

P. Perner (Ed.): MLDM 2016, LNAI 9729, pp. 88–97, 2016.
DOI: 10.1007/978-3-319-41920-6_7

- Analyze historical prices and related volume movements and understand major drivers of discount % and SPF (special pricing form/sales person specific discount) Flag. Since, the prices haven't changed much historically, discount and SPF have been considered as proxies to list price changes.
- To come up with recommended list price along with level of confidence for all SKUs.

2 Solution Framework

2.1 Strategy

A traditional list price optimization would use price changes versus quantity changes, but we never change prices. Discounting practices are an alternative inference method but requires additional steps.

Discount% and SPF requests are useful indicators to infer the customer's assignment of value to a product. Segments will be arrived at for discount % and SPF flag. SPF flag is a discount that can be given by a sales representative on request – mostly given on large order sizes. List price is used to measure VMware's assignment of value to a product. Imbalance between assignments of value then indicate tension in List Price. Following detailed steps were planned –

- Understand segments of the order-SKUs based on discount percentage and also SPF Usage
- Understand the importance of factors that drive variation between segments and within segment
- Incorporate explicit price-volume elasticity algorithms and come up with pseudo elasticity determinant

2.2 Segmentation Approach

Objective was to arrive at segments of order-SKUs based on discount% and SPF usage (proxies to list price changes). Business required that the decision tree be non-binary in nature and in a particular order. Non-Binary decision trees were built to create segments to explain discount% and SPF flag. Product platform, Order Size, Local Currency, Industry Vertical, Partner Tier were used in the same order. Uniqueness of the tree is that non-binary tree implementation that has us determine the order of the input variables based on business perceived improvements.

Working of a normal decision tree – Decision trees [2] in standard software packages decide the list of features and the order of features based on mathematics related to mean squared error, entropy, log-loss, Gini [3] etc.

The customized decision tree made during this study was needed to be built from the leaf node and up using packages in R [4]. It takes the features in the same order required by the business and groups the order-SKUs accordingly. The standard splitting criteria is now replaced by the custom criteria required by the business. The stopping criterion is minimum number of observations. These kind of custom trees are neither available as part of any standard package nor have a custom splitting and stopping criteria. It also uses hand-built mean squared error and log-loss for regression and classification trees respectively. Price elasticity based on above didn't yield significant results and will be trying to arrive at recommendations based on segments of a single product. We will correlate discount % and SPF Usage for segments against their characteristics and use insights from the above segmentation to arrive at list price recommendations at product platform level.

3 Modeling

3.1 Dataset Creation

The data matrix for the model was aggregated at an order-SKU level using Greenplum and Hadoop. This is because our entity is the combination of an order and a SKU. Specific business level judgements were applied like:

- dataset for a time period of FY13Q1 to FY14Q2 with license only transactions for only partner channel.
- Specific products like vCloud Air that had a subscription model/enterprise license agreements (ELAs) that have pre-determined prices transactions while creating this dataset.
- Given large amounts of data, the data was picked from Hadoop, where we took last 5 years of dataset related to site made available by IT in a flat tabular structure at order-SKU level.
- The continuous variables were subjected to coarse classing and fine classing making all the features categorical.

The key independent variables were Product platform, order size group, industry vertical currency code and partner type tier and the dependent variables were Discount percentage (for custom regression tree) and Special Pricing Form (SPF) flag (for custom classification tree).

The goal in our analysis is to arrive at elasticity of list prices by using the following as proxy:

- By using discount %age as a proxy
- Using the special discounting that sales reps give as an indication of the list price elasticity

Subsequently regression trees and classification trees were built to arrive at the above. We developed a unique algorithm that builds non-binary trees and shows all possible combinations of the features to the business. Businesses like white-box models and also control the order when the mathematical uplift generated by a feature is marginal. Our supervised algorithms custom-built allow this.

3.2 Custom Regression Tree for Discount Percentage

Features for the regression tree were chosen in the same order of importance as following: product platform, order size group, industry vertical, currency code, and partner tier.

We found *248* segments with their measures which were used to identify orders where discount % is very different from the segment mean. It was also used to group segments. The tree was a non-binary which used *mean-squared error minimization* [5]. It also can split the dataset using an order of variables/features that business desire. The attributes used to identify segments are mean discount %, proportions of order with SPF flag, # transactions, average booking amount and average unit list price in USD.

Table 1. Mock-up Representation of Regression/Classification tree data

Segment	Order Number	Product SKU	Actual Discount %	Segment Mean Discount %
VCLOUD SUITE UPGRADED	13499196	CLX-VXXX-STD-UG-AXXX	-20%	-34%
VCLOUD SUITE UPGRADED	13471982	CLX-VXXX-STD-UG-BXXX	-20%	-34%

3.3 Custom Classification Tree for SPF Flag

The feature order of importance was same as in the custom regression tree. We found *188* segments with their measures which were used to identify orders where proportion of SPF is high/low. It was also used to group segments. The tree was a non-binary which used *entropy for minimizing classification error*. It also can split the dataset using an order of variables/features that business desire. The attributes used to identify segments are mean discount %, proportions of order with SPF flag, # transactions, average booking amount and average unit list price in USD.

The outputs based on the decision tree approach using classification and regression [6] were joined back to our original dataset. Hence, we have two additional fields in our dataset – Discount Segment and SPF Segment. These were utilized in later part of the project.

3.4 Determining Relative Importance

The relative importance of the factors in the trees were determined by running custom regressions [7] on the predictions from the two decision trees. The relative importance of factors in the decision tree based on discount % was determined using linear regression while logistic regression was used for the decision tree based on SPF usage. The relative importance of the variable were determined for all the depth of the tree and for each splitting rule. Below are the order and magnitude of relative importance of the independent variables at the final depth of the tree:

Table 2. Relative Importance of Factors

Splitting Variable	Average Varlogloss	Relative Importance
Product Platform	0.22312171	30%
Order Size Group	0.22527636	27%
Industry Vertical	0.20068196	15%
Currency Code	0.18544909	14%
Partner Type Tier	0.22435609	14%

3.5 Comparison with Binary Decision Trees

A typical binary decision tree which works on the principle of splitting a node into two child nodes repeatedly, beginning with the root node that contains the whole learning sample uses standard splitting [6] (as an example CART includes least squares and least absolute deviation for regression trees and Gini, entropy, Symgini, twoing, ordered twoing and class probability for classification trees) and stopping criteria which will not give the analyst the ability to choose the features/attributes in the order desired by the business. The customized decision trees which were built for the project used custom splitting criteria which enables the analyst to select the order of the features used for splitting observations. As an example, when the standard binary tree was run on the same dataset, it gave order size group as the most significant variable and was the parent node whereas the business wanted the product platform to be the parent node. The rationale behind this was that discount and SPF are bound to be higher for the SKUs in bigger order size group and the business wanted to evaluate the behavior of the SKUs within the product platform. If the order size group is the parent node then the SKUs will spread across different segments and the discounting behavior cannot be evaluated at product platform level. Also, unlike a standard binary decision tree, the customized decision tree didn't have same number of splits at each level and hence a variable (for example product platform) appeared only once and not more than once as was the case in standard binary decision tree. This applies to all the levels of the customized decision tree and the user has the flexibility to choose desired variable at any level.

Table 3. Overview of comparison of Customized Decision Tree with Standard Binary Tree

Bucket	SI#	Binary Tree	Our custom Non-Binary Tree
Development	1	Only 2 splits at each level	Multiple splits at each level
	2	Same feature might get repeated many times from root to leaf	Same feature cannot get repeated from root to leaf more than once. This explains the business logic better
	3	The features are chosen based on impurity reduction	Full tree is exhaustively built for all combinations and presented as a choice to the business
	4	No overriding of what the algorithm choses	Allows business user to override feature splitting when the mathematical efficiency from split is marginal
	5	Stopping criteria is # of observations OR impurity criteria	Same here
Performance	1	Traditionally binary trees provide better performance than non-binary trees	Our algorithm is able to provide comparable accuracy of binary trees despite being non-binary with the added benefit of business overriding
Implementations	1	Custom implementations exist in R	No such implementation exists in R or in any other software package

The customized decision tree algorithm has been designed to handle large datasets. The algorithm utilizes the following stopping rules in order to control the tree growing process and make the mechanism efficient:

- If the size of a node is less than the user-specified minimum node size (2000), the node will not split.
- If the impurity of the child node is greater than the impurity of the parent node, the node will not split.
- If the current tree depth reaches the user-specified maximum tree depth limit value, the tree growing process will stop.

4 Utilizing Modeling Outputs

4.1 Identifying Anchor Product

A target SKU was selected based on overall bookings and profitability. This SKU was primarily the top selling SKU. *Selecting anchor SKU was a mutual exercise and business judgements were used to identify the SKU.*

4.2 Generating Model Parameters

Transaction Cutoff: The rationale to have a transaction cutoff was to set a minimum number of transactions required for the model to make a recommendation

$$T(Q_i) = \text{mean(transactions)} - \sigma \tag{1}$$

Capped List Price percentage change: Sets a maximum recommendation as % change from original list price. This was set at 15% after due consultation with the pricing team on what is an acceptable limit.

List Price Integer rounding: Rounds recommendations to make value appear more similar to common rounding practice. This was done in alignment to the corporate standards. In this particular case, we rounded to the nearest $5/LC5.

4.3 Calculating Anchor Discount

One of the top selling SKUs was selected as the anchor product. The mean of all the discount % was calculated for the selected SKU. The model was evaluated iteratively using the mean of discount %age. An adjustment was made/error was added to the mean to come up with the anchor discount. Adjustment was made to ensure that the anchor discount satisfies the ranges from the pseudo elasticity determinant.

$$\text{Anchor Discount} = \text{Mean of Discount \%} + \text{Adjustment Amount} \tag{2}$$

4.4 Pseudo Elasticity Determinant

Regression of sum of quantity to gross per unit USD value for different product platform and different pricing and SPF segments. The goal was to arrive at a pseudo elasticity determinant and conform that the change in list price doesn't result in decrease in volume for certain important segments.

4.5 SPF % Intercept and Slope

Expected SPF intercept and slope are used to determine what SPF usage should be expected based on a SKU's list price. The intercept and slope came out of a regression of SPF usage on the list price for selected representative SKUs.

$$\text{SPF Usage} = \alpha + \beta * \text{List Price} \tag{3}$$

$$\text{SPF Intercept} = \alpha \tag{4}$$

$$\text{Slope} = \beta \tag{5}$$

4.6 Expected SPF

This was calculated for each SKU based on the regression equation generated from the regression of SPF usage on the list price for selected representative SKUs.

$$\text{Exp. SPF of SKU} = \alpha + \beta * \text{List Price of SKU} \tag{6}$$

4.7 SPF Flag

This was generated as an indication to whether SPF usage for a SKU is recommended or not. There are three different cases possible for SPF Flag based on following criteria:

$$\text{Flag 'YY'} - \text{if expected SPF} > \text{Actual SPF} + 10\% \tag{7}$$

$$\text{Flag 'Y'} - \text{if expected SPF} > \text{Actual SPF} + \text{Delta } 10\% \tag{8}$$

$$\text{Flag 'N'} - \text{if expected SPF} < \text{Actual SPF} + \text{Delta } 10\% \tag{9}$$

Delta % has to be entered by the business user and is capped at 10%. For current exercise, we took it as 0%.

4.8 Suggested List Price

This is the suggested list price without any adjustment and rounding off.

$$\text{Suggested LP} = \frac{\text{LP x (1-Actual Discount)}}{\text{(1-Anchor Discount)}} \tag{10}$$

4.9 Recommended List Price

The recommended list price was arrived at after applying the transaction cut-off constraint, capped list price percentage change constraint and the list price integer rounding.

Table 4. Model Parameter / Assumptions for Mock-up Calculation of Recommended List Price

Model Parameters/ Assumptions	Values (Product 1)
D_A = Anchor Discount	35%
T_C = Transaction Cut – Off	50
I_{LP} = Capped List Price Increase	15%
LP_R = Rounding to LP to the nearest	$5
SPF_{IN} = SPF % Intercept	0.90%
SPF_S = SPF % Slope (per $1,000)	$0.0000295
SPF_δ = Delta SPF% (capped at 10%)	0%

Table 5. Model Parameter / Assumptions for Mock-up Calculation of Recommended List Price

Model Parameters/ Assumptions	Values (Product 1)
Product Platform	ABC+
Product SKU	VS-ABC-PL-C
LP_{US} = US List Price	$3,495
Rev_{FLAT} = % Platform Revenue	83.40%
Rev_{GROSS} = Gross Revenue	$113,636,152
#tx = Transactions	9,965
Rev_L = List Revenue	$172,292,318
$Disc_E$ = Extended Discount	$58,656,166
Disc % = % Discount = $Disc_E$ / Rev_L	34.04%
SPF_w = Weighted SPF	6.0%

Table 6. Mock-Up Calculation for Estimated Value of List Price

Model Parameters/ Assumptions	Values (Product 1)
SPF_{EXP} = Expected SPF % = LP_{US} x SPF_S+SPF_{IN}	11.21%
SPF Flag	Y
$LP_{Suggested}$ = LP_{US} x (1 – Disc %) / (1-D_A)	$3,546.37
LP_δ = List Price % Delta = ($LP_{Suggested}$ / LP_{US}) - 1	1.47%
LP_{Capped} = Round(If(LP_δ>LP_{US}, LP_{US} x (1 + I_{LP}), $LP_{Suggested}$)	$3,546
$LP_{Recommended}$	$3,545

5 Conclusion

The solution provided for optimizing the list price at VMware Inc. is unique and distinctive because there is hardly any instance of price change at VMware Inc. The custom non-binary decision trees that were built in the first phase of the project are piece of IP creation as there is no existing software/package which builds non-binary decision trees. Also, the implemented approach was very different from standard price optimization methodologies where historical price changes are utilized to arrive at recommended price changes.

From an algorithmic perspective, the approach is unique and there is no such implementation that involves non-binary trees, allows user to override certain decisions in tree building that explain the business processes better. It also allows to see the different combinations of tree building and customized choices. Also, a traditional list price optimization would use price elasticity (price change versus quantity change) but since the list price at VMware Inc. haven't changed significantly, we utilized the segments from the customized decision tree to compute the pseudo elasticity and preclude any decrease in quantity.

The recommendation to change the list price of about ~40% of the SKUs was given to the business. The recommended list price was calculated in local currency for each country and the business was suggested to start with modification to high confidence products. It was also recommended to make the list prices changes independent of other changes as associating a list price change with a version increase will make it harder to directly measure the customer's perception of value. One of the key endorsement was to control the discounting process and enable automated contractual discounts and clear the rules and boundaries. The current method of arriving at the list price utilizes the method of pseudo elasticity determinant. The recommendations from this project will help track the actual list price changes to actual unit quantity changes.

References

1. VMware Inc: Annual Report. http://ir.vmware.com/annuals.cfm
2. Quinlan, J.R.: Induction of Decision Trees. Machine Learning **1**, 81–106 (1986). Kluwer Academic Publishers
3. Shannon, C.E.: A Mathematical Theory of Communication. Bell System Technical Journal **27**(3), 379–423 (1948). doi:10.1002/j.1538-7305.1948.tb01338.x. (PDF)
4. Dowle, M., Srinivasan, A., Short, T., Lianoglou, S.: with contributions from Saporta, R., Antonyan, E. https://cran.r-project.org/web/packages/data.table/data.table.pdf (PDF)
5. Hastie, T., Tibshirani, R., Friedman, J.H.: The elements of statistical learning: Data mining, inference, and prediction. Springer Verlag, New York (2001)
6. Breiman, L., Friedman, J.H., Olshen, R.A., Stone, C.J.: Classification and Regression trees. Wadsworth & Brooks/Cole Advanced Books & Software, Monterey (1984). ISBN 978-0-412-04841-8
7. Cohen, J., Cohen, P., West, S.G., Aiken, L.S.: Applied multiple regression/correlation analysis for the behavioral sciences, 2nd edn. Lawrence Erlbaum Associates, Hillsdale (2003)

AdaMS: Adaptive Mountain Silhouette Extraction from Images

Daniel Braun, Michael Singhof(✉), and Stefan Conrad

Institut für Informatik , Heinrich-Heine-Universität Düsseldorf,
Universitätsstr. 1, 40225 Düsseldorf, Germany
{braun,singhof,conrad}@cs.uni-duesseldorf.de

Abstract. Modern image sharing platforms such as instagram or flickr support an easy publication of photos to the internet, thus leading to great numbers of available photos. However, many of the images are not properly tagged so that there is no notion of what they are showing.

For the example of mountain recognition it is advisable to create reference silhouettes from digital elevation maps. Those are matched with the silhouette extracted from a given image in order to recognise the mountain. It is therefore necessary to obtain a very precise silhouette from the query image.

In this paper, we present AdaMS, an adaptive grid segmentation algorithm, that extracts the silhouette from an image. By the help of an artefact detection method, we find erroneous parts in the silhouette and show how our algorithm uses this information to recalculate the silhouette in the surroundings of the error. We also show that our method yields good results by evaluating our approach on an existing data set of mountain images.

1 Introduction

In times of social media services and a high spread of smart phones, the importance of sharing our experiences with other people rises as part of our today's life. As a result of this, every day the number of publicly accessible photos increases significantly, which can be observed in image sharing platforms like Instagram where users share about 80 million [7] new photos per day. Unfortunately, the majority of these images is not properly tagged and therefore we have no notion of what they are showing, which makes the search for images with specific objects as motif difficult. This leads to a rising need for efficient and precise algorithms for automatic object recognition in images, so that a subsequent tagging, without the need for time consuming human interactions, is possible. Therefore, the significance of this research field increases, what concludes in a high amount of innovations and advances in this area in the last decades.

Our work focuses on the automated landmark recognition for the example of mountain recognition. This task is still challenging, especially because of the problems that are a consequence of the motif itself. These are, for instance, volatile weather conditions, the snowline in combination with clouds,

P. Perner (Ed.): MLDM 2016, LNAI 9729, pp. 98–112, 2016.
DOI: 10.1007/978-3-319-41920-6_8

or vegetation that hides the mountain. As a result, the extraction of meaningful features to describe a mountain, which should be recognized in other images, gets more complex. In addition to this, the appearance of the mountain depends strongly on the viewpoint of the camera, leading to an enormous amount of different feature descriptions for just one mountain to recognize. Therefore, one common method to identify a mountain in an image is to match the silhouette of the mountain with known silhouettes, which for example have been extracted out of a digital elevation map of the mountains to compare with. With the growing spread of devices that have the capability to tag an image with GPS coordinates of the camera position, like smartphones and cameras with an attached GPS unit, this task becomes simpler. This is due to the fact that, with the known position, an algorithm has just to check the surroundings of this location and therefore match the extracted silhouette only against a low number of mountain candidates in this area. However, there are still many images without this advantage, especially in older image collections, so that an algorithm, which can handle images without GPS data, is still a valuable aim. On the downside, we have to check every mountain on the earth, which makes a highly precise silhouette of the mountain in the query image inevitable.

In this work we introduce AdaMS, an adaptive segmentation algorithm, for the purpose of extracting a mountain silhouette out of an image, that tries to reduce the amount of artefacts through segmentation errors or obstacles in the resulting silhouette. For this, we first use an outlier detection algorithm to identify possible artefacts and afterwards classify the encountered outliers to choose the right removal method. If the outlier is classified as a segmentation error, we locally recalculate the segmentation to eliminate it. We therefore use a grid based approach for the segmentation, to get the opportunity to locally modify the segmentation parameters. If, on the contrary, the artefact is caused by an obstacle and therefore the segmentation recalculation would achieve no further improvement, we eliminate this artefact by simply cutting it off.

This paper is structured as follows: In the next section we discuss related work. Afterwards, we introduce our algorithm and give a detailed explanation of its different parts. In chapter four we evaluate the algorithm, before we summarise our work in the last chapter, where we furthermore outline our future work.

2 Related Work

Baatz et al. [2] were the first to target the task of large-scale geo-localisation. For the silhouette extraction they propose an approach which is based on unary data costs for the belonging of a pixel to the sky segment. This approach is combined with a possible user interference, where the user can mark sky or ground pixels for better results. For 49% of the images in their dataset, which they have collected during their research, this user intervention was needed. Thankfully, they have published this dataset, so that it can be used for evaluating our algorithm. For a large scale approach, a fully automated segmentation would be preferable, because every user interference is a bottleneck.

Such an image segmentation is in general a significant technique in many object extraction and recognition tasks. Therefore the research in this field advanced and led to a great number of proposed methods, such as the watershed transform [6], region based algorithms [10] or image clustering [12]. In [10] the authors present a seeded region growing algorithm based on different probability maps calculated for the whole image. Their method relies on homogeneous data and therefore the right probability map for a good segmentation result has to be chosen manually, just as the seed pixel, which determines the region of interest. In contrast to that, our goal is a fully automated parameter determination for the subsequent silhouette extraction. Perner [13] suggests to use case-based-reasoning (CBR) to choose the right parameters for a segmentation task, so that the parameters of the nearest case for an input image are used for the segmentation. In the work of Frucci, Perner and Sanniti di Baja [6], for example, such a CBR system is used for a modified watershed transform algorithm, which reduces the problem of over-segmentation significantly. In comparison to that, we try to choose the parameters locally in the image to automatically reduce the error in respect to the extracted silhouette.

The authors of [12] suggest a k-means clustering for segmentation and introduce an estimation method for the right number of clusters. Therefore, they extract the edges out of the image, by using phase congruency, and merge them as long as the distance of the average colour is below a given threshold. The number of different edges then serves as number of clusters for the subsequent clustering step. For multi-object segmentation both methods can be advisable, but in the case of silhouette extraction we have only two possible classes, namely sky and ground, and therefore the segmentation process has to yield two resulting segments. So, a subsequent region merging step would be necessary to extract the silhouette, but it seems beneficial to use an algorithm, which already takes the requirement of a binarisation for the segmentation step into account and therefore can optimise the result regarding this condition.

This binarisation is, for example, widely used for tasks like character recognition in document images or general foreground/background separation. In the last decades many solutions where presented, like using support vector machines [17], graph cut with Gaussian mixture models [14], global [9,11] and local [16] thresholding, or Markov random fields in combination with the Dempster-Shafer theory [5]. For both tasks, global threshold based algorithms, like Otsu's method [11] or the algorithm from Kittler and Illingworth [9], are often successfully used to partition the image into two groups of pixels [15]. A related technique is proposed in [17]. Their algorithm first clusters the pixels with fuzzy C-means and thereafter the resulting membership values of randomly chosen pixels are used to train a SVM, which serves as a kind of separating threshold, for classifying the remaining pixels. But as a result of the given conditions in mountainous regions an image thresholding algorithm based only on the brightness or colour values of the whole image is not advisable. This is due to the common inhomogeneity of the brightness values, be it the sky, for example through weather conditions, or the ground in general, and a possible fluent passage between the sky and the ground segment. Both lead to a high variation in illumination and contrast which can result in a non-optimal threshold for finding the exact silhouette of the mountain. To overcome similar problems, local thresholding techniques have been proposed. Roman et al. [16], for example, present a local thresholding, which calculates a threshold for a pixel relying on the mean and the mean deviation in a local window around it, by using the integral sum for efficient computation.

With the target of only extracting the border between mountain and sky and the prior knowledge that the sky will be a portion of the upper part of the image, we use a seed growing algorithm with local constraints to find the initial silhouette of the mountain. Having the other target of an unsupervised silhouette extraction process, solutions like GrabCut [14], in which the user gives first a bounding box to locate the foreground object and then possibly interferes to optimise the result, respectively the technique proposed from Chen et al. [5], which needs the user to mark some foreground/background pixels before the segmentation starts, are not feasible. Even so, the GrabCut algorithm can be used for an automated segmentation, if we have some prior knowledge over the background of the image.

Another way to find the mountain silhouette is to analyse all edges in the image in respect to their plausibility to be part of it. The authors of [8] extract the edges in the images with the Canny algorithm [4] and use different filters to reduce the candidate edges. Afterwards they use a measure to find the silhouette with the highest probability to be the skyline. By choosing only one resulting silhouette, this suffers from edges with low contrast, which leads to a fragmented skyline and therefore only a part of the skyline will be found. Ahmad et al. [1] use both, edge-less and egde-based approaches, to find the horizon line as shortest path in a classification map, which values represent the horizon-ness of each pixel, with dynamic programming. A related technique is presented in [3], where the authors use an edge map to find the right terrain alignment by matching this against an edge map created with a digital elevation map. Contrary to our target, this method needs the knowledge of the viewpoint location, through GPS data, and they do not extract the skyline particularly.

3 AdaMS Extraction

As we stated in the first chapter, the use of digital elevation maps to identify the mountain in a query image is beneficial. But without the use of prior knowledge, like GPS data or user pre-selection of possible locations, our algorithm has to check every mountain in the database, which means, that we have to compare the silhouette of the query image with many, through view point relocation, possible silhouettes of these mountains. Therefore we need a highly accurate query silhouette to achieve a good precision in the subsequent matching process. To achieve the desired high accuracy, our AdaMS (Adaptive Mountain Silhouette) extraction algorithm identifies possible artefacts in the extracted silhouette and tries to eliminate them through a refined segmentation, if it is a

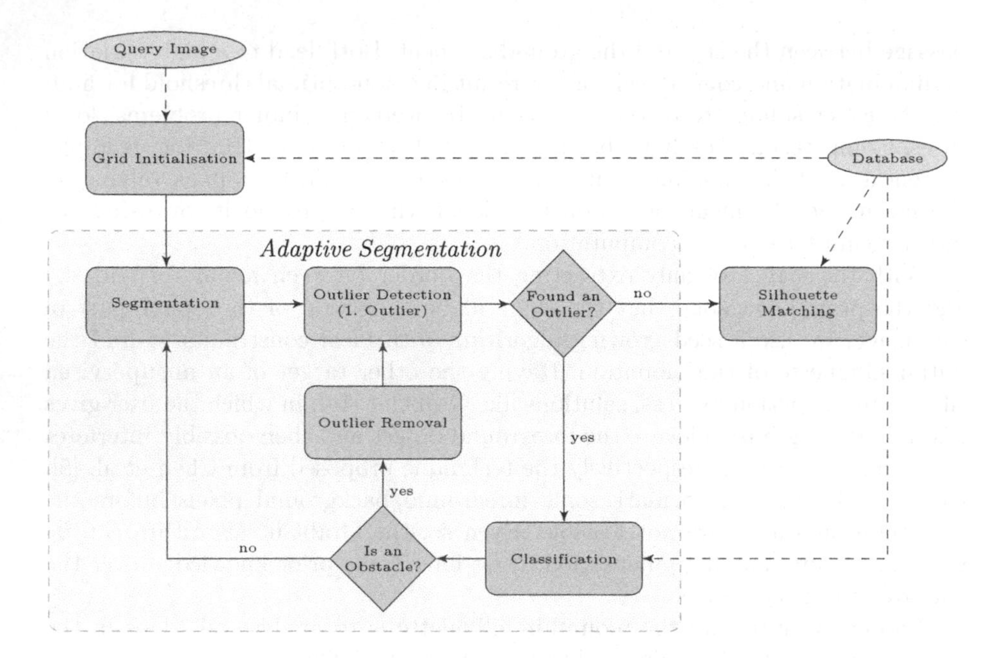

Fig. 1. Flow diagram of AdaMS.

segmentation error, respectively a straight forward removal policy, if the artefact is caused by an obstacle. These different artefact types are described in the following subsection, where we additionally point out some causes for these artefacts. Afterwards, we define the grid which will be used for the segmentation process. Finally, we introduce in the following subsections the five steps of our silhouette extraction algorithm. As can be seen in Figure 1 these steps are grid initialisation, image segmentation, silhouette extraction, outlier detection, and silhouette refinement. The latter one differentiates between outlier removal and recalculation of the segmentation for some parts of the silhouette.

The grid initialisation overlays the image with a grid with predefined grid element spacing. In the segmentation step, in which we use some pixel marked as sky of the image as seed to let the sky segment grow, we finally extract the transition between sky and ground segment as initial silhouette. This silhouette is passed to the outlier detection step, to find all anomalies in its structure which are then classified. For every segmentation error the algorithm tries to recalculate the segmentation in a local area around the artefact.

3.1 Artefact Definition

For the silhouette extraction we have to identify the pixels in the image which are part of the sky. Therefore we have a binary segmentation task to solve, which suffers from possible error sources with the result of possible artefacts in the extracted silhouette. But as a whole we can condense these into the following three main sources: The first one is a general problem with images, where several problems, like noise pollution, background clutter, blur, strong illumination variations, or a reduced contrast of the skyline through over-/underexposure, can occur. As result of such an error, the extracted silhouette can have a saw-tooth-shaped appearance. The second one is a consequence of the motif itself, because of highly volatile weather conditions, clouds or mist can hide parts of the mountain silhouette completely or reduce the contrast of it significantly, especially in combination with snow on mountain peaks, so that a separation of the sky from the ground gets difficult. The last main error source is a result of the natural environment of the mountains. Therefore we have to deal with different possible obstacles, like trees or other vegetation, humans or buildings in the foreground, or cables from a ski-lift, in the image, which partly cover the mountain. Altogether, this problems make the segmentation and therefore the extraction of the real mountain silhouette error prone.

The disparity of these errors leads to another problem, because an algorithm which solves one of it may not be ideal for the other ones. An artefact, which is a result of a low contrast at the mountain silhouette, may be solvable through a modification of the segmentation parameters. But if an obstacle leads to an erroneous silhouette, the segmentation step has worked properly, so that another parameter set will not lead to any improvement. Therefore, if we identify an artefact in the extracted silhouette, we have to classify it, so that we can choose the right method to get rid of it.

3.2 Grid

For the binary segmentation of the image we use a seed based region growing algorithm. This means that the algorithm chooses some start points as sky and then adds every neighbouring pixel that fulfils certain conditions.

For this, we define a grid G as follows:

Definition 1. *Given an image I with the width w_I and height h_I. Then the grid G with step size d_G consists of the pixels*

$$p_{i,j} = (\frac{d_G}{2} + id_G, \frac{d_G}{2} + jd_G)$$

with $0 \leq i < \frac{(w_I - \frac{d_G}{2})}{d_G}$ *and* $0 \leq j < \frac{(h_I - \frac{d_G}{2})}{d_G}$.

This results in a total number of

$$n = \lceil \frac{(w_I - \frac{d_G}{2})}{d_G} \rceil \cdot \lceil \frac{(h_I - \frac{d_G}{2})}{d_G} \rceil$$

grid points g_i with $i \in \{1, \ldots, n\}$ defining n grid cells C overlaying the image.

3.3 Segmentation

For the region growing algorithm we have to specify some seed points, where the algorithm starts to add connected points, if they satisfy a given condition. For this task we can use the assumption, that the sky will be localised at the top of the image. Because of this, we choose the highest pixel row in the image, mark it as sky and finally use this set of pixels as initial sky segment.

Starting at this seed points, the algorithm then adds successively every 8-connected pixel to the sky segment, if for the pixel holds

$$|B_{p_{(x,y)}} - mean^r_{p_{(x,y)}}| < \gamma \cdot \sqrt{V_G}$$

with $B_{p_{(x,y)}}$ the brightness of the pixel $p_{(x,y)}$, $mean^r_{p_{(x,y)}}$ as the mean of the brightness in a neighbourhood of the pixel with the radius of r and γ as a scaling factor. When the algorithm cannot add any further point, the binary segmentation ends. Now the silhouette extraction could start, but there are two drawbacks at this point. First, it is possible, that for instance clouds have been marked as ground and therefore the sky segment is not homogeneous. To get rid of such additional ground patches, we eliminate every patch which has no direct connection to the right or the left border of the image, resulting in only two uniform segments for the sky respectively ground. Finally, through small segmentation errors at the border between sky and ground, the transition can have small parts with the width of one pixel, which would result in a jagged silhouette. To remove this parts, we use erosion as a morphological operation. Therefore, a square with the width w_{struct} centred around each pixel acts as structuring element. Having the class $C_{p_{(x,y)}} \in \{sky, ground\}$ we count every pixel in the hereby defined neighbourhood with the same class and switch $C_{p_{(x,y)}}$ if the ratio of this pixels is below the threshold ζ. After this cleaning step we end with a smoothed border between the two parts of the image.

At this point, the algorithm starts the silhouette extraction by finding a ground pixel in the first pixel column on the left side of the image. If it finds no appropriate point, the algorithm will check the next columns until a ground pixel is found or the right side of the image has been reached. Afterwards it tracks the transition between sky and ground and adds every pixel on the upper border of the ground as vertex $v = (x, y)$, with x and y the pixel coordinates in the image, of the polygonal chain S which constitutes the silhouette of the mountain.

As last step of the extraction, our algorithm eliminates every vertex for which it holds that it is positioned on the line between its predecessor and its ancestor. In this case, the point carries no further information regarding the extracted silhouette and therefore it can be safely removed.

3.4 Outlier Detection

As stated before, the silhouette S extracted in the previous step can contain artefacts and therefore the algorithm now tries to identify the erroneous parts of it. For that, we first convert the silhouette to a representation which is independent from the absolute position of each vertex in the query image. This is necessary, because otherwise equal parts of the different silhouettes could be different through varying positions, which makes pattern recognition difficult. Therefore we define the following representation that is based on relative positions of the vertices and similar to polar coordinates with complex numbers.

Definition 2. *A* relative silhouette $RS = (v_1, \ldots, v_n)$*,* $n > 0$*, is a polygonal chain with* $v_i = (l_i, a_i)$ *for all* $1 \leq i \leq n$*, where* $l_i > 0$ *is the length of a line segment and* $a_i \in (-180°, 180°]$ *is the angle relative to the x-axis.*

Using this representation of the silhouette we now compare parts of a relative silhouette to reference data R, free of outliers, by computing histograms on the silhouettes and computing a distance between the histograms. These histograms contain the same two dimensions as the relative silhouette, namely segment length and angle relative to x-axis. In the following, by H_R we denote the reference histogram, i.e. the histogram computed from the reference data.

Definition 3. *Given a relative silhouette* RS*, then* $H_{RS}(s, l)$ *denotes the histogram consisting of the points* $v_s, \ldots, v_{s+l-1}$ *of* RS *and* H_{RS} *denotes the histogram over all vertices of* RS*.*

Now, for a silhouette $RS = (v_1, \ldots, v_n)$ of a query image we compute the histograms $H_{RS}(1, l)$ to $H_{RS}(n - l, l)$ and their respective distances to the reference histogram $d_i = \text{dist}(H_{RS}(i, l), H_R)$ via a sliding window approach. Then the anomaly score $an(v_j)$ of a vertex v_j is computed as the average of the set of distances $\{d_i | i \leq j \wedge j < i+l\}$, i.e. the distances of histograms that are computed over parts of the silhouette that contain v_j.

Based on the mean μ and the standard deviation σ for the single vertices' distribution of anomaly scores, we introduce two thresholds τ_{in} and τ_{out} similar to the double threshold approach in [4]. These lead to two kinds of anomalies.

Definition 4. *Let* $RS = (v_1, \ldots, v_n)$ *be the silhouette of an image with corresponding anomaly scores* $an(v_i)$ *for the vertex* v_i*, reference outlier score distribution mean* μ *and standard deviation* σ *and the two thresholds* $0 < \tau_{out} < \tau_{in}$*.*

Then we call v_i *a* weak anomaly *if*

$$an(v_i) \geq \mu + \tau_{out} \cdot \sigma$$

and a strong anomaly *if*

$$an(v_i) \geq \mu + \tau_{in} \cdot \sigma.$$

Note, that a strong anomaly always is a weak anomaly, too. The second thing to remember here, is, that for ease of understanding, we do *not* use "anomaly" and "outlier" synonymously. With "anomaly" we refer to a single vertex in a polygonal chain as in the definition above. In contrast to this, with "outlier" we mean a part of a polygonal chain that consists of anomalies as the following definition states.

Definition 5. *Let $l > 0$, and RS, $an(v_i)$, μ, σ, τ_{in} and τ_{out} as in definition 4. We call $o = (v_i, \ldots, v_j)$ an l-outlier if the following is true:*

1. *For all v_k, $i \leq k \leq j$, it holds that v_k is a weak anomaly.*
2. *There exist $m, n \in \{i, \ldots, j\}$ such that $n - m \geq l$ and for all v_k, $m \leq k \leq n$, it holds that v_k is a strong anomaly.*

An outlier $o = (v_i, \ldots, v_j)$ is called a maximum l-outlier *if and only if neither $(v_{i-1}, \ldots, v_j)$ nor $(v_i, \ldots, v_{j+1})$ are l-outliers.*

In the outlier detection step, only maximum l-outliers are of interest, so, whenever we find l consecutive strong anomalies we expand the outlier as long as there are adjacent weak anomalies to that outliers. Given a silhouette RS, we are thus able to find all outliers in linear time relative to the length of the silhouette, since anomaly score computation can be accomplished in linear time as well.

3.5 Outlier Classification

As discussed in section 3.1, different types of outliers exist. For the classification process we use four classes of outliers. All kinds of obstacles are summarised in the class *obstacle*. For segmentation errors, we introduce two classes. If parts of the mountain are recognised as sky, i.e. the silhouette is too far downwards in the image, these outliers are classified as *segUp*. On the other hand, if parts of the sky are recognised as mountain and the silhouette is too far upwards, outliers are classified in *segDown*. Finally, we introduce a class for false positives that contains parts of the silhouette, that get marked as outliers but are not outliers, so we end up with a set of outlier classes

$$OC = \{obstacle, segUp, segDown, falsePos\}.$$

For the outlier classification itself, we choose a weighted k-nearest neighbour approach, using a manually tagged dataset of 80 outliers, 20 of each class, as reference data for the classification. Assuming this module gets an outlier o from the detection module, we compute the distance of its histogram H_o to all reference outlier histograms H_i^{ref} with $1 \leq i \leq 80$, resulting in a set of k nearest outliers, each saved as tuple of type and distance $\{(t_i, d_i)\}$. Finally, for every type $type_j \in OC$ we compute $s_j = \sum_{t_i = type_j} w_i$ and declare o to be of the type with maximum s_j. Here, the weighting w_i for the ith tuple is given by $w_i = \frac{1}{d_i + \varepsilon}$ with ε the smallest positive double in the programming language, since distances can become 0. Having classified the outlier, there are three possible ways to further process it. First, the outlier can be a false positive, which results in deleting it by marking this part of the silhouette as non-outlier. If it is classified as a real outlier, it is, in the case of an obstacle, passed to the artefact elimination tool, described in section 3.6 or, if it is a segmentation error, passed to the segmentation module introduced in section 3.7 for a local refinement.

3.6 Artefact Elimination

In the case of an obstacle hiding the mountain a re-segmentation would be pointless, because the first segmentation has worked properly and a less conservative segmentation can result in a jagged mountain silhouette if the algorithm marks too many pixel as sky. Furthermore, the obstacle hides the background completely, leaving no further information of the mountain contour which could be used for a preliminary processing of this part of the image. Therefore, a good way to handle such an outlier is to cut it off and replace it by a straight line, since, if the width of the outlier is small enough, the mountain will most likely have no greater curves at this place. But even for a greater obstacle such a replacement will be in most cases more natural for the mountain than the obstacle itself. Given the outlier o of length n, with v_1 and v_n as start respectively end vertex, our algorithm removes the obstacle in the following manner: First, we construct the line connecting both endpoints $l = \overline{v_1 v_n}$ and calculate the distance $d(v_i, l)$ for every vertex v_i, resulting in a list of distances D. Then, beginning at the start vertex v_1 and with a threshold parameter δ, the algorithm searches a vertex for which holds

$$d(v_i, l) > \delta \wedge \forall v_j \left(j < i \wedge d(v_j, l) \leq \delta\right).$$

In words, this is the first vertex with a distance exceeding the threshold and consecutive predecessors with distances smaller than δ. If on the one hand no vertex is found, every outlier vertex lies near the connecting line l and therefore the chance for an error, be it through the detection or classification step, rises. Furthermore, with a δ small enough, the impact on the silhouette precision should be negligible, so that the algorithm marks this part as non-outlier. If on the other hand a vertex is found, the algorithm searches such a point in respect to the end vertex and though considering all successors, returning the two vertices v_{low} and v_{high}, whose connecting line $\overline{v_{low} v_{high}}$ is used to remove the obstacle.

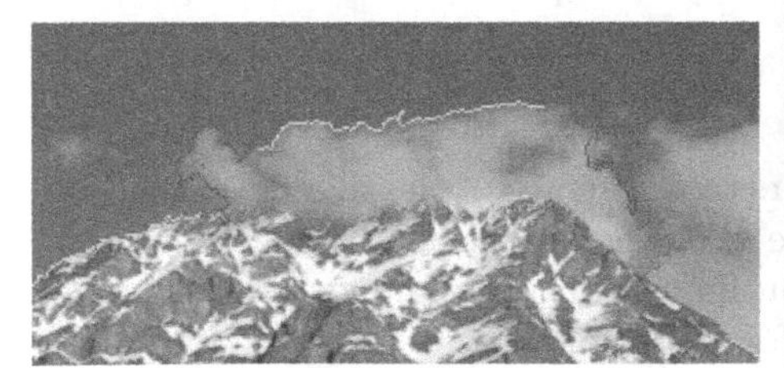

Fig. 2. An example of the local refinement. Left: The initial silhouette with the marked outliers. Right: The resulting silhouette after the local refinement terminated.

3.7 Local Refinement

Segmentation errors can occur when the segmentation is too conservative, which implies that the scaling factor γ is too low, and therefore clouds are not properly marked as sky (*segDown*) or when the contrast of the transition between sky and ground is low, so that the segmentation algorithm starts to add ground pixels to the sky segment (*segUp*), which is a result of a γ too high for this part of the image. With the knowledge of the outlier's class, a recalculation with a changed γ may solve the problem, as long as this refinement is localised around the outlier, because a changed value for the whole image could make the result at other parts of the silhouette even worse. Therefore a local refinement for the outlier removal is the target. For that purpose, we reuse the grid G defined in definition 1. Having the outlier o, the set C_o^{alter} of grid cells containing parts of the outlier and C_o^{update}, the union of C_o^{alter} and all directly connected grid cells, we define the local refinement step as follows.

Given the alteration rate θ, α as the predefined alteration value and the grid cells $C \in C_o^{alter}$ we first compute the new scale factor

$$\gamma_C^* = \alpha \cdot \theta + \gamma_C.$$

Here, the sign of α and therefore the direction of the alteration of γ_C is determined by the class of the segmentation error as follows:

$$\alpha = \begin{cases} 1, & \text{if } o \text{ has been classified as } segDown \\ -1 & \text{else.} \end{cases}$$

Afterwards, we recalculate the segmentation, using

$$|B_{p(x,y)} - mean^r_{p(x,y)}| < \gamma_C^* \cdot \sqrt{V_G},$$

for every point $p(x,y)$ which lies in a grid cell $C \in C_o^{update}$. The resulting silhouette is then extracted and passed back to the outlier detection module. This will be repeated until there are no outliers classified as segmentation errors left or I^{MAX} iterations have been executed. The latter one will result in the use of the originally extracted silhouette by additionally marking it as unsure silhouette. An example for the result after some refinement steps can be seen in Figure 2.

3.8 Global Refinement

If the ratio of detected segmentation errors to the length of the silhouette exceeds a predefined threshold θ, we recalculate the whole segmentation. This is due to the assumption, that such a high outlier ratio is a sign for a messed up segmentation through a wrong initial γ for the whole image. Therefore, we alter each grid point value and restart the segmentation on the whole image. For the selection of the right γ we use a grid search, using a predefined set of possible values, and choose the resulting silhouette with the lowest outlier ratio as start silhouette for the possibly needed local refinement.

4 Evaluation

In the following we evaluate the preciseness of our silhouette extraction approach on the Switzerland dataset from [2]. This dataset consists of 203 images

Table 1. Comparison of extracted silhouettes to ground truth. Values are given in pixels. RG1 and RG2: Region growing with a threshold of 1 respectively 2.

Parameter	GrabCut	Otsu	RG1	RG2	AdaMS
Avg. initial distance					18.80
Avg. final distance	26.03	118.74	27.61	49.31	15.75
Median final distance	2.58	97.44	1.65	1.81	1.43

and corresponding silhouettes as extracted by the authors. We use these silhouettes as ground truth here, since they have been refined manually. However, in some cases there are differences between the approaches. For example, in the ground truth silhouettes, most obstacles have not been removed. Therefore, in our approach, we skip the obstacle removal step in order to get comparable results. We use GrabCut [14], Otsu [11] and a region growing algorithm with a predefined threshold for the maximal allowed divergence in the brightness values of 4-connected pixels in the sky segment, for a comparison of our algorithm with other segmentation methods. GrabCut and the region growing algorithm need some given background pixels and therefore we choose, equally to our approach, the upper row as background. For the latter one, we furthermore choose the values 1 (RG1) and 2 (RG2) as threshold for the maximal brightness difference.

For comparing two silhouettes we use the following distance measure:

Definition 6. *Let S_1 and S_2 be the two silhouettes with n_1 respectively n_2 pixels and $pd(p_i^1, S_2)$ be the pixel distance from the point p_i^1 of S_1 to S_2 defined by*

$$pd(p_i^1, S_2) = \min_{1 \leq j \leq n_2} \{|y_i^1 - y_j^2| \colon x_i^1 = x_j^2\}.$$

Then the distance between S_1 and S_2 is given by

$$\max(\sum_{i=1}^{n_1} pd(p_i^1, S_2), \sum_{i=1}^{n_2} pd(p_i^2, S_1)).$$

Table 1 shows some aggregated results for the comparison. Average initial distance gives the average distance of the silhouettes after the first iteration, i.e. with the initially computed silhouette. Therefore, only our approach has such a value, because for the other algorithms there exists no refinement step. Average final distance gives the distance values after all iterations of the refinement step have been completed. Median final distance is the median of the final distances. As mentioned before, the fact that Otsu finds the threshold by considering the whole image seems to be a great drawback for this task and this can be seen in the results. With an average error of 118.74 pixel per vertex and the fact that for six images the results are too fragmented to extract the silhouette properly, without choosing the bottom pixel row as silhouette candidate, Otsu yields the worst results. On the other side, the region growing algorithm yields relatively good results for a threshold of 1, except for 2 images for which the silhouette extraction fails due to an extremely low contrast. This shows, that the difference

Fig. 3. Huge distance between extracted silhouette and ground truth because we do not cut out obstacles.

in the illumination in the sky segment is very low in many images of the dataset. But increasing the threshold leads to worse results, because the contrast in most images is low and therefore the sky segment grows far to wide, so that, additionally to the higher average distance, the silhouette extraction fails for 19 images when we choose 2 as threshold. This is in contrast to AdaMS and GrabCut. For both, the silhouette extraction has found proper silhouette candidates for all 203 images of the dataset.

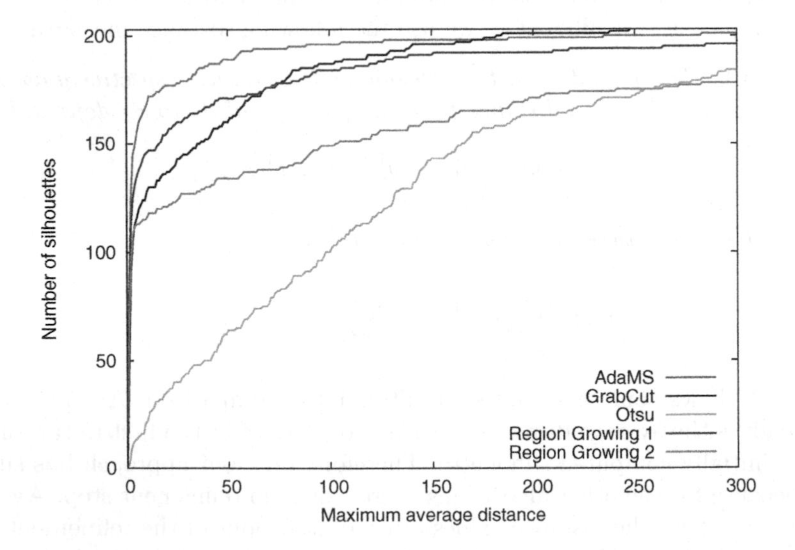

Fig. 4. The cumulative distances of the different algorithms.

The results of GrabCut are nearly equal to RG1 in respect to the overall average distance. But as can be seen in the median value, the region growing algorithm works better on most of the images and furthermore our adaptive approach yields far better results, even for the initial segmentation. In Figure 4 we visualise this by the cumulative sum of the distances. As can be seen, the presented approach can find a good silhouette in significantly more images than the other algorithms. On the other side, GrabCut seems to be a bit more stable in respect to the distance for completely wrong segmentation results, but average distances over 10 pixels are nonetheless not very helpful for a subsequent matching.

Even so, the average initial distance of our algorithm seems to be relatively large with an error of nearly 19 pixels as is the average final distance with a distance of more than 16 pixels. If we compare this, however, to the median final distance, which reaches a good value with 1.43 pixels, it becomes clear, that the high average number is due to only a few images causing problems to our approach. Mostly this is due to omitting the obstacle removal step, as Figure 3 illustrates. It can be seen here, that our approach marks the ropeway's cables as outliers, however, we do not remove those and hence get huge distances in some images.

In comparison to a fixed threshold, our approach regards the global and local appearance of the image to find a better measure for the membership of a pixel to the sky, which leads to a better silhouette for the images with low contrast. In conclusion, we get a great improvement with the precision by the adaptive refinement presented here. Our experiments show that the adaptive refinement improves the quality of the extracted silhouette by 16%.

5 Summary and Future Work

In this work we have proposed a baseline system for an adaptive segmentation approach that is focused on the precise extraction of mountain silhouettes from images of mountains. Our results show that our adaptive solution has great advantages over our initial segmentation and comparable algorithms.

In the future, we aim to make the seed point selection for the grid initialisation more precise by using different feature descriptors combined with a classification step to find pixels with a high probability to be part of the sky, so that we can drop the assumption, that the upper row will be part of the sky. We also want to improve the outlier detection and classification by considering further information such as edge strength and the usage of more than one reference histogram for the description of normal mountain silhouettes.

Furthermore we are currently working on a data set for mountain recognition with ground truth notation that we aim to make publicly available.

References

1. Ahmad, T., Bebis, G., Nicolescu, M., Nefian, A., Fong, T.: Fusion of edge-less and edge-based approaches for horizon line detection. In: 6th IEEE International Conference on Information, Intelligence, Systems and Applications (IISA 2015), Corfu, Greece, July 6–8, 2015. IEEE (2015)

2. Baatz, G., Saurer, O., Köser, K., Pollefeys, M.: Large scale visual geo-localization of images in mountainous terrain. In: Fitzgibbon, A., Lazebnik, S., Perona, P., Sato, Y., Schmid, C. (eds.) ECCV 2012, Part II. LNCS, vol. 7573, pp. 517–530. Springer, Heidelberg (2012)
3. Baboud, L., Čadík, M., Eisemann, E., Seidel, H.P.: Automatic photo-to-terrain alignment for the annotation of mountain pictures. In: 2011 IEEE Conference on Computer Vision and Pattern Recognition (CVPR). IEEE (2011)
4. Canny, J.: A Computational Approach to Edge Detection. IEEE Transactions on Pattern Analysis and Machine Intelligence **PAMI-8**(6) (November 1986)
5. Chen, Y., Cremers, A.B., Cao, Z.: Interactive color image segmentation via iterative evidential labeling. Information Fusion **20** (2014)
6. Frucci, M., Perner, P., Sanniti Di Baja, G.: Case-based-reasoning for image segmentation. International Journal of Pattern Recognition and Artificial Intelligence **22**(05) (2008)
7. Instagram (accessed January 1, 2016). https://instagram.com/press/
8. Kim, B.J., Shin, J.J., Nam, H.J., Kim, J.S.: Skyline extraction using a multistage edge filtering. World Academy of Science, Engineering and Technology **55** (2011)
9. Kittler, J., Illingworth, J.: Minimum error thresholding. Pattern Recognition **19**(1) (1986)
10. Mancas, M., Gosselin, B., Macq, B.: Segmentation using a region-growing thresholding. In: Electronic Imaging 2005. International Society for Optics and Photonics (2005)
11. Otsu, N.: A threshold selection method from gray-level histograms. IEEE Transactions on Systems, Man and Cybernetics **9**(1) (1979)
12. Patil, R., Jondhale, K.: Edge based technique to estimate number of clusters in k-means color image segmentation. In: 2010 3rd IEEE International Conference on Computer Science and Information Technology (ICCSIT), vol. 2 (2010)
13. Perner, P.: An architecture for a CBR image segmentation system. Engineering Applications of Artificial Intelligence **12**(6) (1999)
14. Rother, C., Kolmogorov, V., Blake, A.: "GrabCut": Interactive foreground extraction using iterated graph cuts. In: ACM Transactions on Graphics (SIGGRAPH), vol. 23(3) (2004)
15. Sezgin, M., Sankur, B.: Survey over image thresholding techniques and quantitative performance evaluation. Journal of Electronic Imaging **13**(1) (2004)
16. Singh, T.R., Roy, S., Singh, O.I., Sinam, T., Singh, K.M.: A new local adaptive thresholding technique in binarization. IJCSI International Journal of Computer Science Issues **8**(6) (2011)
17. Wang, X.Y., Zhang, X.J., Yang, H.Y., Bu, J.: A pixel-based color image segmentation using support vector machine and fuzzy -means. Neural Networks **33** (2012)

K-Means over Incomplete Datasets Using Mean Euclidean Distance

Loai AbdAllah[1(✉)] and Ilan Shimshoni[2]

[1] Department of Community Information Systems Zefat Academic College, Department of Mathematics and Computer Science, The College of Sakhnin for Teacher Education, Sakhnin, Israel
loai1984@gmail.com

[2] Department of Information Systems, University of Haifa, Haifa, Israel

Abstract. Missing values in data are common in real world applications. In this research we developed a new version of the well-known k-means clustering algorithm that deals with such incomplete datasets. The k-means algorithm has two basic steps, performed at each iteration: it associates each point with its closest centroid and then it computes the new centroids. So, to run it we need a distance function and a mean computation formula. To measure the similarity between two incomplete points, we use the distribution of the incomplete attributes. We propose several directions for computing the centroids. In the first, incomplete points are dealt with as one point and the centroid is computed according to the developed formula derived in this research. In the second and the third, each incomplete point is replaced with a large number of points according to the data distribution and from these points the centroid is computed. Even so, the runtime complexity of the suggested k-means is the same as the standard k-means over complete datasets. We experimented on six standard numerical datasets from different fields and compared the performance of our proposed k-means to other basic methods. Our experiments show that our suggested k-means algorithms outperform previously published methods.

Keywords: Clustering · k-means · Missing values · Incomplete datasets

1 Introduction

K-Means is the most popular and the simplest partitional clustering algorithm. It has a rich and diverse history as it was independently discovered in different scientific fields [10,11,14]. Ease of implementation, simplicity, efficiency, and empirical success are the main reasons for its popularity. The k-means algorithm (which is an EM type algorithm) has two basic steps, performed at each iteration: (1) It associates each point with its closest centroid. (2) It computes the new centroids.

Developing such an algorithm for datasets with missing values is not a trivial challenge. It is important, since missing values are very common in real world

P. Perner (Ed.): MLDM 2016, LNAI 9729, pp. 113–127, 2016.
DOI: 10.1007/978-3-319-41920-6_9

datasets. They can be caused by human error, equipment failure, system generated errors, and so on. We were introduced to the problem of missing data when we received datasets from Applied Materials (AMAT), a company which develops machines for the semiconductor industry. This data has many missing values.

In general there are two ways to run the k-means clustering algorithm over incomplete datasets: customizing the data or customizing the k-means algorithm. This means that we can preprocess the dataset so that it consists only of complete points and then run the standard k-means, or we can develop a k-means clustering algorithm that can deal with incomplete datasets. Our proposed method is of the latter type.

Based on [2,7–9] there are three basic types of missing data:

1. Missing Completely at Random: Data are said to be MCAR if the failure to observe a value is not related to any other sample.
2. Missing at Random: Data are said to be MAR if the probability that a value is missing does not depend on the other missing values. Thus the conditional probability of missingness may depend on any known values.
3. Not Missing at Random: Data are said to be NMAR if the probability that a known value is missing depends on the value that would have been observed.

Several methods have been proposed to deal with missing data. These methods can be classified into two basic categories: (a) **Case deletion** method, this method assumes that the missing values are missing completely at random (MCAR). It therefore ignores all the instances with missing values and performs the analysis on the rest [16]. (b) Missing data imputation, which replaces each missing value with a known value according to the dataset distribution. A common method that imputes missing data is the **Most Common Attribute Value (MCA)** method. The value of the attribute that occurs most often is selected to be the value for all the unknown values of the attribute [5]. The **Mean Imputation (MI)** method replaces a data point with missing values with the mean of all the instances in the data. A variant of this method is to replace the missing data for a given attribute with the **M**ean of all known values of that **A**ttribute-**MA** (i.e., the mean of each attribute) in the coordinate where the instance with missing data belongs as described by [12]. All these methods assume MCAR since all of them based on the distribution of the whole data and do not take into account the correlations between the observed and the unobserved values. These imputation methods yield complete datasets. As a result the standard k-means clustering algorithm can be run.

There are also methods that run k-means over incomplete datasets without imputation, as described in [4,6]. They estimate a Gaussian Mixture Model-GMM (an extended version of k-means) over datasets with missing values without imputation, but, as we show in this paper, our proposed method is different and yields better results.

AbdAllah and Shimshoni [1] developed a new method to compute the distance function between incomplete points. Their distance is not only efficient but also takes into account the distribution of each attribute. In the computation procedure they take into account all the possible values with their probabilities, which are computed according to the attribute's distribution. This is in contrast to the MCA and the MA methods, which replace each missing value only with the mode or the mean of each attribute.

In a recent paper, Eirola et. al., [3] estimated the pairwise distance between incomplete samples using the Gaussian mixture model with the algorithm described in [4,6]. Mixture models of Gaussians have been studied extensively to describe the distributions of data sets.

In this research we also developed a k-means algorithm that can run over incomplete datasets without a preprocessing procedure. To do so, we first need (1) a distance function to measure the similarity between incomplete points and each centroid in order to associate each point with the closest centroid; and (2) a formula for computing the new centroids of the clusters where each cluster may contain incomplete points.

As a result, in this research we decided to work with the *mean Euclidian distance* (MD_E) presented in [1] to measure the dissimilarity between the incomplete points. The MD_E distance is not only efficient but also takes into account the distribution of each attribute. This distance assumes that the missing values are randomly distributed across all the samples. But, in real world datasets, the missing values may depend on information from the known values of the data. Thus, in this research we generalized this distance function to deal with other types of missing values.

We suggest three variants of k-means that can deal with incomplete datasets. All use the MD_E distance to associate the points with the closest centers. It is important to note that by using this distance we are able to associate points with the centroids without knowing their exact geometric locations. The three directions differ in how to compute the new centroid for each cluster, and more specifically, in how to include the incomplete points within the mean computation procedure. The first direction assumes that each incomplete point represents one point and then it computes the mean according the developed formula for computing the mean. The other two directions assume that each incomplete point represents a set of complete points according to the data distribution, so they replace each incomplete point with a set of points and then compute the *mean* according to the new dataset. It is important to note that even though we replace each incomplete point with a large number of points, we use the histograms of the data distribution in order to make the suggested algorithm more efficient. As a result, the runtime complexity of the suggested k-means algorithms is the same as the standard k-means over complete datasets.

The proposed methods yield better results than previously published methods, as can be seen in the experiments. We experimented on six standard numerical datasets from different fields from the Speech and Image Processing Unit [15]. Our experiments show that the performance of the k-means algorithm using MD_E distance function and the proposed *mean* and the k-means that use the histogram of the data were superior to k-means using other methods.

The paper is organized as follows. A review of the incomplete data distance function measure developed by [1] is described in Section 2. The mean computation is presented in Section 3. Section 4 describes several directions for integrating the (MD_E) distance and the computed *mean* within the k-means clustering algorithm. Experimental results of running several variants of the k-means clustering algorithm on the Speech and Image Processing Unit [15] datasets are presented in Section 5. Finally, our conclusions are presented in Section 6.

2 Incomplete Data Distance Measure

In this section we describe the method for measuring the distance between pairs of points when they may contain missing values developed by [1].

Let $A \subseteq \mathbb{R}^K$ be a set of points. For the ith attribute A^i, the conditional probability for A_i will be computed according to the known values for this attribute from A (i.e., $P(A^i) \sim \chi^i$), where χ^i is the distribution of the ith coordinate.

Given two sample points X and Y from A, the goal is to compute the distance between them. Let x^i and y^i be the ith coordinate values from points X, Y respectively. There are three possible cases for the values of x^i and y^i:

1. Two values are known: When the values of x^i and y^i are given, the distance between them will be defined as the Euclidian distance:

$$D_E(x^i, y^i) = (x^i - y^i)^2. \tag{1}$$

2. One value is missing: Suppose that x^i is missing and the value y^i is given. Since the value of x^i is unknown, we cannot compute its Euclidian distance. Instead we model the distance as a random selection of a point from the distribution of its attribute χ^i and compute its distance. The expectation of this computation is our distance.
   As a result, we approximate the mean Euclidian distance (MD_E) between y^i and the missing value m^i as:

$$MD_E(m^i, y^i) = E[(x - y^i)^2] = \int p(x)(x - y^i)^2 dx = \left((y^i - \mu^i)^2 + (\sigma^i)^2\right).$$

   This metric measures the distance between y^i and each suggested value of x^i and takes into account the probability $p(x)$ for this value according to the evaluated probability distribution. It is important to note that in this computation the probability was computed according to the whole dataset. The authors did not take into account the possible correlations between the missing values and the other known values. It means that they assumed MCAR (missing completely at random) missing data type. The resulting mean Euclidian distance will be:

$$MD_E(m^i, y^i) = \left((y^i - \mu^i)^2 + (\sigma^i)^2\right), \tag{2}$$

   where μ^i and $(\sigma^i)^2$ are the *mean* and the *variance* for all the known values of the attribute.

3. The two values are missing: In this case, in order to estimate the mean Euclidian distance, we have to randomly select values for both x^i and y^i. Both these values are selected from distribution χ^i.
   We compute the expectation of the Euclidean distance between each selected value as we did for the one missing value. As a result the distance is:

$$MD_E(x_i, y_i) = \int \int p(x)p(y)(x-y)^2 dxdy = \left((E[x] - E[y])^2 + \sigma_x^2 + \sigma_y^2 \right).$$

   As x and y belong to the same attribute, $E[x] = E[y] := \mu^i$ and $\sigma_x = \sigma_y := \sigma^i$. Thus:

$$MD_E(x^i, y^i) = 2(\sigma^i)^2. \tag{3}$$

Studying the equation described above, we conclude that the main limitation of this distance is its assumption that the missing data is MCAR. However, many real world datasets are not MCAR. So, if the missing are MAR then the probability $p(x)$ depends on the other observed values and then the distance will be computed as:

$$MD_E(m^i, y^i) = \int p(x|x_{obs})(x - y^i)^2 dx = \left((y^i - \mu^i_{x|x_{obs}})^2 + (\sigma^i_{x|x_{obs}})^2 \right),$$

where x_{obs} denotes the observed attributes of point X, and $\mu^i_{x|x_{obs}}$ and $(\sigma^i_{x|x_{obs}})^2$ are the conditional *mean* and *variance*, respectively.

On the other hand, if the missing values are of type NMAR, then the probability $p(x)$ that was used in Equation 2 will be computed according to this information and then the distance will be:

$$MD_E(m^i, y^i) = \int p(x|m^i)(x - y^i)^2 dx = \left((y^i - \mu^i_{x|m^i})^2 + (\sigma^i_{x|m^i})^2 \right),$$

where $p(x|m^i)$ is the distribution of x when x is missing.

3 Mean Computation

In order to develop a k-means clustering algorithm over datasets with missing values, we still need a formula that uses the MD_E distance to compute the *mean* of a given set that contains missing values.

Let $A \subseteq \mathbb{R}^K$ be a set of n points that may contain points with missing values. Then the *mean* of this dataset is defined as:

$$\bar{x} = \arg\min_{x \in \mathbb{R}} \sum_{i=1}^{n} (distance(x, p_i))^2,$$

for any $x \in \mathbb{R}^K$, where $p_i \in A$ denotes each point from the set A, and *distance*() is a distance function.

Let $f(x)$ be a multidimensional function: $f : \mathbb{R}^K :\rightarrow \mathbb{R}$ which is defined as:

$$f(x) = \sum_{i=1}^{n} (distance(x, p_i))^2 ,$$

In our case, the $distance() = MD_E$. Thus,

$$f(x) = \sum_{i=1}^{n} (distance(x,p_i))^2 = \sum_{i=1}^{n} \Bigg(\underbrace{\sqrt{\sum_{j=1}^{K} MD_E(x^j, p_i^j)}}_{\text{The } MD_E() \text{ distance}} \Bigg)^2 = \sum_{i=1}^{n} \sum_{j=1}^{K} MD_E(x^j, p_i^j),$$

where x^j is the coordinate j and p_i^j is the coordinate j in point p_i. Since each point p_i may contain missing attributes, and according to the definition of the MD_E distance in the previous section, $f(x)$ will be:

$$f(x) = \sum_{j=1}^{K} \Bigg[\underbrace{\sum_{i=1}^{n_j} (x^j - p_i^j)^2}_{\text{there are } n_j \text{ known coordinates}} + \underbrace{\sum_{i=1}^{m_j} \Big((x^j - \mu^j)^2 + (\sigma^j)^2 \Big)}_{\text{there are } m_j \text{ missing coordinates}} \Bigg].$$

$\bar{x}$ is the solution of $f'(x) = 0$, but since $f(x)$ is a multidimensional function then $\bar{x}$ is the solution of $\nabla f = \overrightarrow{0}$, where

$$\nabla f = \Big(f'_{x^1}, f'_{x^2}, ..., f'_{x^k} \Big) = 0,$$

is the gradient of function f. In our derivation we will first deal with one coordinate and then we will generalize it for all the other coordinates.

$$\Rightarrow f'_{x^l} = 2 \sum_{i=1}^{n_l} (x^l - p_i^l) + 2 \sum_{i=1}^{m_l} (x^l - \mu^l) = 0$$

$$\Rightarrow n x^l = \sum_{i=1}^{n_l} p_i^l + m_l \mu^l \Rightarrow x^l = \frac{\sum_{i=1}^{n_l} p_i^l}{n} + \frac{m_l \mu^l}{n}$$

$$\Rightarrow x^l = \frac{n_l}{n} \frac{\sum_{i=1}^{n_l} p_i^l}{n_l} + \frac{n - n_l}{n} \mu^l = \mu^l.$$

Thus, we simply get:

$$x^l = \mu^l. \tag{4}$$

Repeating this for all the coordinates yields $\bar{x} = (\mu^1, \mu^2, ..., \mu^k)$. In other words, each coordinate of the mean is the mean of the known values of that coordinate.

In the same way, we derive a formula for computing the weighted mean for each coordinate l, yielding:

$$\overline{x}_w^l = \frac{\sum_{i=1}^{n_l} w_i x_i^l + \sum_{i=1}^{n_l} w_i \mu^l}{\sum_{i=1}^{n} w_i},$$

where w_i is the weight of point x_i. Thus, in order to compute the *weighted mean* of an attribute l, where some of its values are missing, we must distinguish between two cases: known and missing values. If the value is known we multiply it with its weight. When, however, the value is missing, we replace it with the mean of the known values of the attribute l and then multiply it by the matching weight. We then sum the terms and divide them by the sum of all the weights.

4 K-Means Clustering Using the MD_E Distance

The aim of this research is to develop k-means clustering algorithms for incomplete datasets. The MD_E distance and the *mean* are general and can be used within any algorithm that computes distances and means. It is not, however, clear how it can be integrated into such an algorithm. In this section we describe our proposed method for doing so for the k-means clustering algorithm. For illustration purposes we assume that all the points are from $\mathbb{R}^2$. We propose three different versions for k-means. It is important to note that the first version is similar to the GMM algorithm described [4,6], where each incomplete point is considered a single point. However, the other two versions are different and replace each incomplete point with a set of points according to the data distribution. As will be shown in our experiments, they outperform the first algorithm.

Given dataset D that may contain points with missing values, the k-means algorithm has two basic steps, performed at each iteration: (1) It associates each point with its closest centroid. (2) It computes the new centroids. In the first step the MD_E distance is used to compute the distances between the points and the k centroids in order to associate each point with the closest centroid. There are several possible ways to then compute the new centroids of the clusters. These possibilities will be illustrated using Figure 1(a). According to this example, there are two clusters (i.e., $C1$ is assigned to the yellow cluster and $C2$ is assigned to the brown cluster). The goal is to compute the centroid of each cluster. For simplicity we will deal with $C1$. If none of the instances contain missing values, the centroid will be computed according to the Euclidian *mean* formula, resulting in the magenta star.

When the dataset contains points with missing values, it is not clear how to compute the mean. In the given example, let $(x_0, ?)$ (i.e., the red star) be a point with a missing y value and $x = x_0$. Although the exact geometric location for this point is unknown, we still can associate it with $C1$'s cluster using the MD_E distance as follows:

$$distance^2((x,?),(x_c,y_c)) = MD_E^2((x,?),(x_c,y_c)) = (x_c - x_0)^2 + (y_c - \mu_y)^2 + \sigma_y^2.$$

Using the MA-method, on the other hand, the point $(x_0, ?)$ will be replaced with (x_0, μ_y). It is clear that the difference between the two methods is only in σ_y^2, a fixed value that does not influence the association result.

Let the $Att_{possible}$ group denote all the possible values for each attribute. Thus, in our case:

$$Y_{possible} = \{y \in \mathbb{R} | \exists (x,y) \in D\}.$$

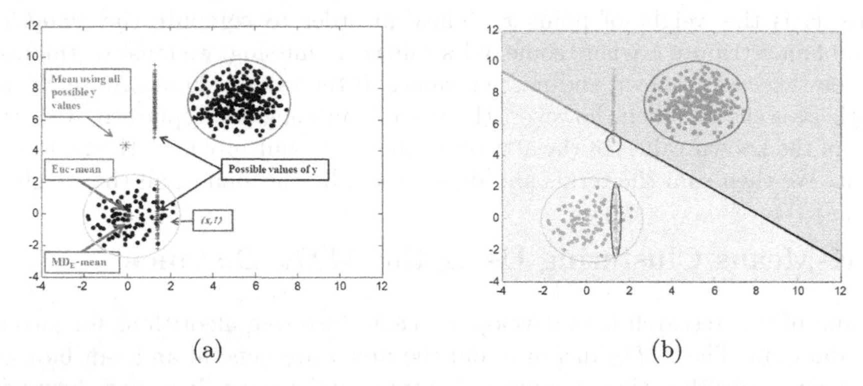

Fig. 1. An example for computing the centroids for two clusters in a dataset with missing values. (a) Shows the results of the different methods of computing the *mean*. (b) Shows the Voronoi diagram.

Let

$$(x_0)_{possible} = \{(x_0, y_p) \big| y_p \in Y_{possible}\}$$

be the set of all the possible points that satisfy $x = x_0$, assume y is missing (the blue "+"), and let

$$C1_{real} = \{(x, y) \in D \big| (x, y) \in C1\}$$

be the set of all the data points without missing values that are associated with the $C1$ cluster.

The naïve method to compute the new centroid is by replacing the point with the missing value with all the possible points $(x_0)_{possible}$, and then computing the *mean* according to these points ($C1_{real}$ and $(x_0)_{possible}$), where each point from $C1_{real}$ has weight one and each point from $(x_0)_{possible}$ has weight $\frac{1}{|Y_{possible}|}$. As a result, the weighted *mean* of $C1$ is:

$$mean(C1) = \frac{\sum_{(x,y)\in C1_{real}}(x, y) + (x_0, \mu_y)}{|C1_{real}| + \sum \frac{1}{|Y_{possible}|}}. \tag{5}$$

This is identical to the Euclidian *mean* when the missing point is replaced with (x_0, μ_y) and is equivalent to the MA method when (x_0, μ_y) is associated with $C1$. As a result, the real centroid of the cluster (the magenta star) moves to the green star. That is because this computation assumes that all the possible points $(x_0)_{possible}$ are represented as (x_0, μ_y), and ignores the other possible points, which in general is suboptimal, as can be seen in Figure 1 (b), where not all the blue "+" marks are associated with $C1$.

Thus, we must take into account the association of each possible point. There are two possible ways to do that. The first (which we name ***k-mean-MD**$_E$*) is to include within the *mean* computation, in addition to the real points within the yellow circle, all the possible points $(x_0, y) \in (x_0)_{possible}$ such that their y coordinates are the y coordinates of the real data points from the yellow cluster (i.e., the blue "+" belonging to the yellow circle). Formally, the mean will be computed according to all the real points $C1_{real}$ and

$$C1_{(x_0)_{possible}} = \{(x_0, y_p) \in (x_0)_{possible} | \exists (x, y) \in C1_{real} \wedge y = y_p\}.$$

The weight for each point from $C1_{real}$ is 1 and the weight for each point from $C1_{possible}$ is: $\frac{1}{|C1_{possible}|}$.

This means that all the weights are computed only according to the distribution of the points associated with $C1$. So, using only this set of points, the *mean* can be computed directly using (4). As a result, the centroid will be preserved, yielding the magenta star in the given example.

Computing the new centroid using (4) yields not only the same centroid as using the Euclidian distance, but also preserves the runtime of the standard k-means using the Euclidian distance.

Another method (which we name ***k-mean-HistMD$_E$***) is to use all the points from $(x_0)_{possible}$ that are associated with the $C1$ cluster and not only the points from $(x_0)_{possible}$ whose y coordinates are from the real points associated with that cluster. Thus, we first associate each of the points from $(x_0)_{Y_{possible}}$ with its closest centroid (the blue "+" in the red circles in Figure 1 (b)), and then compute a weighted *mean*. Formally, the *mean* will be computed according to all the real points $C1_{real}$, and

$$PC1_{possible} = \{(x_0, y_p) \in (x_0)_{possible} | (x_0, y_p) \in C1\}.$$

In this case we cannot use (4), because the weights are computed using the entire dataset. We therefore suggest three methods for implementing the *mean* computation:

1. Simple weighted *mean*: Simply replace each point with a missing value with the $|Y_{possible}|$ points, each with a weight $\frac{1}{|Y_{possible}|}$, and run weighted k-means on the new dataset. Thus, the weighted *mean* of $C1$ is:

$$mean_1(C1) = \frac{\sum_{(x,y) \in C1_{real}} (x, y) + \frac{1}{|Y_{possible}|} \cdot \sum_{(x_0,y_p) \in PC1_{possible}} (x_0, y_p)}{|C1_{real}| + \frac{1}{|Y_{possible}|} \cdot |PC1_{possible}|}. \tag{6}$$

   This method is simple to implement, but its runtime is high, since each point with, for example, a missing y value will be replaced with $|Y_{possible}|$ points. As a result, the size of the dataset will be:

$$|D_{real}| + \left(|D| - |D_{real}|\right) \cdot |Att_{possible}|,$$

   where D_{real} is the set of all the data points that do not contain missing values. We therefore chose to implement more efficient methods to compute the *mean* of the cluster.

2. Voronoi Diagram method: Using the Voronoi diagram, the data space is partitioned to k subspaces (as can be seen in Figure 1 (b)). Each point is associated with the subspace of the cluster in which it lies. We can use the intersection between the Voronoi edges and the $x = x_0$ line to identify all the possible points associated with cluster $C1$, that is, the $PC1_{possible}$ points. Then all the possible points within this cluster will be represented by their Euclidian *mean* (assigned as (x_{m_1}, y_{m_1})). Its weight is:

$$w_{m_1} = \frac{|PC1_{possible}|}{|Y_{possible}|},$$

and the weight for each point within $C1_{real}$ is one. Thus, the *mean* formula is:

$$mean_2(C1) = \frac{\sum_{(x,y)\in C1_{real}}(x,y) + w_{m_1}\cdot(x_{m_1}, y_{m_1})}{|C1_{real}| + w_{m_1}}.$$

It is easy to see that this equation is identical to (6). However, this method is more efficient. Another possible and efficient approximation method is described below.

3. Histogram method: Instead of including each possible point in the computation procedure, we divide the y value space to several disjoint intervals. Each interval will be represented by its mean, and the weight of each interval will be the ratio between the number of points in the interval to the number of all possible points. Formally, let $\Delta_i = \{y_p \in [y_{i-1}, y_i] | y_p \in Y_{possible}\}$ be interval i. Then its representative point is: $r_i = (x_0, mean(\Delta_i))$ and its weight will be $w_i = \frac{|\Delta_i|}{|Y_{possible}|}$. The weight for each point within $C1_{real}$ is one. Thus, the formula for the *mean* is:

$$mean_3(C1) = \frac{\sum_{(x,y)\in C1_{real}}(x,y) + \sum_{r_i\in C1} w_i\cdot r_i}{|C1_{real}| + \sum_{r_i\in C1} w_i}. \quad (7)$$

This method approximates the method that computes the weighted *mean*; the only difference is for the intervals that intersect the Voronoi edges. For all the other intervals the two methods are identical. In the experiments we conclude that this method does not strongly depend on the number of intervals. This method is called ***k-mean-HistMD$_E$***. Consider the following two special cases. In the first case each interval contains only one point (i.e., $\Delta_i = \{y_p\}$, where $y_p \in Y_{possible}$). Then

$$r_i = (x_0, y_p) \in PC1_{possible} \quad , \quad w_i = \frac{1}{|Y_{possible}|},$$

and the *mean* will be:

$$mean(C1) = \frac{\sum_{(x,y)\in C1_{real}}(x,y) + \sum_{r_i\in C1} w_i\cdot r_i}{|C1_{real}| + \sum_{r_i\in C1} w_i}.$$

As a result, this *mean* is identical to $mean_1$, described in (6). In the second case, when there is only one interval, then

$$\Delta_i = Y_{possible}, r_i = (x_0, mean(Y_{possible})) = (x_0, \mu_y), \text{and} w_i = \frac{|\Delta_i|}{|Y_{possible}|} = 1.$$

In this case the *mean* will be:

$$mean(C1) = \frac{\sum_{(x,y)\in C1_{real}}(x,y) + (x_0, \mu_y)}{|C1_{real}| + 1}.$$

Then, this method will be equivalent to the MA imputation method described in (5).

Discussion
There are several differences between k-means-MD_E and k-means-$HistMD_E$. They differ in their performance, efficiency and the way that they work. The k-means-MD_E performs better when there are correlations between the attributes, while the k-means-$HistMD_E$ performs better when there are no strong correlations between the attributes. This is due to the different ways in which they associate points generated to represent the points with the missing values to the clusters.

Moreover, the k-means-MD_E method is more efficient and its runtime is equivalent to the standard k-means as described above. In addition, the results of the two methods are different, as illustrated in Figure 1(b), where in the first method only the blue "+" signs in the yellow circle are included, while in the k-means-$HistMD_E$ method all the points within the red ellipses are included.

4.1 Algorithm Convergence

In the previous section we described our new suggested methods of k-means. Now we will prove that all these new methods converge to a local minimum of the cost function. This has been proven for the original k-means algorithm. In the first method (naïve k-means), as can be seen in the previous section, each missing value is replaced with the mean of all the known values of that coordinate. Then the standard k-means clustering algorithm is run on the modified data set. The algorithm therefore has the same properties as the standard k-means and it therefore also converges. A similar analysis is performed for the k-means-$HistMD_E$ algorithm. In this case each point that has a missing value is replaced with several points using the data distribution. Then again the algorithm can considered as if it were the standard weighted k-means algorithm which is run over this modified dataset. Thus, this method also has the standard k-means properties.

Since a simple reduction is not possible for the k-means-MD_E algorithm we will now prove that this method also converges after a finite number of iterations. Let $c^{(k)}, k = 1..K$ be the centroids of each cluster C_k, where

$$C_k = \{x \in X : \textit{ the closest representative of } x \textit{ is } c^{(k)}\}.$$

The goal of the k- means algorithm is to find the K centroids that minimize the cost function:

$$\{c^{(1)}, c^{(2)}, ..., c^{(K)}\} = \underset{\{c^{(1)}, c^{(2)}, ..., c^{(K)}\}}{\arg\min} \; cost(C_1, C_2, ..., C_K, c^{(1)}, c^{(2)}, ..., c^{(K)})$$

$$= \underset{\{c^{(1)}, c^{(2)}, ..., c^{(K)}\}}{\arg\min} \sum_{k=1}^{K} \sum_{x \in C_k} distance^2(x, c^{(k)}).$$

The standard k-means algorithm converges due to the fact that the value of the cost function monotonically decreases because in each iteration the new centroids $c'^{(i)}$ satisfy $c^{(i)} = \arg\min \sum_{x \in C_i} distance^2(x, c)$ and each point is associated with its closest centroid. We will now show that in the k-means-MD_E algorithm the value of the cost function also decreases monotonically in the same manner.

In the previous section we showed that in the association step that using the MD_E distance function each point will be associated to the closest centroid. This is true for regular points as well as for incomplete points. For each cluster, considering the points associated with it, the mean is computed using the derivation performed in Section 3, which satisfies:

$$c^{(k)} = \underset{c \in \mathbb{R}^d}{\arg\min} \sum_{x \in C_k} distance^2(x, c) = \underset{c \in \mathbb{R}^d}{\arg\min} \sum_{x \in C_i} \sum MD_E^2(x, c),$$

which is what is required.

As a result we conclude that the k-means-MD_E clustering algorithm also converges to a local minimum like the standard k-means.

5 Experiments on Numerical Datasets

In order to measure the ability of k-means-MD_E and k-means-$HistMD_E$ to cluster the incomplete datasets, we compare the performance of the k-means (k is fixed for each dataset) clustering algorithm on complete data (i.e., without missing values) to its performance on data with missing values, using the MD_E distance measure (k-means-MD_E and k-means-$HistMD_E$) and then again using k-means-(MCA, MA, MI), where each missing value in each attribute is replaced using the MCA, MA or MI method respectively on which a standard k-means is run. In our experiments, k-means-$HistMD_E$ was run with 20 intervals. Later (in Section 5.1) we can see that the number of intervals is not critical for the algorithm's performance.

We use the Rand index [13], which is a measure of similarity between two data clusterings, to compare how similar the results of the standard k-means clustering algorithm were to the results of the other algorithms for datasets with missing values.

We ran our experiments on six standard numerical datasets from the Speech and Image Processing Unit [15] from different fields: the Flame dataset, the Jain dataset, the Pathbased dataset, the Spiral dataset, the Compound dataset, and the Aggregation dataset; dataset characteristics are shown in Table 1.

Table 1. Speech and Image Processing Unit Dataset Properties

Dataset	Dataset size	Clusters
Flame	240×2	2
Jain	373×2	2
Pathbased	300×2	3
Spiral	312×2	3
Compound	399×2	6
Aggregation	788×2	7

As can be seen in Figure 2, the two versions of k-means that use the MD_E distance outperformed the other algorithm on all the datasets. Our intuitive explanation for the performance of our algorithms is as follows. In the MA MCA methods, the whole distribution of values is replaced by a single value (the mean or the mode of the distribution of known values). In our two algorithms we use the distribution of the observed values in all the computation stages. This additional information is probably the reason for the improved performance of our methods compared to the known heuristics. Moreover, we can see that, in most cases, k-means-$HistMD_E$ outperformed k-means-MD_E. The difference in performance is due to the data distribution as mentioned in discussion in Section 4.

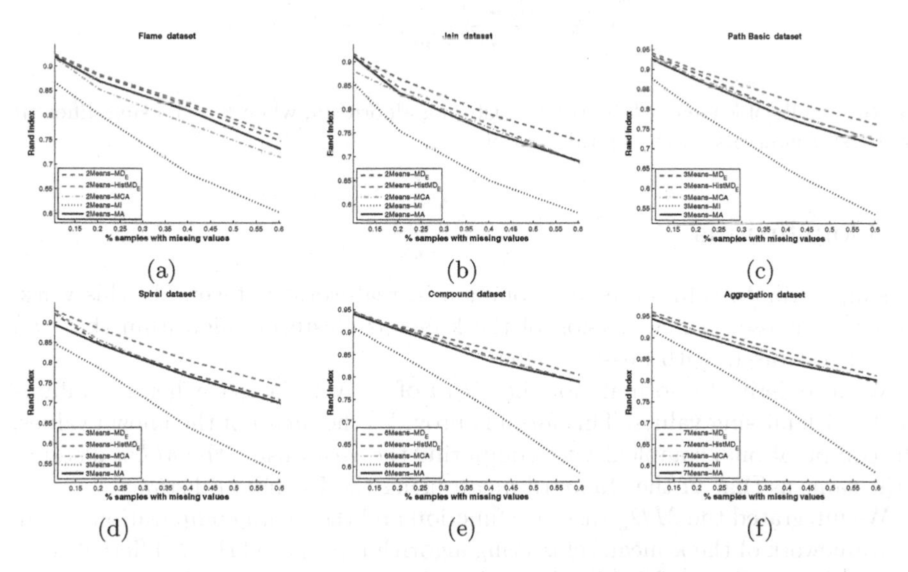

Fig. 2. Results of k-means clustering algorithm using the different distance functions on the six datasets from the Speech and Image Processing Unit.

5.1 The Effect of the Number of Intervals on the Performance of k-means-$HistMD_E$

We performed yet another experiment to determine the extent to which the algorithm depends on the number of intervals. In this experiment we evaluated the performance of the algorithm on the Spiral dataset. We ran k-means-$HistMD_E$ using six different numbers of intervals (1, 5, 10, 15, 20, 50, 100) and compared the performance to that of the standard k-means. As Figure 3 clearly shows, the performance of k-means-$HistMD_E$ does not critically depend on the number of intervals. When k-means-$HistMD_E$ was run with one interval, its performance was identical to that of k-means-MA, and when k-means-$HistMD_E$ was run with five intervals, its performance improved. At 10 intervals, its performance converged and did not change for larger numbers of intervals, as can be seen in the resulting curves. In our experiments we chose therefore to work with 20 disjoint intervals.

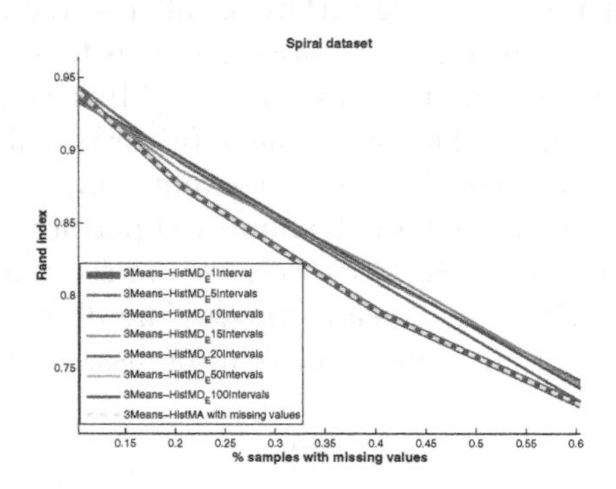

Fig. 3. Results of k-means-$HistMD_E$ clustering algorithm, when $k = 3$, using different numbers of intervals on the Spiral dataset.

6 Conclusions

Missing attribute values are very common in real-world datasets. In this work, we have proposed a new version of the k-means clustering algorithm that can deal with datasets with missing values.

We also derived a formula for the *mean* of a given dataset when it contains points with missing values. This formula provides the *mean* of the known values. The computational complexity for computing the *mean* using the MD_E distance is the same as that of the standard *mean* using the Euclidian distance.

We integrated the MD_E distance function and the *mean* computation within the framework of the k-means clustering algorithm proposed three different ways to compute the centroids while taking into account the associations between the data points and the centroids. This is in contrast to the other basic methods for dealing with missing values, which do not take into account these associations.

We conducted experiments using the k-means clustering algorithm on six standard datasets. Our k-means methods outperformed the standard methods for datasets with missing values.

These proposed methods are general and can be used as part of any algorithm that computes the distance between data points or *means*. Moreover, they can be used for different datasets in different application areas. So, our future vision is devoted to incorporating this distance function within other clustering and classification algorithms.

Acknowledgements. This researh was supported by the Metro 450mm consortium of the Ministry of Economics.

References

1. AbdAllah, L., Shimshoni, I.: Mean shift clustering algorithm for data with missing values. In: Bellatreche, L., Mohania, M.K. (eds.) DaWaK 2014. LNCS, vol. 8646, pp. 426–438. Springer, Heidelberg (2014)
2. Donders, A.R.T., van der Heijden, G.J., Stijnen, T., Moons, K.G.: Review: a gentle introduction to imputation of missing values. Journal of Clinical Epidemiology **59**(10), 1087–1091 (2006)
3. Emil, E., Amaury, L., Vincent, V., Christophe, B.: Mixture of gaussians for distance estimation with missing data. Neurocomputing **131**, 32–42 (2014)
4. Ghahramani, Z., Jordan, M.: Learning from incomplete data. Technical Report, MIT AI Lab Memo, (1509) (1995)
5. Grzymała-Busse, J.W., Hu, M.: A comparison of several approaches to missing attribute values in data mining. In: Ziarko, W.P., Yao, Y. (eds.) RSCTC 2000. LNCS (LNAI), vol. 2005, pp. 378–385. Springer, Heidelberg (2001)
6. Hunt, L., Jorgensen, M.: Mixture model clustering for mixed data with missing information. Computational Statistics & Data Analysis **41**(3), 429–440 (2003)
7. Ibrahim, J.G., Chen, M.H., Lipsitz, S.R., Herring, A.H.: Missing-data methods for generalized linear models: A comparative review. Journal of the American Statistical Association **100**(469), 332–346 (2005)
8. Little, R.J.A.: Missing-data adjustments in large surveys. Journal of Business & Economic Statistics **6**(3), 287–296 (1988)
9. Little, R.J.A., Rubin, D.B.: Statistical Analysis with Missing Data. John Wiley & Sons (2014)
10. Lloyd, S.: Least squares quantization in PCM. IEEE Trans. Information Theory **28**, 129–137 (1982)
11. MacQueen, J.B.: Some methods for classification and analysis of multivariate observations. In: Proceedings of the 5th Symposium on Math, Statistics, and Probability, pp. 281–297 (1967)
12. Magnani, M.: Techniques for dealing with missing data in knowledge discovery tasks. Obtido **15**(01), 2007 (2004). http://magnanim.web.cs.unibo.it/index.html
13. Rand, W.M.: Objective criteria for the evaluation of clustering methods. Journal of the American Statistical Association **66**(336), 846–850 (1971)
14. Steinhaus, H.: Sur la division des corp materiels en parties. Bull. Acad. Polon. Science **4**(3), 801–804 (1956)
15. Speech University of Eastern Finland and Image Processing Unit. Clustering dataset. (2008). http://cs.joensuu.fi/sipu/datasets/
16. Zhang, S., Qin, Z., Ling, C.X., Sheng, S.: "Missing is useful": missing values in cost-sensitive decision trees. IEEE Trans. on KDE **17**(12), 1689–1693 (2005)

A Time Series Model of the Writing Process

Zeev Volkovich(✉)

Software Engineering Department, ORT Braude College of Engineering, 21982 Karmiel, Israel
vlvolkov@braude.ac.il

Abstract. The necessity to operate with the huge number of anonymous documents abounding on the Internet is initiating the study of new methods for authorship recognition. The principal weakness of the methods used in this area is that they assess the similarity of text styles without any regard to their surroundings. This paper proposes a novel mathematical model of the writing process striving to quantify this dependency. A text is divided into a series of sequential sub-documents, which are represented via term histograms. The histograms proximity is estimated through a simple probability distance. Intending to typify the text writing style, a new characteristic representing the mean distance between a current sub-document and numerous earlier ones is advanced. An empirical distribution over the whole document of this feature specifies the writing style. So, dissimilarity of such distributions indicates a difference in the writing styles, and their coincidence implies the styles' identity. Numerical experiments demonstrate high potential ability of the proposed approach.

Keywords: Learning of internal representations and models · Mining text documents · Text Mining · Time series and sequential pattern mining · Authorship recognition

1 Introduction

The necessity to operate with the huge numbers of anonymous documents abounding on the Internet is initiating the study of new methods for authorship recognition. These techniques are intensively used in many modern applications such as author verification (i.e. to settle whether a given text was written by a certain author) [20], plagiarism detection (i.e. finding similarities between two texts) [8], author profiling or characterization (i.e. mining for information concerning such features as the age, education, or sex of the author of a given text) [19], detection of stylistic inconsistencies (as may happen in collaborative writing) [5], and computer forensics (e.g. identifying the authors of the source code of malicious software) [10].

The principal weakness of the methods used in this area (see review in Section 2) is that they assess the similarity of text styles without any regard to their surroundings. One of the widespread outlooks on the human writing

P. Perner (Ed.): MLDM 2016, LNAI 9729, pp. 128–142, 2016.
DOI: 10.1007/978-3-319-41920-6_10

process (see [12]), proposes that the writing process is composed of four main components: planning, drafting, editing, and writing the final draft that obviously leads to intrinsic connection of the current book's part to previously written ones. It is logical to presume that dependence between sequentially composed texts is preserved at the almost identical level if they are written by the same author, and may diminish otherwise.

This paper proposes a novel mathematical model of the writing process striving to quantify this dependency. A text is divided into a series of sequential sub-documents, which are represented via term histograms. The histograms proximity is estimated through a simple probability distance. Intending to typify the text writing style, a new characteristic representing the mean distance between a current sub-document and numerous earlier ones is advanced. An empirical distribution over the whole document of this feature specifies, consistent with our perception, the primary writing style. So, dissimilarity of such distributions indicates a difference in the writing styles, and their coincidence implies the styles' identity.

The rest of the paper is organized as follows. Section 2 is devoted to a review of related works. Section 3 presents the proposed methodology. The numerical experiments are exhibited in Section 4, and Section 5 contains the conclusion.

2 Related Works

Systems of writing style verification are regularly employed to separate texts written by different authors. This task can be studied as a multi-class text-categorization problem where the author of an unnamed text is outlined using a training set of manuscripts with a known authorship. An overview of different techniques with respect to chronological order is presented in [33], [35], and [15]. A discussion of statistical based methods and machine-learning approaches is proposed in [9]. The Impostors Method proposed in [23] is a notable approach converting a one-class text classification task to a two-class task. Initially, a suggested pool of works is different from the studied one in order to train a binary classifiersay, $SVM$to build a model for separation texts, only some of which are written by the author. In the next step the document chunks are categorized using this model, and the majority criterion indicates the authors style.

Many algorithms use word-based features via three types of methods. The first type characterizes documents using a set of functional words, ignoring the content words because they tend to be highly correlated with the document topics [39]. This method is affected by the size of documents. The second type of methods applies the traditional bag-of-words representation and uses single content words as document features [7]. In the bag-of-words approach, texts are represented by the frequency of words, disregarding their relation to grammar. Algorithms based on bag-of-words make the assumption that the style of an author is basically described by the probability distribution of the appearance of certain words, phrases, or any other relevant structure [25]. It is effective when there is a relationship between authors and topics.

A third type of methods considers word $N-$gram features where features consist of sequences of N words. This method captures the style of a text through simple word sequences [27]. Hybrid methods combine several types of features. A method that uses both stylistic and topic features is described in [6]. There are over a thousand different known features, which means that there is no single (universal) feature that robustly separates between different authors [29]. The best results come from analysis of an extremely broad set of features covering many different approaches [15]. The writing style is the most important information source when resolving the problem of author verification that answers the question whether given documents have the same author [17], [35]. The task of authorship verification is restricted by the assumption that there is only one candidate author for the text sample [24]. But since any given author identification problem can be decomposed into a set of authorship verification problems, the problem of authorship verification can be treated as fundamental [22], [36].

The problem of author verification has being discussed since 2000 [38]. In [38] stylometric features were used and the response function for a given author was produced based on multiple regression. The approach was tested on a text corpus generated from Greek newspaper articles. The unmasking method that uses an SVM classifier to separate the documents is presented in [21]. It was shown in [32] that this method is not appropriate for short documents but is effective for long ones. Several modifications of the method are described in [17]. In [24] the author verification problem is considered as a binary classification task where texts of other authors are represented as negative examples. Modern approaches for the authorship verification problem are also reviewed in [37].

Most papers assume that all texts in a verification class match for both genre and topic. In this case a style is affected by genre together with the personal style of each author [22], [24] and [38]. The cross-topic and cross-genre authorship verification is a more complex problem [36].

3 Methodology

In this section, the proposed replica is presented. A new measure is advanced, named the Mean Dependency, for assessment of the association between a given text chunk with several earlier ones. Using this notion we present a new method for modeling a writing style.

3.1 Mean Dependency

Let us consider $\mathbb{D}$ as a collection of finite-length documents such that each sub-document of any document belonging to $\mathbb{D}$ also belongs to $\mathbb{D}$, and take a distance function Dis defined on $\mathbb{D} \times \mathbb{D}$. It is not suggested that $Dis(\mathcal{D}_1, \mathcal{D}_2) = 0$ implies equality of $\mathcal{D}_1$ and $\mathcal{D}_2$. In the framework of our model, we consider for a document $\mathcal{D} \in \mathbb{D}$ a series of m sequential documents having the same size L: $\mathcal{D} = \{\widehat{\mathcal{D}}_1, ..., \widehat{\mathcal{D}}_m\}$. So, in the formal language theory terminology, $\mathcal{D}$ is the concatenation of $\widehat{\mathcal{D}}_1, ..., \widehat{\mathcal{D}}_m$. Further, we introduce the Mean Dependence characterizing the mean relationship between a chunk $\widehat{\mathcal{D}}_i$, $i = T+1, ...m$ and the set of its T "precursors"

$$ZV_{T,Dis,L}(\widehat{\mathcal{D}}_i) = \frac{1}{T}\sum_{j=1}^{j=T} Dis(\widehat{\mathcal{D}}_i, \widehat{\mathcal{D}}_{i-j}), \tag{1}$$

Our perception suggests that a document $\mathcal{D}$ is considered as an outcome provided by "a random number generator" reflecting the writing style of the authors. If a document $\mathcal{D}$ is composed using the same individual writing style then by an appropriate choice of the delay parameter T, a distance function Dis and sub-document size L, the sequence$\{X_i = ZV_{T,Dis,L}(\widehat{\mathcal{D}}_i),\ i = T + 1, ...m\}$ is a strictly stationary sequence of random variables. So, we suggest that each X_i has the same probability distribution and, moreover, each finite set $(X_{i_1+h}, X_{i_2+h}, ..., X_{i_n+h})$ has a joint distribution that does not depend on h. According to the Birkhoff ergodic theorem(see, for example, [11] Section 28.3) there exists a random variable Y such that

$$\lim_{T\to\infty} X_i = Y \tag{2}$$

almost sure. When this sequence is ergodic, for instance having a summable covariance, Y is a constant.

The discussed model makes it possible to offer the following authorship verification procedure. Let us suppose that we have two texts $\mathcal{D}_1$ and $\mathcal{D}_2$, which may have been written by the one author.

Algorithm 1. *Verification if two documents are written by the same author (AV)*

Input:

- $\mathcal{D}_1 \in \mathbb{D}$, $\mathcal{D}_2 \in \mathbb{D}$ - *Two texts to compare.*
- Dis - *Distance function defined on* $\mathbb{D} \times \mathbb{D}$.
- T_0 - *Initial value of the delay parameter* T.
- T^* -*Maximal value of the delay parameter* T.
- ΔT -*Increment step of the delay parameter* T.
- TST - *Two sample test procedure.*
- L -*Chunk size.*

Procedure:

1. *Divide document* D_1 *into* m_1 *chunks of size* L:

$$D_1 = \{\widehat{\mathcal{D}_1^{(1)}}, ..., \widehat{\mathcal{D}_{m_1}^{(1)}}\}.$$

2. Divide document D_2 into m_2 chunks of size L:

$$D_2 = \{\widehat{\mathcal{D}_1^{(2)}, ..., \mathcal{D}_{m_2}^{(2)}}\}.$$

3. Concatenate the partitions:

$$D_0 = \{\widehat{\mathcal{D}_1^{(1)}, ..., \mathcal{D}_{m_1}^{(1)}}, \widehat{\mathcal{D}_1^{(2)}, ..., \mathcal{D}_{m_2}^{(2)}}\} = \left\{\widehat{\mathcal{D}_1^{(0)}, ..., \mathcal{D}_{m_1+m_2}^{(0)}}\right\}.$$

4. For $T = T_0$ to T^* with step ΔT
5. Calculate according to 1:

$$\mathbb{Z}_{T,Dis,L} = \left\{ZV_{T,Dis,L}(\widehat{\mathcal{D}_i^{(0)}}), i = T+1, ..., m_1 + m_2\right\}$$

6. Calculate

$$h = TST\left(\mathbb{Z}_1, \mathbb{Z}_2\right),$$

where $\mathbb{Z}_1 = \{ZV_{T,Dis,L}(\widehat{\mathcal{D}_i^{(0)}}), i = T+1, ..., m_1\}$,
and $\mathbb{Z}_2 = \{ZV_{T,Dis,L}(\widehat{\mathcal{D}_i^{(0)}}), i = m1+1, ..., m_1 + m_2\}$.

7. $if(h = 0)$ then
(a) $\mathcal{D}_1$ and $\mathcal{D}_2$, are written by the same author.
(b) Stop
8. END
9. $if(h = 1)$ then
$\mathcal{D}_1$ and $\mathcal{D}_2$, are not written by the same author.

Comments Regarding the Algorithm
Note that the key step of the procedure is a conversion of the considered texts into "time series" sequences performed in steps 1, 2, 3 and 5 of the algorithm. After this transformation the problem is actually transformed into the change point problem, which strives to recognize times once the distribution of a time series alters. The sample $\mathbb{Z} = (Z_1, \ldots, Z_n)$ is composed of two concatenated parts $\mathbb{X} = (X_1, \ldots, X_k)$ and $\mathbb{Y} = (Y_1, \ldots, Y_m)$, where $m = n - k$, so that $Z_i = X_i$ for $1 \le i \le k$ and $Z_{k+j} = Y_j$ for $1 \le j \le m$. The samples X and Y are created autonomously by two different stationary sources with unknown distributions. In this context, the point k is intended to detect the border between two documents by comparing the distributions of the sets

$$\mathbb{Z}_{1,T,Dis,L} = \left\{ZV_{T,Dis,L}(\widehat{\mathcal{D}_i^{(0)}}), i = T+1, ..., m_1\right\}$$

and

$$\mathbb{Z}_{2,T,Dis,L} = \left\{ZV_{T,Dis,L}(\widehat{\mathcal{D}_i^{(0)}}), i = m1+1, ..., m_1 + m_2\right\}.$$

Random variables within the series are not independent. So, a procedure in the spirit of the statistical inference for ergodic processes (see, e.g. [30]) has to be formally applied. However, assuming in the framework of the model that the process is ergodic, we conclude

$$\lim_{T\to\infty} cov(Z_i, Z_j) = 0,$$

where $Z_i, Z_j \in \mathbb{Z}_{1,T,Dis,L}$ or $Z_i, Z_j \in \mathbb{Z}_{2,T,Dis,L}$ due to 2, where Y is a constant for an ergodic process. Thus, it appears to be reasonable to neglect the dependence between variables inside the series $\mathbb{Z}_{1,T,Dis,L}$ and $\mathbb{Z}_{2,T,Dis,L}$ for sufficiently large values of T and to compare distributions of $\mathbb{Z}_{1,T,Dis,L}$ and $\mathbb{Z}_{2,T,Dis,L}$ by means of a robust two-sample test.

Two-sample hypothesis testing is a statistical analysis approach designed to examine if two samples of independent random elements, drawn from the Euclidean space R^d, have the same probability distribution function. Formally speaking, let $X = X_1, X_2, .., X_k$ and $Y = Y_1, Y_2, .., Y_n$ be two independent random variables whose distribution functions F and G are unknown. A two-sample problem consists of testing the null hypothesis

$$H_0 : F(x) = G(x)$$

against the alternative

$$H_1 : F(x) \neq G(x).$$

The Kolmogorov-Smirnov test, the Cramer-von Mises test, the Friedman's non-parametric ANOVA test, the Wald-Wolfowitz test, the Student t-test and the Z-test have to be mentioned. For our purpose, a TST procedure in an algorithm returns 1 if H_0 is rejected and 0 otherwise.

The most popular method is the Kolmogorov- Smirnov test (the KS-test) [18] [34], which is a nonparametric test of the equality of continuous, one-dimensional probability distributions. The Kolmogorov–Smirnov statistic measures the maximal distance between the empirical distribution functions $\tilde{F}(x)$ and $\tilde{G}(x)$ of two samples. The test is asymptotically distribution-free. We use this test in our study.

The algorithm can be naturally generalized to provide a solution for the author recognition task. Let us suppose that there are several documents

$$\{\mathcal{D}_1, ..., \mathcal{D}n\} \in \mathbb{D},$$

and we we want to assign the document in question $\mathcal{D}_0$ to one of the authors.

Algorithm 2. *Author Determination (AD)*

Input:

- $\mathcal{D}_0$- *the document in question to be assigned.*
- $\{\mathcal{D}_1, \mathcal{D}_2, ..., \mathcal{D}n\} \in \mathbb{D}$- *documents with known authorship;*
- *Dis - Distance function defined on* $\mathbb{D} \times \mathbb{D}$.

– T_0- *Initial value of the delay parameter* T.
– T^*-*Maximal value of the delay parameter* T.
– ΔT-*Increment step of the delay parameter* T.
– TST- *Two sample test procedure.*
– L-*Chunk size.*

Procedure:

1. *For* $T = T_0$ *to* T^* *with step* ΔT
2. *For* i *to* n
3. $h(i) = AV(\mathcal{D}_i, \mathcal{D}_0, Dis, T, T, TST, L)$
4. $if(h(i) = 0)$ *then*
   (a) $\mathcal{D}_0$ *is written by author* i.
   (b) *Stop*
5. *end*
6. *end*
7. $if(h = 1)$ *then*
   $\mathcal{D}_0$ *is not written by any given author.*

3.2 Distance Construction

Distance function choice is essential in the proposed approach. A relevant distance function may be extracted to reflect writing style attributes. It is obviously a challenging and non-trivial task, which has been recently studied as a crucial subject. Formally, measures such as the Levenshtein distance (or the edit distance) can be applied. However, in the text mining domain it is more acceptable to convert texts into a probability distribution and afterwards to utilize a distance between them. In our context, we suggest that there is a transformation $\mathcal{F}$, which maps all the documents belonging to $\mathbb{D}$ into the set $\mathbb{P}$ of all probability distributions on $[0, 1, 2, ...]$:

$$\boldsymbol{P} = \{p_i, i = 0, 1, 2, ...\},\ p_i \geq 0,\ \sum_{i=0}^{\infty} p_i = 1.$$

and

$$Dis(\mathcal{D}_1, \mathcal{D}_2) = dis\left(\mathcal{F}\left(\mathcal{D}_1\right), \mathcal{F}\left(\mathcal{D}_2\right)\right),$$

where dis is a distance function (a simple probability distance) defined on $\mathbb{P}$. The probability metrics theory is stated in [40] and [28]. A comprehensive survey of distance/similarity measures between probability density is presented in [4]. In this paper we compare the following three known distances:

1. The Spearman's correlation distance

   $$S(\boldsymbol{P}, \boldsymbol{Q}) = 1 - \rho(\boldsymbol{P}, \boldsymbol{Q}),$$

   where ρ is the Spearman's ρ (see, e.g, [16]), which is calculated for distributions $\boldsymbol{P}$ and $\boldsymbol{Q}$ treated as a kind of ordinal data such that the frequency values are regarded as the rank positioning. This method has been successively applied to visual word histogram relationship evolution (see, for example [14]), and for clustering genomes within the compositional spectra approach [2].

2. The Euclidean distance:

$$ED^2(\boldsymbol{P}, \boldsymbol{Q}) = \sum_{i=0}^{\infty} (p_i - q_i)^2 .$$

3. The Canberra distance

$$C(\boldsymbol{P}, \boldsymbol{Q}) = \sum_{i=0}^{\infty} \frac{|p_i - q_i|}{(p_i + q_i)}.$$

As usual, a transformation $\mathcal{F}$ is constructed by means of the common Vector Space Model [31]. This model disregards grammar and the order of terms but keeps the collection of terms. Each document is described via a terms frequency table in contradiction of the vocabulary containing all the words (or "terms") in all documents in the corpus. The tables are considered as vectors in a linear space having a dimensionality equal to the vocabulary size. In the bag of words model a document is represented as the distribution of its words. To reduce the space dimensionality, the stop-words are commonly removed. The Keywords Model is an offshoot of the previously discussed model, where a document is represented as a bag not of all terms in the corpus but a bag of selected words. In the N-grams model (for example, see [26], [3]) the vocabulary consists of all N-grams in the corpus, recalling that an N-gram is a connecting sequence of N characters from a text occurring in a slide window of length N. The N-gram approaches are widely applied in the text retrieval area.

In this paper, a Vector Space Model is built, resting upon the content-free words and the stop words. The joint occurrences of the content-free words can provide valuable stylistic evidence for authorship [15], [1]. They also were successfully used in the analysis of quantitative patterns of stylistic influence [13]. Stop words are words that are usually omitted at the preprocessing step, since they are the most common words in a language. In our opinion these terms can play a role similar to the role of the content-free words. Namely, they "glue" together the words in a text. In our study we use a combined set including the content free words and the stop words taken from *http : //www.ranks.nl/stopwords*.

4 Numerical Experiments

Any uppercase characters in the texts involved in the experiments are converted to the corresponding lowercase characters, and all other characters are unchanged. As mentioned earlier, we use the Kolmogorov-Smirnov test with the traditional significance level threshold (0.05) for testing the statistical hypothesis. The results are presented in the tables of the statistical hypothesis testing: '1' indicates that the hypothesis of the same style is rejected, and '0' indicates otherwise.

4.1 Comparison of Book Collections

In the first stage several experiments are provided on large book collections to study the performances of combinations of the stated distance functions with several chunk sizes.

- The first collection consists of the six novels in the Rama series. The first book, "Rendezvous with Rama" (denoted as $R1$), was written by Arthur C. Clarke. He paired up with Gentry Lee for the following three novels "Rama II", "The Garden of Rama", and "Rama Revealed" ($R2-R4$). Gentry Lee also individually authored two novels for the Rama Universe: "Bright Messengers" and "Double Full Moon Night" ($R5 - R6$).
- The second collection consists of the seven novels in Isaac Asimovs Foundation series: "Prelude to Foundation", "Forward the Foundation", "Foundation", "Foundation and Empire", "Second Foundation", "Foundation's Edge", and "Foundation and Earth", which are denoted as $F1, .., F7$.
- The third collection consists of the six fantasy Harry Potter novels written by J. K. Rowling (all books of the series except for "Harry Potter and the Deathly Hallows"), which are denoted in our experiment as $HP1, \dots, HP6$.

The books belonging to each one of the collections are combined in three files matching the collections and denoted $R0$, $F0$ and $H0$. These collections are compared one against one through the chunk size (L) running from 2000 to 10000 with an increment of 1000. The values of T grow from 10 to 100 with an increment 10. Graphs of the Mean Dependency calculated for $T = 10$, $L = 2000, 10000$ and the distance measure S for all the collections are presented in Fig 1.

The charts present realizations of a stationary stochastic process. The corresponding histograms are given in Fig 2.

For the Spearman's and the Canberra distances the expected result stating the collections' differences occur for all values of L and T (Table 1.).

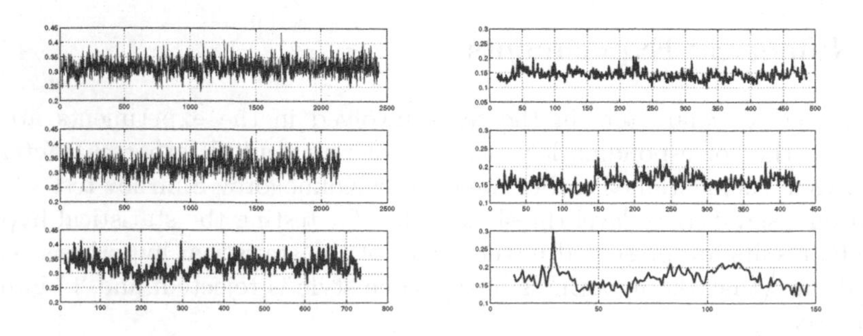

Fig. 1. Graphs of $ZV - value$.

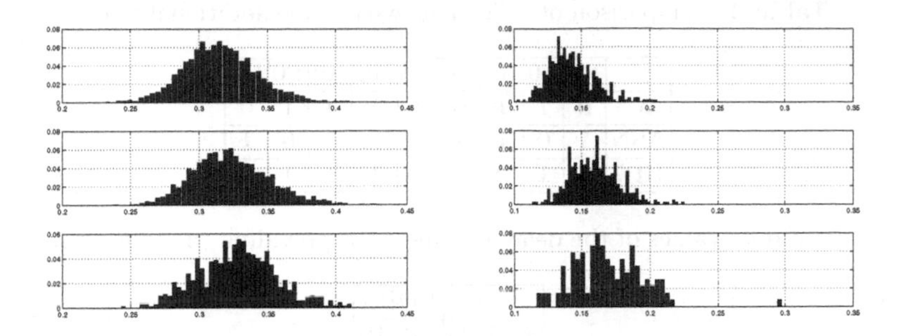

Fig. 2. Histograms of $ZV - value$.

Table 1. Comparison of collections using the Spearman's and the Canberra distances.

	R0	F0	H0
R0	0	1	1
F0	1	0	1
H0	1	1	0

Table 2. Comparison of collections using the Euclidean distance.

	R0	F0	H0	R0	F0	H0
R0	0	1	1	0	1	0
F0	1	0	1	1	0	1
H0	1	1	0	0	1	0

The Euclidean distance demonstrates less stable results presented in Table 2, such that for L in the interval $[2000 - 7000]$ the collections are properly distinguished (the left block in the table), but for the remaining values of L the first and the third collections are not separated (the right block in the table).

In the next step several experiments are provided to evaluate the methods ability to discover authorship. To this aim we compare the following documents with the examined collections:

- 2010: Odyssey two by Arthur C. Clarke (denoted as $AC1$).
- Nemesis by Isaac Asimov (denoted as ASN).
- Harry Potter and the Deathly Hallows by J. K. Rowling (denoted as $HP7$).

The parameter L is 7000 in these experiments because, as can be seen from Tables 1 and 2, this value is a trade-off between a chunk size and the ability to discriminate the collections for all considered distance functions.

The results consist of the ground true, where a book that is assigned only to a collection written by the books author is marked in bold. The first three columns correspond to the Spearmans distance, the next three correspond to the Euclidean distance, and the next three provide the results found for the

Table 3. Comparison of collections with three additional books.

	R0	F0	H0	R0	F0	H0	R0	F0	H0
AC1	**0**	**1**	**1**	**0**	**1**	**1**	1	1	0
ASN	**1**	**0**	**1**	1	1	1	**1**	**0**	**1**
HP7	**1**	**1**	**0**	1	0	1	**1**	**1**	**0**

Table 4. Values of the delay parameter T providing the solutions.

	S	ED	C
AC1	20	50	10
ASN	10	100	10
HP7	80	80	10

Canberra distance. The values of the delay parameter T providing the solutions are presented in Table 4. The Spearmans distance demonstrates the best performance, successfully assigning each single book to its inherent source. On the other hand, applying the Euclidean distance does not provide sufficiently trustworthy results. One of the reasons that can cause such a situation is a possible discrepancy between the styles of the books in a collection. In next section we discuss a semi-supervised approach intended to improve the classification results.

4.2 Comparison of Single Books

The main idea behind the proposed methodology consists of a tentative refinement of the underlying collections aiming to compare a given text with books that are undoubtedly recognized as belonging just to one collection. To specify them, we evaluate a book from a collection against each book from the second collection and categorize them as "undoubtedly" different from the second collection if the majority of the comparisons reject the hypothesis stating a coincidence of the books' styles. In the next stage, all ordered pairs of kind (D_1, D_2), where D_1 is an "undoubtedly" different book selected from collection $i = 1, 2$, are constructed and entered in the tournament, wherever the style of the questionably document is compared with the styles of a pair components. A pair element winners if the style of the questionably document is found identical to the element style but different from the style of another pair element. The collection with the majority of the winners supposes the document classification.

In order to demonstrate the approach ability, experiments through a combination of the Euclidean distance and parameter $L = 10000$, which previously fails, are performed. In the refinement step, the books composing collections $R0$ and $F0$ are compared, and the files $R2$, $R6$, $F1$, and $F4$ are properly assigned to their own collections. In the next phase the book ASN is compared with four pairs of these files. It is pointed out that two pairs vote for the $F0$ collection, and one pair selects the $R0$ collection with a tie in the last tournament, so ASN is correctly assigned.

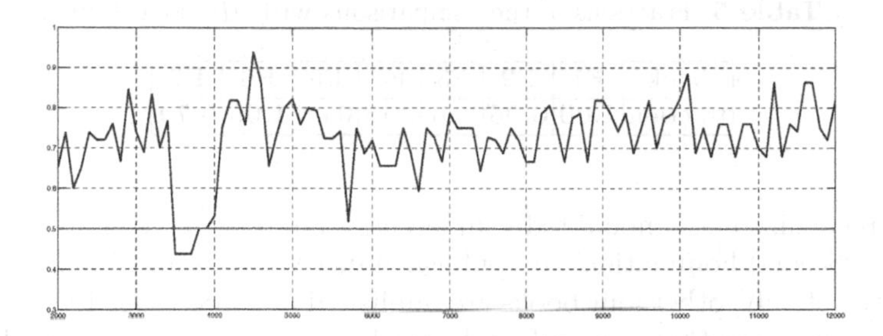

Fig. 3. Graph of the fractions of the winners belonging to HP.

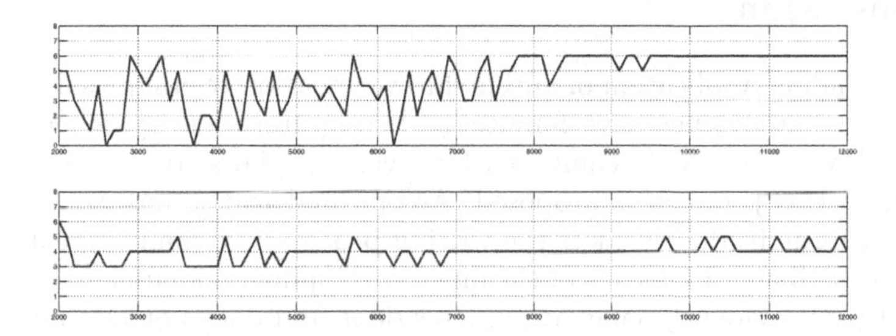

Fig. 4. The quantities of the "undoubtedly " different books.

An analogues attempt to check the ASN style against HP and $F0$ fails for $L = 10000$, because any document from the $F0$ collection survived the first step. However for $L = 12000$ three books from HP and one book from $F0$ are selected with one tie and two successes of $F0$. ASN is again correctly assigned. In the next experiment we compare the $HP7$ style against HP and $F0$ using the Spearman's distance and L running fro 2000 to 12000 with an increment 100. The fractions of the winners belonging to HP are presented in the following figure. The quantities of the "undoubtedly " different books presented in Fig. 5 de facto saturate after $L = 8000$ such that the first collection is turning to a completely undoubtedly different collection (the top panel). The quantities of the "undoubtedly " different books presented in Fig. 5 de facto saturate after $L = 8000$ such that the first collection is turning to a completely "undoubtedly " different collection. Actually, merely four books of the second collection are recognized as "undoubtedly " different in the saturation area. To describe them we consider a table presenting the fractions of the mentioned earlier comparisons with the first collection performed for all considered values of L, where these books are identified as having a style different of the style of $H0$. The undoubtedly different books corresponding to the fractions, which are bigger of 0.5, are marked in bold.

It is very interesting to note that the obtained dichotomy of the Foundation Universe perfectly fits the writing periods of the books. Three novels, which

Table 5. Fractions of the comparisons with $H0$ collection.

book	**F1**	**F2**	F3	F4	F5	**F6**	**F7**
fraction	**1.00**	**0.96**	0.07	0.13	0.05	**1.00**	**0.74**

are not recognized as undoubtedly different, "Foundation", "Foundation and Empire", "Second Foundation" are published in 1951,1952 and 1953 correspondingly. Respectively, other four books are published in 1988, 1993, 1982, and 1986. It makes impression that a considerable gap between the writing periods (about 30 years) leads to this significant difference in the styles.

5 Conclusion

Tasks demanding application of authorship verification techniques arise in many fields such as intelligence, law, finance, and computer security, where there is a need to individualize texts composed by several authors. In this article, a new approach to this problem is proposed. As opposed to the most of the existing methods, this attitude considers the writing process as dynamic creating procedure and describes it by means of the inner developing of documents. The offered methodology is evidently language-independent and does not use any linguistic resource such as ontologies, thesauruses, and language models. In the implementation a document chunks are represented via the histograms of the content free and stop words. A novel characteristic, the mean distance, quantifying association between the sequentially written text chunks by a distance between the histograms is introduced. We consider a series of this feature calculated during a concatenated books collection as a realization of a strictly stationary ergodic process, so that to different writing styles correspond different distributions of the introduced characteristic. We examine several distance functions and found that the Spearmans correlation distance provides the most robust and reliable outcomes. Our method is capable of distinguishing between manuscripts written by different authors. In the future we are going to extend our study to other fields such as plagiarism detection and to compare it to other approaches.

References

1. Binongo, J.: Who wrote the 15th book of Oz? An application of multivariate analysis to authorship attribution. Chance **16**(C), 9–17 (2003)
2. Bolshoy, A., Volkovich, Z., Kirzhner, V., Barzily, Z.: Genome clustering: from linguistic models to classification of genetic texts, vol. 286. Springer Science & Business Media (2010)
3. Brown, P.F., Pietra, V.J.D., deSouza, P.V., Lai, J.C., Mercer, R.L.: Class-based n-gram models of natural language. Computational Linguistics **18**(4), 467–479 (1992)
4. Cha, S.-H.: Comprehensive survey on distance/similarity measures between probability density functions. International Journal of Mathematical Models and Methods in Applied Sciences **1**(4), 300–307 (2007)

5. Collins, J., Kaufer, D., Vlachos, P., Butler, B., Ishizaki, S.: Detecting collaborations in text: Comparing the authors' rhetorical language choices in the federalist papers. Computers and the Humanities **38**, 15–36 (2004)
6. Coyotl-Morales, R.M., Villaseñor-Pineda, L., Montes-y-Gómez, M., Rosso, P.: Authorship attribution using word sequences. In: Martínez-Trinidad, J.F., Carrasco Ochoa, J.A., Kittler, J. (eds.) CIARP 2006. LNCS, vol. 4225, pp. 844–853. Springer, Heidelberg (2006)
7. Diederich, J., Kindermann, J., Leopold, E., Paas, G.: Authorship attribution with support vector machines. Applied Intelligence **19**(1), 109–123 (2003)
8. Eissen, S.M., Stein, B., Kulig, M.: Plagiarism detection without reference collections. Springer, Berlin (2007)
9. Forsyth, R.: New directions in text categorization. Springer, Heidelberg (1999)
10. Frantzeskou, G., Stamatatos, E., Gritzalis, S., Katsikas, S.: Effective identification of source code authors using byte-level information. In: Proceedings of the 28th International Conference on Software Engineering, pp. 893–896. ACM Press, NewYork (2006)
11. Fristedt, B.E., Gray, L.F.: A Modern Approach to Probability Theory. Probability and Its Applications. Birkhäuser, Boston (1996)
12. Harmer, J.: How to Teach Writing. Pearson Education (2006)
13. Hughes, J.M., Foti, N.J., Krakauer, D.C., Rockmore, D.N.: Quantitative patterns of stylistic influence in the evolution of literature. Proc. Natl. Acad. Sci. USA **109**(20), 7682–7686 (2012)
14. Ionescu, R.T., Popescu, M.: Pq kernel. Pattern Recogn. Lett. **55**(C), 51–57 (2015)
15. Juola, P.: Authorship attribution. Foundations and Trends in Information Retrieval **1**(3), 233–334 (2006)
16. Kendall, M.G., Gibbons, J.D.: Rank Correlation Methods. Edward Arnold, London (1990)
17. Kestemont, M., Luyckx, K., Daelemans, W., Crombez, T.: Cross-genre authorship verification using unmasking. English Studies **93**(3), 340–356 (2012)
18. Kolmogorov, A.: Sulla determinazione empirica di una legge di distribuzione. G. Ist. Ital. Attuari **4** (1933)
19. Koppel, M., Argamon, S., Shimoni, A.R.: Automatically categorizing written texts by author gender. Literary and Linguistic Computing **17**(4), 401–412 (2002)
20. Koppel, M., Schler, J.: Authorship verification as a one-class classification problem. In: Proceedings of the 21st International Conferenceon Machine Learning. Press (2004)
21. Koppel, M., Schler, J., Bonchek-Dokow, E.: Measuring differentiability: Unmasking pseudonymous authors. Journal of Machine Learning Research **8**, 1261–1276 (2007)
22. Koppel, M., Winter, Y.: Determining if two documents are written by the same author. Journal of the American Society for Information Science and Technology **65**(1), 178–187 (2014)
23. Koppel, M., Schler, J., Argamon, S.: Computational methods in authorship attribution. JASIST **60**(1), 9–26 (2009)
24. Luyckx, K., Daelemans, W.: Authorship attribution and verification with many authors and limited data. In: Proceedings of the Twenty-Second International Conference on Computational Linguistics (COLING 2008), pp. 513–520 (2008)
25. Manning, C., Schutze, H.: Foundations of Statistical Natural Language Processing. MIT Press, Cambridge (2003)

26. Miao, Y., Kešelj, V., Milios, E.: Document clustering using character n-grams: a comparative evaluation with term-based and word-based clustering. In: Proceedings of the 14th ACM International Conference on Information and Knowledge Management, CIKM 2005, pp. 357–358. ACM, New York (2005)
27. Peng, F., Schuurmans, D., Keselj, V., Wang, S.: Augmenting naive bayes classifiers with statistical languages models. Information Retrieval **7**, 317–345 (2004)
28. Rachev, S.T.: Probability metrics and the stability of stochastic models. Wiley series in probability and mathematical statistics: Applied probability and statistics. Wiley (1991)
29. Rudman, J.: The state of authorship attribution studies: Some problems and solutions. Computers and the Humanities **31**, 351–365 (1998)
30. Ryabko, D., Ryabko, B.: Nonparametric statistical inference for ergodic processes. IEEE Transactions on Information Theory **56**(3), 1430–1435 (2010)
31. Salton, G., Wong, A., Yang, C.S.: A vector space model for automatic indexing. Communications of the ACM **18**(11), 613–620 (1975)
32. Sanderson, C., Guenter, S.: Short text authorship attribution via sequence kernels, markov chains and author unmasking: an investigation. In: Proceedings of the International Conference on Empirical Methods in Natural Language Processing, pp. 482–491 (2006)
33. Sebastiani, F.: Machine learning in automated text categorization. ACM Computing Surveys **34**(1), 1–47 (2002)
34. Smirnov, N.: Table for estimating the goodness of fit of empirical distributions. Annals of Mathematical Statistics **19** (1948)
35. Stamatatos, E.: A survey of modern authorship attribution methods. Journal of the American Society for Information Science and Technology **60**(3), 538–556 (2009)
36. Stamatatos, E., Daelemans, W., Verhoeven, B., Juola, P., Lopez Lopez, A., Potthast, M., Stein, B.: Overview of the author identification task at pan 2015. In: Cappellato, L., Ferro, N., Gareth, J., San Juan, E. (eds.) Working Notes Papers of the CLEF 2015 Evaluation Labs (2015)
37. Stamatatos, E., Daelemans, W., Verhoeven, B., Stein, B., Potthast, M., Juolaand, P., Sanchez-Perez, M.A., Barron-Cedeno, A.: Overview of the author identification task at pan 2014. In: Working Notes for CLEF 2014 Conference, Sheffield, UK, pp. 877–897 (2014)
38. Stamatatos, E., Fakotakis, N., Kokkinakis, G.: Automatic text categorization in terms of genre and author. Computational Linguistics **26**(4), 461–485 (2000)
39. Zhao, Y., Zobel, J.: Effective and scalable authorship attribution using function words. In: Lee, G.G., Yamada, A., Meng, H., Myaeng, S.-H. (eds.) AIRS 2005. LNCS, vol. 3689, pp. 174–189. Springer, Heidelberg (2005)
40. Zolotarev, V.M.: Modern Theory of Summation of Random Variables. Modern Probability & Statistics Series. VSP (1997)

Semantic Aware Bayesian Network Model for Actionable Knowledge Discovery in Linked Data

Hasanein Alharbi(✉) and Mohamad Saraee

The University of Salford, The Crescent, Salford M5 4WT, UK
H.Y.M.AlHarbi@edu.Salford.ac.uk, m.saraee@salford.ac.uk

Abstract. The majority of the conventional mining algorithms treat the mining process as an isolated data-driven procedure and overlook the semantic of the targeted data. As a result, the generated patterns are abundant and end users cannot act upon them seamlessly. Furthermore, interdisciplinary knowledge can not be obtained from domain-specific silo of data.

The emergence of Linked Data (LD) as a new model for knowledge representation, which intertwines data with its semantics, has introduced new opportunities for data miners. Accordingly, this paper proposes an ontology-based Semantic-Aware Bayesian network (BN) model.

In contraxt to the exisiting mining algorithms, the proposed model does nto transorm the original format of the LD set. Therefore, it not only accomodates the sematnic aspects in LD, but also caters to the need of connectign different data-sets from different domains. We evaluate the proposed model on a Bone Dysplasia dataset, Experimental results show promising perfomance.

Keywords: Linked Data (LD) · Actionable Knowledge Discovery (AKD) · Bayesian Network (BN)

1 Introduction

The term Data Mining (DM) refers to methods that aim to extract useful information and knowledge from data. Fayyad et al. have defined these methods as the non-trivial process of identifying valid, novel, potentially useful and ultimately understandable patterns in a database [1, 2].

Despite the fact that the ultimate goal of DM is to identify useful and understandable patterns, the existing mining algorithms are confined to generate frequent patterns and do not illustrate how to act upon them. Accordingly, the concept of Actionable Knowledge Discovery (AKD) has been introduced to overcome the shortages on traditional mining algorithms. The goal of AKD techniques is to bridge the gap between the output of the current mining algorithms and the needs of the real life applications [3,4,5].

This gap has appeared as a result of two major drawbacks, namely, quantity and quality, the former states that the generated patterns are abundant while the later indicates that they cannot be integrated seamlessly into the business domain [3][6]. Upon further investigation, it appears these drawbacks have been caused as a result of viewing the mining process as data driven trial and error practices and ignoring the

P. Perner (Ed.): MLDM 2016, LNAI 9729, pp. 143–154, 2016.
DOI: 10.1007/978-3-319-41920-6_11

surrounding knowledge [3],[7]. Consequently, the mining philosophy has faced a paradigm shift from data-centered into knowledge-centered process, which aims to integrate the surrounding knowledge such as data intelligence in the mining process [3],[8]. Even though the data intelligence could be represented using various techniques, recently LD introduced a new technique to intertwine the data and its semantics in one package. Coupling the data with its semantics not only brings new opportunities for data miner but also raised some challenges; for example, identifying the interesting transaction in heterogeneous data set which have been build based on the description logic and used the triple (subject-predicate-object) format [9,10]. To this end, this paper propos a semantic-aware Bayesian network model which exploits the semantics nature of LD and implicitly accommodates the data intelligence in the mining process. The proposed approach consists of the following steps.

1. Convert the original LD file into BN, which preserves the semantics of the LD file.
2. Initialize the Conditional Probability Tables (CPT's) with default values.
3. Calculate a set of probabilistic constraints using the concept of Maximum a Posterior estimation (MAP) in such a way that it reflects the semantic relation between nodes in the constructed BN.
4. Approximate the CPT's initial values to comply with the set of constraints calculated in the second step using the concept of Iterative proportional Fitting Procedure (IPFP).

The contributions of this paper are twofold. Firstly, the model integrates six semantic relations in the mining process, namely, sub-superclass, equivalent to, complement of, disjoint with, intersection of and union of. Secondly, it does not change the original format of LD set and consequently, it caters for the need of linking various data-sets from multidisciplinary domains using the design principle of LD. The proposed model has been tested using a semi synthetic data-set and the initial results are promising.

The remainder of this paper is organized as following. In section 2 the notion of converting LD file into BN is explained in details while the probabilistic constraints estimation methods are illustrated in section 3. IPFP briefly discussed in section 4, and section 5 discusses in detail the empirical implementation and the initial results. Finally, the paper concluded in section 6 with a brief discussion of planned future work.

2 Bayesian Network Topology Construction

BayesOWL consists of set of construction rules which convert ontology file into BN directed acyclic graph (DAG) thus preserve the semantic of the original ontology file [11,12]. Likewise, the model proposed in this paper follows the same rules to convert the given LD file into BN DAG.

The construction of BayesOWL graph has two main phases. In the first phase, the BayesOWL graph structure is build and the associated CPT's are initialized with default values. The integration of the given probabilistic constrains are then implemented in the second phase [13,14].

The process of construction BN DAG from given ontology file is governed by set of rules. The conventions underpinning these rules can be summarized in the following points [12],[15].

1. Every primitive or defined class is mapped into binary variable.
2. Each parent superclass is connected with its child subclass by an arc.
3. For each concept class C defined as the intersection of set of classes $C_i = \{ C_1, \ldots, C_n \}$ a subnet is created in such a way that there is a link from C and each class in the set C_i toward the class C. Furthermore, there is a link from C and each class in the set C_i toward a logical node called LNodeIntersection. Figure 1.a depicts the creation of the intersection subnet.
4. For each concept class C defined as the union of classes $C_i = \{ C_1, \ldots, C_n \}$; a subnet is created in such a way that there is a link from C to each class in the set C_i. Furthermore, there is a link from C and each class in the set C_i moves toward a logical node called LNodeUnion. Figure 1.b illustrates the creation of the union subnet.

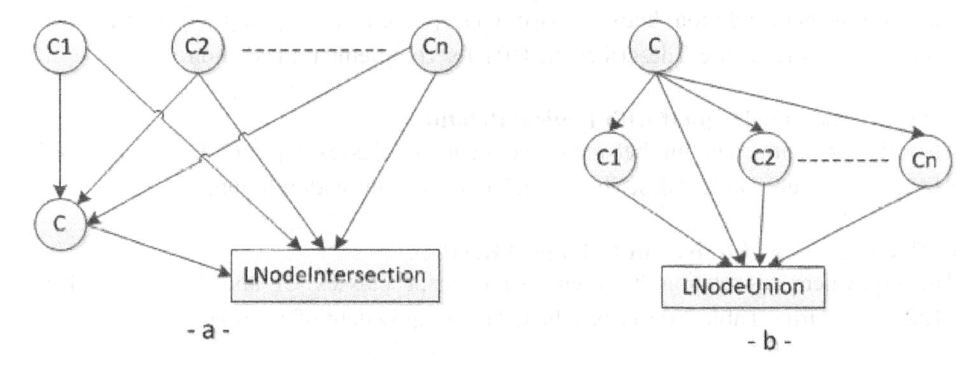

Fig. 1. LNodeIntersection & LNodeUnion [11]

5. For each two concept classes C_1 and C_2 defined as complement of, equivalent to, disjoint with each other a logical node (LNodeComplement, LNodeEquivalent, LNodeDisjoint) are created, which take two input links from C_1 and C_2. Figure 2.a, 2.b and 2.c depicts the creation process for LNodeComplement, LNodeEquivalent and LNodeDisjoint respectively.

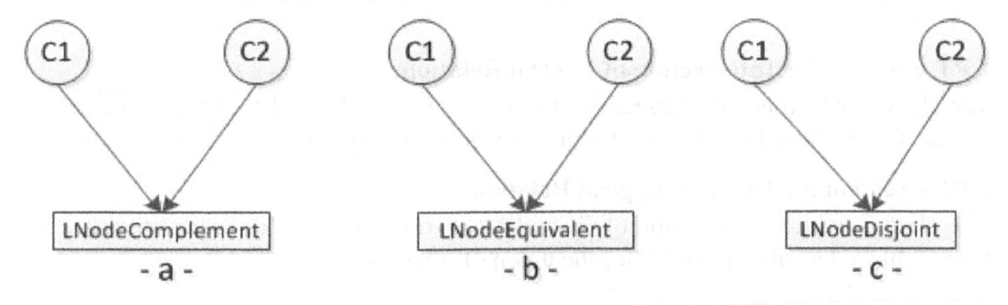

Fig. 2. LNodeComplement & LNodeEquivalent & LNodeDisjoint [11]

It can be seen that the generated DAG contains two types of nodes, namely, regular nodes, which represent classes and logical nodes, which show the logical relation among classes [11],[15]. The combination of these two types of nodes forms the structure of the BN.

So far this section has focused on constructing the BN topology. The following subsection explains in detail the CPT's calculation process for regular and logical nodes [11],[15,16].

2.1 CPT's Calculation for Logical Nodes

It has been stated that the generated DAG cater for five different types of logical nodes, which associated with five logical operations in ontology. The CPT for each logical node is determined by its logical relations. The next subsections explain the CPT creation process for each logical node.

CPT Creation for Complement of Logical Relation
The complement relation between two concepts classes C_1 and C_2 is true IFF $c1\overline{c2c1}c2$ is true. Table 1 describes the CPT for complement of relation.

CPT Creation for Disjoint with Logical Relation
The disjoint with relation between two concept classes C_1 and C_2 is true IFF $c1c2\overline{c1c2}$ is true. Table 2 describes the CPT for disjoint with relation.

CPT Creation for Equivalent to Logical Relation
The equivalent to relation between two concept classes C_1 and C_2 is true IFF $c1c2\overline{c1c2}$ is true. Table 3 describes the CPT for equivalent of relation.

Table 1. CPT for LnodeComplementOF [11]

C_1	C_2	True	False
T	T	0	1
T	F	1	0
F	T	1	0
F	F	0	1

Table 2. CPT for LnodeDisjointWith [11]

C_1	C_2	True	False
T	T	0	1
T	F	1	0
F	T	1	0
F	F	1	0

Table 3. CPT for LnodeEquivalentTo [11]

C_1	C_2	True	False
T	T	1	0
T	F	0	1
F	T	0	1
F	F	1	0

CPT Creation for Intersection of Logical Relation
The class C, which is the intersection of C_1 and C_2 is true IFF $cc1c2\overline{cc1}c2\overline{c}c1\overline{c2cc1c2}$ is true. Table 4 describes the CPT for the intersection of relation.

CPT Creation for Union of Logical Relation
The class C, which is the union of C1 and C2 is true IFF $cc1c2c\overline{c1}c2cc1\overline{c2cc1c2}$ is true. Table 5 describes the CPT for the union of relation.

Table 4. CPT for IntersectionOf [11]

C	C_1	C_2	True	False
T	T	T	1	0
T	T	F	0	1
T	F	T	0	1
T	F	F	1	0
F	T	T	0	1
F	T	F	1	0
F	F	T	0	1
F	F	F	1	0

Table 5. CPT for UnionOF [11]

C	C_1	C_2	True	False
T	T	T	1	0
T	T	F	0	1
T	F	T	1	0
T	F	F	0	1
F	T	T	1	0
F	T	F	0	1
F	F	T	0	1
F	F	F	1	0

2.2 CPT Calculation for Regular Nodes

The CPT's for regular nodes are computed by applying the Bayesian theorem as follows. $P(C|\Pi_C)$ where C is the class of the regular node and Π_C is the set of its parents. The $P(C=\text{True}\ |\Pi_C) = 0$ if any of its parents is false. Hence, the probability for any regular class C is calculated only when all of its parents in true status. This scenario is denoted as $P(C|\ \Pi_C^+)$ where Π_C^+ represent the set of parents classes in true status. This method is used to calculate the probability when probabilistic data is available. Otherwise, a default value (0.5) is assigned based on the following equation [11],[13].

$$P\ (C=\text{True}|\ \Pi_c^+) = P(C=\text{False}|\Pi_c^+) = 0.5. \quad (1)$$

3 Probabilistic Constraints Estimation

The process of converting OWL (i.e. LD) file into BN consists of two phases. In the first phase the BN structure is constructed and the associated CPT's are initialized with default values. Then, the probabilistic constraints are integrated in the second phase. Hereafter, the process of probabilistic constraints estimation is covered in details.

It has been argued that the two main approaches for probabilistic estimation in BN are the Maximum Likelihood Estimation (MLE) and Bayesian estimation [17,18,19]. Thus, these two approached discussed in the next subsections.

3.1 Maximum Likelihood Estimation (MLE)

MLE aims to find the value of ϴ which quantifies the maximum probability of the incoming event. In a data-set D, which consist of n instances of binominal random variable X, MLE aims to estimate the maximum likelihood of occurrence for n+1 incoming event [17],[20].

Let assume that the random variable X represents the event of flipping a thumbtack, which has two possible outcomes, Head and Tail. Furthermore, the observed data consist of 5 observations, 3 of which are Heads and 2 Tails. Therefore, the MLE for the incoming event n+1 for X = Head is:

$$\text{MLE}_{x=\text{Head}} = \frac{\text{X=Head}}{\text{X=Head+ X=Tail}} + \frac{3}{3+2} = 0.6 \quad (2)$$

It can be seen that the likelihood function is maximized by dividing the number of correct trials over the total number of trials. Although MLE has various advantages, it also has some limitations. For example, the size of the observed data-set has no effect in the estimation process. Additionally, MLE does not take the prior knowledge into consideration and relies entirely on the observed data. Therefore, Bayesian method, which integrates the prior knowledge in the estimation process, has been introduced [17,18],[20].

3.2 Maximum a Posterior Estimation (MAP)

An alternative method for parameters estimation, which injects the prior knowledge in form of prior distribution in the estimation process, is MAP. MLE aims to maximize the likelihood function. Likewise, MAP aims to maximize the posterior of Θ given the observed data. This hypothesis can be formalized in the following equation [17],[20].

$$\hat{\theta}\, MAP = \text{argmax } \text{P}(\Theta|\text{d}) \quad (3)$$

Equation (3) could be reformulated using Bayes rule.

$$\hat{\theta}\, MAP = argmax\, \frac{\text{p(d}|\Theta)p(\Theta)}{\text{p(d)}} \text{ Where p(d)} \neq f(\Theta) \quad (4)$$

$$\hat{\theta}\, MAP = argmax\, (\log p(d|\Theta) + \log p(\Theta)) \quad (5)$$

Equation (5) shows that the posterior probability $p(\Theta|d)$ is Beta distribution, which obtained by summing up the likelihood in form of Bernoulli distribution and prior knowledge in form of Beta distribution. Hence, the posterior probability is Beta distribution with $(\alpha + r)$ correct trails out of $(\alpha+\beta+ n)$ total number of trials. Accordingly, the prior and posterior statistics for Beta distribution could be summarized in the following table [19],[21].

Table 6. Prior and posterior statistic for beta distribution [19]

Statistics	Prior	Posterior
Law	Beta(α,β)	Beta(α+r,β+(n-r))
Mean	α/α+β	α+r/α+β+n
Mode	α-1/α+β-2	α+r-1/α+β+n-2
variance	α β/(α+β)2 +(α+β+1)	(α+r)(β+n-r)/(α+β+n)2 +(α+β+n+1)

4 Iterative Proportional Fitting Procedure (IPFP)

The concept of Iterative Proportional Fitting Procedure (IPFP) was first introduced by Deming and Stephan on 1940. It used to estimate the probability in contingency tables, which is subject to given marginal constraints. In 2000, Cramer proposed an extension to the traditional IPFP to accommodate the conditional probability constraints. In fact, the statistical application for probability models with marginal and conditional distributions are comprehensive, such as, Bayesian statistic, contingency table, long-linear models etc. [22,23].

This paper is concerned is the capability of IPFP to approximate a set of probability table according to given set of marginal and conditional probabilistic constraints. The full mathematical and theoretical background of IPFP is beyond the scope of this paper therefore, the reader may refer to the following reference [24] for comprehensive studies on it.

5 Empirical Implementation and Initial Results

5.1 Bone Dysplasia Case Study

Bone Dysplasia (BD) has been defined as a rare and heterogeneous group of genetic based disorders effecting the skeletal development, growth and homeostasis. The genotypes and phenotypes associated with BD are comprehensive. Furthermore, the data and expertise in BD domain are sparse. Therefore, its diagnosis remains a challenge [25,26].

Various researchers in the fields of data mining and machine learning have respond to this challenge using different techniques. For example, Paul et.al. have proposed the use of techniques such as association rule mining, semantic similarity measures and depaster-shafer theory [6],[27].

In addition to that an effort has been spent to develop a comprehensive ontology, namely, the Bone Dysplasia Ontology (BDO), which provides the mean for formal representation of the BD domain and the associated genotypes and phenotypes. It describes over 450 types of bone disorders clustered under 40 groups [25].

One of the most common type of the bone disorder is the Thanatophoric Dysplasia (TD) which characterized by the following physical symptoms: Macrocephaly, Large anterior fontanel, Frontal bossing, Micromyelia, Trident hand with brachydactyly, Redundant skin folds, Narrow bell-shaped thorax with short ribs and protuberant abdomen, Relatively normal trunk length, and Generalized hypotonia [28].

The case study in the next subsections investigated the advantages of applying the proposed model (i.e. Semantically Aware Bayesian Network) to the BDO in order to classify patients with potential skeletal disorder symptoms. The Thanatophoric Dysplasia Type 1 (TD1) has been used as a target class and its associated symptoms as a prediction attributes.

5.2 The Bone Dysplasia Ontology Structure

It has been widely agreed that ontology provides a means for a formal representation of a domain of interest in machine understandable format, which improve knowledge extraction and reasoning over that domain. Hence, the BDO has been developed as a collaborative effort between experts in the skeletal disorders domain and ontology engineers. Consequently, a formal and comprehensive representation, which covers various relationships and axioms between concepts in the domain, has been generated. This case study focused on the "characterized_by" relation, which establishes a link between BD and the associated phenotypic information. Figure (3) illustrates the relations between the top level concepts in the BDO [25].

An effort has been spent to extract the Thanatophoric Dysplasia Type1 and its associated symptoms. Our aim is to construct a Bayesian Network topology that preserves the semantic relations between the disease class and its symptoms. The construction rules proposed by BayesOWL have been used as a guideline to construct the Bayesian Network topology. Figure (4) shows the generated topology. Although, at the current stage of our research the creation of the BN topology has been done manually, there is an ongoing effort to automate this process.

5.3 Data Creation and Initial Results

Access to medical data-set is restricted to medical staff and safe guarded by various polices and regulations due to its sensitivity and confidentiality. Although a great effort has been spent to gain access to the BD data-set, permission has not been granted yet. Alternatively, we have developed codes that randomly generate the required data-set. One thousand records have been generated which cover the target class (the root class in figure 4) and the ten symptoms classes which represented by terminal nodes in figure (4). Additionally, the values for the intermediate nodes, namely, Skeletal Morpholoyg, Abnormality of the Skull and Abnormality of Limb Bone generated based on the truth path rules which indicate that a super-class is true if and only if any of its sub-classes is true.
Finally, the generated data-set has been classified using the model constructed in the previous sections and different evaluation measures have been calculated. Table (7) summarized the obtained results.

5.4 Evaluation and Discussion

The performance of the proposed model has been evaluated using the ten folds cross validation methods and various performance criterias such as F1 measure, accuracy and Receiver Operating Characteristics (ROC) graph (Fig. 5). Furthermore, the performance of the proposed model has been compared to three existing algorithms, namely, Conditional Independences ICSS algorithm, Local score based K2 algorithm and Global score based TAN algorithm. Table (7) summarized the initial results.

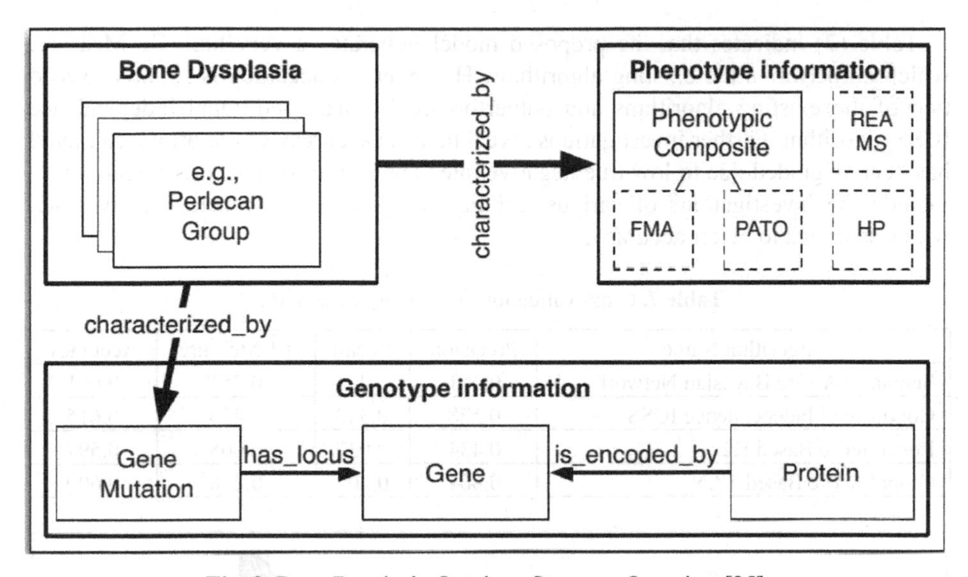

Fig. 3. Bone Dysplasia Ontology Structure Overview [25]

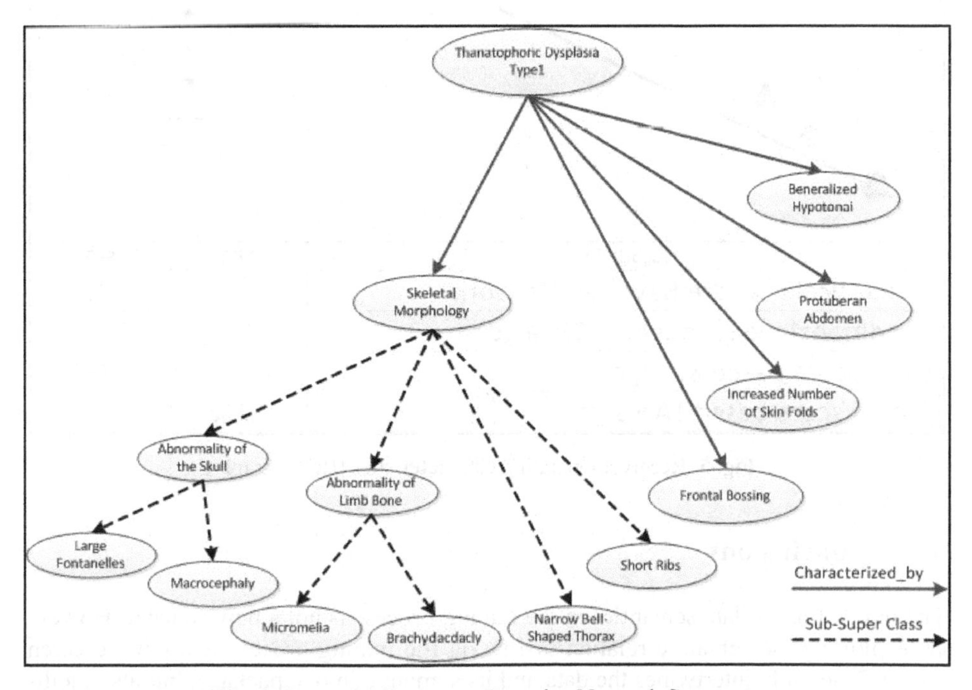

Fig. 4. Semantically Aware Bayesian Network Structure

Table (7) indicates that the proposed model generates a very high F1 Measure, which outperforms the existing algorithms. However, its accuracy is slightly exceed two of the existing algorithms and outperformed by the Conditional Independence ICSS algorithm. Further investigations reveal that the accuracy of the proposed model has been degraded due to low true negative rate. The future work of this research will include the investigations of various techniques to improve the true negative rate, which will lead to better accuracy.

Table 7. Cross validation performance measures

Algorithm Name	Precision	Recall	F1 Measure	Accuracy
Semantic Aware Bayesian Network	0.601	1	0.751	0.601
Conditional Independence ICSS	0.528	0.340	0.414	0.615
Local Score Based K2	0.434	0.037	0.068	0.597
Global Score Based TAN	0.504	0.202	0.288	0.600

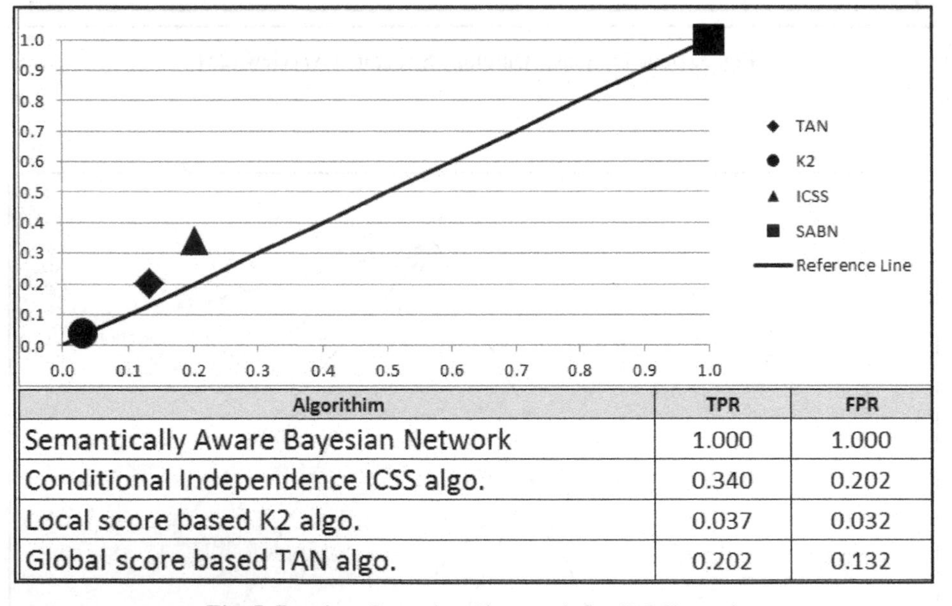

Algorithim	TPR	FPR
Semantically Aware Bayesian Network	1.000	1.000
Conditional Independence ICSS algo.	0.340	0.202
Local score based K2 algo.	0.037	0.032
Global score based TAN algo.	0.202	0.132

Fig. 5. Receiver Operating Characteristics (ROC) graph

6 Conclusions

The integration of data semantics in the mining process is not a new concept. However, exploiting the semantic relation in LD via the mining process is a new research area. LD not only intertwines the data and its semantic in one package, but also facilitates the integration of various data-sets from multiple domains.

One major drawback is that the original format of the LD needs to be transformed into a format which is understood by the conventional mining algorithms. Currently they cannot benefit from the semantical and structural features of the LD. To this end, this paper proposes a semantic-aware Bayesian network model which exploits the semantic nature of LD and implicitly accommodates the data intelligence in the mining process. Furthermore, it does not change the original format of the LD and consequently, linking multiple data-sets is easily achievable.

The proposed model has been tested using semi synthesized data-set, which highlight the significant of the sub-super class relation. Initial results show that injecting the semantically aware probabilistic information into the Bayesian inference algorithm generates more realistic results.

Taking into account the fast accumulation of LD, this paper has investigated the suitability of Bayesian network for LD mining. Although the semi synthesized data-set shows some promising results, the proposed model needs to be tested on a real life data-set.

The finding of this paper suggests that the integration of the semantic relations in the knowledge discovery process is advantageous. Therefore, the plan for the future will be to investigate the integration of other user-defined domain-specific semantic relation. Furthermore, different inference algorithms will be explored.

References

1. Fayyad, U., Piatetsky-Shapiro, G., Smyth, P.: From data mining to knowledge discovery in databases. AI Mag. 37–54 (1996)
2. Zhang, C., Zhang, S.: Association Rule Mining: Models and Algorithms. Springer-Verlag Berlin Heidelberg. XII, p. 244 (2002)
3. Cao, L., Yu, P.S., Zhang, C., Zhao, Y.: Domain driven data mining. Springer US. XVI, p. 248 (2010)
4. Cao, L.: Domain-driven data mining: Challenges and prospects. IEEE Trans. Knowl. Data Eng. **22**, 755–769 (2010)
5. Sexton, M., Lu, S.: The challenges of creating actionable knowledge: an action research perspective. Constr. Manag. Econ. **2**, 683–694 (2009)
6. Paul, R., Groza, T., Hunter, J., Zankl, A.: Semantic interestingness measures for discovering association rules in the skeletal dysplasia domain. J. Biomed. Semantics. **5**, 8 (2014)
7. Dahan, H., Cohen, S., Rokach, L., Maimon, O.: Proactive Data Mining with Decision Trees. Springer New York (2014)
8. Antunes, C., Silva, A.: New trends in knowledge driven data mining a position paper. In: Proc. 16th Int. Conf. Enterp. Inf. Syst., pp. 346–351 (2014)
9. Bizer, C., Heath, T., Berners-Lee, T.: Linked data-the story so far. Int. J. Semant. Web Inf. Syst. **5**, 1–22 (2009)
10. Quboa, Q.K., Saraee, M.: A State-of-the-Art Survey on Semantic Web Mining. Intell. Inf. Manag. **05**, 10–17 (2013)
11. Ding, Z., Peng, Y., Pan, R.: BayesOWL: Uncertainty Modelling in Semantic Web Ontologies. Soft Comput. Ontol. Semant. Web. **204**, 3–29 (2006)
12. Ma, Z.: Soft Computing in Ontologies and Semantic Web. Springer Sci. Bus. Media (2007)
13. Sun, Y.: A Prototype Implementation of BayesOWL. University of Mayryland Baltimore County, Diss (2009)
14. Ding, Z.: BayesOWL. http://www.csee.umbc.edu/~ypeng/BayesOWL/index.html

15. Ding, Z., Peng, Y.: A Bayesian approach to uncertainty modelling in OWL ontology. Maryland Univ Baltimore Dept. of Computer Science and Electrical Engineering (2006). http://oai.dtic.mil/oai/oai?verb=getRecord&metadataPrefix=html&identifier=ADA444453
16. Zhang, S., Sun, Y., Peng, Y., Wang, X.: A practical tool for uncertainty in OWL ontologies. In: Proc. 10th IASTED Int. Conf., vol. 674, pp. 235
17. Koller, D., Friedman, N.: Probabilistic Graphical Models Principles and Techniques. MIT press (2009)
18. Jensen, F.V., Nielsen, T.D.: Bayesian Networks and Decision Graphs. Springer Science & Business Media (2009)
19. Almond, R.G., Mislevy, R.J., Steinberg, L.S., Yan, D., Williamson, D.M.: Learning in models with fixed structure. In: Bayesian Networks Educ. Assessment, pp. 279–330. Springer New York (2015)
20. Heinrich, G.: Parameter estimation for text analysis. Tech. Report, Fraunhofer IGD, Darmstadt, Ger. (2005)
21. Levy, R.: Probabilistic Models in the Study of Language. University of California, San Diego (2012)
22. Fienberg, S.E.: An iterative procedure for estimation in contingency tables. Ann. Math. Statisitics, 907–917 (1970)
23. Cramer, E.: Probability measures with given marginals and conditionals: I-projections and conditional iterative proportional fitting. Stat. Decis. J. Stoch. Methods Model. PhD Thesis, Czech Tech. Univ. Fac. Electr. Eng., 311–330 (2000)
24. Vomlel, J.: Methods of probabilistic knowledge integration. PhD Thesis, Czech Technical University, Faculty Of Electrical Engineering (1999)
25. Groza, T., Hunter, J., Zankl, A.: The Bone Dysplasia Ontology: integrating genotype and phenotype information in the skeletal dysplasia domain. BMC Bioinformatics **13**, 50 (2012)
26. Warman, M.L., Cormier-Daire, V., Hall, C., Krakow, D., Lachman, R., Lemerrer, M., Mortier, G., Mundlos, S., Nishimura, G., Rimoin, D.L., Robertson, S., Savarirayan, R., Sillence, D., Spranger, J., Unger, S., Zabel, B., Superti-Furga, A.: Nosology and classification of genetic skeletal disorders: 2010 revision. Am. J. Med. Genet. Part A **155**, 943–968 (2011)
27. Paul, R., Groza, T., Hunter, J., Zankl, A.: Decision Support Methods for Finding Phenotype - Disorder Associations in the Bone Dysplasia Domain. PLoS One **7** (2012)
28. Liboi, E., Lievens, P.M.J.: Thanatophoric Dysplasia. Dostupné z (2004). http://www.orpha.net/data/patho/GB/uk-Thanatophoric-dysplasia.pdf

Driving Style Identification with Unsupervised Learning

Vladimir Nikulin(✉)

Department of Mathematical Methods in Economy,
Vyatka State University, Kirov, Russia
vniknlin.uq@gmail.com

Abstract. One way to optimise insurance prices and policies is to collect and to analyse driving trajectories: sequences of 2D-points, where time distance between any two consequitive points is a constant. Suppose that most of the drivers have safe driving style with similar statistical characteristics. Using above assumption as a main ground, we shall go through the list of all drivers (available in the database) assuming that the current driver is "bad". We shall add to the training database several randomly selected drivers assuming that they are "good". By comparing the current driver with a few randomly selected "good" drivers, we estimate the probability that the current driver is bad (or has significant deviations from usual statistical characteristics). Note as a distinguished particular feature of the presented method: it does not require availability of the training labels. The database includes 2736 drivers with 200 variable length driving trajectories each. We tested our model (with competitive results) online during Kaggle-based AXA Drivers Telematics Challenge in 2015.

Keywords: Motion mining · Unsupervised learning · Similarity measures · Learning of similarity · Big data · Sequential data

1 Introduction

From the point of view of optimizing of insurance pricing accuracy, it is desirable to be able to identify a driver by his driving style. Using inertial sensor data we can analyse and understand driving style. Consequently, such analysis can help reduce dangerous driving [1].

In the world today, there are over one billion cars with different drivers interacting with each other on the roads. Each driver has their own driving style, which could impact safety, fuel economy, and road congestion, among many things. The precise relationships between driving style and their effects have not been well characterized, although there is some general consensus that "aggressive" driving (e.g. speeding, hard braking or sharp turning) has a mostly negative impact [2].

In the traditional system, a consumer calls an insurance company, provide some basic information, and get a quote for auto insurance based on type of car,

P. Perner (Ed.): MLDM 2016, LNAI 9729, pp. 155–169, 2016.
DOI: 10.1007/978-3-319-41920-6_12

age, gender, marital status, location, driving history, and credit history of the customer. These attributes act as proxies for your auto risk, and the riskier you are the more likely you are to file a claim. These attributes allow an insurance firm to stack you up against the rest of population and see where you are likely to fall in terms of risk. They are playing a guessing game based on averages.

In the last years, there have been active research toward developing systems that make driving safer. In the current insurance markets, consumers have rejected the so-called Pay-As-You-Drive due to two main reasons: the required installation of "black-boxes" in vehicles makes drivers perceive the monitoring as intrusive, and the installation and operation of these units incurs additional costs to insurers and consumers. An alternative approach is presented in [3] to use a smartphone application that is operated at the users discretion, emphasizing that it is more a driving support tool than a "black-box" monitoring device.

In terms of driving cycles characteristics, the research has started since as early as 1978 when Kuhler and Karstens [4] introduced 10 aggregate driving behaviour parameters: average speed, average speed excluding stop, average acceleration, average deceleration, mean length of a driving period, average number of acceleration deceleration changes within one driving period, proportion of standstill time, proportion of acceleration time, proportion of deceleration time, and proportion of time at constant speed.

1.1 Usage-based Insurance

Usage-based insurance (UBI) has been around for a while it began with Pay-As-You-Drive programs that gave drivers discounts on their insurance premiums for driving under a set number of miles. These soon developed into Pay-How-You-Drive programs, pioneered by Progressive, which track your driving habits and give you discounts for 'safe' driving.

With UBI, the need for these best-guess proxies is gone. UBI allows a firm to snap a picture of an individual's specific risk profile, based on that individual's actual driving habits. UBI condenses the period of time under inspection to a few months, guaranteeing a much more relevant pool of information. In addition, it allows consumers to have control over their premiums.

Potentially aggressive driving behavior is currently a leading cause of traffic fatalities in the United States. More often than not, drivers are unaware that they commit potentially-aggressive actions daily. Recently, auto insurance companies have started placing cameras in vehicles to lower insurance rates and monitor driver safety. Similar to vehicle fleet monitoring, insurance companies have also observed that people drive better when being monitored [5].

Being able to dynamically recognize the driving style is invaluable information for modern Intelligent Transport Systems and Road Operators [6]. More specifically, being able to collect contextual information about the driving style coupled with specific traffic or road characteristics, allows the road operator to perform reasoning about safety characteristics of road usage and react upon that information: for example, assess vehicle dynamic information and perform a categorization of drivers acceleration patterns in specific road curves.

If this information is compared with speed limit information and accident statistics in the specific road segment under investigation, specific adjustments of the road infrastructure can be proposed in order to minimize the possibility of bad driving behaviour.

1.2 Global Positioning System

Interesting study [7] on a large real-world Global Positioning System (GPS) dataset of about 23,000 taxis in Shenzhen, China to find anomalities based on trajectory patterns in a road network. The subject of a road segment-based anomaly detection problem is to investigate how anomality on a road segment will cause traffic deviation.

Another application for GPS-data was considered in [8]: we can investigate how optimal is trajectory of the trip, connecting two different points. Further, we can consider the same task with more flexible and advanced structure of data stream [9] considering continuous data arrival online.

The wide use of GPS sensors in smart phones encourages people to record their personal trajectories and share them with others on the Internet. A recommendation service is needed to help people process the large quantity of trajectories and select potentially interesting ones [10].

In this paper we consider just basic driving trajectories. The subjects of more advanced studies are selection of path and the control method for an intelligent car following roads with obstacles. For example, [11] presents a control method to avoid obstacles for an intelligent car based on rough sets and neighborhood systems.

1.3 Unsupervised Learning

One of the advantages of supervised learning is that the final error metric is available during training. For classifiers, the algorithm can directly reduce the number of misclassifications in the training set. Unfortunately, when modeling human learning or constructing classifiers for autonomous robots, supervisory labels are often not available or too expensive [12]. There are many examples of unsupervised methodology. For example, hard clustering with classical algorithm *k-means*, or soft clustering with *EM*-algorithm [13].

Matrix factorization is another popular unsupervised method. In fact, it is rather a framework with a large variety of modifications. In principle, it would be better to describe the high-dimensional data in terms of a small number of meta-features, derived as a result of matrix factorization, which could reduce noise while still capturing the essential features of the data [14]. Framework of neural networks represents another approach to generate meta-features. However, in order to activate backpropagation algorithm we need labels. It is a technical issue: just labels, but not necessarily any specific labels. How to generate those labels is a subject of special consideration [15].

Much recent research has been devoted to learning algorithms for deep learning architectures such as Deep Belief Networks and Convolutional Neural Networks [16], [17], with impressive results obtained in several areas, mostly on vision and language data sets. The best results obtained on supervised learning tasks involve an unsupervised learning component, usually in an unsupervised pre-training phase [18].

Suppose we have a few predictors. How to estimate quality of those predictors without labels represents an important task of a higher level [19].

Unsupervised learning is a difficult problem. It is more difficult when we have to simultaneously find the relevant features as well. A key element here to define precisely the problem [20]. The goal of feature selection for unsupervised learning is to find the smallest feature subset that best uncovers "interesting natural" groupings (clusters) from data according to the chosen criterion.

1.4 Structure of the Paper

This paper is structured as follows. In Section 2 we describe the structure of data and the problem. In Section 2.2 we explain feature engineering as the most essential part of our method. In Section 3 we present the most important experimental results. The main idea of our method is given in Section 3.1. In Section 3.3 we describe some alternative approaches. Finally, Section 4 concludes the paper.

2 AXA Drivers Telematics Challenge

AXA (the orginiser of the competition) is a French multinational insurance firm headquartered in Paris that engages in global insurance, investment management, and other financial services. The company was originally founded in 1816. The AXA Group operates primarily in Western Europe, North America, the Asia Pacific region, and the Middle East; with presence also in Africa. AXA is a conglomerate of independently run businesses, operated according to the laws and regulations of many different countries. The company is a component of the Euro Stoxx 50 stock market index.

For automobile insurers, telematics represents a growing and valuable way to quantify driver risk. Instead of pricing decisions on vehicle and driver characteristics, telematics gives the opportunity to measure the quantity and quality of a driver's behavior. This can lead to savings for careful or infrequent drivers, and transition the burden to policies that represent increased liability.

AXA has provided a dataset of 547200 anonymized driver trips (200 trips per any particular driver). The intent of this competition was to develop an algorithmic signature of driving type. Using available data we can answer on the following important questions. Does a driver drive long trips? Short trips? Highway trips? Back roads? Do they accelerate hard from stops? Do they take turns at high speed? The answers to these questions combine to form an aggregate profile that potentially makes each driver unique.

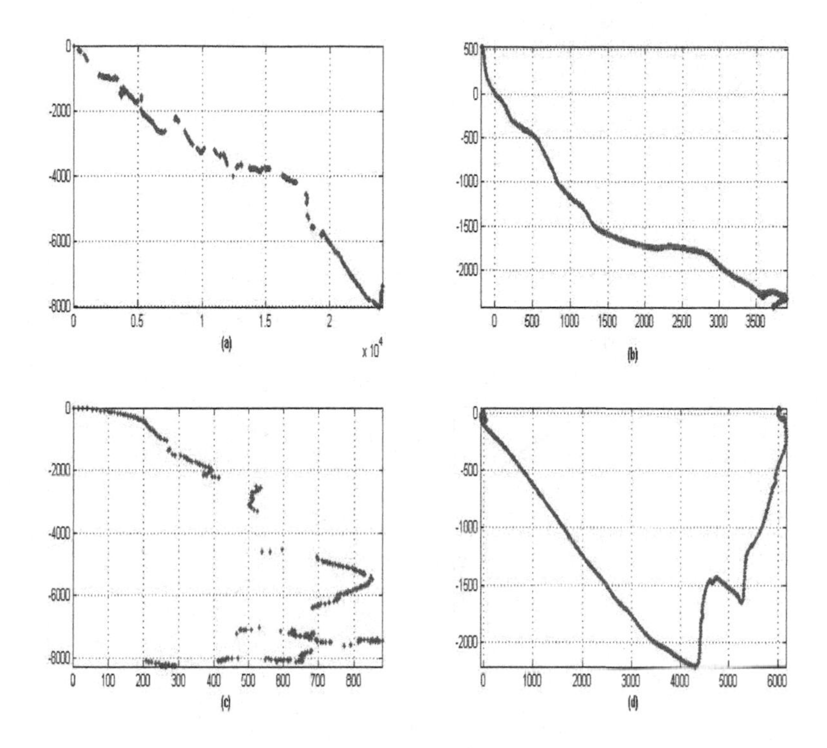

Fig. 1. Driving trajectories: left column - "bad" or risky, right column - "good" or safe.

The duration of the International contest was 91 days: from 15th December 2014 to 16th March 2015. The competition attracted 1528 teams. The AUC - Area under receiver operating curve was used as a criterion. Our result was 0.93606 or 40th place in the contest. Result of the winner was 0.97984. Above results indicate high level of quality in recognition of the driving style and behavior.

2.1 Data

The data includes 2736 folders corresponding to the particular drivers. Any folder contains 200 trips of variable length. The total size of data is about 5.5GB. This competition is particularly interesting because of the following reason: there are no trditional split train/test, and there are no any labels available. The time-difference (a few seconds) between two sequential 2D-point is fixed.

In order to protect the privacy of the drivers' location, the trips were centered to start at the origin (0,0), randomly rotated, and short lengths of trip data were removed from the start/end of the trip, see Figures 1 and 2.

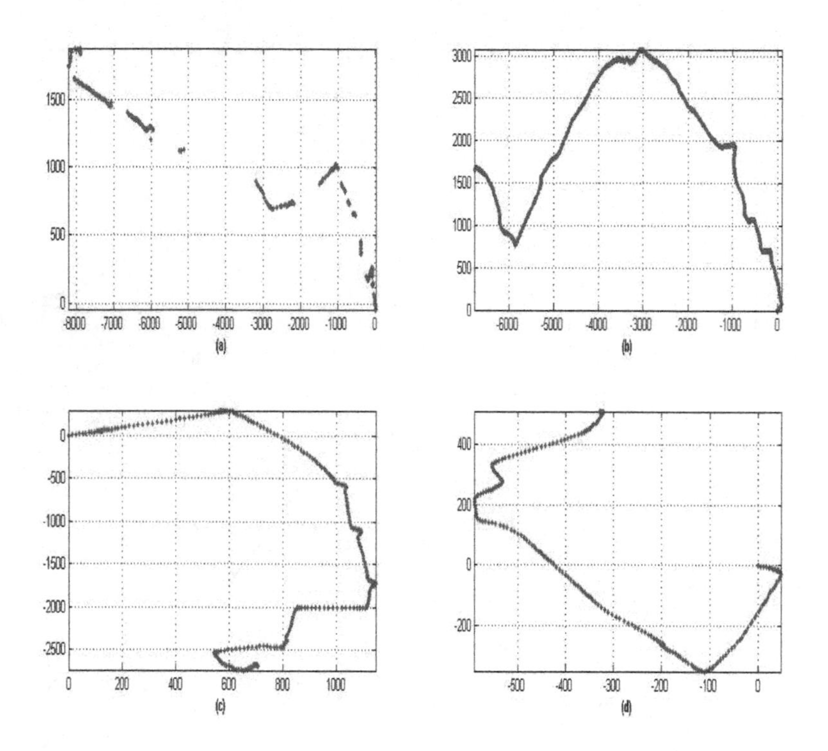

Fig. 2. Driving trajectories: left column - "bad" or risky, right column - "good" or safe.

A small and random number of false trips (trips that were not driven by the driver of interest) are planted in each driver's folder. These false trips were sourced from drivers not included in the competition data, in order to prevent similarity analysis between the included drivers.

The challenge of this competition was to identify "bad" or risky trips based on their telematic features. More specifically, the task was to predict a probability for any particular trip that it is unsafe or risky. The final solution represents a vector of 547200 probabilities.

2.2 Feature Engineering

In mathematical terms we can describe data in the following way:

$$\{x_{d,t,i}, y_{d,t,i}\}, \tag{1}$$

where x and y represent horizontal and vertical coordinates, $1 \leq d \leq n_d$ - index of the driver, $n_d = 2736$; $1 \leq t \leq n_t$ - index of the trip, $n_t = 200$, $1 \leq i \leq n_{d,t}$ - sequential index of the point within particular trip, where $200 \leq n_{d,t} \leq 2000$ - variable length.

In order to simplify notations we shall assume that the driver and trip are fixed. Accordingly, we shall drop indexes d and t.

It is very essential here to note that the time difference any two sequential points in (1) is fixed. Using this fact as a main ground, we can approximate speed of the driver

$$v_i = \sqrt{(x_{i+sv} - x_i)^2 + (y_{i+sv} - y_i)^2}, \tag{2}$$

where $1 \leq i \leq n_{d,t} - sv, sv \geq 1,$ - shift parameter for speed.

Additionally, we can take into account acceleration

$$a_i = v_{i+sa} - v_i, \tag{3}$$

where $1 \leq i \leq n_{d,t} - sv - sa, sa \geq 1,$ - shift parameter for acceleration.

It follows from Figures 1 and 2 that the driving trajectories are very far from straight. Therefore, it is very important to consider how drivers are passing turning points. Note, also, that trajectories in the left column of Figures 1 and 2 correspond to "bad" drivers and trajectories in the right column correspond to "good" drivers. We can see some empty spaces between consequitive points in the left column. Apparently, those spaces indicate very fast movements or high speed.

Generally, shift parameters for speed sv and for accelerations sa are not necessarily the same and maybe different. Let us consider 2D-vectors of speed

$$\boldsymbol{v}_i = \{x_{i+sv} - x_i, y_{i+sv} - y_i\}. \tag{4}$$

It is very essential that vector of speed (4) maybe used as a measurement of the direction of the movement. In the case if driver is changing direction, it will be very important to measure and to take into account how fast is this change. To do this we shall consider the following difference of speed vectors:

$$\boldsymbol{dv}_i = \boldsymbol{v}_{i+sa} - \boldsymbol{v}_i. \tag{5}$$

Consequently, we can consider three new features

$$L(\boldsymbol{dv}_i) = length(\boldsymbol{dv}_i); \tag{6a}$$

$$Lv_i = L(\boldsymbol{dv}_i) \cdot v_i; \tag{6b}$$

$$La_i = L(\boldsymbol{dv}_i) \cdot a_i, \tag{6c}$$

where $L(\boldsymbol{dv})$ is length of the vector $\boldsymbol{dv}$, which is measured according to Euclidean distance.

2.3 Markov Chains

Using one of the above features (speed or acceleration, for example) we can design more advanced features based on the concepts of Markov chains.

Let us consider time-triplet $\{i, i+s_1, i+s_2\}$, where $s_2 > s_1 \geq 1, i = 1, \ldots, n_2,$ $n_2 = n_1 - s_2, n_1 = n_{d,t} - sv.$

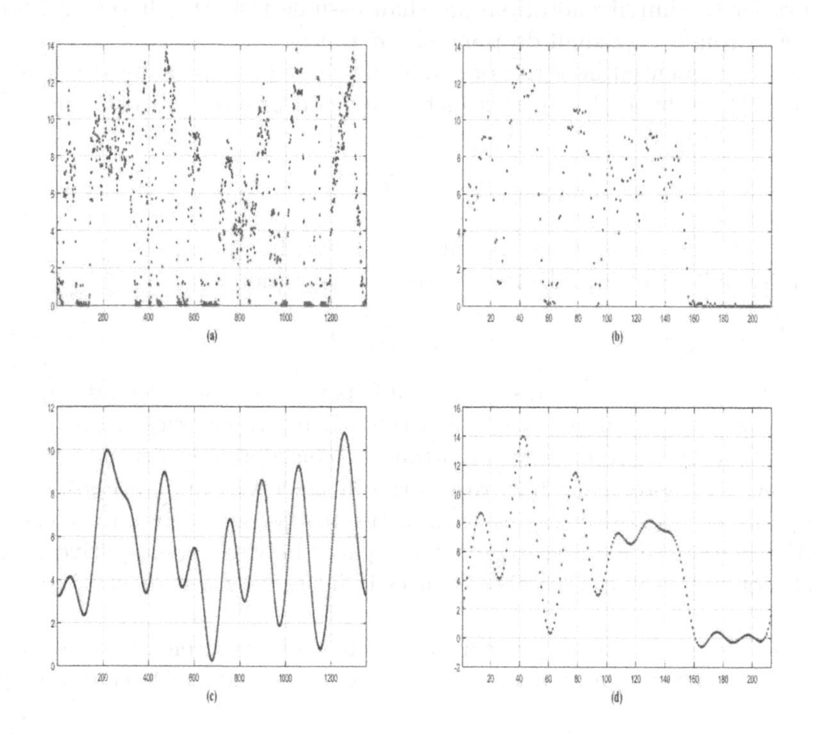

Fig. 3. Upper row: original speeds; second row: smoothed speeds using FFT, see Section 2.4.

Based on the relations between v_{i+s_1} and v_i, we can define the following 5 "enter" conditions: 1) if $v_{i+s_1} - v_i < -\beta \cdot \Delta$, 2) elseif $v_{i+s_1} - v_i < -\alpha \cdot \Delta$, 3) elseif $v_{i+s_1} - v_i < \alpha \cdot \Delta$, 4) elseif $v_{i+s_1} - v_i < \beta \cdot \Delta$, 5) othewise, where $\Delta = \max\{v_i, \delta\}$,

$$\delta = \gamma \cdot \frac{1}{n_1} \sum_{i=1}^{n_1} v_i,$$

where $\beta > \alpha > 0, \gamma > 0$ - regulation parameters.

Similarly, we can define 5 "exit" conditions: 1) if $v_{i+s_2} - v_{i+s1} < -\beta \cdot \Delta$, 2) elseif $v_{i+s_2} - v_{i+s_1} < -\alpha \cdot \Delta$, 3) elseif $v_{i+s_2} - v_{i+s_1} < \alpha \cdot \Delta$, 4) elseif $v_{i+s_2} - v_{i+s_1} < \beta \cdot \Delta$, 5) otherwise, where $\Delta = \max\{v_{i+s_1}, \delta\}$, and remaining parameters are the same as above. For example, we can use the following values: $s_1 = 2, s_2 = 3, \alpha = 0.15, \beta = 0.35, \gamma = 0.01$.

We can go through any driving trajectory and accumulate the 5×5 matrix X of $\{enter, exit\}$ conditions. Then, we can normilise matrix X by dividing any row by the sum of the corresponding elements. As a result, we shall obtain matrix of transactional probabilities. With those 25 probabilities as a secondary features we observed quite significant improvement: from 0.88 to 0.9 in terms of AUC.

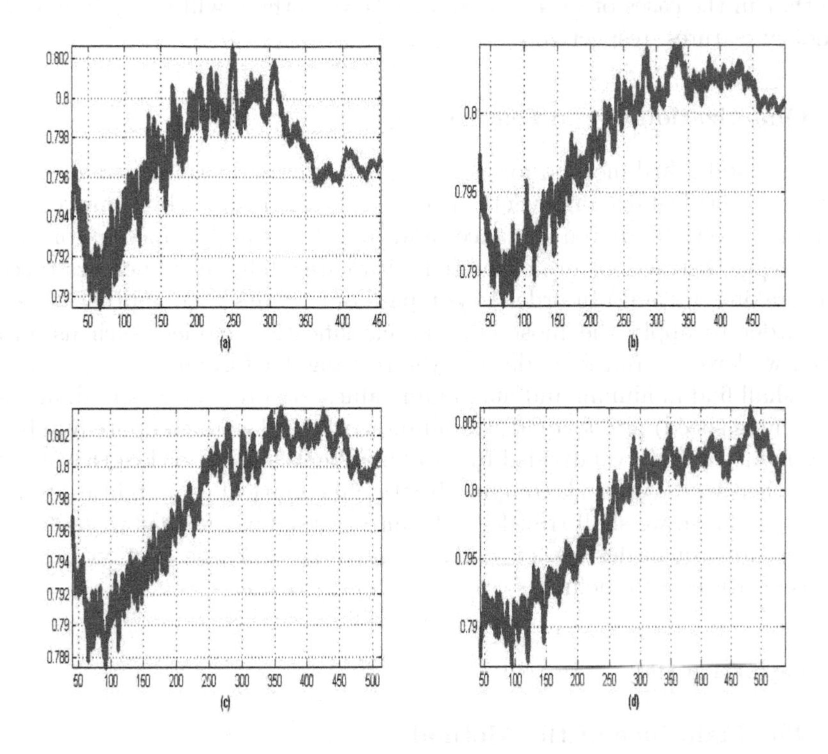

Fig. 4. Moving averages with $\Delta = 5000$, and coefficients: $\{9, 4, 1, 1, 4, 9\}$, see Section 3.2.

2.4 Fast Fourier Transforms

Main idea: driving trajectory maybe represented or approximated as a sum of two components 1) systematic, which corresponds to low frequencies and 2) random noise, which corresponds to high frequencies of the FFT. We are interested to keep first and to filter second component.

Figure 3 illustrates similarities between first and second rows, where upper row represents vectors of original speeds for the particular trips. Smoothed trajectories on the second row were produced using the following MATLAB code:

1) $m = 10$;
2) $h = \mathbf{fft}(speed)$;
3) $n = length(speed)$;
4) $g = zeros(n, 1)$;
5) $g(1 : m) = h(1 : m)$;
6) $g(n - m + 1 : n) = h(n - m + 1 : n)$;
7) $smoothed.speed = real(\mathbf{ifft}(g))$.

Consequently, we can use 39 numbers (FFT-coefficients) as a secondary features for any particular variable-length trajectory: 20 - real and 19 - imaginary.

Note that in the cases of 1) $m = 5$ and 2) $m = 15$ there will be 1) 19 and 2) 59 secondary features, respectively.

2.5 Quantization of the Features

The most established method for assessing driving distractions is to analyze the frequency of critical driving events. They can be aggregated by summation and normalized over the driven distance, and are thus suitable metrics of driving behavior [3]. Distractions are evaluated with three different indicators (acceleration, braking, turning) in order to compare our results with another system.

In order to apply the most efficient classification models such as *xgboost* or *gbm*, we have to transform data to the rectangular format.

We shall find minimum and maximum values for any vector (speed, acceleration, turning speed) into k equal size subintervals, and shall compute numbers of entries in any subinterval divided by total number of entries or length of the vector. In other words, we shall compute fractional empirical probabilities. Suppose that $k = 6$. Then, we shall transfer data into rectangular format with $30 = 5 \cdot k$ novel features (quantiles): $q_{ij}, i = 1, \ldots, 5, j = 1, \ldots, k$, see Section 2.2, where five base features were defined.

3 Experiments

3.1 The Main Idea of the Method

We shall go through the list of all drivers, assuming that the current driver is "bad". We shall add to the database 5 randomly selected drivers assuming that they are "good". Consequently, we shall form a training data table with 1200 rows, where 200 are labelled as one (bad) and remaining 1000 are labelled as zero (good). Further, we shall apply *xgboost* model. Consequently, we shall produce vector of 200 scores, corresponding to the current driver. After completion of the global cycle around all 2736 drivers we shall calculate required vector of all 547200 scores. This vector represents a solution to the given task, and will be called as a pseudo-target in the following section.

3.2 Computation of the Moving Averages

The objective of this section is to explain in full details the computation procedure, which was employed in order to prepare important illustrations Figures 4–6 (similar methodology was used in [21]). We considered the following base features: (a) - speed (2), (b) - acceleration (3), (c) - Lv (6b) and (d) - La (6c).

Let us consider linear combinations of the quantiles:

$$h_i = \sum_{j=1}^{k} c_j \cdot q_{ij}. \tag{7}$$

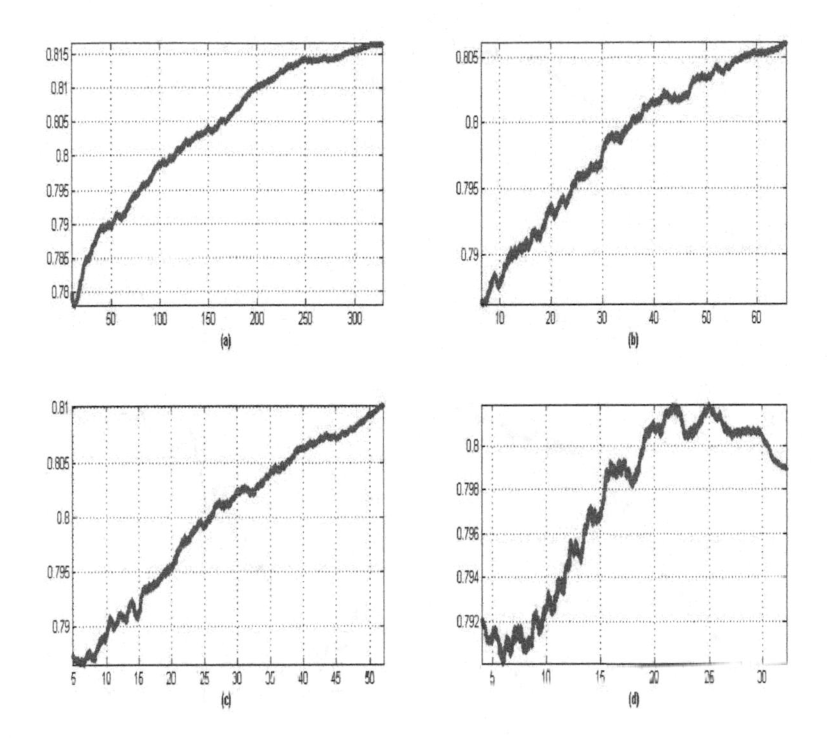

Fig. 5. Moving averages with $\Delta = 20000$, and coefficients: $\{1, 4, 9, 16, 25, 36\}$, see Section 3.2.

Remark 1. Particular values of the coefficients $c_j, j = 1, \ldots, k$, are given in the captions for Figures 4–6.

Let us consider matrix A with 2 columns: 1) h and 2) y - pseudo-target variable. We sort A according to the first column in an increasing order. As a result, we shall obtain matrix B as a sorted matrix A.

Figures 4–6 illustrate moving averages, which are defined by the following formula

$$MovAv_x(t) = \frac{1}{\Delta} \sum_{i=t}^{t+\Delta-1} B_{i,1}, t = 1, \ldots, n - \Delta + 1; \tag{8a}$$

$$MovAv_y(t) = \frac{1}{\Delta} \sum_{i=t}^{t+\Delta-1} B_{i,2}, t = 1, \ldots, n - \Delta + 1, \tag{8b}$$

where n is the number of rows in matrix A; Δ is smoothing parameter; $MovAv_x$ - horizontal axis and $MovAv_y$ - vertical axis.

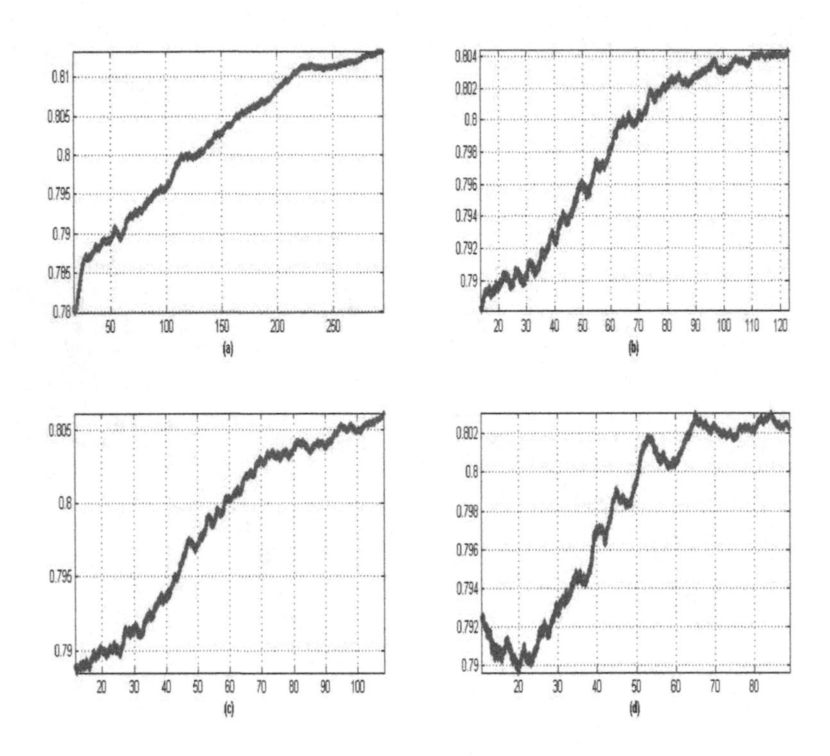

Fig. 6. Moving averages with $\Delta = 20000$, and coefficients: $\{1, 2, 3, 4, 5, 6\}$, see Section 3.2.

Remark 2. We used h as an example, any other feature is to be considered similarly.

Remark 3. We can see that lines on Figure 4 are more noisy compared to the lines on Figures 5 and 6. This property maybe explained by the smaller value of smoothing parameter: $\Delta = 5000$ in the case of Figure 4, and $\Delta = 20000$ in the cases of Figures 5 and 6.

3.3 Some Alternative Methods

The RamerDouglasPeucker algorithm (RDP) is an algorithm for reducing the number of points in a curve that is approximated by a series of points. The starting curve is an ordered set of points combined with distance dimension ϵ. The algorithm recursively divides the line. Initially, it is given all the points between the first and last point. It automatically marks the first and last point to be kept. It then finds the point that is furthest from the line segment with the first and last points as end points (this point is obviously furthest on the curve from the approximating line segment between the end points). If the point is closer than ϵ to the line segment then any points not currently marked to be kept can be discarded without the simplified curve being worse than ϵ. If the point furthest from the line segment is greater than ϵ from the approximation, then that point must be kept. The algorithm recursively calls itself with the first point and the worst point and then with the worst point and the last point (which includes marking the worst point being marked as kept). When the recursion is completed a new output curve can be generated consisting of all (and only) those points that have been marked as kept.

In difference to [7], [8] and [10], AXA data invariant against operations of shift and rotation. Accordingly, we can consider trips as 2D-curves, and can rotate and align the trips to make them closer to each other. The greater number of matches based on some threshold means the higher probability that the trip is normal. As a next step, we can use the results from trip matching as features for the machine learning algorithm. For example, we can consider a sequence of 5 turns were matched (distance between turns and turn size) with other trips.

When comparing trip parts, the distance metric took into account both the distances between turns and the angles (left/right are taken into account). We can compare the match score to the scores obtained when comparing the trip to negative samples (i.e., trips from other drivers), and compute a corresponding probability (p-value).

4 Concluding Remarks

Presented is a model for learning the driving characteristics of a particular driver that could be used to identify him from amongst a pool of drivers. Using speed, features of acceleration/deceleration profiles and turning speed we can detect with high quality bad drivers.

The procedure to discover abnormalities, which is described in Section 3.1, is totally independent on the methods to construct features, see Section 2.2. Based on Figures 4–6, we can see very strong correspondence between features and predicted variable (named in this paper as a pseudo-target variable). This finding may be useful as a key to understand the nature of abnormalities.

References

1. Ly, M., Martin, S., Trivedi, M.: Driver classification and driving style recognition using inertial sensors. In: IEEE Intelligent Vehicles Symposium (IV), June 23–26, Gold Coast, Australia, pp. 1040–1045 (2013)
2. Quek, Z., Ng, E.: Driver Identification by Driving Style, 4 pages (2013). http://cs229.stanford.edu/proj2013/DriverIdentification.pdf
3. Bergasa, L., Almeria, D., Almazan, J., Yebes, J., Arroyo, R.: DriveSafe: an app. for alerting inattentive drivers and scoring driving behaviors. In: IEEE Intelligent Vehicles Symposium Proceedings, pp. 240–245 (2014)

4. Kuhler, M., Karstens, D.: Improved driving cycle for testing automotive exhaust emissions. In: SAE Technical Paper Series 780650 (1978). doi:10.4271/780650
5. Johnson, D., Trivedi, M.: Driving style recognition using a smartphone as a sensor platform. In: 14th International IEEE Conference on Intelligent Transportation Systems Washington, DC, USA, October 5–7, pp. 1609–1615 (2011)
6. Bolovinou, A., Amditis, A., Bellotti, F., Tarkiainen, M.: Driving style recognition for co-operative driving: a survey. In: ADAPTIVE 2014: The Sixth International Conference on Adaptive and Self-Adaptive Systems and Applications, pp. 73–78 (2014)
7. Lan, J., Long, C., Wong, R., Chen, Y., Fu, Y., Guo, D., Liu, S., Ge, Y., Zhou, Y., Li, J.: A new framework for traffic anomaly detection. In: Proceedings of the SIAM International Conference on Data Mining, pp. 875–883 (2014)
8. Zhang, D., Li, N., Zhou, Z., Chen, C., Sun, L., Li, S.: iBAT: detecting anomalous taxi trajectories from GPS traces. In: Proceedings of the 13th International Conference on Ubiquitous Computing, pp. 99–108 (2011)
9. Bu, Y., Chen, L., Fu, A., Liu, D.: Efficient anomaly monitoring over moving object trajectory streams. In: KDD Proceedings of the 15th ACM SIGKDD International Conference on Knowledge Discovery and Data Mining, Paris, France, June 28–July 1, pp. 159–168 (2009)
10. Yin, P., Ye, M., Lee, W.-C., Li, Z.: Mining GPS data for trajectory recommendation. In: Tseng, V.S., Ho, T.B., Zhou, Z.-H., Chen, A.L.P., Kao, H.-Y. (eds.) PAKDD 2014, Part II. LNCS, vol. 8444, pp. 50–61. Springer, Heidelberg (2014)
11. Jiang, Y., Zhao, H., Fu, H.: newblock A control method to avoid obstacles for an intelligent car based on rough sets and neighborhood systems. In: IEEE International Conference on Intelligent Systems and Knowledge Engineering (ISKE), Taipei, Taiwan, November 24–27, pp. 66–70 (2015)
12. de Sa, V.: Learning classification with unlabeled data. In: Advances in Neural Information Processing Systems 6 (NIPS) (1993)
13. Kirshner, S., Cadez, I., Smyth, P., Kamath, C.: Learning to classify galaxy shapes using the EM algorithm. In: Advances in Neural Information Processing Systems 15 (NIPS) (2002)
14. Nikulin, V., Huang, T.H.: Unsupervised dimensionality reduction via gradient-based matrix factorization with two learning rates and their automatic updates. In: Journal of Machine Learning Research, Workshop and Conference Proceedings, vol. 27, pp. 181–195 (2012)
15. Dosovitskiy, A., Springenberg, J., Riedmiller, M., Brox, T.: Discriminative unsupervised feature learning with convolutional neural networks. In: Advances in Neural Information Processing Systems 27 (NIPS) (2014)
16. Le, Q.V., Ngiam, J., Chen, Z., Chia, D., Koh, P.W., Ng, A.Y.: Tiled convolutional neural networks. In: Advances in Neural Information Processing Systems 23 (NIPS), pp. 1279–1287 (2010)
17. Krizhevsky, A., Sutskever, I., Hinton, G.E.: ImageNet classification with deep convolutional neural networks. In: Advances in Neural Information Processing Systems 25 (NIPS), pp. 1106–1114 (2012)
18. Erhan, D., Bengio, Y., Courville, A., Manzagol, P.A., Vincent, P., Bengio, S.: Why does unsupervised pre-training help deep learning? Journal of Machine Learning Research **11**, 625–660 (2010)

19. Donmez, P., Lebanon, G., Balasubramanian, K.: Unsupervised supervised learning i: Estimating classification and regression errors without labels. Journal of Machine Learning Research **11**, 1323–1351 (2010)
20. Dy, J., Brodley, C.: Feature selection for unsupervised learning. Journal of Machine Learning Research **5**, 845–889 (2004)
21. Nikulin, V., Huang, T.H., Lu, J.D.: Mining shoppers data streams to predict customers loyalty. In: IEEE International Conference on Intelligent Systems and Knowledge Engineering (ISKE), Taipei, Taiwan, November 24–27, pp. 27–33 (2015)

A Hybrid Framework for News Clustering Based on the DBSCAN-Martingale and LDA

Ilias Gialampoukidis(✉), Stefanos Vrochidis, and Ioannis Kompatsiaris

Information Technologies Institute, CERTH, Thessaloniki, Greece
heliasgj@iti.gr

Abstract. Nowadays there is an important need by journalists and media monitoring companies to cluster news in large amounts of web articles, in order to ensure fast access to their topics or events of interest. Our aim in this work is to identify groups of news articles that share a common topic or event, without a priori knowledge of the number of clusters. The estimation of the correct number of topics is a challenging issue, due to the existence of "noise", i.e. news articles which are irrelevant to all other topics. In this context, we introduce a novel density-based news clustering framework, in which the assignment of news articles to topics is done by the well-established Latent Dirichlet Allocation, but the estimation of the number of clusters is performed by our novel method, called "DBSCAN-Martingale", which allows for extracting noise from the dataset and progressively extracts clusters from an OPTICS reachability plot. We evaluate our framework and the DBSCAN-Martingale on the 20newsgroups-mini dataset and on 220 web news articles, which are references to specific Wikipedia pages. Among twenty methods for news clustering, without knowing the number of clusters k, the framework of DBSCAN-Martingale provides the correct number of clusters and the highest Normalized Mutual Information.

Keywords: Clustering news articles · Latent Dirichlet Allocation · DBSCAN-Martingale

1 Introduction

Clustering news articles is a very important problem for journalists and media monitoring companies, because of their need to quickly detect interesting articles. This problem becomes also very challenging and complex, given the relatively large amount of news articles produced on a daily basis. The challenges of the aforementioned problem are summarized into two main directions: (a) discover the correct number of news clusters and (b) group the most similar news articles into news clusters. We face these challenges under the following assumptions. Firstly, we take into account that real data is highly noisy and the number of clusters is not known. Secondly, we assume that there is a lower bound for the minimum number of documents per cluster. Thirdly, we consider the names/labels of the clusters unknown.

P. Perner (Ed.): MLDM 2016, LNAI 9729, pp. 170–184, 2016.
DOI: 10.1007/978-3-319-41920-6_13

Towards addressing this problem, we introduce a novel hybrid clustering framework for news clustering, which combines automatic estimation of the number of clusters and assignment of news articles into topics of interest. The estimation of the number of clusters is done by our novel "DBSCAN-Martingale", which can deal with the aforementioned assumptions. The main idea is to progressively extract all clusters (extracted by a density-based algorithm) by applying Doob's martingale and then apply a well-established method for the assignment of news articles to topics, such as Latent Dirichlet Allocation (LDA). The proposed hybrid framework does not consider known the number of news clusters, but requires only the more intuitive parameter $minPts$, as a lower bound for the number of documents per topic. Each realization of the DBSCAN-Martingale provides the number of detected topics and, due to randomness, this number is a random variable. As the final number of detected topics, we use the majority vote over 10 realizations of the DBSCAN-Martingale. Our contribution is summarized as follows:

- We present our novel DBSCAN-Martingale process, which progressively estimates the number of clusters in a dataset.
- We introduce a novel hybrid news clustering framework, which combines our DBSCAN-Martingale with Latent Dirichlet Allocation.

In the following, we present, in Section 2, existing approaches for news clustering and density-based clustering. In Section 3, we propose a new hybrid framework for news clustering, where the number of news clusters is estimated by our "DBSCAN-Martingale", which is presented in Section 4. Finally, in Section 5, we test both our novel method for estimating the number of clusters and our news clustering framework in four datasets of various sizes.

2 Related Work

News clustering is tackled as a text clustering problem [1], which usually involves feature selection [25], spectral clustering [21] and k-means oriented [1] techniques, assuming mainly that the number of news clusters is known. We consider the more general and realistic case, where the number of clusters is unknown and it is possible to have news articles which do not belong to any of the clusters.

Latent Dirichlet Allocation (LDA) [4] is a popular model for topic modeling, given the number of topics k. LDA has been generalized to nonparametric Bayesian approaches, such as the hierarchical Dirichlet process [29] and DP-means [20], which predict the number of topics k. The extraction of the correct number of topics is equivalent to the estimation of the correct number of clusters in a dataset. The majority vote among 30 clustering indices has recently been proposed in [7] as an indicator for the number of clusters in a dataset. In contrast, we propose an alternative majority vote among 10 realizations of the "DBSCAN-Martingale", which is a modification of the DBSCAN algorithm [12] and has three main advantages and characteristics: (a) they discover clusters with not-necessarily regular shapes, (b) they do not require the number of clusters and (c) they extract noise. The parameters of DBSCAN are the density level ϵ and a lower bound for the minimum number of points per cluster: $minPts$.

Other approaches for clustering that could be applied to news clustering, without knowing the number of clusters, are based on density based clustering algorithms. The graph-analogue of DBSCAN has been presented in [5] and dynamically adjusting the density level ϵ, the nested hierarchical sequence of clusterings results to the HDBSCAN algorithm [5]. OPTICS [2] allows for determining the number of clusters in a dataset by counting the "dents" of the OPTICS reachability plot. F-OPTICS [28] has reduced the computational cost of the OPTICS algorithm using a probabilistic approach of the reachability distance, without significant accuracy reduction. The OPTICS-ξ algorithm [2] requires an extra parameter ξ, which has to be manually set in order to find "dents" in the OPTICS reachability plot. The automatic extraction of clusters from the OPTICS reachability plot, as an extension of the OPTICS-ξ algorithm, has been presented in [27] and has been outperformed by HDBSCAN-EOM [5] in several datasets. We will examine whether some of these density based algorithms perform well on the news clustering problem and we shall compare them with our DBSCAN-Martingale, which is a modification of DBSCAN, where the density level ϵ is a random variable and the clusters are progressively extracted.

3 The DBSCAN-Martingale Framework for News Clustering

We propose a novel framework for news clustering, where the number of clusters k is estimated using the DBSCAN-Martingale and documents are assigned to k topics using Latent Dirichlet Allocation (LDA).

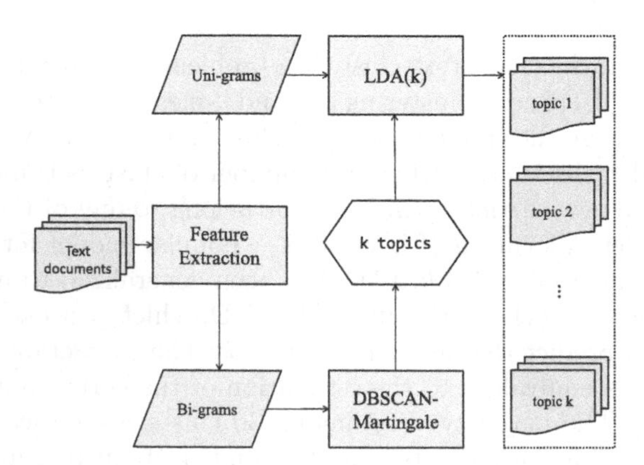

Fig. 1. Our hybrid framework for news clustering, using the DBSCAN-Martingale and Latent Dirichlet Allocation.

We combine DBSCAN and LDA because LDA performs well on text clustering but requires the number of clusters. On the other hand, density-based algorithms do not require the number of clusters, but their performance in text clustering is limited, when compared to LDA.

LDA [4] is a probabilistic topic model, which assumes a Bag-of-Words representation of the collection of documents. Each topic is a distribution over terms in a fixed vocabulary, which assigns probabilities to words. Moreover, LDA assumes that documents exhibit multiple topics and assigns a probability distribution on the set of documents. Finally, LDA assumes that the order of words does not matter and, therefore, is not applicable to word n-grams for $n \geq 2$.

We refer to word n-grams as "uni-grams" for $n = 1$ and as "bi-grams" for $n = 2$. The DBSCAN-Martingale performs well on the bi-grams, following the concept of "phrase extraction" [1]. We restrict our study on textual features (n-grams) in the present work and spatiotemporal features are not used.

In Figure 1 the estimation of the number of clusters is done by DBSCAN-Martingale and LDA follows for the assignment of text documents to clusters.

4 DBSCAN-Martingale

In this Section, we show the construction of the DBSCAN-Martingale. In Section 4.1 we provide the necessary background in density-based clustering and the notation which we adopt. In Section 4.2, we progressively estimate the number of clusters in a dataset by defining a stochastic process, which is then shown (Section 4.3) to be a Martingale process.

4.1 Notation and Preliminaries on DBSCAN

Given a dataset of n-points, density-based clustering algorithms provide as output the clustering vector C. Assuming there are k clusters in the dataset, some of the points are assigned to a cluster and some points do not belong to any of the k clusters. When a point $j = 1, 2, \ldots, n$ is assigned to one of the k clusters, the j-th element of the clustering vector C, denoted by $C[j]$ takes the value of the cluster ID from the set $\{1, 2, \ldots, k\}$. Otherwise, the j-th point does not belong to any cluster, it is marked as "noise" and the corresponding value in the clustering vector becomes zero, i.e. $C[j] = 0$. Therefore, the clustering vector C is a n-dimensional vector with values in $\{0, 1, 2, \ldots, k\}$.

The algorithm DBSCAN [12] is a density-based algorithm, which provides one clustering vector, given two parameters, the density level ϵ and the parameter $minPts$. We denote the clustering vector provided by the algorithm DBSCAN by $C_{DBSCAN(\epsilon, minPts)}$ or simply $C_{DBSCAN(\epsilon)}$ because the parameter $minPts$ is considered as a pre-defined fixed parameter. For low values of ϵ, $C_{DBSCAN(\epsilon)}$ is a vector of zeros (all points are marked as noise). On the other hand, for high values of ϵ, $C_{DBSCAN(\epsilon)}$ is a column vector of ones. Apparently, if a clustering vector has only zeros and ones, only one cluster has been detected and the partitioning is trivial.

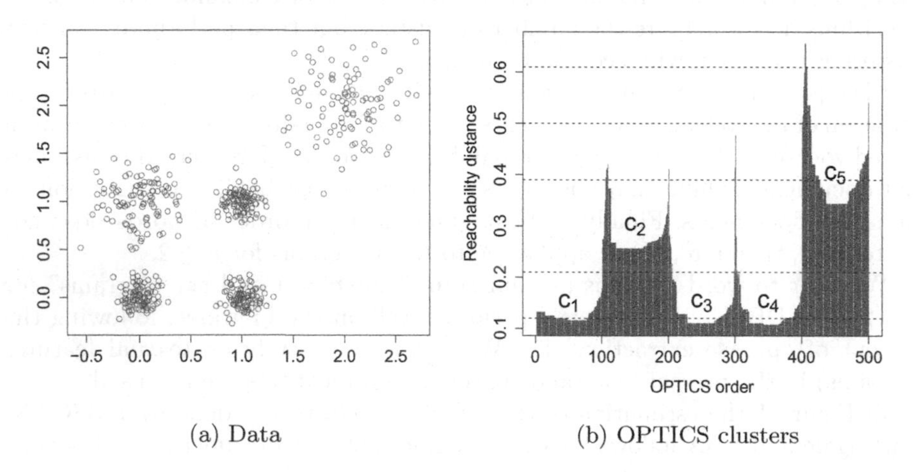

(a) Data (b) OPTICS clusters

Fig. 2. OPTICS reachability plot and randomly generated density levels

Clusters detected by DBSCAN strongly depend on the density level ϵ. An indicative example is shown in Figure 2(a), where the 5 clusters do not have the same density, and it is evident that there is no single value of ϵ that can output all the clusters. In Figure 2(b), we illustrate the corresponding OPTICS reachability plot with 5 randomly selected density levels (horizontal dashed lines) and none of them is able to extract all clusters $C_1, C_2, \ldots, C_5$ into one clustering vector C.

In order to deal with this problem we introduce (in Sections 4.2 and 4.3) an extension of DBSCAN based on Doob's Martingale, which allows for introducing a random variable ϵ and involves the construction of a Martingale process, which progressively approaches the clustering vector which contains all clusters so as to determine the number of clusters.

4.2 Estimation of the Number of Clusters with the DBSCAN-Martingale

We introduce a probabilistic method to estimate the number of clusters, by constructing a martingale stochastic process [11], which is able to progressively extract all clusters for all density levels. The martingale construction is, in general, based on Doob's martingale [11], in which we progressively gain knowledge about the result of a random variable. In the present work, the random variable that needs to be known is the vector of cluster IDs, which is a combination of T clustering vectors $C_{DBSCAN(\epsilon_t)}, t = 1, 2, \ldots, T$.

First, we generate a sample of size T with random numbers $\epsilon_t, t = 1, 2, \ldots, T$ uniformly in $[0, \epsilon_{max}]$, where ϵ_{max} is an upper bound for the density levels. The sample of $\epsilon_t, t = 1, 2, \ldots, T$ is sorted in increasing order and the values of ϵ_t

can be demonstrated on an OPTICS reachability plot, as shown in Figure 2 ($T = 5$). For each density level ϵ_t we find the corresponding clustering vectors $C_{DBSCAN(\epsilon_t)}$ for all stages $t = 1, 2, \ldots, T$.

In the beginning of the algorithm, there are no clusters detected. In the first stage ($t = 1$), all clusters detected by $C_{DBSCAN(\epsilon_1)}$ are kept, corresponding to the lowest density level ϵ_1. In the second stage ($t = 2$), some of the detected clusters by $C_{DBSCAN(\epsilon_2)}$ are new and some of them have also been detected at previous stage ($t = 1$). In order to keep only the newly detected clusters of the second stage ($t = 2$), we keep only groups of numbers of the same cluster ID with size greater than $minPts$.

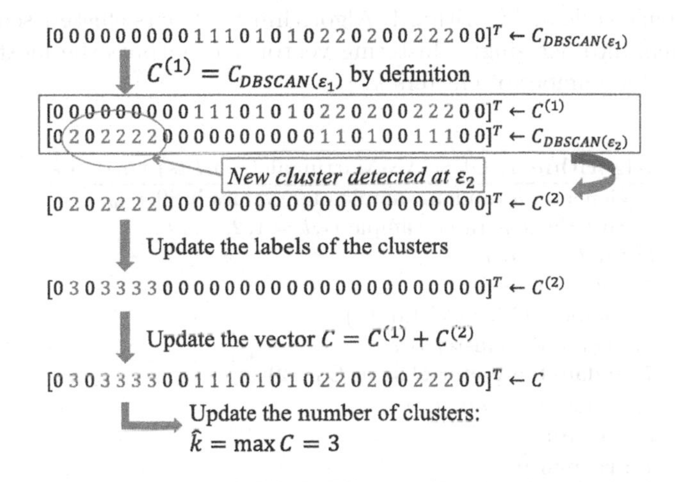

Fig. 3. One realization of the DBSCAN-Martingale with $T = 2$ iterations. The points with cluster label **2** in $C^{(1)}$ are re-discovered as a cluster by $C_{DBSCAN(\epsilon_2)}$ but the update rule keeps only the newly detected cluster.

Formally, we define the sequence of vectors $C^{(t)}, t = 1, 2, \ldots, T$, where $C^{(1)} = C_{DBSCAN(\epsilon_1)}$ and:

$$C^{(t)}[j] := \begin{cases} 0 & \text{if point } j \text{ belongs to a previously extracted cluster} \\ C_{DBSCAN(\epsilon_t)}[j] & \text{otherwise} \end{cases} \quad (1)$$

Since the stochastic process $C^{(t)}, t = 1, 2, \ldots, T$ is a martingale, as shown in Section 4.3, and $C_{DBSCAN(\epsilon_t)}$ is the output of DBSCAN for the density level ϵ_t, the proposed method is called "DBSCAN-Martingale".

Finally, we relabel the cluster IDs. Assuming that r clusters have been detected for the first time at stage t, we update the cluster labels of $C^{(t)}$ starting from $1 + \max_j C^{(t-1)}[j]$ to $r + \max_j C^{(t-1)}[j]$. Note that the maximum value of a clustering vector coincides with the number of clusters.

The sum of all vectors $C^{(t)}$ up to stage T is the final clustering vector of our algorithm:

$$C = C^{(1)} + C^{(2)} + \cdots C^{(T)} \tag{2}$$

The estimated number of clusters $\hat{k}$ is the maximum value of the final clustering vector C:

$$\hat{k} = \max_j C[j] \tag{3}$$

In Figure 3, we adopt the notation X^T for the transpose of the matrix or vector X, in order to demonstrate the estimation of the number of clusters after two iterations of the DBSCAN-Martingale.

The process we have formulated, namely the DBSCAN-Martingale, is represented as pseudo code in Algorithm 1. Algorithm 1 extracts clusters sequentially, combines them into one single clustering vector and outputs the most updated estimation of the number of clusters $\hat{k}$.

Algorithm 1. DBSCAN-Martingale($minPts$) **return** $\hat{k}$

1: Generate a random sample of T values in $[0, \epsilon_{max}]$
2: Sort the generated sample $\epsilon_t, t = 1, 2, \ldots, T$
3: **for** $t = 1$ to T
4: find $C_{DBSCAN(\epsilon_t)}$
5: compute $C^{(t)}$ as in Eq. (1)
6: update the cluster IDs
7: update the vector C as in Eq. (2)
8: update $\hat{k} = \max_j C[j]$
9: **end for**
10: **return** $\hat{k}$

The DBSCAN-Martingale requires T iterations of the DBSCAN algorithm, which runs in $\mathcal{O}(n \log n)$ if a tree-based spatial index can be used and in $\mathcal{O}(n^2)$ without tree-based spatial indexing [2]. Therefore, the DBSCAN-Martingale runs in $\mathcal{O}(Tn \log n)$ for tree-based indexed datasets and in $\mathcal{O}(Tn^2)$ without tree-based indexing. Our code is written in R[1], using the dbscan[2] package, which runs DBSCAN in $\mathcal{O}(n \log n)$ with kd-tree data structures for fast nearest neighbor search.

The DBSCAN-Martingale (one execution of Algorithm 1) is illustrated, for example, on the OPTICS reachability plot of Figure 2 (b) where, for the random sample of density levels $\epsilon_t, t = 1, 2, \ldots, 5$ (horizontal dashed lines), we sequentially extract all clusters. In the first density level $\epsilon_1 = 0.12$, DBSCAN-Martingale extracts the clusters C_1, C_3 and C_4, but in the density level $\epsilon_2 = 0.21$ no new clusters are extracted. In the third density level, $\epsilon_3 = 0.39$, the clusters C_2 and C_5 are added to the final clustering vector and in the other density

[1] https://www.r-project.org/
[2] https://cran.r-project.org/web/packages/dbscan/index.html

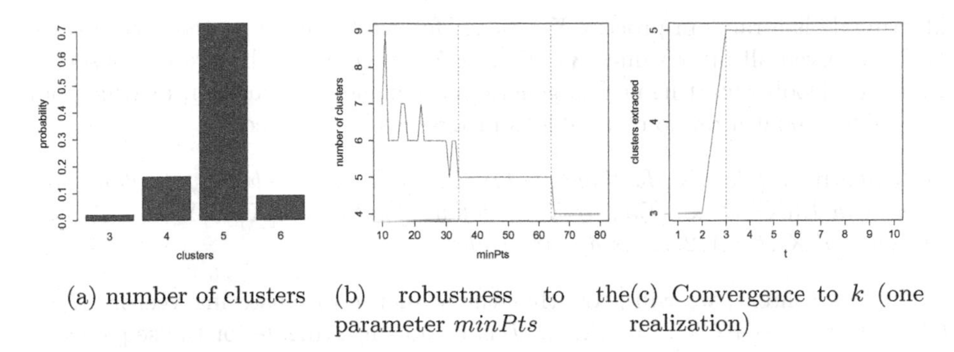

(a) number of clusters (b) robustness to the parameter *minPts* (c) Convergence to k (one realization)

Fig. 4. The number of clusters as generated by DBSCAN-Martingale ($minPts = 50$) after 100 realizations

Algorithm 2. MajorityVote($realizations, minPts$) **return** $\hat{k}$

1: $clusters = \emptyset$, $k = 0$
2: **for** $r = 1$ to $realizations$
3: k =DBSCAN-Martingale($minPts$)
4: $clusters$ = AppendTo($clusters, k$)
5: **end for**
6: $\hat{k}$ = mode($clusters$)
7: **return** $\hat{k}$

levels, ϵ_4 and ϵ_5 there are no new clusters to extract. The number of clusters extracted up to stage t is shown in Figure 4(c). Observe that at $t = 3$ iterations, DBSCAN-Martingale has output $k = 5$ and for all iterations $t > 3$ there are no more clusters to extract. Increasing the total number of iterations T will needlessly introduce additional computational cost in the estimation of the number clusters $\hat{k}$.

The estimation of number of clusters $\hat{k}$ is a random variable, because it inherits the randomness of the density levels $\epsilon_t, t = 1, 2, \ldots, T$. For each execution of Algorithm 1, one realization of the DBSCAN-Martingale generates $\hat{k}$, so we propose as the final estimation of the number of clusters the majority vote over 10 realizations of the DBSCAN-Martingale.

Algorithm 2 outputs the majority vote over a fixed number of realizations of the DBSCAN-Martingale. For each realization, the estimated number of clusters k is added to the list $clusters$ and the majority vote is obtained from the mode of $clusters$, since the mode is defined as the most frequent value in a list. The percentage of realizations where the DBSCAN-Martingale outputs exactly $\hat{k}$ clusters is a probability distribution, such as the one shown in Figure 4(a), which corresponds to the illustrative dataset of Figure 2(a). Finally, we note that the same result ($\hat{k} = 5$) appears for a wide range of the parameter *minPts* (Figure 4(b)), a fact that demonstrates the robustness of our approach.

4.3 The Sequence of Vectors $C^{(t)}$ is a Martingale Process

Martingale is a random process $X_1, X_2, \ldots$ for which the expected future value of X_{t+1}, given all prior values $X_1, X_2, \ldots, X_t$, is equal to the present observed value X_t. Doob's martingale is a generic martingale construction, in which our knowledge about a random variable is progressively obtained:

Definition 1. *(Doob's Martingale) [11]. Let $X, Y_1, Y_2, \ldots$ be any random variables with $E[|X|] < \infty$. Then if X_t is defined by $X_t = E[X|Y_1, Y_2, \ldots, Y_t]$, the sequence of $X_t, t = 1, 2, \ldots$ is a martingale.*

In this context, we will show that the sequence of clustering vectors $X_t = C^{(1)} + C^{(2)} + \cdots + C^{(t)}, t = 1, 2, \ldots, T$ is Doob's martingale for the sequence of random variables $Y_t = C_{DBSCAN(\epsilon_t)}, t = 1, 2, \ldots, T$.

We denote by $< Z_i, Z_l >= \sum_j Z_i[j] \cdot Z_l[j]$ the inner product of any two vectors Z_i and Z_l and we prove the following Lemma:

Lemma 1. *If two clustering vectors Z_i, Z_l are mutually orthogonal, they contain different clusters.*

Proof. The values of the clustering vectors are cluster IDs so they are non-negative integers. Points which do not belong to any of the clusters (noise) are assigned zeros. Since $< Z_i, Z_l >= \sum_j Z_i[j] \cdot Z_l[j] = 0$ and based on the fact that when a sum of non-negative integers is zero, then all integers are zero, we obtain $Z_i[j] = 0$ or $Z_l[j] = 0$ for all $j = 1, 2, \ldots, n$.

For example, the clustering vectors

$Z_i = [0\ 0\ 0\ 0\ 0\ 0\ 0\ 0\ 0\ 0\ 0\ 0\ 0\ 0\ 0\ 1\ 0\ 0\ 0\ 0\ 0\ 1\ 1\ 1\ 0\ 1\ 0\ 1\ 1\ 0\ 0\ 0\ 0\ 0\ 0\ 2\ 2\ 2\ 2\ 2]^T$

$Z_l = [1\ 1\ 0\ 1\ 1\ 1\ 1\ 0\ 1\ 0\ 1\ 0\ 0\ 0\ 0\ 0\ 0\ 0\ 0\ 0\ 0\ 0\ 0\ 0\ 0\ 0\ 0\ 0\ 0\ 0\ 0\ 0\ 0\ 0\ 0\ 0\ 0\ 0\ 0\ 0]^T$

are mutually orthogonal and contain different clusters.

Martingale Construction. Each density level $\epsilon_t, t = 1, 2, \ldots, T$ provides one clustering vector $C_{DBSCAN(\epsilon_t)}$ for all $t = 1, 2, \ldots, T$. As t increases, more clustering vectors are computed and we gain knowledge about the vector C.

In Eq. (1), we constructed a sequence of vectors $C^{(t)}, t = 1, 2, \ldots, T$, where each $C^{(t)}$ is orthogonal to all $C^{(1)}, C^{(2)}, \ldots, C^{(t-1)}$, from Lemma 1. The sum of all clustering vectors $C^{(1)} + C^{(2)} + \ldots + C^{(t-1)}$ has zeros as cluster IDs in the points which belong to the clusters of $C^{(t)}$. Therefore, $C^{(t)}$ is also orthogonal to $C^{(1)} + C^{(2)} + \ldots + C^{(t-1)}$. We use the orthogonality to show that the vector $C^{(1)} + C^{(2)} + \ldots + C^{(t)}$ is our "best prediction" for the final clustering vector C at stage t. The expected final clustering vector at stage t is:
$E[C|C_{DBSCAN(\epsilon_1)}, C_{DBSCAN(\epsilon_2)}, \ldots, C_{DBSCAN(\epsilon_t)}] = C^{(1)} + C^{(2)} + \ldots + C^{(t)}$.

Initially, the final clustering C vector is the zero vector O. Our knowledge about the final clustering vector up to stage t is restricted to $C^{(1)} + C^{(2)} + \ldots + C^{(t)}$ and finally, at stage $t = T$, we have gained all available knowledge about the final clustering vector C, i.e. $C = E[C|C_{DBSCAN(\epsilon_1)}, C_{DBSCAN(\epsilon_2)}, \ldots, C_{DBSCAN(\epsilon_T)}]$.

5 Experiments

5.1 Dataset Description

The proposed methodology is evaluated on the 20newsgroups-mini dataset with 2000 articles, which is available on the UCI repository[3] and on 220 news articles, which are references to specific Wikipedia pages so as to ensure reliable ground-truth: the WikiRef220. We also use two subsets of WikiRef220, namely the WikiRef186 and the WikiRef150, in order to test DBSCAN-Martingale in four datasets of sizes 2000, 220, 150 and 115 documents respectively.

We selected these datasets because we focus on datasets with news clusters which are event-oriented, like "Paris Attacks November 2016" or they discuss about specific topics like "Barack Obama" (rather than "Politics" in general). We would tackle the news categorization problem as a supervised classification problem, because training sets are available, contrary to the news clustering problem where, for example, the topic "Malaysia Airlines Flight 370" had no training set before the 8th of March 2014.

We assume that 2000 news articles is a reasonable upper bound for the number of recent news articles that can be considered for news clustering, in line with other datasets that were used to evaluate similar methods [5,25]. In all datasets (Table 2) we extract uni-grams and bi-grams, assuming a Bag-of-Words representation of text. Before the extraction of uni-grams and bi-grams, we remove the SMART[4] stopwords list and we then stem the words using Porter's algorithm [24]. The uni-grams are filtered out if they occur less than 6 times and the bi-grams if they occur less than 20 times. The final bi-grams are normalized using tf-idf weighting and, in all datasets, the upper bound for the density level is taken $\epsilon_{max} = 3$. We generate a sample of $T = 5$ uniformly distributed numbers using R, for the initialization of Algorithm 1.

Table 1. DBSCAN results without LDA, for the 5 best values of ϵ and $minPts$. The DBSCAN-Martingale requires no tuning for determining ϵ and is able to extract all clusters for datasets (eg. WikiRef220) in which there is no unique density level to extract all clusters.

WikiRef150			**WikiRef186**			**WikiRef220**			**20news**		
ϵ	clusters	NMI	ϵ	clusters	NMI	ϵ	clusters	NMI	ϵ	clusters	NMI
0.8	3	0.3850	0.8	3	0.3662	0.8	3	0.3733	1.6	20	0.0818
0.9	4	0.4750	0.9	3	0.4636	0.9	3	0.4254	1.7	20	0.0818
1.0	3	0.4146	1.0	4	0.4904	1.0	4	0.5140	1.8	20	0.0818
1.1	3	0.4234	1.1	3	0.3959	1.1	3	0.4060	1.9	20	0.0818
1.2	1	0.1706	1.2	2	0.1976	1.2	3	0.4124	2.0	20	0.0818

[3] http://archive.ics.uci.edu/ml/datasets.html

[4] http://jmlr.csail.mit.edu/papers/volume5/lewis04a/a11-smart-stop-list/english.stop

Table 2. Estimated number of topics. The best values are marked in bold. The majority rule for 10 realizations of the DBSCAN-Martingale coincides with the ground truth number of topics.

Index	Ref	WikiRef150	WikiRef186	WikiRef220	20news
CH	[6]	30	29	30	30
Duda	[9]	2	2	2	2
Pseudo t^2	[9]	2	2	2	2
C-index	[17]	27	2	2	2
Ptbiserial	[8]	11	7	6	30
DB	[8]	2	**4**	6	2
Frey	[13]	2	2	2	5
Hartingan	[15]	18	20	16	24
Ratkowsky	[26]	20	24	29	30
Ball	[3]	**3**	3	3	3
McClain	[22]	2	2	2	2
KL	[19]	14	15	17	15
Silhouette	[18]	30	**4**	4	2
Dunn	[10]	2	**4**	**5**	3
SDindex	[14]	4	**4**	6	3
SDbw	[14]	30	7	6	3
NbClust	[7]	2	2	6	2
DP-means	[20]	4	**4**	7	15
HDBSCAN-EOM	[5]	5	5	**5**	36
DBSCAN-Martingale		**3**	**4**	**5**	**20**

5.2 Evaluation

The evaluation of our method is done in two levels. Firstly, we test whether the output of the majority vote over 10 realizations of the DBSCAN-Martingale matches the ground-truth number of clusters. Secondly, we evaluate the overall hybrid news clustering framework, using the number of clusters from Table 2. The index "NbClust", which is computed using the NbClust[5] package, is the majority vote among the 24 indices: CH, Duda, Pseudo t^2, C-index, Beale, CCC, Ptbiserial, DB, Frey, Hartigan, Ratkowsky, Scott, Marriot, Ball, Trcovw, Tracew, Friedman, McClain, Rubin, KL, Silhouette, Dunn, SDindex, SDbw [7]. The Dindex and Hubert's Γ are graphical methods and they are not involved in the majority vote. The indices GAP, Gamma, Gplus and Tau are also not included in the majority vote, due to the high computational cost. The NbClust package requires as a parameter the maximum number of clusters to look for, which is set `max.nc = 30`. For the extraction of clusters from the HDBSCAN hierarchy, we adopt the EOM-optimization [5] and for the nonparametric Bayesian method DP-means, we extended the R-script which is available on GitHub[6].

[5] https://cran.r-project.org/web/packages/NbClust/index.html

[6] https://github.com/johnmyleswhite/bayesian_nonparametrics/tree/master/code/dp-means

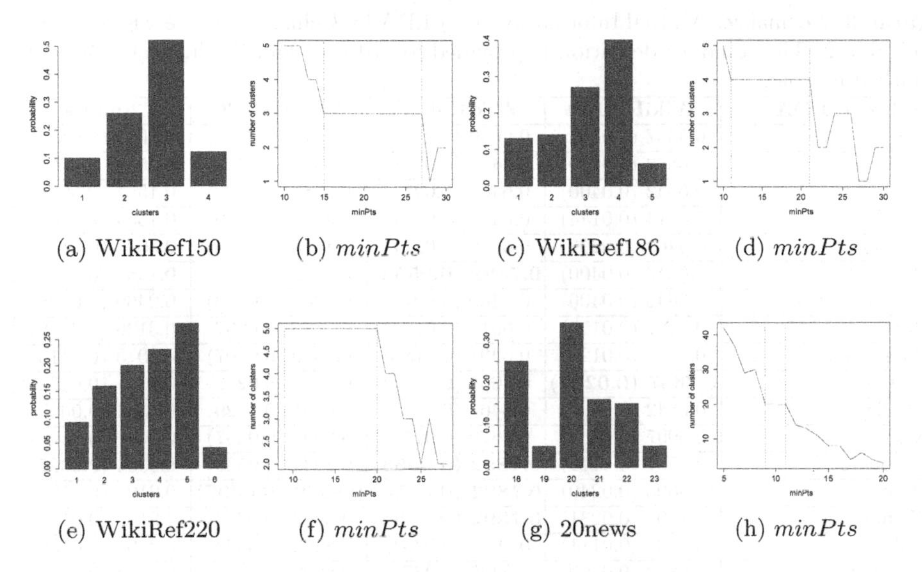

(a) WikiRef150 (b) *minPts* (c) WikiRef186 (d) *minPts*

(e) WikiRef220 (f) *minPts* (g) 20news (h) *minPts*

Fig. 5. The number of clusters as generated by DBSCAN-Martingale

Evaluation of the Number of Clusters: We compare our DBSCAN-Martingale with baseline methods, listed in Table 2, which either estimate the number of clusters directly, or provide a clustering vector without any knowledge of the number of clusters. The Ball index is correct in the WikiRef150 dataset, IIDBSCAN and Dunn is correct in the WikiRef220 dataset and the indices DB, Silhouette, Dunn, SDindex and DP-means are correct in the WikiRef186 datasets. However, in all datasets, the estimation given by the majority vote over 10 realizations of the DBSCAN-Martingale coincides with the ground truth number of clusters. In Figure 5, we present the estimation of the number of clusters for 100 realizations of the DBSCAN-Martingale, in order to show that after 10 runs of 10 realizations the output of Algorithm 1 remains the same. The parameter *minPts* is taken equal to 10 for the 20news dataset and 20 for all other cases.

In all datasets examined, we observe that there are some samples of density levels $\epsilon_t, t = 1, 2, \ldots, T$ which do not provide the correct number of clusters (Figure 5). The "mis-clustered" samples are due to the randomness of the density levels ϵ_t, which are sampled from the uniform distribution. We expect that sampling from another distribution would result to less mis-clustered samples, but searching for the statistical distribution of ϵ_t is beyond the scope of this paper.

We also compared the DBSCAN-Martingale with several methods of Table 2, with respect to the mean processing time. All experiments were performed on an Intel Core i7-4790K CPU at 4.00GHz with 16GB RAM memory, using a single thread and the R statistical software. Given a corpus of 500 news articles, DBSCAN-Martingale run in 0.39 seconds, while the Duda, Pseudo t^2 and Dunn

Table 3. Normalized Mutual Information after LDA by k clusters, where k is estimated in Table 2. The standard deviation is provided for 10 runs and the highest values are marked in bold.

Index + LDA	WikiRef150	WikiRef186	WikiRef220	20news
CH	0.5537 (0.0111)	0.6080 (0.0169)	0.6513 (0.0126)	0.3073 (0.0113)
Duda	0.6842 (0.0400)	0.6469 (0.0271)	0.6381 (0.0429)	0.1554 (0.0067)
Pseudo t^2	0.6842 (0.0400)	0.6469 (0.0271)	0.6381 (0.0429)	0.1554 (0.0067)
C-index	0.5614 (0.0144)	0.6469 (0.0271)	0.6381 (0.0429)	0.1554 (0.0067)
Ptbiserial	0.6469 (0.0283)	0.6469 (0.0271)	0.8262 (0.0324)	0.3073 (0.0113)
DB	0.6842 (0.0400)	**0.7892 (0.0553)**	0.8262 (0.0324)	0.1554 (0.0067)
Frey	0.6842 (0.0400)	0.6469 (0.0271)	0.6381 (0.0429)	0.2460 (0.0198)
Hartingan	0.5887 (0.0157)	0.6513 (0.0184)	0.7156 (0.0237)	0.3126 (0.0098)
Ratkowsky	0.5866 (0.0123)	0.6201 (0.0188)	0.6570 (0.0107)	0.3073 (0.0113)
Ball	**0.7687 (0.0231)**	0.7655 (0.0227)	0.7601 (0.0282)	0.2101 (0.0192)
McClain	0.6842 (0.0400)	0.6469 (0.0271)	0.6381 (0.0429)	0.1554 (0.0067)
KL	0.6097 (0.0232)	0.6670 (0.0156)	0.7091 (0.0257)	0.3077 (0.0094)
Silhouette	0.5537 (0.0111)	**0.7892 (0.0553)**	0.8032 (0.0535)	0.1554 (0.0067)
Dunn	0.5805 (0.0240)	**0.7892 (0.0553)**	**0.8560 (0.0397)**	0.2101 (0.0192)
SDindex	0.7007 (0.0231)	**0.7892 (0.0553)**	0.8262 (0.0324)	0.2101 (0.0192)
SDbw	0.5537 (0.0111)	0.7668 (0.0351)	0.8262 (0.0324)	0.2101 (0.0192)
NbClust	0.6842 (0.0400)	0.6469 (0.0271)	0.8262 (0.0324)	0.1554 (0.0067)
DP-means	0.7007 (0.0231)	**0.7892 (0.0553)**	0.8278 (0.0341)	0.3077 (0.0094)
HDBSCAN-EOM	0.7145 (0.0290)	0.7630 (0.0530)	**0.8560 (0.0397)**	0.3106 (0.0134)
DBSCAN-Martingale	**0.7687 (0.0231)**	**0.7892 (0.0553)**	**0.8560 (0.0397)**	**0.3137 (0.0130)**

in 0.44 seconds, SDindex in 1.06 seconds, HDBSCAN in 1.23 seconds and Silhouette in 1.37 seconds.

Evaluation of News Clustering: The evaluation measure is the popular Normalized Mutual Information (NMI), mainly used for the evaluation of clustering techniques, which allows us to compare results when the number of outputted clusters does not match the number of clusters in the ground truth [20]. For the output k of each method of Table 2, we show the average of 10 runs of LDA (and the corresponding standard deviation) in Table 3. For the WikiRef150 dataset, the combination of Ball index with LDA provides the highest NMI. For the WikiRef220 dataset, the combinations of HDBSCAN with LDA and Dunn index with LDA also provide the highest NMI. For the WikiRef186 dataset, the combinations of LDA with the indices DB, Silhouette, Dunn, SDindex and DP-means perform well. However, in all 4 datasets, our news clustering framework provides the highest NMI score and in the case of 20news dataset, the combination of DBSCAN-Martingale with LDA is the only method which provides the highest NMI score. Without using LDA, the best partition provided by DBSCAN has NMI less than 51.4 % in all WikiRef150, WikiRef186 and WikiRef220, as shown in Table 1. In contrast, we adopt the LDA method which achieves NMI scores up to 85.6 %. Density-based algorithms such as DBSCAN, HDBSCAN and DBSCAN-Martingale assigned too much noise in our datasets, a fact that affected the clustering performance, especially when compared to LDA in news clustering, thus we kept only the estimation $\hat{k}$.

6 Conclusion

We have presented a hybrid framework for news clustering, based on the DBSCAN-Martingale for the estimation of the number of news clusters, followed by the assignment of the news articles to topics using Latent Dirichlet Allocation. We extracted the word n-grams of a news articles collection and we estimated the number of clusters, using the DBSCAN-Martingale which is robust to noise. The extension of the DBSCAN algorithm, based on the martingale theory, allows for introducing a variable density level in the clustering algorithm. Our method outperforms several state-of-the-art methods on 4 corpora, in terms of the number of detected clusters, and the overall news clustering framework shows a good behavior of the proposed martingale approach, as evaluated by the Normalized Mutual Information. In the future, we plan to evaluate our framework using alternatice to LDA text clustering approaches, additional features and content, in order to present the multimodal and multilingual version of our framework.

Acknowledgements. This work was supported by the projects MULTISENSOR (FP7-610411) and KRISTINA (H2020-645012), funded by the European Commission.

References

1. Aggarwal, C.C., Zhai, C.: A survey of text clustering algorithms. In: Mining Text Data, pp. 77–128. Springer, US (2012)
2. Ankerst, M., Breunig, M.M., Kriegel, H.P., Sander, J.: OPTICS: ordering points to identify the clustering structure. In: ACM Sigmod Record, vol. 28(2), pp. 49–60. ACM, June 1999
3. Ball, G.H., Hall, D.J.: ISODATA, a novel method of data analysis and pattern classification. Stanford Research Institute (NTIS No. AD 699616) (1965)
4. Blei, D.M., Ng, A.Y., Jordan, M.I.: Latent Dirichlet Allocation. The Journal of Machine Learning Research **3**, 993–1022 (2003)
5. Campello, R.J.G.B., Moulavi, D., Sander, J.: Density-based clustering based on hierarchical density estimates. In: Pei, J., Tseng, V.S., Cao, L., Motoda, H., Xu, G. (eds.) PAKDD 2013, Part II. LNCS, vol. 7819, pp. 160–172. Springer, Heidelberg (2013)
6. Caliński, T., Harabasz, J.: A dendrite method for cluster analysis. Communications in Statistics-Theory and Methods **3**(1), 1–27 (1974)
7. Charrad, M., Ghazzali, N., Boiteau, V., Niknafs, A.: NbClust: an R package for determining the relevant number of clusters in a data set. Journal of Statistical Software **61**(6), 1–36 (2014)
8. Davies, D.L., Bouldin, D.W.: A cluster separation measure. IEEE Transactions on Pattern Analysis and Machine Intelligence **2**, 224–227 (1979)
9. Duda, R.O., Hart, P.E.: Pattern Classification and Scene Analysis, vol. 3. Wiley, New York (1973)
10. Dunn, J.C.: Well-separated clusters and optimal fuzzy partitions. Journal of cybernetics **4**(1), 95–104 (1974)

11. Doob, J.L.: Stochastic Processes, vol. 101. Wiley, New York (1953)
12. Ester, M., Kriegel, H. P., Sander, J., Xu, X.: A density-based algorithm for discovering clusters in large spatial databases with noise. In: Kdd, vol. 96(34), pp. 226–231, August 1996
13. Frey, T., Van Groenewoud, H.: A cluster analysis of the D2 matrix of white spruce stands in Saskatchewan based on the maximum-minimum principle. The Journal of Ecology, 873–886 (1972)
14. Halkidi, M., Vazirgiannis, M., Batistakis, Y.: Quality scheme assessment in the clustering process. In: Zighed, D.A., Komorowski, J., Żytkow, J.M. (eds.) PKDD 2000. LNCS (LNAI), vol. 1910, pp. 265–276. Springer, Heidelberg (2000)
15. Hartigan, J.A.: Clustering Algorithms. John Wiley Sons, New York (1975)
16. Hubert, L., Arabie, P.: Comparing partitions. Journal of Classification **2**(1), 193–218 (1985)
17. Hubert, L.J., Levin, J.R.: A general statistical framework for assessing categorical clustering in free recall. Psychological Bulletin **83**(6), 1072 (1976)
18. Kaufman, L., Rousseeuw, P.J.: Finding groups in data. An introduction to cluster analysis. Wiley Series in Probability and Mathematical Statistics. Applied Probability and Statistics, 1st edn. Wiley, New York (1990)
19. Krzanowski, W.J., Lai, Y.T.: A criterion for determining the number of groups in a data set using sum-of-squares clustering. Biometrics, 23–34 (1988)
20. Kulis, B., Jordan, M.I.: Revisiting k-means: New algorithms via Bayesian nonparametrics. arXiv preprint (2012). arXiv:1111.0352
21. Kumar, A., Daumé, H.: A co-training approach for multi-view spectral clustering. In: Proceedings of the 28th International Conference on Machine Learning (ICML 2011), pp. 393–400 (2011)
22. McClain, J.O., Rao, V.R.: Clustisz: A program to test for the quality of clustering of a set of objects. Journal of Marketing Research (pre-1986) **12**(000004), 456 (1975)
23. Milligan, G.W., Cooper, M.C.: An examination of procedures for determining the number of clusters in a data set. Psychometrika **50**(2), 159–179 (1985)
24. Porter, M.F.: An algorithm for suffix stripping. Program **14**(3), 130–137 (1980)
25. Qian, M., Zhai, C.: Unsupervised feature selection for multi-view clustering on text-image web news data. In: Proceedings of the 23rd ACM International Conference on Information Knowledge Management, pp. 1963–1966. ACM, November 2014
26. Ratkowsky, D.A., Lance, G.N.: A criterion for determining the number of groups in a classification. Australian Computer Journal **10**(3), 115–117 (1978)
27. Sander, J., Qin, X., Lu, Z., Niu, N., Kovarsky, A.: Automatic extraction of clusters from hierarchical clustering representations. In: Advances in Knowledge Discovery and Data Mining, pp. 75–87. Springer, Heidelberg (2003)
28. Schneider, J., Vlachos, M.: Fast parameterless density-based clustering via random projections. In: Proceedings of the 22nd ACM International Conference on Information Knowledge Management, pp. 861–866. ACM (2013)
29. Teh, Y.W., Jordan, M.I., Beal, M.J., Blei, D.M.: Hierarchical dirichlet processes. Journal of the American Statistical Association **101**(476) (2006)

Diagnosis of Metabolic Syndrome: A Diversity Based Hybrid Model

Nahla Barakat(✉)

Faculty of Informatics and Computer Science,
The British University in Egypt (BUE), Cairo, Egypt
nahla.barakat@bue.edu.eg

Abstract. Metabolic Syndrome (MetS) is a serious disorder, which is mainly characterized by central obesity, abnormal glucose tolerance, hypertension and dyslipidemia. It has been shown that 25% of adults around the world have MetS. The main concern is that those with MetS are more likely to develop type 2 Diabetes, which is found to be the fourth leading cause of global death by disease. Other life threatening complications of MetS include cardiovascular diseases (CVD), heart attack and stroke. It has also been shown that early screening and detection of people at risk may help in preventing or delaying MetS and its further complications. Within this context, data mining and machine learning can be valuable tools for identifying those people, based on their success in diagnosis and prognosis of related diseases like type 2 Diabetes. In this paper, we propose a hybrid diversity based model for diagnosis of Metabolic Syndrome. The proposed model utilizes two learning algorithms only, in particular; a Support Vector Machine (SVM) as the base-level classifier and a different classification algorithm at the meta level. The choice of the meta level classifier is based on different pairwise diversity measures. This is then followed by a final voting stage. Results on real life data set for diagnosis of MetS show that the proposed model is a promising technique, which compares favorably to other well established ensemble methods, and the choice of meta classifiers based on diversity measures was beneficial in this case.

Keywords: Cascade generalization · Classification · Ensemble methods · SVM · Medical diagnosis · Metabolic Syndrome

1 Introduction

Metabolic Syndrome (MetS) is a highly prevalent, serious disorder which may lead to complications like type 2 diabetes, cardiovascular diseases (CVD), heart attack, and/ or stroke [1]. The prevalence of MetS is high and continuously increasing in both developing and developed countries, mainly due to obesity. According to the International Diabetes Federation (IFD), 25% of adults around the world have this disorder. Several studies have also shown that MetS is often associated with type 2 diabetes, which is the fourth leading cause of global death by disease [2-4]. Furthermore, people with MetS, but not diabetic have 5 times higher risk of developing type 2 diabetes

P. Perner (Ed.): MLDM 2016, LNAI 9729, pp. 185–198, 2016.
DOI: 10.1007/978-3-319-41920-6_14

compared to those without MetS [5]. This adds further health challenges worldwide, given that up to 80% of diabetic people die due to CVD, which are known complications of diabetes. Metabolic syndrome also doubles the risk of dying from heart attacks and/or stroke [2,3,5,6]

In addition to the risks of diabetes and CVD, the presence of MetS may also cause memory deficits, affect processing speed, and the overall intellectual functioning [7].

Therefore, finding an intuitive way to screen and diagnose people at risk became a major health concern [6], in order to prevent or delay or prevent MetS and/or its complications [5]. It has been shown that employing computer aided diagnostic systems (CAD) as a "second opinion" has lead to improved diagnostic decisions, in diagnosis of related diseases like diabetes and pre-diabetes [8]. Within this context, machine learning techniques can, also, be valuable tools for identifying subjects with MetS. However, and to the best of the authors' knowledge, no studies have been used machine learning or data mining for the prediction of MetS among Middle East Arab, in particular models based on SVMs and/or ensemble methods.

In this paper, we propose, a new hybrid intelligent model for the diagnosis of metabolic syndrome, using real life dataset. The model is based on cascade generalization concept; in particular CGen-SVM [9] has been utilized, where its meta classifier is chosen based on different diversity measures. CGen-SVM has shown superior performance, compared to other ensemble methods, which have much more base classifiers [9]. For more details about CGen-SVM and diversity measures, please refer to Section 2. Diversity measures were used to choose the meta classifier based on the fact that designing an ensemble with classifiers which have different inductive bias leads to improved performance, compared to those using similar algorithms [10].

As the results section shows, the proposed method has achieved excellent classification accuracy, and Area Under the Operating Receiver Characteristic Curve (AUC) for the diagnosis of MetS. The chosen meta classifier also produced comprehensive rules, which are valid from the medical point of view, and in agreement with the IDF definition of MetS.

The paper is organized as follows: Section 2 provides a brief background of SVMs, cascade generalization, CGen-SVM, diversity measures, MetS definition, and meta-level algorithms utilized in this study. Section 3 summarizes related work, followed by the experimental methodology in section 4. Results are presented in section 5, then discussion and conclusions follow in sections 6, and 7 respectively.

2 Background

In this section we briefly review the learning algorithms utilized in this paper.

2.1 Support Vector Machines (SVMs)

Support Vector Machines are based on the principle of structural risk minimization, which aims to minimize the true error rate. SVMs operate by finding a linear hyperplane that separates the positive and negative examples with a maximum interclass distance or margin. In that sense, SVMs can be considered as stable classifiers as they are robust to noise, and have good generalization performance.

In the case of non-separable data, a soft margin hyper-plane is defined to allow errors ξi (slack variable) in classification. Hence, the optimization problem is formulated as follows [11]:

$$\text{minimize } \frac{1}{2}\|w\|^2 + C\sum_{i=1}^{l}\xi_i \quad \text{subject to } y_i(w.x_i + b) \geq 1 - \xi_i, \quad \xi_i \geq 0$$

where C is a regularization parameter which defines the trade-off between the training error and the margin. In the case of non-linearly separable data, SVMs map input data to be linearly separable in the feature space using kernel functions. Including kernel functions, and Lagrange multiplier α_i, the dual optimization problem is modified as follows [11]:

$$\text{maximize } w(\alpha) = \sum_{i=1}^{l}\alpha_i - \frac{1}{2}\sum_{i=1,j=1}^{l}\alpha_i y_i \alpha_j y_j K(x_i.x_j) \; C \geq \alpha_i \geq 0 \quad \forall_i, \quad \sum_{i=1}^{l}\alpha_i y_i = 0$$

Solving for α, the training examples with non zero α's are called the support vectors (SVs). The separating hyper-plane is completely defined by the SVs and they are the only examples which contribute to the classification decision [11].

The Weka [12] implementation of SVMs, trained using the Sequential Minimal Optimization (SMO) algorithm [13], was utilized in all of our experiments.

2.2 Classifier Ensembles

The idea of classifier ensembles was motivated by the fact that there is no single "best" algorithm generalizes well in all problem domains [10].

In general, designing a classifier ensemble has two phases: the first is the choice of base level individual classifiers. In the second phase, the predictions of base level classifiers are (somehow) combined at the meta level [14]. There are several methods to build base level classifiers. One approach is to apply different machine learning algorithms to a single dataset [10,15]. Another approach is to apply a single algorithm to different versions of a given dataset [16], e.g., Bagging [17] and Boosting [18]. Similarly, different approaches have been proposed to combine the outputs of the base level classifiers [18]. Voting and weighted voting are among the most commonly used methods, which have been used in both Bagging and Boosting. Other effective methods for combining base-level classifiers include Stacking [19] and Cascade Generalization [14].

2.3 CGen-SVM

Cascade Generalization [14] uses the predictions of the base-level classifiers to extend the dimensionality of the input space. This is done by appending the output of each of base level classifiers as a new feature to each training example. Therefore, both base and meta level classifiers utilize the original input features, while the meta-level classifiers also have access to additional features (the base level classifier predictions) [14]. Therefore, and as stated in [14] "Cascade generalization generates a unified theory from the base theories generated earlier".

CGen-SVM [9] extends cascade generalization by utilizing only an SVM as the base-level classifier, and adding a simple majority voting stage at the end, to decide the class of a test example. The voting stage considers three inputs: the output of the meta-classifier at the meta level, the output of the same classifier type at the base-level (trained with the original training data) and the output of the SVM at the base-level (also trained on the original training set). The idea is to try to make use of the bias (predictions) of a specific classifier, before influencing it (relaxing it) [14] at the meta level by the SVM bias, represented by the extended (meta) data.

It should be noted here that only the predictions of the SVM are used to extend the input space and generate the meta data. The prediction of the other classifier is only used at the voting stage.

A description of the proposed method is provided in Figures 1 and 2:

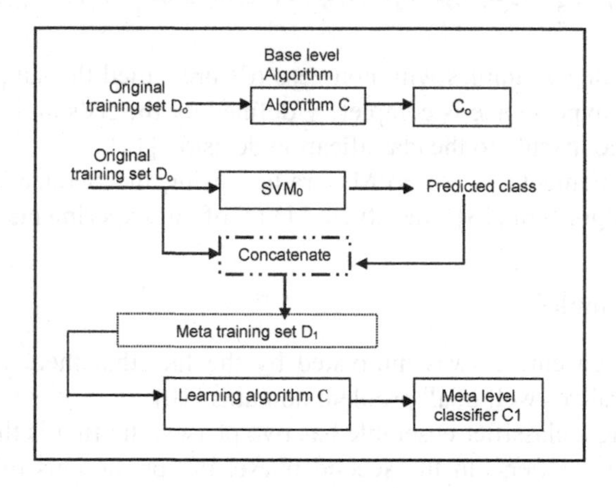

Fig. 1. CGen-SVM training phase

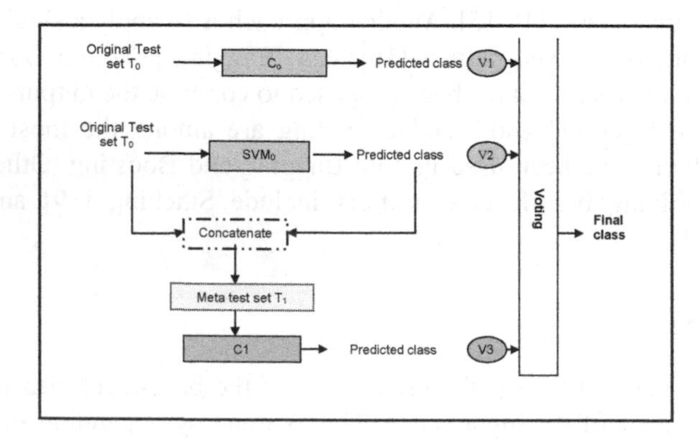

Fig. 2. CGen-SVM testing phase.

2.4 Meta-Level Algorithms

A pool of meta level classifiers has been defined, where the meta level classifier will be chosen from this pool, based on 3different diversity measures. In particular; direct rule learners. C5 [20], CART (Classification and Regression Trees) [21], and RIPPER algorithm [22] rule learners were selected; because in applications such as medical diagnosis, it is preferable to have both good performance (achieved by the SVMs) and transparent classification decisions (facilitated by rule learners).

2.5 Diversity Measures

The diversity among the members of a classifier ensemble is the key to an improved performance of that ensemble [10]. To measure the diversity of the output of binary classifiers, two main categories have been proposed [23]. The averaged pairwise measure; which include the Q statistic, the correlation, the disagreement and the double fault measures [23]. This in addition to other non-pairwise measures which include the Kohavi-Wolpert variance, the Entropy measure, the difficulty index, the generalized diversity, and the coincident failure diversity [23].

In this paper, we use 3 pairwise diversity measures to select the members of the proposed ensemble, in particular, the Q statistic, the disagreement and the double fault measures. The following paragraphs briefly define these diversity measures.

Given labeled data set D of L examples, and two classifiers C1, & C2; assume that the output of a classifier is 1 in case of correctly classified example, and 0 for a misclassified example; the following matrix shows agreements & disagreements between the pair of classifiers C1 & C2:

Table 1. Agreements & disagreements between the pair of classifiers C1 & C2 [23]

	Correctly classified example by C1	Misclassified example by C1
Correctly classified example by C2	L_{11}	L_{10}
Misclassified example by C2	L_{01}	L_{00}

The following are the pairwise diversity measures of the two classifiers C1 & C2 [23].

The Q Statistics

$$Q_{C1,C2}=\frac{L_{11}L_{00}-L_{01}L_{10}}{L_{11}L_{00}+L_{01}L_{10}}$$

Q varies between −1 and 1. Classifiers which agree on correct prediction of the same examples will have positive values of Q, while those disagree on misclassified examples will have negative value of Q [23].

The Disagreement Measure
It is the ratio between the number of examples where the two classifiers disagree in their classification to the total number of examples [23]:

$$Dis_{C1,C2} = \frac{L_{01} + L_{10}}{L_{01} + L_{10} + L_{11} + L_{00}}$$

The Double Fault measure
This measure is defined as the ratio of the number of examples that were misclassified by both classifiers to the total number of examples as follows [23]:

$$DF_{C1,C2} = \frac{L_{00}}{L_{01} + L_{10} + L_{11} + L_{00}}$$

2.6 Definition of Metabolic Syndrome

The underlying risk factors of the MetS are not particularly defined. Therefore, different definitions were suggested by different bodies like the World Health Organization (WHO), International Diabetes Federation (IDF) and other bodies.

In general, obesity, ethnicity of the studied population, abnormal glucose tolerance, and hypertension are the common risk factors among those definitions. Other factors include hormonal changes, age, life style and pro inflammatory state [5].

The IDF defines metabolic syndrome as the presence of central obesity, plus any two of the following factors:

1. Raised triglyceride level>= 1.69 mmol/l
2. Reduced HDL cholesterol
3. Raised blood pressure (systolic BP $\geq$ 130 or diastolic BP $\geq$ 85 mm Hg)
4. Raised fasting plasma glucose (FPG$\geq$100mg/dL , or previously diagnosed as type 2 diabetes) [24].

Central obesity is measured by waist circumference, and it is ethnicity specific [5]. The IDF defined the values for waist circumference to measure central obesity. As there is no standard for the Middle East and Arab countries so far, it is advised to use European cutoff values for central obesity, which is defined as waist circumference>=102 cm for men and, >=88 cm for women.

3 Related Work

Some machine learning and data mining methods have been used for identification of subjects with MetS. In that context, a fuzzy Neural Networks has been used in [25] to search for combinational MetS risk factors. In a similar study, Bayesian network have been utilized for predicting MetS using evolutionary attribute ordering [26]. Association rules have also been used in [27] to find MetS related diseases. Another study used C4 decision tree for identification of MetS in Thailand, which reached accuracy of 99.8 % in diagnosing MetS, with cutoffs close to the ones defined by the IDF [28].

4 Experimental Methodology

The details of the dataset used and the experimental methodology are described in the following subsections:

4.1 The Dataset

The data set utilized in this study is a real-world medical dataset, which was collected for screening and diagnosis of MetS and diabetes. This data set was previously used in several studies for diagnosis of diabetes [8] and pre-diabetes [29]. However, this is the first time to be used for the diagnosis of MetS using machine learning approach.

The dataset has the following input features; Age in years, Sex (male/female), Family history of diabetes (yes/no), Body mass index (BMI),Waist circumference in cm, Hip circumference, Systolic blood pressure, Diastolic blood pressure, Cholesterol, triglyceride, Fasting blood sugar, and the diagnosis of MetS (MetS or None) was the class label. The MetS detection method used in this dataset is the IDF standard MetS definition [5] (please refer to Section 2.7). The data set has 31% with MetS, and 69% without MetS, which resembles the real life prevalence.

4.2 The Experiments

A 10-fold cross validation (CV) was used to select the SVM training parameters (kernel type and the regularization parameter C). The parameters that minimized the error rate over the training set were selected.

To get reliable results, several experiments have been executed with SVM and each of the candidate meta classifiers, to compare their performance, and calculate the diversity measures as follows:

1. 10 fold CV was executed, which were repeated 10 times.
2. The data set was split to two disjoint parts, one for training and the other for testing , with different split ratios for train/test; in particular, 60/40, 65/34, and 70/30.
3. Accuracy, and the AUC were obtained from each run; and the obtained values were then averaged to get the final values.

4.3 The Choice of the Meta Classifier

To choose the best candidate meta classifier for CGen-SVM, we've used three different pairwise diversity measures; namely the Q statistics, the disagreement measure, and the double-fault measure [23], to obtain the diversity between the SVM and other classifiers, namely C5, CART, and JRIP, one at a time.

As explained in section 4.2, several runs have been have been executed as per steps 1 & 2, then the diversity measures have been calculated as follows:

1. For each run in steps 1 & 2, the agreements and disagreement measures between the SVM and each of the candidate meta classifiers, then the Q statistics, disagreement and double fault measures were calculated.
2. The values obtained in 1 were averaged, to get the final diversity measures.
3. To get more rigorous results, the previous steps were repeated at *three different values of misclassification costs*, on false negative and false positive.
4. The final diversity measures were obtained by averaging the values obtained in 2 & 3.

Based on the diversity measures results, the meta classifier for CGen-SVM are chosen.

To further evaluate the performance of the diversity based CGen-SVM, its performance was benchmarked against other well established ensemble methods like Stacking, Bagging, Boosting and Voting, using the same experimental settings in section 4.2. The same candidate meta classifiers tested, and the same data set, were utilized as follows:

- Bagging

  For this method, several experiments have been conducted utilizing SVM, C5, CART and JRIP.

- Boosting

  Again, several experiments have been conducted utilizing, SVM, C5, CART and JRIP.

- Voting

  For this experiment, four base-classifiers, namely; SVM, CART, JRIP and C5 are used as the base classifiers.

- Stacking

  Different meta classifiers were tested to choose the best combination of base and meta classifiers

  The Weka [12] software was used in our experiments. The training parameters that gave the best performance on the inner 10-fold CV were utilized.

5 Results

5.1 Results of SVM and Base Classifiers

Results of training base classifiers are shown in Table 2. From the table, it can be seen that SVM obtained the best accuracy and AUC.

Table 2. Accuracies and AUC of individual base classifiers

	SVM	**JRIP**	**CART**	**C5**
Accuracy	**0.973**	0.950	0.890	0.937
AUC	**0.976**	0.956	0.878	0.95

5.2 Results of Diversity Measures

Table 3 shows the averaged diversity measures for each candidate meta classifiers over all the runs:

Table 3. Pairwise diversity measures for each candidate meta classifiers with SVM

Diversity measure	**JRIP**	**CART**	**C5**
Q statistics	0.960	0.940	0.950
Disagreement measure	0.120	0.150	0.147
Double Fault Measure	0.190	0.210	0.206

5.3 Results of the Meta Classifier

As shown in Table 3, the classifier which has most diverse performance compared to SVM was CART, considering the Q statistics and the disagreement measure, followed by C5. Results of CGen-SVM with CART as met classifier are shown in Table 4. The obtained results are better than the results of individual base classifiers in Table 3.

Table 4. Results of CGen-SVM with CART as met classifier

	CGen-SVM With CART
Accuracy	0.984
AUC	0.988

5.4 Results of CGen-SVM Compared to Other Meta-Classifiers

For the set of experiments with meta classifiers explained in section 4.2, the following was the outcome:

- For Boosting, SVM obtained the best performance followed by JRIP, CART, then C5.
- For stacking, it was found that JRIP as a meta classifier, with SVM, CART and C5 as base classifiers obtained the best performance.
- For Bagging, C5 obtained the best performance followed by, CART, SVM, then JRIP.

Results of CGen-SVM compared to the best results obtained by other meta classifiers are shown in Table 5.

Table 5. Performance of CGen-SVM compared to other methods on same data.

	CGen-SVM With CART	Stacking JRIP	Bagging C5	Boosting SVM	Voting
Accuracy	0.984	0.960	0.958	0.970	0.973
AUC	0.988	0.976	0.962	0.975	0.988

From Table 5, it can be seen that CGen-SVM with CART obtained the best results followed by voting (SVM, CART, JRIP and C5), boosting with SVM, then stacking with JRIP as meta classifier. However, voting and CGen-SVM obtained the same AUC.

5.5 Similar Results on Benchmark Data Sets

In previous work with CGen-SVM [9], the experiments were executed without considering the diversity of candidate meta classifiers and SVM. Four benchmark datasets namely, Pima Indians, Breast cancer, Heart diseases, and Australian were considered. Results of 10 fold CV with same candidate meta classifiers considered here showed that the best average results have been obtained by CART as meta classifier followed by JRIP [9]. Results on validation data set also showed similar pattern, where JRIP obtained best average accuracy followed by C5 then CART.
Going back and calculating the diversity measures for those meta classifiers with SVMs; it was found that the highest values for diversity measures were between C5 and SVM, where the Q statistics value was 0.56, the disagreement measure was 0.21, and the double fault measure was 0.056 on Pima Indian data set. This was followed by CART, where values of 0.83, 0.19, and 0.042 were obtained for Q statistics, disagreement measure, and double fault measure respectively.

Similar calculations revealed that C5 decision tree learner has the highest average values for diversity measures with SVMs as compared to CART, and JRIP, where the values of Q statistics, disagreement measure for the rest of the data sets ranged from 0.93 to 0.81 for Q statistics and from 0.11 to 0.19 for disagreement measure respectively. Therefore, the C5 decision tree could have been selected as meta classifier, without doing all the other experiments with different meta classifiers.

6 Discussion

In this paper, a hybrid, diversity-based ensemble with one base classifier, and an additional voting stage was proposed. Based on different diversity measures, it has been shown that utilizing an SVM as the only base-level classifier, with a decision tree learner at the meta-level, leads to better performance on MetS data set and other benchmark datasets to varying degrees. The reason is that, “In the Cascade framework lower level learners delay the decisions to the high level learners” [14].

Support Vector Machines belong to the class of wide margin classifiers, and are considered “stable” classifiers, due to their excellent generalization performance. Therefore, it can be argued that the good performance of SVMs has positively influenced the performance of CGen-SVM. It has also been noted by [30] and others, that decision trees are considered "unstable" classifiers, as small variations in the training set often result in significant changes in their rules and performance. This explains why the best performance of CGen-SVM was obtained using the CART and C5 tree learner. Our results also agree with those in [14], where it has been shown that the most promising ensemble classifiers use a decision tree as high-level classifier, -as they have low bias and their performance is sensitive to small changes in datasets- and a classifier with strong bias at the lower level [14].

The obtained results are also consistent with the findings of earlier studies [30], where it has been shown that "the combined error rate depends on the error rate of individual classifiers and the correlation between them" [30], and the higher the diversity between classifiers, the better the ensemble's performance [30]. This was evident by the results obtained by CART and C5 as meta classifiers, which had the highest diversity measures with SVMs. Therefore, it can be argued that the SVM and decision tree learners make errors in different regions of the instance space for this data set.

Other studies also argue that if the performance is somehow equal, an ensemble classifier with fewer components will be preferred [10]. This gives an advantage to our proposed algorithm, where we only use SVMs as the base level classifiers and at the same time, make use of three different predictions for the ensemble at the voting stage.

In addition to the excellent performance, the rules learned by CART as a meta classifier; are valid from the medical point of view. Considering the following rules, it can be seen that all the risk factors appear in the rules' antecedents, with the right cutoff values. However, in the last two rules, only two risk factors appear in the antecedents. This can be attributed to the strong correlation between some risk factors like waist circumference as the obesity measure, and others like Triglyceride and hypertension. Figures 3 and 4 show that waist circumference can be a good predictor for high Triglyceride and hypertension respectively, especially for males. Figure 5 also shows high prevalence of 3 metabolic syndrome risk factors with high waist circumference for both mails and females. The rules are also simple and can be a valuable and cheap tool for screening and defining people at risk of MetS, who can be referred for further investigation and follow-up.

– Rules obtained from CGen-SVM with CART

1- BPSYS >= 129.0
WAIST >= 80.5
Triglyceride >= 1.69
Then MetS
2- GLUCFAST >= 99.9
SEXCODE = Female
Then MetS
3- GLUCFAST >= 99.9
WAIST >= 93.5
Then MetS

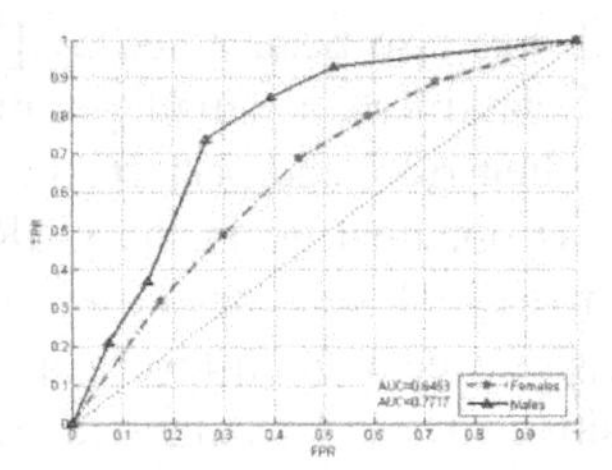

Fig. 3. ROC curve for waist circumference as predictor for high Triglyceride

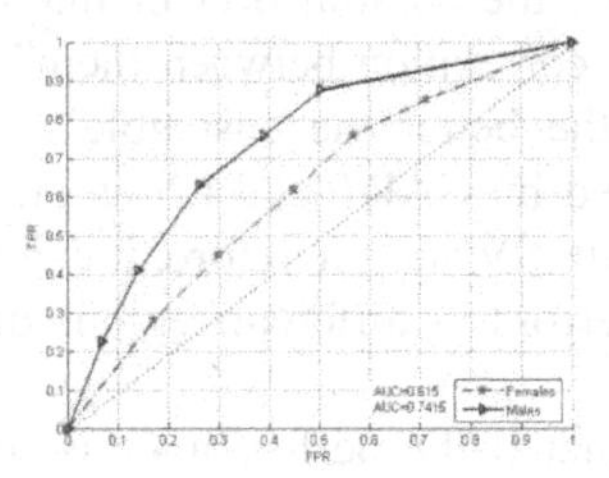

Fig. 4. ROC curve for waist circumference as predictor for Hypertension

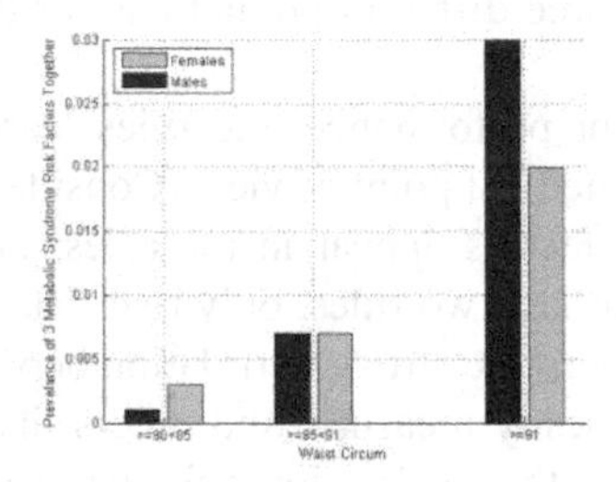

Fig. 5. Prevalence of 3 metabolic syndrome risk factors in different waist circumferences

7 Conclusions

In this paper, a simple diversity based ensemble classifier was proposed. CGen-SVM extends cascade generalization, where an SVM has been utilized as the base level classifier with an additional, different, classifier is chosen based on diversity measures at the meta level.

It has been shown that CGen-SVM with CART as meta classifier is a promising tool for screening of people with or at risk of metabolic syndrome. It has been also shown that CGen-SVM with tree learners compares well to some of other well established ensemble methods. Our results also confirm the findings of previous studies which have shown that the performance of an ensemble classifier is correlated to the performance of the individual classifiers, and the diversity between them. Therefore, an improved performance can be obtained using different families of classifiers at the base and meta-levels. It can be argued also that the difference between a specific classifier's performance at the base-level and the same classifier type performance at the meta level of CGen-SVM can be used as an indicative measure for the effect of the SVM's inductive bias on the input space and therefore, the predictions of the meta-classifiers.

As future research, posterior class probabilities instead of class predictions at the voting stage may be utilized, which can lead to better results after the majority voting stage.

References

1. IDF Worldwide Definition of the Metabolic Syndrome (2015)
2. Lakka, H., Laaksonen, D., Lakka, T., Niskanen, L., Kumpusalo, E., Tumilehto, J., Salonen, J.: The metabolic syndrome and total cardiovascular disease mortality in middle-aged men. JAMA **288**, 2709–2716 (2002)
3. Grundy, S.: Obesity, metabolic syndrome, and cardiovascular disease. Journal of Clinical Endocrinology & Metabolism **89**, 2595–2600 (2004)
4. Goldenberg, R., Punthakee, Z.: Clinical Practice Guidelines: Definition, Classification and Diagnosis of Diabetes, Prediabetes and Metabolic Syndrome. Canadian Journal of Diabetes **37**, 8–11 (2013)
5. IDF Worldwide Definition of the Metabolic Syndrome (2008)
6. Kassi, E., Pervanidou, P., Kaltsas, G., Chrousos, G.: Metabolic syndrome: definitions and controversies. BMC Medicine **9**, 1–13 (2011)
7. Yates, K.F., Sweat, V., Yau, P.L., Turchiano, M.M., Convit, A.: Impact of Metabolic Syndrome on Cognition and Brain A Selected Review of the Literature. Arterioscler. Thromb. Vasc. Biol. **32**, 2060–2067 (2012)
8. Barakat, N., Bradley, A., Barakat, M.N.: Intelligible Support Vector Machines for Diagnosis of Diabetes Mellitus. IEEE Transactions on Information Technology in Biomedicine **14**, 1114–1120 (2010)
9. Barakat, N.: Cascade generalization: is SVMs' inductive bias useful? In: IEEE International Conference on Systems, Man, and Cybernetics: (SMC 2010), Istanbul, Turkey, pp. 1393–1399. IEEE (2010)
10. Galar, M., Fernandez, A., Barrenechea, E., Bustince, H., Herrera, F.: A review on Ensembles for the Class Imbalance problem: Bagging-, Boosting-, and Hybrid-Based Approaches. IEEE Transactions on System, Man and Cybernetics Part C **42**, 463–484 (2012)
11. Cristianini, N., Taylor, J.S.: An introduction to support vector machines and other kernel-based learning methods. Cambridge University Press, Cambridge (2000)
12. Witten, I., Frank, E.: Data Mining: Practical machine learning tools and techniques, 2nd edn. Morgan Kaufmann, San Francisco (2005)
13. Platt, J.: Fast training of support vector machines using sequential minimal optimization. In: Scholkopf, B., Burges, C., Smola, A. (eds.) Advances in Kernel Methods - Support Vector Learning. MIT Press (1998)
14. Gama, J., Brazdil, P.: Cascade Generalization. Machine Learning **41**, 315–343 (2000)
15. Todorovski, L., Dzeroski, S.: Combining classifiers with meta decision trees. Machine Learning **50**, 223–249 (2003)
16. Merz, C.J.: Using correspondence analysis to combine classifiers. Machine Learning **36**, 33–58 (1999)
17. Breiman, L.: Bagging Predictors. Machine Learning **24**, 123–140 (1996)

18. Freund, Y., Schapire, R.E.: Experiments with a new boosting algorithm. In: Saitta, L. (ed.) The 13th International Conference on Machine Learning. Morgan Kaufmann (1996)
19. Wolpert, D.H.: Stacked Generalization. Neural Networks **5**, 241–259 (1992)
20. Quinlan, J.R.: C4.5: Programs for Machine Learning. Morgan Kaufmann, San Mateo (1993)
21. Breiman, L., Friedman, J., Olshen, R., Stone, C.: Classification and regression trees. Wadsworth and Brooks, Monterrey (1984)
22. Cohen, W.W.: Fast effective rule induction. In: Prieditis, A., Russell, S. (eds.) The 12th International Conference on Machine Learning (ML95), pp. 115–123 (1995)
23. Kuncheva, L.I., Skurichina, M., Duni, R.P.W.: An experimental study on diversity for bagging and boosting with linear classifiers. Information Fusion **3**, 245–258 (2002)
24. The IDF consensus worldwide definition of the metabolic syndrome (2006)
25. Ushida, Y., Kato, R., Niwa, K., Tanimura, D., Izawa, H., Yasui, K., Takase, T., Yoshida, Y., Kawase, M., Yoshida, T., Murohara, T., Honda, H.: Combinational risk factors of metabolic syndrome identified by fuzzy neural network analysis of health-check data. BMC Medical Informatics and Decision Making **12**, 1–9 (2012)
26. Park, H.-S., Cho, S.-B.: Evolutionary attribute ordering in Bayesian networks for predicting the metabolic syndrome. Expert Systems with Applications **39**, 4240–4249 (2012)
27. Chan, C.-L.: Discovery of association rules in metabolic syndrome related diseases. In: IEEE International Joint Conference on Computational Intelligence, pp. 856–862. IEEE, Hong Kong (2008)
28. Wan, A.W., Nantasenamat, C., Isarankura-Na-Ayudhya, C., Pidetcha, P., Prachayasittikul, V.: Identification of Metabolic Syndrome using Decision Tree Analysis. Diabetes Research and Clinical Practice **90**, 15–18 (2010)
29. Barakat, N., Barakat, M.N.: Hybrid intelligent model for the diagnosis of impaired glucose tolerance. In: The European Conference on Machine Learning and Principles and Practice of Knowledge Discovery in Databases (ECML PKDD 2011), KD-HCM 2011, Athens, Greece (2011)
30. Ali, K.M., Pazzani, M.J.: Error reduction through learning multiple descriptions. Machine Learning **24**, 173–202 (1996)

EFIM-Closed: Fast and Memory Efficient Discovery of Closed High-Utility Itemsets

Philippe Fournier-Viger[1(✉)], Souleymane Zida[2], Jerry Chun-Wei Lin[3], Cheng-Wei Wu[4], and Vincent S. Tseng[4]

[1] School of Natural Sciences and Humanities, Harbin Institute of Technology Shenzhen Graduate School, Shenzhen, China
philfv@hitsz.edu.cn

[2] Department of Computer Science, University of Moncton, Moncton, NB, Canada
esz2233@umoncton.ca

[3] School of Computer Science and Technology, Harbin Institute of Technology Shenzhen Graduate School, Shenzhen, China
jerrylin@ieee.org

[4] Department of Computer Science, National Chiao Tung University, Hsinchu, Taiwan, People's Republic of China
silvemoonfox@gmail.com, tsengsm@mail.ncku.edu.tw

Abstract. Discovering high-utility temsets in transaction databases is a popular data mining task. A limitation of traditional algorithms is that a huge amount of high-utility itemsets may be presented to the user. To provide a concise and lossless representation of results to the user, the concept of closed high-utility itemsets was proposed. However, mining closed high-utility itemsets is computationally expensive. To address this issue, we present a novel algorithm for discovering closed high-utility itemsets, named EFIM-Closed. This algorithm includes novel pruning strategies named *closure jumping*, *forward closure checking* and *backward closure checking* to prune non-closed high-utility itemsets. Furthermore, it also introduces novel utility upper-bounds and a transaction merging mechanism. Experimental results shows that EFIM-Closed can be more than an order of magnitude faster and consumes more than an order of magnitude less memory than the previous state-of-art CHUD algorithm.

Keywords: Pattern mining · High-utility itemset · Closed itemset

1 Introduction

High Utility Itemset Mining (*HUIM*) [2,3,6–10,14] is a popular data mining task for discovering useful patterns in customer transaction databases. It consists of discovering itemsets that yield a high utility (e.g. high profit), that is *High Utility Itemsets* (*HUIs*). Besides customer transaction analysis, HUIM also has applications in other domains such as click stream analysis and biomedicine [2,3,8,10]. HUIM can be viewed as an extension of the problem of *Frequent itemset Mining* (*FIM*) [1], where a weight (e.g unit profit) may be assigned to each item, and

P. Perner (Ed.): MLDM 2016, LNAI 9729, pp. 199–213, 2016.
DOI: 10.1007/978-3-319-41920-6_15

where purchase quantities of items in transactions are not restricted to binary values. HUIM is generally viewed as a difficult problem, because the *utility* measure used in HUIM is neither *monotonic* or *anti-monotonic*, unlike the *support* measure in FIM. That is, the utility of an itemset may be greater, smaller or equal to the utility of its subsets. For this reason, efficient search space pruning techniques developped in FIM cannot be used in HUIM.

Several algorithms have been proposed for HUIM [2,3,6–10,14]. However, an important limitation of traditional HUIM algorithms is that they often produce a huge amount of high-utility itemsets. Hence, it can be very time-consuming for users to analyze the output of these algorithms. Moreover, this makes HUIM algorithms suffer from long execution times and even fail to run due to huge memory consumption or lack of storage space. To address this issue, it was recently proposed to mine a concise and lossless representation of all HUIs named closed high-utility itemsets (CHUIs) [11]. The concept of CHUI extends the concept of *closed patterns* [12,13] from FIM. A CHUI is a HUI having no proper supersets that are HUIs and appear in the same number of transactions [11]. This latter representation is interesting since it is lossless (it allows deriving all HUIs). Furthermore, it is also meaningful for real applications since it only discovers the largest HUIs that are common to groups of customers. However, CHUI mining can be very computationally expensive [11].

In this paper, we address the need for a more efficient CHUI mining algorithm by proposing an algorithm named EFIM-Closed (EFficient high-utility Itemset Mining - Closed), based on the strict constraint that for each itemset in the search space, all operations for that itemset should be performed in linear time and space. EFIM-Closed propose three strategies to discover CHUIs efficiently: closure jumping (CJU), forward closure checking (FCC) and backward closure checking (BCC). To reduce the cost of database scans, EFIM-Closed relies on two efficient techniques named *High-utility Database Projection* (HDP) and *High-utility Transaction Merging* (HTM). Also, the proposed EFIM-Closed algorithm includes two new upper-bounds on the utility of itemsets named *sub-tree utility* and *local utility* to effectively prune the search space, and an efficient *Fast Utility Counting* (FAC) technique to compute them. An experimental study show that EFIM-Closed is up to 71 times faster and consumes up to 18 times less memory than the state-of-the-art CHUD algorithm, and has excellent scalability.

The rest of this paper is organized as follows. Sections 2, 3, 4, 5 and 6 respectively presents the problem of HUIM, the related work, the EFIM-Closed algorithm, the experimental evaluation and the conclusion.

2 Problem Statement

This section introduces the problem of closed high-utility itemset mining. Let I be a finite set of items (symbols). An itemset X is a finite set of items such that $X \subseteq I$. A *transaction database* is a multiset of transactions $D = \{T_1, T_2, ..., T_n\}$ such that for each transaction T_c, $T_c \subseteq I$ and T_c has a unique identifier c called its TID (Transaction ID). Each item $i \in I$ is associated with a positive number $p(i)$, called its *external utility* (e.g. unit profit). Every item i appearing in a transaction T_c has a positive number $q(i, T_c)$, called its *internal utility* (e.g. purchase quantity). For example, consider the database in Table 1, which will be used as the running example. It contains five transactions ($T_1, T_2..., T_5$). Transaction T_2 indicates that items a, c, e and g appear in this transaction with an internal utility of respectively 2, 6, 2 and 5. Table 2 indicates that the external utility of these items are respectively 5, 1, 3 and 1.

Table 1. A transaction database

TID	Transaction
T_1	$(a,1)(c,1)(d,1)$
T_2	$(a,2)(c,6)(e,2)(g,5)$
T_3	$(a,1)(b,2)(c,1)(d,6)(e,1)(f,5)$
T_4	$(b,4)(c,3)(d,3)(e,1)$
T_5	$(b,2)(c,2)(e,1)(g,2)$

Table 2. External utility values

Item	a	b	c	d	e	f	g
Profit	5	2	1	2	3	1	1

The utility of an item i in a transaction T_c is denoted as $u(i, T_c)$ and defined as $p(i) \times q(i, T_c)$. The *utility of an itemset X in a transaction T_c* is denoted as $u(X, T_c)$ and defined as $u(X, T_c) = \sum_{i \in X} u(i, T_c)$ if $X \subseteq T_c$. The *utility of an itemset* X in a database is denoted as $u(X)$ and defined as $u(X) = \sum_{T_c \in g(X)} u(X, T_c)$, where $g(X)$ is the set of transactions containing X. The *support* of an itemset X in a database D is denoted as $sup(X)$ and defined as $|g(X)|$. For example, the utility of item a in T_2 is $u(a, T_2) = 5 \times 2 = 10$, and its support is 1. The utility of itemset $\{a, c\}$ is $u(\{a, c\}) = u(\{a, c\}, T_1) + u(\{a, c\}, T_2) + u(\{a, c\}, T_3) = u(a, T_1) + u(c, T_1) + u(a, T_2) + u(c, T_2) + u(a, T_3) + u(c, T_3) = 5 + 1 + 10 + 6 + 5 + 1 = 28$.

An itemset X is a *high-utility itemset* if its utility $u(X)$ is no less than a user-specified minimum utility threshold *minutil* given by the user (i.e. $u(X) \geq minutil$). Otherwise, X is a *low-utility itemset.* A HUI X is a *closed high-utility itemset* (*CHUI*) [11] iff there exists no HUI Y such that $X \subset Y$ and $sup(X) = sup(Y)$. The *problem of (closed) high-utility itemset mining* is to discover all (closed) high-utility itemsets, given a threshold *minutil*, set by the user [11]. For example, if $minutil = 30$, the high-utility itemsets in the database of the running example are $\{b, d\}$, $\{a, c, e\}$, $\{b, c, d\}$, $\{b, c, e\}$, $\{b, d, e\}$, $\{b, c, d, e\}$, $\{a, b, c, d, e, f\}$ with respectively a utility of 30, 31, 34, 31, 36, 40 and 30. Among those, the closed high-utility itemsets are $\{a, b, c, d, e, f\}$,{b,c,d,e}, $\{b, c, e\}$ and $\{a, c, e\}$.

3 Related Work

A key challenge in HUIM is that search space prune techniques used in FIM cannot be used in HUIM because the utility measure is neither monotonic nor anti-monotonic [2,9,10]. Several HUIM algorithms circumvent this problem by overestimating the utility of itemsets using the concept of *Transaction-Weighted Utilization* (TWU) measure [2,6,7,9,10,14], defined as follows.

Definition 1 (Transaction Weighted Utilization). The *transaction utility* of a transaction T_c is the sum of the utilities of items from T_c in that transaction. i.e. $TU(T_c) = \sum_{x \in T_c} u(x, T_c)$. The *transaction weighted utilization* (TWU) of an itemset X is defined as $TWU(X) = \sum_{T_c \in g(X)} TU(T_c)$.

For example, The TU of transactions T_1, T_2, T_3, T_4 and T_5 for our running example are respectively 8, 27, 30, 20 and 11. The TWU of single items a, b, c, d, e, f and g are respectively 65, 61, 96, 58, 88, 30 and 38. The following property of the TWU is commonly used in HUIM to prune the search space.

Property 1 (Pruning using the TWU). Let be an itemset X. If $TWU(X) < minutil$, then X is a low-utility itemset as well as all its supersets [9].

Many HUIM algorithms [2,6,7,9–11,14] utilize Property 1 to prune the search space. They operate in two phases. In the first phase, they identify candidate high-utility itemsets by calculating their TWUs. In the second phase, they scan the database to calculate the exact utility of all candidates to filter low-utility itemsets. Recently, algorithms that mine high-utility itemsets using a single phase were proposed to avoid the problem of candidate generation [3,8], and were shown to outperform previous algorithms. One-phase algorithms rely mainly on the concept of *remaining utility* to prune the search space.

Definition 2 (Remaining Utility). Let $\succ$ be a total order on items from I, and X be an itemset. The *remaining utility* of X in a transaction T_c is defined as $re(X, T_c) = \sum_{i \in T_c \wedge i \succ x \forall x \in X} u(i, T_c)$. The *remaining utility* of X in a database is defined as $re(X) = \sum_{T_c \in D} re(X, T_c)$.

Property 2 (Pruning using remaining utility). Let X be an itemset. Let the *extensions* of X be the itemsets that can be obtained by appending an item i to X such that $i \succ x$, $\forall x \in X$. The *remaining utility upper-bound* of an itemset X is defined as $reu(X) = u(X) + re(X)$. If $u(X) + reu(X) < minutil$, then X is a low-utility itemset as well as all its extensions [3,8].

A crucial problem in HUIM is that the set of HUIs is often very large. To address this issue, it was proposed to mine the concise and representative subset of *closed HUIs* [11]. But mining CHUIs can be very computationally expensive. To address this issue, we next introduce a novel more efficient algorithm.

4 The EFIM-Closed Algorithm

The proposed EFIM-Closed algorithm is a highly efficient algorithm for closed HUI mining. It is a one phase algorithm designed using the strict design constraint that for each itemset in the search space, all operations for that itemset should be performed in linear time and space. This section is organized as follows. Subsection 4.1 introduces preliminary definitions related to the depth-first search of itemsets. Subsection 4.2 explains how EFIM-Closed reduces the cost of database scans. Subsection 4.3 explains how EFIM-closed prune low-utility itemsets in the search space. Subsection 4.4 explains how EFIM-Closed prunes non closed HUIs. Finally, subsection 4.5 put all the pieces together, and gives the full pseudocode of EFIM-Closed.

4.1 The Search Space

Let $\succ$ be any total order on items from I. According to this order, the search space of all itemsets 2^I can be represented as a *set-enumeration tree.* For example, the set-enumeration tree of $I = \{a, b, c, d\}$ for the lexicographical order is shown in Fig. 1. The EFIM-Closed algorithm explores this search space using a depth-first search starting from the root (the empty set). During this depth-first search, for any itemset α, EFIM-Closed recursively appends one item at a time to α according to the $\succ$ order, to generate larger itemsets. In our implementation, the $\succ$ order is defined as the order of increasing TWU because it generally reduces the search space for HUIM [2,3,8,10]. However, we henceforth assume that $\succ$ is the lexicographical order, for the sake of simplicity. We next introduce definitions related to the depth-first search exploration of itemsets.

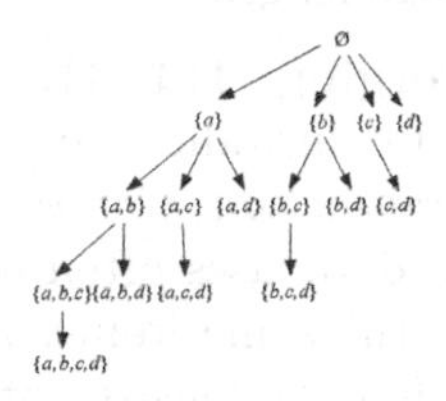

Fig. 1. Set-enumeration tree for $I = \{a, b, c, d\}$

Let be an itemset α. Let $E(\alpha)$ denote the *set of all items that can be used to extend* α according to the depth-first search, that is $E(\alpha) = \{z|z \in I \wedge z \succ x, \forall x \in \alpha\}$. An itemset Z is an *extension* of α (appears in a sub-tree of α in the set-enumeration tree) if $Z = \alpha \cup W$ for an itemset $W \in 2^{E(\alpha)}$ such that $W \neq \emptyset$. An itemset Z is a *single-item extension* of α (is a child of α in the set-enumeration tree) if $Z = \alpha \cup \{z\}$ for an item $z \in E(\alpha)$. For example, consider the database of our running example and $\alpha = \{d\}$. The set $E(\alpha)$ is $\{e, f, g\}$. Single-item extensions of α are $\{d, e\}$, $\{d, f\}$ and $\{d, g\}$. Other extensions of α are $\{d, e, f\}$, $\{d, f, g\}$ and $\{d, e, f, g\}$.

4.2 Scanning the Database Efficiently

As we will later explain, EFIM-Closed performs database scans to calculate the utility of itemsets and upper-bounds on their utility. To reduce the cost of database scans, it is desirable to reduce the database size. In EFIM-Closed this is performed by two techniques.

High-Utility Database Projection (HDP). This technique is based on the observation that when an itemset α is considered during the depth-first search, all items $x \notin E(\alpha)$ can be ignored when scanning the database to calculate the utility of itemsets in the sub-tree of α, or upper-bounds on their utility. A database without these items is called a *projected database.*

Definition 3 (Projected Database). The *projection of a transaction* T *using an itemset* α is denoted as $\alpha\text{-}T$ and defined as $\alpha\text{-}T = \{i | i \in T \wedge i \in E(\alpha)\}$. The *projection of a database* D *using an itemset* α is denoted as $\alpha\text{-}D$ and defined as the multiset $\alpha\text{-}D = \{\alpha\text{-}T | T \in D \wedge \alpha\text{-}T \neq \emptyset\}$.

For example, consider database D of the running example and $\alpha = \{b\}$. The projected database $\alpha\text{-}D$ contains three transactions: $\alpha\text{-}T_3 = \{c, d, e, f\}$, $\alpha\text{-}T_4 = \{c, d, e\}$ and $\alpha\text{-}T_5 = \{c, e, g\}$. Database projections generally greatly reduce the cost of database scans since transactions become smaller as larger itemsets are explored. To implement database projection efficiently, each transaction in the original database is sorted beforehand according to the $\succ$ total order. Then, each projection is performed as a *pseudo-projection*, that is each projected transaction is represented by an offset pointer on the corresponding original transaction. The complexity of performing a projection is $o(nw)$, where n is the number of transactions and w is their average length.

High-Utility Transaction Merging (HTM). To further reduce the cost of database scans, EFIM-Closed also introduce an efficient transaction merging technique named *High-utilty Transaction Merging* (HTM). HTM is based on the observation that transaction databases often contain identical transactions (transactions containing exactly the same items, but not necessarily the same internal utility values). The technique consists of replacing a set of identical transactions $Tr_1, Tr_2, ...Tr_m$ in a (projected) database $\alpha\text{-}D$ by a single new transaction $T_M = Tr_1 = Tr_2 = ... = Tr_m$ where the quantity of each item $i \in T_M$ is defined as $q(i, T_M) = \sum_{k=1...m} q(i, Tr_k)$. For example, consider database D of our running example and $\alpha = \{c\}$. The projected database $\alpha\text{-}D$ contains transactions $\alpha\text{-}T_1 = \{d\}$, $\alpha\text{-}T_2 = \{e, g\}$, $\alpha\text{-}T_3 = \{d, e, f\}$, $\alpha\text{-}T_4 = \{d, e\}$ and $\alpha\text{-}T_5 = \{e, g\}$. Transactions $\alpha\text{-}T_2$ and $\alpha\text{-}T_5$ can be replaced by a new transaction $T_M = \{e, g\}$ where $q(e, T_M) = 3$ and $q(g, T_M) = 7$.

Transaction merging is obviously desirable. However, a key problem is to implement it efficiently. To find identical transactions in $O(nw)$ time, we initially sort the original database according to a new total order $\succ_T$ on transactions defined as the $\succ$ order when the transactions are read backwards. For example, let be transactions $T_x = \{b, c\}$, $T_y = \{a, b, c\}$ and $T_z = \{a, b, e\}$. We have that $T_z \succ_T T_y \succ_T T_x$. Sorting is achieved in $O(nw\ log(nw))$ time. This cost is negligible because it is performed only once.

A database sorted according to the $\succ_T$ order provides the following property. For a database D or any projected database $\alpha\text{-}D$, identical transactions always appear consecutively in the projected database $\alpha\text{-}D$. This property holds because (1) transactions are sorted according to the $\succ$ order when read backwards and (2) projections always remove the smallest items of a transactions according to the $\succ$ order. Using the above property, all identical transactions in a (projected) database can be identified by only comparing each transaction with the next transaction in the database. Thus, using this scheme, transaction merging can be done very efficiently by scanning a (projected) database only once (linear time). It is interesting to note that transaction merging as proposed in EFIM-Closed is not performed in any other one-phase HUIM algorithms.

4.3 Pruning Low-Utility Itemsets

To propose an efficient CHUI mining algorithm, a key problem is to design an effective mechanism for pruning low-utility itemsets in the search space. For this purpose, we introduce two new upper-bounds on the utility of itemsets.

The Subtree-Utility and Local Utility Upper-Bounds. The first upper-bound is defined as follows.

Definition 4 (Sub-Tree Utility). Let be an itemset α and an item z such that $z \in E(\alpha)$. The *Sub-tree Utility* of z w.r.t. α is $su(\alpha, z) = \sum_{T \in g(\alpha \cup \{z\})} [u(\alpha, T) + u(z, T) + \sum_{i \in T \wedge i \in E(\alpha \cup \{z\})} u(i, T)]$.

For example, if $\alpha = \{a\}$, we have that $su(\alpha, c) = (5 + 1 + 2) + (10 + 6 + 11) + (5 + 1 + 20) = 61$, $su(\alpha, d) = 25$ and $su(\alpha, e) = 34$. The following theorem of the sub-tree utility allows EFIM-Closed to prune the search space (proof omitted due to space limitation).

Theorem 1 (Pruning Using Sub-Tree Utility). Let be an itemset α and an item $z \in E(\alpha)$. If $su(\alpha, z) < minutil$, then the single item extension $\alpha \cup \{z\}$ and its extensions are low-utility. In other words, the sub-tree of $\alpha \cup \{z\}$ in the set-enumeration tree can be pruned.

Using Theorem 1, we can prune some sub-trees of an itemset α. To further reduce the search space, we also identify items that should not be explored in any sub-trees of an itemset α.

Definition 5 (Local Utility). Let be an itemset α and an item $z \in E(\alpha)$. The *Local Utility* of z w.r.t. α is $lu(\alpha, z) = \sum_{T \in g(\alpha \cup \{z\})} [u(\alpha, T) + re(\alpha, T)]$.

For example, if $\alpha = \{a\}$, we have that $lu(\alpha, c) = (8+27+30) = 65$, $lu(\alpha, d) = 30$ and $lu(\alpha, e) = 57$. The following property can be used for pruning low-utility itemsets (proof ommitted due to space limitation).

Theorem 2 (Pruning Using the Local Utility). Let be an itemset α and an item $z \in E(\alpha)$. If $lu(\alpha, z) < minutil$, all extensions of α containing z are low-utility. i.e., item z can be ignored when exploring all sub-trees of α.

The relationships between the proposed upper-bounds and the main ones used in previous work are the following. Let be an itemset α, an item z and an itemset $Y = \alpha \cup \{z\}$. It can be demonstrated easily that the relationship $TWU(Y) \geq lu(\alpha, z) \geq reu(Y) = su(\alpha, z)$ holds. Thus, the local utility upper-bound is a tighter upper-bound on the utility of Y and its extensions compared to the TWU, which is commonly used in two-phase HUIM algorithms such as CHUD. About the su upper-bound, one can ask what is the difference between this upper-bound and the reu upper-bound used by some HUIM algorithms since they are mathematically equivalent. The major difference is that su is calculated when the depth-first search is at itemset α in the search tree rather than at the child itemset Y. Thus, if $su(\alpha, z) < minutil$, EFIM-Closed prunes the whole sub-tree of z including node Y rather than only pruning the descendant nodes of Y. Thus, using su instead of reu is more effective for pruning the search space.

In the rest of the paper, for a given itemset α, we respectively refer to items having a sub-tree utility and local-utility no less than *minutil* as *primary* and *secondary items*. Formally, the *primary items of an itemset* α is the set of items defined as $Primary(\alpha) = \{z|z \in E(\alpha) \wedge su(\alpha, z) \geq minutil\}$. The *secondary items of* α is the set of items defined as $Secondary(\alpha) = \{z|z \in E(\alpha) \wedge lu(\alpha, z) \geq minutil\}$. Because $lu(\alpha, z) \geq su(\alpha, z)$, $Primary(\alpha) \subseteq Secondary(\alpha)$. For instance, consider that $\alpha = \{a\}$. $Primary(\alpha) = \{c, e\}$. $Secondary(\alpha) = \{c, d, e\}$.

Calculating Upper-Bounds and Support Efficiently using Fast Utility Counting (FUC). In the previous paragraphs, we introduced two new upper-bounds on the utility of itemsets to prune the search space. We now present a novel efficient array-based approach to compute these upper-bounds in linear time and space that we call Fast Utility Counting (FUC). It relies on a novel array structure called utility-bin.

Definition 6 (Utility-Bin). Let be the set of items I appearing in a database D. A *utility-bin array* U for a database D is an array of length $|I|$, having an entry denoted as $U[z]$ for each item $z \in I$. Each entry is called a *utility-bin* and stores a utility value (an integer in our implementation, initialized to 0).

A utility-bin array can be used to efficiently calculate the following upper-bounds and the support in $O(n)$ time (recall that n is the number of transactions), as follows.

Calculating the TWU of all Items. A utility-bin array U is initialized. Then, for each transaction T of the database, the utility-bin $U[z]$ for each item $z \in T$ is updated as $U[z] = U[z] + TU(T)$. At the end of the database scan, for each item $k \in I$, the utility-bin $U[k]$ contains $TWU(k)$.

Calculating the Sub-Tree Utility w.r.t. an Itemset α. A utility-bin array U is initialized. Then, for each transaction T of the database, the utility-bin $U[z]$ for each item $z \in T \cap E(\alpha)$ is updated as $U[z] = U[z] + u(\alpha, T) + u(z, T) + \sum_{i \in T \wedge i \succ z} u(i, T)$. Thereafter, $U[k] = su(\alpha, k)\ \forall k \in E(\alpha)$.

Calculating the Local Utility w.r.t. an Itemset α. A utility-bin array U is initialized. Then, for each transaction T of the database, the utility-bin $U[z]$ for each item $z \in T \cap E(\alpha)$ is updated as $U[z] = U[z] + u(\alpha, T) + re(\alpha, T)$. Thereafter, we have $U[k] = lu(\alpha, k)\ \forall k \in E(\alpha)$.

Calculating the Support w.r.t. an Itemset α. A utility-bin array U is initialized. Then, for each transaction T of the database, the utility-bin $U[z]$ for each item $z \in T \cap E(\alpha)$ is updated as $U[z] = U[z] + 1$. Thereafter, we have $U[k] = sup(\alpha \cup \{k\})\ \forall k \in E(\alpha)$.

This approach for calculating upper-bounds and the support is highly efficient. For an itemset α, this approach allows to calculate the three upper-bounds and the support for all single extensions of α in linear time by performing a single (projected) database scan. In terms of memory, it can be observed that utility-bins are a very compact data structure ($O(|I|)$ size). To utilize utility-bins more efficiently, we propose three optimizations. First, all items in the database are renamed as consecutive integers. Then, in a utility-bin array U, the utility-bin $U[i]$ for an item i is stored in the i-th position of the array. This allows to access the utility-bin of an item in $O(1)$ time. Second, it is possible to reuse the same utility-bin array multiple times by reinitializing it with zero values before each use. This avoids creating multiple arrays and thus greatly reduces memory usage. In our implementation, only four utility-bin arrays are created, to respectively calculate the TWU, sub-tree utility, local utility and support. This is a reason why the memory usage of EFIM-Closed is very low compared to the state-of-the-art CHUD algorithm, as it will be shown in the experimental section. Third, when reinitializing a utility-bin array to calculate the sub-tree utility or the local utility of single-item extensions of an itemset α, only utility-bins corresponding to items in $E(\alpha)$ are reset to 0, for faster reinitialization of the utility-bin array.

4.4 Pruning Non Closed HUIs

We now explain the techniques used by EFIM-closed to prune non closed HUIs. To find only CHUIs, a naive approach would be to keep all HUIs found until now into memory. Then, every time that a new HUI is found, the algorithm would compare the HUI with previously found HUIs to determine if (1) the new HUI is included in a previously found HUI or (2) if some previously found HUI(s) are included in the new HUI. The drawback of this approach is that it can consume a large amount of memory if the number of patterns is large, and it becomes very time consuming if a very large number of HUIs is found, because a very large number of comparisons would have to be performed. In this paper, we present new checking mechanisms that can determine if a HUI is closed without having to compare a new pattern with previously found patterns. It is inspired by a similar mechanism used in sequential pattern mining [13]. The mechanism is based on two separate checks, which we respectively name *backward extension checking* check and *forward-extension checking*, and are defined as follows.

Definition 7 (Forward/Backward Extensions). *Let be an itemset $\beta = \alpha \cup \{i\}$. The itemset β is said to have a forward extension if there exists an item $z \succ i$ such that $z \in E(\beta)$ and $sup(\alpha \cup \{z\}) = sup(\beta)$. The itemset β is said to have a* backward extension *if there exists an item $z \prec i$ such that $z \notin \beta$ and $sup(\alpha \cup \{z\}) = sup(\beta)$.*

The EFIM-Closed algorithms determine if an itemset is closed as follows.

Property 3 (Identifying non closed itemsets). An itemset $\beta = \alpha \cup \{i\}$ is a CHUI if it is a HUI and it has no backward and forward extension. **Rationale.** By definition, an itemset is not closed if it has a superset $Y = \beta \cup \{z\}$ with the same support. The additional item z can respect either $z \succ i$ or $z \prec i$, which correspond respectively to the cases checked by forward and backward extensions.

The above property only allows to decide if a HUI is closed or not. To also prune the search space of non closed HUIs, the following property is used.

Property 4 (Backward extension pruning). The whole subtree of an itemset $\beta = \alpha \cup \{i\}$ can be pruned during the depth-first search if β has a backward extension. **Rationale.** Because there exists a backward extension with an item z, and z thus appear in all transactions where β appears, it follows that all itemsets in the sub-tree of β also have a backward extension with z, and thus are not CHUIs.

Furthermore, we also introduce a second property for pruning the search space that we name *closure jumping*.

Property 5 (Closure jumping property). Let be an itemset β and a projected database β-D. If $sup(\beta) = sup(\beta \cup \{z\})$ for all item $z \in E(\beta)$, then the itemset $\beta \cup E(\beta)$ is the only closed itemset in the sub-tree of β. The whole sub-tree of β can thus be pruned and $\beta \cup E(\beta)$ can be output if it is a HUI.

This property can be easily demonstrated, and is very powerful. It allows to go directly from an itemset β to its closure and prune the rest of its sub-tree.

4.5 The Algorithm

In this subsection, we present the proposed EFIM-Closed algorithm, which combines all the ideas presented in the previous subsections. The main procedure (Algorithm 1) takes as input a transaction database and the *minutil* threshold. The algorithm initially considers that the current itemset α is the empty set. The algorithm then scans the database once to calculate the local utility of each item w.r.t. α, using a utility-bin array. Then, the local utility of each item is compared with *minutil* to obtain the secondary items w.r.t to α, that is items that should be considered in extensions of α. Then, these items are sorted by ascending order of TWU and that order is thereafter used as the $\succ$ order (as suggested in [2,3,8,11]). The database is then scanned once to remove all items that are not secondary items w.r.t to α since they cannot be part of any high-utility itemsets by Theorem 2. If a transaction becomes empty, it is removed from the database. Then, the database is scanned again to sort transactions by the $\succ_T$ order to allow $O(nw)$ transaction merging, thereafter. Then, the algorithm scans the database again to calculate the sub-tree utility of each secondary item w.r.t. α, using a utility-bin array. Thereafter, the algorithm calls the recursive procedure *Search* to perform the depth first search starting from α.

The *Search* procedure (Algorithm 2) takes as parameters the current itemset to be extended α, the α projected database, the primary and secondary items w.r.t α and the *minutil* threshold. The procedure performs a loop to consider each single-item extension of α of the form $\beta = \alpha \cup \{i\}$, where i is a primary item w.r.t α (since only these single-item extensions of α should be explored according to Theorem 1). For each such extension β, a database scan is performed to calculate the utility of β and at the same time construct the β projected database. Note that transaction merging is performed whilst the β projected

Algorithm 1. The EFIM-Closed algorithm

input : D: a transaction database, *minutil*: a user-specified threshold
output: the set of high-utility itemsets

1 $\alpha = \emptyset$;
2 Calculate $lu(\alpha, i)$ for all items $i \in I$ by scanning D, using a utility-bin array;
3 $Secondary(\alpha) = \{i | i \in I \wedge lu(\alpha, i) \geq minutil\}$;
4 Let $\succ$ be the total order of TWU ascending values on $Secondary(\alpha)$;
5 Scan D to remove each item $i \notin Secondary(\alpha)$ from the transactions, and delete empty transactions;
6 Sort transactions in D according to $\succ_T$;
7 Calculate the sub-tree utility $su(\alpha, i)$ of each item $i \in Secondary(\alpha)$ by scanning D, using a utility-bin array;
8 $Primary(\alpha) = \{i | i \in Secondary(\alpha) \wedge su(\alpha, i) \geq minutil\}$;
9 `Search` (α, D, $Primary(\alpha)$, $Secondary(\alpha)$, *minutil*);

database is constructed. If β has a backward extension, no extensions of β will be explored (by Property 4). Otherwise, the projected database of β is scanned to calculate the support, sub-tree and local utility w.r.t β of each item z that could extend β (the secondary items w.r.t to α), using three utility-bin arrays. This allows determining the primary and secondary items of β. If all items that can extend β have the same support as β, the closure jumping optimization is performed to directly output $\beta \cup \bigcup_{z \succ i \wedge z \in E(\alpha)} \{z\}$ if it is a HUI and prune the subtree of β. Otherwise, the *Search* procedure is recursively called with β to continue the depth-first search by extending β. If no extension of β have the same support as β and the utility of β is no less than *minutil*, β is output as a CHUI (by Property 3). Based on properties and theorems presented in previous sections, it can be seen that when EFIM-Closed terminates, all and only the CHUIs have been output.

Complexity. The complexity of EFIM-Closed is briefly analyzed as follows. In terms of time, a $O(nw\ log(nw))$ sort is performed. This cost is negligible since it is performed only once. Then, to process each primary itemset α encountered during the depth-first search, EFIM-Closed performs database projection, transaction merging, backward/forward extension checking and upper-bound calculation in linear time and space ($O(nw)$). Thus, the time complexity of EFIM-Closed is proportional to the number of itemsets in the search space, and it is linear for each itemset.

5 Experimental Results

We performed experiments to evaluate the performance of the proposed EFIM-Closed algorithm. Experiments were carried out on a computer with a fourth generation 64 bit core i7 processor running Windows 8.1 and 16 GB of RAM. The performance of EFIM-Closed was compared with the state-of-the-art CHUD

Algorithm 2. The *Search* procedure

input : α: an itemset, α-D: the α projected database, $Primary(\alpha)$: the primary items of α, $Secondary(\alpha)$: the secondary items of α, the *minutil* threshold

output: the set of high-utility itemsets that are extensions of α

1 **foreach** *item* $i \in Primary(\alpha)$ **do**
2    $\beta = \alpha \cup \{i\}$;
3   Scan α-D to calculate $u(\beta)$ and create β-D; `// with transaction merging`
4   **if** *β has no backward extension* **then**
5   Calculate $sup(\beta, z)$, $su(\beta, z)$ and $lu(\beta, z)$ for all item $z \in Secondary(\alpha)$ by scanning β-D once, using three utility-bin arrays;
6   **if** $sup(\beta) = sup(\alpha \cup \{z\}) \forall z \succ i \wedge z \in E(\alpha)$ **then**
7   Output $\beta \cup \bigcup_{z \succ i \wedge z \in E(\alpha)} \{z\}$ if it is a HUI; `// closure jumping`
8   **else**
9        $Primary(\beta) = \{z \in Secondary(\alpha) | su(\beta, z) \geq minutil\}$;
10        $Secondary(\beta) = \{z \in Secondary(\alpha) | lu(\beta, z) \geq minutil\}$;
11   `Search` (β, β-D, $Primary(\beta)$, $Secondary(\beta)$, $minutil$);
12   **if** *β has no forward extension and* $u(\beta) \geq minutil$ **then** output β;
13   **end**
14   **end**
15 **end**

algorithm. Moreover, to evaluate the influence of the design decisions in EFIM-Closed, we also compared with a version of EFIM-Closed named EFIM(nop), where transaction merging (HTM), closure jumping, and search space pruning using the sub-tree utility were respectively deactivated, and a version named EFIM(lu), where only the sub-tree utility is desactivated. Algorithms were implemented in Java. Experiments were performed using standard datasets used in the HUIM literature for evaluating HUIM algorithms, namely *Accident*, *BMS*, *Chess*, *Connect*, *Foodmart* and *Mushroom*. These datasets have varied characteristics representing the main types of databases (sparse, dense, long transactions). For these datasets, the number of transactions/number of distinct items/average transaction length are: *Accident* (340,183 / 468 / 33.8), *BMS* (59,601 / 497 / 4.8), *Chess* (3,196 / 75 / 37.0), *Connect* (67,557 / 129 / 43.0), *Foodmart* (4,141 / 1,559 / 4.4), *Mushroom* (8,124 / 119 / 23.0). *Foodmart* contains real external/internal utility values. For other datasets, external and internal utility values have been generated in the $[1, 000]$ and $[1, 5]$ intervals using a log-normal distribution, as done in previous work [2,3,8,10,11]. The datasets and the source code of the compared algorithms can be downloaded at https://goo.gl/ZaeD60.

Influence of the *minutil* Threshold on Execution Time. We ran the algorithms on each dataset while decreasing the *minutil* threshold until algorithms were too slow, ran out of memory or a clear winner was observed. Results are shown in Fig. 2. It can be seen that EFIM-Closed clearly outperforms CHUD on all datasets. For *Accident*, *BMS*, *Chess*, *Connect*, *Foodmart* and *Mushroom*,

EFIM-Closed is respectively up to 71 times, 3 times, 36 times, 2 times, 69 times and 9 times faster than CHUD. The main reasons why EFIM-Closed performs so well are, as we will show in the following experiments, that (1) the proposed sub-tree utility and local-utility upper-bounds allows EFIM-Closed to prune a larger part of the search space compared to CHUD, and that (2) the proposed HTM transaction merging technique greatly reduce the cost of database scans. Beside, the efficient calculation of the proposed upper-bounds and support in linear time using utility-bins also contribute to the time efficiency of EFIM-Closed. A reason why EFIM-Closed is so memory efficient is that it uses a simple database representation, which does not requires to maintain much information in memory (only pointers for pseudo-projections). Moreover, EFIM-Closed is also more efficient because it is a one-phase algorithm (it does not need to maintain candidates in memory), while CHUD is a two-phase algorithm. Lastly, another important characteristic of EFIM-Closed in terms of memory efficiency is that it reuses some of its data structures. As explained in section 4.3, EFIM-Closed uses a very efficient mechanism called Fast Array Counting for calculating upper-bounds. FAC only requires to create four arrays that are then reused to calculate the upper-bounds and support of each itemset considered during the depth-first search.

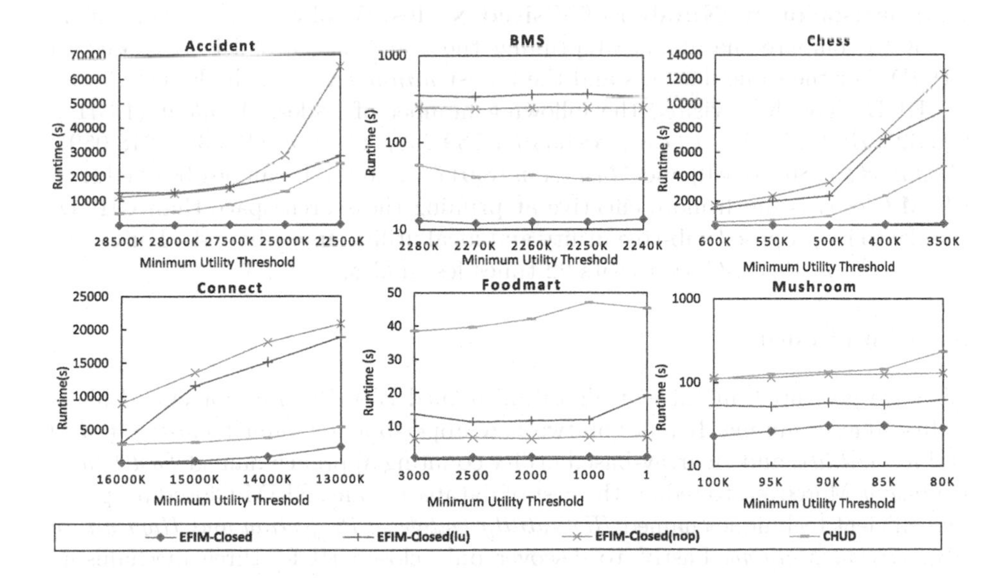

Fig. 2. Execution times on different datasets

Influence of the *minutil* Threshold on Memory Usage. In terms of memory usage, EFIM-Closed clearly outperforms CHUD. The maximum memory usage for EFIM-Closed/CHUD on each dataset are (in megabytes): *Accident* (895 / 2,603), *BMS* (64 / 707), *Chess* (65 / 327), *Connect* (385 / 1,504),

Foodmart (64 / 215) and *Mushroom* (71 / 1,308). It is also interesting that EFIM-Closed utilizes less than 100 MB on four out of the six datasets, and never more than 1 GB, while CHUD often exceeds 1 GB.

Influence of Transaction Merging on Execution Time. In terms of optimizations, the proposed transaction merging and closure jumping techniques used in EFIM-Closed sometimes greatly increases its performance in terms of execution time. This allows EFIM-Closed to perform very well on dense or large datasets such as *Accidents*, *Chess,Chess*, *Connect* and *Mushroom*). For example, for *Accidents* and *minutil* = 22500K, EFIM-Closed terminates in 6 minutes while CHUD terminates in almost 7 hours. On dense datasets or datasets having long transactions, transaction merging and closure jumping is very effective as projected transactions are more likely to be identical. This can be clearly seen by comparing the runtime of EFIM-Closed and EFIM(nop). On *Accidents*, *Chess*, *Connect* and *Mushroom*, EFIM-Closed is up to 183, 90, 9 and 5 times faster than EFIM(nop). For other datasets, transaction merging also reduces execution times but usually by a lesser amount (EFIM-Closed is up to 19, 10 times faster than EFIM(nop) on *BMS* and *Foodmart*). It is also interesting to note that transaction merging could not be implemented efficiently in CHUD because it uses a vertical database representation.

Comparison of the Number of Visited Nodes. We also performed an experiment to compare the ability at pruning the search space of EFIM-Closed to CHUD. For the same datasets and the lowest *minutil* values, the EFIM-Closed / CHUD algorithms visited the following number of nodes: *Accident* (1,341 / 29,932), *BMS* (7 / 27), *Chess* (348,633 / 7,759,252), *Connect* (19,336 / 218,059), *Foodmart* (6,680 / 6,680) and *Mushroom* (8,017 / 17,621). It can be observed that EFIM-Closed is much more effective at pruning the search space than CHUD, thanks to its proposed sub-tree utility and local utility upper-bounds. For example, on *Chess*, EFIM-Closed visits 22 times less nodes.

6 Conclusion

We have presented an efficient algorithm named EFIM-Closed for closed high-utility itemset mining. It relies on two new upper-bounds named *sub-tree utility* and *local utility*, and an array-based utility counting approach named *Fast Utility Counting*. Moreover, to reduce the cost of database scans, EFIM-Closed proposes two efficient techniques named *High-utility Database Projection* and *High-utility Transaction Merging*. Lastly, to discover only closed HUIs, three mechanisms are proposed: (1) forward closure checking, (2) backward closure checking, and (3) closure jumping. Experimental results shows that EFIM-Closed can be up to 71 times faster and consumes up to 18 times less memory than the state-of-the-art CHUD algorithm. Source code and datasets are available as part of the SPMF data mining library [5] at http://www.philippe-fournier-viger.com/spmf/. For future work, we will consider extending ideas introduced in EFIM-closed for top-k HUI mining [4], and high-utility sequent pattern and sequential rule mining [15].

References

1. Agrawal, R., Srikant, R.: Fast algorithms for mining association rules in large databases. In: Proc. Int. Conf. Very Large Databases, pp. 487–499 (1994)
2. Ahmed, C.F., Tanbeer, S.K., Jeong, B.-S., Lee, Y.-K.: Efficient tree structures for high-utility pattern mining in incremental databases. IEEE Transactions on Knowledge and Data Engineering **21**(12), 1708–1721 (2009)
3. Fournier-Viger, P., Wu, C.-W., Zida, S., Tseng, V.S.: FHM: Faster high-utility itemset mining using estimated utility co-occurrence pruning. In: Proc. 21st Intern. Symp. on Methodologies for Intell. Syst., pp. 83–92 (2014)
4. Fournier-Viger, P., Gomariz, A., Gueniche, T., Mwamikazi, E., Thomas, R.: TKS: Efficient mining of top-K sequential patterns. In: Motoda, H., Wu, Z., Cao, L., Zaiane, O., Yao, M., Wang, W. (eds.) ADMA 2013, Part I. LNCS, vol. 8346, pp. 109–120. Springer, Heidelberg (2013)
5. Fournier-Viger, P., Gomariz, A., Gueniche, T., Soltani, A., Wu., C., Tseng, V. S.: SPMF: a Java Open-Source Pattern Mining Library. Journal of Machine Learning Research (JMLR) **15**, 3389–3393 (2014)
6. Lan, G.C., Hong, T.P., Tseng, V.S.: An efficient projection-based indexing approach for mining high utility itemsets. IEEE Transactions on Knowledge and Data Engineering **38**(1), 85–107 (2014)
7. Song, W., Liu, Y., Li, J.: BAHUI: Fast and memory efficient mining of high utility itemsets based on bitmap. Intern. Journal of Data Warehousing and Mining **10**(1), 1–15 (2014)
8. Liu, M., Qu, J.: Mining high utility itemsets without candidate generation. In: Proc. 22nd ACM Intern. Conf. Info. and Know. Management, pp. 55–64 (2012)
9. Liu, Y., Liao, W., Choudhary, A.: A Two-Phase Algorithm for Fast Discovery of High Utility Itemsets. In: Ho, T.B., Cheung, D., Liu, H. (eds.) PAKDD 2005. LNCS (LNAI), vol. 3518, pp. 689–695. Springer, Heidelberg (2005)
10. Tseng, V.S., Shie, B.-E., Wu, C.-W.: Yu., P. S.: Efficient algorithms for mining high utility itemsets from transactional databases. IEEE Transactions on Knowledge and Data Engineering **25**(8), 1772–1786 (2013)
11. Tseng, V., Wu, C., Fournier-Viger, P., Yu, P.: Efficient algorithms for mining the concise and lossless representation of closed+ high utility itemsets. IEEE Transactions on Knowledge and Data Engineering **27**(3), 726–739 (2015)
12. Uno, T., Kiyomi, M., Arimura, H.: LCM ver. 2: Efficient mining algorithms for frequent/closed/maximal itemsets. In: Proc. ICDM 2004 Workshop on Frequent Itemset Mining Implementations. CEUR (2004)
13. Wang, J., Han, J., Li, C.: Frequent closed sequence mining without candidate maintenance. IEEE Transactions on Knowledge and Data Engineering **19**(8), 1042–1056 (2007)
14. Yun, U., Ryang, H., Ryu, K.H.: High utility itemset mining with techniques for reducing overestimated utilities and pruning candidates. IEEE Transactions on Knowledge and Data Engineering **41**(8), 3861–3878 (2014)
15. Zida, S., Fournier-Viger, P., Wu, C.-W., Lin, J.C.-W., Tseng, V.S.: Efficient mining of high-utility sequential rules. In: Perner, P. (ed.) MLDM 2015. LNCS, vol. 9166, pp. 157–171. Springer, Heidelberg (2015)

Fast Detection of Block Boundaries in Block-Wise Constant Matrices

Vincent Brault(✉), Julien Chiquet, and Céline Lévy-Leduc

UMR MIA-Paris, AgroParisTech, INRA, Université Paris-Saclay, 75005 Paris, France
vincentbrault@agroparistech.fr

Abstract. We propose a novel approach for estimating the location of block boundaries (change-points) in a random matrix consisting of a block wise constant matrix observed in white noise. Our method consists in rephrasing this task as a variable selection issue. We use a penalized least-squares criterion with an ℓ_1-type penalty for dealing with this problem. We first provide some theoretical results ensuring the consistency of our change-point estimators. Then, we explain how to implement our method in a very efficient way. Finally, we provide some empirical evidence to support our claims and apply our approach to data coming from molecular biology which can be used for better understanding the structure of the chromatin.

Keywords: Change-points · High-dimensional sparse linear model · HiC experiments

1 Introduction

Detecting automatically the block boundaries in a block wise constant matrix corrupted with noise is a very important issue which may have several applications. One of the main situations in which this problem occurs is the detection of chromosomal regions having close spatial location in the nucleus. Detecting such regions will improve our understanding of the influence of the chromosomal conformation on the cells functioning. The data provided by the most recent technology called HiC consist of a list of pairs of locations along the chromosome which are often summarized as a square matrix such that each entry corresponds to the number of interactions between two positions along the chromosome, see [3]. Since this matrix can be modeled as a block wise matrix corrupted by some additional noise, it is of particular interest to design an efficient and fully automated method to find the block boundaries of large matrices, which may typically have several thousands of rows and columns, in order to identify the interacting chromosomal regions.

A large literature is dedicated to the change-point detection issue for one-dimensional data. This problem can be addressed from a sequential (online) [13] or from a retrospective (off-line) [2] point of view. Many off-line approaches are based on the dynamic programming algorithm which retrieves K change-points

P. Perner (Ed.): MLDM 2016, LNAI 9729, pp. 214–228, 2016.
DOI: 10.1007/978-3-319-41920-6_16

within n observations of a one-dimensional signal with a complexity of $O(Kn^2)$ in time [7]. Such a complexity is however prohibitive for dealing with very large data sets. In this situation, [5] proposed to rephrase the change-point estimation issue as a variable selection problem. This approach has also been extended by [15] to find shared change-points between several signals. To the best of our knowledge no method has been proposed for addressing the case of two-dimensional data where the number of rows and columns may be very large ($n \times n \approx 5000 \times 5000$, namely 2×10^7 observations). The only statistical approach proposed for retrieving the change-point positions in the two-dimensional framework is the one devised by [8] but it is limited to the case where the blockwise matrix is assumed to be blockwise constant on the diagonal and constant outside the diagonal blocks.

It has first to be noticed that the classical dynamic programming algorithm cannot be applied in such a framework since the Markov property does not hold anymore. Moreover, the group-lars approach of [15] cannot be used in this framework since it would only provide change-points in columns and not in rows. As for the generalized Lasso recently devised by [14] or the two dimensional fused Lasso of [6], they are very helpful for image denoising but do not give access to the change-point positions since they are not derived to provide a partitioning of a matrix in rectangular blocks.

The paper is organized as follows. In Section 2, we first describe how to rephrase the problem of two-dimensional change-point estimation as a high dimensional sparse linear model and give some theoretical results which prove the consistency of our change-point estimators. In Section 3, we describe how to efficiently implement our method. In Section 4, we provide experimental evidence of the relevance of our approach on synthetic and real data coming from molecular biology.

2 Statistical Framework

2.1 Statistical Modeling

In this section, we explain how the two-dimensional retrospective change-point estimation issue can be seen as a variable selection problem. Our goal is to estimate $\mathbf{t}_1^\star = (t_{1,1}^\star, \ldots, t_{1,K_1^\star}^\star)$ and $\mathbf{t}_2^\star = (t_{2,1}^\star, \ldots, t_{2,K_2^\star}^\star)$ from the random matrix $\mathbf{Y} = (Y_{i,j})_{1 \leq i,j \leq n}$ defined by

$$\mathbf{Y} = \mathbf{U} + \mathbf{E}, \tag{1}$$

where $\mathbf{U} = (U_{i,j})$ is a blockwise constant matrix such that

$$U_{i,j} = \mu_{k,\ell}^\star \quad \text{if} \quad t_{1,k-1}^\star \leq i \leq t_{1,k}^\star - 1 \quad \text{and } t_{2,\ell-1}^\star \leq j \leq t_{2,\ell}^\star - 1,$$

with the convention $t_{1,0}^\star = t_{2,0}^\star = 1$ and $t_{1,K_1^\star+1}^\star = t_{2,K_2^\star+1}^\star = n+1$. An example of such a matrix $\mathbf{U}$ is displayed in Figure 1 (left). The entries $E_{i,j}$ of the matrix $\mathbf{E} = (E_{i,j})_{1 \leq i,j \leq n}$ are iid zero-mean random variables. With such a definition the

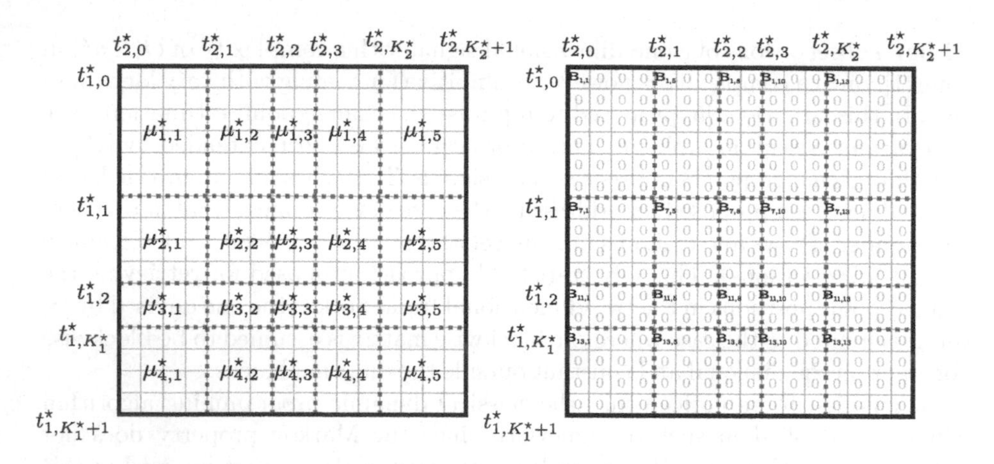

Fig. 1. Left: An example of a matrix $\mathbf{U}$ with $n = 16$, $K_1^\star = 3$ and $K_2^\star = 4$. Right: The matrix $\mathbf{B}$ associated to this matrix $\mathbf{U}$.

$Y_{i,j}$ are assumed to be independent random variables with a blockwise constant mean.

Let $\mathbf{T}$ be a $n \times n$ lower triangular matrix with nonzero elements equal to one and $\mathbf{B}$ a sparse matrix containing null entries except for the $\mathbf{B}_{i,j}$ such that $(i, j) \in \{t^\star_{1,0}, \ldots, t^\star_{1,K_1^\star}\} \times \{t^\star_{2,0}, \ldots, t^\star_{2,K_2^\star}\}$. Then, (1) can be rewritten as follows:

$$\mathbf{Y} = \mathbf{T}\mathbf{B}\mathbf{T}^\top + \mathbf{E}, \tag{2}$$

where $\mathbf{T}^\top$ denotes the transpose of the matrix $\mathbf{T}$. For an example of a matrix $\mathbf{B}$, see Figure 1 (right). Let $\text{Vec}(\mathbf{X})$ denotes the vectorization of the matrix $\mathbf{X}$ formed by stacking the columns of $\mathbf{X}$ into a single column vector then $\text{Vec}(\mathbf{Y}) = \text{Vec}(\mathbf{T}\mathbf{B}\mathbf{T}^\top) + \text{Vec}(\mathbf{E})$. Hence, by using that $\text{Vec}(\mathbf{A}\mathbf{X}\mathbf{C}) = (\mathbf{C}^\top \otimes \mathbf{A})\text{Vec}(\mathbf{X})$, where $\otimes$ denotes the Kronecker product, (2) can be rewritten as:

$$\mathcal{Y} = \mathcal{X}\mathcal{B} + \mathcal{E}, \tag{3}$$

where $\mathcal{Y} = \text{Vec}(\mathbf{Y})$, $\mathcal{X} = \mathbf{T} \otimes \mathbf{T}$, $\mathcal{B} = \text{Vec}(\mathbf{B})$ and $\mathcal{E} = \text{Vec}(\mathbf{E})$. Thanks to these transformations, Model (1) has thus been rephrased as a sparse high dimensional linear model where $\mathcal{Y}$ and $\mathcal{E}$ are $n^2 \times 1$ column vectors, $\mathcal{X}$ is a $n^2 \times n^2$ matrix and $\mathcal{B}$ is $n^2 \times 1$ sparse column vectors. Multiple change-point estimation Problem (1) can thus be addressed as a variable selection problem:

$$\widehat{\mathcal{B}}(\lambda_n) = \underset{\mathcal{B} \in \mathbb{R}^{n^2}}{\text{Argmin}} \left\{ \|\mathcal{Y} - \mathcal{X}\mathcal{B}\|_2^2 + \lambda_n \|\mathcal{B}\|_1 \right\}, \tag{4}$$

where $\|u\|_2^2$ and $\|u\|_1$ are defined for a vector u in $\mathbb{R}^N$ by $\|u\|_2^2 = \sum_{i=1}^N u_i^2$ and $\|u\|_1 = \sum_{i=1}^N |u_i|$. Criterion (4) is related to the popular Least Absolute Shrinkage and Selection Operator (LASSO) in least-square regression. Thanks to the sparsity enforcing property of the ℓ_1-norm, the estimator $\widehat{\mathcal{B}}$ of $\mathcal{B}$ is expected to be sparse and to have non-zero elements matching with those of $\mathcal{B}$. Hence, retrieving the positions of the non zero elements of $\widehat{\mathcal{B}}$ thus provides estimators of $(t^\star_{1,k})_{1\leq k\leq K^\star_1}$ and of $(t^\star_{2,k})_{1\leq k\leq K^\star_2}$. More precisely, let us define by $\widehat{\mathcal{A}}(\lambda_n)$ the set of active variables:

$$\widehat{\mathcal{A}}(\lambda_n) = \left\{ j \in \{1,\ldots,n^2\} : \widehat{\mathcal{B}}_j(\lambda_n) \neq 0 \right\}.$$

For each j in $\widehat{\mathcal{A}}(\lambda_n)$, consider the Euclidean division of $(j-1)$ by n, namely $(j-1) = nq_j + r_j$ then

$$\begin{aligned} \widehat{\mathbf{t}}_1 = (\widehat{t}_{1,k})_{1\leq k\leq |\widehat{\mathcal{A}}_1(\lambda_n)|} &\in \{r_j + 1 : j \in \widehat{\mathcal{A}}(\lambda_n)\}, \\ \widehat{\mathbf{t}}_2 = (\widehat{t}_{2,\ell})_{1\leq \ell\leq |\widehat{\mathcal{A}}_2(\lambda_n)|} &\in \{q_j + 1 : j \in \widehat{\mathcal{A}}(\lambda_n)\} \\ \text{where } \widehat{t}_{1,1} < \widehat{t}_{1,2} < \cdots < \widehat{t}_{1,|\widehat{\mathcal{A}}_1(\lambda_n)|}, \widehat{t}_{2,1} &< \widehat{t}_{2,2} < \cdots < \widehat{t}_{2,|\widehat{\mathcal{A}}_2(\lambda_n)|}. \end{aligned} \quad (5)$$

In (5), $|\widehat{\mathcal{A}}_1(\lambda_n)|$ and $|\widehat{\mathcal{A}}_2(\lambda_n)|$ correspond to the number of distinct elements in $\{r_j : j \in \widehat{\mathcal{A}}(\lambda_n)\}$ and $\{q_j : j \in \widehat{\mathcal{A}}(\lambda_n)\}$, respectively.

As far as we know, neither thorough practical implementation nor theoretical grounding have been given so far to support such an approach for change-point estimation in the two-dimensional case. In the following section, we give theoretical results supporting the use of such an approach.

2.2 Theoretical Results

In order to establish the consistency of the estimators $\widehat{\mathbf{t}}_1$ and $\widehat{\mathbf{t}}_2$ defined in (5), we shall use assumptions (**A1**–**A4**). These assumptions involve the two following quantities

$$\begin{aligned} I^\star_{\min} &= \min_{0\leq k\leq K^\star_1} |t^\star_{1,k+1} - t^\star_{1,k}| \wedge \min_{0\leq k\leq K^\star_2} |t^\star_{2,k+1} - t^\star_{2,k}|, \\ J^\star_{\min} &= \min_{1\leq k\leq K^\star_1, 1\leq \ell\leq K^\star_2+1} |\mu^\star_{k+1,\ell} - \mu^\star_{k,\ell}| \wedge \min_{1\leq k\leq K^\star_1+1, 1\leq \ell\leq K^\star_2} |\mu^\star_{k,\ell+1} - \mu^\star_{k,\ell}|, \end{aligned}$$

which corresponds to the smallest length between two consecutive change-points and to the smallest jump size between two consecutive blocks, respectively.

(**A1**) The random variables $(E_{i,j})_{1\leq i,j\leq n}$ are iid zero mean random variables such that there exists a positive constant β such that for all ν in $\mathbb{R}$, $\mathbb{E}[\exp(\nu E_{1,1})] \leq \exp(\beta\nu^2)$.

(**A2**) The sequence (λ_n) appearing in (4) is such that $(n\delta_n J^\star_{\min})^{-1}\lambda_n \to 0$, as n tends to infinity.

(**A3**) The sequence (δ_n) is a non increasing and positive sequence tending to zero such that $n\delta_n {J^\star_{\min}}^2/\log(n) \to \infty$, as n tends to infinity.

(**A4**) $I^\star_{\min} \geq n\delta_n$.

Proposition 1. *Let $(Y_{i,j})_{1\le i,j\le n}$ be defined by (1) and $\widehat{t}_{1,k}$, $\widehat{t}_{2,k}$ be defined by (5). Assume that $|\widehat{\mathcal{A}}_1(\lambda_n)| = K_1^\star$ and that $|\widehat{\mathcal{A}}_2(\lambda_n)| = K_2^\star$, with probabilty tending to one, then,*

$$\mathbb{P}\left(\left\{\max_{1\le k\le K_1^\star} \left|\widehat{t}_{1,k} - t_{1,k}^\star\right| \le n\delta_n\right\} \cap \left\{\max_{1\le k\le K_2^\star} \left|\widehat{t}_{2,k} - t_{2,k}^\star\right| \le n\delta_n\right\}\right) \underset{n\to\infty}{\longrightarrow} 1. \quad (6)$$

The proof of Proposition 1 is based on the two following lemmas. The first one comes from the Karush-Kuhn-Tucker conditions of the optimization problem stated in (4). The second one allows us to control the supremum of the empirical mean of the noise.

Lemma 1. *Let $(Y_{i,j})_{1\le i,j\le n}$ be defined by (1). Then, $\widehat{\mathcal{U}} = \mathcal{X}\widehat{\mathcal{B}}$, where $\mathcal{X}$ and $\widehat{\mathcal{B}}$ are defined in (3) and (4) respectively, is such that*

$$\sum_{k=r_j+1}^{n} \sum_{\ell=q_j+1}^{n} Y_{k,\ell} - \sum_{k=r_j+1}^{n} \sum_{\ell=q_j+1}^{n} \widehat{\mathcal{U}}_{k,\ell} = \frac{\lambda_n}{2} sign(\widehat{\mathcal{B}}_j), \text{ if } \widehat{\mathcal{B}}_j \neq 0, \quad (7)$$

$$\left|\sum_{k=r_j+1}^{n} \sum_{\ell=q_j+1}^{n} Y_{k,\ell} - \sum_{k=r_j+1}^{n} \sum_{\ell=q_j+1}^{n} \widehat{\mathcal{U}}_{k,\ell}\right| \le \frac{\lambda_n}{2}, \text{ if } \widehat{\mathcal{B}}_j = 0, \quad (8)$$

where q_j and r_j are the quotient and the remainder of the euclidean division of $(j-1)$ by n, respectively, that is $(j-1) = nq_j + r_j$. In (7), sign denotes the function which is defined by $sign(x) = 1$, if $x > 0$, -1, if $x < 0$ and 0 if $x = 0$. Moreover, the matrix $\widehat{\mathbf{U}}$, which is such that $\widehat{\mathcal{U}} = Vec(\widehat{\mathbf{U}})$, is blockwise constant and satisfies $\widehat{U}_{i,j} = \widehat{\mu}_{k,\ell}$, if $\widehat{t}_{1,k-1} \le i \le \widehat{t}_{1,k} - 1$ and $\widehat{t}_{2,\ell-1} \le j \le \widehat{t}_{2,\ell} - 1$, $k \in \{1, \ldots, |\widehat{\mathcal{A}}_1(\lambda_n)|\}$, $\ell \in \{1, \ldots, |\widehat{\mathcal{A}}_2(\lambda_n)|\}$, where the $\widehat{t}_{1,k}$, $\widehat{t}_{2,k}$, $\widehat{\mathcal{A}}_1(\lambda_n)$ and $\widehat{\mathcal{A}}_2(\lambda_n)$ are defined in (5).

Lemma 2. *Let $(E_{i,j})_{1\le i,j\le n}$ be random variables satisfying (A1). Let also (v_n) and (x_n) be two positive sequences such that $v_n x_n^2/\log(n) \to \infty$, then*

$$\mathbb{P}\left(\max_{\substack{1\le r_n < s_n \le n\\ |r_n - s_n| \ge v_n}} \left|(s_n - r_n)^{-1} \sum_{j=r_n}^{s_n-1} E_{n,j}\right| \ge x_n\right) \underset{n\to\infty}{\longrightarrow} 0,$$

the result remaining valid if $E_{n,j}$ is replaced by $E_{j,n}$.

The proofs of Proposition 1, Lemmas 1 and 2 can be seen as a natural extension of the results of [5].

3 Implementation

In order to identify a series of change-points we look for the whole path of solutions in (4), *i.e.*, $\{\hat{\mathcal{B}}(\lambda), \lambda_{\min} < \lambda < \lambda_{\max}\}$ such that $|\hat{\mathcal{A}}(\lambda_{\max})| = 0$ and

$|\hat{\mathcal{A}}(\lambda_{\min})| = s$ with s a predefined maximal number of activated variables. To this end it is natural to adopt the famous homotopy/LARS strategy [4,10]. Such an algorithm identifies in Problem (4) the successive values of λ that correspond to the activation of a new variable, or the deletion of one that became irrelevant. However, the existing implementations do not apply here since the size of the design matrix $\mathcal{X}$ – even for reasonable n – is challenging both in terms of memory requirement and computational burden. To overcome these limitations, we need to take advantage of the particular structure of the problem. In the following lemmas, we show that the most involving computations in the LARS can be made extremely efficiently thanks to the particular structure of $\mathcal{X}$.

Lemma 3. *For any vector* $\mathbf{v} \in \mathbb{R}^{n^2}$, *computing* $\mathcal{X}\mathbf{v}$ *and* $\mathcal{X}^\top \mathbf{v}$ *requires at worse* $2n^2$ *operations.*

Lemma 4. *Let* $\mathcal{A} = \{a_1, \ldots, a_K\}$ *and for each* j *in* $\mathcal{A}$ *let us consider the Euclidean division of* $j-1$ *by* n *given by* $j - 1 = nq_j + r_j$, *then*

$$\left(\left(\mathcal{X}^\top \mathcal{X}\right)_{\mathcal{A},\mathcal{A}}\right)_{1 \le k,\ell \le K} = \left(\left(n - (q_{a_k} \vee q_{a_\ell})\right) \times \left(n - (r_{a_k} \vee r_{a_\ell})\right)\right)_{1 \le k,\ell \le K}. \tag{9}$$

Moreover, for any non empty subset $\mathcal{A}$ *of distinct indices in* $\{1, \ldots, n^2\}$, *the matrix* $\mathcal{X}_\mathcal{A}^\top \mathcal{X}_\mathcal{A}$ *is invertible.*

Lemma 5. *Assume that we have at our disposal the Cholesky factorization of* $\mathcal{X}_\mathcal{A}^\top \mathcal{X}_\mathcal{A}$. *The updated factorization on the extended set* $\mathcal{A} \cup \{j\}$ *only requires solving an* $|\mathcal{A}|$*-size triangular system, with complexity* $\mathcal{O}(|\mathcal{A}|^2)$. *Moreover, the downdated factorization on the restricted set* $\mathcal{A} \setminus \{j\}$ *requires a rotation with negligible cost to preserve the triangular form of the Cholesky factorization after a column deletion.*

Remark 1. We were able to obtain a closed-form expression of the inverse $(\mathcal{X}_\mathcal{A}^\top \mathcal{X}_\mathcal{A})^{-1}$ for some special cases of the subset $\mathcal{A}$, namely, when the quotients/ratios associated with the Euclidean divisions of the elements of $\mathcal{A}$ are endowed with a particular ordering. For addressing any general problem though, we rather solve system involving $\mathcal{X}_\mathcal{A}^\top \mathcal{X}_\mathcal{A}$ by means of a Cholesky factorization which is updated along the homotopy algorithm. These updates correspond to adding or removing an element at a time in $\mathcal{A}$ and are performed efficiently as stated in Lemma 5.

These lemmas are the building blocks for our LARS implementation given in Algorithm 1, where we detail the leading complexity associated with each part. The global complexity is in $\mathcal{O}(mn^2 + ms^2)$ where m is the final number of steps in the while loop. These steps include all the successive additions and deletions needed to reach s, the final targeted number of active variables. At the end of day, we have m block-wise prediction $\hat{\mathbf{Y}}$ associated with the series of m estimations of $\hat{\mathcal{B}}(\lambda)$.

Algorithm 1. Fast LARS for two-dimensional change-point detection

Input: data matrix $\mathbf{Y}$, maximal number of active variables s.

`// Initialization`

Start with no change-point $\mathcal{A} \leftarrow \emptyset$, $\hat{\mathcal{B}} = \mathbf{0}$;

Compute current correlations $\hat{\mathbf{c}} = \mathcal{X}^\top \mathcal{Y}$ with Lemma 3 ; // $\mathcal{O}(n^2)$

while $\lambda > 0$ or $|\mathcal{A}| < s$ **do**

    `// Update the set of active variables`

    Determine next change-point(s) by setting $\lambda \leftarrow \|\hat{\mathbf{c}}\|_\infty$ and $\mathcal{A} \leftarrow \{j : \hat{\mathbf{c}}_j = \lambda\}$;

    Update the Cholesky factorization of $\mathcal{X}_\mathcal{A}^\top \mathcal{X}_\mathcal{A}$ with Lemma 4; // $\mathcal{O}(|\mathcal{A}|^2)$

    `// Compute the direction of descent`

    Get the unormalized direction $\tilde{w}_\mathcal{A} \leftarrow \left(\mathcal{X}_{\cdot\mathcal{A}}^\top \mathcal{X}_{\cdot\mathcal{A}}\right)^{-1} \mathrm{sign}(\hat{c}_\mathcal{A})$; // $\mathcal{O}(|\mathcal{A}|^2)$

    Normalize $w_\mathcal{A} \leftarrow \alpha \tilde{w}_\mathcal{A}$ with $\alpha \leftarrow 1/\sqrt{\tilde{w}_\mathcal{A}^\top \mathrm{sign}(\hat{c}_\mathcal{A})}$;

    Compute the equiangular vector $u_\mathcal{A} = \mathcal{X}_\mathcal{A} w_\mathcal{A}$ and $\mathbf{a} = \mathcal{X}^\top u_\mathcal{A}$ with Lemma 3; // $\mathcal{O}(n^2)$

    `// Compute the direction step`

    Find the maximal step preserving equicorrelation $\gamma_{\mathrm{in}} \leftarrow \min^+_{j \in \mathcal{A}^c} \left\{ \frac{\lambda - \mathbf{c}_j}{\alpha - a_j}, \frac{\lambda + \mathbf{c}_j}{\alpha + a_j} \right\}$;

    Find the maximal step preserving the signs $\gamma_{\mathrm{out}} \leftarrow \min^+_{j \in \mathcal{A}} \left\{ -\hat{\mathcal{B}}_\mathcal{A} / w_\mathcal{A} \right\}$;

    The direction step that preserves both is $\hat{\gamma} \leftarrow \min(\gamma_{\mathrm{in}}, \gamma_{\mathrm{out}})$;

    Update the correlations $\hat{\mathbf{c}} \leftarrow \hat{\mathbf{c}} - \hat{\gamma}\mathbf{a}$ and $\hat{\mathcal{B}}_\mathcal{A} \leftarrow \hat{\mathcal{B}}_\mathcal{A} + \hat{\gamma} w_\mathcal{A}$ accordingly ; // $\mathcal{O}(n)$

    `// Drop variable crossing the zero line`

    if $\gamma_{\mathrm{out}} < \gamma_{\mathrm{in}}$ **then**

        Remove existing change-point(s) $\mathcal{A} \leftarrow \mathcal{A} \backslash \left\{ j \in \mathcal{A} : \hat{\mathcal{B}}_j = 0 \right\}$;

        Downdate the Cholesky factorization of $\mathcal{X}_\mathcal{A}^\top \mathcal{X}_\mathcal{A}$; // $\mathcal{O}(|\mathcal{A}|)$

Output: Sequence of triplet $(\mathcal{A}, \lambda, \hat{\mathcal{B}})$ recorded at each iteration.

Concerning the memory requirements, we only need to store the $n \times n$ data matrix $\mathbf{Y}$ once. Indeed, since we have at our disposal the analytic form of any sub matrix extracted from $\mathcal{X}^\top \mathcal{X}$, we never need to compute neither store this large $n^2 \times n^2$ matrix. This paves the way for quickly processing data with thousands of rows and columns.

4 Numerical Experiments

4.1 Synthetic Data

The goal of this section is to assess the statistical and numerical performances of our methodology. We generated observations according to Model (1) where $\mathbf{U}$ is a symmetric blockwise constant matrix defined by

$$\left(\mu^\star_{k,\ell}\right)_{k \in \{1,\dots K_1^\star + 1\}, \ell \in \{1,\dots K_2^\star + 1\}} = \begin{pmatrix} 1\,0\,1\,0\,1 \\ 0\,1\,0\,1\,0 \\ 1\,0\,1\,0\,1 \\ 0\,1\,0\,1\,0 \\ 1\,0\,1\,0\,1 \end{pmatrix}, \tag{10}$$

and the $E_{i,j}$ are zero mean i.i.d. Gaussian random variables of variance σ^2 where σ is in $\{1, 2, 5\}$. Some examples of data generated from this model with $n = 500$, $K_1^\star = K_2^\star = 4$ can be found in Figure 2 (Top).

Statistical Performances. For each value of σ in $\{1, 2, 5\}$, we generated 1000 matrices following the model described in Section 4.1 with $(t^{\star}_{1,k})_{1\leq k\leq K^{\star}_1} = ([nk/(K^{\star}_1+1)]+1)_{1\leq k\leq K^{\star}_1}$ and $(t^{\star}_{2,k})_{1\leq k\leq K^{\star}_2} = ([nk/(K^{\star}_2+1)]+1)_{1\leq k\leq K^{\star}_2}$, where $[x]$ denotes the integer part of x and $K^{\star}_1 = K^{\star}_2 = 4$ and $n = 500$. Figure 2 (middle) displays the mean square error $n^{-2}\|\mathcal{B} - \widehat{\mathcal{B}}\|_2^2$ for the different samples (in gray) and the median of the mean square errors in thick as a function of the number of active variables s defined in Algorithm 1. We can see from this figure that even in the high noise level case, the mean square error is small. Moreover, the ROC curves displayed in the bottom part of Figure 2 ensure that the change-points in rows: $(t^{\star}_{1,k})_{1\leq k\leq K^{\star}_1}$ are properly retrieved with a very small error rate even in high noise level frameworks. The same results hold for the change-points in columns: $(t^{\star}_{2,k})_{1\leq k\leq K^{\star}_2}$ but are not displayed in order to save space.

Since, to the best of our knowledge, no two-dimensional method are available, we propose to compare our approach to an adaptation of the CART procedure of [1] and to an adaptation of [5] (HL) dedicated to univariate observations. We adapt the CART methodology by using the successive boundaries provided by CART as change-points for the two-dimensional data. The associated ROC curve is displayed with '•' in Figure 3. For adapting the HL methodology, we apply it to each row of the data matrix and for each λ, we obtain the change-points of each row. The change-points appearing in the different rows are claimed to be change-points for the two-dimensional data either if they appear at least in one row (the associated ROC curve for this approach called HL1 is displayed with '+' in Figure 3) or if they appear in $([n/2] + 1)$ rows (the associated ROC curve for this approach called HL2 is displayed with '□' in Figure 3). Since the procedures HL1 and HL2 are much slower than ours, the ROC curves are displayed for matrices of size 250×250. We can see from Figure 3 that our method outperforms the other ones.

Numerical Performances. We implemented Algorithm 1 in C++ using the library **armadillo** for linear algebra [12] and also provide an interface to the R platform [11] through the R package **blockseg** which is available from the Comprehensive R Archive Network (CRAN). All experiments were conducted on Linux workstation with Intel Xeon 2.4 GHz processor and 8 GB of memory.

We generated data as in Model (10) for different values of n: $n \in \{100, 250, 500, 1000, 2500, 5000\}$ and different values of the maximal number of activated variables: $s \in \{50, 100, 250, 500, 750\}$. The median runtimes obtained from 4 replications (+ 2 for warm-up) are reported in Figures 4. Left of Figure 4 (resp. right) gives the runtimes in seconds as a function of s (resp. of n). These results give experimental evidence for the theoretical complexity $\mathcal{O}(mn^2 + ms^2)$ that we established in Section 3 and thus for the computational efficiency of our approach: applying **blockseg** to matrices containing 10^7 entries takes less than 2 minutes.

Model Selection. In practice, we take $s = K^2_{\max}$ where $K_{\max}$ is an upper bound for $K^{\star}_1$ and $K^{\star}_2$. For choosing the final change-points we shall adapt the well-known *stability selection* approach devised by [9]. More precisely, we randomly

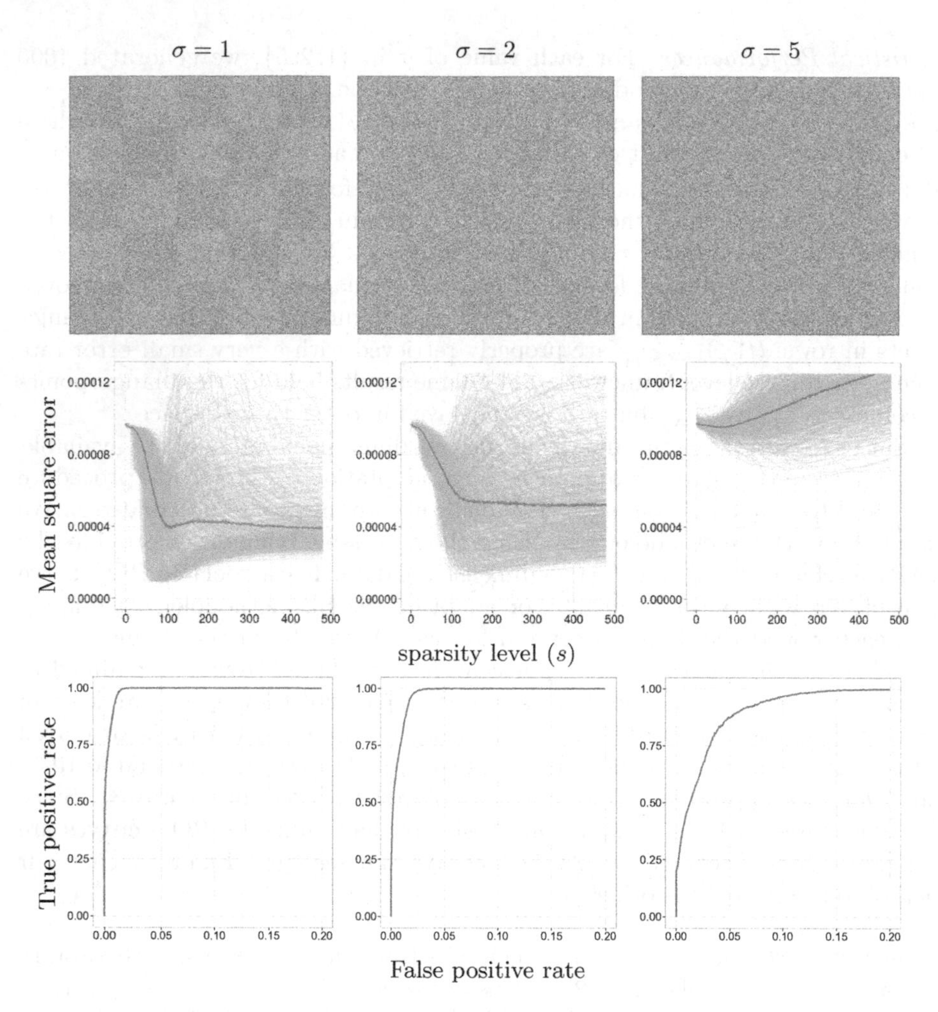

Fig. 2. Top: Examples of 500 × 500 matrices $\mathbf{Y}$ generated from the model described in Section 4.1. Middle: Mean square errors $n^{-2}\|\mathcal{B}-\widehat{\mathcal{B}}\|_2^2$ for the different realizations in gray and the median of the mean square errors in thick line as a function of the number of nonzero elements in $\widehat{\mathcal{B}}$ for each scenario. Bottom: ROC curves for the estimated change-points in rows.

choose M times $n/2$ columns and $n/2$ rows of the matrix $\mathbf{Y}$ and for each set of observations thus generated we select $s = K_{\max}^2$ active variables. Finally, after the M data resamplings, we keep the change-points which appear a number of times larger than a given threshold. By the definition of the change-points given in (5), a change-point $\widehat{t}_{1,k}$ or $\widehat{t}_{2,\ell}$ may appear several times in a given set of resampled observations. Hence, the score associated with each change-point corresponds to the sum of the number of times it appears in each of the M resamplings.

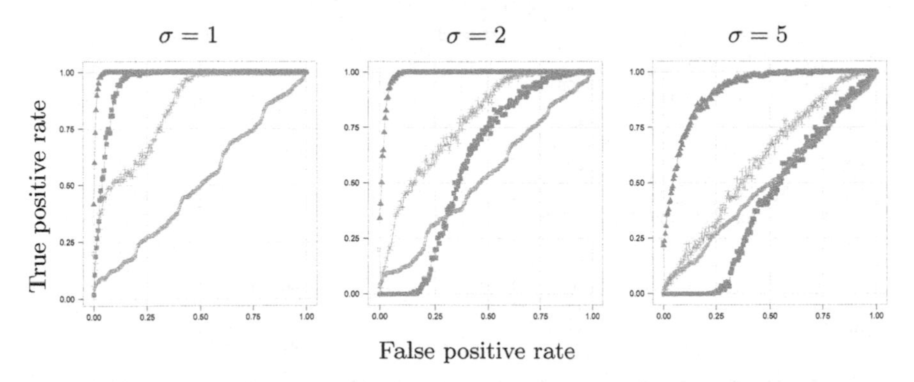

Fig. 3. ROC curves for the estimated change-points in rows for our method ('△'), HL1 ('+'), HL2 ('□') and CART ('•').

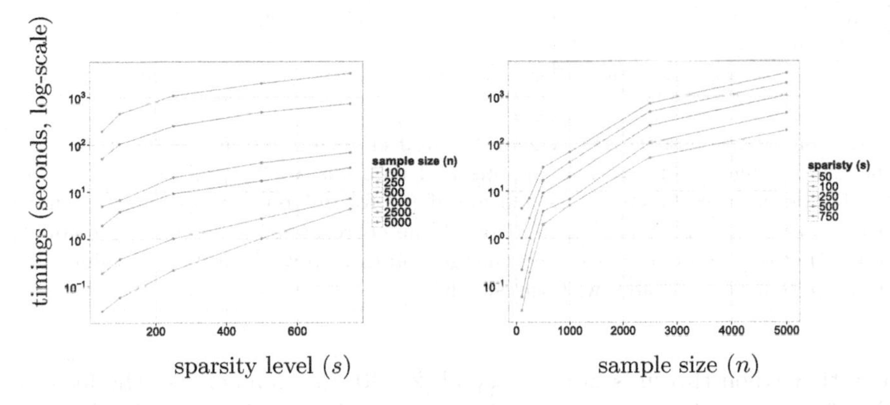

Fig. 4. Left: Computation time (in seconds) for various value of n as a function of the sparsity level $s = |\mathcal{A}|$ reached at the end of the algorithm. The cuvres for $n = 100$ to 5000 are displayed from bottom to top. Right: Computation time (in seconds) as a function of sample size n. The curves for $s = 50$ to 750 are displayed from bottom to top.

To evaluate the performances of this methodology, we generated observations according to the model defined in Section 4.1 with $s = 225$ and $M = 100$. The results are given in Figure 5 which displays the score associated to each change-point for a given matrix $\mathbf{Y}$ (top). We can see from the top part of Figure 5 some spurious change-points appearing near from the true change-point positions. In order to identify the most representative change-point in a given neighborhood, we keep the one with the largest score among a set of contiguous candidates. The result of such a post-processing is displayed in the second and third rows of Figure 5. More precisely the boxplots associated to the estimation of $K_1^\star$ (resp. the histograms of the estimated change-points in rows) are displayed in the middle (resp. bottom) part of Figure 5 for different threshold (resp. when the threshold is equal to $T = 30\%$ of the largest score). We can see from these

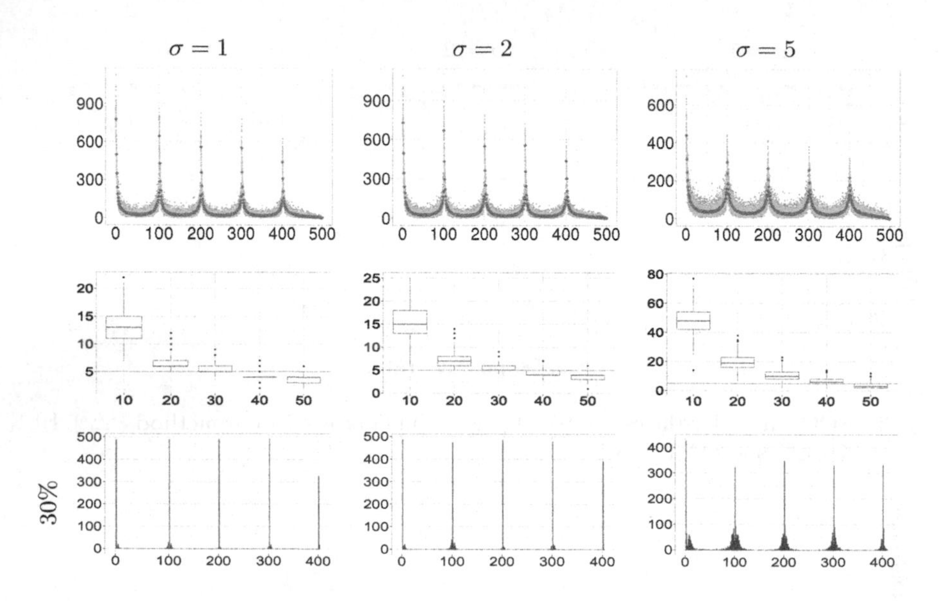

Fig. 5. Top: Scores associated to each estimated change-points for different values of σ; the true change-point positions in rows and columns are located at 101, 201, 301 and 401. Middle: Boxplots of the estimation of $K_1^\star$ for different values of σ and thresh after the post-processing step. The horizontal line corresponds to the true value of $K_1^\star$. Bottom: Histograms of the estimated change-points in rows for different values of σ after the post-processing step with thresh=30%.

figures that when thresh is in the interval $[20, 40]$ the number and the location of the change-points are very well estimated even in the high noise level case.

4.2 Application to HiC Data

In this section, we applied our methodology to publicly available data (http://chromosome.sdsc.edu/mouse/hi-c/download.html) already studied by [3]. More precisely, we studied the interaction matrices of Chromosomes 1 and 19 of the mouse cortex at a resolution 40 kb and we compared the number and the location of the estimated change-points found with our approach with those obtained by [3] on the same data since no ground truth is available. The matrices of these interaction matrices are displayed in Figure 6. We can see from this figure that modeling these matrices as block wise constant matrices corrupted with white seems to be relevant.

We display in Figure 7 the number of change-points in rows found by our approach as a function of the threshold thresh used in our adaptation of the stability selection approach presented in the previous section. We also display in this figure a red line corresponding to the number of change-points found by [3].

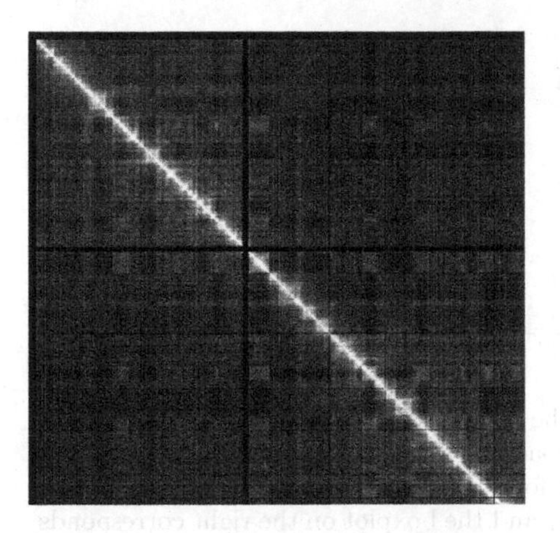
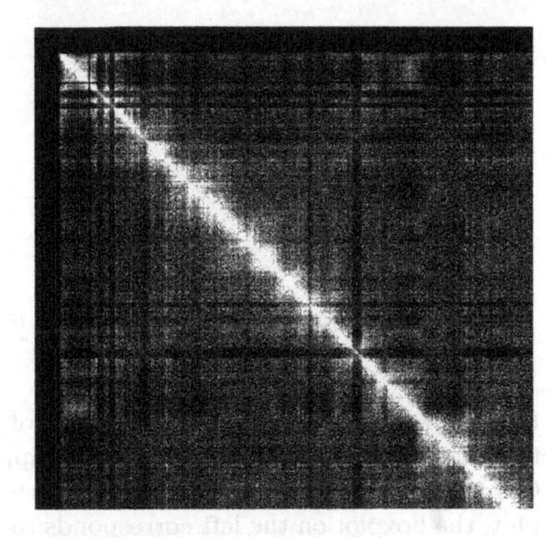

Fig. 6. Raw interaction matrices of Chromosome 1 (left) and Chromosome 19 (right) for the mouse cortex. The darkest entries correspond to the lowest values.

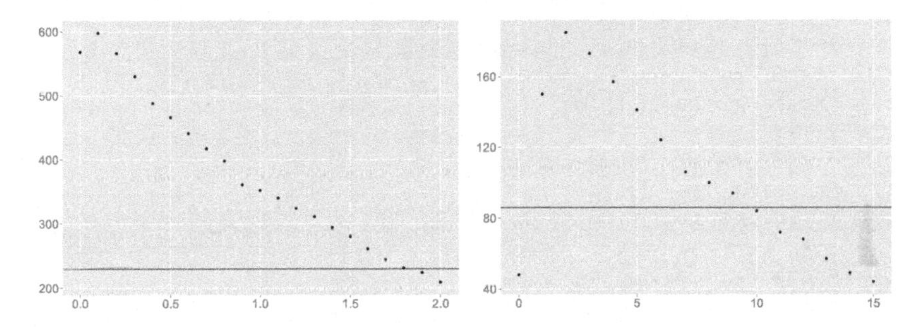

Fig. 7. Number of change-points in rows found by our approach as a function of the threshold ('•') in % for the interaction matrices of Chromosome 1 (left) and Chromosome 19 (right) of the mouse cortex. The horizontal line corresponds to the number of change-points found by [3].

We also compute the two parts of the Hausdorff distance for the change-points in rows which is defined by

$$d\left(\widehat{\boldsymbol{t}}_B, \widehat{\boldsymbol{t}}\right) = \max\left(d_1\left(\widehat{\boldsymbol{t}}_B, \widehat{\boldsymbol{t}}\right), d_2\left(\widehat{\boldsymbol{t}}_B, \widehat{\boldsymbol{t}}\right)\right), \tag{11}$$

where $\widehat{\boldsymbol{t}}$ and $\widehat{\boldsymbol{t}}_B$ are the change-points in rows found by our approach and [3], respectively. In (11),

$$d_1(\mathbf{a}, \mathbf{b}) = \sup_{b \in \mathbf{b}} \inf_{a \in \mathbf{a}} |a - b|, \tag{12}$$

$$d_2(\mathbf{a}, \mathbf{b}) = d_1(\mathbf{b}, \mathbf{a}). \tag{13}$$

More precisely, Figure 8 displays the boxplots of the d_1 and d_2 parts of the Hausdorff distance without taking the supremum in orange and blue, respectively.

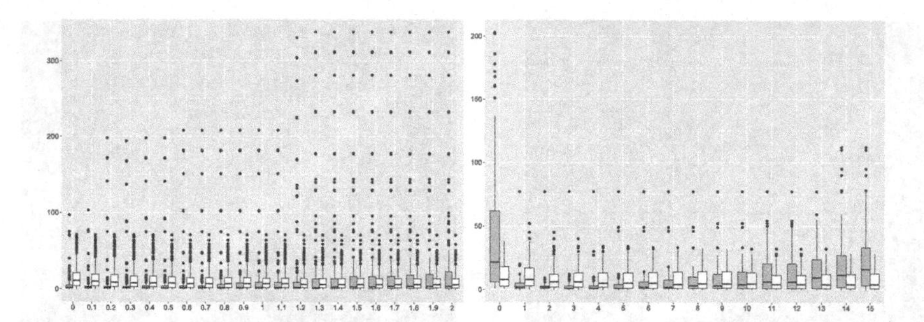

Fig. 8. Boxplots for the infimum parts of the Hausdorff distances d_1 and d_2 between the change-points found by [3] and our approach for the Chromosome 1 (left) and the Chromosome 19 (right) of the mouse cortex for the different thresholds in %. In each plot, the boxplot on the left corresponds to d_1 and the boxplot on the right corresponds to d_2.

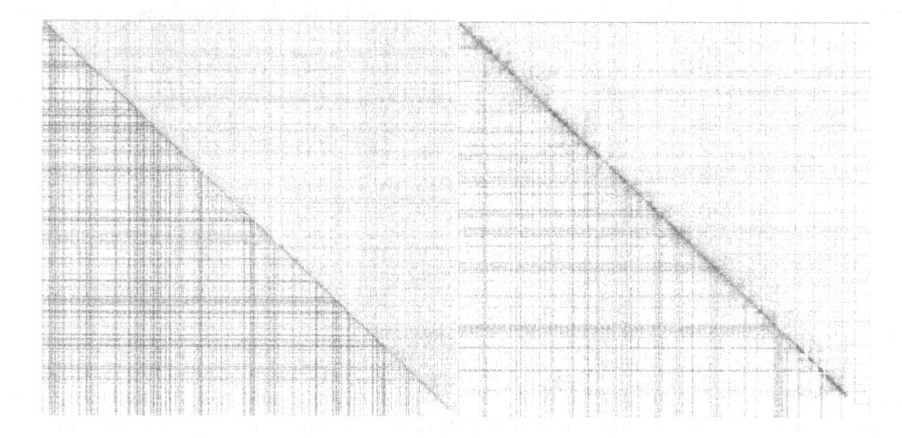

Fig. 9. Topological domains detected by [3] (upper triangular part of the matrix) and by our method (lower triangular part of the matrix) from the interaction matrix of Chromosome 1 (left) and Chromosome 19 (right) of the mouse cortex with a threshold giving 232 (resp. 85) estimated change-points in rows and columns.

We can observe from Figure 8 that some differences indeed exist between the segmentations produced by the two approaches but that the boundaries of the blocks are quite close when the number of estimated change-points are the same, which is the case when thresh = 1.8% (left) and 10% (right).

In the case where the number of estimated change-points are on a par with those of [3], we can see from Figure 9 that the change-points found with our strategy present a lot of similarities with those found by the HMM based approach of [3].

Our method also gives access to the estimated change-point positions for different values of the thresholds. Figure 10 displays the different change-point locations that can be obtained for these different values of the threshold.

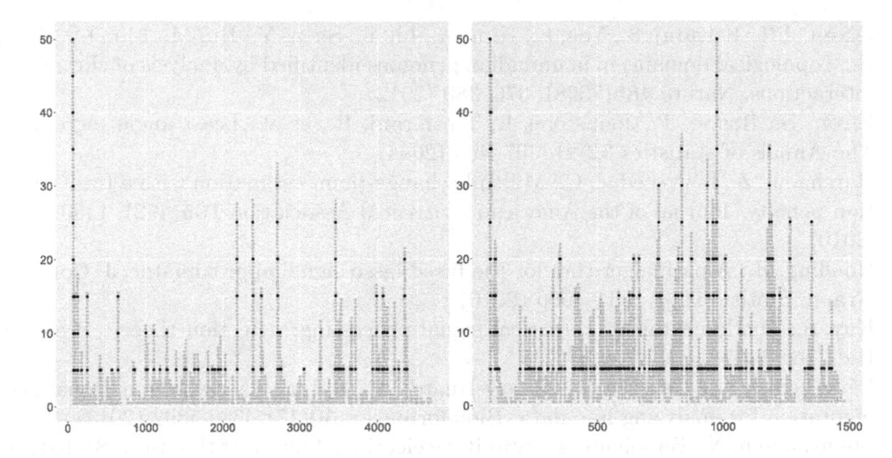

Fig. 10. Plots of the estimated change-points locations (x-axis) for different thresholds (y-axis) from 0.5% to 50% by 0.5%. The estimated change-point locations associated to threshold which are multiples of 5% are displayed with black points.

However, contrary to our method, the approach of [3] can only deal with binned data at the resolution of several kilobases of nucleotides. The very low computational burden of our strategy paves the way for processing data collected at a very high resolution, namely at the nucleotide resolution, which is one of the main current challenges of molecular biology.

5 Conclusion

In this paper, we proposed a novel approach for retrieving the boundaries of a block wise constant matrix corrupted with noise by rephrasing this problem as a variable selection issue. Our approach is implemented in the R package **blockseg** which will be available from the Comprehensive R Archive Network (CRAN). In the course of this study, we have shown that our method has two main features which make it very attractive. Firstly, it is very efficient both from the theoretical and practical point of view. Secondly, its very low computational burden makes its use possible on very large data sets coming from molecular biology.

References

1. Breiman, L., Friedman, J.H., Olshen, R.A., Stone, C.J.: Classification and Regression Trees. Statistics/Probability Series. Wadsworth Publishing Company, Belmont (1984)
2. Brodsky, B., Darkhovsky, B.: Non-parametric statistical diagnosis: problems and methods. Kluwer Academic Publishers (2000)

3. Dixon, J.R., Selvaraj, S., Yue, F., Kim, A., Li, Y., Shen, Y., Hu, M., Liu, J.S., Ren, B.: Topological domains in mammalian genomes identified by analysis of chromatin interactions. Nature **485**(7398), 376–380 (2012)
4. Efron, B., Hastie, T., Johnstone, I., Tibshirani, R., et al.: Least angle regression. The Annals of statistics **32**(2), 407–499 (2004)
5. Harchaoui, Z., Lévy-Leduc, C.: Multiple change-point estimation with a total variation penalty. Journal of the American Statistical Association **105**(492), 1480–1493 (2010)
6. Hoefling, H.: A path algorithm for the fused lasso signal approximator. J. Comput. Graph. Statist. **19**(4), 984–1006 (2010)
7. Kay, S.: Fundamentals of statistical signal processing: detection theory. Prentice-Hall, Inc. (1993)
8. Lévy-Leduc, C., Delattre, M., Mary-Huard, T., Robin, S.: Two-dimensional segmentation for analyzing hi-c data. Bioinformatics **30**(17), i386–i392 (2014)
9. Meinshausen, N., Bühlmann, P.: Stability selection. Journal of the Royal Statistical Society: Series B (Statistical Methodology) **72**(4), 417–473 (2010)
10. Osborne, M.R., Presnell, B., Turlach, B.A.: A new approach to variable selection in least squares problems. IMA Journal of Numerical Analysis **20**(3), 389–403 (2000)
11. R Core Team: R: A Language and Environment for Statistical Computing. R Foundation for Statistical Computing, Vienna, Austria (2015). http://www.R-project.org/
12. Sanderson, C.: Armadillo: An open source C++ linear algebra library for fast prototyping and computationally intensive experiments. Tech. rep, NICTA (2010)
13. Tartakovsky, A., Nikiforov, I., Basseville, M.: Sequential Analysis: Hypothesis Testing and Changepoint Detection. CRC Press, Taylor & Francis Group (2014)
14. Tibshirani, R.J., Taylor, J.: The solution path of the generalized lasso. Ann. Statist. **39**(3), 1335–1371 (2011)
15. Vert, J.P., Bleakley, K.: Fast detection of multiple change-points shared by many signals using group lars. In: Advances in Neural Information Processing Systems, pp. 2343–2351 (2010)

Inferring Censored Geo-Information with Non-Representative Data

Yu Zhang[1(✉)], Tse-Chuan Yang[2], and Stephen A. Matthews[3]

[1] Department of Computer Science, University at Albany, State University of New York, 1400 Washington Avenue, Albany, NY 12222, USA
yzhang20@albany.edu

[2] Department of Sociology, University at Albany, State University of New York, 1400 Washington Avenue, Albany, NY 12222, USA
tyang3@albany.edu

[3] Departments of Sociology and Anthropology, Pennsylvania State University, University Park, Pennsylvania, PA 16802, USA
matthews@psu.edu

Abstract. The goal of this study is to develop a method that is capable of inferring geo-locations for non-representative data. In order to protect privacy of surveyed individuals, most data collectors release coarse geo-information (e.g., tract), rather than detailed geo-information (e.g., street, apt number) when sharing surveyed data. Without the exact locations, many point-based analyses cannot be performed. While several scholars have developed new methods to address this issue, little attention has been paid to how to correct this issue when data are not representative. To fill this knowledge gap, we propose a bias correction method that adjusts for the bias using a bias factor approach. Applying our method to an empirical data set with a known bias associated with gender, we found that our method could generate reliable results despite the non-representativeness of the sample.

Keywords: Bias correction · Non-representative data · Censored geo-location

1 Introduction

Individual information such as home addresses and sociodemographic characteristics are commonly collected by governments and researchers. With an increasing emphasis on data analysis and rapid growth of spatial analysis techniques [1], the demand for individual location information has been growing. However, individual locations have been regarded as sensitive information and cannot be accessed easily. One common solution is to replace detailed locations (e.g., exact address) with a coarse geographic identifier (e.g., tract), which is known as censored location information. There is a growing interest in exploring how to fully utilize the censored geo-information to maintain the statistical utility. However, without actual locations, many point-based spatial analysis techniques cannot

P. Perner (Ed.): MLDM 2016, LNAI 9729, pp. 229–235, 2016.
DOI: 10.1007/978-3-319-41920-6_17

be applied in exploring the spatial associations embedded in the data. To address this issue, a recent study [2] proposed three probability-based allocation algorithms to impute the coordinates by using a Gaussian distribution, Bayesian function, and multinomial models. The imputed coordinates enable researchers to obtain spatial analysis results that are comparable with those based on the actual geocodes. One shortcoming of their study is to overlook the fact that many surveys may not be representative and this study aims to filled this gap. Building upon their algorithms, we proposed an approach that can increase utility of the censored geo-information and tackle the known bias in the data. After correcting the bias, the statistical findings can be generalized to the population.

Assume that we are given a survey data set where individual sociodemographic information is available and individual addresses have been censored with tract identifiers. We propose that we can link the individual information (e.g., gender) to other data from external sources to help us predict in which block group an individual may reside. The external sources are the aggregate information of tracts and block groups (e.g., the percentage of male and female). It should be noted that surveyed individuals could be biased or at least non-representative due to sampling designs or research purposes. When survey data are not representative, the overall population in block groups will have different attribute distributions compared with surveyed individual data. For example, in the surveyed data, the percentage of female in a certain block group could be 80%, while females in fact accounts for only 50% of the actual population in this block group. Without addressing the potential problem of the selection bias, making inference about individuals from surveyed data may be unreliable and misleading. We propose an innovative method to fill the gap. The experiments show that the method can generate meaningful results without sacrificing individual privacy.

2 Methodology

The following information is required by our methodology:

Survey data: (a) A group of N surveyed individuals, I_n, $n = 1, \ldots, N$. (b) Individual attributes $a_1, \ldots, a_k$ (e.g., gender and race/ethnicity), and for individual I_n attribute values are $a_{n1}, \ldots, a_{nK}$ (e.g., female, non-Hispanic Black).

External sources: (c) The censored location unit (e.g., tract), say G_m, $m = 1, \ldots, M$. The aggregate information of G_m (e.g., the percentage of female in G_m). (d) The subunits (e.g., block groups) that each censored location G_m contains, S_{ms}, $s = 1, \ldots, S$, and population aggregate information of S_{ms} (e.g., the percentage of female in S_{ms}).

Suppose there is a bias factor β such that it is correlated with a_{n1}, β alters (either increases or decreases) the odds of a_{n1} and affects how attribute a_1 is sampled from S_{ms} in surveyed data. Assuming β is independent of other covariates $a_{n2}, \ldots, a_{nK}$, the entire surveyed data set is biased by conditioning on β. Based on G_m and S_{ms} aggregate information, biased attribute β, and individual attributes, our objective is to calculate the probability $P(S_{ms}|I_n, \beta)$

of an individual living in a certain subunit. A cumulative probability strategy is used to select the subunit for I_n, and the method will randomly select individuals from group without repetition. Random geographic coordinates will be generated within the chosen subunit. The posterior distribution is designed as:

$$P(S_{ms}|I_n,\beta) = P(S_{ms}|a_{n1}, a_{n2}, ..., a_{nK}, \beta) \tag{1}$$

$$= \frac{P(a_{n1}|\beta, S_{ms}) * P(S_{ms}) \prod_{k=2}^{K} P(a_{nk}|S_{ms}))}{\sum_{s=1}^{S}(P(a_{n1}|\beta, S_{ms}) * P(S_{ms}) * \prod_{k=2}^{K} P(a_{nk}|S_{ms}))}$$

Where:

The probability of subunit S_{ms} is chosen from all the subunits in G_m:

$$P(S_{ms}) = \frac{1}{S} \tag{2}$$

The probability of biased attribute value in subunit S_{ms}:

$$P(a_{n1}|\beta, S_{ms}) = \frac{\beta * P(a_{n1}|S_{ms})}{1 + (1-\beta) * P(a_{n1}|S_{ms})} \tag{3}$$

The probability of attribute value in subunit S_{ms} (external sources aggregate information):

For $k \in 1, ..., K$:

$$P(a_{nk}|S_{ms}) = \frac{number\ of\ population\ in\ subunit\ S_{ms}\ with\ attribute\ a_{nk}}{number\ of\ population\ in\ subunit\ S_{ms}} \tag{4}$$

Without loss of generality, assume there are T possible categorical values for a_1, and a_{n1} indicates I_n's observed category of a_1. We define $\overline{a_{n1}}$ as the set which contains other $(T-1)$ categories attribute values of a_1 excepting a_{n1}. For example, if a_{n1} is non-Hispanic Black, then $\overline{a_{n1}}$ is the set of any other race/ethnicity groups except non-Hispanic Black. For I_n, we know it is within the unit G_m (i.e., the known censored geo-information), but which subunit I_n lives in is unknown. So is the subunit aggregate information from survey data. β is introduced when calculating the probability for attribute value a_{n1}, so comparing with the calculation of non-baised attributes values $a_{n2}, \ldots, a_{nK}$, we can ease the bias from macroscopic view. Thus, for G_m, the bias factor β is calculated in function (5). By using β, more attentions will be paid to the information obtained from the unit G_m rather than the subunit, and it reflects the bias degree of a_{n1} in survey data and censored data, because β is monotonic increasing with the difference of the percentage of a_{n1} between the survey data and the censored data (the more biases of the survey data, the bigger of the β). Hence, comparing with the function (2), function (3) can not only obtain information from the external sources, but also can integrate the biased influence of the a_{n1} from the survey data.

$$\beta = \frac{\left(\frac{number\ of\ I_n\ in\ survey\ data\ from\ G_m\ with\ a_{n1}}{number\ of\ I_n\ in\ survey\ data\ from\ G_m\ with\ \overline{a_{n1}}}\right)}{\left(\frac{number\ of\ I_n\ in\ G_m\ aggregate\ information\ with\ a_{n1}}{number\ of\ I_n\ in\ G_m\ aggregate\ information\ with\ \overline{a_{n1}}}\right)} \tag{5}$$

3 Experiments

To evaluate the effectiveness of our framework, we conducted experiments on the tasks of generating coordinates for 1,167 individuals. They are from two sites of the Philadelphia metropolitan area: Site 1 contains 11 census tracts and 65 block groups; Site 2 contains 8 census tracts and 44 block groups. Multiple imputation strategy and the standard errors method proposed by Rubin [3] are used to test whether the generated coordinates maintain statistical utility. Specifically, suppose that there are V imputations, Q_j (e.g., the proportion of males) is the estimated quantity of interest obtain from j^{th} imputation and U_j is the estimated variance associated with Q_j. $\overline{Q}$ is the average estimated quantity of interest and the average within-imputation variance will be $\overline{U} = (1/V)\sum_{j=1}^{V} U_j$. The between-imputation variance can be expressed as $B = 1/(V-1)\sum_{j=1}^{V}(Q_j - \overline{Q})^2$. The total stand error is $\overline{U} + (1 + 1/V)B$. We dichotomize gender into two categories (male and female), age into two categories (age above or equal to 55, and age below 55), and grouped educational attainment into five groups (less than high school, high school graduate, some college, college graduate and post-college degree). For this experiment, we assume that the three variables are mutually independent.

For each block group, the empirical aggregate attribute percentage and the imputed attributes proportion are used to calculate the overall standard errors. We use the standard errors to obtain the 95% confidence intervals. If the true proportions from the Census Bureau are covered by the confidence intervals, it

Table 1. Results of standard error comparison for 3 variables (gender, education and age).

	Site 1				Site 2			
Attribute	Accurate rate	Bias correction	Non-Bias correction	Improved percentage	Accurate rate	Bias correction	Non-Bias correction	Improved percentage
Male	100%	0.0241	0.0251	62.5%	97.72%	0.0416	0.0417	52.2%
Female	100%	0.0243	0.0255	62.5%	97.72%	0.0416	0.0417	52.2%
Less than High School	100%	0.0306	0.0317	43.7%	100%	0.0225	0.0226	47.7%
High School	100%	0.0191	0.0197	53.1%	100%	0.0144	0.0146	50.0%
Some College	100%	0.0205	0.0209	46.8%	100%	0.0152	0.0154	52.4%
College	100%	0.0115	0.0115	44.4%	100%	0.0107	0.0112	52.2%
Post-college	95.3%	0.0058	0.0075	53.1%	100%	0.0062	0.0067	45.4%
Age Below 55	100%	0.0244	0.0261	60.9%	100%	0.0181	0.0184	61.3%
Age Above 55	100%	0.0242	0.0249	57.8%	100%	0.0181	0.0184	61.3%

Table 2. Results of standard error comparison for 2 variables (gender, education).

	Site 1			Site 2		
Attributes	Bias correction	Non-Bias correction	Improved percentage	Bias correction	Non-Bias correction	Improved percentage
Male	0.0240	0.0251	57.8%	0.0416	0.0417	50.0%
Female	0.0241	0.0254	57.8%	0.0416	0.0417	50.0%
Less than High School	0.0306	0.0316	53.1%	0.0226	0.0227	59.0%
High School	0.0192	0.0198	50.0%	0.0147	0.0146	38.6%
Some College	0.0208	0.0212	45.3%	0.0154	0.0157	47.7%
College	0.0111	0.0112	40.6%	0.0107	0.0115	59.0%
Post-college	0.0059	0.0075	48.4%	0.0062	0.0067	45.4%

Table 3. Results of standard error comparison for 2 variables (gender, age).

	Site 1			Site 2		
Attributes	Bias correction	Non-Bias correction	Improved percentage	Bias correction	Non-Bias correction	Improved percentage
Male	0.0241	0.0253	56.2%	0.0416	0.0417	52.2%
Female	0.0241	0.0255	56.2%	0.0416	0.0417	52.2%
Age Below 55	0.0237	0.0256	70.3%	0.0179	0.0183	56.8%
Age Above 55	0.0237	0.0246	67.1%	0.0179	0.0182	56.8%

is coded 1, otherwise 0. The testing statistical utility for site 1 and site 2 are presented in Table 1 column "Accurate rate" (i.e., 95% confidence intervals cover the true values) among our block groups. Based on the results, we can conclude that our algorithm attain valuable results.

We also compare the bias-corrected results with Yang's [2] non bias-corrected results (note that we use the same data set). We used three attribute combinations to conduct experiments for the two sites. Table 1 shows the results based on all three attributes: gender, education and age. Table 2 considers two attributes, namely gender and education, whereas Table 3 focuses on gender and age.

Note that gender is the bias source and that is why it is included in all three tables. Columns of "Bias correction" and "Non-Bias correction" indicate the average standard errors among subunits for all imputations. For 97.5% attribute cases, the bias correction method attains lower average standard error than Non-Bias correction algorithm. The "Improved percentage" column shows the percentage of block groups, for which our method provides lower standard error than the Non-Bias correction algorithm. In this case, our method is better for

72.5% of the attributes values. Therefore, the bias correction method outperforms Yang's algorithm and achieves our goal to preserve the statistical utility in non-representative data.

4 Related Work

Most of the previous work has attempted to generate synthetic data by either adding random noise or replacing variables with values simulated from a probability distribution [4]. However these approaches cannot guarantee that the synthetic data could maintain the statistical utility. The difference-in-difference approach [5] used non-representative data to analyzed trends in mobility and migration flows. When correcting the bias for the input data, they simply evaluates the influence from external sources, authors should pay more attention to the variables within the input data, and since they do not have 'ground truth' (i.e., empirical data) to calibrate the model, it is hard to evaluate the accuracy of the model and trends results. The calibration analytical approach of Emilio et al. [6] only works in some circumstances; it has a limitation to extract signal from various noisy and messy data. Other methods aiming to handle non-representative data like Wei et al. [7] required rich information in order to forecast. They used the multilevel regression and post-stratification strategy to correct the bias. The computation cost for calculating more than ten thousands cases remains huge.

5 Discussion

For survey data owners, censoring geo-information when sharing data with others is a common practice but doing so does not allow others to effectively use the censored data, particularly when data may not be representative. Our method links the survey data to other external data sources and imputes the potential locations. A possible future direction is to use the information-theoretic to develop a new model to extract useful data from non-representative data and the researchers can obtain useful data without compromising individual privacy.

6 Conclusion

In this paper, we proposed a framework that fills the gap of inferring censored geo-location for non-representative data. The results suggest that our approach carries promising implications for adjusting for the known bias sources. In the future, we plan to conduct a more detailed study with more complicated data sets. The current bias correction method is developed for the biases associated with categorical variables and we will extend the scope to ordinal variables, e.g., income and age groups.

Acknowledgements. The findings in this report are part of the project entitled "Utility for Private Data Sharing in Social Science", which is supported by the National Science Foundation (NSF, 1228669).

References

1. Schabenber, O., Gotwa, C.A.: Statistical Methods for Spatial Data Analysis. Chapman and Hall/CRC (2005)
2. Yang, T.C., Kifer, D., Matthews, S.A., Zhang, Y., Bacon, R.: Multiple Imputations of Subregions with Censored Location Data and Statistical Utility Examination (2015) (under review)
3. Rubin, D.B.: Multiple Imputation for Nonresponse in Surveys. John Wiley&Sons (1987)
4. Abowd, J.M., Vilhuber, L.: How Protective are Synthetic Data? Privacy in Statistical DataBases, 239–246 (2008)
5. Emilio, Z., Venkata, R.K., Bogda, S.: Inferring International and Internal Migration Patterns from Twitter Data. In: WWW (2014)
6. Emilio, Z., Ingmar, W.: Demographic Research with Non-representative Internet Data. In: IJM (2014)
7. Wang, W., Rothschild, D., Goel, S., Gelman, A.: Forecasting Elections with Non-representative Pools. In: International Institute of Forecasters (2014)

Efficient Mining of Weighted Frequent Itemsets in Uncertain Databases

Jerry Chun-Wei Lin[1(✉)], Wensheng Gan[1], Philippe Fournier-Viger[2], and Tzung-Pei Hong[3,4]

[1] School of Computer Science and Technology, Harbin Institute of Technology Shenzhen Graduate School, Shenzhen, China
jerrylin@ieee.org, wsgan001@gmail.com

[2] School of Natural Sciences and Humanities, Harbin Institute of Technology Shenzhen Graduate School, Shenzhen, China
philfv@hitsz.edu.cn

[3] Department of Computer Science and Information Engineering, National University of Kaohsiung, Kaohsiung, Taiwan
tphong@nuk.edu.tw

[4] Department of Computer Science and Engineering, National Sun Yat-sen University, Kaohsiung, Taiwan

Abstract. Frequent itemset mining (FIM) is a fundamental set of techniques used to discover useful and meaningful relationships between items in transaction databases. Recently, extensions of FIM such as weighted frequent itemset mining (WFIM) and frequent itemset mining in uncertain databases (UFIM) have been proposed. WFIM considers that items may have different weight/importance, and the UFIM takes into account that data collected in a real-life environment may often be inaccurate, imprecise, or incomplete. Recently, a two-phase Apriori-based approach called HEWI-Uapriori was proposed to consider both item weight and uncertainty to mine the high expected weighted itemsets (HEWIs), while it generates a large amount of candidates and is too time-consuming. In this paper, a more efficient algorithm named HEWI-Utree is developed to efficiently mine HEWIs without performing multiple database scans and without generating enormous candidates. It relies on three novel structures named element (E)-table, weighted-probability (WP)-table and WP-tree to maintain the information required for identifying and pruning unpromising itemsets early. Experimental results show that the proposed algorithm is efficient than traditional methods of WFIM and UFIM, as well as the HEWI-Uapriori algorithm, in terms of runtime, memory usage, and scalability.

Keywords: Frequent itemsets · Uncertain databases · Weighted frequent itemsets · WP-table

1 Introduction

Knowledge Discovery in Database (KDD) aims at finding the meaningful and useful information from the amounts of mass data; frequent itemset mining (FIM) and

P. Perner (Ed.): MLDM 2016, LNAI 9729, pp. 236–250, 2016.
DOI: 10.1007/978-3-319-41920-6_18

association rule mining (ARM) [4,8] are the important and fundamental issues in KDD. However, the limitations on most FIM algorithms are stated below.

The first limitation is that previous studies generally ignore the fact that data collected in real-life applications such as wireless sensor network or location based-service, in which the data may be inaccurate, imprecise or incomplete [1,2,5]. In recent years, discovering patterns in uncertain databases has become a critical issue. Numerous algorithms have been designed for the discovery of frequent patterns in uncertain databases [1,2,5,7,15], which can be generally classified into two categories as the expected support model [1,7] and the probabilistic frequent itemset mining model [5,15].

The second limitation is that only the presence or absence of items in transactions is considered in most studies, and all items are assumed to have the same importance. However, in real-life situations, all items are not equally important to the user. To address this problem, the prominent solution is to assign weights to the items according to users preferences, which can be used to indicate the importance of items or other factors (e.g. interestingness, profit or risk). Mining weighted frequent itemsets has been extensively studied [6,9,11,12,16–20], but the proposed approaches cannot directly discover WFIs in uncertain databases. In the past, the concept of high expected weighted itemset (HEWI) was proposed by Lin et al. [10] as part of the high expected weighted itemset (HEWIM) framework for extracting HEWIs by considering both the weight and data uncertainty constraints. However, the generation-and-test mechanism to discover HEWIs in a level-wise manner may have a poor performance when the database has many long transactions or the minimum threshold is set quite low.

Motivated by the aforementioned real-world applications and the drawback of the HEWI-Uapriori approach, an enumeration tree-based algorithm called HEWI-Utree is proposed in this paper to discover the HEWI in uncertain databases. Major contributions are summarized as follows:

- Based on the concept of HEWI, an efficient HEWI-Utree algorithm is proposed to directly discover HEWIs with only two database scans. The pruning strategies of the developed downward closure property of HEWIs are also designed to speed up computation.
- Three novel structures: the element (E)-table, the weighted-probability (WP)-table and the WP-tree, are designed to improve the efficiency of mining process. These structures are used in the proposed HEWI-Utree algorithm to directly mine HEWIs without relying on candidate generate-and-test approach that requires to repeatedly scan the database.
- Based on the designed WP-tree, two efficient pruning strategies are further proposed for reducing the number of operations in the "constructing during mining process" and thus increasing mining efficiency.

2 Related Work

In real-life applications, a huge amount of data is collected and the data is often inaccurate, imprecise or incomplete. For example, this is often the case for data collected from wireless sensor networks or location-based services [2,7]. Discovering frequent itemsets in uncertain databases (UFIM) is an emerging topic in recent years [1,2,5,7,15]. In general, algorithms of UFIM can be classified into two categories: those based on the expected support model [1,7] and those based on the probabilistic frequent itemset mining model [5,15].

The weight constraint is an intrinsic attribute of objects in real-world applications. According to users preferences, a weight can be assigned to each item, which indicates its relative importance or interestingness. To improve the usefulness of mining results in real-world applications, the problem of weighted frequent pattern mining has been extensively studied in recent years, such as weighted frequent itemset mining [9,11,18,20], weighted association rule mining [6,16,17,19], and weighted sequential pattern mining [12].

Cai et al. [6] defined the weighted support measure, which is calculated by multiplying the support of an item by its average weight, and further designed the k-support bound to maintain the anti-monotone property in mining association rules with weighted items. Wang et al. [19] proposed an efficient mining methodology to mine Weighted Association Rules (WARs). Tao et al. [17] then proposed the Weighted Association Rule Mining (WARM) algorithm to mine WARs by considering weights during the iterative process of frequent itemset generation.

The Weighted Frequent Itemset Mining (WFIM) algorithm [20] was proposed to consider the weight constraint when mining itemsets using a pattern-growth approach. Lan et al. [11] then proposed the Projection-based Weighted itemset mining Approach (named PWA) to mine WFIs, an upper-bound model also was developed to avoid information loss in any WFI mining case. Extending the idea of using weights from finding WFIs to weighted sequential patterns (WSPs) [12]. Up to now, many related techniques for mining weighted patterns are still developed. However, only the HEWI-Uapriori approach [10] addressed the problem of mining weighted frequent itemsets in uncertain databases.

3 Preliminaries and Problem Statement

3.1 Preliminaries

Let $I = \{i_1, i_2, \ldots, i_m\}$ be a finite set of m distinct items appearing. An uncertain database is a set of transactions $D = \{T_1, T_2, \ldots, T_n\}$, where each transaction $T_q \in D$ is a subset of I, and has a unique identifier called its *TID*. Based on the attribute uncertainty database model [5,7], a unique existence probability $p(i_j, T_q)$ is assigned to each item i_j in each transaction T_q. It indicates that an item i_j exists in T_q with a probability $p(i_j, T_q)$. A unique existence weight $w(i_j)$ is assigned to each item $i_j \in I$ in a weight table (*w-table* $= \{w(i_1), w(i_2), \ldots, w(i_m)\}$), which represents its importance (e.g. profit, interest, risk). An itemset $X \in I$ with k distinct items $\{i_1, i_2, \ldots, i_k\}$ is of length k and is referred to as a k-itemset. A predefined minimum expected weighted-support threshold ε is defined such that $\varepsilon \in (0, 100\%]$. As a running example in Table 1, it shows an uncertainty database containing 10 transactions. We also assume that the

Table 1. An uncertain database

TID	Transaction (*item, prob.*)
T_1	B:0.5, C:0.45, F:1.0
T_2	A:0.7, B:0.82, D:0.3, F:0.75
T_3	C:0.9, D:1.0, E:0.7
T_4	A:0.48, B:0.8, C:0.6, D:1.0
T_5	B:0.7, D:0.3, E:1.0
T_6	B:0.65, C:1.0, D:0.8
T_7	C:0.9, D:0.5, F:1.0
T_8	A:0.4, E:0.4
T_9	A:0.8, B:1.0, D:0.8, F:0.7
T_{10}	B:0.4, C:0.9, D:1.0

Table 2. Derived HEWIs

Itemset	*Weight*	*Fre.*	*expSup*	*expWSup*
(B)	0.700	7	4.870	3.409
(C)	1.000	6	4.750	4.750
(D)	0.550	8	5.700	3.135
(E)	0.850	3	2.100	1.785
(F)	0.300	4	3.450	1.035
(BC)	0.850	4	1.715	1.458
(BD)	0.625	6	2.976	1.860
(CD)	0.775	5	3.650	2.829
(BCD)	0.750	3	1.360	1.020

w-table is defined as: $\{w(A)$:0.40, $w(B)$:0.70, $w(C)$:1.00, $w(D)$:0.55, $w(E)$:0.85, $w(F)$:0.30$\}$.

Definition 1 (Item weight). The weight of an item i_j in D is a value in the (0, 1] interval denoted as $w(i_j)$. It represents the importance of the item i_j according to users' preferences.

In the given example of *w-table*, the weight of (B) and (D) are respectively $w(B)$ $(= 0.7)$ and $w(D)$ $(= 0.55)$.

Definition 2 (Itemset weight in a transaction). The weight of an itemset X in T_q is denoted as $w(X, T_q)$, where $w(X, T_q)$ is the sum of the weights of all items in X divided by the number of items in X, which is formally defined as: $w(X, T_q) = \frac{\sum_{i_j \in X} w(i_j, T_q)}{|k|}$, where $|k|$ is the number of items in X, and $w(i_j, T_q)$ equals to $w(i_j)$.

For example, the weight of (BCD) in transaction T_4 is calculated as $w(BCD, T_4) = (w(B, T_4) + w(C, T_4) + w(D, T_4))/3 = (0.7 + 1.0 + 0.55)/3$ $(= 0.75)$.

Definition 3 (Itemset weight in D). The weight of an itemset X in a database D is denoted as $w(X)$ and is the same for all transactions in D, that is:

$$w(X) = \frac{\sum_{i_j \in X} w(i_j)}{|k|} = w(X, T_q). \tag{1}$$

For example, the weight of itemset (BCD) in D is calculated as $w(BCD) = w(BCD, T_4) = w(BCD, T_6) = w(BCD, T_{10}) = (0.7 + 1.0 + 0.55)/3(= 0.75)$.

Definition 4 (Itemset probability in a transaction). The probability of an itemset X in a transaction t_q is denoted as $p(X, T_q)$, and defined as:

$$p(X, T_q) = \prod_{i_j \in X} p(i_j, T_q), \tag{2}$$

where j is the j-th item in X.

Definition 5 (Expected support of an itemset in D). The expected support of an itemset X in an uncertain database D is denoted as $expSup(X)$, and defined as:

$$expSup(X) = \sum_{X \subseteq T_q \wedge T_q \in D} p(X, T_q) = \sum_{X \subseteq T_q \wedge T_q \in D} (\prod_{i_j \in X} p(i_j, T_q)). \quad (3)$$

Definition 6 (Expected weighted support of an itemset in D). The expected weighted support of an itemset X in an uncertain database D is denoted as $expWSup(X)$, and defined as the weight of X multiplied by the expected support of X, that is:

$$expWSup(X) = w(X) \times \sum_{X \subseteq T_q \wedge T_q \in D} p(X, T_q) = w(X) \times expSup(X). \quad (4)$$

Definition 7 (High expected weighted itemset, HEWI). An itemset X in a database D is called high expected weighted support itemset (HEWI) if it satisfies $expWSup(X) \geq \varepsilon \times |D|$, in which ε is the user-specified minimum expected weighted support threshold.

For example, the expected weighted supports of itemset (B) and (BCD) are respectively calculated as $expWSup(B) = w(B) \times expSup(B) = 0.7 \times 4.87$ (= 3.409) and $expWSup(BCD) = w(BCD) \times expSup(BCD) = 0.75 \times 1.36$ (= 1.02).The results of discovered HEWIs are shown in Table 2.

3.2 Problem Statement

Given an uncertain database D, a user-specified weight table *w-table*, and a user-specified minimum expected weighted-support threshold ε. The problem of mining high expected weighted itemsets (HEWIs) in an uncertain database D is to discover the set of expected weighted frequent k-itemsets while considering both the weight and the existential probability constraints.

4 Proposed HEWI-Utree Algorithm

4.1 Element (E)-table and Weighted-Probability (WP)-table

In the proposed HEWI-Utree algorithm, the element-table (E-table) and the weighted-probability table (WP-table) are designed to keep essential information about itemsets of an uncertain database.

Definition 8 (E-table). The element-table (E-table) of an itemset X consists of two elements; the list of transaction IDs containing X (***tid***) and the existential probability of X in each transaction T_q (***Pro***).

Since $w(i_j, T_q)$ is equal to $w(i_j)$, the weight of X in D can be directly calculated as $w(X) = w(X, T_q)$, thanks to the E-table. In fact, three important types of information are obtained from the designed E-table: 1) the name of X, 2) the weight of X in D, and 3) the expected support of X in D. The weight of X can also be directly calculated from the defined *w-table*, and the expected support of an itemset X can be calculated as the sum of the existential probabilities of X in T_q. The WP-table is thus built to keep the necessary information.

Definition 9 (WP-table). A weighted-probability (WP)-table of an itemset X consists of three elements; the itemset X, the expected existential probability of X (*expSup*) and the weight of X (*weight*).

Notice that it is necessary to initially construct the E-tables and WP-tables of the discovered $HUBEWI^1$ [10] by scanning the original database. These tables can be used to later construct the tables of larger itemsets without scanning the database (for itemsets that are within the search space of the proposed algorithm). The E-table construction procedure is illustrated in Algorithm 1.

```
   Input: X, X_a, X_b (a ≠ b).
   Output: X_ab which having the E-table denoted as X_ab.ET.
1  set X_ab.ET ← ∅;
2  for each element E_a ∈ X_a.ET do
3      if ∃E_a ∈ X_b.ET ∧ E_a.tid = E_b.tid then
4          if X.ET ≠ ∅ then
5              Search for element E ⊂ X.ET, E.tid = E_a.tid;
6              E_ab ←< E_a.tid, E_a.Pro × E_b.Pro/E.Pro >;
7          else
8              E_ab ←< E_a.tid, E_a.Pro × E_b.Pro >;
9          X_ab.ET ← X_ab.ET ∪ E_ab;
10 return X_ab;
```

Algorithm 1. E-table construction procedure

Before constructing the E-table and WP-table, it is necessary to discover the high upper-bound expected weighted-frequent 1-itemsets ($HUBEWI^1$) with the value of item upper-bound weighted probability of an itemset (abbreviated as *iubwp*). Details of $HUBEWI$ and *iubwp* can be found in [10]. In this running example, the discovered $HUBEWI^1$ are: *iubwp*(A): 2.614, *iubwp*(B): 6.124, *iubwp*(C): 6.0, *iubwp*(D): 7.124, *iubwp*(E): 2.19, *iubwp*(F): 3.274. In the proposed algorithm, the E-tables and WP-tables are built in weight-descending order of 1-items. From Table 2, the weight-descending order of six items is $\{w(C) > w(E) > w(B) > w(D) > w(A) > w(F)\}$. The built E-tables for all 1-items are shown in Fig. 1. The constructed WP-tables for the $HUBEWI^1$ are shown in Fig. 2.

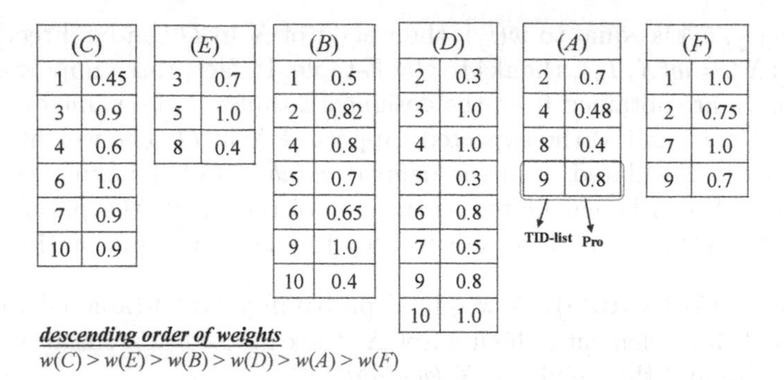

(C)	
1	0.45
3	0.9
4	0.6
6	1.0
7	0.9
10	0.9

(E)	
3	0.7
5	1.0
8	0.4

(B)	
1	0.5
2	0.82
4	0.8
5	0.7
6	0.65
9	1.0
10	0.4

(D)	
2	0.3
3	1.0
4	1.0
5	0.3
6	0.8
7	0.5
9	0.8
10	1.0

(A)	
2	0.7
4	0.48
8	0.4
9	0.8

TID-list Pro

(F)	
1	1.0
2	0.75
7	1.0
9	0.7

descending order of weights
$w(C) > w(E) > w(B) > w(D) > w(A) > w(F)$

Fig. 1. Constructed E-tables of 1-items.

WP-table		
itemset	*weight*	*expSup*
(A)	0.400	2.380

WP-table		
itemset	*weight*	*expSup*
(B)	0.700	4.870

WP-table		
itemset	*weight*	*expSup*
(C)	1.000	4.750

WP-table		
itemset	*weight*	*expSup*
(D)	0.550	5.700

WP-table		
itemset	*weight*	*expSup*
(E)	0.850	2.100

WP-table		
itemset	*weight*	*expSup*
(F)	0.300	3.450

Fig. 2. Constructed WP-tables of $HUBEWI^1$.

4.2 Search Space of the HEWI-Utree Algorithm: WP-tree

Definition 10. Assume that the total order $\prec$ on items in the designed weighted-probability (WP)-tree is the weight-descending order of 1-items.

Definition 11. The extensions of an itemset w.r.t. node X can be obtained by appending an item y to X such that y is greater than all items already in X according to the total order $\prec$.

The search space of the proposed HEWI-Utree algorithm can be represented by a WP-tree in the total order $\prec$, as shown in Fig. 3. Based on the WP-tree, the following lemmas are used to efficiently prune the search space and without performing additional database scans

Lemma 1. *The search space of the proposed HEWI-Utree algorithm can be represented by a WP-tree where items are sorted according to the weight-descending order of the discovered* $HUBEWI^1$.

Proof. According to the definition of a Set-enumeration tree [13], the complete search space of I (where m is the number of items in I) contains 2^m patterns, by systematically enumerating all subsets of I, w.r.t. all the possible patterns. As it can be seen in Fig. 3, either breadth-first or depth-first search may be used to traverse nodes of a WP-tree. Thus, all the supersets of the root node can be enumerated according to the weight-descending order of 1-items. Hence, the developed WP-tree can represent the complete search space of the proposed HEWI-Utree algorithm.

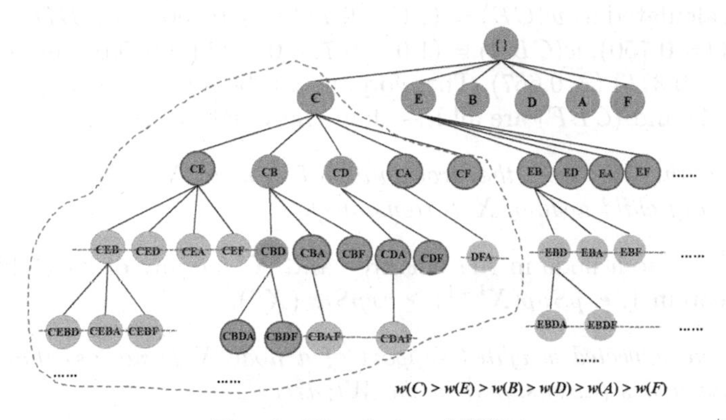

Fig. 3. The designed WP-tree.

Lemma 2. *In the WP-tree, the weight of a node is no less than any of its child nodes (extension nodes).*

Proof. Let X^k be a node (w.r.t k-itemset) in the WP-tree, as shown in Fig. 3, then X^{k-1} be its parent node and X^k be any of child nodes of X^{k-1}. With the same assumption, all the 1-items are sorted in the total order $\prec$ of item weights, i.e. $w(X_1) \geq w(X_2) \geq \cdots \geq w(X_k) > 0$. Since $w(X_1) \geq w(X_2)$;the weight of a 2-itemset can be calculated as:
$2 \times w(X_1) \geq w(X_1) + w(X_2) \Longrightarrow \dfrac{w(X_1)}{1} \geq \dfrac{w(X_1) + w(X_2)}{2}$;
Thus, the weight of a 3-itemset can be calculated as:
$w(X_1) \geq w(X_2) \geq w(X_3)$
$\Longrightarrow w(X_1) + w(X_2) \geq w(X_3) + w(X_3)$
$\Longrightarrow 2 \times (w(X_1) + w(X_2)) + w(X_1) + w(X_2) \geq 2 \times (w(X_1) + w(X_2)) + 2 \times w(X_3)$
$\Longrightarrow 3 \times (w(X_1) + w(X_2)) \geq 2 \times (w(X_1) + w(X_2) + w(X_3))$
$\Longrightarrow \dfrac{w(X_1) + w(X_2)}{2} \geq \dfrac{w(X_1) + w(X_2) + w(X_3)}{3}$.
The following relationships can thus be obtained: for each transaction T_q in D,
$w(X_1) \geq w(X_2) \geq \cdots \geq w(X_{k-1}) \geq w(X_k) > 0$
$\Longrightarrow (w(X_1) + w(X_2) + \cdots + w(X_{k-1})) \geq (k-1) \times w(X_k)$
$\Longrightarrow (k-1) \times (w(X_1) + w(X_2) + \cdots + w(X_{k-1})) + (w(X_1) + w(X_2) + \cdots + w(X_{k-1})) \geq (k-1) \times (w(X_1) + w(X_2) + \cdots + w(X_{k-1})) + (k-1) \times w(X_k)$
$\Longrightarrow (k) \times (w(X_1) + w(X_2) + \cdots + w(X_{k-1})) \geq (k-1) \times (w(X_1) + w(X_2) + \cdots + w(X_{k-1}) + w(X_k))$
$\Longrightarrow \dfrac{w(X_1) + \cdots + w(X_{k-1})}{k-1} \geq \dfrac{w(X_1) + \cdots + w(X_{k-1}) + w(X_k)}{k}$.
Thus, for each T_q in D, $w(X^k, T_q) \leq w(X^{k-1}, T_q)$. Based on definition 3, the lemma $w(X^k) \leq w(X^{k-1})$ holds.

For example, the WP-tree is sorted in weight-descending order of the discovered $HUBEWI^1$; the weight of (CB) is always no less than that of its child nodes (w.r.t. extensions) (CBD), (CBA) and (CBF) by Lemma 2, which are respectively calculated as $w(CB) = (1.0 + 0.7)/2$ (= 0.850), $w(CBD) = (1.0 + 0.7 + 0.55)/3$ (= 0.750), $w(CBA) = (1.0 + 0.7 + 0.4)/3$ (= 0.700), and $w(CBF) = (1.0 + 0.7 + 0.3)/3$ (= 0.667). Therefore, it can be found that the weights of (CBD), (CBA) and (CBF) are all less than the weight of (CB).

Lemma 3. *In the WP-tree, the probability of a node X is no less than the probability of any child node of X (extensions).*

Proof. Let X^{k-1} be a node in the WP-tree and X^k be any of its child nodes. Based on Theorem 1, $expSup(X^{k-1}) \geq expSup(X^k)$.

Lemma 4. *The expected weighted support of a node X is no less than that of any of its child nodes (extensions) in the WP-tree.*

Proof. Let X^k denotes any child node of a (k-1)-itemset X^{k-1}. Based on Lemmas 2 and 3, the two following conditions are obtained: (1) $w(X^{k-1}) \geq w(X^k)$; (2) $expSup(X^{k-1}) \geq expSup(X^k)$. Thus,

$$w(X^{k-1}) \times expSup(X^{k-1}) \geq w(X^k) \times expSup(X^k) \quad (5)$$

Therefore, $expWSup(X^{k-1}) \geq expWSup(X^k)$, this lemma holds.

Pruning Strategy. *If a node is not a HEWI, any of its child nodes is not a HEWI either. This strategy can be directly applied to eliminate unpromising nodes when a depth-first search is performed in the WP-tree.*

Rationale. According to Lemma 4, it can be obtained that if an itemset X is not a HEWI, any of its child nodes will not be a HEWI either. Thus, unpromising itemsets in the developed WP-tree, can be directly pruned by using the built WP-table of X.

Fig. 3 illustrates the visited nodes of the WP-tree for those itemsets having (C) as prefix. Notice that the traversed nodes are orange; the orange nodes with blue edges represents nodes can be avoided of exploration using the pruning strategy of the traversed node; gray nodes are unpromising nodes which were not explored. Hence, the designed pruning strategy is very powerful to early prune the unpromising itemsets and speed up computation.

4.3 Procedure of HEWI-Utree Algorithm

As shown in the main procedure of HEWI-Utree (Algorithm 2), the proposed HEWI-Utree first initially sets $D.ET$, $D.WPT$, $i.ET$, $i.WPT$ as an empty set (Line 1), then scans the database to calculate the $iubwp(i)$ value of each item $i \in I$ (Line 2). The set of I^* w.r.t. $HUBEWI^1$ (Line 3) are discovered. The discovered $HUBEWI^1$ are then sorted in weight-descending order to construct their E-tables by scanning database again, thus forming the final set of WP-tables as $D.WPT$ (Lines 4 to 7). Afterwards each 1-itemset X in the set of $HUBEWI^1$ is processed in weight-descending order to find HEWIs from the WP-tree, using the constructed WP-tables without rescanning the database (Line 8, HEWI-Search procedure).

Input: $D(n = |D|)$, *w-table*, ε
Output: The set of high expected weighted itemsets (HEWIs).
1 $D.ET \leftarrow \emptyset, D.WPT \leftarrow \emptyset, i.ET \leftarrow \emptyset, i.WPT \leftarrow \emptyset$;
2 scan D to calculate the *iubwp* value of each single item;
3 find $I^* \leftarrow \{i \in I | iubwp(i) \geq \varepsilon \times |D|\}$, w.r.t. $HUBEWI^1$;
4 sort I^* in the designed total order $\prec$ (weight-descending order);
5 scan D once to build the EI-table ($i.ET$) for each 1-item $i \in I^*$;
6 $D.ET \leftarrow \bigcup i.ET$;
7 $i.WPT \leftarrow i.ET$, and put it into the set of $D.WPT$;
8 **call HEWI-Search**(ϕ, I^*, ε);
9 return $HEWIs$.

Algorithm 2. HEWI-Utree algorithm

As shown in Algorithm 3, the main idea of the **HEWI-Search** procedure is that for each 1-itemset X, extensions of X w.r.t. $extenETOfX$ are recursively explored using a depth-first search. Moreover, each itemset encountered during that search is evaluated to determine if it is a HEWIs (Lines 2 to 5). Notice that the depth search is only performed for an itemset X if the $expWSup$ of X (which is directly obtained from the relevant WP-tables by calculating its weight and probabilities) is larger than or equal to $\varepsilon \times |D|$ (Lines 3 to 4). Simultaneously, the E-table construction procedure is used to construct a series of extensions of X (Lines 7 to 10). The above process is recursively executed until all the 1-itemsets in $HUBEWI^1$ have been processed. After that, the final set of HEWIs has been discovered. The algorithm has "constructing during mining" property; the HEWIs are discovered by exploiting the constructed WP-tables. The time-consuming database scans can be thus avoided (Line 11).

Input: X, an itemset; $extenETOfX$, a set of E-tables of all 1-extensions of X; ε.
Output: The set of HEWIs.
1 **for** *each itemset* $X_a \in extenETOfX$ **do**
2 obtain the $w(X_a)$ and $expSup(X_a)$ values from the built $Xa.WPT$;
3 calculate $expWSup(X_a) := w(X_a) \times expSup(X_a)$;
4 **if** $expWSup(X_a) \geq \varepsilon \times |D|$ **then**
5   $HEWIs \leftarrow HEWIs \cup X_a$;
6   $extenETOfX_a \leftarrow \emptyset$;
7   **for** *each itemset* X_b *after* X_a *in* $extenETOfX$ **do**
8    $X_{ab} \leftarrow X_a \cup X_b$;
9    $X_{ab}.ET \leftarrow construct(X, X_a, X_b)$;
10    $extenETOfX_a \leftarrow extenETOfX_a \cup X_{ab}.ET$;
11   **call HEWI-Search**$(X_a, extenETOfX_a, \varepsilon)$;
12 **return** $HEWIs$;

Algorithm 3. HEWI-Search procedure

Lemma 5. *The HEWI-Utree algorithm is a correct and complete algorithm to discover the set of HEWIs.*

Proof. Given an uncertain database, the developed HEWI-Utree algorithm can extract the set $HUBEWI^1$. The E-tables of these itemsets can be used to recursively build the E-tables and WP-tables of any itemset in the WP-tree using the proposed construction process. Based on Lemmas 2, 3, and 4, the HEWI-Utree algorithm can ensure that no HEWIs will be discarded (**completeness**), and that the information about each remaining itemset can be exactly obtained using the WP-table (**correctness**). Therefore, the algorithm is correct and complete for discovering the HEWIs from the WP-tree.

5 Experiments

We performed extensive experiments to evaluate the proposed algorithm. Note that the HEWI-Uapriori algorithm [10] (for mining high expected weighted frequent itemsets in an uncertain dataset), the existing Uapriori algorithm [7] (for mining expected support frequent itemsets in an uncertain dataset) and the existing PWA algorithm [11] (for mining weighted frequent itemsets in a precise dataset) are used as baseline algorithms to evaluate the improvements afforded by the proposed HEWI-Utree algorithm. In order to perform a fair comparison, all algorithms used in the experiments are also implemented in Java. Experiments were performed on a computer having an Intel Core 2 Duo 2.8 GHz processor with 4 GB of main memory, running the 32 bit Microsoft Windows 7 operating system. Besides, the same four datasets in [10], including both real-world datasets (foodmart, retail, mushroom, the foodmart dataset was acquired from Microsoft foodmart 2000) [14] and synthetic dataset (T10I4D100K) [3], are conducted on our experiments. We assessed execution time, number of patterns and scalability. The weight of each item was randomly selected in (0,1].

In the experiments, the size of the test dataset is fixed and the user-specified minimum expected weighted-support threshold $minEWSup$ is varied. Notice that here $minEWSup$ has different meanings for the different compared algorithms. It refers to the minimum expected support threshold ($minESup$) as used in Uapriori algorithm; the minimum weighted-support threshold ($minWSup$) as used in PWA algorithm; and the minimum expected weighted-support threshold (ε) as used in HEWI-Uapriori algorithm and the proposed HEWI-Utree algorithm.

5.1 Execution Time

To evaluate the efficiency of the proposed algorithm, a runtime analysis is first described and shown in Fig. 4.

It can be observed that the proposed HEWI-Utree algorithm always considerably outperforms the state-of-the-art HEWI-Uapriori algorithm in terms of runtime. Moreover, HEWI-Utree is the fastest algorithm among the compared algorithms for the four datasets, and is generally about one to two orders

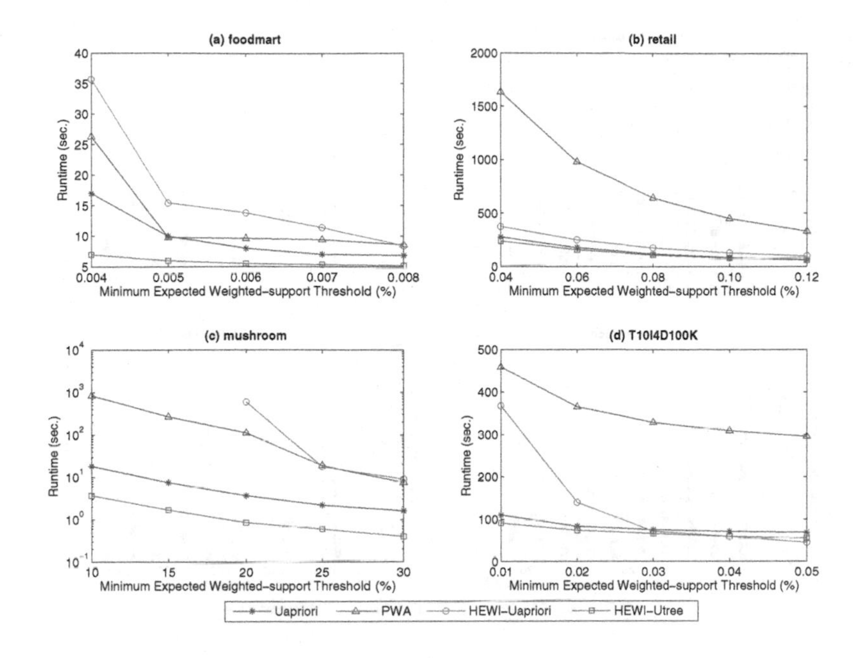

Fig. 4. Runtime performance.

of magnitude faster than HEWI-Uapriori. It is reasonable since HEWI-Utree finds HEWIs using the proposed E-table and WP-table structures, which are constructed using only two database scans to reduce runtime. Second, the necessary information for discovering HEWIs can be easily obtained from the built E-tables and WP-tables. These structures allow HEWI-Utree to directly mine HEWIs without generate-and-test mechanism for generating a large amount of candidates. Third, HEWI-Utree also considers both the probability and weight constraints; thus it significantly outperforms the Uapriori and PWA algorithms. Moreover, it can be observed that the HEWI-Uapriori algorithm is slower than the Uapriori algorithm, but it is faster than the projection-based weighted frequent itemset mining algorithm PWA in some cases. HEWI-Uapriori may spend additional time compared to Uapriori, which only considers data uncertainty.

5.2 Memory Usage

In this section, the peak memory usage are compared under the same parameters as before. Fig. 5 shows that the proposed HEWI-Utree algorithm always requires less memory than HEWI-Uapriori, as well as the other compared algorithms in four datasets. In particular, the HEWI-Utree algorithm requires nearly constant memory under various $minEWSup$, while the HEWI-Uapriori algorithm requires the most memory among the compared algorithms in four datasets. Only extensions of a HUBEWI are processed by using a depth-first search. In other

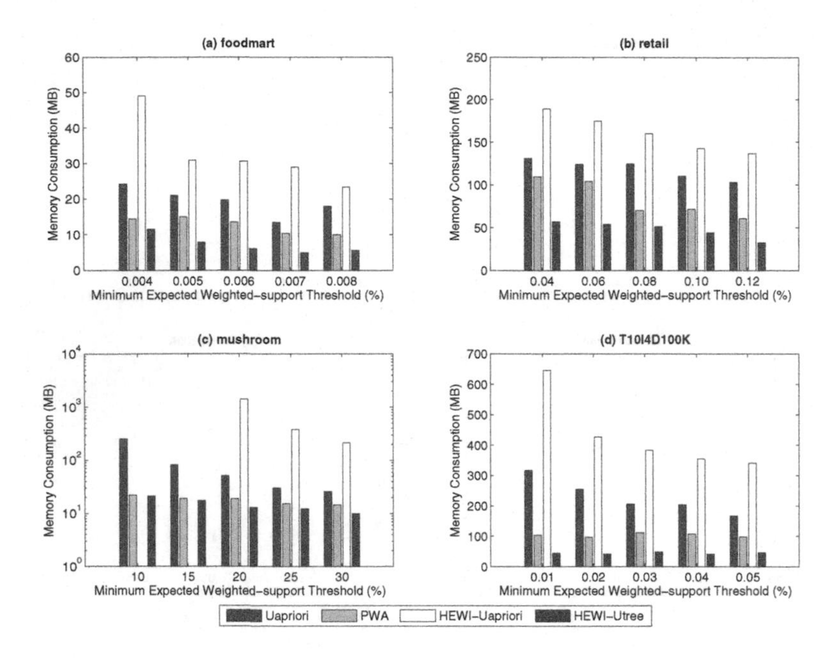

Fig. 5. Memory usage performance.

words, the designed E-table and WP-table data structures are not constructed for unpromising itemsets. Therefore, the E-tables and WP-tables constructed from the databases are very small and the intersection operation of E-tables can be quickly performed by HEWI-Utree. Hence, HEWI-Utree considerably outperforms HEWI-Uapriori. Moreover, the performance gap between the HEWI-Utree and HEWI-Uapriori algorithms becomes larger as $minEWSup$ decreases, but the memory usage of HEWI-Utree does not considerably change when $minEWSup$ is varied.

5.3 Scalability Analysis

Fig. 6 shows the scalability of the compared algorithms under varied dataset sizes T10I4N4KD$|X|K$, from 100K to 500K, increments of 100K each time. The runtime, memory usage, and the number of derived patterns (EFIs, WFIs, and HEWIs) were evaluated when $minEWSup$ is set to 0.03%. It can be seen that both the HEWI-Uapriori and HEWI-Utree algorithms have good scalability in terms of runtime and memory usage, and that the HEWI-Utree algorithm performs better than the others. The runtimes of all compared algorithms significantly increase when the dataset size is increased, and these results showed that HEWI-Utree is much more scalable than HEWI-Uapriori. In particular, the larger dataset size is, the bigger difference between HEWI-Utree and the compared algorithms is, in terms of runtime and memory usage. The aforementioned

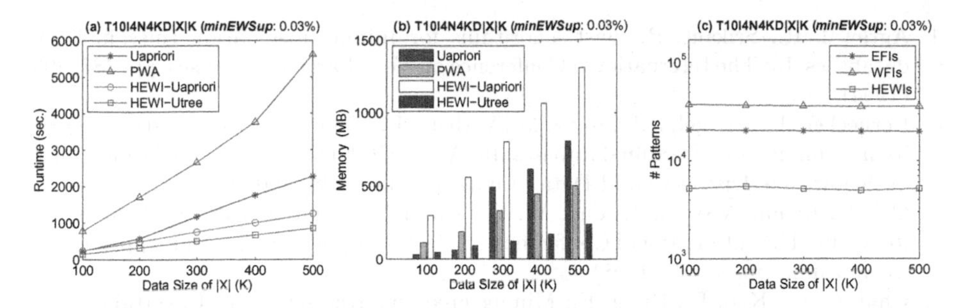

Fig. 6. Scalability performance.

reasons can also be used to explain the results in terms of number of patterns depicted in Fig. 6(c). In summary, we can conclude that the proposed HEWI-Utree algorithm is more scalable for large datasets and it runs much faster and consumes less memory than the compared algorithms. The above experimental results show that the effectiveness and efficiency of the HEWI-Utree algorithm is acceptable for real-world applications.

6 Conclusion

In this paper, an efficient algorithm named HEWI-Utree is proposed. It relies on three novel structures: the element-table, the weighted-probability (WP)-table and the WP-tree. These structures allow HEWI-Utree to directly mine the HEWIs without performing multiple database scans. A pruning strategy is also developed to effectively filter unpromising patterns early, thus reducing execution time compared to the HEWI-Uapriori algorithm. An extensive experimental study shows that the proposed algorithm significantly outperforms the HEWI-Uapriori algorithm in terms of execution time, memory usage, and scalability.

Acknowledgment. This research was partially supported by National Natural Science Foundation of China (NSFC) under grant no.61503092, and by the Tencent Project under grant CCF-TencentRAGR20140114.

References

1. Aggarwal, C.C., Li, Y., Wang, J., Wang, J.: Frequent pattern mining with uncertain data. In: ACM SIGKDD International Conference on Knowledge Discovery and Data Mining, pp. 29–38 (2009)
2. Aggarwal, C.C., Yu, P.S.: A survey of uncertain data algorithms and applications. IEEE Transactions on Knowledge and Data Engineering. **21**(5), 609–623 (2009)
3. Agrawal, R., Srikant, R.: Quest synthetic data generator. http://www.Almaden.ibm.com/cs/quest/syndata.html

4. Agrawal, R., Srikant, R.: Fast algorithms for mining association rules in large databases. In: The International Conference on Very Large Data Bases, pp. 487–499 (1994)
5. Bernecker, T., Kriegel, H.P., Renz, M., Verhein, F., Zuefl, A.: Probabilistic frequent itemset mining in uncertain databases. In: ACM SIGKDD International Conference on Knowledge Discovery and Data Mining, pp. 119–128 (2009)
6. Cai, C.H., Fu, A.W.C., Kwong, W.W.: Mining association rules with weighted items. In: The International Conference on Database Engineering and Applications Symposium, pp. 68–77 (1998)
7. Chui, C.-K., Kao, B., Hung, E.: Mining Frequent Itemsets from Uncertain Data. In: Zhou, Z.-H., Li, H., Yang, Q. (eds.) PAKDD 2007. LNCS (LNAI), vol. 4426, pp. 47–58. Springer, Heidelberg (2007)
8. Han, J., Pei, J., Yin, Y., Mao, R.: Mining frequent patterns without candidate generation: A frequent-pattern tree approach. Data Mining and Knowledge Discovery. **8**(1), 53–97 (2004)
9. Lin, J.C.W., Gan, W., Fournier-Viger, P., Hong, T.P.: RWFIM: Recent weighted-frequent itemsets mining. Engineering Applications of Artificial Intelligence. **45**, 18–32 (2015)
10. Lin, J.C.W., Gan, W., Fournier-Viger, P., Hong, T.P., Tseng, V.S.: Weighted frequent itemset mining over uncertain databases. Applied Intelligence. **44**(1), 166–178 (2016)
11. Lan, G.C., Hong, T.P., Lee, H.Y., Lin, C.W.: Mining weighted frequent itemsets. The Workshop on Combinatorial Mathematics and Computation Theory, pp. 85–89 (2013)
12. Lan, G.C., Hong, T.P., Lee, H.Y.: An efficient approach for finding weighted sequential patterns from sequence databases. Applied Intelligence. **41**, 439–452 (2014)
13. Rymon, R.: Search through systematic set enumeration. In: The International Conference on Principles of Knowledge Representation and Reasoning, pp. 539–550 (1992)
14. SPMF: A Java Open-Source Data Mining Library. http://www.philippe-fournier-viger.com/spmf/
15. Sun, L., Cheng, R., Cheung, D.W., Cheng, J.: Mining uncertain data with probabilistic guarantees. In: ACM SIGKDD International Conference on Knowledge Discovery and Data Mining, pp. 273–282 (2010)
16. Sun, K., Bai, F.: Mining weighted association rules without preassigned weights. IEEE Transactions on Knowledge and Data Engineering. **20**, 489–495 (2008)
17. Tao, F., Murtagh, F., Farid, M.: Weighted association rule mining using weighted support and significance framework. In: ACM SIGKDD International Conference on Knowledge Discovery and Data Mining, pp. 661–666 (2003)
18. Vo, B., Coenen, F., Le, B.: A new method for mining frequent weighted itemsets based on wit-trees. Expert Systems with Applications. **40**, 1256–1264 (2013)
19. Wang, W., Yang, J., Yu, P.S.: Efficient mining of weighted association rules (WAR). In: ACM SIGKDD International Conference on Knowledge Discovery and Data Mining, pp. 270–274 (2000)
20. Yun, U., Leggett, J.: WFIM: Weighted frequent itemset mining with a weight range and a minimum weight. In: SIAM International Conference on Data Mining, pp. 636–640 (2005)

Automatic Detection of Latent Common Clusters of Groups in MultiGroup Regression

Minhazul Islam Sk(✉) and Arunava Banerjee

University of Florida, Gainesville, Florida, USA
smislam@cise.ufl.edu

Abstract. We present a flexible non-parametric generative model for multigroup regression that detects latent common clusters of groups. The model is founded on techniques that are now considered standard in the statistical parameter estimation literature, namely, Dirichlet process(DP) and Generalized Linear Model (GLM), and therefore, we name it "Infinite MultiGroup Generalized Linear Models" (iMG-GLM). We present two versions of the core model. First, in iMG-GLM-1, we demonstrate how the use of a DP prior on the groups while modeling the response-covariate densities via GLM, allows the model to capture latent clusters of groups by noting similar densities. The model ensures different densities for different clusters of groups in the multigroup setting. Secondly, in iMG-GLM-2, we model the posterior density of a new group using the latent densities of the clusters inferred from previous groups as prior. This spares the model from needing to memorize the entire data of previous groups. The posterior inference for iMG-GLM-1 is done using Variational Inference and that for iMG-GLM-2 using a simple Metropolis Hastings Algorithm. We demonstrate iMG-GLM's superior accuracy in comparison to well known competing methods like Generalized Linear Mixed Model (GLMM), Random Forest, Linear Regression etc. on two real world problems.

1 Introduction

Multigroup Regression is the method of choice for research design whenever response-covariate data is collected across multiple groups. When a common regressor is learned on the amalgamated data, the resultant model fails to identify effects for the responses specific to individual groups because the underlying assumption is that the response-covariate pairs are drawn from a single global distribution, when the reality might be that the groups are not statistically identical, making the joining of them inappropriate. Modeling separate groups via separate regressors results in a model that is devoid of common latent effects across the groups. Such a model does not exploit the patterns common among the groups ensuring in turn the transferability of information among groups in the regression setting. This is of particular importance when the training set is very small for many of the groups. Joint learning, by sharing knowledge between the statistically similar groups, strengthens the model for each group, and the resulting generalization in the regression setting is vastly improved.

P. Perner (Ed.): MLDM 2016, LNAI 9729, pp. 251–266, 2016.
DOI: 10.1007/978-3-319-41920-6_19

The complexities that underlie the utilization of the information transfer between the groups are best motivated through examples. In Clinical Trials, for example, a group of people are prescribed either a new drug or a placebo to estimate the efficacy of the drug for the treatment of a certain disease. At a population level, this efficacy may be modeled using a single Normal or Poisson mixed model distribution with mean set as a (linear or otherwise) function of the covariates of the individuals in the population. A closer inspection might however disclose potential factors that explain the efficacy results better. For example, there might be regularities at the group level—Caucasians as a whole might react differently to the drug than, say, Asians, who might, furthermore, comprise many groups. Identifying this across group information would therefore improve the accuracy of the regressor. Similarly in the Stock Market, future values and trends for a group of stocks are predicted for various sectors such as Energy, Materials, Consumer discretionary, Financials, Telecomm., Technology, etc. Within each sector, various stocks share trends and therefore predicting them together (modeling them with the same time series via autoregressive density) is usually much more accurate than predicting and capturing individual trends. Modeling the latent common clustering effects of cross-cutting subgroups is therefore an important problem to solve. We present a framework here that accomplishes this.

We begin with a brief description of the weaknesses of the most popular multilevel regression techniques, namely, Generalized Linear Models [19] and Mixed model [7]. In regression theory, Generalized Linear Model (GLM), proposed in [19], brings erstwhile disparate techniques such as, Linear regression, Logistic regression, and Poisson regression, under a unified framework. GLM is formally defined as:

$$f(y;\theta,\psi) = exp\left\{\frac{y\theta - b(\theta)}{a(\psi)} + c(y;\psi)\right\} \tag{1}$$

Here, ψ is a dispersion parameter. exp denotes the exponential family density. The mean response is $E[Y|\mathbf{X}] = b(\theta) = \mu = g^{-1}\left(X^T\beta\right)$, where g is the link function, $X^T\beta$ is the linear predictor. For multigroup regression, Generalized Linear Mixed Model (GLMM) [7] and Hierarchical Generalized Linear Mixed Model [14] have been developed where similarities between groups is captured though a Fixed effect and variation across groups is captured through random effects. Statistically, these models are very rigid since every group is forced to manifest the same fixed effect, while the random effect only represents the intercept parameter of the linear predictors. Cluster of groups may have significantly different properties from other clusters of groups, a feature that is not captured in these traditional GLM based models. Furthermore, various clusters of groups may have different uncertainties with respect to the covariates which we denote as heteroscedasticity. In recent progress, [3] has proposed a Bayesian Hierarchical model, where a prior is used for the mixture of groups. Nevertheless, individual groups are given weights as opposed to jointly learning various groups. Also, the number of mixtures are fixed in advance.

Before, presenting our algorithm, we describe our basis for identifying group-correlation. First, two groups are correlated if their responses follow the same distribution. Second, two groups that have the same response variance with respect to the covariates are deemed to be correlated. This is achieved via a Dirichlet Process prior on the groups and the covariate co-efficients (β). The posterior is obtained by appropriately combining the prior and the data likelihood from the given groups. The prior helps cluster the groups and the likelihood from the individual groups help in the sharing of trends between groups to create the single posterior density between the many potential groups, thereby leading to group-correlation.

We now present an overview of our iMG-GLM framework. Our objective is to achieve (a) shared learning of various groups in a regression setting, where data may vary in terms of temporal, geographical or other modalities and (b) automatic clustering of groups which display correlation. iMG-GLM-1 solves this task. In iMG-GLM-2, we model a completely new group after modeling previous groups through parameters learned in iMG-GLM-1. In the first part, the regression parameters are given a Dirichlet Process prior, that is, they are drawn from a DP with the base distributions set as the density of the regression parameters. Since a draw from a DP is an atomic density, to begin, one group will be assigned one density of the regression parameters which signifies the response density with respect to its covariates. As the drawn probability weight from the DP increases, the cluster starts to consume more and more groups in this mutigroup setting. We employ a variational Bayes algorithm for the inference procedure in iMG-GLM-1 for computational efficiency. iMG-GLM-1 is then extended to iMG-GLM-2 for modeling a completely new group. Here we transfer the information (covariate coefficients) obtained in the first part, to learning a new group. In essence, the cluster parameters (covariate coefficients for the whole group) are used as a prior distribution for the model parameters of the new group's response density. This therefore leads to a mixture model where the weights are given by the number of groups that one cluster consumed in the first part and the mixture components are the regression parameters obtained for that specific cluster. The likelihood comes from the data of the new group. We use a simple accept-reject based Metropolis Hastings algorithm to generate samples from the posterior for the new group regression parameter density. For both iMG-GLM-1 and iMG-GLM-2, we use Monte Carlo integration for evaluating the predictive density of the new test samples.

We evaluate both iMG-GLM-1 and iMG-GLM-2 Normal models in two real world problems. The first is the prediction and finding of trends in the Stock Market. We show how information transfer between groups help our model to effectively predict future stock values by varying the number of training samples in both previous and new groups. In the second, we show the efficacy of i-MG-GLM-1 and 2 Poisson model against its competitors in a very important Clinical Trial Problem Setting.

2 Mathematical Background

2.1 Models Related to iMG-GLM

After its introduction, Generalized Linear Model was extended to Hierarchical Generalized Linear Model (HGLM) [14]. Then it included structured dispersion in [15] and models for spatio-temporal co-relation in [16]. Generalized Linear Mixed Models (GLMMs) were proposed in [7]. The random effects in HGLM were specified by both mean and dispersion in [17]. Mixture of Linear Regression was proposed in [22]. Hierarchical Mixture of Regression was done in [13]. Varying co-efficient models were proposed in [11]. Multi-tasking Model for classification in Non-parametric Bayesian scenario was introduced in [23]. Sharing Hidden Nodes in Neural Networks was introduced in [4,5]. General Multi-Task learning was described first in [8]. Common prior in hierarchical Bayesian model was used in [24,25]. Common structure sharing in the predictor space was presented in [1]. All of these models suffer the shortcomings of not identifying the latent clustering effect across groups as well as varying uncertainty with respect to covariates across groups, which the iMG-GLM inherently models.

2.2 Dirichlet Process and its Stick-Breaking Representation

A Dirichlet Process [10], $D(\alpha, G_0)$ is defined as a probability distribution over a sample space of probability distributions, $G \sim DP(\alpha, G_0)$ and $\eta_j | G \sim G$. Here, α is the concentration parameter and G_0 is the base distribution.

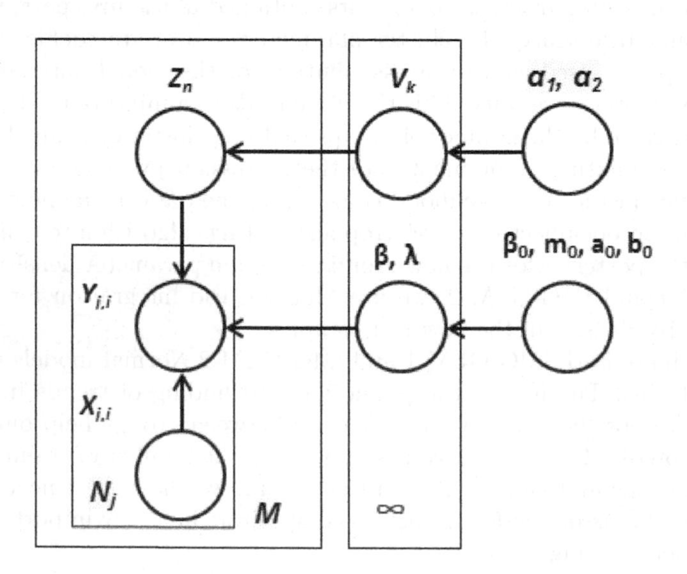

Fig. 1. Graphical Representation of iMG-GLM-1 Model.

When we integrate over G, the conditional density of η_j, given previous $\eta_{1:j-1}$ is given by the *Chinese Restaurant process* [2]. $\eta_j|\theta_{1:j-1}, \alpha, G_0 \sim \frac{\alpha}{\alpha+j-1}G_0 + \frac{1}{\alpha+j-1}\sum_{k=1}^{j-1} n_{-j,k}\delta_{\eta_k^*}$. Here, $n_{-m,k}$ denotes the number of η's equal to η_k^* (From K distinct values) excluding η_j.

According to the stick-breaking construction [21] of DP, G, which is a sample from DP, is an atomic distribution with countably infinite atoms drawn from G_0.

$$\begin{aligned} &v_k|\alpha, G_0 \sim Beta(1, \alpha), \quad \theta_k|\alpha, G_0 \sim G_0, \\ &\pi_i = v_k \prod_{p=1}^{k-1}(1 - v_p), \quad G = \sum_{k=1}^{\infty} \pi_k.\delta_{\theta_k} \end{aligned} \tag{2}$$

In the DP mixture model [9], DP is used as a non-parametric prior over parameters of an Infinite Mixture model.

$$\begin{aligned} &z_n|\{v_1, v_2, ...\} \sim Categorical\{\Pi_1, \Pi_2, \Pi_3....\}, \\ &X_n|z_n, (\theta_k)_{i=1}^{\infty} \sim F(\theta_{z_n}) \end{aligned} \tag{3}$$

Here, F is a distribution parametrized by θ_{z_n}. $\{\pi_1, \pi_2, \pi_3, ...\}$ is defined by Eq. 2.2.

3 iMG-GLM Model Formulation

We consider M groups indexed by $j = 1, ..., M$ and the complete data as $\mathcal{D} = \{x_{j,i}, y_{j,i}\}$ s.t. $i = 1, ...N_j$. $\{x_{j,i}, y_{j,i}\}$ are covariate-response pairs and are drawn i.i.d. from an underlying density which differs along with the nature of $\{x_{j,i}, y_{j,i}\}$ among various models.

3.1 Normal iMG-GLM-1 Model

In the Normal iMG-GLM-1 model, the generative model of the covariate-response pair is given by the following set of equations. Here, X_{ji} and Y_{ji} represent the ith continuous covariate-response pairs of the jth group. The distribution of $Y_{j,i}|X_{j,i}$ is normal parametrized by $\beta_{0:D}$ and λ. The distribution, $\{\beta_{kd}, \lambda_k\}$ (Normal-Gamma) is the prior distribution on the covariate coefficient β. This distribution is the base distribution (G) of the Dirichlet Process. The set $\{m_0, \beta_0, a_0, b_0\}$ constitute the hyper-parameters for the covariate coefficients (β) distribution. The graphical representation of the normal model is given in Figure 1.

$$\begin{aligned} &v_k \sim Beta(\alpha_1, \alpha_2), \quad \pi_k = v_k \Pi_{n=1}^{k-1}(1 - v_n) \\ &\mathrm{N}\left(\beta_{kd}|m_0, (\beta_0, \lambda_k)^{-1}\right) \mathrm{Gamma}\left(\lambda_k|a_0, b_0\right) \\ &Z_j|v_k \sim Categorical\left(\pi_1, ...\pi_\infty\right) \\ &Y_{ji}|X_{ji} \sim \mathcal{N}\left(Y_{ji}|\sum_{d=0}^{D}\beta_{Z_j d}X_{jid}, \lambda_{Z_j}^{-1}\right) \end{aligned} \tag{4}$$

3.2 Logistic Multinomial iMG-GLM-1 Model

In the Logistic Multinomial iMG-GLM-1 model, a Multinomial Logistic Framework is used for a Categorical response, Y_{ji}, for a continuous covariate, X_{ji}, in the case of ith data point of the jth group. t is the index of the category. The distribution of $Y_{j,i}|X_{j,i}$ is Categorical parametrized by $\beta_{0:D,0:T}$. The distribution, $\{\beta_{ktd}\}$ (Normal) is the prior distribution on the covariate coefficient β which is the base distribution (G) of the Dirichlet Process. The set $\{m_0, s_0\}$ constitute the hyper-parameters for the covariate coefficients (β) distribution.

$$\begin{aligned} &v_k \sim Beta(\alpha_1, \alpha_2), \quad \pi_k = v_k \Pi_{n=1}^{k-1}(1 - v_n) \\ &\beta_{ktd} \sim \mathcal{N}\left(\beta_{ktd}|m_0, s_0^2\right), \quad Z_j|v_k \sim Categorical\,(\pi_1, ...\pi_\infty) \\ &Y_{ji} = t|X_{ji}, Z_j \sim \frac{\exp\left(\sum_{d=0}^{D} \beta_{Z_j td} X_{jid}\right)}{\sum_{t=1}^{T} \exp\left(\sum_{d=0}^{D} \beta_{Z_j td} X_{jid}\right)} \end{aligned} \tag{5}$$

3.3 Poisson iMG-GLM-1 Model

In the Poisson iMG-GLM model, a Poisson distribution is used for the count response. Here, X_{ji} and Y_{ji} represent the ith continuous/ordinal covariate and categorical response pair of the jth group. The distribution of $Y_{j,i}|X_{j,i}$ is Poisson parametrized by $\beta_{0:D,0:T}$. The distribution, $\{\beta_{kd}\}$ (Normal) is the prior distribution on the covariate coefficient β which is the base distribution (G) of the Dirichlet Process. The set $\{m_0, s_0\}$ constitute the hyper-parameters for the covariate coefficients (β) distribution.

$$\begin{aligned} &v_k \sim Beta(\alpha_1, \alpha_2), \quad \pi_k = v_k \Pi_{n=1}^{k-1}(1 - v_n), \\ &\{\beta_{k,d}\} \sim \mathcal{N}\left(\beta_{kd}|m_0, s_0^2\right) \\ &Y_{ji}|X_{ji}, Z_j \sim Poisson\left(y_{ji}|\exp\left(\sum_{d=0}^{D} \beta_{Z_j d} X_{jid}\right)\right) \end{aligned} \tag{6}$$

4 Variational Inference

The inter-coupling between Y_{ji}, X_{ji} and z_j in all three models described above makes computing the posterior of the latent parameters analytically intractable. We therefore introduce the following fully factorized and decoupled variational distributions as surrogates.

4.1 Normal iMG-GLM-1 Model

The variational distribution for the Normal model is defined formally as:

$$\begin{aligned} &q\,(z, v, \beta_{kd}, \lambda_k) = \prod_{k=1}^{K} Beta\left(v_k|\gamma_k^1, \gamma_k^2\right) \prod_{j=1}^{M} Multinomial\,(z_j|\phi_j) \\ &\prod_{k=1}^{K} \prod_{d=0}^{D} \mathcal{N}\left(\beta_{kd}|m_{kd}, (\beta_k, \lambda_k)^{-1}\right) \text{Gamma}\,(\lambda_k|a_k, b_k) \end{aligned} \tag{7}$$

Firstly, each v_k follows a Beta distribution. As in [6], we have truncated the infinite series of $v_k^{'}$s into a finite one by making the assumption $p\,(v_K = 1) = 1$ and $\pi_k = 0 \forall k > K$. Note that this truncation applies to the variational surrogate distribution and *not* the actual posterior distribution that we approximate. Secondly, z_j follows a variational multinomial distribution. Thirdly, $\{\beta_{kd}, \lambda_k\}$ follows a Normal-Gamma distribution.

4.2 Logistic Multinomial iMG-GLM-1 Model

The variational distribution for the Logistic Multinomial model is given by:

$$q(z, v, \beta_{kd}, \lambda_k) = \prod_{k=1}^{K} Beta\left(v_k|\gamma_k^1, \gamma_k^2\right) \prod_{j=1}^{M} Multinomial\left(z_j|\phi_j\right) \\ \prod_{k=1}^{K} \prod_{t=1}^{T} \prod_{d=0}^{D} \left\{\mathcal{N}\left(\beta_{ktd}|m_{ktd}, s_{ktd}^2\right)\right\} \tag{8}$$

Here, v_k and z_j represent the same distributions as described in the Normal iMG-GLM-1 model above. $\{\beta_{ktd}\}$ follows a variational Normal Model.

4.3 Poisson iMG-GLM-1 Model

The variational distribution for the Poisson iMG-GLM-1 model is given by:

$$q(z, v, \beta_{kd}, \lambda_k) = \prod_{k=1}^{K} Beta\left(v_k|\gamma_k^1, \gamma_k^2\right) \\ \prod_{j=1}^{M} Multinomial\left(z_j|\phi_j\right) \prod_{k=1}^{K} \prod_{d=0}^{D} \left\{\mathcal{N}\left(\beta_{ktd}|m_{ktd}, s_{ktd}^2\right)\right\} \tag{9}$$

Here, v_k and z_j represent the same distributions as described in the Normal iMG-GLM-1 model above. $\{\beta_{kd}\}$ follows a variational Normal Model.

5 Parameter Estimation for Variational Distribution

We bound the log likelihood of the observations in the generalized form of iMG-GLM-1 (same for all the models) using Jensen's inequality, $\phi(E[X]) \geq \mathrm{E}[\phi(X)]$, where, ϕ is a concave function and X is a random variable. In this section, we differentiate the individually derived bounds with respect to the variational parameters of the specific models to obtain their respective estimates.

5.1 Parameter Estimation of iMG-GLM-1 Normal Model

The parameter estimation of the Normal Model is as follows:

$$\begin{aligned}
&\gamma_k^1 = 1 + \sum_{i=1}^{M} \phi_{ik}, \quad \gamma_k^2 = \alpha + \sum_{i=1}^{M} \sum_{p=k+1}^{K} \phi_{n,p} \\
&\phi_{jk} = \frac{exp(S_{jk})}{\sum_{k=1}^{K} exp(S_{jk})} \quad s.t. \\
&S_{jk} = \sum_{j=1}^{k} \left\{\Psi\left(\gamma_j^1\right) - \Psi\left(\gamma_j^1 + \gamma_j^2\right)\right\} + P_{jk} \quad s.t. \\
&P_{jk} = \frac{1}{2} \sum_{j=1}^{M} \sum_{i=1}^{N_j} \phi_{jk} \{log\left(\tfrac{1}{2\pi}\right) + \Psi(a_k) - log(b_k) \\
&-\beta_k \left(1 + \sum_{d=1}^{D} X_{jid}^2\right) - \tfrac{a_k}{b_k}\left(Y_{ji} - m_{k0} - \sum_{d=1}^{D} m_{kd} X_{jid}\right)^2\} \\
&\beta_k = \frac{(D+1)\beta_0 + \sum_{j=1}^{M} \sum_{i=1}^{N_j} \phi_{jk}\left(1 + \sum_{d=1}^{D} X_{jid}^2\right)}{D+1} \\
&a_k = \sum_{d=0}^{D} a_0 + \tfrac{1}{2} \sum_{j=1}^{M} \sum_{i=1}^{N_j} \phi_{jk} \\
&b_k = \tfrac{1}{2}\{\sum_{d=0}^{D} \beta_0 (m_{kd} - m_0)^2 + 2b_0 \\
&+ \sum_{j=1}^{M} \sum_{i=1}^{N_j} \phi_{jk} \left(Y_{ji} - m_{k0} - \sum_{d=1}^{D} m_{kd} X_{jid}\right)^2\} \\
&m_{k0} = \frac{m_0\beta_0 + \sum_{j=1}^{M} \sum_{i=1}^{N_j} \phi_{ji}\left(Y_{ji} - \sum_{d=1}^{D} m_{kd} X_{jid}\right)}{\beta_0 + \sum_{j=1}^{M} \sum_{i=1}^{N_j} \phi_{jk}} \\
&m_{kd} = \frac{m_0\beta_0 + \sum_{j=1}^{M} \sum_{i=1}^{N_j} \phi_{ji}\left(Y_{ji} - m_{k0} - \sum_{d=1}^{D-(d)} m_{kd} X_{jid}\right) X_{jid}}{\beta_0 + \sum_{j=1}^{M} \sum_{i=1}^{N_j} \phi_{jk} X_{jid}^2}
\end{aligned} \tag{10}$$

5.2 Parameter Estimation of iMG-GLM-1 Multinomial Model

For the Logistic Multinomial Model, the estimation of $\gamma_i^1, \gamma_i^2, \phi_{jk}$ and are identical to the Normal model with the only difference being that P_{jk} is given as,

$$
\begin{aligned}
&P_{jk} = \tfrac{1}{2} \textstyle\sum_{j=1}^{M} \sum_{i=1}^{N_j} \phi_{jk} \{log\left(\tfrac{1}{2\pi}\right) + \\
&\textstyle\sum_{t=1}^{T} Y_{jit} \left(m_{k0t} + \sum_{d=1}^{D} X_{jid} m_{kdt}\right) \\
&m_{kdt} = m_0 s_0^2 + s_{kdt}^2 \textstyle\sum_{j=1}^{M} \phi_{jk} \sum_{j=1}^{N_j} Y_{jit} X_{jid}, \quad s_{kdt}^2 = s_0^2 + \\
&\textstyle\sum_{j=1}^{M} \phi_{jk} \sum_{j=1}^{N_j} \left(\sum_{d=0}^{D} X_{jid}^2 \exp\left(\sum_{d=0}^{D} X_{jid} m_{kdt}\right)\right)
\end{aligned}
\tag{11}
$$

5.3 Parameter Estimation of Poisson iMG-GLM-1 Model

Again, in the Poisson Model, estimation of $\gamma_i^1, \gamma_i^2, \phi_{jk}$, are similar to the Normal model with the only difference being that the term P_{jk} is given as,

$$
\begin{aligned}
&P_{jk} = \tfrac{1}{2} \textstyle\sum_{j=1}^{M} \sum_{i=1}^{N_j} \phi_{jk} \{- \sum_{d=0}^{D} \exp\left(\frac{s^{kd}}{2} + \frac{m_{kd} X_{jid}}{s^{kd}}\right) + \\
&Y_{ji} \left(\textstyle\sum_{d=0}^{D} X_{jid} m_{kd}\right) - \log\left(Y_{ji}\right) \\
&\tfrac{m_{kd}}{s_{kd}^2} + \exp\left(m_{kd}\right) + \textstyle\sum_{j=1}^{M} \phi_{jk} \sum_{i=1}^{N_j} \frac{X_{jid}}{s_{kd}^2} \\
&= \textstyle\sum_{j=1}^{M} \sum_{i=1} N_j \phi_{jk} Y_{ji} X_{jid}
\end{aligned}
\tag{12}
$$

For, m_{kd} and s_{kd}, does not have a close form solution. However, it can be solved quickly via any iterative root-finding method.

5.4 Predictive Distribution

Finally, we define the predictive distribution for a new response given a new covariate and the set of previous covariate-response pairs for the trained groups.

$$
\begin{aligned}
&p\left(Y_{j,new} | X_{j,new}, Z_j, \beta_{k=1:K,d=0:D}\right) = \\
&\textstyle\sum_{k=1}^{K} \int Z_{jk} p\left(Y_{j,new} | X_{j,new}, \beta_{k,d=0}^{D}\right) q\left(z, v, \beta_{kd}, \lambda_k\right)
\end{aligned}
\tag{13}
$$

Table 1. Algorithm: Variational Inference Algorithm for iMG-GLM-1 Normal Model.

1. **Initialize Generative Model Latent Parameters** $q\left(z, v, \beta_{kd}, \lambda_k\right)$ **Randomly in its State Space.**
Repeat
2. **Estimate** γ_k^1 **and** γ_k^2 **according to Eq.5.10. for** $k = 1\ to\ K$.
3. **Estimate** ϕ_{jk} **according to Eq.5.10. for** $j = 1\ to\ M$ **and for** $k = 1\ to\ K$. 4. **Estimate the model density parameters,** $\{m_{kd}, \beta_k, a_k, b_k\}$ **according to Eq.5.10. for** $k = 1\ to\ K$ **and** $d = 0\ to\ D$. **until** converged
6. **Evaluate** $E[Y_{j,new}]$ **for a new covariate,** $X_{j,new}$, **according to Eq.5.14 and Eq.5.15.**

Integrating out the $q(z, v, \beta_{kd}, \lambda_k)$, we get the following equation for the Normal model.

$$p(Y_{j,new}|X_{j,new}) = \sum_{k=1}^{K} \phi_{jk} St\left(Y_{j,new}| \left(\sum_{d=0}^{D} m_{kd} X_{j,new,,d}, L_k, B_k\right)\right) \tag{14}$$

Here, $L_k = \frac{(2a_k - D)\beta_k}{2(1+\beta_k)b_k}$, which is the precision parameter of the Student's t-distribution and $B_i = 2a_{y,i} - D$ is the degrees of freedom. For the Poisson and Multinomial Models, the integration of the densities is not analytically tractable. Therefore, we use Monte Carlo integration to obtain,

$$\begin{aligned} E[Y_{j,new}|\mathbf{X_{j,new}}, \mathbf{X}, \mathbf{Y}] &= E[E[Y_{j,new}|\mathbf{X_{j,new}}, \mathbf{q}(\beta_{\mathbf{kd}})]|\mathbf{X}, \mathbf{Y}] \\ &= \frac{1}{S}\sum_{s=1}^{S} E[Y_{j,new}|\mathbf{X_{j,new}}, \mathbf{q}(\beta_{\mathbf{kd}})] \end{aligned} \tag{15}$$

In all experiments presented in this paper, we collected 100 i.i.d. samples (S=100) from the density of β to evaluate the expected value of $Y_{j,new}$. The complete Variational Inference Algorithm for iMG-GLM-1 Normal Model is given Table 1.

6 iMG-GLM-2 Model

We can now learn a new group $M+1$, after all of the first M groups have been trained. For this process, we memorize the learned latent parameters from the previously learned data.

6.1 Information Transfer From Prior Groups

First, we write down the latent parameter conditional distribution given all the parameters in the previous groups. We define the set of latent parameters (Z, v, β, λ) as η. From the description of Dirichlet Process we write down the probability for the latent parameters for the $(M+1)^{th}$ group given previous ones,

$$p(\eta_{M+1}|\eta_{1:M}, \alpha, G_0) = \frac{\alpha}{M+\alpha} G_0 + \frac{1}{M+\alpha}\sum_{k=1}^{K} n_k \delta_{\eta_k^*} \tag{16}$$

Where, $n_k = \sum_{j=1}^{M} Z_{jk}$, represents count where $\eta_j = \eta_k^*$. If we substitute $\eta_k^* = E[\eta_k^*]$, which we define by $\Omega = \{\phi_{jk}, \gamma_k, m_{dk}, \lambda_k, s_{dk}\}$, we get,

$$p(\eta_{M+1}|\eta_k^*, \alpha, G_0) = \frac{\alpha}{M+\alpha} G_0 + \frac{1}{M+\alpha}\sum_{k=1}^{K} n_k \delta_{\eta_k^*} \tag{17}$$

Where, $n_k = \sum_{j=1}^{M} index_{jk}$ and $index_{jk} = \delta_{argmax(\phi_{jk})}$. This distribution represents the prior belief about the new group latent parameters in the Bayesian setting. Now our goal is to compute the posterior distribution of the new group latent parameters after we view the likelihood with the data in $(M+1)^{th}$ group.

$$p(\eta_{M+1}|\Omega, \alpha, D_{M+1}) = \frac{p(D_{M+1}|\eta_{M+1})p(\eta_{M+1}|\Omega, G_0)}{p(D_{M+1}|\Omega, G_0)} \tag{18}$$

Here, $p(D_{M+1}|\eta_{M+1}) = \Pi_{i=1}^{N_{M+1}} p(Y_{M+1,i}|\eta_{M+1}, X_{M+1,i})$.

6.2 Posterior Sampling

The posterior of Eq. 6.18 does not have a closed form solution apart from the Normal Model. So, we apply a Metropolis Hastings Algorithm [18,20] for the Logistic Multinomial and Poisson Model. For the Normal model, $p(\eta_{M+1}|\Omega, \alpha, D_{M+1})$ turns out to be a mixture of Normal-Gamma density, $Normal-Gamma\left(\eta_{M+1}|m_k^{'}, \beta_k^{'}, a_k^{'}, b_k^{'}\right)$ with following parameters,

$$\begin{aligned} m_k^{'} &= \left\{X_{M+1}^T X_{M+1} + (\beta_k) I\right\}^{-1} \left\{X_{M+1}^T Y_{M+1} + \beta_k I m_k\right\} \\ \beta_k^{'} &= \left(X_{M+1}^T X_{M+1} + \beta_k I\right), \quad a_k^{'} = a_k + N_{M+1}/2 \\ b_k^{'} &= b_k + \tfrac{1}{2}\left\{Y_{M+1}^T Y_{M+1} + m_k^T \beta_k m_k - m_k^{'T} \beta_k^{'} m_k^{'}\right\} \end{aligned} \quad (19)$$

For the Poisson and Logistic Multinomial Model, The Metropolis Hastings Algorithm has the following steps. First, we draw a sample $\acute{\eta}$ from Eq. 6.17. Then we draw a candidate sample η, Next, we compute the acceptance probability, $\left[min\left[1, \frac{p(D_{M+1}|\eta)}{p(D_{M+1}|\acute{\eta})}\right]\right]$. We set the new $\acute{\eta}$ to η with this acceptance probability. Otherwise, it remains the old value. We repeat the above 4 steps until enough samples has been collected. This yields the approximation of the posterior.

6.3 Prediction for New Group Test Samples

We seek to predict the future $Y_{M+1,new}|X_{M+1,new}, \Omega$, by the following equation with the previous collection of posterior samples $\eta_{t=1:T}$. T is the number of samples.

$$\begin{aligned} & p\left(Y_{M+1,new}|X_{M+1,new}, \Omega\right) \\ & = \tfrac{1}{T}\sum_{t=1}^{T} p\left(Y_{M+1,new}|X_{M+1,new}, \eta_t\right) \end{aligned} \quad (20)$$

7 Experimental Results

We present empirical studies on two realworld applications: (a) a Stock Market Accuracy and Trend Detection problem and (b) a Clinical Trial problem on the efficacy of a new drug.

7.1 Trends in Stock Market

We propose iMG-GLM-1 and iMG-GLM-2 as a trend spotter in Financial Markets where we have chosen daily close out stock prices over 51 stocks from NYSE and Nasdaq in various sectors, such as, Financials (BAC, WFC, JPM, GS, MS, Citi, BRK-B, AXP), Technology (AAPl, MSFT, FB, GOOG, CSCO, IBM, VZ), Consumer Discretionary (AMZN, DIS, HD, MCD, SBUX, NKE, LOW), Energy (XOM, CVX, SLB, KMI, EOG), Health Care (JNJ, PFE, GILD, MRK, UNH, AMGN, AGN), Industrials (GE, MMM, BA, UNP, HON, UTX, UPS), Materials (DOW, DD, MON, LYB) and Consumer Staples (PG, KO, PEP, PM, CVS, WMT). The task is to predict future stock prices given past stock value for all these stocks and spot general trends in the cluster of the stocks which might be

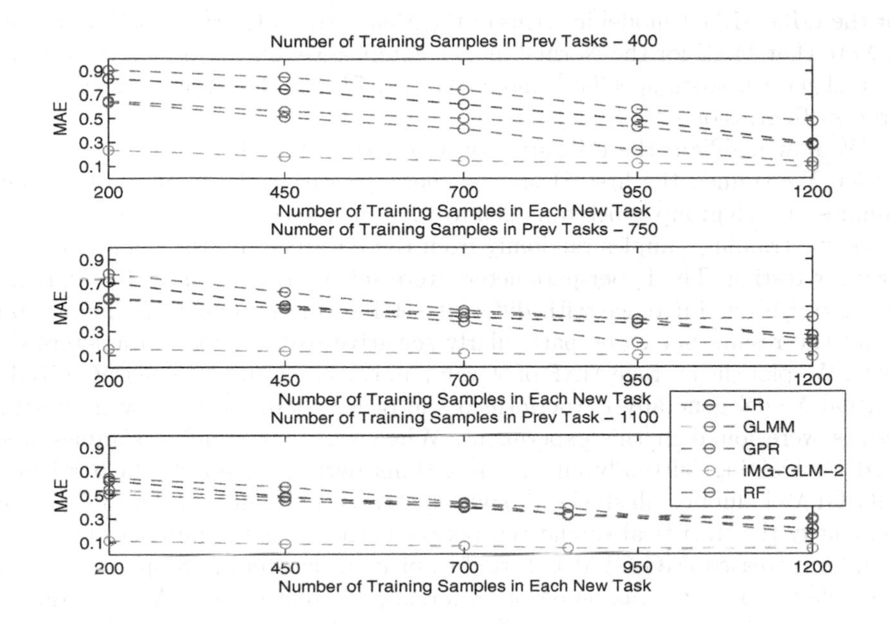

Fig. 2. The Average Mean Absolute Error for 10 New Stocks for 50 random runs for iMG-GLM-2 Model with varying number of training samples in both previous and New Groups

helpful in finding a far more powerful model for prediction. The general setting is a auto-regressive process via the Normal iMG-GLM-1 model with lags representing the predictor variables and response being the current stock price. The lag-length was determined to be 3 by trial and error with 50-50 training-testing split. Data was collected from September 13th, 2010 to September 13th, 2015 with 1250 data points, from Google Finance.

Some very interesting trends were noteworthy. After the clustering was accomplished for the Normal model, the stocks became grouped almost entirely by the sectors they came from. Specifically, we witnessed a total of 9 clusters of stocks, close in makeup to the 8 sectors chosen originally consolidating all the stocks sectors such as, financial, healthcare etc. For example, Apple, Microsoft Verizon, Google, Cisco and AMZN were clubbed together in one cluster. This signifies that all of these stocks share the same auto-regressive density with the same variance. In comparison, single and separate modeling of the stocks resulted in a much inferior model. Joint modeling was particularly useful because we had only 625 data points per stocks for training purposes over the past 5 years. As a result, transfer of stock data points from one stock to another helped mitigate the problem of over-fitting the individual stocks while ensuring a much improved model for density estimation for a cluster of stocks. We report the clustering of the stocks in Table 2. We also show the accuracy of the prediction for the iMG-GLM-1 model in terms of the Mean Absolute error (MAE) in Table 3. Note that MAE for the Normal model significantly outperformed the GLMM normal model, stock specific Random Forest, Linear Regression and Gaussian Process Regression.

We now highlight the utilization of information transfer in the iMG-GLM-1 model. We trained the first 51 stocks where we varied the number of training samples in each group/stock from 200 to 1200 in steps of 250. For each group we chose the training samples randomly from the datasets and the remaining were used for testing. The hyper-parameters were set as, $\{m_0, \beta_0, a_0, b_0\} = 0, 1, 2, 2$. We also ran our inference with different settings of the hyper-parameters but found the results not to be particularly sensitive to the hyper-parameters settings. We plot the average MAE of 50 random runs in Figure 3. The iMG-GLM-1 Normal Model generally outperformed the other competitors. Few interesting results were found in this experiment. When very few training samples were used for training, virtually all the algorithms performed poorly. In particular, iMG-GLM-1 clubbed all stocks into one cluster as sufficient data was not present to identify the statistical similarities between stocks. As the number of training samples increased iMG-GLM-1 started to pick out cluster of groups/stocks as it was able find latent common densities among different groups. As, the training samples got closer to the number of data points (1200), all other models started to perform close to the iMG-GLM-1 model, because they managed to learn each stock well in isolation, indicating that further data from other groups became less useful.

We now proceed to iMG-GLM-2, where we trained 10 new stocks from different sectors (CMCSA, PCLN, WBA, COST, KMI, AIG, GS, HON, LMT, T). Two features which influenced the learning were considered. First, we varied the number of training samples from 400 to 750 to 1100 for each previous groups that were used to further train β_{M+1}. Then, we changed the number of training samples for the new groups from 200 to 1200 in steps of 250. We plot the MAE results for 50 random runs in Figure 2. The prior belief is that the new groups are similar in response density to the previous groups. iMG-GLM-2 efficiently transfers this information from a previous groups to new groups. The iMG-GLM-1 model learns an informative prior for new groups when the number of training samples for each previous group is very small (as seen in the first part in Figure 2). The accuracy increases very slightly as the number of training samples increases in each group. But, with the number of training samples for the new groups increasing, iMG-GLM-2 does not improve at all. This is due to the flexible information transfer from the previous groups. The model does not require more training samples for its own group to model its density, because it has already obtained sufficient information as prior from the previous groups.

7.2 Clinical Trial Problem Modeled by Poisson iMG-GLM Model

Finally, we explored a Clinical Trial problem [12] for testing whether a new anticonvulsant drug reduces a patient's rate of epileptic seizures. Patients were assigned the new drug or the placebo and the number of seizures were recorded over a six

Table 2. Clusters of Stocks from Various Sectors. We note 9 clusters of stocks consolidating all the pre-chosen sectors such as, financials, materials etc.

Group No.	1	2	3	4	5	6	7	8	9
	AAPL, MSFT, VZ, GOOG, CSCO, AMZN	BAC, WFC, JPM, AXP, PG, CITI, GS,MS	DIS, HD, LOW, SBUX, MCD	XOM, CVX, SLB, EOG, KMI	GILD, MRK, UNH, AMGN, AGN	GE, MMM, BA, UNP, HON	DOW, DD, MON, LYB, JNJ, PFE	KO, PEP, PM, CVS, WMT	BRK-B, IBM, FB, NKE, UTX, UPS

Table 3. Mean Absolute Error (MAE) for All Stocks. iMG-GLM has Much Higher Accuracy than Other Competitors.

	AAPL	MSFT	VZ	GOOG	CSCO	AMZN	BAC	WFC	JPM	AXP	PG	CITI	GS	MS	DIS	HD	LOW
GPR	.023	.004	.087	.078	.093	.189	.452	.265	.176	.190	.378	.018	.037	.098	.278	.038	.011
RF	.278	.903	.370	.256	.290	.570	.159	.262	.329	.592	.746	.894	.956	.239	.934	.189	.045
LR	.381	.865	.280	.038	.801	.706	.589	.491	.391	.467	.135	.728	.578	.891	.389	.790	.624
GLMM	.378	.489	.389	.208	.972	.786	.289	.768	.189	.389	.590	.673	.901	.490	.209	.391	.991
iMG-GLM	.012	.002	.009	.011	.018	.028	.047	.038	.035	.079	.069	.087	.019	.030	.139	.189	.213
	SBUX	MCD	XOM	CVX	SLB	EOG	KMI	GILD	MRK	UNH	AMGN	AGN	GE	MMM	BA	UNP	HON
GPR	.837	.289	.849	.583	.185	.810	.473	.362	.539	.289	.306	.438	.769	.848	.940	.829	.691
RF	.884	.321	.895	.843	.774	.863	.973	.729	.894	.794	.695	.549	.603	.738	.481	.482	.482
LR	.380	.391	.940	.995	.175	.398	.539	.786	.591	.320	.793	.839	.991	.839	.698	.389	.298
GLMM	.649	.720	.364	.920	.529	.369	.837	.630	.729	.481	.289	.970	.740	.649	.375	.439	.539
iMG-GLM	.003	.018	.128	.291	.005	.060	.052	.017	.014	.078	.009	.067	.191	.034	.098	.145	.238
	DOW	DD	LYB	JNJ	PFE	KO	PEP	PM	CVS	WMT	BRK-B	IBM	FB	NKE	UTX	UPS	MON
GPR	.689	.890	.745	.907	.678	.378	.867	.945	.361	934	.580	.845	.901	.310	.483	.828	.748
RF	.181	.098	.489	.237	.692	.827	.490	.295	.749	.692	.957	.295	.478	.694	.747	.806	.945
LR	.67	.386	.984	.982	.749	.294	.256	.567	.345	.767	.893	.956	.294	.389	.694	.921	.702
GLMM	.727	.389	.288	.592	.402	.734	.923	.900	.571	.312	.839	.956	.638	.490	.390	.372	.512
iMG-GLM	.038	.078	.063	.019	.024	.007	.089	.192	.138	.111	.289	.390	.289	.218	.200	.149	.087

week period. A measurement was made before the trial as a baseline. The objective was to model the number of seizures, which being a count datum, is modeled using a Poisson distribution with a Log link. The covariates are: Treatment Center size (ordinal), number of weeks of treatment (ordinal), type of treatment–new drug or placebo (nominal) and gender (nominal). A Poisson distribution with log link was used for the count of seizures. Here, X_{ji} and Y_{ji} represent the ith covariate and count response pair of the jth group. The distribution, $\{\beta_{kd}\}$ (Normal) is the prior distribution on the covariate coefficient β.

We found that a patient's number of seizures are clustered (they form the groups) in multiple collections. This signifies that a majority of the patients across groups show the same response to the treatment. We obtained 8 clusters from 300 out of 565 patients for the iMG-GLM-1 model (the remaining 265 were set aside for modeling through the iMG-GLM-2 model). Among them 5 clusters showed that the new drug reduces the number of epileptic seizures with increasing number of weeks of treatment while the remaining 3 clusters did not show any improvement. We also report the forecast error of the number of epileptic seizures of the remaining 265 patients in Table 4. Our recommendation for the usage of the new drug would be a cluster based solution. For a specific patient, if she falls in one of those clusters with decreasing trend in the number of seizures

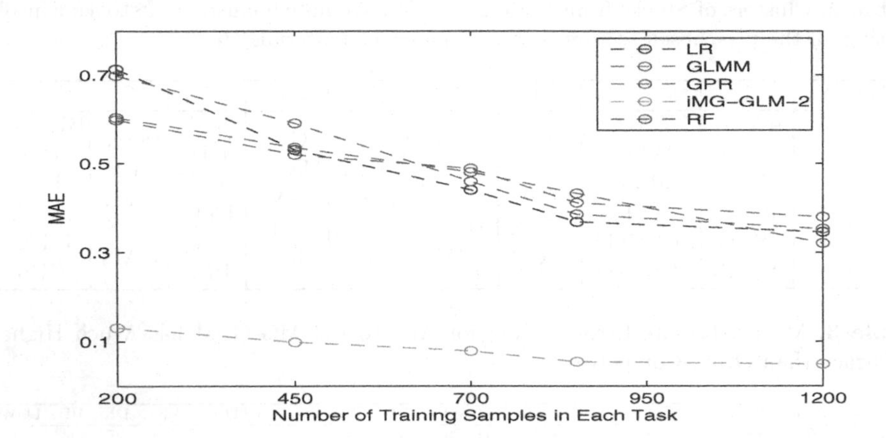

Fig. 3. The Average Mean Absolute Error for 51 Stocks for 50 random runs for iMG-GLM-1 Model with varying number of training samples.

Table 4. MSE and MAE of the Algorithms for the Clinical Trial Dataset and Number of Patients in Clusters for iMG-GLM-1 and iMG-GLM-2 Model.

<table>
<tr><td colspan="8">Patient Number in Clusters for iMG-GLM-1 Model</td></tr>
<tr><td colspan="4">Positive</td><td colspan="4">Negative</td></tr>
<tr><td>46</td><td>30</td><td>40</td><td>27</td><td>33</td><td>24</td><td>37</td><td>24</td></tr>
<tr><td colspan="8">Patient Number in Clusters for iMG-GLM-2 Model</td></tr>
<tr><td colspan="4">Positive</td><td colspan="4">Negative</td></tr>
<tr><td>33</td><td>24</td><td>41</td><td>29</td><td>30</td><td>31</td><td>34</td><td>43</td></tr>
<tr><td colspan="2">iMG-GLM</td><td colspan="2">Poisson GLMM</td><td colspan="2">Poisson Regression</td><td colspan="2">RForest</td></tr>
<tr><td colspan="8">Mean Square Root Error(L2 Error) fpr iMG-GLM-2 Model</td></tr>
<tr><td colspan="2">1.53</td><td colspan="2">1.58</td><td colspan="2">1.92</td><td colspan="2">1.75</td></tr>
<tr><td colspan="8">Mean Absolute Error Root Error(L1 Error) for iMG-GLM-2 Model</td></tr>
<tr><td colspan="2">1.14</td><td colspan="2">1.34</td><td colspan="2">1.51</td><td colspan="2">1.62</td></tr>
</table>

with time, we would recommend the new drug, and otherwise not. Out of 265 test case patients modeled through iMG-GLM-2, 180 showed signs of improvements while 85 did not. We kept all the weeks as training for the iMG-GLM-1 model and the first five weeks as training and the last week as testing data for the iMG-GLM-2 model. Traditional Poisson GLMM cannot infer these findings since the densities are not shared at the patient group level. Moreover, only the Poisson iMG-GLM-1/2 based prediction is formally equipped to recommend a patient cluster based solution for the new drug, whereas all traditional mixed models predict a global recommendation for all patients.

8 Conclusion

In this paper, we have formulated an infinite multigroup Generalized Linear Model (iMG-GLM), a flexible model for shared learning among groups in grouped regression. The model clusters groups by identifying identical response-covariate densities for different groups. It also models heteroscedasticity among groups by modeling different uncertainty among groups. We experimentally evaluated the model on a wide range of problems where traditional mixed effect models and group specific regression models fail to capture structure in the grouped data. Although the Metropolis Hastings algorithm turned out to be fairly accurate for the iMG-GLM-2 model, developing a variational inference alternative would be an interesting topic for future research. Finally, the number of groups in each cluster depends on the scale factors α_1 and α_2 (scale parameters of the DP) of the model, and at times grows large in specific cluster. This occurs mostly when any cluster has a large number of groups which becomes representative of the whole data. In most cases, beyond a few primary clusters, the remaining clusters represent outliers. Although, careful tuning of scale parameters can mitigate these problems, a theoretical understanding of the dependence of the model on scale parameters could lead to better modeling and application.

References

1. Ando, R.K., Zhang, T.: A framework for learning predictive structures from multiple tasks and unlabeled data. Journal of Machine Learning Research **6**, 1817–1853 (2005)
2. Antoniak, C.: Mixtures of dirichlet processes with applications to bayesian nonparametric problems. Annals of Statistics **2**(6), 1152–1174 (1974)
3. Bakker, B., Heskes, T.: Task clustering and gating for bayesian multitask learning. Journal of Machine Learning Research **4**, 83–99 (2003)
4. Baxter, J.: Learning internal representations. In: International Conference on Computational Learning Theory, pp. 311–320 (1995)
5. Baxter, J.: A model of inductive bias learning. Journal of Artificial Intelligence Research **12**, 149–198 (2000)
6. Blei, D.: Variational inference for dirichlet process mixtures. Bayesian Analysis **1**, 121–144 (2006)
7. Breslow, N.E., Clayton, D.G.: Approximate inference in generalized linear mixed models. Journal of the American Statistical Association **88**(421), 9–25 (1993)
8. Caruana, R.: Multitask learning. Machine Learning **28**(1), 41–75 (1997)
9. Escobar, M.D., West, M.: Bayesian density estimation and inference using mixtures. Journal of the American Statistical Association **90**, 577–588 (1994)
10. Ferguson, T.: A bayesian analysis of some nonparametric problems. Annals of Statistics **1**, 209–230 (1973)
11. Hastie, T., Tibshirani, R.: Varying-coefficient models. Journal of the Royal Statistical Society. Series B (Methodological) **55**(4), 757–796 (1993)
12. IBM: Ibm spss version 20. IBM SPSS SOFTWARE (2011)
13. Jordan, M., Jacobs, R.: Hierarchical mixtures of experts and the EM algorithm. In: International Joint Conference on Neural Networks (1993)
14. Lee, Y., Nelder, J.A.: Hierarchical generalized linear models. Journal of the Royal Statistical Society. Series B (Methodological) **58**(4), 619–678 (1996)
15. Lee, Y., Nelder, J.A.: Hierarchical generalised linear models: A synthesis of generalised linear models, random-effect models and structured dispersions. Biometrika **88**(4), 987–1006 (2001)

16. Lee, Y., Nelder, J.A.: Modelling and analysing correlated non-normal data. Statistical Modelling **1**(1), 3–16 (2001)
17. Lee, Y., Nelder, J.A.: Double hierarchical generalized linear models (with discussion). Journal of the Royal Statistical Society: Series C (Applied Statistics) **55**(2), 139–185 (2006)
18. Neal, R.M.: Markov chain sampling methods for dirichlet process mixture models. Journal of Computational and Graphical Statistics **9**(2), 249–265 (2000)
19. Nelder, J.A., Wedderburn, R.W.M.: Generalized linear models. Journal of the Royal Statistical Society, Series A (General) **135**(3), 370–384 (1972)
20. Robert, C., Casella, G.: Monte Carlo Statistical Methods (Springer Texts in Statistics). Springer-Verlag New York, Inc. (2005)
21. Sethuraman, J.: A constructive definition of dirichlet priors. Statistica Sinica **4**, 639–650 (1994)
22. Viele, K., Tong, B.: Modeling with mixtures of linear regressions. Statistics and Computing **12**(4), 315–330 (2002)
23. Xue, Y., Liao, X., Carin, L.: Multi-task learning for classification with dirichlet process priors. Journal of Machine Learning Research **8**, 35–63 (2007)
24. Yu, K., Tresp, V., Schwaighofer, A.: Learning gaussian processes from multiple tasks. International Conference on Machine Learning, pp. 1012–1019 (2005)
25. Zhang, J., Ghahramani, Z., Yang, Y.: Learning multiple related tasks using latent independent component analysis. Advances in Neural Information Processing Systems, pp. 1585–1592 (2005)

Content Based Identification of Talk Show Videos Using Audio Visual Features

Atta Muhammad(✉) and Sher Muhammad Daudpota

Department of Computer Science, Sukkur IBA, Sukkur, Sindh, Pakistan
{a_muhammad,sher}@iba-suk.edu.pk
http://www.iba-suk.edu.pk

Abstract. TV Talk Shows are used for exchanging views among participants on a particular topic. Huge audiences, among all age groups, follow the talk shows worldwide to acquire knowledge on current affairs and other topics of their interest. A major portion of these audiences use online video databases to search talk shows of their interest. Online video databases contain videos of different genres like movie, drama, talk shows, animations, sports, horror, music etc. Searching a particular talk show, in presence of many other video files, is a tedious task, especially when the uploader has not used proper naming convention while assigning caption to the file as search in the video databases is still text based. Different contents based classification techniques have been proposed in literature to efficiently index video contents on Internet. Literature also includes few attempts at identifying if a video clip is representing a talk show; however, this identification already found in literature uses a long list of features which makes its processing slow. This paper proposes a solution based on audio visual features and employing basic grammar of talk shows recording to automatically identify if a video recently uploaded on video database is containing a talk show. We have performed experiments on different genres of videos collected from YouTube, Dialymotion, movies from Bollywood and Hollywood. The proposed system classifies a video file in TalkShow and OtherVideo classes with precision and recall of 93% and 100% respectively.

1 Introduction

A talk show is a chat show in which discussants, listener, viewers or studio audience are invited to participate in the discussion on a particular topic. Recording videos of these talk shows or broadcasting them live on different TV channels is a serious business worldwide. These recorded videos are then usually stored in online databases like Youtube, Dailymotion, etc for more viewership. Nowadays, the Internet has become an effective tool to share the knowledge. People share whatever they want on social networking sites and video databases. Video databases on Internet are increasing in numbers and size every day. By 2018, 84% of the Internet traffic would be based on video contents [1], which is increase of 6% over the current video chunk on online Internet pie. According to Cisco, video traffic on the Internet is 37 Exabyte (EB) per month. There are numerous

P. Perner (Ed.): MLDM 2016, LNAI 9729, pp. 267–283, 2016.
DOI: 10.1007/978-3-319-41920-6_20

video databases available online like YouTube, Dailymotion, WIRE, etc. According YouTube statistics [2], millions of people watch YouTube, generating trillions of views daily.

The video databases contain videos of different genres like movies, dramas, talk-shows, animations, sports, horror, adult and many more. IMDb [3] enlist 321,409 different talk-shows. According to [4] there are 31.49 million people only in the USA who watch daytime talk-shows.

Those contents are viewed by the community who surf on the Internet. Sometimes, people do not get what they desire to see on the Internet. Our focus is only on video contents uploaded on video databases by the Internet community. Uploading a video clip on a video database is very simple. All one needs is just to give the correct directory path and the desired title to the video. When a user uploads a video, sometimes title of the contents does not match with genre type. This lead to exlusion of video when another user wants to query the same input video with its genre because querying a video database is still caption based. The proposed system would classify the input video into TalkShow or OtherVideo genres.

In talk-shows, there are limited numbers of cameras to record the program. Each camera records the video from a particular angle called a shot. The shots recorded from various angles are juxtaposed to create a scene. In talk-shows, due to limited number of cameras, the repetition of shots is very high. Due to repetition of same shots again and again in same environment, mostly there is only one scene in the talk-shows. So, to detect the genre of input video as talk-show, first we have to identify the shots and scenes in input video.

In literature, many approaches have been proposed to identify the scenes in the video based on high or low level features. Shirahama [6] categorizes these features as low and high according to their very nature. An object or person or anything which has some semantic meaning is categorized as a high level object in a scene. Whereas color histogram, edges, movement, hue, contrast, etc. are categorized as low level features. For instance, in Figure 1, an actor and a fan are high level features whereas its color histogram, edges position, red shirt, and white background are low level features. These features may be utilized to detect

Fig. 1. An image taken from a popular Bollywood movie

scenes in a video. However, [8][9] argued on detection of the scene through high level features may be challenging and proposed a scene detection through low level features.

Based on low level features many approaches are proposed in the literature [11][12][13][14] for event detection. In these approaches, an event as input is given to the system. System queries the entire video and find similar events in the video in hand and results the related events based on the similarity of their low level features e.g. color histogram, edges and motion etc. More or less these approaches are related to each other as they performed the same task, which is named as "query by shot" in the literature. The accuracy of [6] is better than other approaches as they use multiple query rather than a single one like others. Li et al. in [24] utilized visual as well audio cues to index and analyze movies. Their approach incorporates audio features with visual features to classify high level semantic events. Those semantic events mainly comprises of two speakers dialogue, multiple speakers dialogue detection, and hybrid events. Visual features in action scene change rapidly and are more importance than audio features because there are fewer speech dialogues during an action scene.

1.1 SCENE

As discussed, the scene is a collection of shots. Scenes are mainly caterogrised as action and dialogue scenes. Before discussing action and dialogue scenes in detail, we present here some of the important rules for dialogue and action scenes recording.

Film Making Rules. There are some norms set by the movie industry, which are followed when the recording of a movie takes place. One rule is that the duration of a dialogue scene is always kept larger than that of an action scene because in that scene the expression of the actor involved must be captured. While in contrast to that, for increasing the suspense in the movie the duration of the action scene is usually kept low. Table 1 gives a glance of various parameter values found in a dialogue and action scene. A viewer may find it problematic, if those norms would not be followed in a movie or video clip. For example, what if an action scene is played with no background music or there is a dialogue scene with background noise and music?

Action and Dailogue Scene Detection Using Film Making Convention. There are many proposed approaches to classify the movie scene as dialogue or

Table 1. Film making rules for action and dialogue scene [5]

Scene Type	Shot Duration	Actor Movement	Background Music	Envirnoment Sound	Camera Motion
Dialogue Scene	High	Low	Low	Low or Zero	Low
Action Scene	Low	High	High	High	High

action. Sundaram in [15] proposed a method to classify a scene as action or dialogue. According to them a scene has three common properties that must be same, which are chromaticity means color, lightning, and ambient sound. They reported 100% precision in scene change. While recall for scene detection and dialogue detection they reported were 94% and 90% respectively. For dialogue and action scene detection, lei chen et al. Chen and Rizvi in [17] proposed a top-down approach. Action Scene and dialogue scene were detected by audio cues and video editing rules. Like our approach they used audio cues to train a support vector machine (SVM). The audio features includes zero-crossing rate, harmonic ratio, and silence ratio. They performed experiments on two movies. In Gladiator for dialogue detection they achieved precision and recall of 89.47% and 96.6% respectively. On the other hand, action scenes precision and recall calculated was 84%. Likewise, In Crouching Tiger and Hidden Dragon, precision and recall for dialogue scene detection were 76.5% and 81.6% respectively. Zhai et al. in [19] proposed a model to categorize the movie scene into three categories: Conversation, Suspense, and Action. A Finite Stat Machine (FSM) was constructed based on mid and low level features for different scenes. Mid-level feature used in FSM was body, whereas low level features utilized were motion and audio energy. This system was tested on 80 different clips from different movies. For dialogue scene, they achieved 94.3% precision and 97.1% recall. Whereas, for suspense scene 100% precision and 97% recall was reported. For action scenes recall and precision achieved were 91.4% and 94.0% respectively. Lehane et al. [5] proposed an audio-visual based model with film making convention to detect action scene in a movie. They proposed a bottom-up approach with low, mid and high level features.

Figure 2 shows the system proposed by Lehance et al. [5]. The system takes a video clip as input and at low level, shot boundary is calculated. Mid level Motion activity analysis and similar shots are clustered. At higher level, they used film making rules to detect an aciton scene. As defined in table2.1, action scene has high motion in shots. Thus motion in the shot can be principle discriminator between action and dialogue scene. In addition to that in 2007, Lehane et al.[7] proposed another model which was based on audio-visual features for detecting dialogue, action and montage. Musical events like funeral, songs, sad moment are categorized as montage. The author proposed many FSM to categorize the scene to the categories mentioned above. Movie Browser, a complete system was proposed. The system achieved precision and recall of 94% and 81% respectively.Hence significant work has been done to categorize scenes in movies as action, dialogue or montage.
Sher and Guha [25] proposed a model to automatically extract song from a movie based on audio-visual cues. This model constructs the timed automaton to differentiate between the songs and other scenes in a movie.

Lu et al. [20] proposed a model based on KNN classifier to discriminate music and non-music. The audio features used were shot time energy, spectrum flux, and zero-crossing rate. Figure 3 shows the proposed model, which takes a digital audio stream as input, extract requisite audio features from the audio stream.

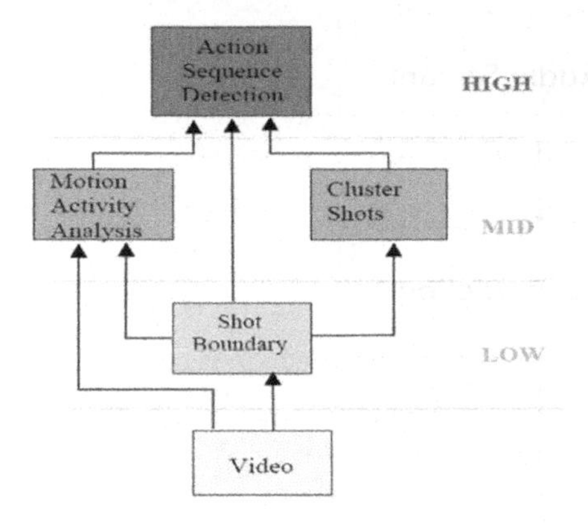

Fig. 2. Abstarct Lehane Proposed Model

Based on audio features extract KNN, classifier classifies the input audio stream as speech or non-speech. For further classification as Speaker segmentation , Music, silence or environment sound, proposed model utilizes band periodicity, spectral flux, and the noise frame ratio.

Panagiotakis and Tziritas in [32] proposed an approach to classify the input audio stream into music and non-music class. They used zero-crossing rate (ZC) and root mean square (RMS) as audio cues. The main objective of this proposed model was to discriminate audio as music or speech at runtime. A window size of 20ms was selected for feature calculation. They acheived an accuracy of 95% in classification and 97% in segmentation. The last part of our research is to classify the input video as talk-show or another genre.

Roach [10] classified the input video into cartoon, sports, news based on foreground motion object, and background camera movement. They tested this system on 30 seconds video. Based on object motion and camera motion they could classify a video into 3 categories with error rate of 17%, 18% and 6% respectively. Hazim et al. [23] proposed a classification mechanism to categorize the videos to their genre based on low and high level features. In low level visual features, they used HSV color histogram, color moments, autocorrelogram. For low level audio features, they experimented on Mel-frequency cepstral coefficients, fundamental frequency, signal energy, zero-crossing rate. For cognitive and structural features they used face detection. Figure 4 shows the architecture of this proposed model. The input video stream is applied to system, which extract its audio visual features, each audio and visual feature is recorded and checked on its relevant SVM model. At last majority voting classifies the video into its relevant genre.They used RAI TV, French TV and Youtube video with accuracy of 99.6%, 99% and 92.4% correct classification respectively.

The rest of the paper is organized as follows. Section 2 discusses the methodology used in the proposed system and section 3 presents results and finally section 4 concludes the paper.

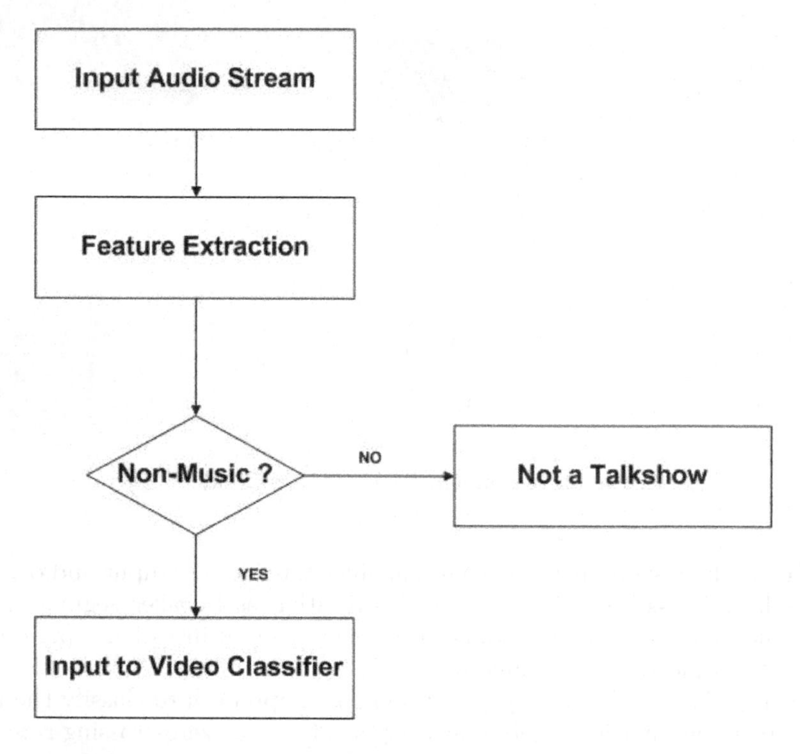

Fig. 3. Audio Classifier [17]

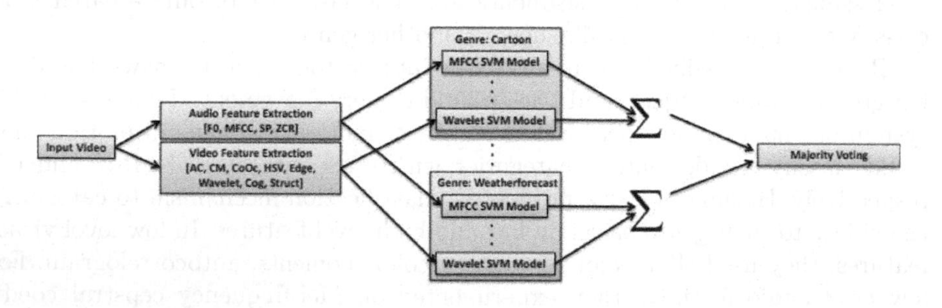

Fig. 4. Audio Classifier [23]

2 Automated Video Content Genre Based Classification

This section would look how to detects shots based on shot boundary detection. We would see how accurately proposed model detect shots in a video. Secondly, the average motion in a video clip, which is also a key factor for analysis. Third, Scene detection algorithm to detect the number of scenes in the video. Finally, we will put everything together to see how proposed system works in classifying videos based on their genres.

2.1 System Overview

The abstract model of the system proposed to classify an input video into talk-show or OtherVideo genre is shown in Figure 5. The system is logically divided into two components, Audio Classifier and Video Classifier. Audio Classifier's role is to classify the input video as music or non-music. The purpose behind classifying input into music and non-music classes is that talk-shows usually contain dialog conversation between participants with a very small portion of music elements in it, thus classification based on music and non-music would certainly help us differentiating music videos, movies, dramas, and other video genres from talk-shows significantly.

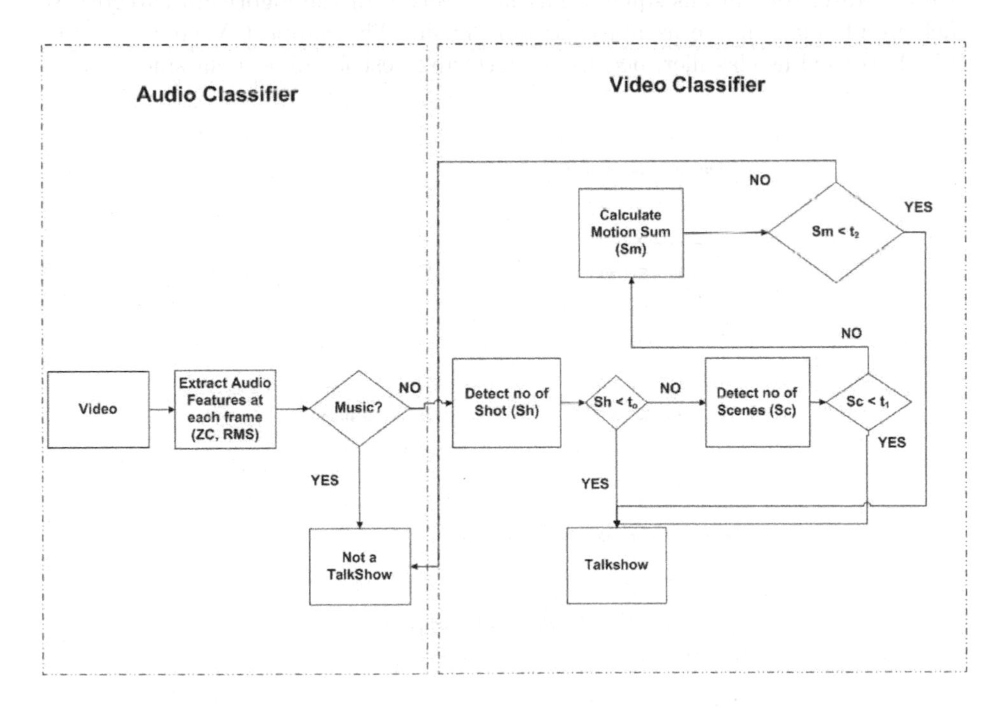

Fig. 5. System Architecture

If the input video is classified as a total music file, it would not be further processed in video classifier as the basic grammar of talk-shows recording does not include music element. If the Audio Classifier, classify the input video as non-music, the input video would be passed to the second logical component as video classifier. The role of video classifier is to classify the input video as talk-show or OtherVideo based on the number of scenes and average motion in an input video. The rest of this section explains working of each of these components in detail.

2.2 Audio Classifier

Talk-shows are mainly based on dialogue rather than music. Therefore, audio classifier is the first part of the proposed model. The audio classifier classifies the input video file as a music or non-music file. There are various attempts in the literature to classify the audio file into music and non-music classes [28] [29] [30] [31] [22] [33] [21] [34] [35]. Zhang et al. [21] proposed model is used in our case. To classify the input video as music or non-music, algorithm used is in Figure 6. The first step of algorithm is to divide the entire input video into frames with a window of 2 seconds. Audio features in the classifier are key factors to the performance of the classifier. The next step of the algorithm is to extract zero-crossing rate, root means square. The last section of the algorithm categorizes each input audio frame as music or non-music. The Support Vector Machine (SVM) is used as classifier, because it performs best for binary classification.

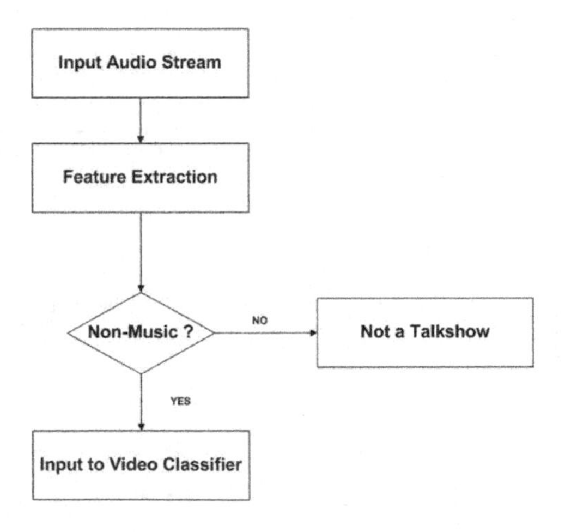

Fig. 6. Audio Classifier

Audio Features. The audio features used in audio classification are zero-crossing rate and root mean square. In this section we will give a little description about these features.

1. **Root Mean Square** measures amplitude of input signal. The signal Amplitude A, is defined by

$$A = \sum_{m=1}^{N-1} x^2(n)$$

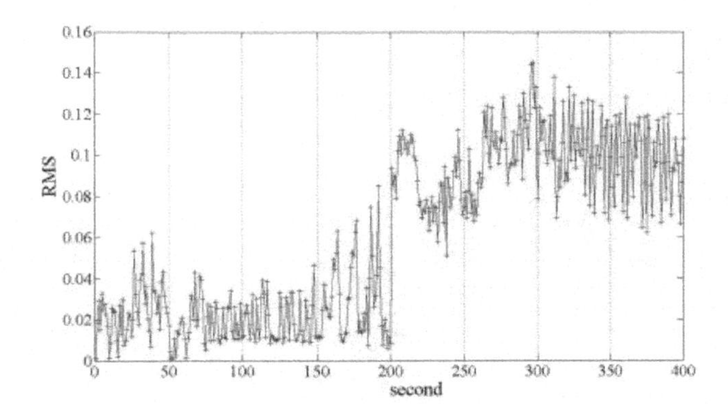

Fig. 7. Root Mean Squared [25]

Figure 7 show that variation in RMS values of music files are typically higher than non-music file. From 1 to 200 seconds shows RMS values of a non-music file. From 201 to 400 seconds show the RMS values for music file.

2. **Zero-crossing Rate** is the number of time-domain zero-crossing within a frame of an input audio file.

$$ZCR = \frac{1}{2(N-1)} \sum_{m=1}^{N-1} |sgn[x(m+1)] - sgn[x(m)]|$$

Where $x(m)$ is a discrete input audio signal and $sgn[]$ is a sign function (Lu et al., 2003). The music part of the input video has a lower level of zero-crossing rate than non-music part. This is because the non-music part contains speech. Speech has higher zero-crossing rate than music. Therefore, the zero-crossing rate can be an effective discriminator between the music and non-music components.
Figure 8 show ZCR of non-music and music part of an input video. 1 to 200 seconds demonstrate ZCR of a speech signal. Whereas, non-music ZCR are shown from 201 to 400. One can visualize that zero-crossing rate (ZCR) during speech part is much greater than music file.

2.3 Video Classification

If the input video is categorized into the music class that means it is not a talk-show. If the input video is classified into non-music class, the next step is to extract possible number of shots.

Shot Detection. There are multiple approaches to detect shot boundary in literature, we have used similar approach as proposed in [26] [27][14] based on

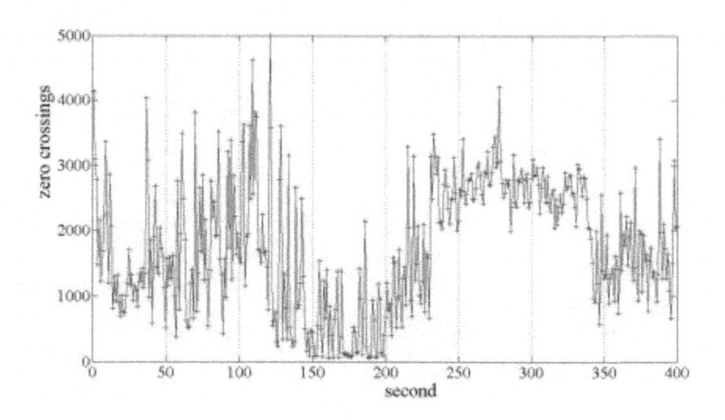

Fig. 8. Zero-crossing Variance [25]

color histogram difference. Figure 3.1 show the model followed to detect the shot boundary in a video clip. The color histogram (RGB) of current frame (fc) is calculated and compared with the previous shot boundary (fp). If the color difference between previous boundaries (fp) and current frame (fc) is greater than the threshold defined, there is a shot change in the video otherwise shot is still continue and check for the next frame. The threshold value is set by performing experiments on 105 videos of different genres. The performance of shot detection algorithm is demonstrated in result section.

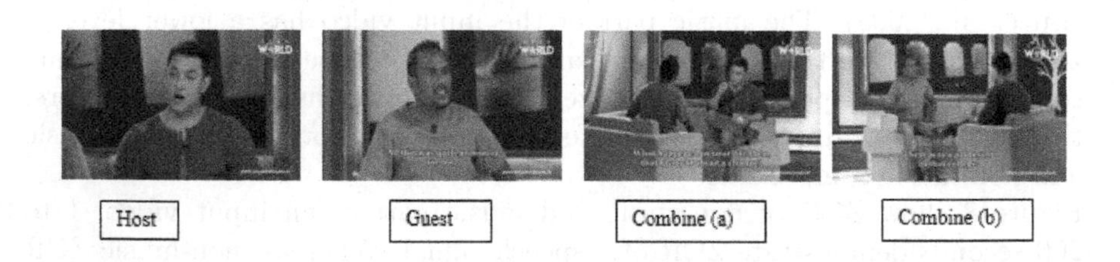

Fig. 9. Major Shots in Talk-show

Scene Detection. Once shots are detected, the next task is to detect the scene within those shots. In talk-shows, mostly people sit together and discuss on a topic. There are limited possible shots. Those possible shots would be repeated over time. Figure 9 shows the major shots in a talk-show, Satyamev Jayate, a popular talk-show with 10/10 rating on IMDb. The right most frame is the host of the show, adjacent to host is the guest, and two possible combine shots.

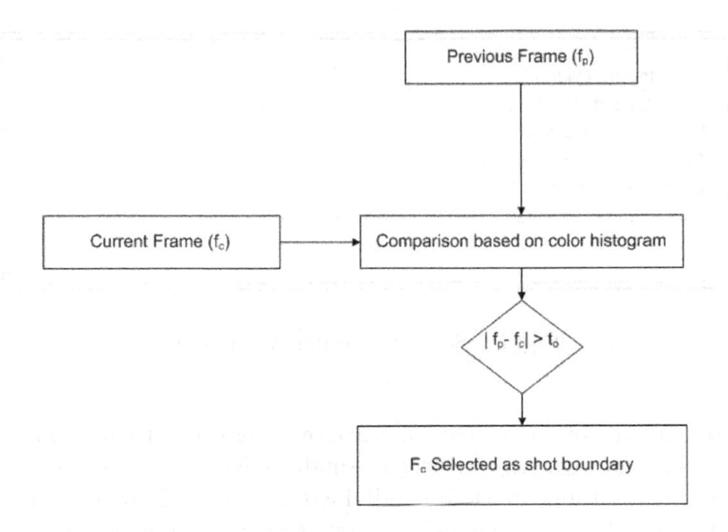

Fig. 10. Shot Detection [17]

Hence, those shots are continuously repeated. If same shots are repeated continuously that means, the scene is continued. In talk-shows, mostly there is only one scene.

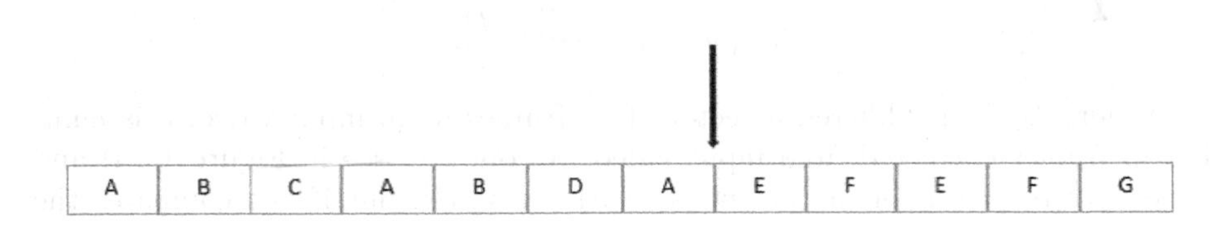

Fig. 11. Scene Cut

The reason is there are the limited number of people in a confined environment with the limited possible number of shots. On the other hand, in movies, there are multiple scenes.

Figure 12 shows, the sequence of shots (A, B, C, D, E, F and G). As one can see that, the shot A, B, C, and D are repeated for a limited period of time in the scene A. After sometime, repetition of shots A, B, C, D stops and E, F, G shots in scene B start to appear. Black Arrow facing downward indicates a scene change.

Actor Movement. In talk-shows the actors usually sit on a chair and communicate with the other actors available. On the other hand, in movies, sports,

Input: A video File
Output: No of Scenes
1 Segment an input video to m frames
2 Calculate shot Boundaries based on Difference between adjacent frames.
3 Detect Repeated Shots.
4 Identify the Scene cuts based on Repetition of Shots.
5 Return no of Scene

Fig. 12. Scene Detection Algorithm

dramas, etc. actor got more freedom to move here and there. Thus, in talk-shows, the actor movement is relatively small as compare to other genres. For Actor Movement, a frame image is subdivided into in a grid of 48 (8 x 6) cells as shown in Figure 13. These cells between two consecutive images are subtracted to find out the difference. Given two images,$image_i$ and $image_j$, their absolute difference D can be calculated as

$$D = \sum_{r=1}^{48} |image_i(r) - image_j(r)|$$

Where r is the region in the image. The Average Actor movement can also be given by

$$A_v ActorM = \frac{\sum_{i=1}^{n} D_i}{n}$$

Where D_i is the difference vector of all frames in an input video. n is number of frames extracted from input video. As one can see in Figure 13(a) and Figure 13(b) the actor movement is relatively small, but if one compares the distance vector of Figure 13(b) and Figure 13(c) the different would be much higher.

a

b

c

Fig. 13. Actor Movement

3 Results

The results produced by shot detection algorithm are demonstrated in 2 which shows the input video clip name, number of shots detected by the shot detection algorithm and human calculated shots. One can see that the number of shots in talk show are mostly less than other genre videos.

Table 2. Shot Detection Results

Video Clip Name	Genre	System Detected Shots	Actual Shots
Penguins-1	animation	71	77
Penguins-2	animation	95	91
Game of Thrones ep 1	drama	64	60
Game of Thrones ep 2	drama	12	35
Spider Man 1	movie	99	88
Warrior 2	movie	37	37
The Dead Lands	movie	68	54
BBC news	news	46	49
BBC news 2	news	40	38
The Daily Show	talkshow	16	17
Mubashir Lukman	talkshow	1	1
The Daily Show-2	talkshow	25	25

Like shot detection algorithm, the results are also extracted by scene detection algorithm for different genre videos. Table 3 summarizes the results. It can be clearly scene that the scene in talk show is usually one unlike otherVideo genres.

Table 3. Scene Detection Results

Video Name	Genre	System Detected Scenes	Actual Scenes
Penguins-1	animation	9	6
Penguins-2	animation	13	10
House of Cards ep 1	drama	4	7
House of Cards ep 2	drama	4	8
Spider Man 1	movie	14	3
Warrior 2	movie	4	5
BBC news	news	6	7
BBC news2	news	6	9
Football 1	sport	2	2
Football 2	sport	6	3
Mere Mutabik-2	talkshow	1	1
Mere Mutabik-3	talkshow	1	2

Actor Movement is also an efficient parameter to classify the input video into Talk Show or OtherVideo. Table 4 shows the results calculated for different genres. As one can see that average Actor movement in talk shows are very less than other genres.

Table 4. Actor Movement

Video Name	Type	Actor Movement (Pixels)	Avg Movement
Penguins-1	animation	131887	231.38
Penguins-2	animation	200580	351.89
House of Cards ep 1	drama	84704	149.126
House of Cards ep 2	drama	83167	145.9
Spider Man 1	movie	224829	394.4
Warrior 2	movie	81744	143
BBC news 1	news	112352	187.25
BBC news 2	news	61762	102.93
Lost Desperation	song	145115	296.153
Dont believe me	song	188626	348.66
Football 1	sport	147489	245.815
Football 2	sport	164875	274.791
The Daily Show	talkshow	33990	59.63
Mubashir Lukman	talkshow	18956	33.25614035

Finally, based on all features, the system classify the input video into talk show and OtherVideo genre. Table 5 summarizes the results and accuracy of the model proposed.

Table 5. Video Genre Classification Results

S.No	Name of Video	Actual Genre	Automated Genre	Remarks
1	Penguins-1	animation	Not a Talks Show	correct
2	Penguins-2	animation	Not a Talk Show	correct
15	House of Cards ep 1	drama	Not a Talk Show	correct
16	House of Cards ep 2	drama	Not a Talk Show	correct
30	Spider Man 1	movie	Not a Talk Show	correct
31	Warrior 2	movie	Not a Talk Show	correct
45	BBC News1	news	Not a Talk Show	correct
46	BBC News2	news	Not a Talk Show	correct
57	Football 1	sport	Not a Talk Show	correct
58	Football 2	sport	Not a Talk Show	correct
70	The Daily Show	talkshow	Talk Show	correct
71	Mubashir Lukman	talkshow	Talk Show	correct

The system is tested on different genres of videos including Bollywood and Hollywood movies. Table 6 demonstrates the accuracy of the proposed model in terms of percision and recall respectively.

Table 6. Precision and Recall of Algorithms

Algorithm Name	Percision	Recall
Shot Detection Algorithm	94.7%	94.4%
Scene Detection Algorithm	96%	85%
Talk-show Detection Algorithm	93%	100%

4 Conclusion

Video databases on the Internet are increasing in numbers and size every day. According to [1] by 2018, 84% of the Internet traffic would be video. An increase of 6% over the current video chunk on online Internet pie. According to Cisco, video traffic on the Internet is 37-Exabyte (EB) per month traffic. There are numerous video databases available online like YouTube, daily Motion, WIRE, etc. According YouTube [2] statistics millions of people who access this video database, generating trillions of views daily. Our proposed model is a semantic approach to detect the uploaded video as talk-show or any other genre video file.The recorded precision and recall of the proposed system is 100% and 89% respectively.

References

1. REELSO, 2018-internet-traffic-video (2015). http://www.reelseo.com
2. YouTube Statistics (2015). https://www.youtube.com/yt/press/statistics.html
3. IMDb Statistics (2015). http://www.imdb.com/stats
4. Statistica, TV program type Number of TV viewers who typically watch daytime talk shows (2015). http://www.statista.com/statistics/229097/tv-viewers-who-typically-watch-daytime-talk-shows-usa
5. Lehane, B., O'Connor, N.E., Murphy, N.: Action sequence detection in motion pictures. In: European Workshop on the Integration of Knowledge, Semantic and Digital Media Technologies (2004)
6. Shirahama, K., Uehara, K.: Query by shots, retrieving meaningful events using multiple queries and rough set theory. Eurasip Journal on Image and Video Processing (2008)
7. Lehane, B., O'Connor, N.E., Lee, H., Smeaton, A.F.: Indexing of fictional video content for event detection and summarisation. Eurasip Journal on Image and Video Processing (2007)
8. Naphade, M.R., Smith, J.R., Tesic, J., Chang, S.F., Hsu, W.H., Kennedy, L.S., et al.: Large-Scale Concept Ontology for Multimedia. IEEE Multimedia (2006)

9. Haering, N., Qian, R.J., Ibrahim, M.: A semantic event-detection approach and its application to detecting hunts in wildlife video. IEEE Transactions on Circuits and Systems for Video Technology (1999)
10. Raoch, M.: Video genre classification using dynamics, acoustics, speech, and signal processing. In: IEEE Proceedings (2001)
11. Kim, Y.T., Chua, T.S.: Retrieval of news video using video sequence matching. In: Multimedia Modeling (2005)
12. Peng, Y., Ngo, C.W.: Emd-Based Video Clip Retrieval by Many-to-Many Matching (2005)
13. Liu, X., Zhuang, Y., Pan, Y.: A new approach to retrieve video by example video clip. In: ACM Multimedia Conference (1999)
14. Jain, A.K., Vailaya, A., Xiong, W.: Query by video clip. Multimedia Systems (1999)
15. Sundaram, H., Chang, S.F.: Computable scenes and structures in films. IEEE Transactions on Multimedia (2002)
16. Lienhart, R., Pfeiffer, S., Effelsberg, W.: Scene determination based on video and audio features. Multimedia Tools and Applications (1998)
17. Chen, L., Rizvi, S.J., Ozsu, M.T.: Incorporating audio cues into dialog and action scene extraction. In: Storage and Retrieval for Image and Video Databases (2003)
18. Kotti, M., Kotropoulos, C., Ziółko, B., Pitas, I., Moschou, V.: A framework for dialogue detection in movies. In: Gunsel, B., Jain, A.K., Tekalp, A.M., Sankur, B. (eds.) MRCS 2006. LNCS, vol. 4105, pp. 371–378. Springer, Heidelberg (2006)
19. Zhai, Y., Rasheed, Z., Shah, M.: A framework for semantic classification of scenes using finite state machines. In: Enser, P.G.B., Kompatsiaris, Y., O'Connor, N.E., Smeaton, A.F., Smeulders, A.W.M. (eds.) CIVR 2004. LNCS, vol. 3115, pp. 279–288. Springer, Heidelberg (2004)
20. Lu, L., Zhang, H.J.: Content analysis for audio classification and segmentation. IEEE Transactions on Audio, Speech and Language Processing (2002)
21. Zhang, H.J., Li, S.Z.: Content-based audio classification and segmentation by using support vector machines. Multimedia Systems (2003)
22. Panagiotakis, C., Tziritas, G.: IEEE Transactions on Multimedia (2005)
23. Ekenel, H.K., Semela, T., Stiefelhagen, R.: Content-based video genre classification using multiple cues. In: AIEMPro (2004)
24. Li, Y., Narayanan, S., Kuo, C.J.: Content-based movie analysis and indexing based on audiovisual cues. IEEE Transactions on Circuits and Systems for Video Technology (2004)
25. Doudpota, S.M., Guha, S.: Mining movies to extract song sequences. In: MDMKDD 2011 (2011)
26. Zhang, H., Low, C.Y., Smoliar, S.W.: Video parsing and browsing using compressed data. Multimedia Tools and Applications (1995)
27. Zhang, H., Low, C.Y., Smoliar, S.W.: ACM Multimedia Conference (1995)
28. Scheirer, E., Slaney, M.: Construction and evaluation of a robust multifeature speech/music discriminator. In: ACM Multimedia Conference (1997)
29. Li, D., Sethi, I.K., Dimitrova, N., Mcgee, T.: Classification of general audio data for content-based retrieval. Pattern Recognition Letters (2001)
30. Harb, H., Chen, L., Auloge, J.Y.: Speech/music/silence and gender detection algorithm. In: Proceedings of the 7th International Conference on Distributed Multimedia Systems, DMS 2001 (2001)
31. Aggelos Pikrakis, T.G., Theodoridis, S.: A speech/music discriminator of radio recordings based on dynamic programming and bayesian networks. IEEE Transactions on Multimedia (2008)

32. Panagiotakis, C., Tziritas, G.: A speech/music discriminator based on rms and zero-crossings. IEEE Transactions on Multimedia (2005)
33. Lu, L., Li, S.Z., Zhang, H.J.: Content-based audio segmentation using support vector machines. In: IEEE International Conference on Multimedia and Expo (2001)
34. Saunders, J.: Real time discrimination of broadcast speech/music. In: International Conference on Acoustics, Speech, and Signal Processing (1996)
35. Zhang, T., Jay Kuo, C.C.: Real time discrimination of broadcast speech/music. In: International Conference on Acoustics, Speech, and Signal Processing (1999)
36. Zhang, H., Low, C.Y., Smoliar, S.W.: Video parsing, retrieval and browsing: an integrated and content-based solution. In: ACM Multimedia Conference (1995)
37. Oger, S., Linares, G., Matrouf, D.: Audio-Based Video Genre Identification (2015)
38. Martins, G.B., Almeida, J., Papa, J.P.: Supervised video genre classification using optimum-path forest. In: Pardo, A., Kittler, J. (eds.) Progress in Pattern Recognition, Image Analysis, Computer Vision, and Applications. LNCS, vol. 9423, pp. 735–742. Springer, Switzerland (2015)

IncMSTS-PP: An Algorithm to the Incremental Extraction of Significant Sequences from Environmental Sensor Data

Carlos Roberto Silveira Junior(✉), Marcela Xavier Ribeiro, and Marilde Terezinha Prado Santos

Federal University of São Carlos, Washington Luís km 235, Sao Carlos, Brazil
carlos_jr@dc.ufscar.br
http://www.ufscar.br

Abstract. The mining of sequential patterns from environmental sensor data is a challenging task. The data can present noises and contain sparse patterns hide in a huge amount of information. The knowledge extracted from environmental sensor data can have many applications: indicate climate changes, risk of ecologic catastrophes and help to determine environment degradation in face of humans actions. However, there is a lack of methods that can handle this kind of data. Based on that, we proposed IncMSTS-PP: an incremental algorithm that finds sequential sparse patterns and enhances them semantically facilitating the interpretation. IncMSTS-PP implements STW-method to extract stretchy patterns (patterns with time gaps) in data with noises. The enhancement use post-processing method that generalizes the patterns using the fuzzy ontology knowledge. Our experiment shows that IncMSTS-PP extracts 2.3 times more relevant sequences than traditional algorithms in sensor domain. The post-processing summarizes the patterns reducing to 22.47% of the original number of patterns. In conclusion, IncMSTS-PP is efficient and reliable in the extraction of significant sequences from sensor data.

Keywords: Data mining · Fuzzy ontology · Real data · Sequence generalization

1 Introduction

The patterns extraction from environment sensor data is a challenging task. In this domain, the data is spatio-temporal: it is organized by the location and the moment in which the event was measured. Another characteristic of the environment sensor domain is that the data can present noises and/or errors originated from devices or humans errors (in the case where data collection have been made manually). Normally, the collection process has many *ad hoc* procedures [BAJ+10] making the data be stored without standardization. Furthermore, environment sensor data are incremental and, in most of the time, the

P. Perner (Ed.): MLDM 2016, LNAI 9729, pp. 284–293, 2016.
DOI: 10.1007/978-3-319-41920-6_21

incremental is irregular (the measures are conditioned to a random event occurrence). And also, the domain can present time sparse patterns (that present time interval between events) making harder to be mined by the traditional algorithms of data mining.

The state of art to the mining sequence from the environment sensor data is the algorithm IncMSTS [SJCSR15]. IncMSTS is an incremental version of MSTS [JSR13] that implements the incremental technique of IncSpan [CYH04]. MSTS solves the problem of mining time sparse patterns and the problem of database with noises. IncMSTS solves the problem of the irregular increment of data. However, the mined patterns are not semantically expressive as it is expected to be. Based on that, we propose a Post-Processing (PP) method that uses the knowledge stored on fuzzy ontology. PP method is added in the final steps of IncMSTS algorithm creating the IncMSTS-PP. Our experiments shows that IncMSTS-PP is a good alternative to extract sparse patterns from noise database with irregular data increment.

This paper is organized as follows: Section 2 presents important concepts and related works. Section 3 presents IncMSTS-PP, Section 4 presents the experiments, results and a discussion about the results. Section 5 finishes the paper presenting the conclusions and future works.

2 Background and Related Work

A sequential pattern is a sequence of events (itemsets)[1] that keep an order among their occurrences. Consider a sequence $s =< i_1 \dots i_n >$, where i_k is an itemset (a nonempty set of items that happen together), $n \geq 2$ and i_{k-1} precedes i_k for $1 < k \leq n$. s' is a sub-sequence of s ($s \prec s'$), if and only if, for all itemset $i_k \in s$, s' has an itemset i' that is sub-itemset of i_k or an empty set keeping the sequential order of s.

The sequential mining algorithms usually separate the sequences into frequent and non-frequent ones. The frequency of a sequence is called support, it can be calculated as $support(s) = \frac{|number\ of\ occurrence\ of\ s|}{|number\ of\ sequences\ on\ the\ database|}$ and it has a value between zero and one ([0; 1]) [Hua09]. If the $support(s)$ is equal or greater than a minimum value of support ($minSup$) set by the user, then the sequence s is considered frequent.

The sequential pattern mining was introduced with the algorithm AprioriAll [AS95]. Based on AprioriAll, the GSP algorithm was proposed [SA96]. AprioriAll and GSP work with vertical databases where all the tuples are organized by time of event occurrence. Others algorithms, such as PrefixSpan [PHMa+01], use the horizontal mode. In horizontal database, all tuples are an identification (e.g. site of collection of samples) followed by a list of events ordered by time of occurrence.

[1] The word "events" is sometimes used instead of "itemsets" because "event" is more semantic to explain.

Incremental Miner of Stretchy Time Sequences, proposed by [SJCSR15], is an incremental algorithm that aims to find Stretchy Time Patterns. Stretchy Time Pattern is a pattern that presents time gaps in-between its events (sparse pattern) and it is defined by Equation (A), where $i_{1...n}$ are itemsets (not necessarily disjoint) and $\Delta t_{1...n-1}$ are time gaps [JSR13].

$$s = < i_1 \; \Delta t_1 \; i_2 \; \ldots \; \Delta t_{n-1} \; i_n > \quad (A)$$

A frequent sequence s has many occurrences, for each occurrence o the sum of its time gaps will never be greater than μ (Maximum Time Gap Size – parameter set by user). Since the pattern's time gaps variate between occurrences of sequence s, the patterns are called stretchy time. The strategy IncMSTS uses to find the Stretchy Time Patterns is called Sliding Time Window introduced in [JSR13]. Sliding Time Window is a adaptive window that changes its size based on pattern's occurrences.

The incremental strategy of IncMSTS is based on the semi–frequent strategy, as IncSpan [CYH04]: IncMSTS tries to find the frequent patterns by processing only the data increment d database D has received, since the database D has already been mined. To do so, IncMSTS and IncSpan store semi–frequent patterns between the executions. A semi–frequent pattern is almost frequent that can became frequent with the data increment d. To a pattern be considered semi–frequent, its support must be in-between $[minsup \times \delta; minsup)$. The δ-parameter is set by the user and it dictates how far a support can be to be considered semi–frequent.

Based on IncSpan strategy, in the first time a dataset is processed, the there is no frequent and semi–frequent patterns to be processed. So IncMSTS processes the data using the same strategy MSTS does but storing also the semi–frequent ones. After a data increment d, IncMSTS finds the frequent and semi–frequent Stretchy Time Sequences by processing d and in some scenarios $D+d$. IncMSTS starts updating the support of the frequent and semi–frequent patterns, with the update, the algorithm sorts the patterns in frequents, semi–frequent and non-frequent (that is discarded). The data increment d is processed locking for frequent patterns that are not in the updated ones. Those patterns demand a update considering the whole database $(D+d)$ to calculate their support.

Ontology is an formal and explicit specification of a shared contextualization [Gua98]. In post-processing, the knowledge stored by ontologies can be used to generalize found patterns. The generalized patterns are semantically richer than the usual extracted patterns. These patterns usually simplifies the understanding of the patterns [LT10].

NARFO, proposed by [MYSB09], is an algorithm of association rules extraction based on Apriori [AS94]. NARFO uses ontologies in two of the processing steps: (i) to catch the similarity between items that allows extracting rule like $x \; y \rightarrow z$ (item x similar to y are associated to item z). And (ii), in the post-processing: similar rules are generalized using ontologies. Therefore, if the items a and b have the same father-item f in the ontology, the rules $a \rightarrow c$ and $b \rightarrow c$ can became $f \rightarrow c$.

NARFO* is a NARFO's extension, proposed by [MYSF10], where the parameter *MinGen* is proposed. *MinGen* affects step (ii) of NARFO allowing just the generalization of the rules containing a minimum percentage of "children" in the ontology. Both NARFO and NARFO* do not extract sequential patterns. Furthermore, they are not prepared to handle database that is either incomplete or with noises.

3 The Proposed Algorithm: *Incremental Miner of Stretchy Time Sequences with Post-Processing*

Figure 1 presents IncMSTS-PP diagram. As previously explained, the Post-Processing module and the Extraction of Stretchy Time Patterns module (IncMSTS) are separated. It makes the generalization (done by the post-processing modulo) completely independent of extraction algorithm.

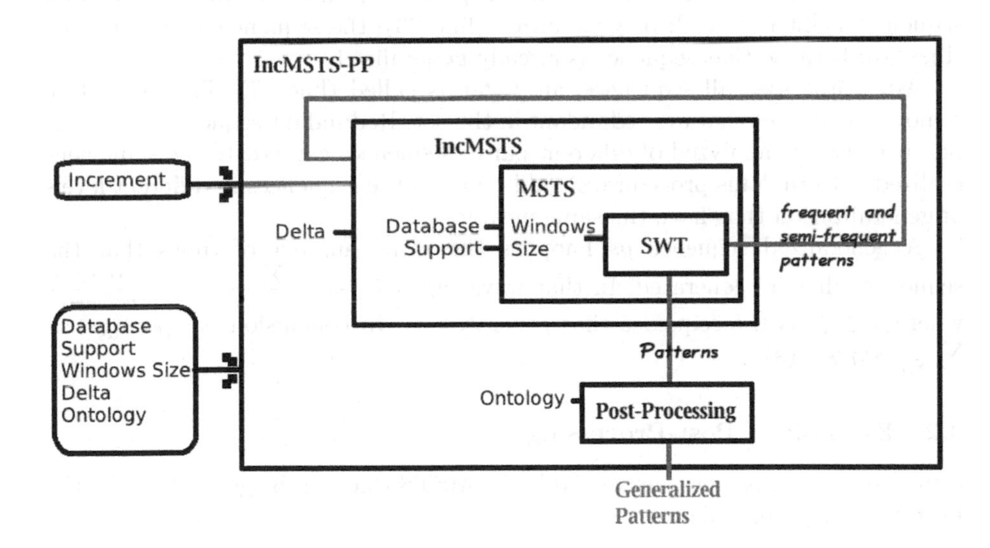

Fig. 1. IncMSTS–PP Diagram.

The IncMSTS-PP explanation is organized as follow: Subsection 3.1 presents the Post-Processing module. Subsection 3.2 presents an example of the Post-Processing.

3.1 Post-Processing Module

The Post-Processing module generalizes the sequences. That reduces the number of sequences presented to the user and also gives more semantic significance to the patterns. Algorithm 1 presents the generalization process.

The algorithm has as input the set of found patterns *sp* and the domain ontology Ω. The output is the set of generalized patterns, *gp*. If a pattern cannot be generalized it is added in *gp* because it is considered generalized.

The algorithm tries to combine each sequence (lines 1 and 2, Algorithm 1). Two sequences can be candidates to generalization, if they have the same size and itemsets different (line 3). The generalization process tries the generalization (lines 5 to 17) looking each different itemsets in both sequences at the same position (lines 8 to 16). Then the algorithm tries finding a common "father" to the different itemsets (line 9) using the knowledge in the ontology Ω. If it might find the common "father" (line 10), the father is put in the generalized sequence that is forming *temp* (line 11). If it has not find the common "father" (line 12), the generalized sequence is throwed out (line 13) and the sequences processing ends (line 13). In the case where the itemsets analyzed in both sequences are the same (line 6), the itemset is concatenated in *temp* (line 7).

If the processing of an sequence ends creating an generalized sequence (line 18). It is added to the set of generalized sequence, *gp* (line 19). In the case of an sequence might not at all be generalized (line 23), the sequence is added to *gp* (line 24): because that sequence is already generalized.

After processed all sequences, an *filter* is called (line 27). This procedure removes sequences that are redundant in the set. Redundant sequences can happen because the analyzed of different par of sequences can create the same generalized pattern. This procedure should removes the sequences in different levels of generalization that have the same meaning.

A generalized sequence *gs* happens the same number of times that the sequences that its generated. In that way, $support(gs) = \frac{\sum_{s \in gs} |occurrence\ of\ s|}{|sequences \in dataset|}$ where $s \in gs$ is the sequence that generates gs. In conclusion, $support(gs) = \sum_{s \in gs} support(s)$.

3.2 Example of Post-Processing

Consider the two patterns extracted by IncMSTS that can be generalized by the Post-Processing module:

$$Patterns \begin{cases} s_1 =< a\ b\ c > support : sup_{s_1} \\ s_2 =< a\ d\ e > support : sup_{s_2} \end{cases}$$

And, consider the ontology Ω:

$$\Omega \begin{cases} W \rightarrow_{fatherOf} X \quad W \rightarrow_{fatherOf} Y \\ X \rightarrow_{fatherOf} b \quad Y \rightarrow_{fatherOf} d \\ X \rightarrow_{fatherOf} c \quad Y \rightarrow_{fatherOf} e \end{cases}$$

The generalization process tries to combine the sequences s_1 and s_2. The first step is to check if they have the same size (number of itemset). In this case both are 2-sequence. After that, the post-processing tries the generalization. The algorithm analyses the first item of each sequences and detects that are the same item. Then item a goes to the generalized sequences *sg* that has been created,

Algorithm 1. Post-Processing Module: generalization process.

```
   Input: set of patterns sp, ontology Ω
   Output: set of generalized patterns gp
1  foreach pattern p1 ∈ sp do
2      foreach pattern p2 ∈ sp do
           /* condition to generalize two sequences                    */
3          if p1 ≠ p2 and sizeOf(p1) = sizeOf(p2) then
4              pattern temp ← ∅ ;
               /* trying the generalization.                            */
5              for i ← 1 to sizeOf(p1) do
                   /* same itemsets at the same position.               */
6                  if itemsetAt(i, p1) = itemsetAt(i, p2) then
7                      concat(itemsetAt(i, p1), temp) ;
8                  else
                       /* different itemsets at the same position.      */
9                      itemset
                       ι ← comunFather(Ω, itemsetAt(i, p1), itemsetAt(i, p2));
10                     if ι ≠ ∅ then
                           /* different itemset that can be generalized. */
11                         concat(ι, temp) ;
12                     else
                           /* sequences that cannot be generalized.     */
13                         temp ← ∅ ;
14                         break ;
15                     end
16                 end
17             end
18             if temp ≠ ∅ then
19                 add(temp, gp);
20             end
21         end
22     end
       /* sequence that cannot be generalized are generalized.          */
23     if couldNotBeGeneralized(p1) then
24         add(p1, gp);
25     end
26 end
27 filter(gp);
```

so $sg =< a$. After that, the algorithm looks the second item in both sequences, they are different although the ontology Ω says that item b and item d have the same "father", X. So X goes to the generalized sequences $sg =< a\ X$. The third items are processed following the same algorithm, than $sg =< a\ X\ Y >$. It quites the processing and sg is the result. Now the sequences s_1 and s_2 can be unconsidered. The support of sg is $sup_{s_1} + sup_{s_2}$ because sg happens $s_{1,2}$-time.

4 Experiment and Results

In order to validate IncMSTS-PP approach, we performed several experiments. The data we used comes from environmental sensors installed at *Feijão* River (in São Carlos city, São Paulo state, Brazil), the same data used in [JSR13, SJCSR15]. The data are composed by measurements of discharge rate of *Feijão* River and the rainfall rate. The information starts at 1977 and goes until 2004, its granularity is in days and the data is sorted by geographic location of the collection. A real example of the data from *Feijão* River dataset is presented by Table 1: the discharge rate (in cubic meters per second) of *Feijão* River and the rainfall rate (in millimeters per hour).

To use IncMSTS-PP, the dataset pass through a pre–processing step that discrete the data. It is the same process applied in [JSR13, SJCSR15] that uses Omega algorithm [RTT08]. Table 2 presents the same data in Table 1 after Omega processes.

Table 1. Example of tuples in the original *Feijão* River database.

Date	Discharge	Rainfall
1979.03.19	$2.41m^3/s$	$1.2mm/h$
1979.03.10	$2.25m^3/s$	$0mm/h$
1979.03.21	$2.1m^3/s$	$0.6mm/h$
1979.03.22	$3.68m^3/s$	$22.5mm/h$

Table 2. Same tuples after being processed by Omega Algorithm.

Tuple	Rainfall	Discharge
703	$Rainfall_4$	$Discharge_1$
704	$Rainfall_0$	$Discharge_1$
705	$Rainfall_2$	$Discharge_0$
706	$Rainfall_{17}$	$Discharge_3$

After the pre-processing is done, the data are grouped for weeks that aims to reduce the granularity of the data. These grouping are made by calculating the average value for the rainfall and discharge of *Feijão* river. That also reduced the number of records to be mined.

After this process, the information in the data could be organized on domain ontologies. This organization was done using the works [AMS12, Jeb07, Cha96, NK96]. The result is shown by Figure 2 and Figure 3. These ontologies can be joined because Rain and Discharge refer to Characteristic of River Point.

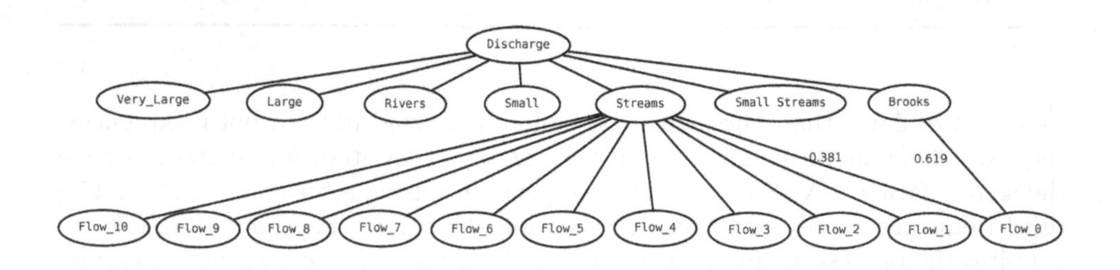

Fig. 2. Ontology created to the discharge after the pre-processing.

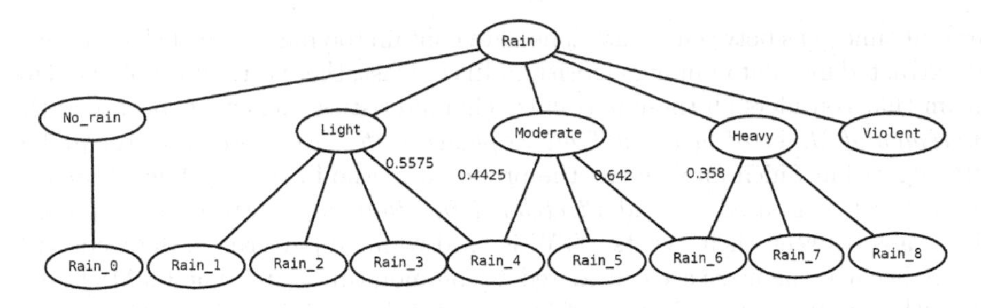

Fig. 3. Ontology created to the rain after the pre-processing.

The Stretchy Time Patterns mined, presented in this subsection, were extracted using $\mu = 15$ weeks, $minSup = 0.05$ and $\delta = 0.5$ (i.e. a semi-frequent pattern has its support greater or equal 0.025). Table 3 presents examples of generalized patterns extracted by IncMSTS-PP and the original patterns (patterns that have been generalized). Both examples presented are 2-sequences.

Table 3. Examples of generalized patterns and their original patterns.

Generalized Patterns	Support	Original Patterns	Support
$< Streams\ No\_Rain >$	12%	$< Flow_0\ Rain_0 >$	6%
		$< Flow_1\ Rain_0 >$	6%
$< No\_Rain\ Streams >$	17.5%	$< Rain_0\ Flow_0 >$	5%
		$< Rain_0\ Flow_1 >$	12.5%

The first pattern, $< Streams\ No\_Rain >$, came from the combination between $< Flow_0\ Rain_0 >$ and $< Flow_1\ Rain_0 >$. The ontology brings the information that $Flow_0$ and $Flow_1$ have the same father $Streams$, as it is shown by Figure 2. $Rain_0$, in both original patterns, is generalized to No_Rain because $Rain_0$ is the unique son of No_Rain, as it is shown by Figure 3. The support of $< Streams\ No\_Rain >$ is 12% because its number of occurrence is the sum each original pattern number of occurrence.

The second pattern is $< No\_Rain\ Streams >$. It presents the same itemsets of the first generalized pattern, but in different order. And the explanation of generalization process is almost the same of the first example. Also, the support of the second pattern if 17.5% by the same reason.

5 Conclusion

The extraction of significant sequence in environmental sensor data is a challenge task. The domain presents many characteristics to be considered, for instance: spacial–temporal data, periodic increments of data, noises, and patterns that can present time gaps between events. The data contain too much knowledge that can be extracted by a data mining algorithm. In contrast, there is no state-of-art algorithm that considers all these particular characteristics. Hence, we proposed the *Incremental Miner of Stretchy Time Sequence with Post-Processing* (IncMSTS-PP) algorithm which implements the proposed method *Stretchy Time Windows* (STW). STW is used to find *Stretchy Time Patterns*: patterns that present time gaps between their events. STW is used to mine patterns in datasets with noises because the noise are considered as common time gaps. The IncMSTS-PP algorithm implements technique of incremental data mining which stores semi-frequent patterns. The incremental technique improve the performance in incremental dataset. Moreover, IncMSTS-PP applies a generalization process in set of found patterns. The generalization summarizes patterns and attributes to them greater semantic value that simplifies the interpretation process of knowledge extracted. Our experiments have shown that IncMSTS-PP returns sequences up to 5 times larger and up to 2.3 times in higher number than GSP algorithm. Furthermore, the incremental module presents good results on incremental dataset. Also, the post-processing reduces in 22.47% the number of sequences showed to the user.

Acknowledgments. The authors thank the Brazilian agencies CAPES, CNPq, FAPESP and FINEP for the financial support and Federal University of Itabubá for the access on *Feijão* River data.

References

[AMS12] American-Meteorological-Society. Glossary of meteorology. Digital, novembro 2012. amsglossary.allenpress.com/glossary/search?id=rain1

[AS94] Agrawal, R., Srikant, R.: Fast algorithms for mining association rules in large databases. In: Bocca, J.B., Jarke, M., Zaniolo, C. (eds.) VLDB 1994, Proceedings of 20th International Conference on Very Large Data Bases, Santiago de Chile, Chile, September 12–15, 1994, pp. 487–499. Morgan Kaufmann (1994). DBLP: conf/vldb/94

[AS95] Agrawal, R., Srikant, R.: Mining sequential patterns. In: Proceedings of the Eleventh International Conference on Data Engineering, Taipei, Taiwan, pp. 3–14, March 1995

[BAJ+10] Barseghian, D., Altintas, I., Jones, M.B., Crawl, D., Potter, N., Gallagher, J., Cornillon, P., Schildhauer, M., Borer, E.T., Seabloom, E.W., Hosseini, P.R.: Workflows and extensions to the kepler scientific workflow system to support environmental sensor data access and analysis. Ecological Informatics **5**(1), 42–50 (2010)

[Cha96] Chapman, D.: Water Quality Assessments: A guide to the use of biota, sediments and water in environmental monitoring. Behalf of WHO by F & FN Spon 11 New Fetter Lane London EC4, 2nd edn. (1996). ISBN 0 419 21590 5 (HB) 0 419 21600 6 (PB)

[CYH04] Cheng, H., Yan, X., Han, J.: Incspan: incremental mining of sequential patterns in large database. In: Proceedings of the Tenth ACM SIGKDD International Conference on Knowledge Discovery and Data Mining, KDD 2004, pp. 527–532. ACM, New York (2004)

[Gua98] Guarino, N.: Formal Ontology and Information Systems, pp. 3–15. IOS Press, Amsterdam (1998)

[Hua09] Huang, T.C.-K.: Developing an efficient knowledge discovering model for mining fuzzy multi-level sequential patterns in sequence databases. In: International Conference on New Trends in Information and Service Science, NISS 2009, pp. 362–371, June 30, 2009–July 2, 2009

[Jeb07] Jebson, S.: Fact sheet number 3: Water in the atmosphere. Met Office, p. 13, August 2007. http://cedadocs.badc.rl.ac.uk/255/1/factsheet03.pdf

[JSR13] Silveira Junior, C.R., Santos, M.T.P., Ribeiro, M.X.: Stretchy time pattern mining: a deeper analysis of environment sensordata. In: FLAIRS Conference (2013)

[LT10] Loh, B.C.S., Then, P.H.H.: Ontology-Enhanced Interactive Anonymization in Domain-Driven Data Mining Outsourcing, pp. 9–14 (2010) (cited By (since 1996) 0)

[MYSB09] Miani, R.G., Yaguinuma, C.A., Santos, M.T.P., Biajiz, M.: NARFO algorithm: mining non-redundant and generalized association rules based on fuzzy ontologies. In: Filipe, J., Cordeiro, J. (eds.) Enterprise Information Systems. LNBIP, vol. 24, pp. 415–426. Springer, Heidelberg (2009)

[MYSF10] Miani, R.G., Yaguinuma, C.A., Santos, M.T.P., Ferraz, V.R.T.: Narfo* Algorithm: Optimizing the Process of Obtaining Non-Redundant and Generalized Semantic Association Rules, vol. 2. AIDSS, pp. 320–325 (2010) (cited By (since 1996) 0)

[NK96] Nanson, G.C., David Knighton, A.: Anabranching rivers: Their cause, character and classification. Earth Surface Processes and Landforms **21**(3), 217–239 (1996)

[PHMa+01] Pei, J., Han, J., Mortazavi-asl, B., Pinto, H., Chen, Q., Dayal, U., Hsu, M.C.: Prefixspan: mining sequential patterns efficiently by prefix-projected pattern growth. In: Proceedings of the 17th International Conference on Data Engineering, ICDE 2001, pp. 215–224. IEEE Computer Society, Washington (2001)

[RTT08] Ribeiro, M.X., Traina, A.J.M., Traina Jr., C.: A new algorithm for data discretization and feature selection. In: Proceedings of the 2008 ACM symposium on Applied computing, SAC 2008, pp. 953–954. ACM, New York (2008)

[SA96] Srikant, R., Agrawal, R.: Mining sequential patterns: generalizations and performance improvements. In: Apers, P., Bouzeghoub, M., Gardarin, G. (eds.) Advances in Database Technology — EDBT 1996. LNCS, vol. 1057, pp. 1–17. Springer, Heidelberg (1996)

[SJCSR15] Silveira Junior, C.R., Carvalho, D.C., Santos, M.T.P., Ribeiro, M.X.: Incremental mining of frequent sequences in environmental sensor data. In: FLAIRS Conference, pp. 452–456 (2015)

DSCo: A Language Modeling Approach for Time Series Classification

Daoyuan Li[✉], Li Li, Tegawendé F. Bissyandé, Jacques Klein, and Yves Le Traon

University of Luxembourg, Luxembourg, Luxembourg
{daoyuan.li,li.li,tegawende.bissyande, jacques.klein,yves.letraon}@uni.lu

Abstract. Time series data are abundant in various domains and are often characterized as large in size and high in dimensionality, leading to storage and processing challenges. Symbolic representation of time series – which transforms numeric time series data into texts – is a promising technique to address these challenges. However, these techniques are essentially lossy compression functions and information are partially lost during transformation. To that end, we bring up a novel approach named Domain Series Corpus (DSCo), which builds per-class language models from the symbolized texts. To classify unlabeled samples, we compute the fitness of each symbolized sample against all per-class models and choose the class represented by the model with the best fitness score. Our work innovatively takes advantage of mature techniques from both time series mining and NLP communities. Through extensive experiments on an open dataset archive, we demonstrate that it performs similarly to approaches working with original uncompressed numeric data.

1 Introduction

Time series data refers to a sequence of data that is ordered either temporally, spatially or in other defined order. Such data are abundant in various domains including health-care, finance, energy and industry applications. Furthermore, time series data are often characterized as large in size and high in dimensionality [7]. These characteristics of time series data – together with its abundance – has led to various challenges in both storage and analytics. For example, the BLUED non-intrusive load monitoring dataset [1] records voltage and current measurements in a single household for one week with a sampling rate of 12 kHz, leading to a total of tens of billions of numeric readings and making it extremely difficult to mine meaningful patterns in real-time using these raw numeric data. Researchers from EPFL also argue that while smart meter technologies make it possible for utility companies to analyze household energy consumption data in real-time, data acquired by smart meters are often so large that analytic tasks become extremely expensive [28].

P. Perner (Ed.): MLDM 2016, LNAI 9729, pp. 294–310, 2016.
DOI: 10.1007/978-3-319-41920-6_22

Traditionally, researchers have proposed various methodologies to represent time series more efficiently, including dimensionality reduction [9] and numerosity reduction [29] techniques. Another line of research on time series representation focuses on converting numeric values into symbolic form [7], while one of the most prominent approaches is Symbolic Aggregate approXimation (SAX) [14], which adapts both dimensionality and numerosity reduction techniques to transform numeric time series into symbolic representations. Although SAX comes with a distance measure that can be used for nearest neighbor classification, it is still unclear how classification performance will be affected when classifying SAX's symbolic representations of time series.

In this paper, we set to investigate how symbolic representations of time series can tackle time series classification (TSC) challenges. Specifically, we propose a novel TSC approach named Domain Series Corpus (DSCo, pronounced as `disco`), which firstly transforms numeric values into texts and then builds per-class language models from these texts. To classify unlabeled samples, we compute the fitness of each symbolized sample against all per-class models and choose the class represented by the model with the best fitness score. Our work innovatively takes advantage of mature techniques from both time series mining and Natural Language Processing (NLP) communities. Through extensive experiments on an open dataset archive, we demonstrate that our approach not only performs similarly or better than state-of-the-art approaches (which works with original data that possesses much more information), but can also work on reduced data, an essential property to ensure scalability in TSC.

Overall, the contributions of this paper are summarized as follows:

- We bring up a novel method for TSC by leveraging mature techniques from the NLP community. By taking advantage of language modeling techniques we are able to consider both *local* and *global* similarities among time series.
- We have tested our approach extensively on an open archive which contains datasets from various domains, demonstrating by comparison with state-of-the-art approaches that DSCo is performant, efficient and can be generalized.
- We prove that although our approach works with approximated data, DSCo can perform similarly to approaches that work with original uncompressed numeric data.
- We propose a new perspective for TSC: we view time series data as *sentences* (as in natural languages), where some *words* and their *combinations* will define different classes. In this way, we approximate a TSC task to a pseudo language detection problem.

The remainder of this paper is organized as follows. Section 2 briefly surveys related research work to ours. Section 3 provides our intuition and the necessary background information on time series classification as well as preliminaries on language modeling. Section 4 presents the details of our approach, while experiments and evaluation results compared with related work are described in Section 5. Section 6 concludes the paper with directions for future work.

2 Related Work

TSC is a major task in time series mining thanks to its wide application scenarios. As a consequence, there are a plethora of classification algorithms for TSC. Fu [7] has surveyed extensively on time series mining and TSC in his review paper. Due to space limitations, here we only introduce works that are closely related to ours.

Classically, machine learning classification is built by defining two elements: a *distance measure* to compare samples and a *classification algorithm* which implements the method of comparison. For example, in 1NN classification, a given sample is directly compared with samples from the training set. The tested sample will be assigned the class label of the sample from the training set which is the closest to it following a distance measure, for instance, the Euclidean distance. This approach can be expensive if the training dataset is large and thus will lead to a high number of pairwise comparisons.

As stated by Battista et al. "all of the current empirical evidence suggests that simple nearest neighbor classification is very difficult to beat" [2]. The core of nearest neighbor classifiers lies in defining an accurate distance measure. To perform best, kNN classifiers leverage the Dynamic Time Warping (DTW) distance which mitigates problems caused by distortion in the time axis [4, 20]. Thanks to its maturity and performance, DTW based 1NN classification has become one of the most prominent TSC approaches. Although recent empirical comparison [23] reveals that the Time Warp Edit Distance (TWED) [15] performs statistically more accurate than DTW. Other common distance measures include Euclidean distance, Edit Distance on Real sequence (EDR) [5] and Minimum Jump Cost (MJC) [24].

One issue with DTW, however, is that it focuses on finding *global* similarities, i.e., the overall curve *shape* of two time series in the time dimension. As a result, it requires applications to specify a proper warping window size or to properly align data samples. To solve this issue, *shapelets*-based algorithms [8, 16, 19, 30] try to find phase-independent defining subsequences in time series based on *local* similarities, so that such subsequences can be used as discriminatory features (or *primitives*) for classification. Our approach differs from *shapelets* since we consider both *global* and *local* similarities using language modeling techniques.

Some other approaches also borrow paradigms from the text mining community. One of the most prominent ones is the *bag-of-words* approach, which inspired the *bag-of-features* [3, 27] approach for TSC. SAX-VSM [22] also takes advantage of bag-of-words approach and builds term frequency-inverse document frequency (*tf*idf*) vectors in its training phase. It defines a similarity measure of two vectors (that are constructed from original series) based on their inner product.

The closest work related to DSCo are probably compression-based TSC approaches. Xi et. al. proposes numerosity reduction techniques [29] in order to speed up DTW distance calculation, while our approach takes advantage of both numerosity reduction and dimensionality reduction techniques to reduce the size of time series data. Furthermore, DSCo is not based on DTW and it is computationally more efficient than DTW, although there has been efforts to optimize DTW's time complexity [10,25]. Finally, this approach is inspired by our previous work [12], where we consider TSC in the household appliance profiling domain and draw an analogy between dialects in natural languages and different patterns of electricity signals each type of electrical appliance emits.

3 Background and Key Intuition

In this section, we briefly introduce the mechanism behind SAX and how SAX is traditionally used for TSC tasks. Then we present language modeling and how it can be used in combination with *SAX-ified* strings. In the meanwhile, we introduce our intuitions on tackling TSC with symbolic representation and language modeling.

3.1 Text Representation with SAX

Time series classification often involves learning what patterns are associated to a specific class of data readings. Consider the case of two samples from the `BirdChicken` [6] dataset illustrated in Fig. 1, where bird/chicken images have been transformed into time series (illustrated in gray curves). By observing the readings in different segments of the time series and which segment succeeds another, one can immediately summarize the characteristics of each class.

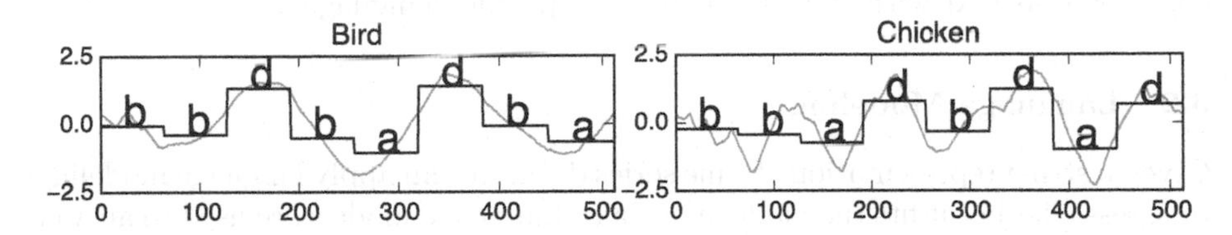

Fig. 1. Illustrative examples of how SAX works. The gray curves represent time series samples from the `BirdChicken` dataset. The black curves illustrate PAA representations of two time series samples where dimension is reduced from 512 to 8 ($n = 512$ and $s = 8$). Alphabets on top of the PAA curves are SAX representations of two time series samples with an alphabet size of four.

However, to better visualize the significant patterns, one must first reduce the dimensionality of the time series by considering only relevant variations. To that end, we leverage Piecewise Aggregate Approximation (PAA) [11] which can reduce the time series from n dimensions to s dimensions by dividing the data into s segments of equal size. The data reduction representation is then a vector of the mean values of the data readings per segment. Let $\bar{V} = \bar{v}_1, ..., \bar{v}_s$ be this vector where each $\bar{v}_i$ is computed by equation 1 (for more information please refer to [11,14]).

$$\bar{v}_i = \frac{s}{n} \sum_{k=\frac{n}{s}(i-1)+1}^{\frac{n}{s}i} v_k \tag{1}$$

The black curves in Fig. 1 illustrate the PAA representation of the original time series readings. Although very simple, the PAA dimensionality reduction enables to make an analogy with languages: the succession of segments (at different levels) in a time series is comparable to a succession of words and expressions which may define the vocabulary and phrases of a language. The intuition becomes straightforward: if one can collect a dictionary of the segments and their co-occurrence frequencies, pairwise comparison between samples can be effective. In order to benefit from the plethora of algorithms that exist in the NLP field, we must transform the PAA representation into a more symbolic representation with alphabets. To that end, we build on the Symbolic Aggregate approXimation (SAX) [14].

As an example of symbolic representation that fits with our intuition for language modeling using PAA values of time series segments, we consider the Symbolic Aggregate approXimation (SAX) [14]. SAX was initially brought up to transform real valued time series data into a sequence of alphabets, i.e., a string. It has then been proven especially efficient for motif (repeated patterns) discovery tasks. For example, it is advantageous to use SAX in order to find variable-length motifs [13,21]. Fig. 1 illustrates the SAX representations of samples from the two classes in the `BirdChicken` dataset. The string sequences yielded can now be considered as text with expressions from a specific "language".

3.2 Language Modeling

Given a string representation of time series data, we can apply language modeling to assess whether it fits the model of a class. Language models are used to answer questions such as *"How likely a string of words from a language vocabulary is good phrasing in this language?"*. A statistical language model is a probability distribution over strings of a corpus [18]. Thus, any sequence of words W has a probability score $P(W) = P(w_1, ..., w_n)$ in the language model, indicative of its relative validity within a language.

N-gram language models are common means of language modeling. In the simplest case of *unigram* models (1-gram models), the probability score of the sequence of words W is approximated to the product of the probabilities of each word. Equation 2 provides the formula for computing this score.

$$P(W) = P(w_1, ..., w_n) \approx \prod_{i=1}^{n} P(w_i) \tag{2}$$

Bigram models put conditions on the previous word to account for the likelihood of co-occurrence between two words (for instance, *beer drinkers* appears more often than *beer eaters*). The probability score of the sequence W is then approximately the product of conditional probabilities of words with their previous peer. It is computed by the formula in Equation 3.

$$P(W) = P(w_1, ..., w_n) \approx \prod_{i=1}^{n} P(w_i|w_{i-1}) \tag{3}$$

Since in a given corpus it is possible that certain bigrams are never observed beforehand, it is reasonable to use a back-off mechanism to take into account only their unigram probabilities. Although N-grams models can be theoretically insufficient because language has long-distance dependencies (for instance, some words may co-occur in a sentence but not directly following each other in the sequence: *"the computer which I just bought and setup in my room crashed"*), these models have been shown to be efficient in practice and used in various fields including speech recognition, author attribution and malware detection.

4 Domain Series Corpora for TSC

Since symbolic representation of time series data is a promising mechanism to tackle the numeriosity and high dimensionality issue in the era of big data, we investigate how symbolic representation and language modeling can be used for TSC. Recall that DSCo builds on the simple intuition that time series patterns in a specific class of a given domain can be differentiated from other class patterns, which is similar to NLP methods that distinguish texts from different languages or dialects. The assumption is thus that the language model extracted from the samples of a specific class will be descriptive and discriminative enough to differentiate it from another language model within the same domain. DSCo therefore consists of building a corpus of *words* representative of time series subsequences (or *segments*) for a given domain and the associated language models for its classes.

Figure 2 illustrates the steps for building per-class Domain Series Corpora for a specific domain. First of all, data readings of time series samples from each class are transformed into texts. Next, language modeling is applied on these texts to extract the corresponding language models, so that afterwards these models can be used to test and classify unlabeled samples.

4.1 Data Representation as Texts

As described in Section 3, we create symbolic representations for real-valued samples. In DSCo we have leveraged SAX for this task. It is nonetheless possible to leverage another symbolic representation algorithms for time series. The output of this step is a string representation for each time series sample.

4.2 Language Model Inference

DSCo explores a training set of time series to extract meaningful patterns of segments by studying their occurrence frequencies. Once time series are represented as texts, a language model can be built to summarize each time series class. To build a language model for a time series class, DSCo generates its corresponding dictionary by collecting *words* that appear in the training set. A large body of work have been proposed in the NLP literature on how to obtain such dictionaries.However, since the symbolic representations generated by SAX have no word boundaries, we need to break them into smaller pieces first by employing a corpus acquisition mechanism. This is common procedure for some natural languages such as Chinese, which has no obvious word boundaries.

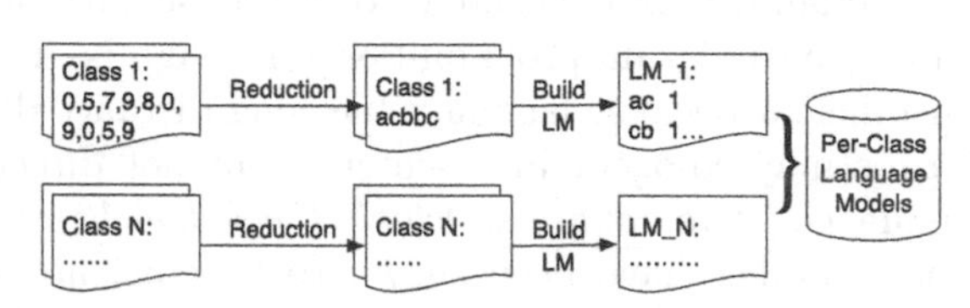

Fig. 2. Process for building language models in DSCo.

One approach to dictionary acquisition is to break the sequences using an annealing algorithm, which is a probabilistic and non-deterministic algorithm that randomly permutes the possible segmentations and searches the whole solution space with the best segmentation until thresholds are met according to an objective function. Such an approach is able to find word boundaries with reasonable accuracy given a large training set. Unfortunately, this algorithm is highly time-consuming. Besides, time series training sets are seldom sufficiently large to accommodate this algorithm. As a result, we extract *words* from the symbolic representations using a naïve sliding window method described in Algorithm 1. This algorithm collects all possible sub-strings of length w within a string, so that no descriptive segment is left uncaptured from the original time series. For example, we can break string `abccc` into the following 2-alphabet segments: `ab`, `bc`, `cc` and `cc`.

Algorithm 1. Extract *words* from a string (S) using a sliding window (of length l).

1: **procedure** ExtractWords(S, l)
2: $\quad words \leftarrow \varnothing$
3: $\quad$ **for** $i \leftarrow 0, GetLength(S) - l + 1$ **do**
4: $\quad\quad word \leftarrow SubString(S, i, l)$ ▷ Sub-string of size l
5: $\quad\quad words \leftarrow words \cup \{word\}$
6: $\quad$ **return** *words*

Next, in order to better preserve the descriptive information that time series generally come together in a sequence, we also compute the frequencies of n-grams. We build n-gram language models for each time series class in our training set. This process is illustrated in Algorithm 2. In order to be generic, we define a minimum (`minWL`) word length and a maximum word length (`maxWL`): The intuition behind this is that 2-alphabet segments may be generic but not descriptive, while segments with larger length may be descriptive but not generic enough. These extreme values can be easily determined with enough domain knowledge. For instance, if we assume that electrocardiogram (ECG) patterns generally have similar lengths, the `minWL` and `maxWL` values can be predefined to avoid producing noisy segments.

Algorithm 2. Build language models (*LMs*) from a list (*SL*) of (*string*, *label*) pairs.

1: **procedure** BUILDLM($SL, minWL, maxWL$)
2: $LMs \leftarrow \varnothing$
3: **for all** $(string, label) \in SL$ **do**
4: **if** $NGrams_{label} \notin LMs$ **then**
5: $NGrams_{label} \leftarrow \varnothing$
6: **for** $wl \leftarrow minWL$, $maxWL$ **do**
7: $words \leftarrow ExtractWords(string, wl))$
8: **for all** $ngram \in GetNGrams(words)$ **do**
9: InsertOrIncreaseFreq($NGrams_{label}, ngram$)
10: $LMs \leftarrow LMs \cup NGrams_{label}$
11: ConvertFreqToProbability(LMs)
12: **return** LMs

When all n-grams in every per-class language model are counted, we convert their frequencies into probabilities within each language model. Note that frequencies may need normalization when there are different number of instances in each class. Otherwise n-gram probabilities will be biased and lead to classification errors in the next step.

4.3 Classification

In DSCo, classification is performed by checking which language model is the best fit for the tested sample, as shown in Fig. 3. First, similar to the training phase, each test sample is reduced to a string using the same algorithms and parameters. Then, in order to test model fitness, each language model – which summarizes the characteristics of all samples from a given class – is used to segment the time series' text representation. To be computationally efficient, we consider a probabilistic language modeling approach with bigrams as introduced in Section 3. We compute the segmentation score as the product of conditional probability of all segmented words following Equation 3. If a bigram is not known in the language model, we just back-off to the unigram probability values for this specific segment. Since there are different ways to segment the text according to a specific language model, we only consider the best segmentation score that can be

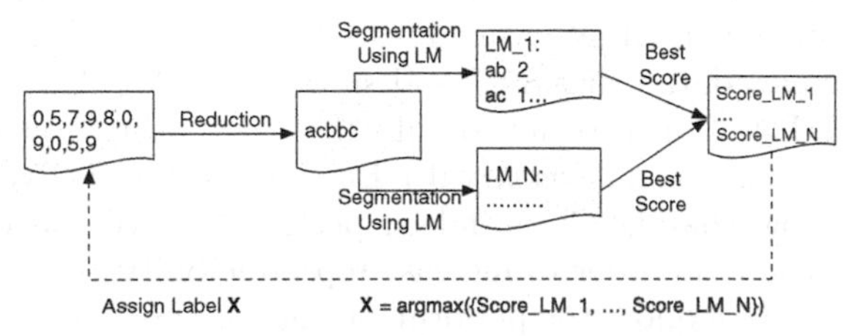

Fig. 3. Illustration of DSCo's classification process.

Algorithm 3. Given language models, find the best way (with the maximum probability) to segment a string (S).

Require: global $minWL$ and $maxWL$
1: **procedure** SPLITSTRING(S)
2: $P \leftarrow \varnothing$
3: $sl \leftarrow GetLength(S)$
4: **for** $l \leftarrow minWL, min(sl, maxWL)$ **do**
5: **if** $sl - l \geq minWL$ **then**
6: $P \leftarrow P \cup \{(Slice(S, 0, l), Slice(S, l, sl))\}$
7: **return** P
8: **procedure** SEGMENT(S, prev)
9: UpdateViterbiTable() ▷ Dynamic programming to avoid repetitive computing
10: **if** $GetLength(S) < minWL$ **then**
11: **return** 0
12: $Sgs \leftarrow \varnothing$
13: **for all** $(h, t) \in SplitString(S)$ **do**
14: $Sgs \leftarrow Sgs \cup \{P(S|prev) * Segment(t, h)\}$
15: **return** $max(Sgs)$

obtained with each language model. That is, each language model segments the sample and keeps its best segmentation score, then all language models compare their scores and the winning language model's class label will be applied to the sample.

5 Evaluation

In order to evaluate the feasibility and performance of our approach, we have implemented DSCo and tested on an open dataset archive. We first reduce time series data using SAX and show that SAX distance-based 1NN under-performs DTW. Then we investigate the value added by DSCo on top of SAX. Finally, through comparison with Euclidean- and DTW-based 1NN classification, we show that DSCo is indeed performant. We have open sourced our implementation[1] in order to increase reproducibility.

To explore the performance of the DSCo approach and investigate the extent of its applicability, we consider the Time Series Classification Archive [6] from University of California, Riverside. The datasets contained in this archive are popular within the TSC community, allowing for a reliable comparison baseline. Besides, the archive includes error rates for DTW- and Euclidean-based nearest neighbor classification as a performance benchmark for TSC. The UCR archive is composed of two sub-archives: `Pre_Summer_2015_Datasets` and `Newly Added Datasets` which include datasets from various fields, ranging from electrocardiograms (ECG) to intra-species image recognition data. We tested on the

[1] https://github.com/serval-snt-uni-lu/dsco

latter (which contains 39 different datasets) because its file format and internal data structures are consistent, making it possible to conduct batch processing in a content-agnostic manner. Furthermore, both sub-archives have similar dataset diversity and some datasets for specific domains (for example, ECG data) appear in both sub-archives. Specifically, these 39 datasets have various number of classes from 2 to 60 and different number of training and testing instances from 20 to 8,926, with time series lengths varying from 80 to 2,079.

5.1 Reducing Data Using SAX

In order to validate SAX's data reduction performance, we take those `Newly Added Datasets` from UCR and transform the numeric data records into symbols. We have set the maximum time series length to 100 and varied SAX's alphabet size from 3 to 20, so that we can take advantage of SAX's dimensionality and numeriosity reduction mechanism. Since SAX can be viewed as a lossy compression function, we evaluate how well the original information are kept for TSC using 1NN classification and compare the classification performance between DTW distance and SAX's internal distance measure as defined in [14]. SAX's distance measure is essentially a variance of Euclidean distance except that the distance between two alphabets are predefined in SAX's look-up table. For instance, for an alphabet size of 4, $dist(a, b) = 0$ and $dist(a, c) = 0.67$. Fig. 4 presents the 1NN classification comparison, where the solid lines indicate SAX distance-based 1NN classification accuracy for each of the 39 datasets, and the dashed lines shows DTW distance-based 1NN classification performance.

Here we show the classification accuracy of DTW-based 1NN because it is the most mature and widely used distance measure among the research community. Furthermore, classification performance using DTW is readily available from the UCR archive. Finally, using DTW requires one to explicitly setting its warping size parameter. In Fig. 4 we show DTW's performance with the best warping size (parameter space is 100: warping size varies from 0 to 100 percent of original time series length) that is found in [6]. When comparing SAX distance to DTW's best performance, we believe it is fair to present SAX's best accuracy results as well (parameter space is 18: alphabet size varies from 3 to 20). In this case DTW-based 1NN outperforms SAX in 69.2%(27/39) datasets indicating that SAX distance-based 1NN classification indeed under-performs DTW, possibly due to two major facts, namely SAX's lossy reduction of the original numeric data and SAX distance's inability to consider time series' global similarities. As a result, we take advantage of DSCo and take into account both global and local similarities of time series to counter SAX's lossy reduction. And as we shall demonstrate later, DSCo indeed classifies time series more accurately than SAX distance-based 1NN classification.

5.2 Implementation and Setup

Normally, DSCo's classification process can be extremely expensive due to the need of recursively dividing strings in to smaller pieces and segmenting these

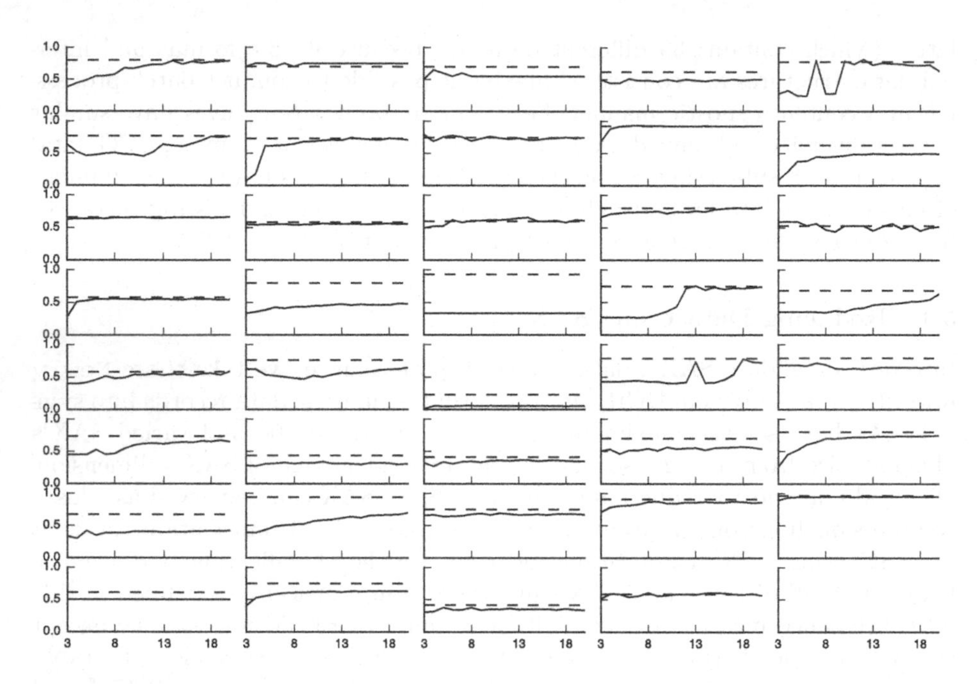

Fig. 4. 1NN classification accuracy comparison between DTW (dashed) and SAX (solid) distance.

sub-strings. However, a good implementation may take advantage of the Viterbi algorithm [26] – which is in essence a dynamic programming approach – to avoid redundant computation. In addition, the classification process calculates best segmentation scores based on the text segmentation algorithm provided in [17], which works exceptionally well for NLP text segmentation tasks and has a computational complexity of $O(nL^2)$, where n is the length of a testing string, and L is the maximum word length. Finally, DSCo computes segmentation scores using mainly bigram probabilities, only falling back to unigram probabilities when a specific bigram is not found.

Recall that time series longer than 100 have been arbitrarily reduced to 100-alphabet strings during dimensionality reduction, in order to speed up the classification process; and we have varied SAX's alphabet size from 3 to 20 to search for the best text representation. Since DSCo requires two parameters: a $(minWL, maxWL)$ tuple, we experiment with three sets of parameter settings: short segments $(minWL = 2, maxWL = 10)$, long segments $(minWL = 11, maxWL = 20)$ and short-long combined $(minWL = 2, maxWL = 20)$. Results illustrated in Fig. 5 show that DSCo's performance are relatively consistent regardless short or long segments (*words*) are configured. As a result, in the following comparisons, we fix $minWL$ to 2 and $maxWL$ to 20.

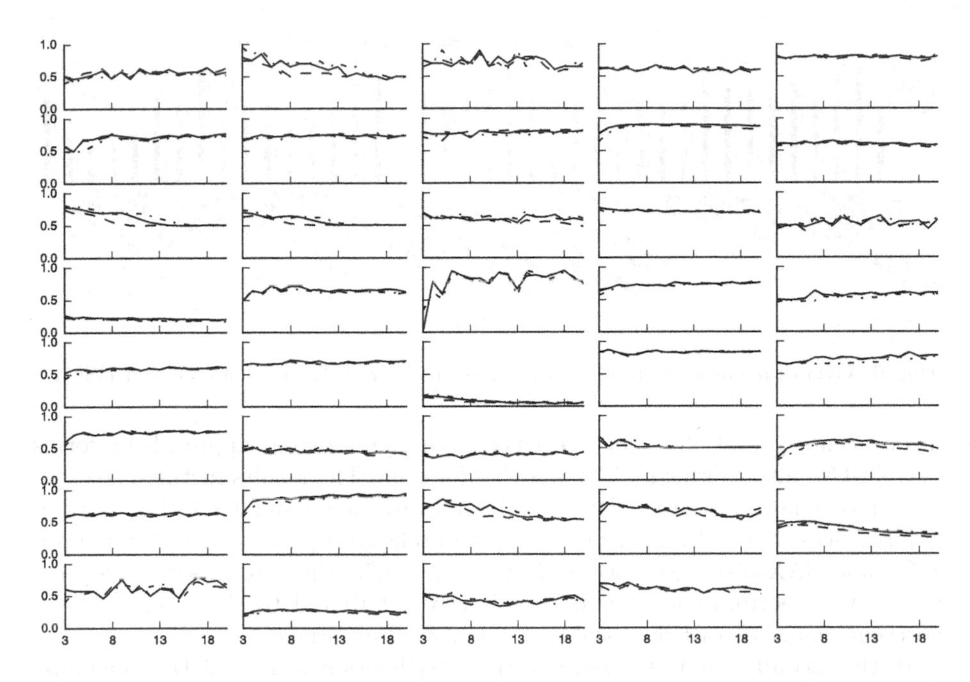

Fig. 5. DSCo's classification accuracy with different parameter settings: short segments (dashed-dotted lines), long segments (dashed lines) and combined (solid lines).

5.3 Comparison of Classification Performance

In the first round of experiments, we investigate the value added by the language modeling to the use of the SAX representations. Since SAX already comes with a distance metric which was demonstrated to yield better performance than the Euclidean distance on real-valued time series data [14], we compare DSCo against SAX-distance-based 1NN classification. Table 1 shows respectively the classifi-

Table 1. Classification accuracy comparison between the best performance of DSCo and 1NN with SAX distance, where $|\alpha|$ is the alphabet size when best performance is achieved.

Data-	SAX's Best		DSCo's Best		Data-	SAX's Best		DSCo's Best		Data-	SAX's Best		DSCo's Best	
set	$\|\alpha\|$	acc.	$\|\alpha\|$	acc.	set	$\|\alpha\|$	acc.	$\|\alpha\|$	acc.	set	$\|\alpha\|$	acc.	$\|\alpha\|$	acc.
1	14	**0.79**	11	0.62	14	20	**0.81**	3	0.77	27	11	0.34	4	**0.53**
2	4	0.75	5	**0.95**	15	3	0.59	14	**0.64**	28	14	0.37	17	**0.45**
3	4	0.65	9	**0.90**	16	8	**0.56**	3	0.27	29	18	0.57	3	**0.68**
4	6	0.51	9	**0.67**	17	19	0.49	6	**0.72**	30	20	**0.75**	11	0.64
5	12	0.81	12	**0.83**	18	3	0.33	6	**0.95**	31	20	0.42	19	**0.66**
6	20	**0.77**	20	**0.77**	19	13	0.76	17	**0.77**	32	20	0.70	15	**0.94**
7	14	0.74	20	**0.76**	20	20	**0.63**	7	**0.63**	33	11	0.69	5	**0.85**
8	3	**0.79**	3	0.78	21	20	0.59	20	**0.62**	34	9	**0.88**	5	0.78
9	7	**0.93**	11	0.91	22	20	0.64	8	**0.72**	35	14	**0.95**	6	0.50
10	19	0.50	9	**0.65**	23	8	0.06	4	**0.18**	36	3	0.50	17	**0.80**
11	20	0.66	3	**0.78**	24	18	0.78	4	**0.87**	37	17	**0.63**	7	0.30
12	11	0.56	3	**0.71**	25	7	0.71	17	**0.83**	38	5	0.38	6	**0.54**
13	14	0.66	3	**0.70**	26	19	0.67	18	**0.78**	39	8	0.61	6	**0.71**

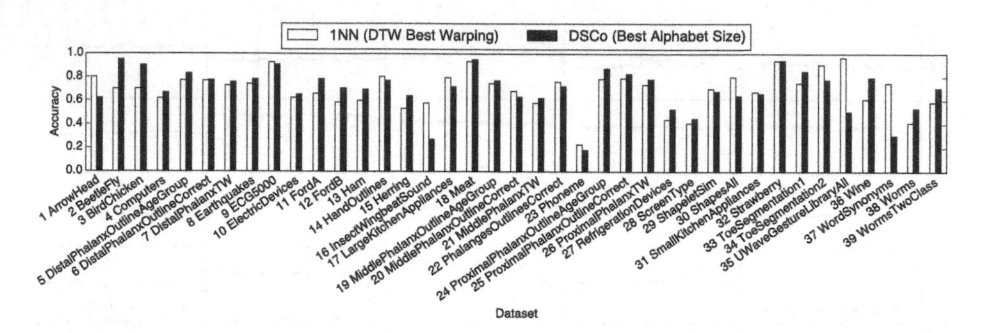

Fig. 6. Overall accuracy comparison between 1NN with DTW distance and DSCo.

cation accuracy and alphabet size when each classification approach performs best. DSCo outperforms in *74% (29/39)* datasets. The results further shows no explicit correlation between DSCo and SAX-distance-based 1NN classification performances, both obtained mostly with unrelated alphabet sizes. Since both DSCo and SAX-distance-based 1NN leverage SAX, these results thus suggest that DSCo's performance cannot be directly attributed to the usage of SAX representation, but rather to the language modeling process.

In the second round, we compare the classification results of DSCo against the benchmark 1NN with Euclidean distance. In *74% (29/39)* of the datasets, DSCo performs better in terms of classification accuracy. We also compare the improvement brought by DTW – the state-of-the-art approach – over Euclidean distance: DTW-based 1NN beats 1NN with Euclidean distance in *64% (25/39)* datasets. We further compare directly DSCo with DTW-based 1NN classification. Fig. 6 illustrates the results where we consider the best performance with DTW (i.e., with the best warping window size) and the best performance of DSCo (i.e., with the best SAX alphabet size). As shown, DSCo performs similarly to DTW in most datasets. DSCo appears to have good performance in image recognition tasks (for example, `BeetleFly` and `BirdChicken`), while it performs badly for some datasets where the training set has unbalanced distribution of different classes (e.g., `WordSynonyms`), making it difficult for DSCo to extract discriminatory n-grams. Overall, DSCo performs better than DTW in *64% (25/39)* datasets, indicating good classification performance.

Finally, we investigate if pruning bigrams affect classification accuracy. Fig. 7 illustrates the case with the `ECG5000` dataset, where we removed all bigrams that has frequencies lower than the mean value from each language model and use the pruned language models for classification. In some cases the pruned lan-

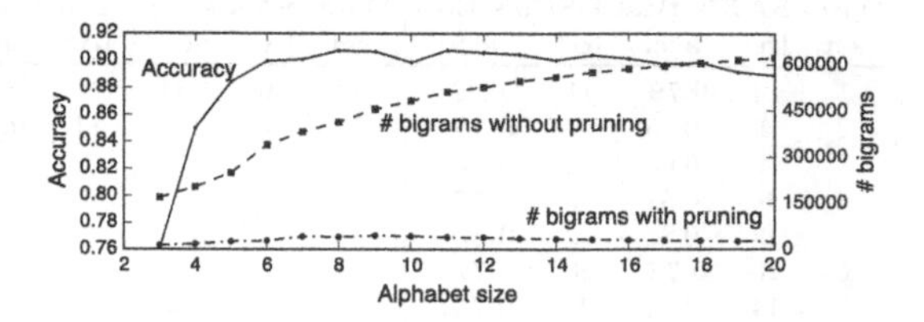

Fig. 7. For the `ECG5000` dataset, classification accuracy remains the same after pruning up to 95.8% bigrams.

guage models has less than 5% bigrams from the original ones. Yet surprisingly, the classification accuracy remains exactly the same. This indicates high redundancy in the language models and pruning techniques can be used in order to reduce memory footprints and speed up the classification process.

5.4 Time and Space Complexity

SAX has a linear time and space complexity when transforming real-valued time series into PAA and then to text representation. DSCo's language model inference step processes each training sample constant times and stores the models to external storage, resulting an $O(n)$ time and space complexity. We have shown that the fitness metric algorithm for classification in DSCo has a computational complexity of $O(nL^2)$ (in our experiments, $n \leq 100$ and $L = 20$). In comparison, the state-of-the-art DTW-based metric has a computational complexity of $O(n^2)$, although with LB_Keogh lower bounding technique this complexity can be reduced to $O(n)$ [25]. However, for complete classification processes, where a tested sample is compared against all samples from the training set, the 1NN approach using DTW with LB_Keogh yields a complexity of of $O(mn)$ where m is the size of training set. DSCo, on the other hand, has a complexity of $O(cnL^2)$ where c is the number of classes ($c \ll m$ for most scenarios).

Next, we compute the space complexity of DSCo. Given a time series sample T_i of length n, Algorithm 1 produces maximum $n-l+1$ unique *words* of length l. We denote T_i's l-sized words as $W_{i,l}$, then

$$1 \leq |W_{i,l}| \leq n - l + 1 \quad (4)$$

We denote the set of *unigrams* from a specific class of time series consisting of m instances ($T = \{T_1, ..., T_m\}$) as W, then

$$m \leq |W| \leq m \sum_{l=minWL}^{maxWL} (n - l + 1) \quad (5)$$

Since we use *bigrams* in our DSCo implementation, it thus has a space complexity of $O(m^2n^2)$, equivalent to that of shapelet-based algorithms, which also have a time complexity of $O(m^2n^2)$ for their full search procedures [8].

5.5 Limitations

We have demonstrated DSCo's performance through extensive experiments and complexity analysis. However, it also has its own limitations. Since DSCo essentially summarizes training samples into models, it performs best when there are a sufficient number of training samples in each class. Furthermore, our current implementation is built on top of SAX, tweaking SAX's parameters – for example, alphabet size – may require users to look inside data samples in order to achieve best performance. Nevertheless, we believe DSCo has offered a new perspective for TSC tasks and its limitations can be tackled in the near future by employing more advanced symbolization and corpus acquisition algorithms.

6 Conclusions and Future Work

In this work, we have brought up a novel approach named DSCo for time series classification. It works on symbolized time series data and builds per-class language models, against which testing samples are fitted in order to predict their corresponding class labels. Through extensive experiments we are able to prove that DSCo performs similarly or better than some state-of-the-art TSC approaches that works with original numeric data, namely 1NN with Euclidean and DTW distance. By taking advantage of mature algorithms from the NLP community, DSCo is able to achieve close-to-linear time complexity, which will be a great advantage for real-time applications.

Our future work will focus on further improving DSCo's performance, including reducing computation overhead and memory consumption by more effectively pruning n-grams in language models, improving classification accuracy and finding key defining subsequences for better user comprehension. In addition, other symbolization techniques can be taken advantage of to make DSCo more generalized and parameter free.

References

1. Anderson, K., Ocneanu, A., Benitez, D., Carlson, D., Rowe, A., Berges, M.: Blued: a fully labeled public dataset for event-based non-intrusive load monitoring research. In: Proceedings of the 2nd KDD Workshop on Data Mining Applications in Sustainability, pp. 1–5 (2012)
2. Batista, G.E., Wang, X., Keogh, E.J.: A complexity-invariant distance measure for time series. SDM **11**, 699–710 (2011)
3. Baydogan, M.G., Runger, G., Tuv, E.: A bag-of-features framework to classify time series. IEEE Transactions on Pattern Analysis and Machine Intelligence **35**(11), 2796–2802 (2013)
4. Berndt, D.J., Clifford, J.: Using dynamic time warping to find patterns in time series. In: KDD Workshop, vol. 10, pp. 359–370 (1994)
5. Chen, L., Özsu, M.T., Oria, V.: Robust and fast similarity search for moving object trajectories. In: Proceedings of the 2005 ACM SIGMOD International Conference on Management of Data, pp. 491–502. ACM (2005)
6. Chen, Y., Keogh, E., Hu, B., Begum, N., Bagnall, A., Mueen, A., Batista, G.: The UCR Time Series Classification Archive, July 2015. www.cs.ucr.edu/~eamonn/time_series_data/
7. Fu, T.C.: A review on time series data mining. Engineering Applications of Artificial Intelligence **24**(1), 164–181 (2011)
8. Hills, J., Lines, J., Baranauskas, E., Mapp, J., Bagnall, A.: Classification of time series by shapelet transformation. Data Mining and Knowledge Discovery **28**(4), 851–881 (2014)
9. Keogh, E., Chakrabarti, K., Pazzani, M., Mehrotra, S.: Dimensionality reduction for fast similarity search in large time series databases. Knowledge and Information Systems **3**(3), 263–286 (2001)
10. Keogh, E., Wei, L., Xi, X., Lee, S.H., Vlachos, M.: Lb_keogh supports exact indexing of shapes under rotation invariance with arbitrary representations and distance measures. In: Proceedings of the 32nd International Conference on Very Large Data Bases, pp. 882–893. VLDB Endowment (2006)

11. Keogh, E.J., Pazzani, M.J.: Scaling up dynamic time warping for datamining applications. In: Proceedings of the 6th ACM SIGKDD International Conference on Knowledge Discovery and Data Mining, pp. 285–289. ACM (2000)
12. Li, D., Bissyandé, T.F., Kubler, S., Klein, J., Le Traon, Y.: Profiling household appliance electricity usage with n-gram language modeling. In: The 2016 IEEE International Conference on Industrial Technology (ICIT 2016). IEEE, Taipei, March 2016
13. Li, Y., Lin, J.: Approximate variable-length time series motif discovery using grammar inference. In: Proceedings of the Tenth International Workshop on Multimedia Data Mining, p. 10 (2010)
14. Lin, J., Keogh, E., Wei, L., Lonardi, S.: Experiencing sax: a novel symbolic representation of time series. Data Mining and Knowledge Discovery **15**(2), 107–144 (2007)
15. Marteau, P.F.: Time warp edit distance with stiffness adjustment for time series matching. IEEE Transactions on Pattern Analysis and Machine Intelligence **31**(2), 306–318 (2009)
16. Mueen, A., Keogh, E., Young, N.: Logical-shapelets: an expressive primitive for time series classification. In: Proceedings of the 17th ACM SIGKDD International Conference on Knowledge Discovery and Data Mining, pp. 1154–1162. ACM (2011)
17. Norvig, P.: Natural language corpus data. In: Segaran, T., Hammerbacher, J. (eds.) Beautiful Data: The Stories Behind Elegant Data Solutions, pp. 219–242. O'Reilly Media, Inc. (2009)
18. Ponte, J.M., Croft, W.B.: A language modeling approach to information retrieval. In: Proceedings of the 21st Annual International ACM SIGIR Conference on Research and Development in Information Retrieval, pp. 275–281 (1998)
19. Rakthanmanon, T., Keogh, E.: Fast shapelets: a scalable algorithm for discovering time series shapelets. In: Proceedings of the Thirteenth SIAM Conference on Data Mining (2013)
20. Ratanamahatana, C.A., Keogh, E.: Three myths about dynamic time warping data mining. In: Proceedings of SIAM International Conference on Data Mining, pp. 506–510 (2005)
21. Senin, P., Lin, J., Wang, X., Oates, T., Gandhi, S., Boedihardjo, A.P., Chen, C., Frankenstein, S., Lerner, M.: GrammarViz 2.0: a tool for grammar-based pattern discovery in time series. In: Calders, T., Esposito, F., Hüllermeier, E., Meo, R. (eds.) ECML PKDD 2014, Part III. LNCS, vol. 8726, pp. 468–472. Springer, Heidelberg (2014)
22. Senin, P., Malinchik, S.: Sax-vsm: interpretable time series classification using sax and vector space model. In: IEEE 13th International Conference on Data Mining, pp. 1175–1180. IEEE (2013)
23. Serrà, J., Arcos, J.L.: An empirical evaluation of similarity measures for time series classification. Knowledge-Based Systems **67**, 305–314 (2014)
24. Serrà, J., Arcos, J.L.: A competitive measure to assess the similarity between two time series. In: Agudo, B.D., Watson, I. (eds.) ICCBR 2012. LNCS, vol. 7466, pp. 414–427. Springer, Heidelberg (2012)
25. Smith, A.A., Craven, M.: Fast multisegment alignments for temporal expression profiles. In: Proceedings of the 7th Annual International Conference on Computational Systems Bioinformatics, vol. 7, pp. 315–326. World Scientific (2008)
26. Viterbi, A.J.: Error bounds for convolutional codes and an asymptotically optimum decoding algorithm. IEEE Transactions on Information Theory **13**(2), 260–269 (1967)

27. Wang, X., Mueen, A., Ding, H., Trajcevski, G., Scheuermann, P., Keogh, E.: Experimental comparison of representation methods and distance measures for time series data. Data Mining and Knowledge Discovery **26**(2), 275–309 (2013)
28. Wijaya, T.K., Eberle, J., Aberer, K.: Symbolic representation of smart meter data. In: Proceedings of the Joint EDBT/ICDT 2013 Workshops, pp. 242–248. ACM (2013)
29. Xi, X., Keogh, E., Shelton, C., Wei, L., Ratanamahatana, C.A.: Fast time series classification using numerosity reduction. In: Proceedings of the 23rd International Conference on Machine Learning, pp. 1033–1040. ACM (2006)
30. Ye, L., Keogh, E.: Time series shapelets: a new primitive for data mining. In: Proceedings of the 15th ACM SIGKDD International Conference on Knowledge Discovery and Data Mining, pp. 947–956. ACM (2009)

Statistical Learning on Manifold-Valued Data

Alexander Kuleshov[1] and Alexander Bernstein[2,3(✉)]

[1] Skolkovo Institute of Science and Technology, Moscow, Russia
A.Kuleshov@skoltech.ru
[2] Institute for Systems Analysis, FRC CSC RAS, Moscow, Russia
a.bernstein@mail.ru
[3] Kharkevich Institute for Information Transmission Problems RAS, Moscow, Russia

Abstract. Regression on manifolds problem is to estimate an unknown smooth function f that maps p-dimensional manifold-valued inputs, whose values lie on unknown Input manifold **M** of lower dimensionality $q < p$ embedded in an ambient high-dimensional input space R^p, to m-dimensional outputs from training sample consisting of given 'input-output' pairs. We consider this problem in which Jacobian $J_f(X)$ of function f and Input manifold **M** should be also estimated. The paper presents a new geometrically motivated method for estimating a triple ($f(X)$, $J_f(X)$, **M**) from given sample. The proposed solution is based on solving a Tangent bundle manifold learning problem for specific unknown Regression manifold embedded in input-output space R^{p+m} and consisting of input-output pairs (X, $f(X)$), $X \in \mathbf{M}$.

Keywords: Regression on manifolds · Regression on features · Input manifold reconstruction · Jacobian estimation · Tangent bundle manifold learning

1 Introduction

Regression is a part of Statistical Learning whose general goal is finding a predictive function based on data [1–3]. Common regression problem is as follows. Let f(X) be an unknown smooth mapping from its domain of definition **M** lying in input space R^p to m-dimensional output space R^m. Given training 'input-output' sample

$$\mathbf{Z}_n = \left\{Z_i = \begin{pmatrix} X_i \\ f(X_i) \end{pmatrix}, i = 1, 2, \ldots, n\right\}, \tag{1}$$

the problem is to construct an estimator (learned function) $f^*(X)$ which maps the inputs $X \in \mathbf{M}$ to outputs $f^*(X) \in R^m$ with small predictive error:

$$f^*(X) \approx f(X), \tag{2}$$

and, thus, can be used to predict output f(X) for new 'previously unseen' input $X \in \mathbf{M}$.

There are various approaches and methods for reconstruction of an unknown function from the training sample, such as least squares techniques (linear and nonlinear),

P. Perner (Ed.): MLDM 2016, LNAI 9729, pp. 311–325, 2016.
DOI: 10.1007/978-3-319-41920-6_23

artificial neural networks, kernel nonparametric regression and kriging, SVM-regression, Gaussian processes, Radial Basic Functions, Gradient Boosting Regression, etc. [4–7].

In many applications it is necessary to estimate not only function f(X) but also its Jacobian matrix $J_f(X) = \nabla_X f(X)$. For example, many design tasks in Engineering are formulated as optimization of given function $f_1(X)$ over design variable $X \in \mathbf{M}$ under constraints defined in terms of other functions $f_2(X)$, $f_3(X)$, … , $f_m(X)$, which together determine m-dimensional vector-function f(X). Various gradient-based algorithms are used usually in solving of such optimization tasks. In general, the vector-function f is unknown and all available information about f is contained only in the sample $\mathbf{Z}_n$ (1). In this case, at first the learned function f*(X), which in engineering applications is referred to as surrogate function (model) or meta-model, is constructed from the sample. Then, an initial optimization task is replaced by 'surrogate' optimization task about the constructed surrogate function f*(X) [8–10]. Therefore, for using gradient-based algorithms in this surrogate optimization task, we need either to provide proximity $J_{f^*}(X) \approx J_f(X)$ between m×p Jacobian matrices $J_f(X)$ and $J_{f^*}(X)$ of unknown and learned functions f(X) and f*(X), respectively, or to construct sample-based m×p matrix J*(X) which accurately reconstructs the Jacobian matrix $J_f(X)$:

$$J^*(X) \approx J_f(X) \tag{3}$$

for all points $X \in \mathbf{M}$; the pair (f*(X), J*(X)) can be then used in optimization procedures. Note that most regression methods don't include Jacobian estimation.

In this paper, we will consider regression task as a sample-based construction of estimators f*(X) and J*(X) for unknown function f(X) and its Jacobian $J_f(X)$.

It is well known that if a dimensionality p of inputs X is large than many regression methods perform poorly due a statistical and computational 'curse of dimensionality': a collinearity or 'near-collinearity' of high-dimensional inputs cause difficulties when doing regression; reconstruction error in (2) cannot achieve a convergence rate faster than $n^{-s/(2s+p)}$ when nonparametric learned function f* estimates at least s times differentiable function f [11].

Fortunately, in many applications, especially in imaging and medical ones, high-dimensional inputs X are 'manifold-valued' data: all input values X lie on or near unknown manifold called Input manifold (IM) of lower dimensionality q < p (usually, q << p) embedded in an ambient high-dimensional input space R^p (Manifold assumption [12] about high-dimensional data); typically, this assumption is satisfied for 'real-world' high-dimensional data obtained from 'natural' sources.

Example. Wing shape optimization is one of important problem in aircraft designing in which design variables include a number of p-dimensional detailed descriptions X of wing airfoils consisting of coordinates of points lying densely on the airfoils' contours (in the airfoils coordinate system) [13]. In practical applications, the dimension p varies in the range from 50 to 200; a specific value of p is selected depending on the required accuracy of airfoil description. But high-dimensional descriptions of 'real' aerodynamic airfoils occupy only a very small part $\mathbf{M} \subset R^p$ of the 'airfoil-description' space R^p [14, 15] whose intrinsic dimension q varies in the range from 5 to 10.

An estimation of unknown function f defined on low-dimensional manifold, which is embedded in high-dimensional input space, is usually referred to as the Regression on manifolds problem; the papers [16–25] contain many other real examples of regression tasks in which inputs are manifold-valued data.

Various Regression on manifolds methods in an explicit or an implicit form use Dimensionality reduction technique for discovering low-dimensional structure of the Input manifold that allows avoiding the curse of dimensionality. In [16, 18], the kernel regression estimator is constructed directly on the manifold, using the geodesic distances on the IM. Another approach [17] is to employ usual Local Linear Regression technique in the ambient space R^p with regularization imposed on the coefficients in the directions perpendicular to certain estimators of tangent spaces to the IM. The derivatives of the unknown function f were estimated in [19]. A Manifold Learning tool called Manifold Adaptive Local Linear Estimator for the Regression [20] allows estimating the unknown function and its gradient. This tool explores the Riemannian geometric structure of the IM and constructs the Local Linear Regression directly on estimated tangent spaces to the IM, without knowing the geodesic distances and manifold structure. Bayesian nonparametric regression method for Regression on manifolds was proposed in [21, 22]. Geodesic regression and Polynomial regression on Riemannian manifolds were proposed in [23] and [24], respectively. A more general case concerning nonparametric regression between Riemannian manifolds (it means that output space R^m has dimension $m > 1$) is studied in [25] where minimization of regularized empirical risk is used for constructing the learned function.

In real examples, the IM $\mathbf{M}$ = Supp(f) being a support of unknown mapping f is also usually unknown, and all available information about $\mathbf{M}$ is contained in Input sample $\mathbf{X}_n = \{X_1, X_2, \ldots, X_n\}$. Manifold dimension q can be also estimated from this sample [26–29]; further we will suppose that the manifold dimension q is known.

In this paper, we will consider Regression on manifolds as problem of estimation of unknown triple (f(X), $J_f(X)$, $\mathbf{M}$) from given sample (1). Note that Regression on manifolds problem in above papers concerns, as a rule, only an estimation of unknown one-dimensional ($m = 1$) function f, without estimation of its Jacobian $J_f(X)$ and domain of definition $\mathbf{M}$ = Supp(f).

The Regression on manifolds problem is considered in Manifold Learning framework [30] in which various Data Analysis tasks (such as Pattern Recognition, Classification, Clustering, Prognosis, and others) are studied under Manifold assumption about the processed data.

We propose a new geometrically motivated method for considered Regression on manifolds which is based on reducing the problem to certain Dimensionality reduction task (namely, Tangent bundle manifold learning) [31, 32]) for specific unknown q-dimensional 'Regression manifold' $\mathbf{M}$(f) embedded in input-output space R^{p+m} and consisting of input-output vectors $(X, f(X))^T$, $X \in \mathbf{M}$; hereinafter, the vectors are written as column-vectors, symbol ^{T} denotes transposition.

The proposed approach gives also a new solution of common 'full-dimensional' Regression problem in which possible input values fill a full-dimensional area in the input space R^p (i.e., an intrinsic dimension q of the IM **M** equals p). The similar approach has already been used in [33] in common regression problem and resulted in new regression algorithm called Manifold Learning Regression which allowed avoiding serious drawbacks that are intrinsic to many regression methods (kernel nonparametric regression, kriging, Gaussian processes) when they are applied to functions with strongly varying gradients [2, 6, 7].

This paper is organized as follows. Section 2 contains strict definition of considered Regression on manifolds problem and describes possible natural approach to its solution; the difficulties arising at use of this approach are discussed also in this section. The proposed solution to the considered problem, which allowed avoiding these difficulties, is described in Sect. 3.

2 Regression on Manifolds: Problem Definition

2.1 Assumptions on the Input Manifold

Consider q-dimensional manifold

$$\mathbf{M} = \{X = \varphi(b) \in R^p: b \in \mathbf{B} \subset R^q\} \quad (4)$$

covered by a single unknown chart φ and embedded in an ambient p-dimensional space R^p, $q < p$. The chart φ is one-to-one mapping from open bounded Coordinate space $\mathbf{B} \subset R^q$ to the manifold $\mathbf{M} = \varphi(\mathbf{B})$ with differentiable inverse map φ^{-1}: $\mathbf{M} \to \mathbf{B}$. The intrinsic manifold dimension q is assumed to be known.

It is supposed that manifold **M** has positive condition number c(**M**) which is the number such that any point $X \in R^p$ distant from **M** by not more than $c^{-1}(\mathbf{M})$ has an unique projection onto the **M** [34], thus, no self-intersections, no 'short-circuit'; this means also that **M** has tubular neighborhood (ε-tube $U_\varepsilon(\mathbf{M})$ of radius $\varepsilon < c^{-1}(\mathbf{M})$).

Let h(X) be an arbitrary one-to-one Embedding mapping from the manifold **M** to the Feature space (FS) $\mathbf{Y}_h = h(\mathbf{M}) \subset R^q$ which gives low-dimensional representations (features) $y = h(X) \subset R^q$ of manifold-valued vectors $X \in \mathbf{M}$. Denote $g = h^{-1}$ an inverse mapping from the FS $\mathbf{Y}_h$ to the **M**. Thus, the manifold **M** is covered by chart g defined on the FS $\mathbf{Y}_h$ considered as Coordinate space and, therefore, can be written as

$$\mathbf{M} = \{X = g(y) \in R^p: y \in \mathbf{Y}_h \subset R^q\}. \quad (5)$$

If the mappings h(X) and g(y) are differentiable (covariant differentiation is used for h(X), $X \in \mathbf{M}$) and $J_h(X)$ and $J_g(y)$ are their q×p and p×q Jacobian matrices, respectively, than q-dimensional linear space

$$L(X) = \mathrm{Span}(J_g(h(X))) \quad (6)$$

in R^p is tangent space to the manifold **M** at the point $X \in \mathbf{M}$; hereinafter, Span(H) is linear space spanned by columns of arbitrary matrix H.

As follows from identity $g(h(X)) \equiv X$ for all $X \in \mathbf{M}$, the Jacobian matrices $J_h(X)$ and $J_g(y)$ satisfy the relations $J_g(h(X)) \times J_h(X) \equiv \pi(X)$ for all $X \in \mathbf{M}$, here $\pi(X)$ is p×p projection matrix onto the tangent space L(X) (6) to manifold **M** at the point $X \in \mathbf{M}$.

2.2 Regression on Manifolds: Problem Definition

Let f(X) be an unknown differentiable mapping from input space R^p to output space R^m whose domain of definition Supp(f) contains ε-tube $U_\varepsilon(\mathbf{M})$ of the Input manifold **M**, $\varepsilon < c^{-1}(\mathbf{M})$, and

$$J_f(X) = \nabla_X f(X) \tag{7}$$

be its m×p Jacobian matrix; covariant differentiation is used in (7) when we consider a restriction $f|_{\mathbf{M}}: X \in \mathbf{M} \to f(X) \in R^m$ of the mapping f on the IM **M**.

Using representation (5), function f(X) of manifold-valued vectors $X \in \mathbf{M}$ can be written as $f(X) = g_f(h(X))$ where function

$$g_f(y) = f(g(y)) = (f \bullet g)(y) \tag{8}$$

is defined on the FS $\mathbf{Y}_h$. Thus, $J_f(X) = J_{f\bullet g}(h(X)) \times J_h(X)$, where $J_{f\bullet g}(y)$ is m×q Jacobian matrix of the function $g_f(y)$, and, so, Jacobian $J_f(X)$ has rank $\min(m, q) \le \min(m, p)$.

The Regression on manifolds is considered in the following setting: Given a sample $\mathbf{Z}_n$ (1), construct the learned function f*(X), m×p matrix J*(X), and q-dimensional manifold **M*** embedded in an ambient space R^p, which is domain of definition of learned function f*, providing approximate equalities (2), (3) for all points $X \in \mathbf{M}$ and proximity

$$\mathbf{M}^* \approx \mathbf{M} \tag{9}$$

that means a small Hausdorff distance $d_H(\mathbf{M}^*, \mathbf{M})$ between the manifolds **M*** and **M**.

2.3 Regression on Manifolds: Possible Approach to Solution

If the IM **M** (5) has been known, a Regression on manifolds problem could be reduced (at least relative to an estimation of unknown function f) to the following 'full-dimensional' regression task on the low-dimensional FS $\mathbf{Y}_h$: based on 'feature' sample

$$\{(y_{i,h} = h(X_i), g_f(y_{i,h}) = f(X_i)), i = 1, 2, \ldots, n\} \tag{10}$$

consisting of known input-output values of an unknown function $g_f(y)$ (8), construct the learned function g*(y) that accurately approximates the unknown function $g_f(y)$: $g^*(y) \approx g_f(y)$ for all $y \in \mathbf{Y}_h$; this reduced task will be referred to as the Regression on Feature space problem (Regression on features, shortly). Then $f^*(X) = g^*(h(X))$ can be taken as an estimator of unknown function f of manifold-valued inputs $X \in \mathbf{M}$.

If the IM **M** is unknown, a possible and natural approach to the Regression on manifolds problem may consist of two solved sequentially by problems:

- Input manifold reconstruction problem [35, 36] consisting in estimation (reconstruction) of unknown IM **M** from given Input sample $\mathbf{X}_n$. Its solution gives a pair (h, g_{in}) consisting of an Embedding mapping h from the IM **M** to the Estimated Feature space (EFS) $\mathbf{Y}_h = h(\mathbf{M})$ and an Input reconstruction mapping g_{in} from the EFS $\mathbf{Y}_h$ to the input space R^p which provide proximity

$$r_{in}(X) \equiv g_{in}(h(X)) \approx X \quad (11)$$

between points $X \in \mathbf{M}$ and their recovered values $r_{in}(X)$ which are results of successively applying the embedding and input reconstruction mappings to the inputs $X \in \mathbf{M}$. The pair (h, g_{in}) determines an estimator

$$\mathbf{M}_{in} = \{X = g_{in}(y) \in R^p : y \in \mathbf{Y}_h \subset R^q\}$$

of the IM **M** which, due proximity (11), provides proximity $\mathbf{M}_{in} \approx \mathbf{M}$;
- followed by Regression on features problem consisting in estimation of an unknown function $g_f(y) = f(g_{in}(y))$ defined on the EFS $\mathbf{Y}_h$ from feature sample (10). Its solution is a function $g_{out}(y)$, $y \in \mathbf{Y}_h$, which provides proximity

$$g_{out}(y) \approx g_f(y) \qquad \text{for all } y \in \mathbf{Y}_h. \quad (12)$$

It follows from (12) that $g_{out}(h(X)) \approx g_f(h(X)) = f(r_{in}(X))$ for all points $X \in \mathbf{M}$. If a solution (h, g_{in}) of the Input manifold reconstruction problem meets, at the same time with proximity (11), an additional 'functional' proximity

$$f(r_{in}(X)) \approx f(X), \quad (13)$$

than the learned function $f^*(X) = g_{out}(h(X))$ provides the required proximity (2) and, thus, this 'two-step' approach really allows avoiding the curse of dimensionality.

The proximity (13) is a requirement to the solution (h, g_{in}), which is obtained on the Input manifold reconstruction step, but this requirement is formulated with use of unknown function f which will be estimated on next, Regression on features, step. Thus, these steps cannot be solved sequentially and independently.

That is why we propose new unified approach in which these interrelated problems are formulated and solved simultaneously and interdependently.

3 Regression on Manifolds: Approach and Solution

3.1 Regression on Manifolds: Proposed Approach

Consider a smooth Regression manifold (RM)

$$\mathbf{M}(f) = \{Z = F(X) \in R^r : X \in \mathbf{M} \subset R^p\} \quad (14)$$

embedded in an ambient input-output space R^r, $r = p + m$, in which

$$F: X \to Z = F(X) = \binom{X}{f(X)} \in R^r \quad (15)$$

is the mapping defined on the ε-tube $U_\varepsilon(\mathbf{M})$ of the IM **M**. The dataset $\mathbf{Z}_n$ (1) can be considered as a sample from the unknown Regression manifold **M**(f).

It follows from (4) that $\mathbf{M}(f) = \{Z = F(\varphi(b)) \in R^r: b \in \mathbf{B} \subset R^q\}$ is q-dimensional manifold embedded in ambient space R^r and covered by a chart $F(\varphi)$.

The mapping F has r×p Jacobian matrix

$$J_F(X) = \begin{pmatrix} I_q \\ J_f(X) \end{pmatrix} \times \pi(X) \tag{16}$$

in which first multiplier (r×p matrix) is split into p×p unit matrix I_p and m×p Jacobian matrix $J_f(X)$ (covariant differentiation is used in the $J_f(X)$ (7) when $X \in \mathbf{M}$). Thus, $\text{rank}(J_F(X)) = q$ and q-dimensional linear space

$$L_f(X) = \text{Span}(J_F(X)) \tag{17}$$

is tangent space to the RM **M**(f) at the point $F(X) \in \mathbf{M}(f)$ (if $X \in \mathbf{M}$).

We propose a new geometrically motivated approach to the Regression on manifolds problem which reduces this problem to Tangent bundle manifold learning problem (TBML) for the RM **M**(f) consisting in estimating both the RM **M**(f) (14) and tangent spaces $\{L_f(X), X \in \mathbf{M}\}$ (17) to the RM **M**(f) from the dataset $\mathbf{Z}_n$ (1) sampled from the RM **M**(f). In this approach, both the above formulated intermediate 'Input manifold reconstruction' and 'Regression on features' problems are solved simultaneously ensuring satisfying the 'simultaneous' requirement (13) and resulting in required estimators of the triple $(f(X), J_f(X), \mathbf{M})$. The approach allowed also estimating the tangent spaces to the Input manifold **M**.

First, we will use the previously proposed [31, 32, 36] Grassmann&Stiefel Eigenmaps (GSE) method which gives a solution to this TBML problem. In particular, the GSE results in Embedding mapping $h_{GSE}(Z)$, which maps the RM **M**(f) to the Regression Feature Space (RFS) $\mathbf{Y}_{GSE} = h_{GSE}(\mathbf{M}(f))$, and reconstruction mapping $g_{GSE}(y)$ from the RFS $\mathbf{Y}_{GSE}$ to the input-output space R^r in such a way that

$$r_{GSE}(F(X)) \equiv g_{GSE}(h_{GSE}(F(X))) \approx F(X) \tag{18}$$

for all points $Z = F(X) \in \mathbf{M}(f)$, or, the same, for all points $X \in \mathbf{M}$. In accordance with splitting of vector $Z \in R^{p+m}$ into two vectors $Z_{in} \in R^p$ and $Z_{out} \in R^m$, the mapping $g_{GSE}(y)$ also can be split into two mappings $g_{GSE,in}$ and $g_{GSE,out}$:

$$g_{GSE}(y) = \begin{pmatrix} g_{GSE,in}(y) \\ g_{GSE,out}(y) \end{pmatrix} \tag{19}$$

from the RFS $\mathbf{Y}_{GSE}$ to the input space R^p and output space R^m, respectively. Denote

$$h_{GSE,f}(X) = h_{GSE}(F(X)), \tag{20}$$

$$f_{GSE}(X) = g_{GSE,out}(h_{GSE,f}(X)), \tag{21}$$

then, as it follows from notation (15), the approximate relations (18) can be written as

$$r_{GSE,in}(X) \equiv g_{GSE,in}(h_{GSE,f}(X)) \approx X, \quad (22)$$

$$f_{GSE}(X) \approx f(X), \quad (23)$$

for all $X \in \mathbf{M}$. Therefore,

- function $f_{GSE}(X)$ (21) accurately reconstructs (23) unknown function f,
- function $h_{GSE,f}(X)$ (20) maps the IM **M** into the RFS $\mathbf{Y}_{GSE} = h_{GSE,f}(\mathbf{M})$,
- q-dimensional Recovered input manifold (RIM)

$$\mathbf{M}_{GSE} = \{g_{GSE,in}(y) \in R^p: y \in \mathbf{Y}_{GSE} \subset R^q\} \quad (24)$$

  embedded in input space R^p and covered by a single chart $g_{GSE,in}$ defined on the RFS $\mathbf{Y}_{GSE}$ meets Manifold proximity

$$\mathbf{M}_{GSE} \approx \mathbf{M}; \quad (25)$$

  the latter follows from inequality $d_H(\mathbf{M}_{GSE}, \mathbf{M}) \leq \sup_{X \in \mathbf{M}} |r_{GSE,in}(F(X) - X|$, see (22).

The GSE allows also constructing m×p matrix $G_{GSE,f}(X)$ that accurately approximates Jacobian $J_f(X)$. The first step is described in details in next Sect. 3.2.

But although the GSE-solution ($f_{GSE}(X)$, $G_{GSE,f}(X)$, $\mathbf{M}_{GSE}$) accurately approximates the unknown triple (f(X), $J_f(X)$, **M**), it cannot considered as solution to the Regression on manifolds problem because it depends on unknown function f.

Because of this, we consider this GSE-solution as preliminary one to the considered problem, based on which we construct in the second step final desired triple (f*(X), J*(X), **M***) which meets the required properties and, thus, solves the Regression on manifolds problem. The details of this step are described in Sect. 3.3.

3.2 Tangent Bundle Manifold Learning for Regression Manifold

Tangent Bundle Manifold Learning Definition. As was shown in [32], a minimization of reconstruction error when solving Manifold reconstruction problem implies additional requirement to the Embedding and Reconstruction mappings: they should ensure not only Manifold proximity, but also Tangent proximity property consisting in proximity between the tangent spaces to the initial and recovered manifolds.

In topology, a set composed of manifold points equipped by tangent spaces at these points is known as Tangent bundle of the manifold. An amplification of the Manifold reconstruction problem consisting in reconstructing of not only manifold points but also of the tangent spaces at these points can be referred to as the Tangent bundle manifold learning problem [31, 32]. The paper [32] contains theoretical justification of the need of accurate recovery of the tangent spaces for providing good generalization ability properties of a solution to the manifold reconstruction problem.

Grassmann&Stiefel Eigenmaps. The GSE being applied to a solving of the TBML problem for the RM $\mathbf{M}(f)$ constructs:

- Embedding mapping $y = h_{GSE}(Z)$ that defines Embedding mapping $h_{GSE,f}(X)$ (20),
- reconstruction mapping $g_{GSE}(y)$ that maps the RFS $\mathbf{Y}_{GSE}$ to input-output space R^r,

which ensure together proximity (18) between the manifold points $Z = F(X) \in \mathbf{M}(f)$ and their recovered values $r_{GSE}(Z) = g_{GSE}(h_{GSE}(Z))$ which implies manifold proximity

$$\mathbf{M}_{GSE}(f) \approx \mathbf{M}(f) \tag{26}$$

between the RM $\mathbf{M}(f)$ and Recovered Regression manifold (RRM)

$$\mathbf{M}_{GSE}(f) = \{Z = g_{GSE}(y) \in R^r: y \in \mathbf{Y}_{GSE} \subset R^q\}. \tag{27}$$

The GSE constructs also r×q matrix $G_{GSE}(y)$ which accurately reconstructs r×q Jacobian matrix $J_{GSE,g}(y)$ of the mapping $g_{GSE}(y)$:

$$G_{GSE}(y) \approx J_{GSE,g}(y), \tag{28}$$

that implies proximity between tangent space $L_f(X)$ (17) and tangent space to the RRM $\mathbf{M}_{GSE}(f)$ at the recovered point $r_{GSE}(X) \in \mathbf{M}_{GSE}(f)$. The tangent spaces are considered as elements of the Grassmann manifold Grass(r, q) consisting of all q-dimensional linear subspaces in R^r; a closeness between the tangent spaces is defined by use of certain metric on the Grassmann manifold.

If sample size n tends to infinity, the reconstruction error $|r_{GSE}(F(X)) - F(X)|$ in approximate equality (18) tends to zero uniformly in points $X \subset \mathbf{M}$ with a rate

$$|r_{GSE}(F(X)) - F(X)| = O(n^{-2/(q+2)}) \tag{29}$$

as $n \to \infty$ with high probability [37], the term 'with high probability' means that relation (29) holds with probability at least $(1 - C_\alpha / n^\alpha)$, where $\alpha \geq 1$ is an arbitrary number and C_α depends only on the number α (not on n).

Therefore, asymptotic relation $d_H(\mathbf{M}_{GSE}(f), \mathbf{M}(f)) = O(n^{-2/(q+2)})$ holds as $n \to \infty$, and this convergence rate coincides with the asymptotically minimax lower bound [38] for Hausdorff distance between the RM $\mathbf{M}(f)$ and RRM $\mathbf{M}_{GSE}(f)$. The same result $d_H(\mathbf{M}_{GSE}, \mathbf{M}) = O(n^{-2/(q+2)})$ as $n \to \infty$ takes a place for the IM $\mathbf{M}$ and RIM $\mathbf{M}_{GSE}$ (24). Thus, the GSE provides asymptotically minimal reconstruction error.

Preliminary GSE-Based Solution. The relation $F_{GSE}(X) \approx F(X)$ (18) admits applying a covariant differentiation with respect to the $X \in \mathbf{M}$ which results in relation

$$\nabla_X F_{GSE}(X) \equiv J_{GSE,g}(h_{GSE,f}(X)) \times J_{GSE,h,f}(X) \approx J_F(X), \tag{30}$$

where $J_{GSE,h,f}(X)$ is q×p Jacobian matrix of the mapping $h_{GSE,f}(X)$ (20).

The r×q matrices $J_{GSE,g}(y)$ and $G_{GSE}(y)$ can be written as

$$J_{GSE,g}(y) = \begin{pmatrix} J_{GSE,g,in}(y) \\ J_{GSE,g,out}(y) \end{pmatrix} \text{ and } G_{GSE}(y) = \begin{pmatrix} G_{GSE,in}(y) \\ G_{GSE,out}(y) \end{pmatrix}$$

by splitting these matrices into p×q matrices $J_{GSE,g,in}(y)$, $G_{GSE,in}(y)$ and m×q matrices $J_{GSE,g,out}(y)$, $G_{GSE,out}(y)$, respectively. Taking into account representation (16) and equalities (28), approximate equalities (30) can be written as

$$G_{GSE,in}(h_{GSE,f}(X)) \times J_{GSE,h,f}(X) \approx \pi(X), \tag{31}$$

$$G_{GSE,out}(h_{GSE,f}(X)) \times J_{GSE,h,f}(X) \approx J_f(X). \tag{32}$$

Write QR-decomposition [39] of the p×q matrix $G_{GSE,in}(y)$:

$$G_{GSE,in}(y) = G_{GSE,in,ort}(y) \times R(y),$$

where $G_{GSE,in,ort}$ is p×q orthogonal matrix and R(y) is q×q nonsingular upper triangular matrix. Then p×p projection matrix

$$\pi_{GSE}(X) = G_{GSE,in,ort}(h_{GSE,f}(X)) \times G^{T}_{GSE,in,ort}(h_{GSE,f}(X)) \tag{33}$$

onto the tangent space to the RIM $\mathbf{M}_{GSE}$ (24) at the point $r_{GSE,in}(X)$ (22) accurately approximates projection matrix π(X) to the tangent space L(X) (6):

$$\pi_{GSE}(X) \approx \pi(X).$$

With use standard Least-squares technique, it follows from the relations (31) that

$$G_{GSE,h}(X) = G^{-}_{GSE,in}(h_{GSE,f}(X)) \times \pi_{GSE}(X) \tag{34}$$

accurately reconstructs the Jacobian $J_{GSE,h,f}(X)$:

$$G_{GSE,h}(X) \approx J_{GSE,h,f}(X) \tag{35}$$

for all points X ∈ **M**, here

$$G^{-}_{GSE,in}(h_f(X)) = (G^{T}_{GSE,in}(h_f(X)) \times G_{GSE,in}(h_f(X)))^{-1} \times G^{T}_{GSE,in}(h_f(X)) \tag{36}$$

is q×p pseudoinverse Moore-Penrose matrix [39]. Finally, consider p×q matrix

$$G_{GSE,f}(X) = G_{GSE,out}(h_{GSE,f}(X)) \times G_{GSE,h}(X) \tag{37}$$

that meets, due to relations (32) and (35), required proximity

$$G_{GSE,f}(X) \approx J_f(X). \tag{38}$$

The triple ($f_{GSE}(X)$, $G_{GSE,f}(X)$, $\mathbf{M}_{GSE}$) defined in formulas (21), (37), and (24) depends on the unknown function f and is considered as 'preliminary' solution to the Regression on manifolds problem.

3.3 Regression on Manifolds: Solution

An Approach. The triple ($f_{GSE}(X)$, $G_{GSE,f}(X)$, $\mathbf{M}_{GSE}$) was constructed with use of the following GSE-based statistics: Embedding mapping $h_{GSE,f}(X) = h_{GSE}(F(X))$ (20), Reconstructing mapping $g_{GSE}(y)$ (19), and estimator $G_{GSE}(y)$ (28) of its Jacobian matrix. Based on these statistics, other required statistics were constructed: the RFS $\mathbf{Y}_{GSE}$, RIM $\mathbf{M}_{GSE}$ (24), estimator $G_{GSE,h}(X)$ (34) of Jacobian of Embedding mapping $h_{GSE,f}(X)$, learned function $f_{GSE}(X)$ (21), and estimator $G_{GSE,f}(X)$ (37) of Jacobian $J_f(X)$. But all these quantities depend on unknown function f and their values are known only at sample points $\{X_i\}$ or their features $\{y_{i,h} = h_{GSE,f}(X_i)\}$, respectively.

Based on these known values, we construct the estimators $h^*(X)$ and $g^*_{GSE}(y)$ for the mappings $h_{GSE,f}(X)$ and $g_{GSE}(y)$. A peculiarity of these regression problems is that we know (already estimated) Jacobians of these mappings at sample points. An unified approach to solving such regression problems, called Known Jacobian Regression (KJR), is proposed in next subsection.

Based on obtained solutions to these KJR-problems, we construct final solution for considered Regression on manifolds problem.

Regression Problems with Known Jacobians. Let $W = T(V)$ be an unknown smooth mapping from its domain of definition $\mathbf{V} \subset R^t$ to s-dimensional Euclidean space R^s with unknown s×t Jacobian matrix $J_T(V)$. Let

$$\{(W_i = T(V_i), H_i = J_T(V_i)), i = 1, 2, \ldots, n\} \tag{39}$$

be given sample consisting of values of both the function $T(V)$ and its Jacobian $J_T(V)$ in sample points $\{V_i \in \mathbf{V}\}$. The problem is to estimate the unknown mapping $T(V)$ and its Jacobian matrix $J_T(V)$ at every point $V \in \mathbf{V}$ from the sample (39).

Standard Kernel Nonparametric Regression (KNR) approach [40] to this problem is as follows. Let $K(V, V')$ be some kernel on the $\mathbf{V}$ reflecting nearness between the points V and V′; $K_\varepsilon(V, V') = I\{|V - V'| < \varepsilon\}$ is simplest often used kernel in which $I\{\cdot\}$ is indicator function and ε is chosen small parameter. The estimator

$$T_{KNR}(V) = \frac{1}{K(V)} \sum_{j=1}^{n} K(V, V_j) \times W_i, \quad K(V) = \sum_{j=1}^{n} K(V, V_j), \tag{40}$$

minimizes the 'first-order' residual $\Delta_1(T) = \sum_{i=1}^{n} K(V, V_i) \times |T - T(V_i)|^2$ over T.

Using the Taylor series expansion $T(V) \approx T(V') + J_T(V') \times (V - V')$ of the function $T(V)$ at near points V′, V, consider the following 'second-order' cost function

$$\Delta_2(T) = \sum_{i=1}^{n} K(V, V_i) \times |T - W_i - H_i \times (V - V_i)|^2$$

depending on vector $T \in R^s$. Then the Known Jacobian Regression estimator

$$T_{KJR}(V) = T_{KNR}(V) + \frac{1}{K(V)} \sum_{i=1}^{n} K(V, V_i) \times H_i \times (V - V_i), \tag{41}$$

minimizes cost function $\Delta_2(T)$ over T.

If some kernel K(V, V′) equals to zero when $|V - V'| > \varepsilon$ (the kernel $K_\varepsilon(V, V')$ satisfies this condition) than the estimators (40) and (41) have asymptotic expansions $|T_{KNR}(X) - T(X)| = O(\varepsilon)$ and $|T_{KJR}(X) - T(X)| = O(\varepsilon^2)$ as $\varepsilon \to 0$.

The KNR-statistic

$$J_{T,KNR}(V) = \frac{1}{K(V)} \sum_{j=1}^{n} K(V, V_i) \times H_i \tag{42}$$

accurately reconstructs s×t Jacobian matrix $J_T(V)$ for all points $V \in \mathbf{V}$.

Used Kernels. Hereinafter, by $\{\varepsilon_i\}$ we will denote algorithm parameters which are small positive numbers. Let set $U_n(X, \varepsilon_1) \subset \mathbf{M}$ consists of chosen point $X \in \mathbf{M}$ and sample points that belong to ε_1-ball in R^p centered at X. An applying the Principal Component Analysis (PCA) [41] to the set $U_n(X, \varepsilon_1)$ results in p×q orthogonal matrix $Q_{PCA}(X)$ whose columns are PCA principal eigenvectors corresponding to q largest PCA eigenvalues. In what follows, we assume that the IM **M** is 'well-sampled' (this means that sample size n is large enough) to ensure a positive value of the qth eigenvalue in the PCA and provide proximities

$$L_{PCA}(X) \approx L(X) \text{ for all } X \in \mathbf{M}, \tag{43}$$

here $L_{PCA}(X) = \mathrm{Span}(Q_{PCA}(X))$. Taking the ball radius ε_1 tending to 0 as $n \to \infty$ with rate $O(n^{-1/(q+2)})$, the PCA-error in (43) is $O(n^{-1/(q+2)})$ with high probability [42, 43].

Let $K_E(X, X') = I\{|X' - X| < \varepsilon_1\} \times \exp\{-\varepsilon_2 \times |X - X'|^2\}$ be Euclidean kernel and

$$K_G(X, X') = I\{d_{BC}(L_{PCA}(X), L_{PCA}(X')) < \varepsilon_3\} \times K_{BC}(L_{PCA}(X), L_{PCA}(X'))$$

be data-based 'Grassmann' kernel on the IM; here

$$d_{BC}(L_{PCA}(X), L_{PCA}(X')) = \{1 - \mathrm{Det}^2[Q_{PCA}^T(X) \times Q_{PCA}(X')]\}^{1/2},$$

$$K_{BC}(L_{PCA}(X), L_{PCA}(X')) = \mathrm{Det}^2[Q_{PCA}^T(X) \times Q_{PCA}(X')]$$

are the Binet-Cauchy metric and Binet-Cauchy kernel, respectively, on the Grassmann manifold Grass(p, q) [44, 45]. Consider new 'aggregate' kernel

$$K(X, X') = K_E(X, X') \times K_G(X, X') \tag{44}$$

reflecting not only Euclidean nearness between X and X′ but also nearness between the spaces $L_{PCA}(X)$ and $L_{PCA}(X')$; this nearness results in nearness between the tangent spaces L(X) and L(X′) to the IM **M**. The p×p projection matrix

$$\pi^*(X) = Q_{PCA}(X) \times Q_{PCA}^T(X) \tag{45}$$

onto the linear space $L_{PCA}(X)$ provides proximity $\pi^*(X) \approx \pi(X)$ and will be used further instead of the matrices $\pi_{GSE}(X)$ (33) and $\pi(X)$.

The GSE also provides feature kernel k(y, y′) defined on the pairs $(y, y_{i,h})$, in which $y \in \mathbf{Y}_{GSE}$ and $y_{i,h}$ are known sample features, that meets proximity

$$k(h_{GSE,f}(X), h_{GSE,f}(X_i)) \approx K(X, X_i)$$

for all $X \in \mathbf{M}$, $i = 1, 2, \ldots, n$. Denote $k(y) = \sum_{i=1}^{n} k(y, y_{i,h})$.

Estimation of an Embedding Mapping. The proposed KJR- estimator (41) of Embedding mapping $h_{GSE,f}(X)$, in which the kernel (44) and known values $\{G_{GSE,h}(X_i)\}$ (34) at sample points are used, results in the estimator

$$h^*(X) = \frac{1}{K(X)} \sum_{i=1}^{n} K(X, X_i) \times \{h_{GSE,f}(X_i) + G_{GSE,h}(X_i) \times (X - X_i)\} \approx h_{GSE,f}(X);$$

the projector π^* (45) in Jacobian matrix $G_{GSE,h}(X)$ (34) is used instead of the projector π_{GSE} (33), This estimator determines Estimated Feature space (EFS) $\mathbf{Y}^* = h^*(\mathbf{M})$.

Estimation of a Reconstruction Mapping. The proposed KJR- estimator (41) of Reconstruction mapping $g_{GSE}(y)$, in which feature kernel $k(y, y_{i,h})$ and known values $\{G_{GSE}(y_{i,h})\}$ at sample feature points are used, results in the estimator

$$g^*(y) = \begin{pmatrix} g_{in}^*(y) \\ g_{out}^*(y) \end{pmatrix} = \frac{1}{k(y)} \sum_{i=1}^{n} k(y, y_{i,h}) \times \{X_i + G_{GSE}(y_{i,h}) \times (y - y_{i,h})\} \approx g_{GSE}(y).$$

Estimation of Jacobians of Embedding and Reconstruction Mappings. The KNR-estimator (40), (42) results in the estimators

$$G_h^*(X) = \frac{1}{K(X)} \sum_{j=1}^{n} K(X, X_i) \times G_{GSE,in}^{-}(y_{i,h}) \times \pi^*(X_i),$$

$$G_g^*(y) = \frac{1}{k(y)} \sum_{i=1}^{n} k(y, y_{i,h}) \times G_{GSE,g}(y_{i,h})$$

of the Jacobians of the mappings $h^*(X)$ and $g^*(y)$, respectively.

Regression on Manifolds Final Solution. The triple ($f^*(X)$, $J^*(X)$, $\mathbf{M}^*$)) consisting of the estimators $f^*(X) = g_{out}^*(h^*(X))$, $J^*(X) = G_g^*(h^*(X)) \times G_h^*(X)$, and

$$\mathbf{M}^* = \{g_{in}^*(y) \in R^p : y \in \mathbf{Y}^* \subset R^q\},$$

meets the required proximities (2), (3) and (9) and gives the final solution of the Regression on manifolds problem.

Acknowledgments. The study was performed in the IITP RAS exclusively by the grant from the Russian Science Foundation (project № 14-50-00150).

References

1. Vapnik, V.: Statistical Learning Theory. John Wiley, New York (1998)
2. Hastie, T., Tibshirani, R., Friedman, J.: The Elements of Statistical Learning: Data Mining, Inference, and Prediction, 2nd edn. Springer, New York (2009)
3. James, G., Witten, D., Hastie, T., Tibshirani, R.: An Introduction to Statistical Learning with Applications in R. Springer Texts in Statistics. Springer, New York (2013)
4. Bishop, C.M.: Pattern Recognition and Machine Learning. Springer, Heidelberg (2007)
5. Friedman, J.H.: Greedy Function Approximation: A Gradient Boosting Machine. Ann. Stat. **29**(5), 1189–1232 (2001)
6. Rasmussen, C.E., Williams, C.: Gaussian Processes for Machine Learning. MIT Press, Cambridge (2006)
7. Loader, C.: Local Regression and Likelihood. Springer, New York (1999)
8. Wang, G.: Gary Shan S.: Review of metamodeling techniques in support of engineering design optimization. J. Mech. Des. **129**(3), 370–381 (2007)
9. Forrester, A.I.J., Sobester, A., Keane, A.J.: Engineering Design via Surrogate Modelling: A Practical Guide. Wiley, New York (2008)
10. Kuleshov, A.P., Bernstein, A.V.: Cognitive technologies in adaptive models of complex plants. Inf. Control Probl. Manuf. **13**(1), 1441–1452 (2009)
11. Stone, C.J.: Optimal global rates of convergence for nonparametric regression. Ann. Stat. **10**, 1040–1053 (1982)
12. Seung, H.S., Lee, D.D.: The Manifold Ways of Perception. Science **290**(5500), 2268–2269 (2000)
13. Rajaram, D., Pant, R.S.: An improved methodology for airfoil shape optimization using surrogate based design optimization. In: Rodrigues, H., et al. (eds.) Engineering Optimization IV, pp. 147–152. CRC Press, Taylor & Francis Group, London (2015)
14. Bernstein, A., Kuleshov, A., Sviridenko, Y., Vyshinsky, V.: Fast aerodynamic model for design technology. In: Proceedings of West-East High Speed Flow Field Conference (WEHSFF-2007), Moscow, Russia (2007). http://wehsff.imamod.ru/pages/s7.htm
15. Zhu, F., Qin, N., Burnaev, E.V., Bernstein, A.V., Chernova, S.S.: Comparison of three geometric parameterization methods and their effect on aerodynamic optimization. In: Poloni, C. (ed.) Eurogen 2011, Optimization and Control with Applications to Industrial and Societal Problems International Conference on Proceedings - Evolutionary and Deterministic Methods for Design, pp. 758–772. Sira, Capua (2011)
16. Pelletier, B.: Nonparametric regression estimation on closed Riemannian manifolds. J. Nonparametric Stat. **18**(1), 57–67 (2006)
17. Loubes, J.-M., Pelletier, B.: A kernel-based classifier on a Riemannian manifold. Statistics and Decisions **26**(1), 35–51 (2008). Verlag, Oldenbourg
18. Bickel, P., Li, B.: Local polynomial regression on unknown manifolds. In: Complex Datasets and Inverse Problems: Tomography, Networks and Beyond . IMS Lecture notes – Monograph Series, vol. 54, pp. 177–186 (2007)
19. Aswani, A., Bickel, P., Tomlin, C.: Regression on manifolds: Estimation of the exterior derivative. Ann. Stat. **39**(1), 48–81 (2011)
20. Cheng, M.-Y., Wu, H.-T.: Local Linear Regression on Manifolds and its Geometric Interpretation. J. Am. Stat. Assoc. **108**(504), 1421–1434 (2013)
21. Yang, Y., Dunson, D.B.: Bayesian manifold regression. In: arXiv:1305.0167v2 [math.ST], pp. 1–40, June 16, 2014
22. Guhaniyogi, R., Dunson, D.B.: Compressed gaussian process. In: arXiv:1406.1916v1 [stat.ML], pp. 1–29, June 7, 2014
23. Fletcher, P.T.: Geodesic regression on Riemannian manifolds. In: Proceedings of International Workshop on Mathematical Foundations of Computational Anatomy (MFCA), pp. 75–86 (2011)

24. Hinkle, J., Muralidharan, P., Fletcher, P., Joshi, S.: Polynomial regression on riemannian manifolds. In: Fitzgibbon, A., Lazebnik, S., Perona, P., Sato, Y., Schmid, C. (eds.) ECCV 2012, Part III. LNCS, vol. 7574, pp. 1–14. Springer, Heidelberg (2012)
25. Steinke, F., Hein, M., Schölkopf, B.: Nonparametric regression between general Riemannian manifolds. SIAM J. Imaging Sci. **3**(3), 527–563 (2010)
26. Levina, E., Bickel, P.J.: Maximum likelihood estimation of intrinsic dimension. In: Saul, L., Weiss, Y., Bottou, L. (eds.) Advances in Neural Information Processing Systems, vol. 17, pp. 777–784. MIT Press, Cambridge (2005)
27. Fan, M., Qiao, H., Zhang, B.: Intrinsic dimension estimation of manifolds by incising balls. Pattern Recogn. **42**, 780–787 (2009)
28. Fan, M., Gu, N., Qiao, H., Zhang, B.: Intrinsic dimension estimation of data by principal component analysis. In: arXiv:1002.2050v1 [cs.CV], pp. 1–8, February 10, 2010
29. Rozza, A., Lombardi, G., Rosa, M., Casiraghi, E., Campadelli, P.: IDEA: intrinsic dimension estimation algorithm. In: Maino, G., Foresti, G.L. (eds.) ICIAP 2011, Part I. LNCS, vol. 6978, pp. 433–442. Springer, Heidelberg (2011)
30. Ma, Y., Fu, Y. (eds.): Manifold Learning Theory and Applications. CRC Press, London (2011)
31. Bernstein, A.V., Kuleshov, A.P.: Tangent bundle manifold learning via Grassmann&Stiefel eigenmaps. In: arXiv:1212.6031v1 [cs.LG], pp. 1–25, December 2012
32. Bernstein, A.V., Kuleshov, A.P.: Manifold Learning: generalizing ability and tangent proximity. Int. J. Softw. Inf. **7**(3), 359–390 (2013)
33. Bernstein, A., Kuleshov, A., Yanovich, Y.: Manifold learning in regression tasks. In: Gammerman, A., Vovk, V., Papadopoulos, H. (eds.) SLDS 2015. LNCS, vol. 9047, pp. 414–423. Springer, Heidelberg (2015)
34. Niyogi, P., Smale, S., Weinberger, S.: Finding the homology of submanifolds with high confidence from random samples. Discrete Comput. Geom. **39**, 419–441 (2008)
35. Bernstein, A.V., Kuleshov, A.P.: Data-based manifold reconstruction via tangent bundle manifold learning. In: ICML-2014, Topological Methods for Machine Learning Workshop, Beijing, June 25, 2014. http://topology.cs.wisc.edu/KuleshovBernstein.pdf
36. Kuleshov, A., Bernstein, A.: Manifold learning in data mining tasks. In: Perner, P. (ed.) MLDM 2014. LNCS, vol. 8556, pp. 119–133. Springer, Heidelberg (2014)
37. Kuleshov, A., Bernstein, A., Yanovich, Yu.: Asymptotically optimal method in manifold estimation. In: Márkus, L., Prokaj, V. (eds.) Abstracts of the XXIX-th European Meeting of Statisticians, July 20–25, 2013, Budapest, p. 325 (2013)
38. Genovese, C.R., Perone-Pacifico, M., Verdinelli, I., Wasserman, L.: Minimax Manifold Estimation. J. Mach. Learn. Res. **13**, 1263–1291 (2012)
39. Golub, G.H., Van Loan, C.F.: Matrix Computation, 3rd edn. Johns Hopkins University Press, Baltimore (1996)
40. Wasserman, L.: All of Nonparametric Statistics. Springer Texts in Statistics, Berlin (2007)
41. Jollie, T.: Principal Component Analysis. Springer, New York (2002)
42. Singer, A., Wu, H.-T.: Vector Diffusion Maps and the Connection Laplacian. Commun. Pure Appl. Math. **65**(8), 1067–1144 (2012)
43. Tyagi, H., Vural, E., Frossard, P.: Tangent space estimation for smooth embeddings of Riemannian manifold. In: arXiv:1208.1065v2 [stat.CO], pp. 1–35, May 17, 2013
44. Hamm, J., Lee, D.D.: Grassmann discriminant analysis: a unifying view on subspace-based learning. In: Proceedings of the 25th International Conference on Machine Learning (ICML 2008), pp. 376–383 (2008)
45. Wolf, L., Shashua, A.: Learning over sets using kernel principal angles. J. Mach. Learn. Res. **4**, 913–931 (2003)

A New Strategy for Case-Based Reasoning Retrieval Using Classification Based on Association

Ahmed Aljuboori(✉), Farid Meziane, and David Parsons

School of Computing Science and Engineering, University of Salford, Salford M5 4WT, UK
a.s.aljuboori@edu.salford.ac.uk,
{f.meziane,d.j.parsons}@salford.ac.uk

Abstract. This paper proposes a novel strategy, Case-Based Reasoning Using Association Rules (CBRAR) to improve the performance of the Similarity base Retrieval SBR, classed frequent pattern trees FP-CAR algorithm, in order to disambiguate wrongly retrieved cases in Case-Based Reasoning (CBR). CBRAR use class association rules (CARs) to generate an optimum FP-tree which holds a value of each node. The possible advantage offered is that more efficient results can be gained when SBR returns uncertain answers. We compare the CBR Query as a pattern with FP-CAR patterns to identify the longest length of the voted class. If the patterns are matched, the proposed strategy can select not just the most similar case but the correct one. Our experimental evaluation on real data from the UCI repository indicates that the proposed CBRAR is a better approach when compared to the accuracy of the CBR systems used in our experiments.

Keywords: Class Association Rules · Frequent pattern trees · Case-Based Reasoning · Retrieval · P-trees

1 Introduction

The basic premise of case-based reasoning (CBR) is that experience in the form of previous cases can be influenced to solve new problems [1]. An individual experience is named a case, and its collection is stored in a case base [2]. Basically, each case is defined by a problem description and its corresponding solution description. Among the four main phases, retrieval is a key stage, with success being heavily reliant on its performance [3]. Its aim is to retrieve similar or useful cases that can be successfully used to solve a target problem. This is of particular importance because if the retrieved cases are not useful, CBR systems may not ultimately produce a suitable solution to the problem [2].

Fundamentally, retrieval is performed through a specific strategy of leveraging similarity knowledge (SK) referred to as 'similarity-based retrieval.' (SBR) [3]. In SBR, SK is utilized to determine the benefit of stored cases with regards to a target problem. SK is typically encoded via similarity measures between the problem and stored cases. In SBR, the measures are used to identify cases ranked by their similarities to the problem. Their solutions are then used to solve the problem.

P. Perner (Ed.): MLDM 2016, LNAI 9729, pp. 326–340, 2016.
DOI: 10.1007/978-3-319-41920-6_24

Association rules mining (ARM) is an important technique in the field of data mining (DM). ARM is used to extract interesting correlations, associations or casual structures among a set of items in a transaction database or other data repositories. It is used in various application areas, such as banking, products relationships and frequent patterns. The class association rule (CAR) technique was first proposed by [4]. It generates classification rules based on association rules (ARs). Other techniques for mining CARs have been suggested in recent years. They include GARC [5], ECR-CARM [6], CBC [7], CAR-Miner [8], CHISC-AC [9] and developed d2O [10]. The methods of classification based on CARs were demonstrated to be more accurate than the classic methods e.g. C4.5 [11]and ILA [12, 13] in their practical results [4].

Frequent pattern mining (FPM) plays a major role in ARM. On its own FPM is concerned with finding frequent patterns (frequently co-occurring sub-sets of attributes) in data. A number of FPM algorithms have been proposed for instance Apriori [14]. With respect to pattern matching the majority of these have been integrated with ARM algorithms. Of these, the best known, and most frequently cited, is the FP-Growth algorithm [15]. FP-growth is constructed on a set enumeration tree structure called the FP-tree. It takes a totally different approach to discovering frequent itemsets. Unlike Apriori, it does not generate and test the paradigm. Instead, FP-growth compacts the data set structure using FP-tree and extracts the frequent pattern directly from this structure [16]. FP-tree is a compressed representation of the input data. It is built by reading the dataset transaction and allocating each transaction to a path in the FP-tree. As various transactions can have many items in common, their paths might overlap. The more the paths overlap with one another, the more can be achieved by using the FP-tree structure. The performance of this process will depend on the amount of memory available on the system being used. If the FP-tree can be held entirely within the available memory, the extraction of frequent itemsets will be faster as it will be possible to avoid repeated passes over stored data.

In this paper, we propose CBRAR a new strategy for enhancing the performance of CBR by using a new more efficient algorithm (FP-CAR) for mining all CARs with FP-tree values for a CBR query *Q*. The proposed algorithm uses an optimum tree derived from the FP-tree and optimized by P-tree concepts to produce a super-pattern that matches the new CBR case. Our initial experimental results show that the CBRAR strategy is able to disambiguate the answers of the retrieval phase compared to those obtained when using Jcolibri [17] and FreeCBR [18] systems for example.

2 Literature Review of CBR and other Types of Knowledge

CBR is a well-studied area in machine learning. In the past decades several researchers have studied CBR methods in real world applications, such as medical diagnosis [19], [20], product recommendation [21] and personal rostering decisions [22]. CBR is a cyclic and integrated process of solving a problem and learning from the experience of experts, which is used to build a knowledge domain which is then recorded to be used to help solve future problems. It can be defined as "to solve a problem, remember a similar problem you have solved in the past and adopt the old solution to solve the new problem" [23]. CBR methods are composed of four steps: retrieve-find the best matching of previous cases, reuse-find what can be reused from old cases, revise-check if the proposed solution may be correct, and retain-learn from the problem solving experience. This decomposition of CBR phases is based on [1] and illustrated in Fig. 1.

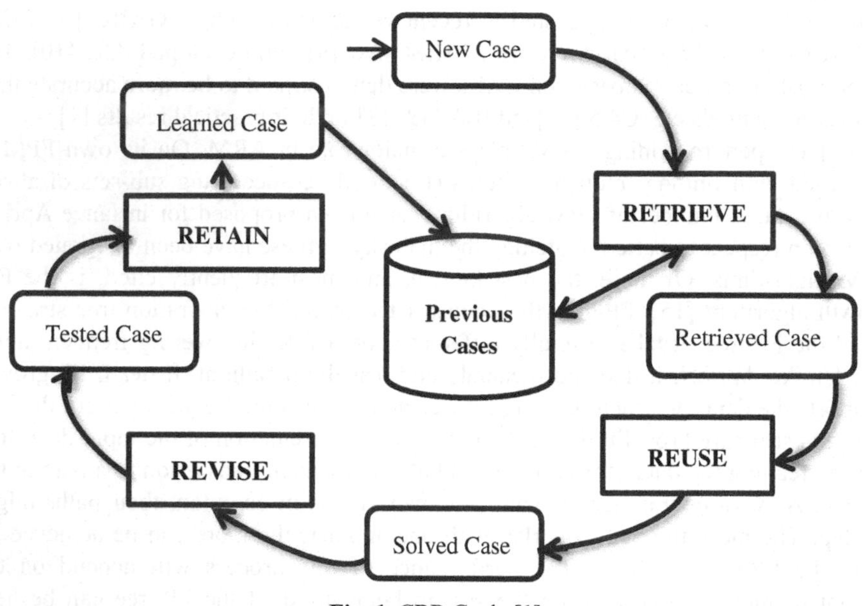

Fig. 1. CBR Cycle [1]

2.1 Machine Learning and Retrieval

The development of machine learning has resulted in retrieval approaches that SBR merges with rule-induction (RI) approaches to enhance SBR. RI systems often learn domain-particular knowledge and represent it as IF–THEN rules. It is suggested that such rules can be utilized for determining the weights of case features in SBR [24]. [25] shows that decision tree algorithms can be used to discover domain-specific rules from a specific case base. From such rules, users select useful rules according to the thresholds set up by experts. The extracted rules are then used to point a target problem to its most similar case set and to calculate the weights of the case features. Such knowledge is finally used to retrieve the most similar case from the case base. A retrieval paradigm in [26] chooses between SBR or a RI method (using decision trees) for the target problem, considering the similarities of cases in a case base.

The CBRAR approach is different from these approaches in that Association Knowledge (AK) is not used to measure the weights of case features, but to refine the cases retrieved by SBR and guide more specific rules to the target problem.

2.2 Data Mining and CBR

Over time, techniques of integrating data mining (DM) and K nearest neighbor (KNN) have often been implemented in CBR research to improve KNN through three main platforms. First, to integrate feature weighting (FW) and feature selection (FS) into KNN. In this framework, FW is used to estimate the optimal weights of the original features of cases [27], [28] , and FS is employed when choosing relevant features of cases [20], [22], or their aggregation is used to leverage their usefulness [19]. Second, to merge data clustering with KNN, where the structure of clustered cases is leveraged to lead to more relevant cases [29], [30]. For case retrieval, the similarity between a target problem and each case is combined with the relevance of the clustered group containing the case considered [31]. Third, to apply both DM and SBR techniques together to discover cases related to the target problem. For instance, [32] displays how to integrate DM with SBR to improve liver diagnosis. Given a target problem, once a DM method (a back-propagation neural network) is applied on the case base, some cases thought to be relevant to the problem are retrieved. These cases are then tested to verify whether these are adequately similar to the target problem through SBR. Similar cases are merely used as a retrieval result for the problem. Unlike this scheme, our approach is based mainly on the use of AK built via CARs.

2.3 Retrieval and CAR

Basically, retrieval is achieved by employing two methods: (AK) and (SK). The retrieval is normally achieved utilizing SBR which is a technique based on SK. In SBR, SK is utilized for estimating the retrieval of similar cases to the target problem. The similarity measure is used between the various cases available and the problem to find those cases that can be selected to solve the target. Nevertheless, defining the SK can be considered as a main disadvantage of SBR because it is reliant on domain experts and is a time consuming process [33]. The similarity standard defined for one domain differs for numerous domains that are helpful for some problems and not for others. Therefore, the performance of SBR varies from problem to problem even within the same domain [26].

Association rules (ARs) aim to find interesting relationships (associations) in a transaction database [14]. The focus is usually on discovering a set of highly co-occurring features shared by a large number of transactions in a database. It is an implication of the form $X \rightarrow Y$, where X and Y are nonintersecting sets of items. For example, {milk, eggs}→{bread} is an association rule that says that when milk and eggs are purchased, bread is likely to be purchased. In the context of CBR, ARs can be employed to determine interesting relationships from a given case base. Furthermore, the transaction of the item can be considered as a case and an attribute as a value pair, respectively. The most traditional algorithm is Apriori [34] which has been used to evaluate and rank a large number of extracted ARs that have support and confidence which is not less than that specified by a user [35]. CAR is a specific subset of ARs whose consequents are restricted to one target class. In the context of CBR, a CAR is considered as an AR whose consequent holds the item formed as a pair of a solution attributes and its value [36]. In a given case base library, AK is encoded to show how a specific problem's features are associated with a certain solution.

3 Related Work

3.1 Soft Matching of ARM (SARM)

A limitation of traditional ARM algorithms for rule $X \rightarrow Y$ e.g. Apriori [34] is that items X and Y are discovered based on the relation of equality. Basically, these algorithms perform poorly when dealing with similar items. For instance, Apriori cannot find rules like 70% of the customers who buy products similar to yogurt (e.g. milk) and products similar to mayonnaise (e.g. egg) also buy baguettes. Soft matching was suggested to address this [37], where the consequents and antecedent of ARs are discovered by similarity valuation. The SARM standard is used to find all rules from $X \rightarrow Y$, where minimum support and minimum confidence of each rule are not less than soft support and soft confidence, respectively. Support and confidence are used to generalize the definition of soft support and soft confidence.

This generalization is performed by allowing elements to match, so long as their similarity exceeds minimum similarity (minsim) as specified by the user. The soft-matching criteria can be employed to model better relationships among features of cases instead of the equality relation, by using the concept of similarity.

3.2 Soft - CAR Algorithm

This algorithm calculates the soft support and finds the frequency of each item soft matching CARs. It also discovers the seed set of rules found in every pass in the corresponding class. For every rule item, the seed set of rules are utilized to generate new rule items known as candidate rule items. The soft support is computed through the set of different cases.

It produces SCARs rules in the last pass after it finds the candidate rule items which are frequent from those frequent items [38]. However, experts are required for calculating and defining the SK domain, making this a time consuming and difficult process.

3.3 USIMCAR Algorithm

This algorithm is an expansion of the retrieval phase to improve the performance of the SBR. It encodes the AK in Soft-CARs together with SK to improve the performance of CBR [38]. USIMCAR is used to enhance the usefulness of cases, retrieved through the SK [36], with regard to a new case Q in addition to including the SCAR, thus meaningfully utilizing the cases with their usefulness [36]. In addition, it leverages the AK by searching and finding those SCARs whose usefulness is greater than others concerning Q, therefore valuably using them with their usefulness. Patel [39] also developed the USIMCAR strategy for hierarchical cases which combines the support-count bit from multilevel and soft-matching criteria (SC-BF) algorithm for the SCARs. Patel also applied the unified knowledge of the AK and similarity to enhance the performance of the SBR. Both strategies [38] and [39] are a simulation of the retrieval phase by providing a percentage value but do not involve providing a CBR system with feedback inputs as part of the original cycle.

In this paper, we propose the FP-CAR algorithm to generate an optimum tree using CARs and FP-tree. The tree is optimized by utilizing various types of association knowledge i.e. P-trees and an equivalence table of implications. FP-CAR is also a part of the suggested CBRAR technique which is an expansion to the SBR. The novel CBRAR is used to disambiguate the wrong retrieved answers as feedback to the CBR.

4 Proposed Algorithm FP-CAR

The FP-CAR (frequent pattern class association rules algorithm) is based on two steps. First, it generates a FP-tree from a set of CARs [40]. Second, the tree is optimized by utilizing the P-tree [41] concepts and equivalence table of implication. These two steps are combined to gain an optimum tree that can be compared with a new case Q of the CBR as a super-pattern to improve the performance of the SBR. The start of the observation is where the options of CARs have been selected as follows (lower support $\xi = 0.1$ and confidence = 0.9, delta = 0.05, number of rules = Maximum), then the existence of rule $X \rightarrow c$ is a subset should make it necessary to consider it as an antecedent of a superset $X,Y \rightarrow c$. Practically, however, we may still find a rule $Y \rightarrow c$, say, where Y is another subset of the same class, where both X and Y form a Superset-Pattern $X,Y \rightarrow c$. In the first case scenario, logical equivalences concepts are utilized to prove the theory behind gaining the equivalence of $((X \rightarrow c) \vee (Y \rightarrow c)) \equiv (X \wedge Y) \rightarrow c$. In other words if X implies c or Y implies c, it is equivalent to X and Y both implying c.

The second case scenario uses the acute inflammation dataset from UCI (see Table 1, Table 2, Fig. 2 and Fig. 3). In this case, as in Coenen [42] we take advantage of the P-tree to gain a superset. We consider the partial total accumulated at *ABCD* which makes a contribution for all the subsets of *ABCD*. In other words, the contribution in respect of the subsets of *ABC* is already included in the interim total for *ABCD*, therefore, when considering the superset *ABCD*, we need to examine only those subsets which include the attribute *D* [43].

In this paper we suggest an alternative explanatory method: If we can identify a generic rule $X \rightarrow c$ which meets the required support and confidence thresholds, then it is necessary to look for other rules whose antecedent is a superset combined with $(X \wedge Y)$ and whose consequent is c which distinguishes our algorithm compared to [40]. The objective of the FP-CAR algorithm is to continue to look for rules that select other classes in order to reduce the risk of overfitting and the number of the considered candidate rules.

FP-CAR uses the concepts of classification based on association and the Total From Partial Classification (TFPC) algorithm [40]. It builds a set-enumeration tree structure of the CARs, where the FP-tree contains an incomplete summation of support-counts for relevant sets and patterns. Using the FP-tree structure to represent all patterns of the CARs, the T-tree [41] concept is used to build an optimum tree that finally contains all the frequent patterns sets (i.e. those that can be compared to the pattern of the CBR query). The FP-CAR is built level by level, the first level comprising all the subsets that contain a value of the attribute under consideration. It compresses the subsets into a prefix tree, where the root *c* holds all frequent items according to their frequency. In the second pass, the unnecessary subsets are removed, from the tree. Candidate-subsets then form a superset from the remaining sets considering the pattern of the CBR. The process continues, with the voting of a length in each class label, until no more candidate sets can be generated. The patterns of subsets will contain a value of each node which can be compared with a CBR query *Q*.

Fig. 2 shows the form of a FP-CAR, for the subsets $\{\{A,B,C,D\},\{A,E,F\},L,c_1\}$, $\{\{A,B,C\},L,c_2\}$ where L is a length identifier, c_1 and c_2 are class identifiers, each node of a subset holds a value i.e. $A=\{yes, no\}$. This tree includes all possible related supersets that are not resolved by SBR, except for those including both c_1 and c_2 which we will assume were pruned. The target of FP-CAR is to find a CBR case problem that caused uncertain answers i.e. $\{A=yes,\ B=yes,\ C=yes,\ D=yes,\ E=no,\ F=40,\ L=6\}$. FP-CAR nodes include a value of each node for a superset Q i.e. $A=\{yes, no\}$. Practically, an actual FP-tree would contain all those nodes representing the frequent subsets where FP-CAR includes the voting length and values. For instance, if the set $\{\{A,B,C\},L\}$ fails to reach the required support threshold, and length identifier e.g. 4 to conform to the case problem pattern, then the class of the subset $\{A,B,C\}$ would be ignored, and the superset would not be created. All the candidates that contain the class-identifier *c1* with required length can be found in the subtree rooted at *c1* ,starts with *A* node descended by $\{B,C,D,E,F\}$ frequency as shown in Table 1. FP-Tree Hash Table Therefore, all the rules that classify to c_1 can be derived from the root *A* (and also for c_2) whereas those subsets which start with other roots will be removed to gain a super-pattern.

Table 1. FP-Tree Hash Table

Item	Frequency	Priority
B	58	2
A	62	1
C	50	3
D	49	4
E	44	5
F	26	6

→

Ordered-Subsets	Length	Class
A,B,C	3	*c1*
A,C	2	*c1*
A,B	2	c2
A,B,C	3	c2
A	1	c2
A,B,C,D	4	*c1*

The algorithm used to build the FP-CAR tree in Fig. 2 is a modification of the original FP-Tree approach using TFPC concepts. As each pass is concluded, we ignore from the tree all those subsets that fail to meet the target pattern to form a superset. The remaining (frequent) sets that are included within the class-identifier subtrees of (c_1), define a possible partial superset if one matches the voted length and other one is a complement. For example, the set $\{ABCDc_1\}$ is a partial one where $X \rightarrow c_1$ of $L=4$

and $AEFc_1$ is a complement that corresponds to rule $Y \rightarrow c_1$. We now build the supersets of all such sets that match the new case Q = *{yes, yes, yes, yes, no and 40.0}*. If the threshold of L of the subsets is greater than or equal to the voted class c_1, we add the subset to our target set considering the nodes values, and ignore the corresponding subset from the tree that occurs in c_2. The complement of the superset will then be completed from the same cluster of c_1 i.e. $\{X \wedge Y\} \rightarrow c_1 \equiv Q \rightarrow c_1$ as shown in Fig. 2. Connecting the tree below to the results in Table 2, proves the theory behind the proposed algorithm.

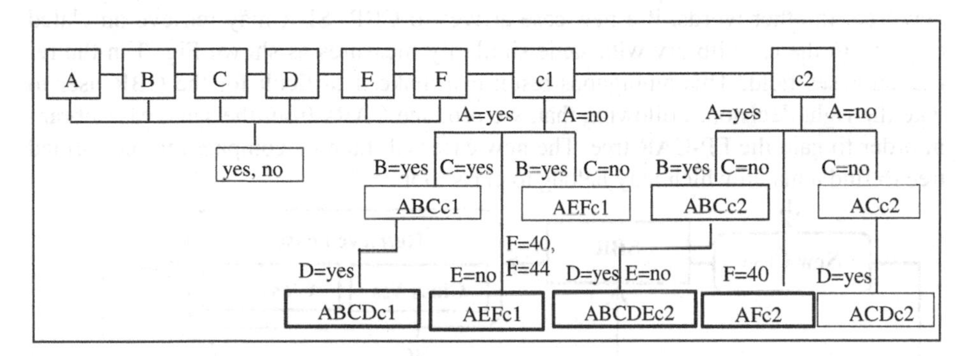

Fig. 2. FP-CAR Algorithm Tree

5 New Strategy CBRAR to Enhance the Performance of SBR

This section presents the proposed new technique CBRAR of integrating CARs into CBR. Basically, there is a possible problem in CBR which is retrieving unrelated cases that cause an incorrect solution. To overcome this problem, CAR is utilized to find the relationship between the case library and a target case. Normally, to achieve the retrieval phase, CBR systems execute similarity SBR. However, SBR tends to depend on similarity knowledge, ignoring other types of knowledge that can benefit and improve retrieval performance. In this research, the challenge is how to retrieve not just the most similar case in CBR but the correct one. Some studies which apply ARs into CBR, for example [38], are much dependent on the experts domain for finding SK. [39] focused on the case representation hierarchically by combining SK and AK depending on the Apriori algorithm when a number of passes are needed to generate new candidates. Both strategies [38], [39] are a simulation of the retrieval phase by providing a percentage value of related cases but do not involve providing a CBR system with feedback, which is part of the original cycle. The new approach CBRAR produces a correct case pattern not just a similar one. It also enables a correct case to be returned back into the retrieval phase to disambiguate any wrong answer produced by CBR.

As shown in Fig. 3, we start to remove one case from the case based library of the CBR until the system retrieves two different classes with the same similarity. The new method adapts the CARs to produce the FP-tree considering a class label, length of subsets and support. This is because in mining association rule algorithms, any associated method does not consider class clusters and length in the process of producing frequent patterns of a specific class. Thus, in experiments to date an attempt has been made to develop a FP-tree to make the frequent rules more effective to one class by using a parent root of each class label. As a consequence of that, every frequent rule will belong to its class. In the experiments, the first step of the FP-tree algorithm is changed to classify subsets according to its frequency before the rules are produced. Hence, considering the new case as a pattern to be compared with the constructed FP-tree will provide a correct match based on the new case built from the new tree. In other words, if a new case arrives to CBR, SBR may retrieve unrelated cases from the case library with same similarity measures as shown Fig. 3 in the retrieved cases field. This ambiguous result can make it difficult for the CBR user to take the right decision. Following that, we produce CARs from the same case library in order to gain the FP-CAR tree. The new case will then be compared to the formed tree to find a match which may belong to class root.

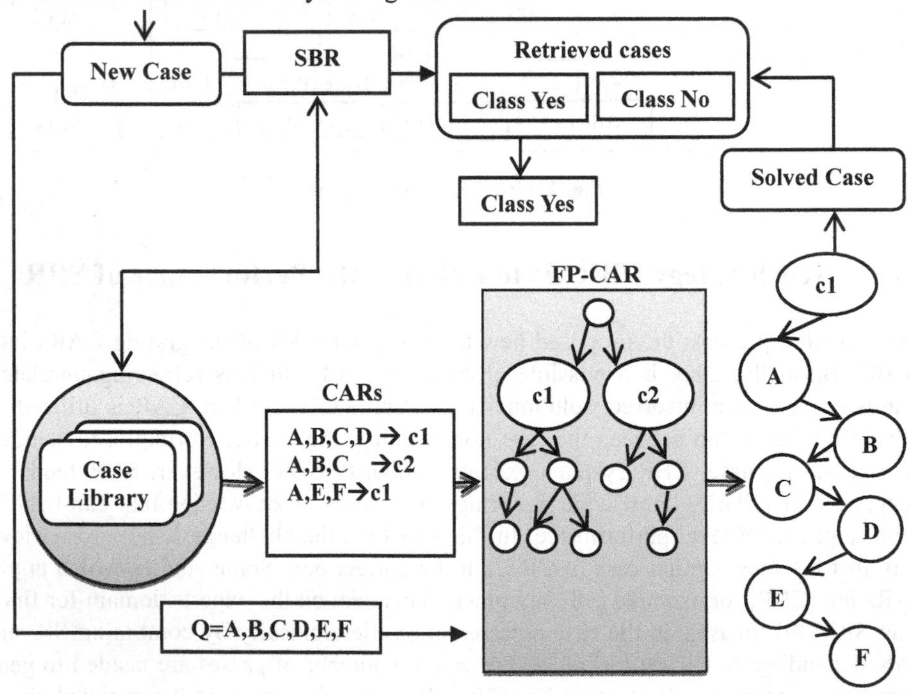

Fig. 3. CBRAR Model

The proposed strategy is compared to existing CBR tools in the following steps:

- Splitting: the new algorithm splits rules into different classes, where each rule represents a subset which belongs to a particular class.
- Comparing: the new algorithm compares a CBR query as a pattern which actually represents a new case; it should match exactly a frequent path FP-tree.

- Voting: the process of voting is performed by considering the longest length of the nodes considering values of the modified FP-tree in terms of finding a partial match.
- P-trees: a P-trees procedure is invoked to complete any missing nodes in the tree if needed to form an equivalent pattern to the CBR query.

In the final step, the result obtained by our new model is compared with the outcomes of the retrieval phase to select a correct answer. We compare the solved case with the result of the retrieved cases to remove unrelated answers as shown in Fig. 3. It can be seen that two different labels i.e. class (yes and no) are retrieved by CBR in the retrieved cases field. By returning the solved case into the retrieved cases phase, the ambiguity of the SBR outcomes was removed.

6 Experimental Results

To investigate the accuracy of CBRAR, we conducted experiments using a dataset taken from the UCI Machine Learning Repository. The implementation of CBRAR used a Java platform Eclipse (4.5.0), and for comparison purposes we have used the Jcolibri framework [17] and FreeCBR [18] as powerful CBR tools. WEKA 3.6 is used as an open source in order to generate the CARs. In the set of experiments, we have removed one case from the CBR case library to be considered as a new case in each run of both Jcolibri and FreeCBR. We used the acute inflammations dataset as the same source to measure the CBR and CBRAR accuracy. By default, SBR returns the 5 most similar answers when using Jcolibri when a new case is applied. However, the pre-determined cases 73,76,85,88 have registered an ambiguity that misleads the decision maker as all retrieved cases have the same percentage of similarity with different labels i.e. (yes, no). When using FreeCBR, more potential cases were identified in addition to those found by Jcolibri.

The results are shown in Table 2; vertically, the first column refers to the new case Q followed by the cases retrieved by the CBR tools i.e. NewCase73 followed by cases (71, 72, 76, 77 and 79, for Jcolibri) and cases (71, 72, 77 until 107 for FreeCBR). The "Attributes" columns start with a temperature attribute F followed by 5 additional attributes A, B, C, D and E. The class label column indicates a diagnosis of Inflammation of the urinary bladder with values (yes and no). The "Accuracy" columns show the comparison between Jcolibri, FreeCBR and CBRAR. In the table, we use symbols TP, TN, FP and FN as follows True Positive, True Negative, False Positive and False Negative. The assumption is made to indicate the four probabilities on the confusion matrix. Table 2 shows that, for each new case applied to CBR, 5 different cases are retrieved by Jcolibri with the same similarity ratio i.e. 0.912. In the first experiment, a NewCase73 applied to the CBR, Jcolibri retrieved 3 TP and 2 FP cases with the same similarity ratio, and this is equal to 60% of accuracy, whereas FreeCBR retrieved 9 TP and 2 FP, and this is equal to 81% of accuracy. CBRAR retrieved 1 TP case from new model. In the second experiment, a NewCase76 applied to the CBR, Jcolibri retrieved 4 TN and 1 FN with same similarity and this is equal to 80% of accuracy, whereas FreeCBR retrieved 6 TN and 1 FN and this is equal to 86% of accuracy.

Table 2. Results of Wrong Retrieved Cases

Cases	Attributes							Accuracy		
	F	A	B	C	D	E	Class	Jcolibri	FreeCBR	CBRAR
NewCase73	**40.0**	**yes**	**yes**	**yes**	**yes**	**no**	**yes**	**0.912**	**59.1751**	**TP**
Case71	40.0	yes	yes	yes	yes	yes	yes	TP	TP	
Case72	40.0	yes	yes	yes	yes	yes	yes	TP	TP	
Case76	40.0	yes	yes	no	yes	no	no	FP	FP	
Case77	40.0	yes	yes	no	yes	no	no	FP	FP	
Case79	40.1	yes	yes	yes	yes	no	yes	TP	TP	
Case85	40.4	yes	yes	yes	yes	no	yes		TP	
Case86	40.4	yes	yes	yes	yes	no	yes		TP	
Case89	40.5	yes	yes	yes	yes	no	yes		TP	
Case94	40.7	yes	yes	yes	yes	no	yes		TP	
Case100	40.9	yes	yes	yes	yes	no	yes		TP	
Case107	41.1	yes	yes	yes	yes	no	yes		TP	
NewCase76	**40.0**	**yes**	**yes**	**no**	**yes**	**no**	**no**	**0.912**	**59.1751**	**TN**
Case73	40.0	yes	yes	yes	yes	no	yes	FN	FN	
Case82	40.2	yes	yes	no	yes	no	no	TN	TN	
Case88	40.4	yes	yes	no	yes	no	no	TN	TN	
Case92	40.6	yes	yes	no	yes	no	no	TN	TN	
Case96	40.7	yes	yes	no	yes	no	no	TN	TN	
Case104	41.0	yes	yes	no	yes	no	no		TN	
Case109	41.1	yes	yes	no	yes	no	no		TN	
NewCase85	**40.4**	**yes**	**yes**	**yes**	**yes**	**no**	**yes**	**0.912**	**55.278**	**TP**
Case73	40.0	yes	yes	yes	yes	no	yes	TP	TP	
Case79	40.1	yes	yes	yes	yes	no	yes	TP	TP	
Case84	40.4	yes	yes	yes	yes	yes	yes	TP	TP	
Case88	40.4	yes	yes	no	yes	no	no	FP	FP	
Case89	40.5	yes	yes	yes	yes	no	yes	TP	TP	
Case94	40.7	yes	yes	yes	yes	no	yes		TP	
Case100	40.9	yes	yes	yes	yes	no	yes		TP	
Case107	41.1	yes	yes	yes	yes	no	yes		TP	
NewCase88	**40.4**	**yes**	**yes**	**no**	**yes**	**no**	**no**	**0.912**	**55.276**	**FN**
Case76	40.0	yes	yes	no	yes	no	no	TN	TN	
Case77	40.0	yes	yes	no	yes	no	no	TN	TN	
Case82	40.2	yes	yes	no	yes	no	no	TN	TN	
Case85	40.4	yes	yes	yes	yes	no	yes	FN	FN	
Case86	40.4	yes	yes	yes	yes	no	yes	FN	FN	
Case92	40.6	yes	yes	yes	yes	no	yes		TN	
Case96	40.7	yes	yes	yes	yes	no	yes		TN	
Case104	41.0	yes	yes	yes	yes	no	yes		TN	
Case109	41.1	yes	yes	yes	yes	no	yes		TN	
Average								70	83	75

CBRAR retrieved 1 TN case from the suggested algorithm. When NewCase85 is applied to the CBR in the third experiment, Jcolibri retrieved 4 TP and 1 FP cases with the same similarity percentage, and this is equal to 80% accuracy whilst FreeCBR retrieved 7 TP and 1 FP and this is equal to 88% of accuracy. CBRAR retrieved 1 TP case from FP-CAR tree. In the fourth experiment, a NewCase88 applied using Jcolibri again retrieved 5 cases with 3 TN and 2 FN with same similarity and this is equal to 60% accuracy. FreeCBR retrieved 9 cases with 7 TN and 2 FN. CBRAR incorrectly retrieved 1 FN as a wrong case.

The results show that 14 out of the 20 Jcolibri retrieved cases are classified as TP and TN giving 70% accuracy. By comparison, 29 of the 35 cases retrieved by FreeCBR are classified as TP and TN giving 83% accuracy. However, both Jcolibri and FreeCBR deliver "confusing" results. Our CBRAR strategy demonstrates advantages over both Jcolibri and FreeCBR by resolving 3 out of 4 cases with 75% accuracy and no confusion. The accuracy of CBRAR was better compared to Jcolibri and FreeCBR. CBRAR resolved the ambiguity of the FP and FN cases without confusion. Cases 73, 76 and 85 in Table 2 can be reworked in Fig. 2 to prove that CBRAR identifies a correct case using a frequent classed tree.

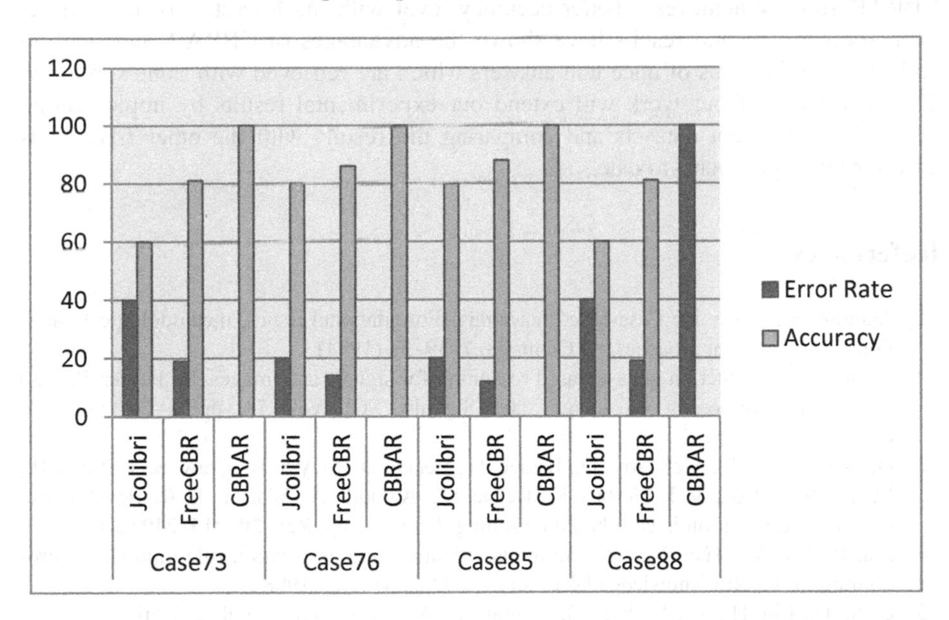

Fig. 4. Error Rate and Accuracy

The bar chart in Fig. 4 illustrates the error rate and accuracy of Jcolibri, FreeCBR and CBRAR. From the chart, it is clear that in Case73, CBRAR registered 0 error rate, which is the lowest among the rates (40, 19) when compared to Jcolibri and FreeCBR. The results also show that the error rate of CBRAR is the lowest on Case76 and Case85 thus giving the highest accuracy, when compared to the other CBR tools used. CBRAR also correctly resolved 3 out of 4 cases. In Case88, it noticeable that the (40, 19) % error rate of Jcolibri and FreeCBR was considerably lower than CBRAR. However, whilst CBRAR did not resolve Case88 neither of the other CBR tools offered any advantage when compared to the new model. In conclusion, we have shown that the other CBR tools used inherit the same problem of error rates, whereas CBRAR has shown a better performance in overall error rate.

7 Conclusion

This paper has presented a new approach, CBRAR, to improve the performance of SBR. The CBRAR approach includes a new algorithm FP-CAR which produces far fewer frequent classed subsets than would be produced from a generic FP-tree. It uses a new method of length voting compared to the TFPC algorithm where a value of nodes is considered whilst building the tree. Moreover, the subsets left on the tree that meet the support, confidence and longest length of pattern can be used to classify subsets when sorted in a hash table. A superset could be derived; to be compared with other new CBR cases when compared with the CBR tools Jcolibri and FreeCBR, the CBRAR strategy achieves a better accuracy level with the lowest error rate. Moreover, the experimental results have shown the advantages of CBRAR over Jcolibri and FreeCBR in terms of uncertain answers which are retrieved with same similarity. The next phase of our work will extend our experimental results by implementing CBRAR on different datasets and comparing the results with the other CBR tools used for our experiments to date.

References

1. Aamodt, A., Plaza, E.: Case-based reasoning: Foundational issues, methodological variations, and system approaches. AI Commun. **7**, 39–59 (1994)
2. Perner, P.: Introduction to case-based reasoning for signals and images. In: Perner, P. (ed.) Case-Based Reasoning on Images and Signals. SCI, vol. 73, pp. 1–24. Springer, Heidelberg (2008)
3. De Mantaras, R.L., McSherry, D., Bridge, D., Leake, D., Smyth, B., Craw, S., Faltings, B., Maher, M.L., Cox, M.T., Forbus, K., Keane, M., Aamodt, A., Watson, I.: Retrieval, reuse, revision and retention in case-based reasoning. Knowl. Eng. Rev. **20**, 215–240 (2005)
4. Ma, B., Liu, W., Hsu, Y.: Integrating classification and association rule mining. In: Proceedings of the 4th Knowledge Discovery and Data Mining (1998)
5. Chen, G., Liu, H., Yu, L., Wei, Q., Zhang, X.: A new approach to classification based on association rule mining. Decis. Support Syst. **42**, 674–689 (2006)
6. Vo, B., Le, B.: A novel classification algorithm based on association rules mining. In: Richards, D., Kang, B.-H. (eds.) PKAW 2008. LNCS, vol. 5465, pp. 61–75. Springer, Heidelberg (2009)
7. Deng, H., Runger, G., Tuv, E., Bannister, W.: CBC: An associative classifier with a small number of rules. Decis. Support Syst. **59**, 163–170 (2014)
8. Nguyen, L.T.T., Vo, B., Hong, T.-P., Thanh, H.C.: CAR-Miner: An efficient algorithm for mining class-association rules. Expert Syst. Appl. **40**, 2305–2311 (2013)

9. Ibrahim, S.P.S., Chandran, K.R., Kanthasamy, C.J.K.: CHISC-AC: Compact Highest Subset Confidence-Based Associative Classification[1]. Data Sci. J. **13**, 127–137 (2014)
10. Nguyen, L.T.T., Nguyen, N.T.: An improved algorithm for mining class association rules using the difference of Obidsets. Expert Syst. Appl. **42**, 4361–4369 (2015)
11. Quinlan, J.R.: C4.5: programs for machine learning. Morgan Kaufmann, San Mateo (1993)
12. Tolun, M.R., Abu-Soud, S.M.: ILA: an inductive learning algorithm for rule extraction. Expert Syst. Appl. **14**, 361–370 (1998)
13. Tolun, M.R., Sever, H., Uludag, M., Abu-Soud, S.M.: ILA-2: An inductive learning algorithm for knowledge discovery. Cybern. Syst. **30**, 609–628 (1999)
14. Agrawal, R., Srikant, R.: Fast algorithms for mining association rules. In: Proceedings of the 20th International Conference Very Large Data Bases, VLDB, pp. 487–499 (1994)
15. Han, J., Pei, J., Yin, Y.: Mining frequent patterns without candidate generation. In: ACM SIGMOD Record, pp. 1–12. ACM (2000)
16. Cagliero, L., Garza, P.: Infrequent weighted itemset mining using frequent pattern growth. Knowl. Data Eng. IEEE Trans. **26**, 903–915 (2014)
17. jCOLIBRI | GAIA – Group of Artificial Intelligence Applications. http://gaia.fdi.ucm.es/research/colibri/jcolibri
18. FreeCBR. http://freecbr.sourceforge.net/index.shtml
19. Ahn, H., Kim, K.: Global optimization of case-based reasoning for breast cytology diagnosis. Expert Syst. Appl. **36**, 724–734 (2009)
20. Pandey, B., Mishra, R.B.: Case-based reasoning and data mining integrated method for the diagnosis of some neuromuscular disease. Int. J. Med. Eng. Inform. **3**, 1–15 (2011)
21. Anand, S.S., Mobasher, B.: Intelligent techniques for web personalization. In: Mobasher, B., Anand, S.S. (eds.) ITWP 2003. LNCS (LNAI), vol. 3169, pp. 1–36. Springer, Heidelberg (2005)
22. Beddoe, G.R., Petrovic, S.: Selecting and weighting features using a genetic algorithm in a case-based reasoning approach to personnel rostering. Eur. J. Oper. Res. **175**, 649–671 (2006)
23. Althof, K.-D., Auriol, E., Barlette, R., Manago, M.: A Review of Industrial Case Based Reasoning. AI Intelligence, Oxford (1995)
24. Cercone, N., An, A., Chan, C.: Rule-induction and case-based reasoning: hybrid architectures appear advantageous. IEEE Trans. Knowl. Data Eng. **11**, 166–174 (1999)
25. Huang, M.-J., Chen, M.-Y., Lee, S.-C.: Integrating data mining with case-based reasoning for chronic diseases prognosis and diagnosis. Expert Syst. Appl. **32**, 856–867 (2007)
26. Park, Y.-J., Choi, E., Park, S.-H.: Two-step filtering datamining method integrating case-based reasoning and rule induction. Expert Syst. Appl. **36**, 861–871 (2009)
27. Bradley, K., Smyth, B.: Personalized information ordering: a case study in online recruitment. Knowledge-Based Syst. **16**, 269–275 (2003)
28. Vong, C.M., Wong, P.K., Ip, W.F.: Case-based classification system with clustering for automotive engine spark ignition diagnosis. In: 2010 IEEE/ACIS 9th International Conference on Computer and Information Science (ICIS), pp. 17–22. IEEE (2010)
29. Azuaje, F., Dubitzky, W., Black, N., Adamson, K.: Discovering relevance knowledge in data: a growing cell structures approach. IEEE Trans. Syst. Man, Cybern. Part B Cybern. **30**, 448–460 (2000)
30. Zhuang, Z.Y., Churilov, L., Burstein, F., Sikaris, K.: Combining data mining and case-based reasoning for intelligent decision support for pathology ordering by general practitioners. Eur. J. Oper. Res. **195**, 662–675 (2009)
31. Perner, P.: Prototype-based classification. App. Intell. **28**(3), 238–246 (2008)

32. Chuang, C.-L.: Case-based reasoning support for liver disease diagnosis. Artif. Intell. Med. **53**, 15–23 (2011)
33. Guo, Y., Hu, J., Peng, Y.: Research on CBR system based on data mining. Appl. Soft Comput. **11**, 5006–5014 (2011)
34. Agrawal, R., Imieliński, T., Swami, A.: Mining association rules between sets of items in large databases. In: ACM SIGMOD Record, pp. 207–216. ACM (1993)
35. Geng, L., Hamilton, H.J.: Interestingness measures for data mining: A survey. ACM Comput. Surv. **38**, 9 (2006)
36. Aparna, V., Ingle, M.: Enriching Retrieval Process for Case Based Reasoning by using Vertical Association Knowledge with Correlation. Int. J. Recent Innov. Trends Comput. Commun. **2**, 4114–4117 (2014)
37. Nahm, U.Y., Mooney, R.J.: Using soft-matching mined rules to improve information extraction. Language (Baltim) **11**, 50 (2004)
38. Kang, Y.-B., Krishnaswamy, S., Zaslavsky, A.: A Retrieval Strategy for Case-Based Reasoning Using Similarity and Association Knowledge. IEEE Trans. Cybern. **44**, 473–487 (2014)
39. Patel, D.: A Retrieval Strategy for Case-Based Reasoning using USIMSCAR for Hierarchical Case. Int. J. Adv. Eng. Res. Technol. **2**, 65–69 (2014)
40. TFPC Classification Association Rule Mining (CARM) Software. https://cgi.csc.liv.ac.uk/~frans/KDD/Software/Apriori-TFPC/Version2/aprioriTFPC.html
41. Coenen, F., Leng, P., Ahmed, S.: Data structure for association rule mining: T-trees and P-trees. IEEE Trans. Knowl. Data Eng., 774–778 (2004)
42. Goulbourne, G., Coenen, F., Leng, P.: Algorithms for computing association rules using a partial-support tree. Knowledge-Based Syst. **13**, 141–149 (2000)
43. Coenen, F., Goulbourne, G., Leng, P.: Tree structures for mining association rules. Data Min. Knowl. Discov. **8**, 25–51 (2004)

Tracking Communities over Time in Dynamic Social Network

Etienne Gael Tajeuna[1(✉)], Mohamed Bouguessa[2], and Shengrui Wang[1]

[1] Department of Computer Science, University of Sherbrooke, Sherbrooke, QC, Canada
{etienne.gael.tajeuna,shengrui.wang}@usherbrooke.ca

[2] Department of Computer Science, University of Quebec at Montreal, Montreal, QC, Canada
bouguessa.mohamed@uqam.ca

Abstract. This poster paper presents an approach for tracking community structures. In contrast to the vast majority of existing methods, which are based on time-to-time consecutive evaluation, the proposed approach uses a similarity measure that involves the global temporal aspect of the network under investigation. A notable feature of our approach is that it is able to preserve the generated content across different time points. To demonstrate the suitability of the proposed method, we conducted experiments on real data extracted from the DBLP.

Keywords: Community evolution · Similarity measure · Topological structure

1 Introduction

To understand the evolution of communities over time, several approaches have been proposed. Most of these approaches investigate the common nodes of two communities at consecutive time stamps t_i and t_{i+1} using a Jaccard or modified Jaccard measure [1], [2]. However, as demonstrated in [3], at the end of lifespan such an approach may yield a community that does not share any nodes with the initially observed community.

In fact, a tracking approach that considers only consecutive time points may not necessarily capture the overall temporal evolution of a community. For purposes of clarification, let's look at the the evolution of community $C_{t_1}^1$ from t_1 to t_4 in two different cases as presented in Fig. 1. In the first case (Fig. 1 (First case)) we have an evolution obtained from a simple one-to-one investigation of nodes, with the corresponding evolution of the bag of topics from $B_{t_1}^1$ to $B_{t_4}^1$. As time evolves from time t_1 to t_4, we can see how nodes initially observed in $C_{t_1}^1$ gradually disappear as the topics are gradually change from the computer science field to the mathematic field. However, we can not say that the main topic of community $C_{t_1}^1$ has gradually changed from social network analysis to boolean algebra due to the fact that individuals found in $C_{t_4}^1$ may not share the same

P. Perner (Ed.): MLDM 2016, LNAI 9729, pp. 341–345, 2016.
DOI: 10.1007/978-3-319-41920-6_25

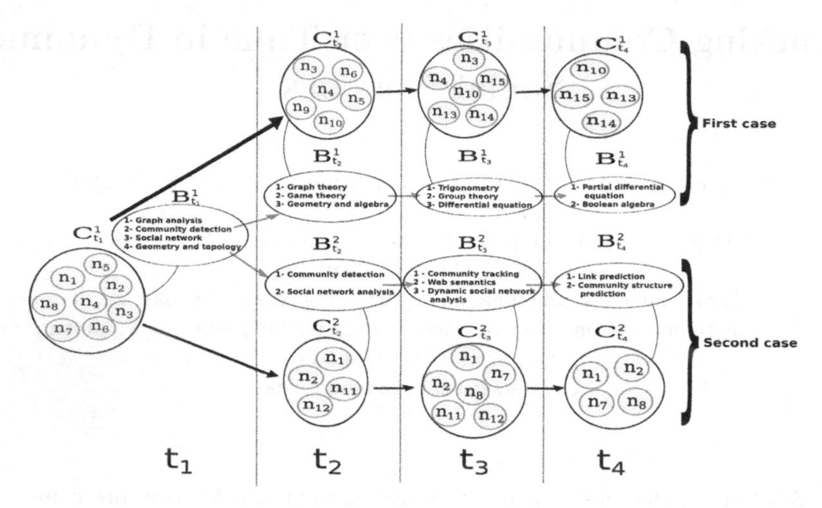

Fig. 1. An example of community and topic evolution.

interest as the individuals initially observed in $C_{t_1}^1$. In the second case (Fig. 1 (Second case)), we have an evolution from $C_{t_1}^1$ to $C_{t_4}^2$, against the corresponding evolution of the bag of topics from $B_{t_1}^1$ to $B_{t_4}^2$. We can see how some nodes initially observed in $C_{t_1}^1$ persist over time, as topics remain in the computer science field due to the fact that individuals found in this evolution are all interested on social network analysis.

The example in Fig. 1 suggests that consecutive tracking yields an inappropriate sequence which may inappropriately reflect the temporal evolution of a community for which the main topics may not be preserved across the time points. To alleviate this, in the next section we present a tracking method that uses a similarity measure which takes into consideration the temporal relation between communities.

2 The Tracking Approach

We denote by $G = \{V_{t_i}, E_{t_i} \,|\, 1 \leq i \leq m\} = (g_{t_i})_{1 \leq i \leq m}$ the dynamic social network over the time period from t_1 to t_m. For each graph g_{t_i}, we define a partition $\{C_{t_i}^1, C_{t_i}^2, ..., C_{t_i}^{q_i}\}$ representing the communities detected at time t_i using an existing community detection algorithm. To capture the temporal relation aspect of a given community C_{t_j} observed at time t_j, we use the framework that we have developed in [3] to extract the corresponding row vector v_j which captures the temporal aspect relation of C_{t_j} with other communities detected in the social network.

Specifically, we build a binary membership matrix $A_{(N_n \times N_c)}$, in which the rows correspond to the nodes found in G while the columns represent the discovered communities across different time points while. In A, the value 1 indicates that a given node is "present in" a specific community at specific time, while the value 0 reflects the opposite case (that is, the value 0 indicated that a given node is "absent in" a specific community at specific time). Next, we define the contingency matrix $B = A^T \times A = (b_{\alpha,\beta})_{1 \leq \alpha, \beta \leq N_c}$, where A^T is the transpose of A . By normalizing each row of B using the relation $p_{\alpha,\beta} = \frac{b_{\alpha,\beta}}{\sum_{\beta=1}^{N_c} b_{\alpha,\beta}}$, we obtain the matrix $B^* = (p_{\alpha,\beta})_{1 \leq \alpha, \beta \leq N_c}$, where each row j corresponds to the vector v_j. Note that each component $p_{\alpha,\beta}$ represents the probability that community C^α change to community C^β. Hence, the individual row vectors $v_j = (p_{j,1}, p_{j,2}, ..., p_{j,N_c})$ reflect the transition probabilities of community C^j over time which, in turn, represent the proportions of nodes found in community C^j over all detected communities.

To track communities over time, we use the following similarity measure:

$$sim(C_{t_i}, C_{t_j}) = \begin{cases} \sum_{\alpha=1}^{N_c} 2 \frac{p_{i,\alpha} \times p_{j,\alpha}}{p_{i,\alpha} + p_{j,\alpha}} \; if \; \sum_{\alpha=1}^{N_c} 2 \frac{p_{i,\alpha} \times p_{j,\alpha}}{p_{i,\alpha} + p_{j,\alpha}} > \lambda \\ 0 \; otherwise \end{cases} \tag{1}$$

where λ is the jonction point between the two Gammas curves estimated from the non-zeros values obtained when scoring the similarity between two transition probabilities vectors; $p_{i,\alpha}$ and $p_{j,\alpha}$ the respective components of vectors v_i and v_j.

By using the similarity measure described in (1), we define the evolution of community C_{t_i}, as the sequence of sorted communities $S_{C_{t_i}} = C_{t_i} \rightarrow C_{t_{i+\eta}} \rightarrow ... \rightarrow C_{t_k}$, $t_i < t_k \leq t_m$ such that all communities in $S_{C_{t_i}}$ are similar. Moreover, all communities C_{t_j} in $S_{C_{t_i}}$ should always share nodes with C_{t_i} such that the Jaccard coefficient exceeds zero.

To evaluate the framework presented above, we analyse communities with lifespans of 3 to 14 years. We compare the performance of our approach to existing methods that track communities in a time-sequential manner using similarity measures proposed in [2], [4] and [5]. For an objective comparison of the sequence communities obtained by competing approaches, we adopt a general criterion based on the resemblance between each pair of selected communities in the evolving communities. Specifically, we look at the proportion of nodes persisting in an evolving community (that is, the proportion of nodes observed at the first time and during the time duration of the evolving community). Formally, the proportion of node persisting in an evolving community $S_{C_{t_i}}$, is expressed as follows:

$$N_p(S_{C_{t_i}}) = \frac{1}{|\bigcup_{V \in S_{V_{t_i}}}|} |V_{t_i} \bigcap (S_{V_{t_i}} - V_{t_i})| \tag{2}$$

where $S_{V_{t_i}} = \{V_{t_i}, V_{t_{i+\eta}}, ..., V_{t_j}\}$ is the set of nodes corresponding to the sequence of community $S_{C_{t_i}}$.

Moreover, we look at the general trend of the top 5 most frequent words used for communities with lifespans of 3 to 14 years. Then, we take the particular case of a community with 14 years lifespan to provide a more detailed illustration of the topological structure and content evolution. For topological structure, as time evolves, we look at the transitivity, the conductance [6] and the community's average power to attract and keep nodes [7].

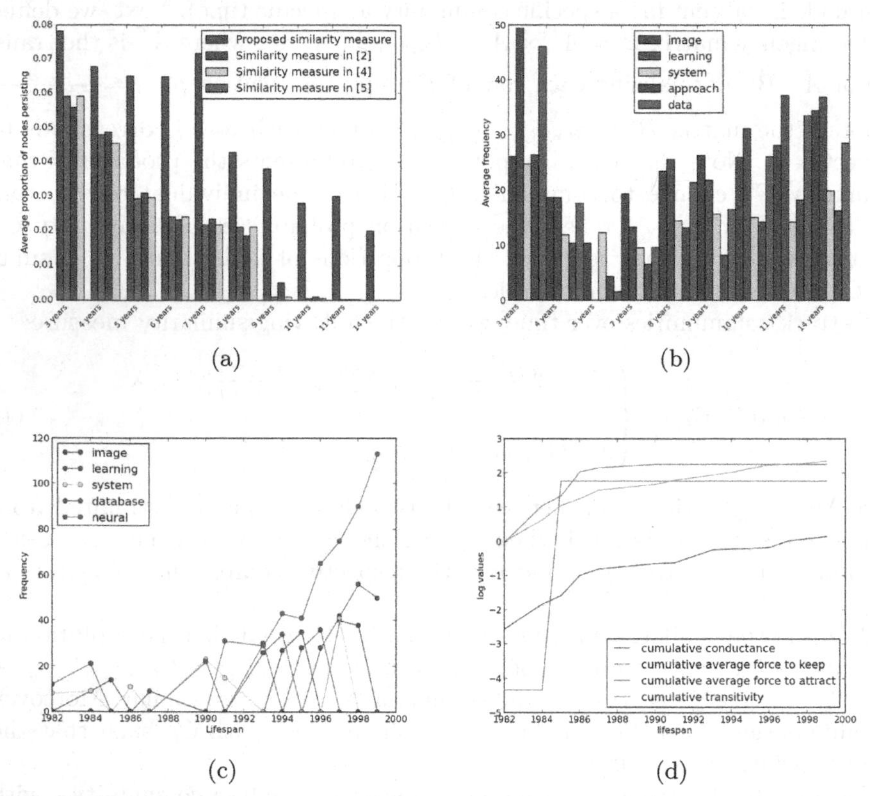

Fig. 2. (*a*) Average proportion of nodes persisting reported by competing approaches. (*b*) General trend of the top 5 most frequent words for communities having lifespans of 3 to 14 years. (*c*) Frequency of the top 5 most used words in $S_{C_{1982}^{18}}$. (*d*) Cumulative topological information.

3 Experiments

We demonstrate the suitability of the proposed approach on the seventh version of the DBLP dataset[1] [8]. The DBLP dataset contains the co-publications of authors. For each paper published it gives the paper title, the authors, the year, the publication venue, the index identification of the paper and the identifications of references to the paper. We built undirected, unweighted graphs between co-authors in the fields of data mining, machine learning and artificial intelligence and databases from year 1977 to 1999, taking each year as a snapshot. Note that in this dataset, nodes correspond to authors and edges reflect co-authorship relations. The total number of nodes is 12,178 while the number of nodes in the different snapshots varies from 80 to 1,709. To identify communities in each snapshot, we use the Infomap algorithm, a parameter-free approach for community detection. For each detected community, we define the bag-of-words obtained from titles of papers written by co-authors.

[1] http://arnetminer.org/citation

The experimental results are given in Fig. 2. Fig. 2(a) gives the histogram of the average proportion of nodes persisting in all competing approaches. As we can see from Fig. 2(a), the average proportion of nodes persisting is higher in our approach. This suggests that the proposed similarity measure is capable to track well communities of co-authors over time than those proposed in [2], [4] and [5]. Moreover, as depicted by Fig. 2(b) which illustrates the general trends of the top 5 most frequent words that persist for communities with lifespans of 3 to 14 years, we can see that, in general, the most frequent words remain present over the different communities' lifespan. The same observation occurs for the community $S_{C_{1982}^{18}}$ with a 14 years lifespan (Fig. 2(c)).

We note that the results observed in Fig. 2(c) can be explained by the cumulative topological structures depicted in Fig. 2(d) where we observe how the capacity of the community to attract and keep a node remains stable from the year 1986, while the conductance and transitivity of their evolution show the strength of the community over time.

4 Conclusion

In this paper, we have presented an approach for tracking community structures in dynamic social networks. Our experiment suggests that the proposed approach can track communities as a sequence and preserve the generated content across different time points. Moreover, the evolution of topological structure explored in the experiment reveals interesting information which may be important to better understand the evolution of communities with their related content. In our continuing research, we are exploring the topological structure to render the approach capable of predicting the future transitions a community may undergo.

Acknowledgments. This work is supported by research grants from the Natural Sciences and Engineering Research Council of Canada (NSERC).

References

1. Asur, S., Parthasarathy, S., Ucar, D.: An event-based framework for characterizing the evolutionary behavior of interaction graphs. In: ACM KDD, pp. 913–921 (2007)
2. Takaffoli, M., Fagnan, J., Sangi, F., Zaiane, O.R.: Tracking changes in dynamic information networks. In: IEEE CASoN, pp. 94–101 (2011)
3. Tajeuna, E.G., Bouguessa, M., Wang, S.: Tracking the evolution of community structures in time-evolving social networks. In: IEEE DSAA, pp. 1–10 (2015)
4. Bourqui, R., Gilbert, F., Simonetto, P., Zaidi, F., Sharan, U., Jourdan, F.: Detecting structural changes and command hierarchies in dynamic social networks. In: IEEE ASONAM, pp. 83–88 (2009)
5. Greene, D., Doyle, D., Cunningham, P.: Tracking the evolution of communities in dynamic social networks. In: ASONAM, pp. 176–183 (2010)
6. Leskovec, J., Lang, K.J., Dasgupta, A., Mahoney, M.W.: Statistical properties of community structure in large social and information networks. In: ACM WWW, pp. 695–704 (2008)
7. Ye, Z., Hu, S., Yu, J.: Adaptive clustering algorithm for community detection in complex networks. Physical Review E **78**, 046115 (2008)
8. Tang, J., Zhang, J., Yao, L., Li, J., Zhang, L., Su, Z.: ArnetMiner: extraction and mining of academic social networks. In: ACM KDD, pp. 990–998 (2008)

Generalized Hand Gesture Recognition for Wearable Devices in IoT: Application and Implementation Challenges

Darshan Iyer, Fahim Mohammad, Yuan Guo, Ebrahim Al Safadi, Benjamin J. Smiley, Zhiqiang Liang, and Nilesh K. Jain(✉)

Intel Corporations, Hillsboro, OR, USA
{darshan.iyer,fahim.mohammad,yuan.guo,ebrahim.al.safadi, benjamin.j.smiley,zhiqiang.liang,nilesh.jain}@intel.com

Abstract. The proliferation of low power and low cost continuous sensing technology is enabling new and innovative applications in wearables and Internet of Things (IoT). At the same time, new applications are creating challenges to maintain real-time response in a resource-constrained device, while maintaining an acceptable performance. In this paper, we describe an IMU (Inertial Measurement Unit) sensor-based generalized hand gesture recognition system, its applications, and the challenges involved in implementing the algorithm in a resource-constrained device. We have implemented a simple algorithm for gesture spotting that substantially reduces the false positives. The gesture recognition model was built using the data collected from 52 unique subjects. The model was mapped onto Intel® Quark™ SE Pattern Matching Engine, and field-tested using 8 additional subjects achieving 92% performance.

Keywords: Gesture recognition · Machine learning · Pattern recognition · Feature engineering · Wearable device · Internet of Things

1 Introduction

Gesture recognition refers to the process of understanding and classifying meaningful movements by human's body parts such as hands, arms, face, and sometimes head. The proliferation in technology, and in microelectronics more specifically, has inspired research in the field of IMU-based gesture recognition. Three-axis accelerometers are increasingly embedded into many personal electronic devices such as smartphones, Wiimote, etc [Akl et al., 2011; Liu et al., 2009; Liu et al., 2010]. A growing number of devices are being equipped with 3-axis gyroscopes in addition a 3-axis accelerometer. Combining the data from these two sensor types provides significantly more accurate motion information compared to only using an accelerometer. Kratz et al. (2013) showed an increase in performance with an addition of gyroscope data, which allows for more complex gestures to be used in mobile applications.

Gesture recognition has wide range of applications in telerobotics [Speeter, 1992], character-recognition in 3D space using inertial sensors [Zhou et al., 2008; Oh, JK et al., 2004], controlling a TV set remotely [Freeman and Weissman, 1995], enabling hand as

P. Perner (Ed.): MLDM 2016, LNAI 9729, pp. 346–355, 2016.
DOI: 10.1007/978-3-319-41920-6_26

a 3D mouse [Bretzner and Lindeberg, 1998], using hand gestures as a control mechanism in virtual reality [Xu, 2006; Liu et al., 2009], understanding the actions of a musical conductor [Je et al., 2007].

Previous studies have used data from smaller number of subjects, though the variety of gestures and number of samples per each gesture were higher. Liu et al. (2010) collected a set of 8 gestures with 150 to 200 samples per gesture from a single subject. Xu et al. (2012) used a set of 7 gestures with 30 samples per gesture with no reference to the number of users. Akl et al. (2011) collected a set of 18 gestures with total 3780 samples collected from 7 participants. Liu et al. (2009) used a set of 8 gestures with over 4000 samples collected from 8 users. Kratz et al. (2013) used a set of 15 gestures with 15 samples per gesture collected from 6 participants. Mace et al. (2013) used 4 gestures with 5 samples per gesture collected from 5 participants. Pylvänäinen (2005) collected 10 gestures with 20 samples per gesture from 7 users. A model constructed with smaller number of subjects does not capture enough inter-subject variability, which makes it difficult a build a generalized system.

Most of the above algorithms, especially the ones implemented on wearable devices, are for personalized application [Liu et al., 2009; Liu and Pan, 2010; Li, 2010; Kratz et al., 2013], applicable only for single user against whom the classifier is trained and tested. Algorithms by Pylvänäinen (2005) based on HMM and Akl et al. (2011) based on DTW were applicable to both user-dependent and user-independent, generalized applications. Being user-dependent limits the applications of the system. Researchers on gesture recognition are envisaging a universal system that, given a dictionary of gestures, can recognize different gesture traces with competitive accuracy and with minimal dependence on the user.

An increasingly large number of devices can capture gestures. A few example of such devices include Nintendo Wiimote, joystick, trackball, touch tablet, smart phone, data glove, TI eZ430-Chronos watch, etc. The two most popular algorithms found in gesture recognition literature are hidden markov models (HMMs) and dynamic time warping (DTW). The computational complexity of statistical methods like HMM is directly proportional to the quantity as well as the dimension of feature vectors. Besides, variations in gestures are not necessarily Gaussian and perhaps other formulations may turn out to be a better fit [Akl et al., 2011]. Another limitation of HMM-based methods is that they often require knowledge of the vocabulary in order to configure the models properly, e.g., the number of states in the model. Therefore, HMM-based methods may suffer when users are allowed to choose gestures freely, or for personalized gesture recognition [Liu et al., 2009]. Naïve template-based techniques such as nearest-neighbor search, though easy to implement, will generally not compensate for variations in gesture execution time. A popular algorithm that does compensate for differences in time series lengths is DTW [Akl et al., 2011; Liu et al., 2009; Liu and Pan, 2010; Mace et al., 2013]. Protractor3D [Kratz et al., 2011] is a template-based technique developed for 3D acceleration data that compensates for rotational derivations between input sequences and templates by finding the optimal registration between them. Compared to DTW, the method does not do any kind of warping in the time domain. Xu et al. (2012) presented an algorithm based on reducing a gesture signal waveform to a code of 8 numbers based on peaks and valleys, and comparing against a database of coded numbers. All these algorithms, though delivering excellent performance in different scenarios, are computationally challenging to implement on a wearable device with extremely lean resources.

Segmenting the data related to gesture amidst the background is challenging and might lead to lot of false positives and consequent drop in recognition performance. To overcome this issue, users of smart phones and handheld devices like Nintendo Wii use push-to-gesture button, wherein recording starts when button is pressed and stops when button is released [Akl et al., 2011; Liu et al., 2009; Liu and Pan, 2010; Kratz et al., 2013]. Ruiz and Li (2011) addressed the segmentation issue by the concept of DoubleFlip. DoubleFlip employs a simple delimiter gesture for entering gesture input mode. The criteria for choosing the "double flip" gesture were (1) the gesture being easy to perform, and (2) the gesture being sufficiently distinct from everyday movement. The segmentation algorithm of Xu et al. (2012) required that the time interval between two gestures be no less than 0.2 seconds for it to work, which was based on checking five conditions on all the data points and picking out the most likely ones as the gesture termination points.

Designing systems for wearable devices is always a challenge primarily because of three limiting factors: small memory, limited processing power, and the limited battery life. The gesture recognition system presented in this paper is a user-independent, generalized model developed keeping industrial workers in mind. Nevertheless, the algorithm is general enough for various applications. The entire sensor subsystem and the recognition system are located in a resource-constrained platform of a wrist-worn wearable device. The model was obtained from training data collected from 52 different male subjects (<=10 samples per user and gesture), and tested on data collected at different times from the same set of users as well on data obtained from 8 independent users that included a mixture of male and female subjects. The proposed gesture recognition system uses data from the Y-axis of the gyroscope for spotting a gesture and from X, Y, and Z-axes of accelerometer for feature extraction and recognition. The algorithm is based on down sampling data to a fixed size to eliminate speed differences, amplitude-normalization to eliminate intensity differences, supervised hierarchical clustering for building gesture templates, followed by simple pattern matching for recognition. The segmentation of gestures is based on a variant of DoubleFlip [Ruiz and Li, 2011], specifically suitable for wearable device that incorporates gesture key at both the beginning and the end of the gesture to better delineate the gesture under noisy conditions. The algorithm is overall much simpler in terms of computational complexity and is suitable for a resource-constrained platform such as wearable device.

The rest of the paper is organized as follows: section 2 on "Data" discusses the details of data collection. Section 3 on "Methods" elaborates on different methods used for gesture data analysis. Section 4 on "Results" presents the performance summary for different test sets. Finally, section 5 on "Discussion" discusses the results and highlights the unique aspects of this work.

2 Data

Data were collected from custom-built wearable device containing BMI 160 (accelerometer and gyroscope) sensor. As shown in Figure 1, the device was placed on the right wrist for data collection. The orientation was pre-determined and fixed across all subjects such that +y axis was the dominant axis for gesture key identification.

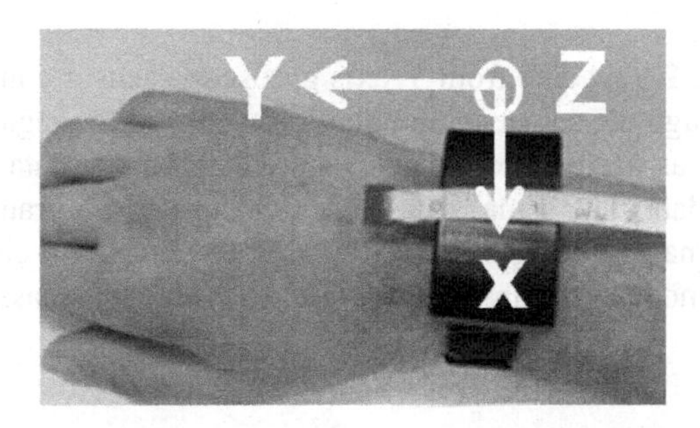

Fig. 1. Placement of device containing sensors and recognition engine on the wrist.

Sampling rate was set at 100 Hz, and accelerometer values were collected with a dynamic range of 8g. The devices were ensured to have the same internal orientations for X, Y, and Z-axes.

2.1 Training Data

Training data were collected in two stages from 52 subjects. In the first stage, gestures were conducted while subjects were standing, while in the second stage, subjects were in prone position to ensure we capture gesture in different positions. In each of these stages, subject repeated each of the five gestures (A, D, M, L and U) five times (therefore a total of 25 gestures overall per stage). The subject would start with five repetitions of first gesture (i.e., letter A), followed by five reps of second gesture (i.e., letter D), and so on.

2.2 Test Data

Two different sets of test data were collected to verify the robustness of gesture model.

2.2.1 Test Dataset A

From the same set of subjects involved in training set, an additional 90 sec of data were captured while the subjects were performing various industrial activities such as climbing, walking, etc. along with gesture M and gesture A. These data were used to test the robustness of the segmentation algorithm, and to test the performance of models for final model selection.

2.2.2 Test Dataset B

Data from a total of eight subjects were collected, of which five subjects were new (3 males and 2 females) with an objective to quantify the robustness of the final model. The subjects followed the same protocol as the one mentioned for training data collection, performing the same sets of gestures under different conditions.

3 Methods

As shown in Figure 2, raw data from Y-axis gyroscope sensor connected to the wrist were streamed through the segmentation algorithm to find the region of interest. The region of interest was used to extract relevant accelerometer data from X, Y, and Z axes called signal data. The signal data were run through the transform functions to generate a feature matrix, which in turn was used to generate a classification model. This classification model was subsequently used to predict the class for the test data.

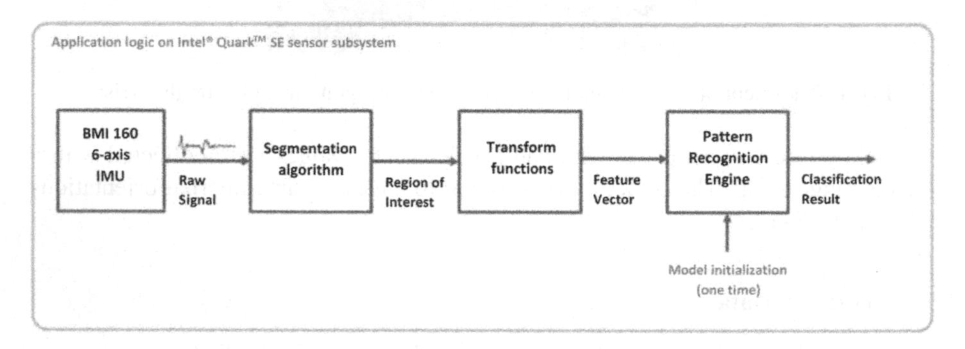

Fig. 2. Overall flow of the data collection and recognition engine.

3.1 Segmentation

Identifying the proper segment containing a gesture in the time-series data is a challenge in wearable devices because push-to-gesture (usually performed on handheld device using extra hardware button) or press-to-record (on mobile phones) functionalities are generally inconvenient or not possible. As per the protocol, for the gesture recognition mode to be activated, the subject must perform a gesture key consisting of twisting the wrist twice, performing the hand gesture, followed by one twist. The gesture-spotting algorithm was designed with two objectives in mind: reducing the number of false positives and false negatives when the gesture key is performed correctly, and dealing with possible deviations from the protocol.

The segmentation algorithm in essence counts twists by detecting peaks with certain characteristics. The algorithm is designed to operate on a small buffer of approximately 5 sec data, typically needed for wearable devices. The algorithm operates in four stages.

1. **Bump Detection:** A stream of data are first demeaned, followed by smoothing, and then computing the first order difference. A bump is detected when a sequence of consecutive positive values (state P) are followed by a sequence of consecutive negative values (state N), followed finally by a positive value (state P). This final transition from state N to state P triggers bump detection.
2. **Peak Detection:** Not all bumps are peaks. Only bumps with amplitudes above a certain threshold are considered to be peaks caused by intentional key-activation twists by the subject. This threshold, however, is not static (although it begins with an initial value) but adapts to the average peak values of a subject. More specifically, if a bump is found to have a peak value above the initial threshold, and the next bump is greater than half of this value, it is also considered a peak. Once these peaks are taken, an average of their values is registered and the new threshold becomes half of the new average, and so on.
3. **Twist State:** The number of peaks is accumulated until we have three. Once this number is reached, the values of these peaks, their locations (i.e., indices), and the location of their shoulders are processed to determine whether a proper gesture key was done. The objective is to ensure that the first two peaks are close to each other (but not too close, else they will be merged as one single peak), while the third peak is farther apart but only within an average maximum duration for a gesture to take place in between the second and third twists.
4. **Final Check:** This final step demands that the algorithm refrain from making a key-activation decision for an extra (fixed) number of samples to ensure a fourth twist is not present. If no peak is detected in the tail region, then spotting can immediately be made, else if a fourth peak is detected, the algorithm checks the amount of time (or samples) between the second and the third twist. Depending on this duration, the algorithm decides whether the subjects simply did the fourth twist by mistake, or whether the subject did two twists, changed his/her mind, and re-initiated key activation (hence the algorithm would drop the first two gestures, and assume the third and the fourth to be the first and the second).

All the statistics related to thresholds and initial values were generated using the training data.

3.2 Feature Generation

We observed that concatenated raw accelerometer signal of X, Y, Z axes had good separation between various gestures, so decided to derive features from the raw signals. This finding really helped us to keep the feature computation cost down. We used down sampling of raw signal to fixed window size, and amplitude normalization to get rid of time and intensity variance across various subjects. The feature vector was finally quantized to make it compatible with the Intel® Quark™ SE hardware.

3.3 Model Construction and Pattern Recognition

A supervised hierarchical clustering algorithm was used to cluster the feature vectors based on similarity characteristics and labels. The centroids were computed for each cluster and labels were assigned to each centroid based on majority. The numbers of cluster centroids were optimized to leverage the pattern recognition capability of Intel® Quark™ SE in Radial Basis Function (RBF) mode, which enables the use of finite surface around each centroid. Combination of centroid and other related information defined a pattern.

During recognition phase, the test vector is assigned a class based on its proximity to the pattern, taking into account the finite surface around the pattern. In case the test vector falls outside the surface of the pattern, it is assigned an unknown classification (UNK). This feature is very useful as it eliminates the need for a negative class to deal with gestures that fall outside the ones that were used during model construction.

3.4 Model Selection

Selecting a model with the desired number of patterns was based on leave-one-subject out performance values. In each iteration, feature vectors from all but one subject constitute the training set and the left out feature vectors constitute the validation set. A new model was generated in each iteration and the performance was tested on the validation set and metrics (recall, precision, and accuracy) were collected. This was repeated for all the subjects and leave-one-subject out performance numbers were calculated as average metric values. Figure 3 depicts the performance values as a function of model size. As one can see, the performance becomes stable at around 100 patterns.

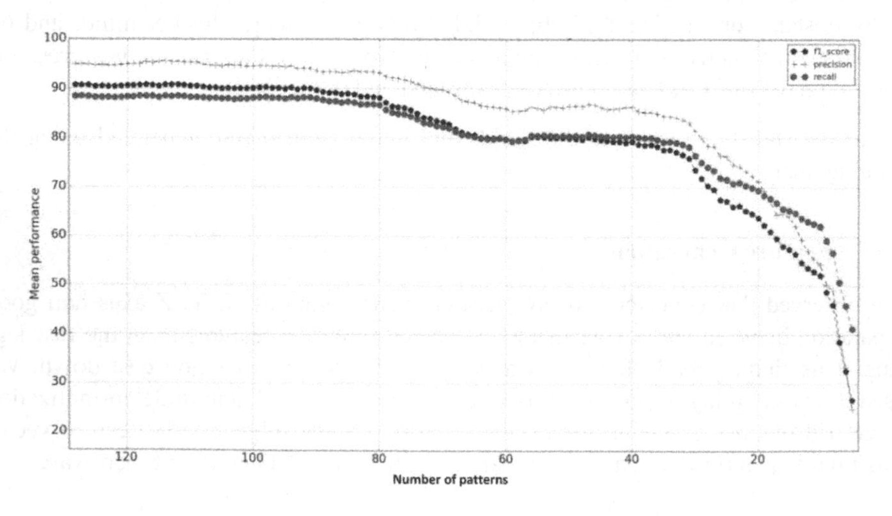

Fig. 3. Selection process of optimum number of patterns.

4 Results

Different models were constructed to evaluate the model performance by varying the number of patterns. We finally selected one model consisting of 100 patterns based on the performance on the validation data. We first tested this model on test dataset A from 52 subjects, consisting of only gestures A and M from each user. These gestures were performed while doing various industrial activities. Table 1 summarizes the overall performance with a precision of 0.96, recall of 0.89 and f1-score of 0.92

Table 1. Performance summary on test dataset A.

	UNK	Precision	Recall	f1-score	Support
Gesture A	3	0.94	0.91	0.92	54
Gesture M	3	1	0.88	0.94	50
Average		0.96	0.89	0.92	104

We tested the performance of thc same model on test dataset B and the results are shown in Table 2. The support column has variable values as some of the gestures were repeated more or less than five times. As is evident from the table, for this dataset, the overall values for precision, recall, and f1-score were 0.92, 0.92, and 0.92, respectively.

Table 2. Performance summary on tcst dataset B.

	UNK	Precision	Recall	f1-score	Support
Gesture A	0	0.94	0.86	0.9	79
Gesture D	1	0.84	0.92	0.88	77
Gesture M	1	1	0.95	0.97	78
Gesture L	0	0.88	0.98	0.92	86
Gesture U	0	0.97	0.86	0.91	80
Average		0.92	0.92	0.92	400

5 Discussion

Gesture recognition using body worn sensor such as IMU is an interesting and challenging problem. The objective of the current work is to showcase the wearable technologies for first responders, industrial workers such as firefighters, mining workers, etc. that will equip them to perform their job efficiently, and at the same time reduce the risk of injuries.

To this end we have developed a model suited for a wearable device for gesture-based communication for the connected worker. It is a wrist-worn device, which allows users to communicate using gestures in situations when other modes of communication such as radio are not available. This could be lifesaving, especially when the radio is cluttered with multiple simultaneous chats, and/or when there is no visibility to fellow firefighters. Gesture devices locally detect the user gesture and send the detected gesture via Bluetooth low energy (BLE) to a mobile hub, which can take further application-specific action.

Most of the published approaches are 1) personalized and thus not user-independent, 2) if generalized, need computationally intensive approaches like HMM and DTW, 3) not applicable for complex movements, 4) gesture delineation requires push button approach, 5) part of handheld device such as smart phones, Wiimote, etc.

To detect the presence of a gesture, the time series data were buffered. The buffer size needs to be sufficient enough to hold the entire gesture including gesture key. Smaller buffers may be insufficient to hold the entire gesture segment, while a larger buffer would need more computation and result in a delay in processing. Once the segmentation algorithm detects a valid gesture segment, the detection algorithm needs to be fast enough to display the result. A delay in processing may miss the gesture performed immediately after the current gesture. The proposed method is a user-independent and computationally simple gesture delineation method based on variant of DoubleFlip, thus suitable for wearable device such as a wrist-worn sensor.

A machine learning model based on large number of features or complex features may be more accurate but inappropriate for wearables. Calculating complex features based on fast Fourier transform (FFT) (O(n.logn) computational complexity) or principal component analysis (PCA) ($O(n^3)$ computational complexity) requires complex matrix manipulations which in turn needs more space, insufficient to fit on the device, and more processing power which leads to more time to process and subsequent faster draining of battery. The proposed method (O(n) computational complexity) eliminates the need for expensive feature computation.

Also, the device once trained using the proposed framework, works for new users as opposed to personalized framework, which works only for the specific user on whom the device is trained. This is great for industrial use as it provides great out-of-the-box experience. This could be further improved over a period of time by adding personalized samples to the training set.

We built machine learning models using the training data collected from 52 subjects performing 5 gestures viz., A, D, M, L, and U. The model was first tested using data that included test gesture samples collected amidst other activities. We got an overall precision of 0.96, recall of 0.89, and f1-Score of 0.92. To further test the generalization performance of the model, we tested it on an independent set of data. We got an overall precision of 0.92, recall of 0.92, and f1-Score of 0.92. Thus, the system is simple and works in different scenarios: male and female, different gestures from same and different users, same gesture at different speeds from the same user. Our method based on robust generalized gesture recognition algorithm can be applied to wide range of applications.

References

1. Akl, A., Feng, C., Valaee, S.: A Novel Accelerometer-Based Gesture Recognition System. IEEE Transcations on Signal Processing **59**(12), December 2011
2. Bretzner, L., Lindeberg, T.: Use your hand as a 3-D mouse, or, relative orientation from extended sequences of sparse point and line correspondences using the affine trifocal tensor. In: Burkhardt, H.-J., Neumann, B. (eds.) ECCV 1998. LNCS, vol. 1406, pp. 141–157. Springer, Heidelberg (1998)

3. Freeman, W.T., Weissman, C.D.: TV control by hand gestures. In: IEEE Int. Workshop on Automatic Face and Gesture Recognition, Zurich, Switzerland (1995)
4. Je, H., Kim, J., Kim, D.: Hand gesture recognition to understand musical conducting action. In: IEEE Int. Conf. Robot & Human Interactive Communication, August 2007
5. Kratz, S., Rohs, M.: Protractor3D: a closed-form solution to rotation-invariant 3D gestures. In: Proc. IUI 2011, pp. 371–374. ACM (2011)
6. Kratz, S., Rohs, M., Essl, G.: Combining acceleration and gyroscope data for motion gesture recognition using classifiers with dimensionality constraints. In: IUI 2013, Santa Monica, CA, USA, pp. 173–178, March 19–22, 2013
7. Liu, J., Zhong, L., Wickramasuriya, J., Vasudevan, V.: uWave: Accelerometer-based personalized gesture recognition and its applications. Pervasive and Mobile Computing **5**, 657–675 (2009)
8. Liu, J., Pan, A., Li, X.: An Accelerometer-Based Gesture Recognition Algorithm and its Application for 3D Interaction. Computer Science and Information Systems **7**(1) (2010)
9. Mace, D., Gao, W., Coskun, A.: Accelerometer-based hand gesture recognition using feature weighted naïve bayesian classifiers and dynamic time warping. In: IUI 2013 Companion, pp. 83–84 (2013)
10. Oh, J.K., Cho, S.J., Bang, W.C., et al.: Inertial sensor based recognition of 3-D character gestures with an ensemble of classifiers. In: 9th Int. Workshop on Frontiers in Handwriting Recognition (2004)
11. Pylvänäinen, T.: Accelerometer based gesture recognition using continuous HMMs. In: Marques, J.S., Pérez de la Blanca, N., Pina, P. (eds.) IbPRIA 2005. LNCS, vol. 3522, pp. 639–646. Springer, Heidelberg (2005)
12. Ruiz, J., Li, Y.: DoubleFlip: a motion gesture delimiter for mobile interaction. In: Proc. CHI 2011, pp. 2717–2720. ACM (2011)
13. Speeter, T.H.: Transformation human hand motion for telemanipulation. Presence **1**(1), 63–79 (1992)
14. Xu, D.: A neural network approach for hand gesture recognition in virtual reality driving training system of SPG. In: 18th Int. Conf. Pattern Recognition (2006)
15. Xu, R., Zhou, S., Li, W.J.: MEMS accelerometer based nonspecific-user hand gesture recognition. Sensors Journal **12**(5), 1166–1173 (2012)
16. Zhou, S., Dong, Z., Li, W.J., Kwong, C.P.: Hand-written character recognition using MEMS motion sensing technology. In: Proc. IEEE/ASME Int. Conf. Advanced Intelligent Mechatronics, pp. 1418–1423 (2008)

A Review on Artificial Intelligence Based Parameter Forecasting for Soil-Water Content

Ferhat Özçep[1(✉)], Eray Yıldırım[2], Okan Tezel[1], Metin Aşçı[3], Savaş Karabulut[1], and Tazegül Özçep[4]

[1] Department of Geophysical Engineering, İstanbul Üniversity, Istanbul, Turkey
{ferozcep,otezel,savask}@istanbul.edu.tr
[2] Department of Geophysical Engineering, Sakarya University, Sakarya, Turkey
[3] Department of Geophysical Engineering, Kocaeli University, Kocaeli, Turkey
[4] Sirinevler Mehmet Sen Okulu, Ministry of National Education, Istanbul, Turkey

Abstract. The purpose of this study, by using an artificial intelligent approaches, is to compare a correlation between geophysical and geotechnical parameters. The input variables for this system are the electrical resistivity reading, the water content laboratory measurements. The output variable is water content of soils. In this study, our data sets are clustered into 120 training sets and 28 testing sets for constructing the fuzzy system and validating the ability of system prediction, respectively. Relationships between soil water content and electrical parameters were obtained by curvilinear models. The ranges of our samples are changed between 1 - 50 ohm.m (for resistivity) and 20 - 60 (%, for water content). An artificial intelligent system (artificial neural networks, Fuzzy logic applications, Mamdani and Sugeno approaches) are based on some comparisons about correlation between electrical resistivity and soil-water content, for Istanbul and Golcuk Soils in Turkey.

Keywords: Fuzzy logic applications · AI · Mamdani and Sugeno approaches · Geophysical and geotechnical data

1 Introduction

Relationships between soil water content and electrical parameters were measured in field and laboratory conditions and mostly curvilinear models were obtained. Curvilinear relationships were also proposed between electrical resistivity and temperature [1,2,3]. But, Ananyan [4] derived and experimentally proved exponential relationship between electrical resistivity, soil temperature, and water content based on a series of experiments. There are several studies related to the electrical resistivity and water content [5, 6, 7, 8, 9, 10,11, 12, 13]. Artificial neural networks (ANNs) are part of a much wider field called artificial intelligence, which can be defined as the study of mental facilities through the use of computational models [14].

P. Perner (Ed.): MLDM 2016, LNAI 9729, pp. 356–361, 2016.
DOI: 10.1007/978-3-319-41920-6_27

2 Artificial Intelligent (AI) Technique

Artificial Intelligent (AI) Technique encompass computer algorithms that solve several types of problems. The problems include classification, parameter estimation, parameter prediction, pattern recognition, completion, association, filtering, and optimization [15]. ANNs are composed of a large number of highly interconnected processing elements, or neurons, usually arranged in layers. These layers generally include an input layer, a number of hidden layers, and an output layer.

Fuzzy logic was first developed by Zadeh [16] in 1960s for representing uncertain and imprecise information. It provides approximate but effective descriptions for highly complex or difficult to analyze mathematical systems. Fuzzy inference system (FIS) is a rule based on system consisting of three conceptual components. These are: (1) a rule-base, containing fuzzy if-then rules, (2) a database, defining the membership functions (MF) and (3) an inference system, combining the fuzzy rules and produces the system results [17]. There are two types of widely used fuzzy inference systems, Takagi- Sugeno FIS and Mamdani FIS [8]. In geophysics there are several applications related to use of artificial Intelligences techniques in general such as Albora et al. [18, 19, 20], and especially electrical resistivity data such as [21, 22, 23].

3 Artificial Intelligent (AI) Technique

Study area is located in Istanbul (Yesilkoy, Florya, Basinkoy) and Golcuk areas). Investigation depth for soil mechanics procedures and geoelectrical measurements is up to 15 m. The ranges of our samples are changed between 1 - 50 ohm.m (for resistivity) and 20 - 60 (%, for water content).

3.1 Artificial Neural Networks Applications

Hidden and output are formed in this study. ANN models was formed by taking i and k values as 1 and j value as 3, 5 and 10. Analysis of 148 data couple with a input vector (ROA) and output vector (Wn) was carried out. To apply the models, all data was normalized between 0.1 and 0.9 in the flowing relation:

$$X\_ = 0{,}8\ .\ (X\_ - XMIN) / (XMAX - XMIN) + 0{,}1 \quad (1)$$

Where, Xi normalized values, XMAX and XMIN maximum and minimum measurement values. By the normalization process, data can be dimensionless. Data are divided in to two categories as training and testing data. Training set includes 120 values, and other 28 testing data were used in the performance evaluation. Performance values obtained by ANN models were given in Table 1. In this study, neuron number of hidden layer after several tests are determined 5 from performance evaluation of test set (as shown in Table 1).

Table 1. Performance Evaluation of Test Data of ANN models

Model	MAEP (Mean Absolute error percent)	MSE (Nean Square Error	R^2
ANN (1 3 1)	17.66	30.50*	0.8754
ANN (1 5 1)	17.76	33.62	08.44*
ANN (1 10 1)	16.53*	34.97	0.8723

Note: best results signed as *

3.2 Fuzzy Logic Applications

3.2.1 Mamdani Method Application and Results

It was formed 5 different sub-sets. In the formation of sub-set, triangle membership function was used. These sub-sets are defined as very lower, (VL), lower (L), middle (M), high (H) and very high (VH). Figure 1 shows the membership functions of sub-sets of ROA input and Wn output were shown. In this study, the rules were defined between ROA and Wn by using sub-sets. For example ''If ROA is VL, then Wn is VH''. Fuzzy logic rules to estimate water content was given in Table 2. 120 data set was used in Mamdani-Fuzzy logic modeling. This formed model was tested by 28 data set that used in testing of artificial neural networks. In evaluation of performance of test group, MAEP and R2 values are calculated respectively 19.99 and 0.8268.

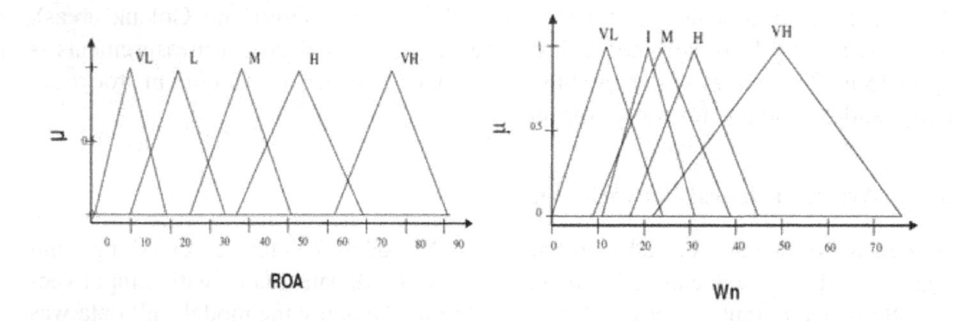

Fig. 1. The membership functions of sub-sets of ROA input and Wn output

Table 2. Fuzzy logic (Mamdani) rules that shows relation between water content (Wn) ile electrical resistivity ROA

If (ROA = VL), Then (Wn=VH)
If (ROA = L), Then (Wn=H)
If (ROA = M), Then (Wn=M)
If (ROA = H), Then (Wn=L)
If (ROA = VH), Then (Wn=VL)

3.2.2 Sugeno Method Application and Results

In this approach, calculations based on mathematical relations each sub-set. In this study, 5 sub-sets was formed by considering data distribution, and curve equation for each sub-set was obtained. Distribution of these sub-sets was shown in Figure 2a to e. Apart from Mamdani approach, sub-set of Sugeno method was formed by using linear relation between input-out. Rule definitions 'of these sub-sets was shown in Table 3. Membership function of sub-set for input and variations of membership grade was given in Figure 3. With the training of 120 input-output, data set was used in Sugeno- Fuzzy logic modeling. In evaluation of performance of test group, MAEP and R2 values are calculated respectively 17.63 and 0.8025. Sugeno-Fuzzy logic approach gives more effective results than Mamdani-Fuzzy logic approach. Moreover, by using Sugeno-Fuzzy logic approach, formation of fuzzy logic model is very easy than others

4 Results and Discussions

The prediction accuracy of AI system was fairly good (predictive ability and for the coefficient of correlation) based on the results of the testing performance, and the

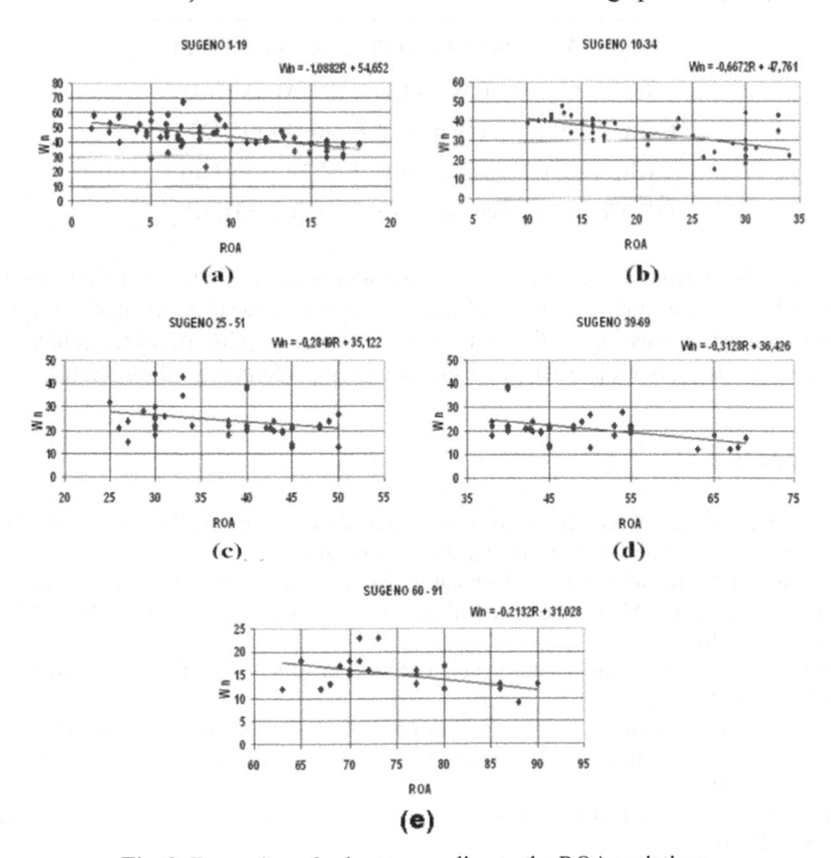

Fig. 2. Formation of sub-set according to the ROA variations.

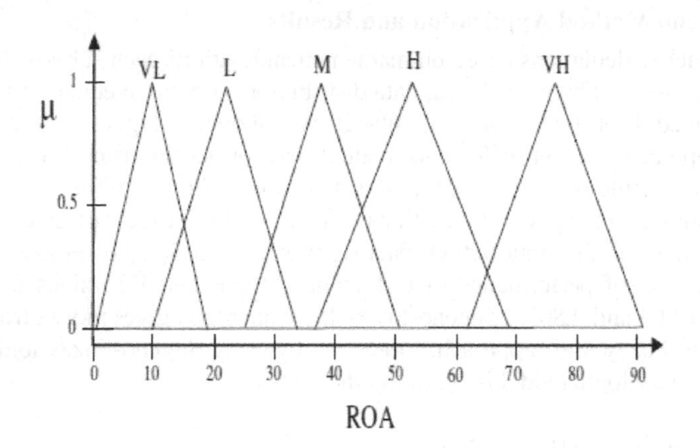

Fig. 3. Membership functions of sub-set for input and variations of membership grade.

Table 3. Fuzzy logic (Sugeno) rules that indicates relation between water content and ROA.

If (ROA = VL), Then (Wn=-1.0882 (ROA) +54.652)
If (ROA = L), Then (Wn=-0.6672 (ROA) +47.761)
If (ROA = M), Then (Wn=-0.2849 (ROA) +35.122)
If (ROA = H), Then (Wn=-0.3128 (ROA) +36.426)
If (ROA = VH), Then (Wn=-0.2132 (ROA) +31.028)

calculated coefficient of correlation of training and testing. - The compared results by classical (regression) analysis, artificial neural networks and fuzzy estimates show that the conventional regression method can easily estimate a value of water content from electrical resistivity, but it is weak in the evaluation of performance of estimations.

References

1. Liu, N.: Soil and Site Characterization Using Electromagnetic Waves, Ph.D. Thesis, Virginia Polytechnic Institute and State University (2007)
2. Raisov, O.Z.: Relationship of electrical resistivity and temperature for sod-serozemic solonchak. Vestnik MGU. Biology, Soil Science. N 3. Moscow University Press, Moscow (1973). (in Russian)
3. Ananyan, A.A.: Permafrost studies, vol. 1. Academy of Science of USSR Press, Moscow (1961). (in Russian)
4. Abidin, M.H.Z., Saad, R., Ahmad, F., Wijeyesekera, D.C., Yahya, A.S.: Soil moisture content and density prediction using laboratory resistivity experiment. Int. J. Eng. Technol. **5**, 731–735 (2013)
5. Pozdnyakova, L.A.: Electrical Properties of Soils. Ph.D.Thesis, University of Wyomins, USA (1999)

6. Ozcep, F., Tezel, O., Asci, M.: Correlation between electrical resistivity and soil-water content: Istanbul and Golcuk. International Journal of Physical Sciences **4**(6), 362–365 (2009)
7. Ball, R.J., Allen, G.J., Carter, M.A., Wilson, A.A., Ince, C., El-Turki, A.: The application of electrical resistance measurements to water transport in lime–masonry systems. Applied, Physics A **106**(3), 669–677 (2012)
8. Jang, J., Sun, C., Mizutani, E.: Neuro-fuzzy and Soft Computing. Prentice Hall (1997)
9. Osman, O., Albora, A.M., Ucan, O.N.: Forward Modeling with Forced Neural Networks for Gravity Anomaly Profile. Mathematical Geology **39**(6), 593–605 (2007)
10. Islam, T., Chik, Z.: Improved near surface soil characterizations using a multilayer soil resistivity model. Geoderma **209**, 136–142 (2013)
11. Hazreek, Z.A., Aziman, M., Azhar, A.T.S., Chitral, W.D., Fauziah, A.M., Rosli, A.S.: The behaviour of laboratory soil electrical resistivity value under basic soil properties influences. In: IOP Conference Series: Earth and Environmental Science, vol. 23(1), p. 012002. IOP Publishing (2015)
12. Islam, S., Chik, Z., Mustafa, M.M., Sanusi, H.: Model with artificial neural network to predict the relationship between the soil resistivity and dry density of compacted soil. Journal of Intelligent & Fuzzy Systems: Applications in Engineering and Technology **25**(2), 351–357 (2013)
13. Abidin, Z., Hazreek, M., Azhar, A.T.S.: The behaviour of laboratory soil electrical resistivity value under basic soil properties influences. Earth and Environmental Science **25**(1) (2015)
14. Charniak, E., McDermott, D.: Introduction to artificial intelligence. Addison (1985)
15. Brown, M.P., Poulton, M.M.: Locating buried objects for environmental site investigations using neural networks. JEEG **1**, 179–188 (1996)
16. Zadeh, L.A.: Fuzzy sets. Information and Control **8**(3), 338–353 (1965)
17. Takagi, T., Sugeno, M.: Fuzzy identification of systems and its application to modeling and control. IEEE Trans. Syst. Man Cybern. **15**, 116–132 (1985)
18. Albora, A.M., Özmen, A., Ucan, O.N.: Residual separation of magnetic fields using a cellular neural network approach. Pure and Applied Geophysics **158**(9–10), 1797–1818 (2001)
19. Albora, A.M., Ucan, O.N., Ozmen, A., Ozkan, T.: Separation of Bouguer anomaly map using cellular neural network. Journal of Applied Geophysics **46**(2), 129–142 (2001)
20. Albora, A.M., Uçan, O., Aydogan, D.: Tectonic modeling of Konya-Beysehir Region (Turkey) using cellular neural networks. Annals of Geophysics **50**(5) (2007)
21. Ozcep, F., Yildirim, E., Tezel, O., Asci, M., Karabulut, S.: Correlation between electrical resistivity and soil-water content based artificial intelligent techniques. International Journal of Physical Sciences **5**(1), 047–056 (2010)
22. Bian, H., Liu, S., Cai, G., Tian, L.: Artificial neural network model for predicting soil electrical resistivity. Journal of Intelligent & Fuzzy Systems **29**(5), 1751–1759 (2015)
23. Chik, Z., Islam, T.: Near surface soil characterizations through soil apparent resistivity: a case study. In: 2013 IEEE 7th International Conference on Intelligent Data Acquisition and Advanced Computing Systems (IDAACS), vol. 1, pp. 57–60. IEEE (2013)

Dictionary Comparison for Anomaly Detection on Aircraft Engine Spectrograms

Mina Abdel-Sayed[1,2,4(✉)], Daniel Duclos[1], Gilles Faÿ[2], Jérôme Lacaille[3], and Mathilde Mougeot[4]

[1] TSI, Safran Tech (Safran Group), Magny-les-Hameaux, France
{mina.abdel-sayed,daniel.duclos}@safran.fr
[2] MICS, Ecole CentraleSupélec, Chatenay-Malabry, France
gilles.fay@centralesupelec.fr
[3] Snecma (Safran Group), Moissy-Cramayel, France
jerome.lacaille@snecma.fr
[4] LPMA, Université Paris Diderot, Paris, France
mathilde.mougeot@univ-paris-diderot.fr

Abstract. To ensure the liability of civil aircrafts, engines have to be tested after their production. Vibrations are one of the most informative measures to diagnose some damages in the engine if any. The representation of these vibrations as spectrograms provides visual signatures related to damages. However, this representation is noisy and high-dimensional. Moreover, the relevant signatures are localized in small parts of the spectrogram and the number of damaged engines in the database is extremely low. These elements disturb the elaboration of detection algorithms. A new adequate representation computed from the spectrograms is needed in order to perform automatic diagnose of the aircraft engines. In this paper, we study two kinds of representations with dictionaries that can be learnt from the data (NMF) or fixed in advance (curvelets). We present some dictionary comparison methods taking into account the low number of damaged engines.

Keywords: Dictionary learning · Curvelets · NMF · Spectrogram · Vibrations · Anomaly detection

1 Introduction

The aeronautic domain is governed by many regulations, controls and certifications. Each engine manufactured has to be tested during its whole life, from production line to the flight operations.

Vibrations are one of the most pertinent information to analyze the engine behavior. Each potential defect in an engine component may induce a source of vibrations visually detectable in vibration spectrograms rather than in the temporal signal. However the representation of vibration with a spectrogram is noisy and its high dimension overwhelms the pertinent information such as abnormal signatures. Nevertheless,

P. Perner (Ed.): MLDM 2016, LNAI 9729, pp. 362–376, 2016.
DOI: 10.1007/978-3-319-41920-6_28

an expert is mostly able to detect some signatures just looking at spectrograms. The large fleet of engines in-flight and the always increasing number of engines cannot be inspected manually and has to be checked automatically. Machine learning methods such as a reduction dimension process [4,5] or the estimation of a spectrogram mask [7] may help to perform anomaly detection. Alternative signal processing methods directly based on the raw signals and the engines' cinematics are also available [6][8].

The representation of an object into a dictionary is a well-known tool to reduce the dimension of the data. We introduce a dictionary to model an object; this model is then defined by the used dictionary. This approach is popular in statistics, machine learning community [1,2,3], signal and image processing, etc. One of the key point is the choice of the model. The different dictionaries are often compared by quantifying the restoration of an image from the dictionary [3] or by the percentage of good classification of the target object [2]. In the anomaly detection domain, the detection percentage of abnormal objects is a way to compare the dictionaries used for decomposition, if the number of abnormal data is sufficient. Concerning the current application, the data correspond to serial engine reception tests at the exit of the production line; the number of damaged engines is extremely low preventing simple comparison.

In this paper, we compare two kinds of models. The first one is defined by a data-driven dictionary called non-negative matrix factorization (NMF) [5][9]. The particularity of such dictionary, as for principal components analysis (PCA), lies in the computation of the dictionary atoms directly from the data. The second is a dictionary based on the curvelet frame [10,11], which atoms are analytically designed and fixed beforehand. Data adaptation is carried on by atom selection within the dictionary. The Fourier and other wavelet bases or frames provide typical representatives of such dictionary. We are interested in the comparison of these models for the purpose of anomaly detection with the constraint of a limited number of abnormal engines.

For a better understanding of our problematic, we describe our data in Section 2. We present the algorithm in Section 3. We explain the NMF in Section 4 and the curvelets frame in Section 5. The different comparison methods are defined and results are presented in Section 6.

2 Vibration Analysis and Data

Vibration measurements contain multiple data concerning the different elements of the engine such as its dynamic health status. The vibrations are recorded as acceleration of accelerometers in function of the time. The investigation of these signals may provide indications to potential failures if any.

In the test bench, the vibrations signals are acquired at 50 kHz by two accelerometers on the engine during different phases such as the acceleration and deceleration. The non-stationary phases are more informative from a vibratory point of view than a steady phase. The studied engines are composed of two shafts, the high pressure (HP) shaft with rotation speed denoted N2 and the low pressure (LP) shaft with rotation speed denoted N1. Several vibration sources are dependent of these shafts which speeds are also acquired by tachometers (Figure 1).

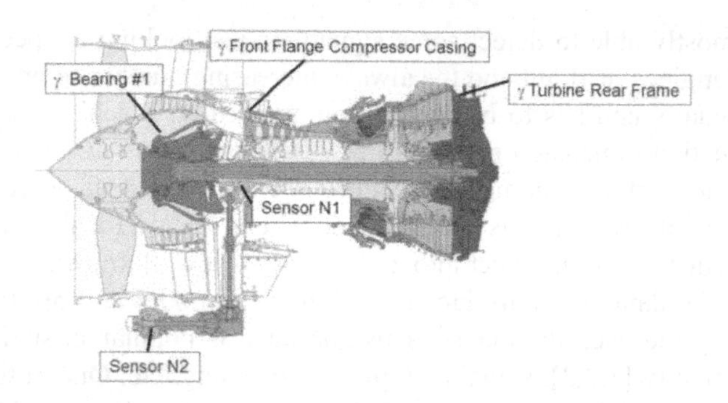

Fig. 1. Position of the accelerometer and tachometer sensors

The vibrations acquired by the system consist in high frequency raw temporal signals which appear to be difficult to visually analyze by a human expert. In order to perform visual inspection of the vibration data, the temporal signals are converted into spectrogram by the following process.

- The temporal signal is decomposed into small temporal buffers with overlap during which the regime may be considered almost stationary.
- A short time Fourier transform is computed on each buffer.
- The square modulus of the spectrum are concatenated to construct the spectrogram.

The output of this process is a temporal spectrogram with the time as the x-axis and frequencies as y-axis. Resampling the spectrogram in one of the two shaft speeds (N1 or N2) leads to a better visualization of the vibration sources. Our data are sampled according to the high-pressure shaft speed N2 at 10 rpm (revolution per minute). Figure 2 shows a computed spectrogram with the shaft speed N2 as x-axis and a zoom on an abnormal part.

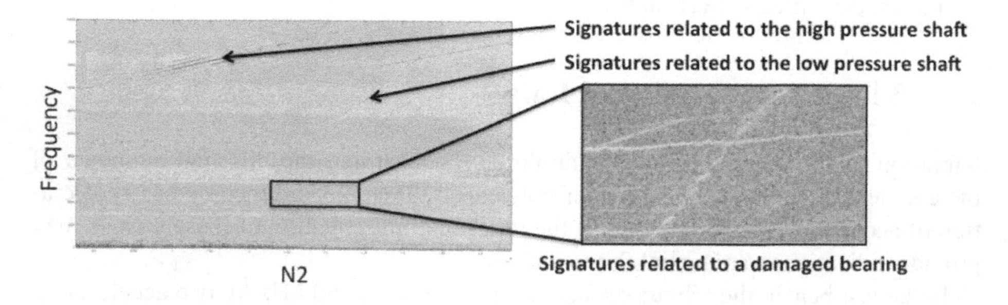

Fig. 2. Abnormal spectrogram sampled in N2

The frequencies of vibrations related to the HP shaft are of the form $\alpha \times N_2, \alpha \in \mathbb{N}$. In this representation (Figure 2), they are then characterized by straight lines. Rays corresponding to the same vibrations source are in the exact same localization in every spectrogram (independently of the engine). Vibrations related on the LP shaft are characterized by the relation $\beta \times N_1, \beta \in \mathbb{N}$ and are represented in the spectrogram by curve homothetic to the curve representing the relation between N1 and N2. This relation is subject to variability depending on external conditions which leads to the shift of these curves in the different spectrograms (depending on the engine).

As far as spectrograms are results of high bandwidth vibration measurements and digital signal processing on amplitude, they are subject to noise. Moreover, relevant signature related to damage represents a tiny part of the high dimensional spectrogram (Figure 2). The relevant information is then overwhelmed by noise and normal signatures. This data acquisition process is also subject to several variabilities in the test bench and in external conditions.

These elements constitute serious obstacles to the elaboration of automatic anomaly detection process. We study a new representation of the spectrograms based on their decomposition on a dictionary in order to discriminate damaged engines.

3 Anomaly Detection Algorithm

Due to the low number of abnormal spectrograms, the detection algorithm is naturally designed to characterize exclusively the normal behaviors of the engine's vibrations. We define the representation of the normal spectrograms as reference in order to compare them with the representation of other engines. These normal engines are used to calibrate the model. In this configuration, an abnormal engine can be defined by the distance between its own representation and the representations of the set of reference normal engines. Another way to define the abnormal engines is to compare the spectrogram with its projection on the normal domain. This projection consists on the reconstruction of the spectrogram from the dictionary learnt on normal data.

Learning the dictionary and representation on non-damaged engines is imposed by the low number of damaged engine. Supervised classification methods cannot be used in this case. The damaged engines should have the highest distance from the reference representation (the normal engines).

3.1 Subdivision into Patches

The ratio between non-pertinent information and the relevant signatures is huge. The signatures of abnormal phenomena, if any, correspond to small-localized parts of the spectrogram. Inspecting the spectrogram as a whole is not efficient. Experts inspect visually the spectrograms by looking at different ranges of frequencies successively. Our approach is inspired from this method; we subdivide the spectrogram into patches P (1) which may overlap. These patches are defined by a range of frequencies and regimes N2 either by experts or iteratively. The representation in a dictionary is then applied on each patch separately.

$$\{[f_i^1, f_i^2] \times [N_{2,i}^1, N_{2,i}^2]\}_{i \in P} \quad (1)$$

with f_i and $N_{2,i}$ respectively the range of frequencies and regimes of patch i.

This approach has two benefits; the first one is algorithmic, since we consider each patch independent from the others, it is possible to parallelize the process. The second one consists in a rough approximation of the localization of the signature in the spectrogram. Only the patch containing the signature may trigger an alarm, the others should behave normally, thus we just have to study the alarming patch.

3.2 Process

The process is composed of two phases, a learning one and a testing one illustrated in Figure 3.

Learning Phase. This phase is characterized by two steps. The first consists in the decomposition of normal spectrogram over the dictionary. The data-driven model (dictionary learning) must be learnt from the data beforehand. For a designed dictionary, the representation of normal data is computed in this step. At the end of this step, both the dictionary and representation of the normal engines considered as reference are computed.

The second step consists in the modeling of the distribution of the 'anomaly' scores of normal engines. Normal engines, that are not been used as reference, are decomposed over the dictionaries and their scores are computed. These scores are defined below (4.4 and 5.2). We propose a Gamma distribution to fit the empirical distribution of each score. The use of Gamma distribution has been confirmed by a statistical test of Kolmogorov-Smirnoff and by a stochastic test [14]. It shall be used as a proxy for the normal behaviors of the engines.

We define the engine database used in this learning phase by $D_{learning}$. This set is composed of two disjoint sets, D_{model} containing the engines employed as reference and D_{fit} used for the estimation of the distribution.

Testing Phase. In the testing phase, new engines not used in the learning phase (either normal or damaged) are represented in the dictionaries and their anomaly scores are computed. The probability of anomaly of these engines is then estimated thanks to the distribution estimated in the learning phase. The probability consists in the p-value of a statistical test accepting or rejecting the hypothesis H0 : "the patch is normal". The p-value is the probability of wrongly rejecting H0 under the probability of this hypothesis. The probability of the H0 hypothesis is the estimated Gamma distribution. The p-value of a spectrogram Y is defined in (2).

$$pvalue(Y) = \mathbb{P}(X > score(Y)) = 1 - F_X(score(Y)) \quad (2)$$

with X following the Gamma distribution corresponding to *score(.)*, and F_X its distribution function. The lower the p-value is, the more the engine is suspected.

We define D_{test} as the database used in this phase disjoint from $D_{learning}$ and containing damaged engines.

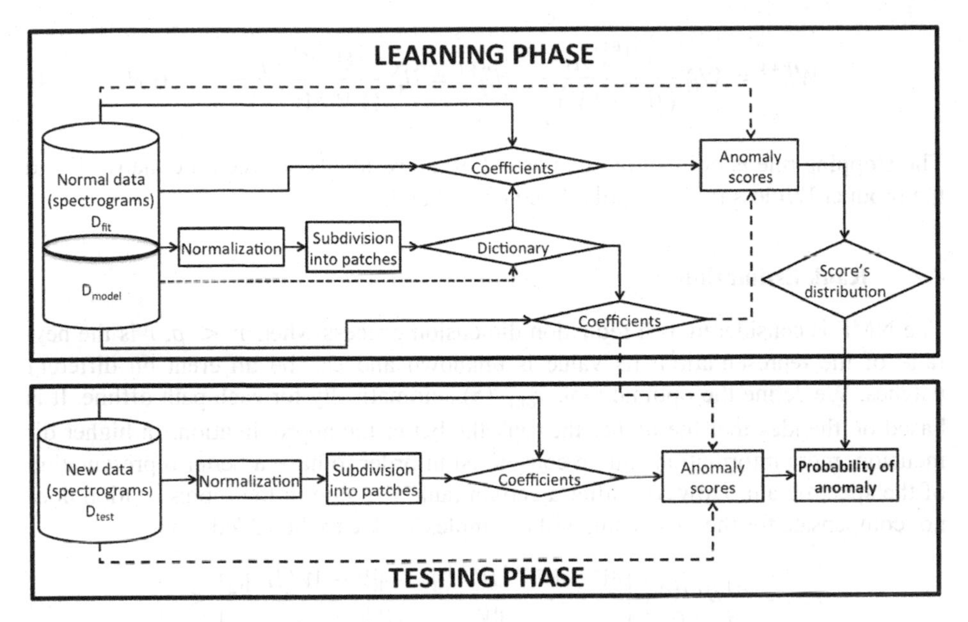

Fig. 3. Process of the algorithm based on representation in a dictionary

4 Data-Driven Dictionary - Non-negative Matrix Factorization (NMF)

This approach relies on the learning of a dictionary directly from the database, being reminiscent of the projection of the data on a principal subspaces in a Principal Component Analysis.

4.1 Problem Formulation

NMF [9] is a matrix decomposition process consisting in the approximation of a positive matrix $V \in \mathbb{R}_+^{p \times n}$ into the product of two positive matrices $W \in \mathbb{R}_+^{p \times r}$ and $H \in \mathbb{R}_+^{r \times n}$. This problem can be summarized by the formulation in (3).

$$(W^*, H^*) = \underset{W \geq 0, H \geq 0}{argmin} \|V - WH\|_2^2 \text{ such that } \forall i, \|W(.,i)\|_2^2 = 1 \quad (3)$$

with $W(.,i)$ the columns of W.

In this case, each column of V represents the spectrogram of a given engine (flatten from a bitmap matrix to a vector); the columns of W constitute the dictionary while the columns of H are the coefficients representing the spectrogram in the dictionary. The second constraint ensures the identifiability of the solution.

This problem (3) is biconvex. A solution can be find by iterative alternating update rules on W and H (4) proposed by Lee and Seung [9][5].

$$W_{ij}^{k+1} = W_{ij}^{k} \frac{(H^k V)_{ij}}{\left(H^{k^T} H^k V\right)_{ij}} \quad H_{ij}^{k+1} = H_{ij}^{k} \frac{(V W^k)_{ij}}{\left(V W^k W^{k^T}\right)_{ij}} \qquad k \in \mathbb{N} \tag{4}$$

The stopping rule is defined by the norm 2 of the residuals between the matrix V and the product WH less than a small ε fixed beforehand.

4.2 Rank Estimation

The NMF is considered as a reduction dimension process when $r < p$, r is the new rank of the representation. Its value is unknown and can be different on different patches. We define the optimal rank r_{opt} (5) automatically for each path offline. It is based on the idea that the higher the rank the better the approximation. In higher dimension, more information can be considered in order to have a better representation of the spectrogram. However, after a certain rank, the gain in the representation does not compensate for the increasing of the complexity due to the added rank.

$$r_{opt} = \underset{r \geq 0}{argmin} \left\{ \left| \frac{\|V - W^{r+1}H^{r+1}\|_2^2 - \|V - W^r H^r\|_2^2}{\|V - W^r H^r\|_2^2} \right| \leq \varepsilon \right\} \tag{5}$$

where W^r and H^r correspond to the approximation results of the NMF at rank r.

4.3 The Dictionary

The NMF allows learning the dictionary along with the coefficients. We have already seen that because of the low number of damaged engine, the model is learnt using no damaged engine in order to code normal spectrograms in the dictionary. However, even among them, normal spectrograms are subject to a lot of variability, for example the different lines related to the low pressure shaft. Thus, the learnt model has to take into account these different behaviors. Figure 4 presents different atoms of a dictionary for a patch.

The atoms in Figure 4 are similar to the spectrogram's patch on which they are estimated. We can see different kind of N1 sources of vibrations in the different atoms.

The dictionary seems well-adapted as it captures the different singularities present in the spectrogram with their variability. In the dictionary, different N1 vibration signatures are represented (due to the variability between N1 and N2) whereas those of N2 are at the exact same localization on each atom.

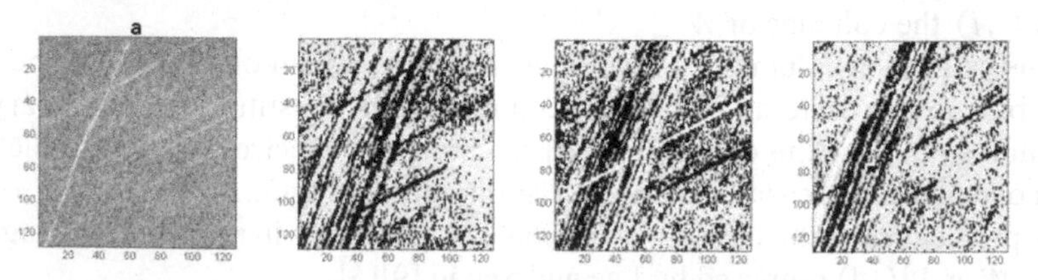

Fig. 4. NMF dictionary atoms learnt from a spectrograms' patch (a)

4.4 Anomaly Scores Based on this Method

Distance to Nearest Neighbor Spectrogram on the Coefficients (NMF DNN). The spectrograms are represented in a lower dimension. Abnormal signatures can alter some coefficients of the representation. An interesting score is then the minimal distance between the representations of the tested engine and the reference engines.

$$score_{dnn}(Y) = \min_{X \in D_{model}} \|H_Y - H_X\|_2^2 \tag{6}$$

where Y and X are respectively a tested spectrogram and those from the learning database, H_Y and H_X refers to the representation in the NMF dictionary (a solution of (3) based on the learnt dictionary W; $Y \cong WH_Y$) of respectively Y and X.

Mahalanobis Distance on the Coefficients Domain (NMF Mahal). In the previous score, by applying the L2 norm, we consider that each coefficient is independent from the others which is an oversimplifying hypothesis. The Mahalanobis distance allows to consider these possible interactions.

$$score_{mahal} = (H_Y - \overline{H})^T \Sigma^{-1} (H_Y - \overline{H}) \tag{7}$$

$\overline{H}$ and Σ represent respectively the mean and the covariance matrix estimated on the coefficients of the reference engines.

Reconstruction Error (NMF RE). The representation in this dictionary learnt on normal engines can be considered as a projection of the spectrogram in the space of regular spectrogram. It is precisely the closest element from the spectrogram in the convex cone generated by regular spectrogram. The abnormal signatures cannot be reconstructed as no close features are available in the NMF atoms. Those signatures are then present in the residuals of the reconstruction.

$$score_{RE}(Y) = \|Y - WH_Y\|_2^2 \tag{8}$$

5 Nonadaptive Dictionary: Curvelets

The curvelet frame [10,11] is chosen as a way to represent the spectrograms. This choice of model is based on the conjunction of two facts, the first one being the perception of vibrations as curves in the spectrogram, the second one the efficiency of the curvelets to represent curves with a small number of coefficients. In this section, we briefly explain the construction of the curvelet transform and its use for anomaly detection.

5.1 The Curvelet Transform

The curvelet transform denoted T in this paper, is built from a process which comes from the ridgelet transform [10,11,12,13] and the idea that any curve can be approximated by a succession of small straight lines.

The Ridgelet Transform. The ridgelet transform consists in a multiscale representation of the data built with ridge function. A ridgelet atom (9) is a function defined by a scale parameter $a > 0$, an orientation parameter $\theta \in \mathbb{R}$, a localization parameter $b \in \mathbb{R}$ and a mother function ψ that is usually a 1D-wavelet.

$$\psi_{a,b,\theta}(x) = a^{-\frac{1}{2}}\psi((x_1 cos\theta + x_2 sin\theta - b)/a) \ with \int \left|\hat{\psi}(\omega)\right|^2 \omega^{-2} d\omega < \infty \quad (9)$$

This function $\psi_{a,b,\theta}$ is then constant along ridges and can be considered as wavelets elsewhere. This transform is well adapted to characterize linear singularities such as straight lines. The interaction between the ridgelet transform and linear singularities is similar to the interaction between wavelets and punctual singularities.

The continuous ridgelet transform is not computable on the whole space $\mathbb{R}_+^* \times \mathbb{R}^2$, for this reason a discrete ridgelet transform (DRT) is defined on a dyadic discretization of this space where the discretization of the orientation and localization is dependent of the scale discretization [12]. This discrete ridgelet transform (as the continuous one) is invertible and respect the Parseval relation.

For the curvelet transform, a specific ridgelet is used which respects the previous statements and defines a new basis for $L^2(\mathbb{R}^2)$, the orthonormal ridgelets [13].

Construction of the Curvelet Transform. The ridgelet transform is not adapted to characterize the spectrogram since some vibrations correspond to non-linear curves in the spectrograms. Still, at fine scale, a curve can be considered as a straight line. The curvelet transform is based on this idea. We give a brief description of the different steps for its construction (Figure 5), for more details we refer to [10,11].

Applying the ridgelet transform on different dyadic squares extracted from the object is computationally intensive and redundant. For this reason, a specific decomposition in dyadic squares is chosen for the object. The decomposition is not applied directly on the object, but on a filtered version at specific dyadic range of frequencies, the scale of the decomposition is dependent of this filter. The extracted dyadic squares are rescaled to form a unitary object on which the orthonormal ridgelet transform is applied. The curvelet transform is invertible and respect the Parseval relation.

Figure 6 presents some atoms at different scales. These atoms are obtained by using the inverse curvelet transform T^{-1} on coefficients set to 0 except for one set to 1.

The atoms are in a way local, each atom is characterized by a position, but some rebounds with lower energy appear in the atoms.

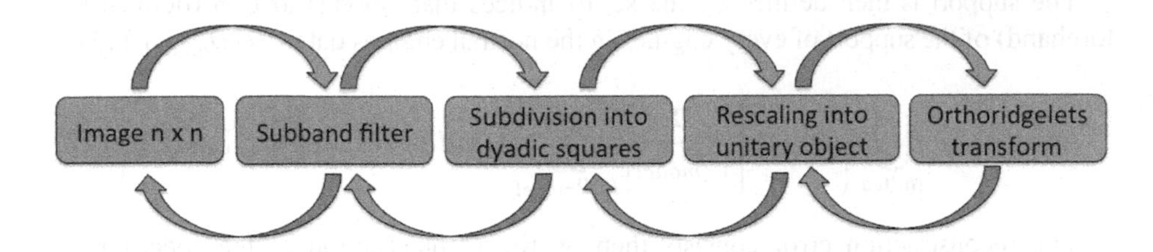

Fig. 5. Construction process of the curvelets transform and its inverse

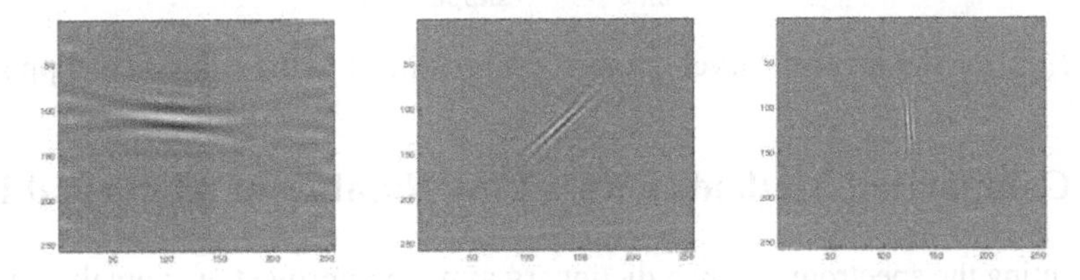

Fig. 6. Atoms of the curvelets from coarse to fine scale (left to right)

5.2 Anomaly Scores Based on this Representation

Nearest Neighbor Spectrogram in the Coefficient Domain (Curve dnn). This score is similar to the one in the NMF representation; it is based on the locality of the curvelets. Indeed, abnormal signatures can activate some atoms of the curvelets which are absent for normal spectrograms.

$$score_{dnn}(Y) = \min_{X \in D_{model}} \|T_X - T_Y\|_2^2 \tag{10}$$

where T_X and T_Y are respectively the curvelet transform of X and Y.

Reconstruction Error (Curve RE). This second score is specific to local dictionaries. The curvelet transform, unlike the NMF, give a perfect reconstruction of the spectrogram, then in this configuration, there are no gain to compute the reconstruction error as no abnormal information is contained in the residuals. We then decided to reconstruct only from the atoms of the dictionary that characterize the normal behaviors of the spectrogram. These atoms are denoted *support*. This support is learnt from the representation of normal data. The support of one spectrogram is defined by the indices of the coefficients where the value is different of 0 (11) after thresholding. The threshold consists in a trade-off between a good reconstruction and a minimal number of coefficients. We define our threshold in order to keep the 10% largest coefficients.

$$support_Y = \{(j,k,l) | \; T_Y(j,k,l) \neq 0\} \tag{11}$$

The support is then defined by the set of indices that belongs to $Q\%$ (defined beforehand) of the support of every engines in the normal engines database D_{model} (12).

$$support = \bigcup_{indice} \left\{ indice \,\middle|\, \frac{1}{|D_{model}|} \sum_{i \in D_{model}} \mathbb{I}\{indice \in support_i\} \geq Q\% \right\} \quad (12)$$

The reconstruction error consists then on the reconstruction of the spectrogram from the coefficients restrained to the learnt support (13).

$$score_{ER}(Y) = T_{support}^{-1}(T_Y) \quad (13)$$

with $T_{support}^{-1}$ the inverse curvelet transform restrained to the computed support.

6 Comparison Methods with a Low Number of Abnormal Data

Representing the spectrograms in a dictionary aims to improve the anomaly detection. Which model is preferable is an open question, even more when the number of damaged engine is very low. The model is learnt on normal engines in order for abnormal engines to be distant from the reference database. It is preferable to compare the different dictionaries via the probability of anomaly (the p-value) since the scores have different order of magnitude.

We dispose of 583 engines among which one engine is damaged. We present the results on the damaged engine; we use 500 normal engines for the learning database $D_{learning}$, with a repartition of 75% for D_{model} and 25% for D_{fit}. The engines in both databases are selected randomly 5 times. The results correspond to the mean results on D_{test} of these 5 runs. D_{test} contains 83 engines among which the damaged engine.

6.1 Basic Comparison of the Detection Process

Anomaly detection is the objective of our work. Consequently, we study which method has the best performance in discriminating the damaged engine in the patch containing abnormal signatures.

Table 1. Mean p-values of the different scores for normal and damaged engines on two patches; one containing abnormal signatures and one without any marks of anomaly.

	Abnormal patch		Normal patch	
	Normal engine	Damaged engine	Normal engine	Damaged engine
NMF DNN	0.55 ± 0.07	**$(2.25 \pm 0.13)\times10^{-9}$**	0.57 ± 0.09	0.5 ± 0.11
NMF Mahal	0.57 ± 0.06	**$(6.03 \pm 0.6)\times10^{-11}$**	0.5 ± 0.09	0.29 ± 0.09
NMF RE	0.52 ± 0,03	**$< 10^{-16}$**	0.49 ± 0.04	0.34 ± 0.05
Curve DNN	0.54 ± 0.03	**$< 10^{-16}$**	0.53 ± 0.03	0.51 ± 0.07
Curve RE	0.57 ± 0.03	**$< 10^{-16}$**	0.44 ± 0.03	0.27 ± 0.04

Table 1 presents the mean p-values of our scores for abnormal and normal patches. These results consist in the mean of the different p-values with random selection of D_{model} and D_{fit}. The damaged engine is quite well detected in the corresponding patch for every score (the p-values are all sufficiently low to discriminate the damaged engine) and not detected in the normal patch. It confirms the algorithm but also suggests that this engine can be discriminated quite easily. Then looking at the p-values is not sufficient to deeply compare the different dictionaries.

6.2 Comparison with a Shift of the Window

Previously, the patch was centered on the abnormal signatures. The different methods may have worse results if the abnormal signatures are located at an extremity of the window. The results on windows translated along the frequency or regime axis can give pertinent comparison of the different methods on the patches containing non-centered abnormal signatures. We consider a collection of windows translated by one pixel steps.

For each window, we perform a bootstrap by selecting randomly the engines in D_{model} and D_{fit} five times.

Figure 7 presents the results of the different scores on the spectrogram of the damaged engine when the window is shifted on frequency (the range of regimes is the same for each window). The results correspond to the mean in the different runs of $-\log_{10}(pvalue)$, in this configuration the higher the value, the more the engine is suspected.

We already know that the damaged engine is well-detected by every method on the windows which abnormal signatures are centered on it. It is more interesting to compare them on the extremity of the graph where the abnormal signatures are low (c&d). The reconstruction error applied to the curvelets (Curvelets RE) seems to detect before the other methods on (c) even when the signatures are quite low, it is equivalent on the other side (d), but the method decreases at the end in spite of a small mark of anomaly. This comparison suggests that the Curvelet RE give the best results.

6.3 Creation of Low Signatures

We have only one damaged engine with a signature of high intensity leading to easy detection on the patch. We can then expect that with lower signatures, some methods will have more difficulty to detect the anomaly. To verify this claim, we first search in our database of normal engines the one with the nearest configuration N1 from the abnormal spectrogram one (14). The reason to find this engine is to not compromise the experience with some variability of the relation between N1 and N2.

$$engine_{selected} = \underset{\substack{X \in normal \\ databse}}{argmin} \left\{ \| N_1(X) - N_1(engine_{ano}) \|_2^2 \right\} \tag{14}$$

with $N_1(X)$ the LP shaft speed of the engine X and $engine_{ano}$ the damaged engine.

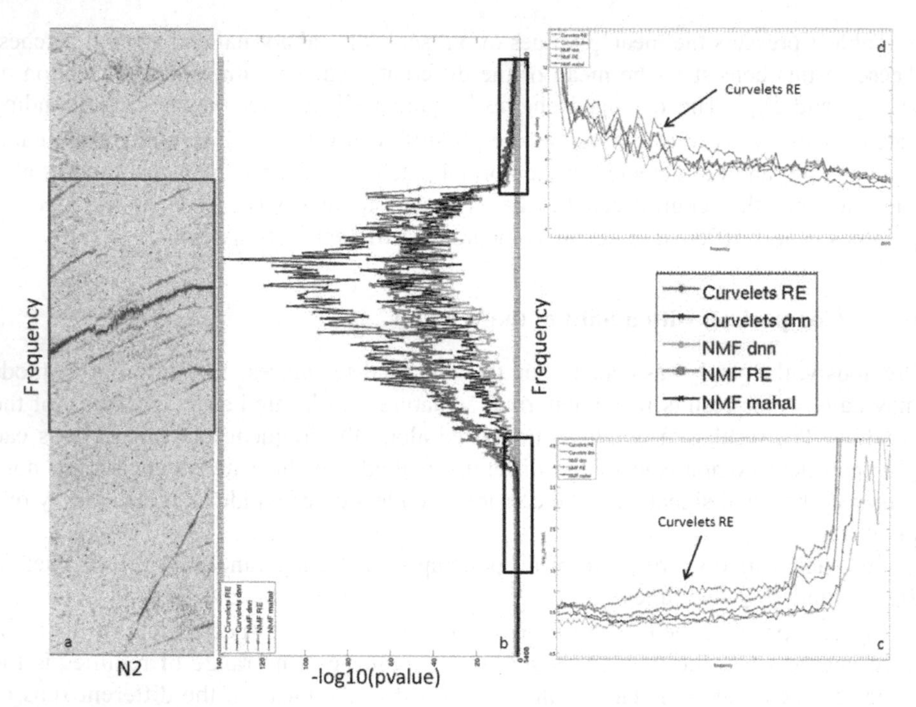

Fig. 7. Results corresponding to –log10(p-value) (b) when the window is shifted in frequency along the abnormal spectrogram zone (a). Zoom of the beginning of the graph (c) and the ending (d) are given. The abnormal signatures are highlighted (red) on (a).

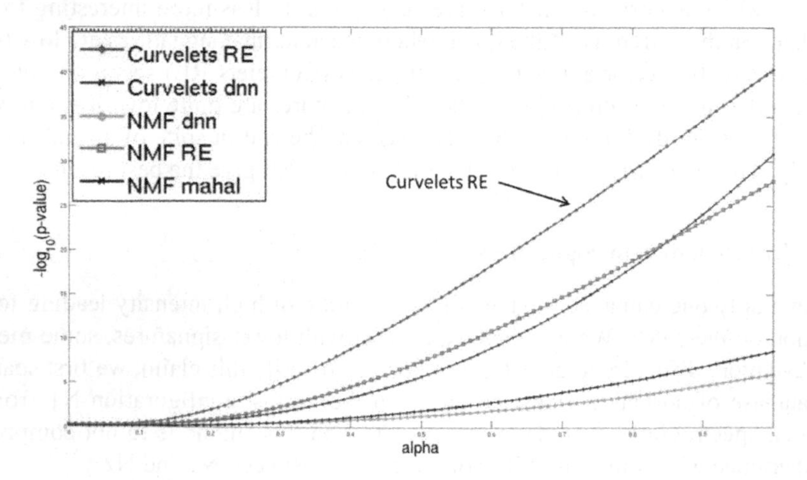

Fig. 8. P-value of the created abnormal spectrogram for every method

With this engine selected, it is possible to create low abnormal signatures by combining the abnormal spectrogram with the selected one with the convex relation (15).

$$engine_{ano}^{new}(\alpha) = \alpha * engine_{ano} + (1-\alpha) * engine_{selected}\ , \alpha \in [0,1] \quad (15)$$

The closer α is to 1, the higher is the intensity of the abnormal signature. We evaluate the different models for different values of α in order to determine which methods can detect a damaged engine with weak signatures. The minimal value of α on which the method detects abnormal signatures is the tool for comparison. The method with the highest performance is the one which detects with the smallest value of α.

Figure 8 presents the results of the created abnormal engines for different values of α from 0 to 1 at step of 0.1. The results correspond to the mean of $-log10(pvalue)$.

The results of this comparison process are quite clear, the reconstruction error applied to the curvelets (Curvelets RE) detect for values of α lower than the others. The curve representing 'Curvelets RE' is always above the other methods.

7 Conclusion

In this work, to avoid the curse of dimensionality we have tested different dictionaries in order to perform anomaly detection on the vibrations spectrograms of aircraft engines. One of them is data-driven, the model is learnt directly on a learning database, the NMF. The second, the curvelets, is designed beforehand. It is chosen for the fact that the curvelets is a good characterization for objects containing curves such as the spectrograms. We introduced different indicators of anomaly based either on the distance between a spectrogram and a reference database of normal spectrogram or on the reconstruction process of the model.

The comparison process of models is simple in the domain of anomaly detection; it consists in the rate of good classification of the engines. However, this process is possible only when the database contains several damaged engines. In order to get rid of this constraint, we have elaborated different strategies to be able to evaluate and compare the dictionaries with their scores. These methods rely on the shifting of the window in order to have the abnormal signatures at its extremity and not in the center or on the creation of low signatures from the damaged engine

These different comparisons have shown that the reconstruction error applied with the curvelet transform enables to discriminate patches containing small hints of abnormal signatures. Even if the dictionary is fixed, the method in itself is data-driven. Indeed, the process to learn the support is dependent of the data. In fact, we select, in an over-complete dictionary, the atoms to activate in order to characterize the normal behaviors of the vibrations spectrogram and reconstruct from these atoms.

One direction of future work concentrates on enlarging the study of different kinds of data-driven models and multiple testing theory will be introduced to take into account the decision of the different patches.

Acknowledgments. We thank Julien Griffaton and Pierre Lalague from Snecma – SAFRAN Group for their contribution in our comprehension of vibration analysis.

This work was supported by ANRT through the CIFRE sponsorship No 63/2014 and by Safran Tech – SAFRAN Group.

References

1. Gribonval, R., Jenatton, R., Bach, F.: Sparse and Spurious: Dictionary Learning With Noise and Outliers. IEEE Transactions on Information Theory **61**(11), 6298–6319 (2015)
2. Mairal, J., Ponce, J., Sapiro, G., Zisserman, A., Bach, F.: Supervised dictionary learning. In: Advances in Neural Information Processing, pp. 1033–1040 (2009)
3. Mairal, J., Elad, M., Sapiro, G.: Sparse representation for color image restoration. Image Processing IEEE Transactions **17**(1), 53–69 (2008)
4. Lacaille, J.: Searching similar vibration patterns on turbofan engines. In: 10th International Conference on Condition Monitoring and Machinery Failure Prevention Technologies (2014)
5. Abdel-Sayed, M., Duclos, D., Faÿ, G., Lacaille, J., Mougeot, M.: NMF-based decomposition for anomaly detection applied to vibration analysis. In: 12th International Conference on Condition Monitoring and Machinery Failure Prevention Technologies (2015)
6. Hazan, A., Lacaille, J., Madani, K.: Extreme value statistics for vibration spectra outlier detection. In: 9th International Conference on Condition Monitoring and Machinery Failure Prevention Technologies (2012)
7. Griffaton, J., Picheral, J., Tenenhaus, A.: Enhanced visual analysis of aircraft engines based on spectrograms. In: International Conference on Noise and Vibration Engineering (ISMA) (2014)
8. Klein, R., Rudyk, E., Masad, E.: Methods for diagnostics of bearings in non-stationary environment. International Journal of Condition Monitoring **2**(1), 2–7 (2012)
9. Lee, D.D., Seung, H.S.: Algorithms for non-negative matrix factorization. In: 13th Neural Information Processing Systems Conference, pp. 556–562 (2001)
10. Candès, S.J., Donoho, D.L.: Curvelets – a surprisingly effective nonadaptive representation for objects with edges. In: Cohen, A., Rabut, C., Schumaker, L.L. (eds.) Curve and Surface Fitting: Saint-Malo, pp. 105–120. Vanderbilt University Press, Nashville (2000)
11. Candès, E.J.: Donoho, D.L: New tight frames of curvelets and optimal representations of objects with piecewise-C^2 singularities. Communications on Pure and Applied Mathematics **57**(2), 219–266 (2004)
12. Candès, S.J., Donoho, D.L.: Ridgelets: a key to higher-dimensional intermittency ? Philosophical Transactions-Royal Society. Mathematical, Physical and Engineering Sciences **357**(1760), 2495–2509 (1999)
13. Donoho, D.L.: Orthonormal ridgelets and linear singularities. SIAM Journal of Mathematical Analysis **31**(5), 1062–1099 (2000)
14. Hastie, T., Tibshirani, R., Friedman, J.: The elements of statistical learning, vol. 2(1). Springer Edition (2009)

Location-Aware Ad Recommendation to Bid for Impressions

Satish Kumar Verma(✉)

SAP Innovation Center, Singapore, Singapore
satish.kumar.verma@sap.com

Abstract. We present our work-in-progress design and development of a location-aware advertisement recommendation system. The ad-delivery platform acts as a combination of a DSP (Demand Side Platform) and an ad exchange. Advertisers act as clients and set up ad-campaigns defining the ad-targeting and budget criterion. The live system connects to SSP (Supply Side Platform) and receives ad bid requests selling ad impressions. The bid request contains information about the device, user, location, impression type, publisher site info, etc. In response to a bid request, the ad delivery platform can decide to bid or not bid for an impression. Upon a successful bid, an impression is show which the user may click or ignore. This process follows an established protocol, the RTB (real-time bidding) and the entire process completes within 100s of milliseconds. The objective of the ad recommender is to pick the most relevant ad that should be shown to the user and this ad form the part of the BidResponse, in case the ad-platform decides to bid for the impression. They key context that we use to pick the recommended ad is user's location. In addition to designing a location-aware recommender, we also describe our approach for identify interesting locations from raw gps trace in the impression bid requests received from SSPs.

1 Introduction

In this paper, we describe our ongoing research efforts towards designing and implementing a location-aware advertisement recommendation system. Advertising is a key component of any company's marketing strategy to bolster its brand perception and ultimately, increase revenues. On one side of the spectrum, we have advertisers who want to reach the right customers and are willing to pay considerable dollars to achieve their goals, such as lifting brand perception, converting ad clicks to sales, etc. On the other side, we have publishers, which are sites where advertisers can buy impressions. Their goal is maximize their revenue as well as attract and retain advertisers and users to their sites. In the middle, we have entities which create a matching between the needs to advertisers and publishers. They can be agencies where sales personal build relationship with advertisers and publishers or they can be automated entities which leverage the newer developments in automated programmatic buying and selling of ad impressions using new technologies and protocols such as RTB (real-time

P. Perner (Ed.): MLDM 2016, LNAI 9729, pp. 377–386, 2016.
DOI: 10.1007/978-3-319-41920-6_29

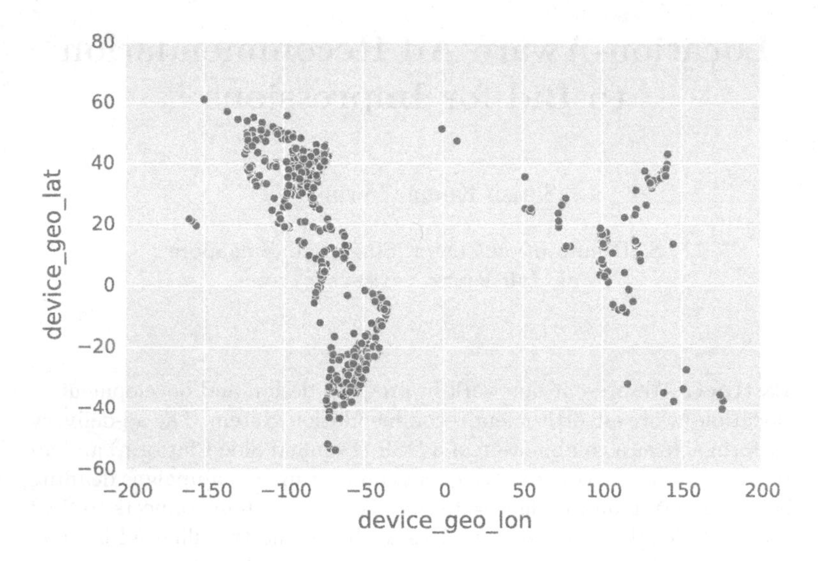

Fig. 1. GPS Distribution for Live Mobile BidRequests

bidding), trading desks, ad-exchanges, DSPs (demand side platforms for advertisers) and SSPs (supply side platform for publishers). These automated entities communicate with each other using well-defined protocols such as the RTB.

Our Ad-serving product is quite generic in terms of functionality. It works in part as a DSP and SSP with premium partnerships with publishers and advertisers. At the same time, it makes use of automated RTB protocol to connect to third party services such as other DSPs and SSPs. It also leverages other third party services for financial aspects such as billings as well as obtaining consumer and geographical data from DMPs. We briefly describe how RTB works and motivate why we think having an Ad Recommender to pick an ad before bidding for an impression can be useful in a real setting.

Real-time bidding (RTB) (as defined on wikipedia) is a means by which advertising inventory is bought and sold on a per-impression basis, via programmatic instantaneous auction, similar to financial markets . With real-time bidding, advertising buyers bid on an impression and, if the bid is won, the buyers ad is instantly displayed on the publishers site. The OpenRTB API (iab.net/media/file/OpenRTB-API-Specification-Version-2-3.pdf) specification describes in detail the protocol transport, security, data format and the specifications of data being exchanged, namely, the BidRequest and BidResponse. Upon receiving a bid request which is potentially an invitation to buy an ad impression, our ad-platform decides to bid for it or not. At this point, it has to choose an ad which optimizes an end goal such as click-through-rate (CTR) or conversions via sale. We believe that using a recommendation algorithm to pick the right ad to send using Bid Response can improve performance in terms advertiser objectives and thus, make the system more attractive for such customers. We are currently in the process of identifying key use-cases for such an

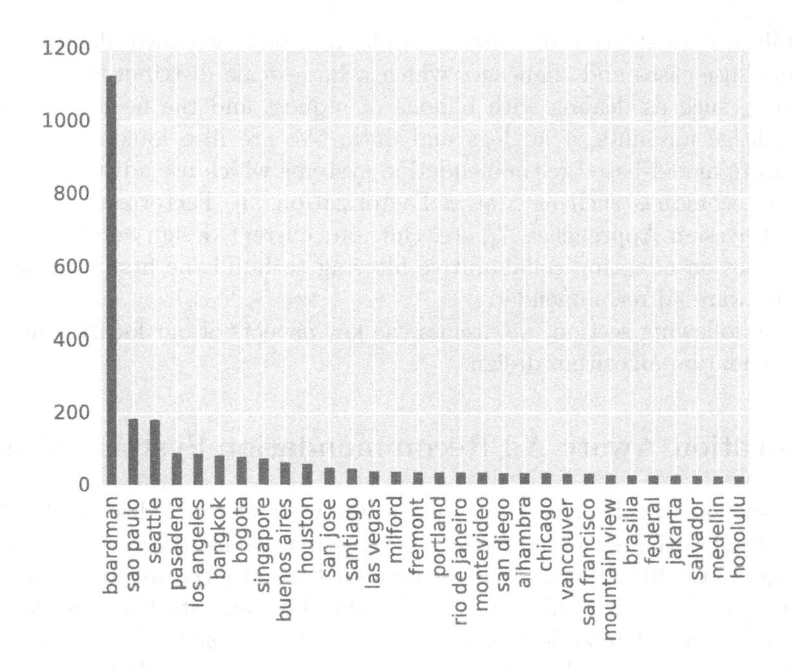

Fig. 2. GPS Distribution for Live Mobile BidRequests for a Single City

ad recommender. For a select set of such use cases, for example the Location-Aware Ad Recommender, we are designing and implementing a recommender module which will fit in the ad-delivery software logic.

2 Related Work

In this section, we present some of the research work which are relevant to our current work. Our current research efforts make use of memory-based collaborative filtering [2]. Sarwar et al. [6] discuss the design of an item-based collaborative filtering using similarity function to compute similar users and items, and a weighted function to compute user-item score. For our ad recommender, we take a similar approach where we build user-ad, user-location and location-ad matrics to capture the relationships between the three entities. We then define similarity and scoring functions to compute the optimally recommended ad.

A large amount of ad and impression bid requests contain raw GPS (latitude and longitude information). One of the key data preprocessing that we need to do is to identify a small set of interesting locations where location-aware ad recommendation will hold relevance. Similar ideas based on mapping raw gps trace to semantically interesting locations are discussed in works such as [8], [1] and [9].

While we are starting off with a simple approach, we have identified many long-term use-cases and challenges which a large-scale distributed ad-platform will bring, such as dealing with billions of request and the need to store and process large amounts of ad logs and data. We are also looking at ideas to incorporate model-based recommendation systems which use advanced machine learning techniques such as Matrix Factorization [3], Factorization Machines [5], [4], Bayesian Approaches [7], etc. But, our current design and implementation focuses on adapting collaborative filtering to build the first version of the location-aware ad recommender.

In the following section, we discuss the key aspects of our location-aware ad recommendation algorithm design.

3 Location Aware Ad Recommendation System Design

In this section, we describe the assumptions we make and discuss the design steps of the location-aware ad recommender. First, we list down some of the key challenges in the first design. Next, we define what our recommendation problem statement is. Following this, we describe the data and its features. Next, we discuss the issues that we foresee and present a first version of recommendation algorithm based on collaborative filtering. Some of the key challenges that we faced at the first step are as follows:

- Dealing with missing data: RTB BidRequests contain a lot of information about device, user, impression, location, publisher/site, content category, etc but most of the information is optional. Thus, the recommender must be able to handle missing information.
- Mapping raw gps location to a small set of interesting locations : Using raw gps locations is not a scalable approach. Hence, we need to cluster and map raw gps locations to a location ontology. This will result is set of interesting locations which can be used to create user-location, location-ad matrices for collaborative filtering.
- Designing a Memory-based Recommender Algorithms: Design a first version of collaborative filtering algorithm by finding similar users as a function of location and recommending an ad suitable for a given location.
- Designing Model-based Recommender Algorithm: Design scalable and more efficient machine learning based model-based recommendation algorithms.

3.1 Recommendation Problem Statement and Data Format

In our ad-delivery platform, advertisers launch campaings with target parameters and ads. For analysis, we assume that we have M ads represented by A_i for $0 \leq i < M$. Similarly, we assume that users are represented by U_i for $0 \leq i < N$ and that, interesting locations are represented by L_i for $0 \leq i < K$. **The goal of our ad recommendation system is to select the highest ranked ad in response to a RTB BidRequest from a device looking to fill an**

impression. This is for a specific user U at a location L. This selected ad is sent as a part of the BidResponse and if the bid wins, the ad is displayed at the site location filling the impression. The BidRequest can be viewed as a bunch of key-value pairs with information about the impression type, device, user, location, site, app, content category, etc. Note that most of this information is actually optional so we need to design our algorithm assuming the absence of many of these fields. The BidResponse contains the bid information such as ad-serving url, the bid price and the ad that will be shown in case of winning the bid.

In response to BidRequest, many ad exchanges or DSP's will bid for the impression. One may or may not choose to bid. Even in case of bid, there is no assurance of winning the bid in which case that ad is shown. The shown ad may or may not be clicked by the user. To generate the data for building a recommender, we assume that BidResponse was made, and was successful in showing the ad, which may or may not be clicked. Thus the data format used to build the recommendation algorithm would be [**BidRequest Data, BidResponse Data, ClickInfo**]. Each of the two items, BidRequest and BidResponse contain many key-values pairs which can be used for feature engineering to extract meanigful features that help optimize a goal (for example CTR, click-through rate or a conversion measure). The first iteration of our recommender will optimize for the probability of clicking an ad, i,e, the CTR.

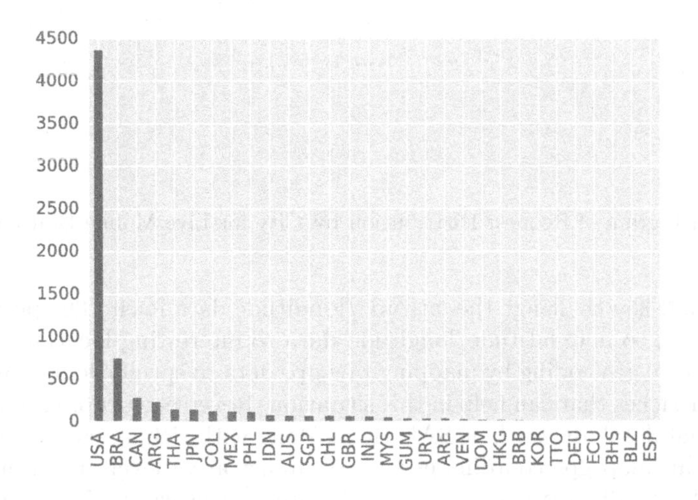

Fig. 3. Histogram of Request Distribution by Country for Live Mobile BidRequests

3.2 Challenges and Recommendation Algorithms

The first key challenge that we face is the nature of data. Recall that for implementing the recommender, we need historical data of the form [**BidRequest**

Data, BidResponse Data, ClickInfo]. However, to recommend an ad, we only need a new BidRequest. Many of the fields in BidRequest are optional, so they may or may not be present. Depending on how granular the location information in a BidRequest is, we can come up with multiple suitable approaches. We received BidRequest dataset from one of our partners and analyzed the nature of data. The location data can be fine-grained with latitude/longitude data. In general, we found that only a small fraction of data contains such gps data. However, the city and country information were frequently present. Thus, our first design idea is to build recommenders at different level of granularity, such as latitude/longitude, city, country and none (assuming that no location data is available).

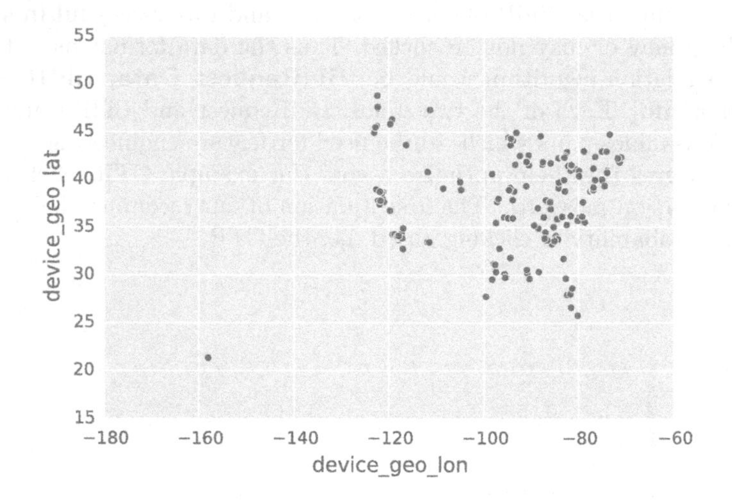

Fig. 4. Histogram of Request Distribution by City for Live Mobile BidRequests

The second key challenge the latitude/longitude data itself. The perprocessing task here it to map latitude/longitude data to interesting locations by clustering or reverse geo-coding by mapping raw gps data to specific locations. There are two approaches that can help in this situation. Reverse-geocoding functionalities in spatial databases such as SAP Hana along with rich maps such as Nokia HereMaps can map gps coordinates to 5 administrative level at various granularity. GPS coordinates can be clustered based on nearness. Locations when ad clicks have happened frequently can be used to identify important locations from a location-aware ad recommender perspective.

Evaluation of a live recommender system is a huge challenge. We propose using click-through-rate of ads as an optimization in the first iteration even though it is well-known that CTR is not be the most reasonable optimization criterion. Later on, we plan to integrate other ideas such as ad conversion and point of sales data to evaluate the recommender. For the first version, we are using baseline recommenders which recommends popular clicked ad for a given location.

Next, we discuss some basic mathematical framework we are using to design and implement the first version of our recommender.

4 Recommendation Algorithms and Experiments

To build the ad recommender prototype, we design the following:

- Recommenders R_i for i in [interesting location set, city, country, none]: Ad Recommenders at various granularity of location information (interesting location obtained from latitude/longitude, city, country, none)
- GeoMap[lat_lon list] $\rightarrow L_i$ for $0 \leq i \leq K$. : Mapping of raw location data to K locations by clustering where $K << $ Number of GPS data points.
- For each R_i, Design three functions. (Similarity Function, User-Ad Score Computation, Selecting the highest ranked ad to recommend).
- Baseline Recommender Systems Rb_i for i in [interesting location set, city, country, none] which select an ad based on popularity or revenue potential grouped by location.

The premise of our recommendation begins with the question: Which ad should we choose to send as BidResponse when a user U_i goes to location L_j. We maintain three sets of matrices similar in the spirit of collaborative filtering, User-Location, Location-Ad and User-Ad. We represent these matrices as UL_m, LA_m, UA_m. These can be used to measure similarity between pairs of objects (such as user-user similarity based on what locations they visit or ads they click on).

4.1 Collaborative Filtering Style Ad-Recommender

To compute an ad to recommend to the U_i when he goes to location L_j, we do the following steps:

- Initialize and maintain the three matrices, UL_m, LA_m, UA_m. For example, depending on the ad campaign parameters, we can initialize LA_m as binary vector of 1s if the campaign has selected the location or 0 if not. This vector is updated with user response to impressions. Even if a campaign has not picked a location, our system can set a non-zero value for an ad if data from similar campaigns suggests so. Thus, location-ad relationships are updated through recommendations learnt by performance of other similar campaigns at a given location.
- For the location L_j, get a candidate list of ads Ad_cand relevant to that location. For every location, ads can be shown multiple times, some of which may have been clicked. Thus, the LA_m matrix contains the fraction of times an ad was clicked in a given location. For a location L_j, we extract ads which have had good performance at that location. We also find locations similar to

L_j by computing location-location similarity from the LA_m using similarity score (cosine or pearson). We can sort these ads by their click percentage value and return the top few as a candidate list for potential recommendation at that given location. The baseline recommenders use the most clicked ads at any location to recommend the ad.
- Similar Users Computation: Using collaborative filtering similarity metrics such as adjusted cosine similarity or pearson r-correlation, we compute users that are similar to U_i in terms of locations that they have visited. For this, we use the UL_m Matrix and a similarity function such a cosine or person correlation. We get an ordered list of similar users sorted by weights equal to similarity score.
- Rank Ads by scoring: Using the similar users list from the previous step, we compute score for pairs of U_i and Ads in the Ad_cand by a weighted sum of user similarity and ad-score for all users in similar user list. Thus, our location aware ad recommender, finds affinity between ads and locations as well as users and locations and finally orders ads based on a weighted score.
- Finally, we recommend the ad with the highest score. If any of the previous steps returns null, i.e., empty Ad_cand, empty similar users list or ranked ad list, we invoke the baseline recommender (location, city or country). The baseline recommender picks an ad based on certain metric such as ad campaign parameters, revenue or popularity.

4.2 Identifying the Interesting Locations

Our goal is to map raw gps locations to a small set of interesting locations L_j. We are currently working on how to efficiently do this but our initial ideas is to combine clustering and location ontology. Using the raw gps data, we first identify a set of interesting locations by using third-party data such a Nokia Here Maps. Locations where ads have been clicked often also give us an indication of useful locations such as airport or malls. These locations are the pivots in our algorithm. We initialize the algorithm using such pivot locations and then merge other gps points to these or cluster and create new interesting locations if gps data is not close enough to these points. We make sure that the number of interesting locations is substantially lower than the raw dataset size.

- Enrich the raw gps locations using reverse geo-coding or mapping them to location ontology if found. This can be done via spatial database such as Hana enriched with Nokia Heremaps which maps gps coordinates to 5 administrative levels from country to neighborhood.
- Locations where ad clicks are high are treated as interesting locations. Use the first two steps to generate pivot locations.
- Using the interesting pivot locations we compute in the first two steps, we cluster the rest of the gps locations to one of the earlier identified locations if they are within a distance threshold. If not, they are clustered to create new locations.

4.3 Experimental Work

In this section, we report our finding from the basic data analysis of live bidrequest data. Our system will go live in the near future, which will enable data collection for recommendation, i.e., [BidRequest, BidResponse, Click or NoClick] We discuss what kind of information is usually present in real data which sort of inspired us to design the recommender system. We received data regarding BidRequests from one of our partners and experimented too see what kind of data is available to build the recommender. We present some statistics about BidRequests coming from mobile devices. The data is in json format and contains information about mobile app, device, impression, site and user. The location information is embedded in the device parent category. For instance, we plot the histogram distribution of device_geo_lat and device_geo_lon in Figure 1 and Figure 2 the whole world and a specific city. In Figures 3 and 4, we plot the distribution of number of BidReqests by country and city (for top 30 cities).

Missing data is an important challenge in our recommendation system. We mapped the BidRequest data information mobile app, device, impression, site and user to a total of 113 column dataset. We noticed that around 24.8% data had latitude and longitude information. Around 99.95% of requests have city and country information. All the data points have an associated ip address. Thus, our conclusions is that if latitute and longitude data are present, we invoke the fine-grain recommender else we recommend a city or country based recommender. If nothing is available, then we invoke a baseline recommender. We computed the fraction of missing data for BidRequests and it was around 71%, showing a bulk of BidRequest fields are missing. These observations indicate that we need to build multiple recommenders which will be triggered under different input data scenario.

The next objective is to collect [BidRequest, BidResponse, Click Info] data from the live system and let the recommender learn and recommend. This would help us tweak, evaluate and improve future versions of the recommendation algorithm, and improve over the baseline recommender algorithms.

5 Future Work and Conclusion

In this paper, we discussed our ongoing efforts to build an Ad-Recommender to the ad-delivery platform with an objective to optimize performance criterion for advertisers and publishers, and in turn, make our ad-platform attractive to customers and increase long-term revenues. In addition to location-aware recommendation, we are also looking in many other use cases which can leverage machine learning and data-driven solutions. Some of these cases are the optimal time for ad delivery, dealing with anonymous users, etc.

References

1. Cao, X., Cong, G., Jensen, C.S.: Mining significant semantic locations from gps data. Proc. VLDB Endow. **3**(1–2), 1009–1020 (2010)
2. Goldberg, D., Nichols, D., Oki, B.M., Terry, D.: Using collaborative filtering to weave an information tapestry. Communications of the ACM **35**, 61–70 (1992)
3. Koren, Y., Bell, R., Volinsky, C.: Matrix factorization techniques for recommender systems. Computer **42**(8), 30–37 (2009)
4. Kula, M.: Metadata embeddings for user and item cold-start recommendations. CoRR abs/1507.08439 (2015)
5. Rendle, S.: Factorization machines with libFM. ACM Trans. Intell. Syst. Technol. **3**(3), 57:1–57:22 (2012)
6. Sarwar, B., Karypis, G., Konstan, J., Riedl, J.: Item-based collaborative filtering recommendation algorithms. In: Proceedings of 10th International Conference on The World Wide Web, pp. 285–295 (2001)
7. Stern, D., Herbrich, R., Graepel, T.: Matchbox: large scale bayesian recommendations. In: Proceedings of the 18th International World Wide Web Conference (2009)
8. Leung, K.W.T, Lee, D.L., Lee, W.C.: Clr: a collaborative location recommendation framework based on co-clustering. In: Proceedings of the 34th International ACM Conference on Research and Development in Information Retrieval (SIGIR), pp. 305–314 (2011)
9. Zhou, C., Bhatnagar, N., Shekhar, S., Terveen, L.: Mining personally important places from gps tracks. In: 2007 IEEE 23rd International Conference on Data Engineering Workshop, pp. 517–526 (2007)

Building FP-Tree on the Fly: Single-Pass Frequent Itemset Mining

Nima Shahbazi(✉), Rohollah Soltani, Jarek Gryz, and Aijun An

Department of Computer Science and Engineering, York University, Toronto, Canada
{nima,rsoltani,jarek,aan}@cse.yorku.ca

Abstract. The FP-Growth algorithm has been studied extensively in the field of frequent pattern mining. The algorithm offers the advantage of avoiding costly database scans in comparison with Apriori-based algorithms. However, since it still requires two database scans, it cannot be used on streaming data. Also, the algorithm is designed for static datasets, where the input transactions are fixed and thus cannot be used for incremental or interactive mining. Existing incremental mining algorithms are not easily adoptable for on-the-fly, fast, and memory efficient FP-tree mining. In this paper we propose a novel SPFP-tree (single pass frequent pattern tree) algorithm that scans the database only once and provides the same tree as FP-Growth. Our algorithm changes the tree structure dynamically to create a highly compact frequency-ordered tree on the fly. With the insertion of each new transaction our algorithm dynamically maintains a tree identical to an FP-tree. Experimental results show the efficiency of the SPFP-tree algorithm in both incremental and interactive mining of frequent patterns.

Keywords: Data mining · Frequent pattern · Incremental mining · Association rule · Interactive mining

1 Introduction

Since its introduction [1], the problem of frequent pattern mining has been an active research area and subject of numerous studies. It plays an essential role in mining for association rules, correlations, emerging patterns, maximal-patterns and frequent closed pattern, multi-dimensional patterns, sequential patterns, classification and clustering. A large number of solutions have been proposed to solve the problem of frequent pattern mining efficiently, either through new algorithms or by improving existing ones.

The Apriori candidate generation and test technique [1] was the first approach to computing frequent itemsets. In order to discover frequent patterns, the algorithm performs n or $n + 1$ database scans, where n is the length of the longest pattern. To reduce the number of scans the FP-Growth algorithm was proposed [2]. This algorithm constructs a highly compact FP-tree (Frequent Pattern tree) which is usually much smaller than the original database and reduces the required number of database scans to two.

P. Perner (Ed.): MLDM 2016, LNAI 9729, pp. 387–400, 2016.
DOI: 10.1007/978-3-319-41920-6_30

In general, if we can scan the database only once to construct the FP-tree, we are able to mine patterns incrementally. Several algorithms such as CanTree [3] and CP-tree [4] have been proposed to capture all the necessary information in a database in one scan. The CanTree construction is based on the lexicographic order (i.e. alphabetical order) while CP-tree is based on the frequency descending order.

The key contribution of this paper is an algorithm that efficiently constructs an FP-tree on the fly, which is thus suitable for incremental and interactive mining. The algorithm constructs an SPFP-tree (single pass frequent pattern) by scanning the database only once and changes dynamically the structure of the tree to maintain the FP-tree structure. Our experimental results show that frequent itemsets mining with the SPFP-tree algorithm is more efficient than existing algorithms for single pass incremental mining.

The rest of the paper is organized as follows. In Section 2 we review the existing algorithms that mine frequent itemsets either statically or incrementally. Section 3 presents the SPFP-tree construction for incremental and interactive mining. Experimental results are described in Section 4 and conclusions are given in Section 5.

2 Related Work

Mining frequent itemsets can be broadly divided into two categories; static mining and incremental/interactive mining. The algorithms in static mining assume that the data set does not change during the mining process. These algorithms can be further divided into two main subgroups: Apriori based algorithms and FP-Growth based algorithms.

Apriori [1] was first proposed in 1993 and was followed by many algorithms based on the same idea. The main weakness of Apriori-based algorithms is the use of multiple database scans and numerous candidate generations. The FP-Growth algorithm, first proposed in [2], eliminates the process of generating candidates and reduces the number of database scans to two. It has three major steps. First, it scans the entire set of transactions to obtain each item's total count. In the second scan of a database, it builds the FP-tree based on frequent items. Finally, it recursively mines (in main memory) the FP-tree to find frequent patterns using a divide-and-conquer strategy. Many other algorithms were also proposed to improve the performance of FP-Growth; recent ones include PrePost [5], PrePost+ [6] and FIN [7]. However, these algorithms only work on static datasets and cannot be used for incremental or interactive mining.

The algorithms for incremental and interactive mining must incrementally update some dedicated data structure and mine frequent patterns from this structure with various support thresholds ("build once, mine many"). The algorithms in this category can also be divided into two subgroups: Apriori based and FP-Growth based. Algorithms in the first group include FUP [8], FUP2 [9] and UWEP [10]. All these algorithms require generating a large number of candidates and multiple database scans. However, within the FP-Growth based framework, the algorithms use a tree structure that captures all the necessary database information and mines the frequent itemsets in two or fewer database scans. Algorithms in this group include AFPIM [11], EFPIN [12], FUFP-tree [13], FELINE with CATS tree [14], CFP-tree [15], CanTree [3] and CP-tree [4]. The FP-Growth based incremental algorithms perform mining by incrementally updating a compact data structure, usually an FP-tree structure. APFIM, EFPIM and FUFP-tree algorithms require two database scans in order to build the corresponding FP-tree. Upon updating the database, if any infrequent patterns become frequent, the APFIM and EFPIM algorithms may require rescanning the updated database and FUFP-tree may require rescanning the original database.

FELINE with the CATS tree is well suited for interactive mining but its efficiency for incremental mining (when the database changes frequently) is unclear, due to its complex tree construction process. CanTree captures all the database information in a prefix tree which is based on lexicographic-order and uses the same mining technique as FP-Growth. The creation of a CanTree is faster than that of an FP-tree because the use of the lexicographic order requires no swapping or reconstruction of the tree. But it suffers from poor mining performance in comparison to FP-Growth due to the potentially large size of the resulting tree. CP-tree does not maintain the FP-tree structure at all times, but intermittently (e.g. after an addition of every 5 transactions) does so. In order to keep the prefix tree the same as FP-tree, this algorithm adjusts the tree using bubble sort. The algorithm needs to check all the branches in the tree to make sure the nodes in a branch are sorted, and rearrange the nodes in the branches to keep them in the sorted order. This will make the overall mining time reasonably long when a large fraction of branches in the tree require reconstruction.

3 SPFP-tree Algorithm

3.1 Tree Construction/Reconstruction

The construction of FP-tree in FP-Growth [2] consists of two major steps. First, the algorithm scans the database and finds the total frequency count of each item. Second, in the second database scan, items in a transaction are sorted in descending order of their frequency and added to the tree with prefix merging.1 The key feature of our algorithm is to maintain the structure of the FP-tree at all times so it can be mined efficiently due to its compact structure. The novelty of our approach lies in minimizing the number of branch comparisons whenever a new transaction is added.

The proposed algorithm consists of three steps: tree construction, tree reconstruction, and mining. Transactions are read, one at a time, and inserted into the tree. Then, the reconstruction phase modifies the tree structure to maintain its FP-tree structure. Finally, the mining algorithm finds the frequent itemsets from the tree. The major advantage of the method is the fact that the first two steps require only one database scan. Since the third step is identical as the tree mining part of the FP-Growth algorithm, we only describe the first two steps.

We maintain two hash tables for our algorithm. $Hash_1$ stores pairs (*item, count*) where *count* is the number of transactions containing the *item*, and $Hash_2$ stores pairs

[1] We are assuming the reader is familiar with the FP-tree structure.

(*count, list of items*). Thus, pairs in $Hash_2$ are reversed pairs from $Hash_1$ with items with identical counts collapsed to a single entry (the use of both tables will be explained below). Pairs in both hash tables are ordered with respect to the value of *count*. Consider a database with transactions: $t_1: (A, B)$, $t_2: (A, C)$ and $t_3: (A, B, C)$. The resulting hash tables are: $Hash_1 = \{A: 3\ ,\ B: 2\ , C: 2\}$ and $Hash_2 = \{3: (A)\ , 2: (B, C)\}$. The frequency count order of the items is: $A > B \geq C$.

Let us walk through the details of the algorithm. We maintain a proper FP-tree at all times updating it on the fly. Whenever a transaction is processed, the following steps are taken:

– Sort items in the transaction based on $Hash_1$ count.
– Add the transaction to the tree in a prefix-tree manner.
– Use $Hash_2$ to determine if any item violate the frequency count order (we describe below what exactly this involves) and reconstruct the tree if necessary.
– Update hash tables

As an example, consider the transactions and corresponding FP-tree in Fig. 1.a and Fig. 1.b respectively (the equivalent hash tables are shown in Fig. 1.c). The frequency count order of the items is: $A > B \geq C \geq D > E > F > K > G > H \geq I \geq J$.[2] we use $Hash_1$ table for ordering items in each transaction based on frequency count, and $Hash_2$ to find out if the frequency count order has changed.

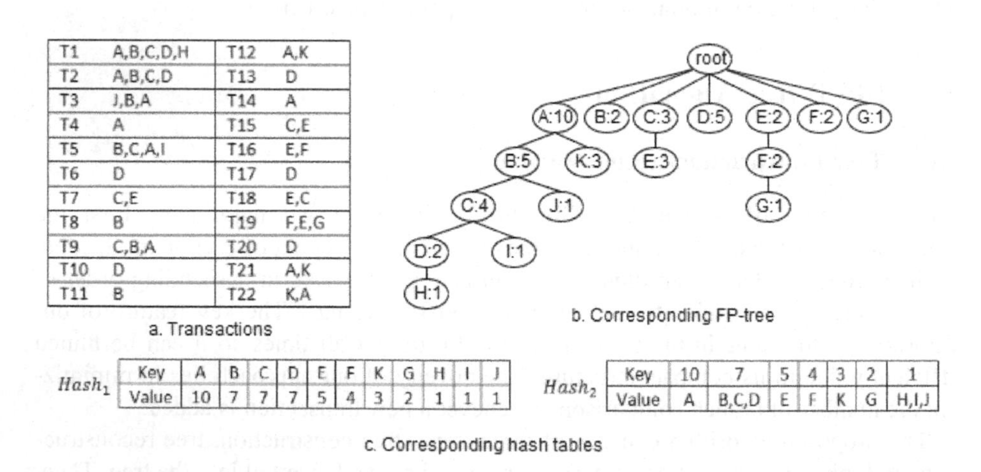

T1	A,B,C,D,H	T12	A,K
T2	A,B,C,D	T13	D
T3	J,B,A	T14	A
T4	A	T15	C,E
T5	B,C,A,I	T16	E,F
T6	D	T17	D
T7	C,E	T18	E,C
T8	B	T19	F,E,G
T9	C,B,A	T20	D
T10	D	T21	A,K
T11	B	T22	K,A

a. Transactions

$Hash_1$

Key	A	B	C	D	E	F	K	G	H	I	J
Value	10	7	7	7	5	4	3	2	1	1	1

$Hash_2$

Key	10	7	5	4	3	2	1
Value	A	B,C,D	E	F	K	G	H,I,J

c. Corresponding hash tables

Fig. 1. Hash tables and corresponding FP-tree

Now suppose transaction $T_{23}: (F, E)$ is to be processed (so far, the count for E and F is 5 and 4 respectively). The items in the transactions are first sorted with respect to the $Hash_1$ table to $T_{23}: (E, F)$ and added to the tree in a prefix-tree manner as shown in Fig. 2 (only the affected branch of the tree is shown). After items E and F are added, we need to verify that there are no items in the $Hash_2$ table with counts equal to

[2] The fact that the items are ordered alphabetically here is purely incidental.

the count of E or F (key 5 has only E as its value and key 4 has only F as its value). This test guarantees that no tree reconstruction is required because adding items F and E does not change the frequency count order of the items (still $A > B \geq C \geq D > E > F > K > G > H \geq I \geq J$). The change in order is possible only when the added item has its current count equal to that of another item. In that case, the addition of a new transaction with that item increases its count above the other one. Finally, the hash tables are updated as shown in Fig. 2.

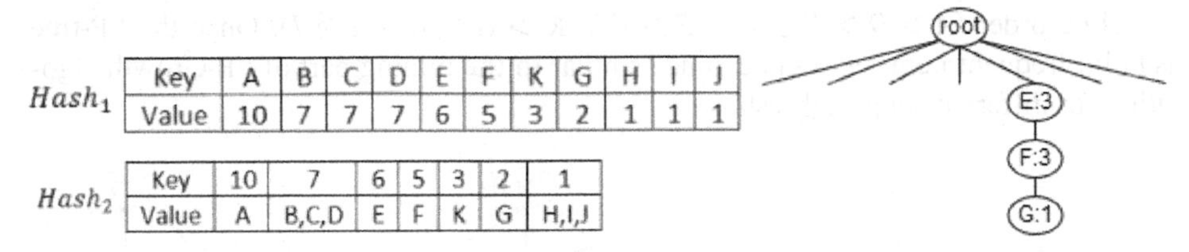

Fig. 2. Transaction $(\boldsymbol{F}, \boldsymbol{E})$ added

To see when the tree reconstruction is necessary, consider transaction T_{24}: (D, G). Again, the items are sorted first within the transaction according to their order in $Hash_1$ table ((D, G) is the correct order) and added to the tree as shown in Fig. 3. $Hash_2$ table (as shown in Fig. 2, that is, before it is updated for T_{24}) shows that there are two other items, B and C, whose counts are equal to item D's count (key 7 has values B, C, and D). In this case, we call node D a winner and nodes B and C losers. If there is a branch in the tree where node D's parent is B or C, it will lead to reconstruction of the tree (this is necessary to retain the FP-tree properties). Finally, hash tables are updated as shown in Fig. 3. Note that the frequency order count has now changed to: $A > D > B \geq C > E > F > K > G > H \geq I \geq J$).

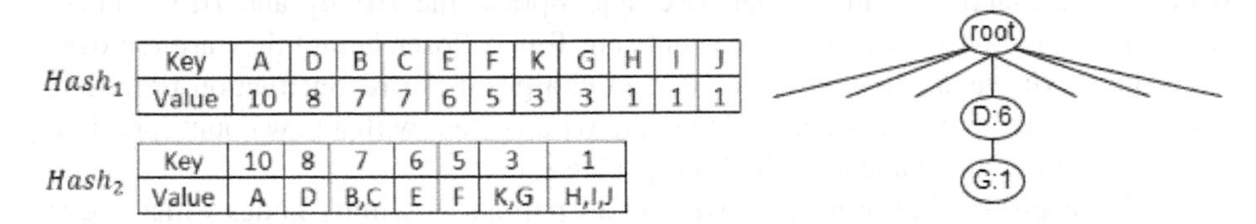

Fig. 3. Transaction $(\boldsymbol{D}, \boldsymbol{G})$ added.

Since there is a branch in the FP-tree where D's parent is C (leftmost branch in Fig. 1.b), the reconstruction phase is called. The winner node and all of its children are removed from its parent node and all of its children are kept in memory to be reassigned as children to other nodes. This is illustrated in Fig. 4.a. Then the loser nodes' local counts in the branch they share with the winner node are decreased by the local count of the winner node as shown in Fig. 4.b. If the local count of the loser node is equal to the local count of the winner node, the loser node is removed as its count becomes 0. The winner node is then added as a child to the immediate parent of the loser nodes, node A in our case, as shown in Fig. 4.b. If node A's children already contain the winner node, then these two nodes can be merged (changing the local count of the two nodes to the sum of the local counts). This is the only case in which two nodes are allowed to be merged. Then the final phase of reconstruction commences as illustrated in Fig. 4.c. In this phase, we first iterate through the list of loser node items (nodes B and C), a copy of the loser node items will be created and added to the winner node as children. The counts of these items however will be set to the winner node count (Node D with count 2). Finally, the child of the winner node (node H) is added as a child to the bottom-most loser node.

The resulting tree is the same as the FP-tree, and all items are in frequency descending order ($A > D > B \geq C > E > F > K > G > H \geq I \geq J$). Once the FP-tree is built, frequent item sets can be mined similar to the mining part of FP-Growth algorithm for different support thresholds.

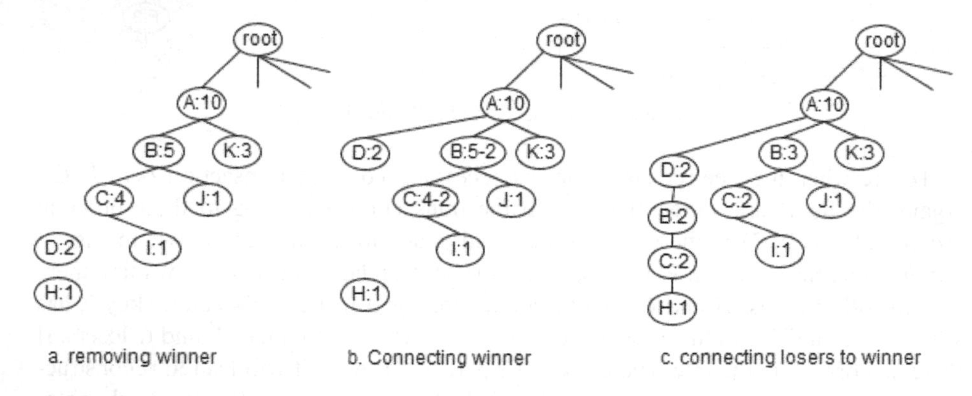

Fig. 4. Reconstruction

Figure 5 shows the corresponding pseudo code of the algorithm. Lines 1 to 17 read each transaction from the database, and update the $Hash_1$ and $Hash_2$ tables respectively. If items with the same counts are found (loser items) they are removed from the $Hash_2$ table (line 6). Line 10 is in charge of the reconstruction procedure given the winner and loser items and then the winner item with a new count (previous count plus one) will be added to the $Hash_2$ table.

In the reconstruction procedure, we find all parents of winner nodes which are in the losers list, and decrease their counts by winner's local count (if the local count for loser's node becomes zero, the node will be removed). Then we add the winner node to the immediate parent of the loser nodes. Finally, starting from the winner node, iterating through the list of loser nodes' items, a copy of loser nodes will be created and their counts will be set the same as winner's count, and added to the winner's children. Also, the children of the winner node will be added to the children of bottom-most loser node (line 23 to 33).

```
Input: Transactional DB
Output: SPFP-Tree
Begin
 1.  For each transaction T in DB
 2.    For each item X in T
 3.      If X is in hash_1
 4.          HithertoCountX := Hash_1[X]
 5.          Increment value of X in Hash_1
 6.          Remove X from Hash_2[HithertoCountX]
 7.          If Hash_2[HithertoCountX] is not empty
 8.            WinnerItem=X
 9.            LosersList=Hash_2[HithertoCountX]
10.          reconstruction(WinnerItem, LosersList)
11.        EndIf
12.        insert X into Hash_2[HithertoCountX+1]
13.      else
14.        insert X into Hash_1 with count 1
15.        insert X into Hash_2[1]
16.      EndIf
17.    EndFor
18.    sort T based on frequent descending based on Hash_1
19.    insert T into SPFP-Tree
20. End For
21. End
22.
23. reconstruction (WinnerItem ,LosersList):
24.   For each node W with value equal to WinnerItem
25.       Find the farthest ancestor A from W, which is
            in LosersList.
26.     If A is not null
27.       Reduce the local  count of all nodes  from W to A by
            the local count of W
28.       Insert W as a child of A's parent
29.       Add path from A to W's parent,  with local count of W
            To the new inserted W.
30.       Assign all children of W to the end of the new inserted
            path.
31.       Delete all node having zero local count
32.     EnIf
33.   EndFor
```

Fig. 5. SPFP-tree Algorithm

3.2 Correctness of the SPFP-tree Algorithm

We prove the correctness of the SPFP-tree algorithm. The objective of the proof is to show that tree maintained by the algorithm is the FP-tree. The key procedure is the tree reconstruction: we show that after the reconstruction the resulting tree is an FP-tree.

Let $A = \{a_1 \ldots a_m\}$ be a set of m distinct items. Each transaction $\mathcal{T}$ in database $\mathcal{D}$ has a unique identifier $\mathcal{T}id$, and contains a set of items (itemsets) such that $itemset \subseteq A$. $\mathcal{D}$ consists of N transactions $\{\mathcal{T}_1, \mathcal{T}_2 \ldots \mathcal{T}_N\}$.

Definition 1: A batch $\mathcal{B}$ is a finite sequence of transactions. Let $\mathcal{B}_k$ be the first k transanctions $\{\mathcal{T}_1, \mathcal{T}_2 \dots \mathcal{T}_k\}$ based on their appearance in $\mathcal{D}$, $1 \leq k \leq N$.

We define two operators: $|\ |_k$ as count operator and $\leq_k$ as ordered relation operator on $\mathcal{B}_k$ respectively:

$$|a_i|_k = count\ of\ a_i\ in\ \mathcal{B}_k \qquad 1 \leq i \leq m,\ 1 \leq k \leq N \tag{1}$$

$$a_i \leq_k a_j \Leftrightarrow \left(|a_i|_k < |a_j|_k \text{ or } \left(|a_i|_k = |a_j|_k \text{ and } i < j\right)\right) \quad 1 \leq i,j \leq m, 1 \leq k \leq N \tag{2}$$

By (2) items are ordered in a batch by their count, and if they have equal count, they ordered by their index. Consider a batch $\mathcal{B}_k$, $1 \leq k \leq N$. We define the following trees:

- An unpacked tree T_k^U $(1 \leq k \leq N)$ created from $\mathcal{B}_k$, is defined as follow:
  - All the transactions in B_k are attached to the root of the tree, $\mathcal{R}$, without prefix merging.
- Packed tree T_k^P $(1 \leq k \leq N)$ created from $\mathcal{B}_k$, is defined as follow:
  - All the transactions in B_k attached to $\mathcal{R}$ with prefix merging (prefix-tree).
- Level k for each tree consists of all nodes that have distance k from the root (the root is at level 0). The depth of the tree is equal to the largest level of the tree.

In order to create the FP-tree, we need to create the $\leq_N$ order first and then the T_N^P tree (prefix merging). Our algorithm works in N steps. In step i $(1 \leq i \leq N)$ an $\leq_i$ order is created from the $\leq_{i-1}$ order and T_i^P tree from T_{i-1}^P tree. In order to show that our tree is the same as FP-tree, it is required to show that the proposed method for reconstructing T_i^P from T_{i-1}^P tree is correct (which means that (3) and (4) produce the same tree).

$$T_i^P \qquad 1 \leq i \leq N \tag{3}$$

$$T_{i-1}^P \text{ and adding } \mathcal{T}_i \qquad 1 \leq i \leq N \tag{4}$$

If adding $\mathcal{T}_i$ does not require tree reconstruction, then (3) and (4) are obviously the same. But suppose that, there are two items a and b with equal total counts in T_{i-1}^P, and $\mathcal{T}_i$ consist of one item a. By adding $\mathcal{T}_i = \{a\}$ the order of items a and b will change (that is, $a \leq_{i-1} b\ and\ b \leq_i a$). Now a branch in T_{i-1}^P needs reconstruction if it has items a and b, as shown in the left branch of Fig. 6.a. In this case for T_i^P node a must have lower level than node b (because, $b \leq_i a$). The algorithm will change the tree structure as follow: First $\mathcal{T}_i = \{a\}$ added to the tree in a prefix-tree manner (changing node $a's$ count in l_1 from w to $w + 1$), see right branch in Fig. 6.b. Then node a in left branch will go to level l_1 in that branch and it's node count k_1 will add up with a node count $w + 1$ in that level (total node count for a is now: $k_1 + w + 1$), then k_1 node count of b will be $a's$ child in left branch and all $a's$ children in left branch will be $b{:}\,k_1$'s child now. New node count for b in l_1 will be $(k_2 - k_1)$, see right branch in Fig. 6.c.

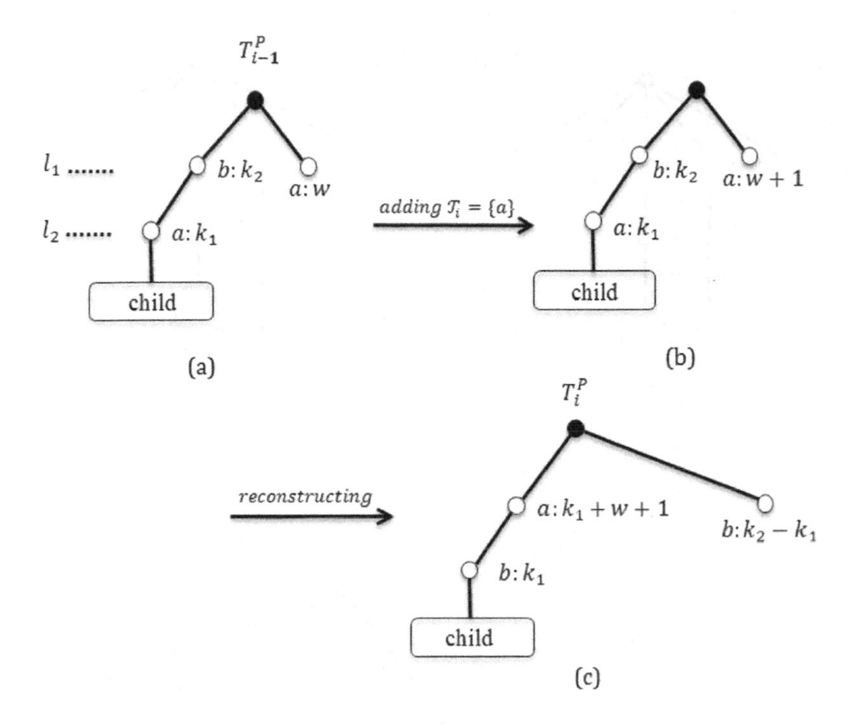

Fig. 6. Constructing T_i^P from T_{i-1}^P

To validate the above reconstruction, consider unpacked tree (unpack version of T_{i-1}^P is shown in Fig. 7 top left). Due to $b \leq_i a$, item a must have lower level than item b. So for two items in the same path *we only change the position of those items one by one* (now $(k_2 - k_1)$ node count of b will remain in l_1 level and remaining k_1 node count of b will be $a's$ child in left branch). In order to get back to T_i^P tree is to pack everything again, after that an exact replica of an FP-tree will be created, see Fig. 7 (the steps in the figure shows that: first by considering unpack tree, and then changing the node position one by one followed by packing back the result, the same T_i^P in Fig. 6.c is created). ■

3.3 Incremental Mining with the SPFP-tree

Items can be stored in a compact FP-tree structure regardless of whether they are frequent or not. It is then straightforward to add or delete transactions from the tree. In order to delete a transaction it needs to be sorted first based on the $Hash_1$table. Then the tree is traversed downwards in order to find the corresponding nodes and their local counts are decreased by one. Afterwards, the $Hash_1$ and $Hash_2$ tables are updated and the reconstruction phase procedure of the tree will be called. To add a transaction one needs to follow the routine described in the SPFP-tree algorithm.

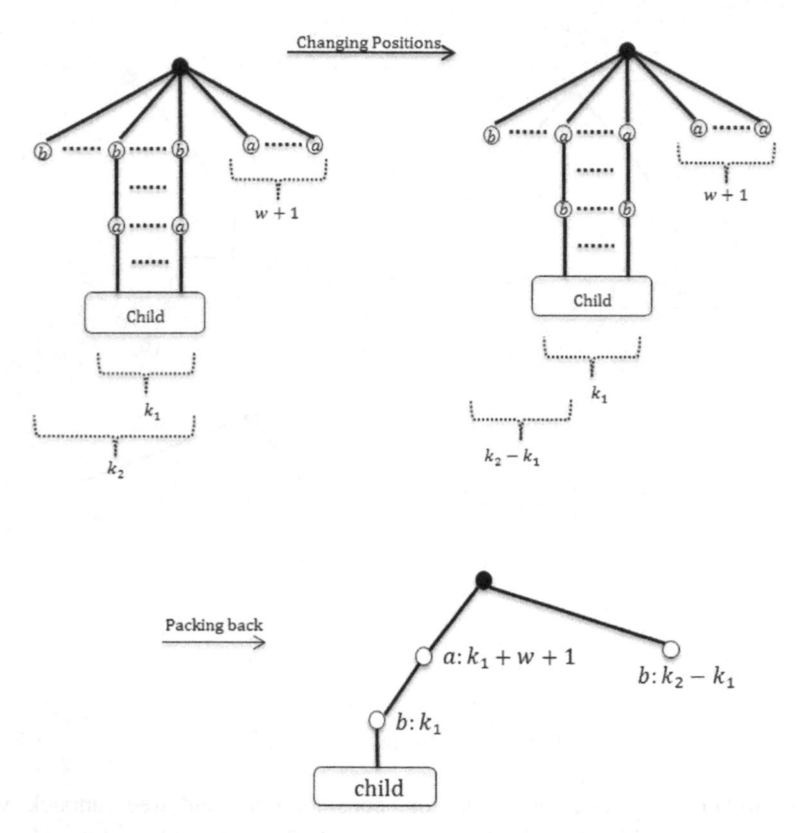

Fig. 7. Changing position in ***unpacked*** tree and then packing back the tree.

3.4 Interactive Mining with SPFP-tree

The original FP-Growth algorithm builds an FP-tree based on a given support threshold, hence the tree can only be used to find patterns satisfying the given threshold. Since the SPFP-tree algorithm builds a tree from all the items in transactions, it supports mining the tree with different support thresholds. Given a minimum support threshold (min_sup), the algorithm traverses the tree upward from nodes that have a min_sup count in $Hash_1$ table, and builds the corresponding FP-tree. The tree is in frequency-descending order and nodes placed below the traversed nodes are not frequent. Similar to other interactive mining methods if the new min_sup is greater the min_sup in the previous round, it is possible to cache the frequent patterns in previous round and reuse them in the next round. This allows for further reducing of the total mining time.

4 Performance Study and Experimental Results

In this section, the performance of the SPFP-tree algorithm is evaluated in comparison with Can-Tree and CP-tree. These two were chosen for comparison since they show good performance among other incremental frequent item set mining algorithms. All programs are implemented in JAVA and run on Linux centos 6 with Intel core 2 duo 3.00 GHz CPU and 4 GB memory. The reported figures are based on the average execution time over multiple runs. The experiments were performed on several datasets from the UCI Machine Learning Repository [16] including Chess, Connect, Mushroom and Accidents.

4.1 Performance Study of Execution Time for Different Threshold Levels

In the first experiment we measure how the minimum support threshold affects the runtime of the algorithms. Fig. 8 shows the runtime for SPFP-tree versus CP-tree and CanTree on different support thresholds for four datasets. The total execution time includes time for the prefix tree construction, and frequent item set mining steps, $(Total\ Time = Construction\ Time + Mining\ Time)$.

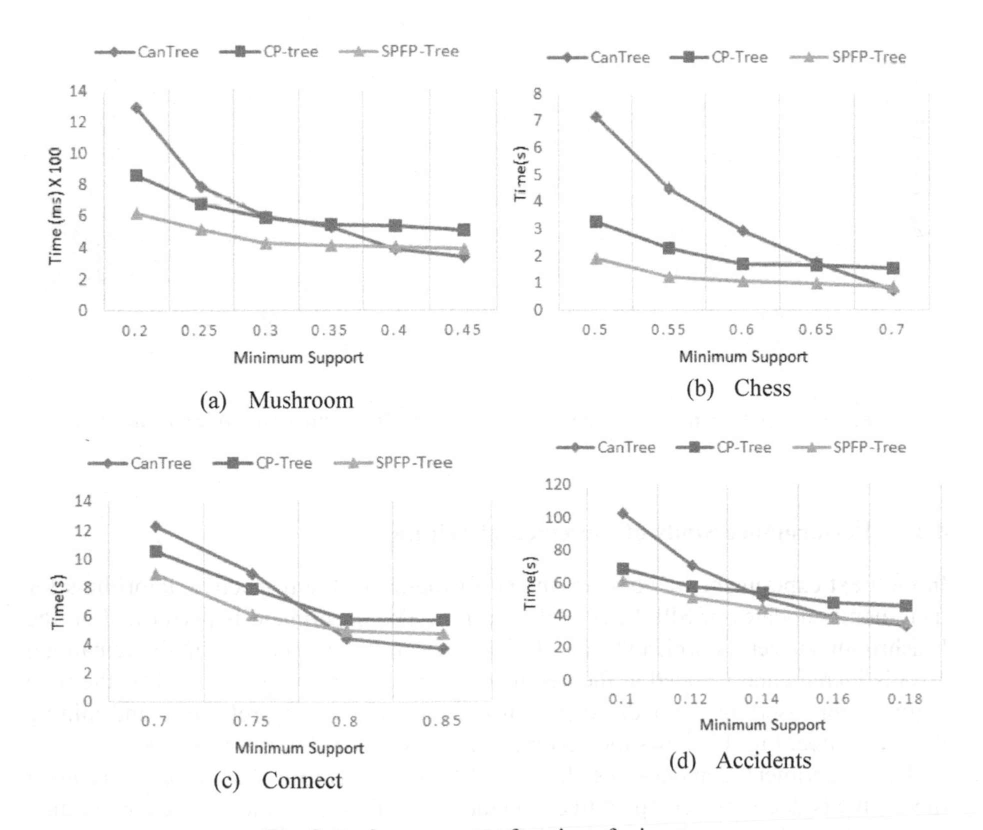

(a) Mushroom (b) Chess

(c) Connect (d) Accidents

Fig. 8. Performance as a function of min_sup

CanTree has smaller *Construction Time* in comparison to CP-tree and SPFP-tree, because it uses alphabetical order to add transactions, and no reconstruction is needed. But the tree generated by the CanTree algorithm is larger than CP-tree or SPFP-tree (see Fig. 10) resulting in larger *Mining Time*. The mining time is highly correlated to the number of nodes in the pruned prefix-tree (pruned based on the minimum support threshold). Therefore, when a low min_sup is used the pruned prefix-tree (either for FP-tree or alphabetical tree) has more nodes than when a higher min_sup is used. This causes a dominance in mining time, whereas when we have min_sup (either for the FP-tree or alphabetical tree), the *Construction Time* will be dominant.

Mining Times for SPFP-tree and CP-tree are the same, since both algorithms have the same final FP-tree structure. Therefore, the difference between them depends on the construction time. The construction time in the proposed algorithm is lower than that of CP-tree (see Fig. 9). Therefore, in all cases it outperforms the CP-tree algorithm.

Comparing SPFP-tree with CanTree, it can be seen that with an increase in the support threshold (for example more than 0.8 in the Connect dataset) the total required time for CanTree becomes smaller than that of the SPFP-tree algorithm. This is due to the fact that a higher support threshold makes *Construction Time* dominant, causing CanTree to have better performance due to its fast prefix-tree creation.

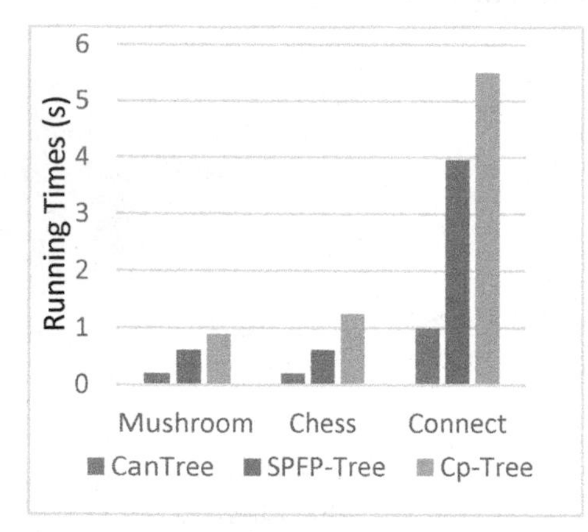

Fig. 9. Tree Construction Time

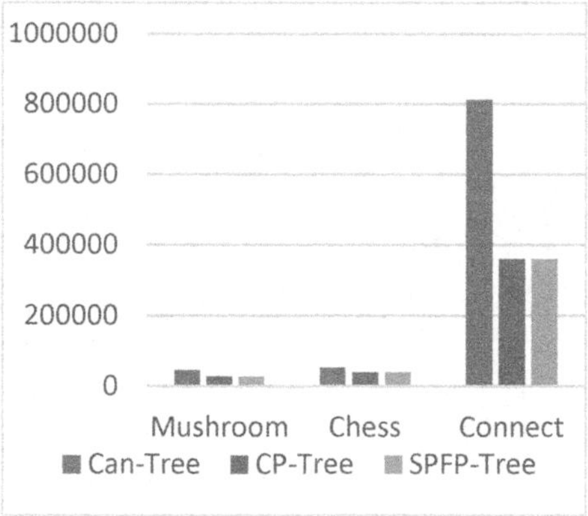

Fig. 10. Number of nodes in each tree

4.2 Performance Study of Incremental Mining

In the next experiment we compare the performance of the respective algorithms for incremental updates of SPFP-tree and CanTree. The experiment is performed on the Mushroom dataset for which 90% of the transactions are preloaded and the remaining 10% is incrementally added to the tree in 10 steps. The running time will be the time required for inserting the updated part of the database to the prefix-tree and mining the prefix-tree. Fig. 11 shows the performance of SPFP-tree versus CanTree.

The experiment demonstrates that SPFP-tree outperforms Can-Tree on running time. This is due to the compact tree structure of SPFP-tree, which makes the mining part of its process much faster than CanTree.

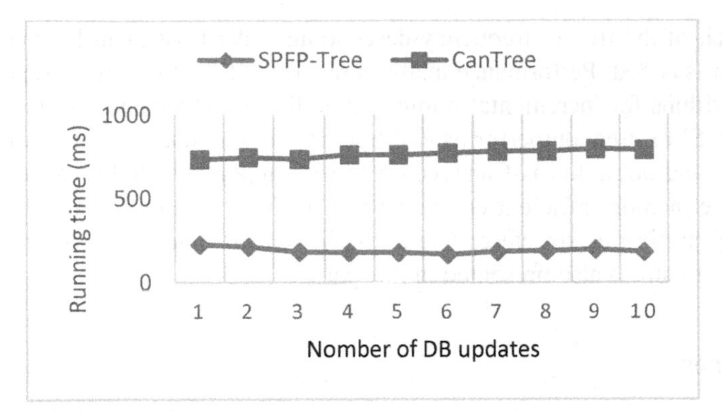

Fig. 11. Incremental mining on mushroom with min_sup=0.1

4.3 Performance Study of Interactive Mining

Interactive mining occurs when the user plans to mine a fixed database with different minimum support thresholds. The results of interactively mining the Mushroom dataset with the proposed SPFP-tree and CanTree are shown in Fig. 12.

Both algorithms need to construct the tree once, and then prune it based on a min_sup interactively. The time reported here covers only the mining time (we assume the tree has already been built). The result shows that the SPFP-tree outperforms CanTree, due to its frequency-descending item ordering and more compact tree structure.

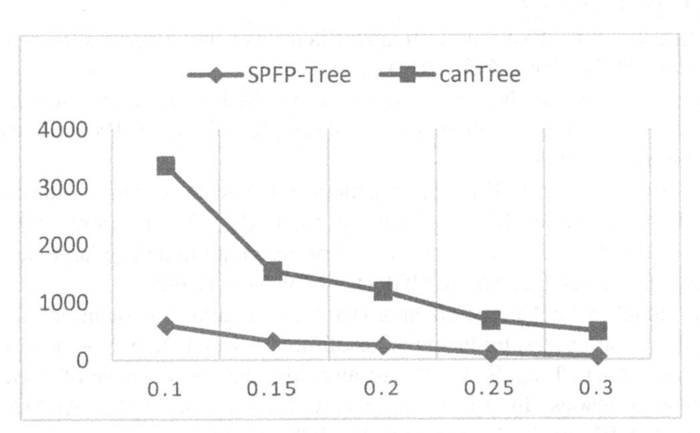

Fig. 12. Interactive Mining for Mushrooms Dataset

5 Conclusion

In this paper, a new method called SPFP-tree (single pass frequent pattern tree) for incrementally constructing FP-tree with a single pass of the data set has been proposed. The proposed algorithm rearranges the tree on the fly, and keeps the items in each branch of the tree in frequency-descending order (just as in FP-tree) after each transaction is added. Performance analysis results of SPFP-tree were reported against other algorithms for incremental mining, including CanTree and CP-tree. Our results show that SPFP-tree outperforms CP-tree in all the datasets on various support thresholds, and outperforms CanTree on lower support thresholds. Moreover, SPFP-tree is more memory efficient compared to CanTree because of its dense frequency-descending prefix-tree structure. The feasibility of the algorithm for incremental and interactive mining is also presented in this paper.

References

1. Agrawal, R., Imielinski, T., Swami, A.: Mining association rules between sets of items in large databases. In: Proceedings of the SIGMOD, New York (1993)
2. Han, J., Pei, J., Yin, Y.: Mining frequent patterns without candidate generation. In: Proceedings of the ACM SIGMOD International Conference on Management of Data, Dallas, Texas (2000)
3. Leung, C.K.-S., Khan, Q.I., Li, Z., Hoque, T.: CanTree: a canonical-order tree for incremental frequent-pattern mining. Knowledge and Information Systems **11**(3), 287–311 (2007)
4. Tanbeer, S.K., Ahmed, C.F., Jeong, B.S., Lee, Y.K.: Efficient single-pass frequent pattern mining using a prefix-tree. Information Sciences **179**(5), 559–583 (2009)
5. Deng, Z.H., Wang, Z., Jiang, J.J.: A new algorithm for fast mining frequent itemsets using N-lists. SCIENCE CHINA Information Sciences **55**(9), 2008–2030 (2012)
6. Deng, Z.H., Lv, S.L.: PrePost+: An efficient N-lists-based algorithm for mining frequent itemsets via Children-Parent Equivalence pruning. Expert Systems with Applications **42**(13), 5424–5432 (2015)
7. Deng, Z.H., Lv, S.L.: Fast mining frequent itemsets using Nodesets. Expert Systems with Applications **41**(7), 3506–3513 (2014)
8. Cheung, D.W., Han, J., Ng, V.T., Wong, C.Y.: Maintenance of discovered association rules in large databases: an incremental updating technique. In: Proceedings of the ICDE, Los Alamitos, CA (1996)
9. Cheung, D.W., Lee, S.D., Kao, B.: A general incremental technique for maintaining discovered association rules. In: Proceedings of the DASFAA, Singapore (1997)
10. Ayan, N.F., Tansel, A.U., Arkun, E.: Efficient algorithm to update large itemsets with early pruning. In: Proceedings of the SIGKDD, New York (1999)
11. Koh, J.L., Shieh, S.F.: An efficient approach for maintaining association rules based on adjusting FP-tree structures. In: Proceedings of the DASFAA, New York (2004)
12. Li, X., Deng, Z.-H., Tang, S.-W.: A fast algorithm for maintenance of association rules in incremental databases. In: Li, X., Zaïane, O.R., Li, Z.-H. (eds.) ADMA 2006. LNCS (LNAI), vol. 4093, pp. 56–63. Springer, Heidelberg (2006)
13. Hong, T.P., Lin, C.W., Wu, Y.L.: Incrementally fast updated frequent pattern trees. Expert Systems with Applications **34**(4), 2424–2435 (2008)
14. Cheung, W., Zaiane, O.R.: Incremental mining of frequent patterns without candidate generation or support constraint. In: Proceedings of the IDEAS, Los Alamitos, CA (2003)
15. Liu, G., Lu, H., Yu, J.X.: CFP-tree: A compact disk-based structure for storing and querying frequent itemsets. Information Systems **32**(2), 295–319 (2007)
16. Blake, C., Merz, C.: UCI repository of machine learning databases. University of California, Irvine (1998)

Energy Disaggregation Based on Semi-Binary NMF

Masako Matsumoto[1(✉)], Yu Fujimoto[2], and Yasuhiro Hayashi[1]

[1] Department of Electrical Engineering and Bioscience, Waseda University, Tokyo, Japan
masako-sakura@toki.waseda.jp, hayashi@waseda.jp
[2] Advanced Collaborative Research Organization for Smart Society, Waseda University, Tokyo, Japan
y.fujimoto@aoni.waseda.jp

Abstract. The large-scale introduction of renewable energy resources will cause instability in the power supply. Residential energy management systems will be even more important in the near future. An important function of such systems is visualization of appliance-wise energy consumption; residents will be able to consciously avoid unnecessary consumption behavior. However, visualization requires sensors to measure appliance-wise energy consumption and is generally a costly task. In this paper, an unsupervised method for non-intrusive appliance load monitoring based on a semi-binary non-negative matrix factorization model is proposed. This framework utilizes the total power consumption patterns measured at the circuit breaker panel in a house, and derives disaggregated appliance-wise energy consumption. In the proposed approach, the energy consumption of individual appliances is estimated by considering the appliance-specific variances based on an aggregated energy consumption data set. The authors implement the proposed method and evaluate disaggregation accuracy using real world data sets.

Keywords: Semi-Binary NMF · Energy disaggregation

1 Introduction

Large-scale introduction of renewable energy is driven by the effect of cutting one's dependence on fossil fuels and as a countermeasure against global warming. However, renewable energy will cause instability of the power balance between supply and demand because the generated power fluctuates greatly. Therefore, residential energy management is expected to contribute to the stabilization of the supply-demand balance. Visualization of appliance-wise energy consumption is a useful approach for an efficient residential energy management. Residents can recognize the relationship between ant power consumption by aware appliance-wise energy consumption; visualization will help them to consciously change their appliance usage to avoid heavy consumption during periods of peak demand. Visualization of appliance-wise energy consumption requires appliance-wise energy consumption sensors; however, setting up these sensors is a costly task in general. Therefore, visualizing appliance-wise energy consumption is not widely prevalent.

P. Perner (Ed.): MLDM 2016, LNAI 9729, pp. 401–414, 2016.
DOI: 10.1007/978-3-319-41920-6_31

An appliance-wise energy disaggregation technique is a resolution for the above mentioned cost issue; a typical framework of this technique comprises a single sensor placed at the circuit breaker panel in a house. This sensor derives appliance-wise energy consumption on the basis of the historical records of the total power consumption curves. This type of energy disaggregation technique is more attractive for realizing services to visualize appliance-wise power consumption without using many sensors.

In this paper, we propose an energy disaggregation method based on Semi-Binary Non-negative Matrix Factorization (SBNMF); in our approach, appliance-wise energy consumption is estimated by using the total power consumption curves. We experimentally show the accuracy of our proposed method using real-world data sets.

The remainder of this paper is organized as follows. In Sec. 2, we explain the basic idea of the matrix factorization approach used in this paper. In Sec. 3, we discuss the problem statement and provide several ideas for improvement. In Sec. 4, we show the simulation results of the proposed approach on real-word data sets for evaluation. Finally, we conclude the study in Sec. 5 and point out the directions for future improvements and generalization.

2 Energy Disaggregation via SBNMF

The estimation task of appliance-wise energy consumption based only on the total power consumption patterns; is called "energy disaggregation" in this paper. An appliance-wise energy disaggregation technique is also known as non-intrusive appliance load monitoring (NIALM), and has been proposed in the early 1990s [1]. The typical setting of energy disaggregation uses a single sensor at the circuit breaker panel in a house and derives appliance-wise energy consumption based on the historical records of the total power consumption patterns. In the literature, applications of hidden Markov model (HMM) [2, 3] and non-negative matrix factorization (NMF) [4] have been proposed for energy disaggregation task. Particularly, NMF approach utilizes semi-supervised learning by means of several appliance-wise energy consumption pattern data sets. However this approach needs auxiliary information of appliance-wise energy consumption for estimation. The authors have previously considered applying the method of modified NMF, so called Semi-Binary NMF (SBNMF) under the unsupervised setting [5]. In this paper, we only use total energy consumption measured at a single sensor, and attempt to improve the accuracy of unsupervised energy disaggregation based on SBNMF.

Let $r \in \{1,\ldots,R\}$ be the index of the load appliance in a household, and $\mathbf{Y}_r = [y_{rtn}] \in \mathbb{R}_+^{T\times N}$ be the matrix composed of power consumed by appliance r in time slice $t \in \{1,\ldots,T\}$ in day $n \in \{1,\ldots,N\}$. Assume that our accessible information is given as the total energy consumption matrix $\mathbf{Y} = \sum_r \mathbf{Y}_r$. The basic idea of energy disaggregation based on SBNMF is given as the following decomposition form

$$\mathbf{Y} \cong \mathbf{XA}, \tag{1}$$

where $\mathbf{X} = [x_{tk}] \in \mathbb{R}_{+}^{T \times K}$ is a non-negative basis matrix whose column vectors express typical appliance-wise energy consumption patterns and $\mathbf{A} = [a_{kn}] \in \{0,1\}^{K \times N}$ is an indicator matrix whose column indicates binary weights for K bases in each day. Under the assumption that total energy consumption $y_{tn} = \sum_r y_{rtn}$ on day n at time t is according to the following Gaussian distribution,

$$p(y_{tn}; \sum_k x_{tk} a_{kn}, \sigma^2) \sim \frac{1}{\sqrt{2\pi\sigma^2}} \exp\left(-\frac{(y_{tn} - \sum_k x_{tk} a_{kn})^2}{2\sigma^2}\right) \quad (\forall\, t, n), \tag{2}$$

the maximum likelihood estimation of matrix $\mathbf{X}, \mathbf{A}$ is reduced to the Frobenius norm minimization problem as follows:

$$\begin{aligned} \mathrm{argmax}_{\mathbf{X},\mathbf{A}} \sum_{t,n} \log p(y_{tn}; \sum_k x_{tk} a_{kn}, \sigma^2) &= \mathrm{argmin}_{\mathbf{X},\mathbf{A}} \sum_{t,n} (y_{tn} - \sum_k x_{tk} a_{kn})^2 \\ &= \mathrm{argmin}_{\mathbf{X},\mathbf{A}} \|\mathbf{Y} - \mathbf{X}\mathbf{A}\|_2^2. \end{aligned} \tag{3}$$

The optimization problem given in Eq.(3) can be solved by the iterative update approach like the ordinary NMF [6]. The update rule of $\mathbf{X}$ is derived based on the auxiliary function technique [7] as follows,

$$\mathbf{X} \leftarrow \mathbf{X} \circledast \mathbf{Y}\mathbf{A}^{\mathrm{T}} \oslash \mathbf{X}\mathbf{A}\mathbf{A}^{\mathrm{T}}, \tag{4}$$

where $\circledast$ is the Hadamard (element-wise) product and $\oslash$ is the element-wise division. There are several approaches to update binary matrix $\mathbf{A}$ [8, 9]; here, we use the greedy random search under the fixed $\mathbf{X}$ so that exhaustive search in 2^K candidates is adopted independently for N days as shown in Algorithm 1. Note that Eq.(3) generally has local minima; therefore, we adopt the algorithm from different initial matrices and select the best result from the viewpoint of the objective function as the final estimate. In this paper, the relationship between K bases (K column vectors in $\mathbf{X}$) and R appliances is given as $y_{rtn} = \sum_{k \in S_r} x_{tk} a_{kn}$ $(r = 1, \ldots, R)$, where $S_r \subseteq \{1, \ldots, K\}$ be the index subset for the appliance r, such that

$$\begin{aligned} &\bigcup_{r \in \{1,\ldots,R\}} S_r = \{1, \ldots, K\}, \\ &S_r \cap S_{r'} = \emptyset \quad (r \neq r'), \end{aligned}$$

holds.

The final estimate $\hat{y}_{rtn}$ for energy consumption y_{rtn} of the appliance r under the given total energy consumption y_{tn} can be derived according to the following maximization problem,

$$\begin{aligned} &\max \prod_{r=1}^{R} p(\hat{y}_{rtn} | y_{tn}; \sum_{k \in S_r} x_{tk} a_{kn}, \sigma^2) \\ &\text{s.t.} \quad \hat{y}_{rtn} \geq 0 \ (r = 1, \ldots, R) \\ &\qquad \sum_{r=1}^{R} \hat{y}_{rtn} = y_{tn}. \end{aligned} \tag{5}$$

Algorithm 1. Energy Disaggregation based on SBNMF

```
Input: total energy consumption matrix Y.
Initialize: X by random non-negative values and A by ran-
dom binary values.
while ||Y − XA||_2^2 is not converged do
  Update X according to Eq. (4).
  for each n ∈ {1,…,N}
    update binary vector [a_kn; k = 1,…,K] to minimize the ob-
     jective function
  end for
end while
Output: matrices X̂, Â.
```

Fig. 1. Algorithm of energy disaggregation based on SBNMF

We assume that the estimate for energy consumption $\hat{y}_{rtn}$ is derived by using the following softmax function,

$$\hat{y}_{rtn} = \begin{cases} y_{tn} \times \frac{\exp(\theta_r)}{1+\sum_{r'=1}^{R-1}\exp(\theta_{r'})} & (r \neq R) \\ y_{tn} \times \frac{1}{1+\sum_{r'=1}^{R-1}\exp(\theta_{r'})} & (r = R), \end{cases} \tag{6}$$

and numerically optimize Eq. (5) in terms of $\theta_r \in \mathbb{R}$ $(r = 1, \ldots, R)$ for each pair (t, n). Note that the optimal number R is derived based on the cross-validation from the viewpoint of reconstruction error $\|\mathbf{Y} - \mathbf{XA}\|_2^2$.

To evaluate the disaggregation result of the appliance me, focus on the following appliance-wise match rate,

$$MR_r = \frac{\sum_{t,n}\min\{y_{rtn}, \hat{y}_{rtn}\}}{\sum_{t,n}\max\{y_{rtn}, \hat{y}_{rtn}\}}. \tag{7}$$

Equation (7) indicates the goodness of the disaggregation result $\hat{\mathbf{Y}}_r = [\hat{y}_{rtn}]$ for appliance r by comparing it with $\mathbf{Y}_r$; $MR_r = 1$ holds if and only if $y_{rtn} = \hat{y}_{rtn}$ holds for all t and n

Figure 2 shows examples of the daily energy consumption; Fig. 2(a) shows a real-world appliance-wise energy consumption and 2(b) shows estimation results. This example implies that energy consumption of the refrigerator is nearly constant in the real-world, however its estimation result is not similar at all. And, appliance-wise energy consumption differs in magnitude. In the observation, energy consumption of the air conditioner is the largest, but that of the TV is small. The result shows that the magnitude of their consumption is not accurately estimated. The figure also shows that there are several appliances operated only for short periods of time, like the nothing machine; the energy consumption of these appliance are not appropriately estimated in the naive implementation of SBNMF.

3 Improvement of Energy Disaggregation Based on SBNMF

3.1 Basic Property of Energy Consumption

To improve accuracy of SBNMF based energy disaggregation, we investigate half-hourly energy consumption of the following major electric appliances, i.e. refrigerators, washing machines, TVs and air conditioners, and focus on several appliance-specific properties of energy consumption.

Figure 3 shows the histograms of appliance-wise energy consumption, and dashed lines indicate the Gaussian approximation of their estimates derived with SBNMF introduced in the previous section. Energy consumption varies greatly depending on the appliance types; e.g., energy consumption of the refrigerator is almost constant, though that of the air conditioner varies. Table 1 shows the daily frequency of the washing machine operation in 28 days. Figure 4 also shows the running time-zone of the washing machine in a day. Figure 4 indicates that the washing machine is frequently used in the morning, especially during 6:30–7:30, and operated only for a short period in a day.

In SBNMF, the above mentioned appliance-specific properties are not considered. However, it is important to consider such appliance-specific properties for achievement of the accurate energy disaggregation. In this study, we propose the following three assumptions to achieve plausible energy disaggregation.

1. Existence of an appliance with stable energy consumption (base load).
2. Appliance-specific variance of energy consumption.
3. Existence of appliances running for a short time.

To introduce the first assumption, we focus on the K-th basis in $\mathbf{X}$ as "base load" to express the appliance with the constant energy consumption;

$$x_{tK} = c \quad (t = 1, \ldots, T), \tag{8}$$

where c is the load energy consumption of the base load and estimate it by expressing stable energy consumption appliance and upload measure of energy consumption stable value. The update rule of $\mathbf{X}$ is derived based on Eq. (4), however K-th basis in $\mathbf{X}$ have stable elements. The binary indicator matrix $\mathbf{A}$ is updated according to the greedy random search exactly the same as that of ordinary SBNMF. In the following subsections, we introduce ideas for applying assumptions 2 and 3 to our disaggregation framework.

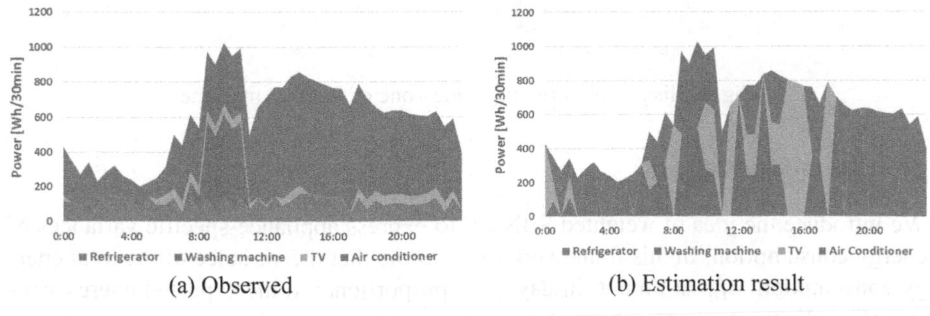

(a) Observed (b) Estimation result

Fig. 2. Example of daily energy consumption

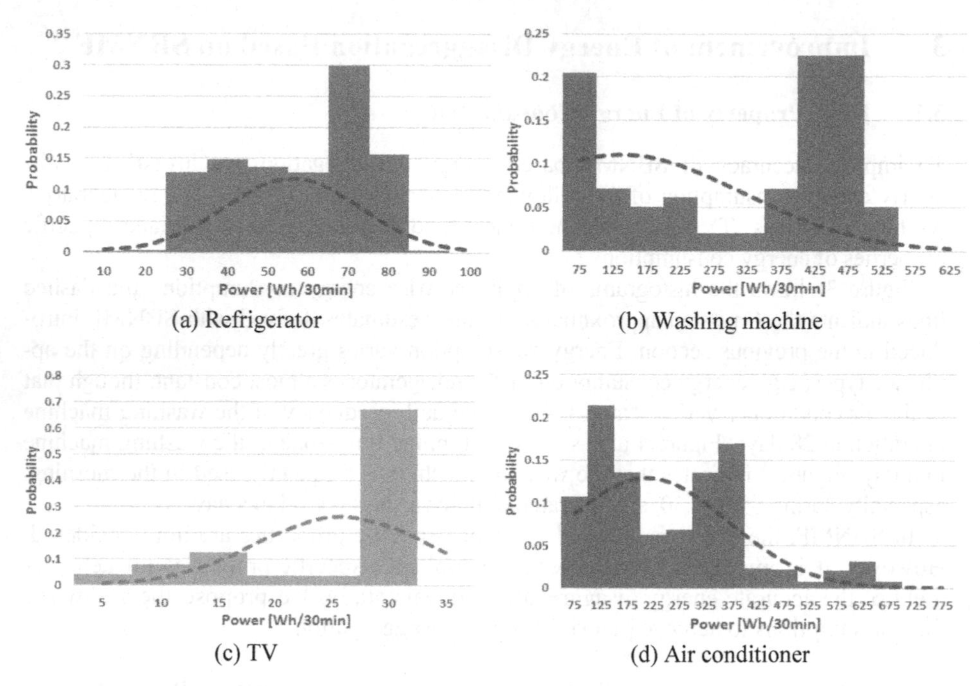

Fig. 3. Histograms and Gaussian distribution of appliance-wise energy consumption

Table 1. Operation frequency of washing machine in a day

Frequency of usage in a day	1	2	3
Days	12	13	3

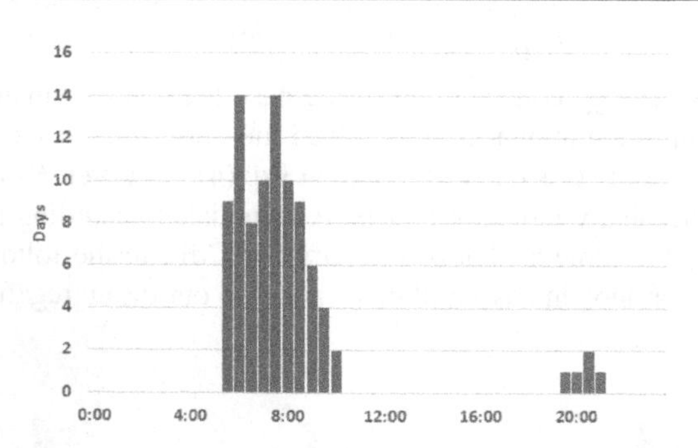

Fig. 4. Histogram of running time-zone of washing machine

3.2 Weighted SBNMF

We introduce the idea of weighted SBNMF to express appliance-specific variances of energy consumption. In this framework, we assume that the variance of the total energy consumption σ_{tn}^2 at time t in day n is proportional to the expected energy consumption $\sum_k x_{tk} a_{kn}$, i.e,

$$\sigma_{tn}^2 = \gamma \sum_k x_{tk} a_{kn}, \tag{9}$$

where γ is a positive coefficient which should be estimated. Equation (9) suggests that the total energy consumption can vary according to the running appliances. The optimization problem for disaggregation is slightly modified as

$$\begin{aligned} &\mathrm{argmax}_{\mathbf{X},\mathbf{A},\gamma} \sum_{t,n} \log p(y_{tn}; \sum_k x_{tk} a_{kn}, \ \gamma \sum_k x_{tk} a_{kn}) \\ &= \mathrm{argmin}_{\mathbf{X},\mathbf{A},\gamma} \left\{ \sum_{t,n} \log\left(\gamma \sum_k x_{tk} a_{kn}\right) + \sum_{t,n} \frac{(y_{tn} - \sum_k x_{tk} a_{kn})^2}{\gamma \sum_k x_{tk} a_{kn}} \right\}. \end{aligned} \tag{10}$$

We solve the minimization problem Eq. (10) by using an iterative optimization approach for estimate $\mathbf{X}, \mathbf{A}$, and γ. The update rule of matrix $\mathbf{X}$ is derived from the estimation procedure of the weighted NMF [10], which is given as

$$\mathbf{X} \leftarrow \mathbf{X} \circledast (\mathbf{W} \circledast \mathbf{Y})\mathbf{A}^{\mathrm{T}} \oslash \left(\mathbf{W} \circledast (\mathbf{X}\mathbf{A})\right)\mathbf{A}^{\mathrm{T}}, \tag{11}$$

where $\mathbf{W}$ is the variance weight matrix defined as:

$$\mathbf{W} = \left[w_{tn} = \frac{1}{\sqrt{\gamma \sum_k x_{tk} a_{kn}}} \right] \in \mathbb{R}_+^{T \times N}. \tag{12}$$

The binary indicator matrix $\mathbf{A}$ is updated according to the greedy random search exactly the same as that of ordinary SBNMF. The coefficient γ is numerically estimated on the basis of the quasi-Newton method under fixed $\mathbf{X}$ and $\mathbf{A}$. The update of weighted SBNMF is shown in Algorithm 2.

Algorithm 2. Energy Disaggregation based on Weighted SBNMF

Input: total energy consumption matrix $\mathbf{Y}$.
Initialize: $\mathbf{X}$ by random non-negative values, $\mathbf{A}$ by random binary values, and γ by random positive value.
Calculate $\mathbf{W}$ by using $\mathbf{X}, \mathbf{A}$, and γ on the basis of Eq. (12).
while γ is not converged **do**
  while $\sum_{t,n} \log(\gamma \sum_k x_{tk} a_{kn}) + \sum_{t,n} \frac{(y_{tn} - \sum_k x_{tk} a_{kn})^2}{\gamma \sum_k x_{tk} a_{kn}}$ is not converged **do**
    Update $\mathbf{X}$ according to Eq. (11).
    for each $n \in \{1, \ldots, N\}$
      update binary vector $[a_{kn}; k = 1, \ldots, K]$ to minimize the objective function
    end for
  end while
  Update γ on the basis of the quasi-Newton algorithm to minimize the objective function under current $\mathbf{X}$ and $\mathbf{A}$.
end while
Output: matrices $\hat{\mathbf{X}}$, and $\hat{\mathbf{A}}$.

Fig. 5. Algorithm of energy disaggregation based on Weighted SBNMF

3.3 Shift-invariant Weighted SBNMF

We focus on a specific appliance which is operated for a short period at the various timing in a day. In this framework, we prepare a short-time basis and consider to use it by admitting time shift. This method is called shift-invariant weighted SBNMF in this paper. Let vector $\mathbf{x}^0$ be the ancestral basis corresponding to the appliance running only for a period of P,

$$\mathbf{x}^0 = \left[x_1^0, \dots, x_P^0, \overbrace{0, \dots, 0}^{T-P}\right]^{\mathrm{T}}, \tag{13}$$

and $\mathbf{x}^m$ $(m \in 0, \dots, M)$ be the time-shifted vector of $\mathbf{x}^0$,

$$\mathbf{x}^m = \left[\overbrace{0, \dots, 0}^{m}, x_1^0, \dots, x_P^0, \overbrace{0, \dots, 0}^{T-m-P}\right]^{\mathrm{T}}, \tag{14}$$

where $M = T - P$. We compose the base matrix for this approach by using Eqs. (13) and (14) as

$$\mathbf{X}^0 = [\mathbf{x}^0, \mathbf{x}^1, \dots, \mathbf{x}^M] \left(\in \mathbb{R}_+^{T \times (M+1)}\right). \tag{15}$$

Algorithm 3. Energy Disaggregation based on Shift-invariant Weighted SBNMF

Input: total energy consumption matrix $\mathbf{Y}$.
Initialize: $\mathbf{X}$ and $\mathbf{A}$ by substituting estimation results of SBNMF, $\mathbf{X}^0$ by random non-negative values, $\mathbf{A}^0$ by random binary values, and γ by random non-negative value.
Calculate $\mathbf{W}$ by using $\mathbf{X}' = [\mathbf{X}, \mathbf{X}^0], \mathbf{A}' = [\mathbf{A}, \mathbf{A}^0]$ and γ on the basis of Eq. (12).
while γ is not converged **do**
  while $\sum_{t,n} \log(\gamma \sum_k x'_{tk} a'_{kn}) + \sum_{t,n} \frac{(y_{tn} - \sum_k x'_{tk} a'_{kn})^2}{\gamma \sum_k x'_{tk} a'_{kn}}$ is not converged **do**
    Update $\mathbf{X}'$ according to Eq. (19).
    for each $n \in \{1, \dots, N\}$
      update binary vector $\left[[a_{kn}; k = 1, \dots, K], [a_{mn}^0; m = 1, \dots, (M+1)]\right]^{\mathrm{T}}$ to minimize the objective function subject to $\sum_m a_{mn}^0 \leq 2$
    end for
  end while
  Update γ on the basis of the quasi-Newton algorithm to minimize the objective function under current $\mathbf{X}$ and $\mathbf{A}$.
end while
Output: matrices $\widehat{\mathbf{X}'}$ and $\widehat{\mathbf{A}}'$.

Fig. 6. Algorithm of energy disaggregation based on shift-invariant weighted SBNMF

Similarly, we let the binary matrix $\mathbf{A}^0 = [a_{mn}^0] \in \{0,1\}^{(M+1)\times N}$ be the usage indicator corresponding to $\mathbf{X}^0$. We also let $\mathbf{X}'$and $\mathbf{A}'$ be the complete basis and indicator matrices composed of ordinary matrices $\mathbf{X}, \mathbf{A}$ and the time-shifted matrices $\mathbf{X}^0, \mathbf{A}^0$ as follows,

$$\mathbf{X}' = [\mathbf{X}, \mathbf{X}^0] \in \mathbb{R}_+^{T\times(K+M+1)}, \tag{16}$$

$$\mathbf{A}' = \left[\mathbf{A}^{\mathrm{T}}, \mathbf{A}^{0^T}\right]^T \in \{0,1\}^{(K+M+1)\times N}. \tag{17}$$

In this framework, we consider the following approximation

$$\mathbf{Y} \cong \mathbf{X}'\mathbf{A}', \tag{18}$$

instead of Eq. (1). The estimation of matrices $\mathbf{X}'$and $\mathbf{A}'$ is achieved on the basis of the following iterative update procedure. The update rule of $\mathbf{X}'$ is given as follows,

$$\mathbf{X}' \leftarrow \mathbf{X}' \circledast (\mathbf{W} \circledast \mathbf{Y})\mathbf{A}'^{\mathrm{T}} \oslash \left(\mathbf{W} \circledast (\mathbf{X}'\mathbf{A}')\right)\mathbf{A}'^{\mathrm{T}}. \tag{19}$$

The elements of $\mathbf{X}^0$ are modified after updating $\mathbf{X}'$ as follows

$$x_p^0 = \frac{1}{M+1}\sum_{m=0}^{M} x_p^m \quad (p = 1, \ldots, P), \tag{20}$$

where x_p^m is an element of $\mathbf{X}^0$ which is updated according to Eq. (19). The whole procedure for estimation of shift-invariant weighted SBNMF is shown in Algorithm 3. We should stress that a typical appliance with a short running period is basically used only a few times in a day as shown in Table 1. Therefore, we assume the following heuristic constraint in the updating process for the time-shift basis $\mathbf{A}^0$,

$$\sum_m a_{mn}^0 \leq 2 \quad (\forall\, n). \tag{21}$$

4 Numerical Experiments

In this section, we experimentally shows the effectiveness of the proposed energy disaggregation methods by using real-world data sets acquired at nine houses every 30 and 15 min. The data sets are composed of the sum of energy consumption collected from the refrigerator, washing machine, TV, and air conditioner in the following period;

- Learning data: 6/29/2014 – 7/26/2014 (28 days)
- Evaluation data: 7/27/2014 – 8/3/2014 (7 days).

In this experiment, we compare the following four factorization methods,

- ◆ SBNMF (SB NMF(w/o base))
- ◆ SBNMF with base load (SBNMF)

- Weighted SBNMF with base load (Weighted SBNMF)
- Shift-invariant weighted SBNMF with base load (Shift-invariant weighted SBNMF)

Disaggregation results are evaluated from the viewpoint of the appliance-wise match rate defined in Eq. (7).

Figures 7 and 8 compare the disaggregation results of the nine houses in terms of the appliance-wise match rate based on 30 min and 15 min observations, respectively. Result of weighted SBNMF shows a great improvement by comparing it with that of SBNMF; especially in the case of the 15 min data sets. The results indicate that considering appliance-specific variance is effective for energy disaggregation; particularly shift-invariant weighted SBNMF shows the best result for most houses.

Figure 9 shows the averages of appliance-wise match rates based on 15 and 30 min data sets, respectively. In this figure, results of shift-invariant weighted SBNMF show the best among all methods. The results imply that the introduction of assumptions of appliance-specific variances and existence of an appliance with short running time works appropriately. Meanwhile, result of SBNMF shows little improvement; introduction of base load seems to have poor efficacy in energy disaggregation. The results imply that the refrigerator is easy to estimate since it is operated most of the time, and similarly the air conditioner is easy to estimate since it consumer prominently large energy as shown in figure 1. However, results of the TV and the washing machine show low match rates because these appliances have various operation pattern, show the experimental results indicate that the estimation of appliance-wise energy consumption for these appliances are notably difficult, but the proposed approach achieves to improve then.

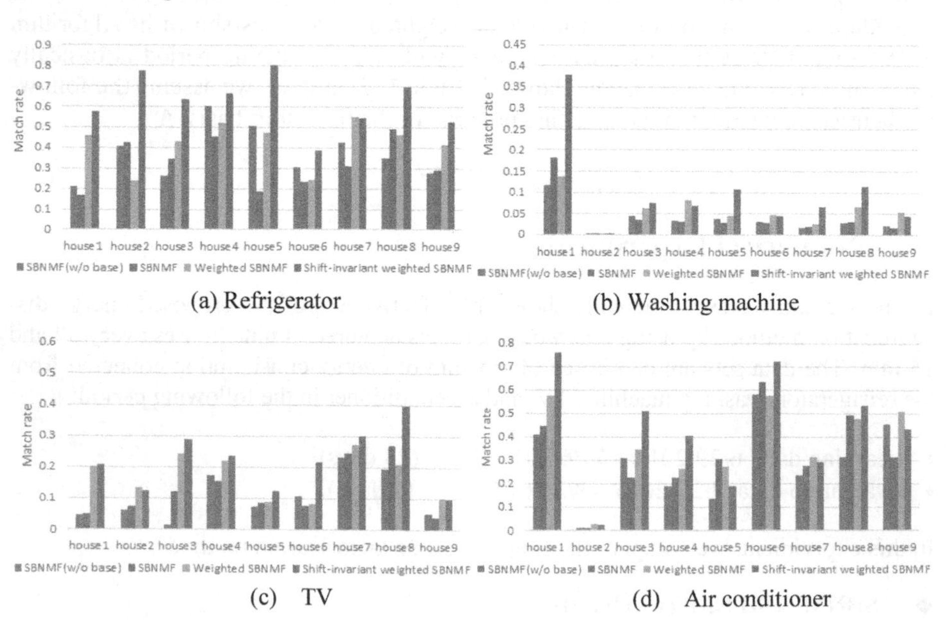

(a) Refrigerator (b) Washing machine

(c) TV (d) Air conditioner

Fig. 7. The match rates by using 30min data sets

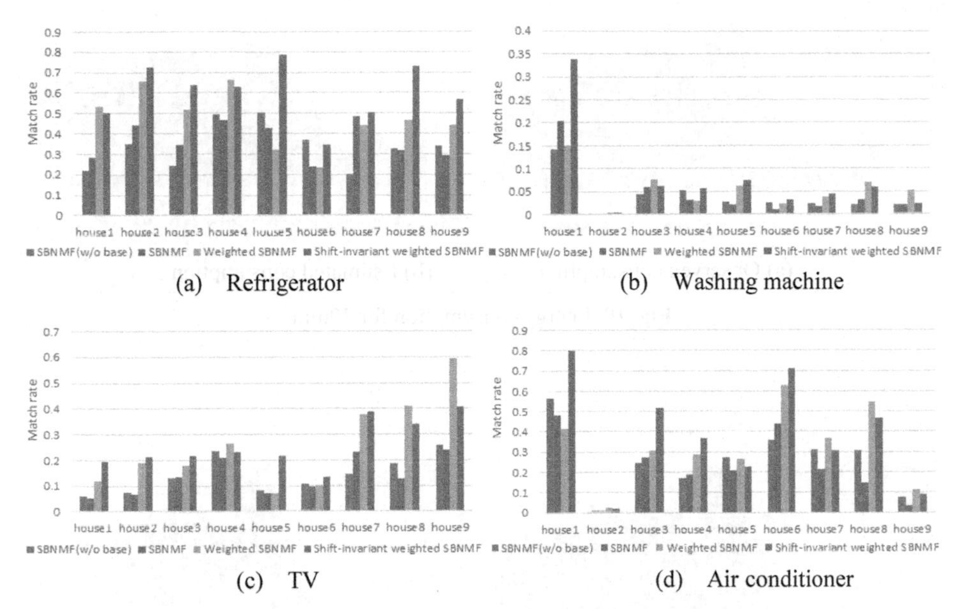

Fig. 8. The match rates by using 15 min data sets

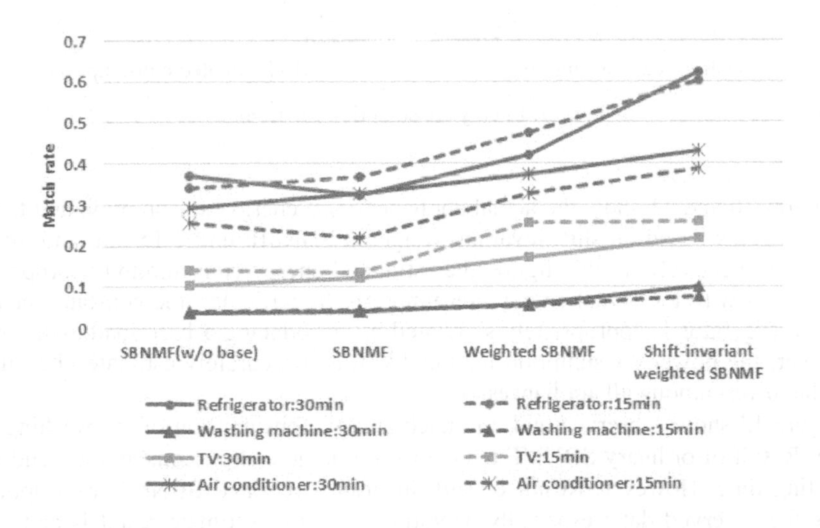

Fig. 9. Average appliance-wise match rates of nine houses

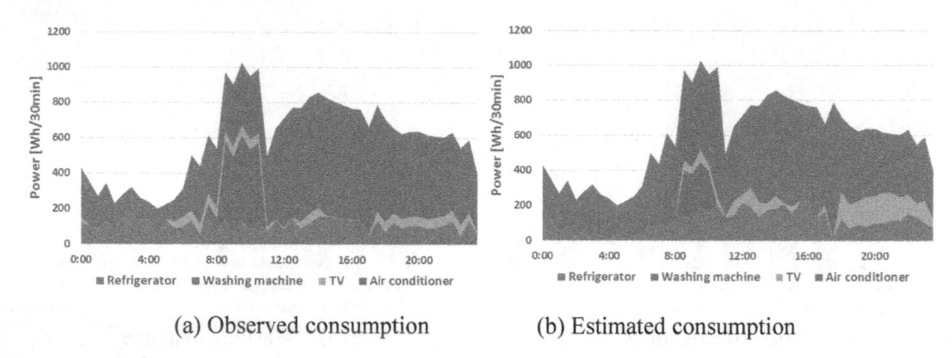

(a) Observed consumption (b) Estimated consumption

Fig. 10. Energy consumption for 30min

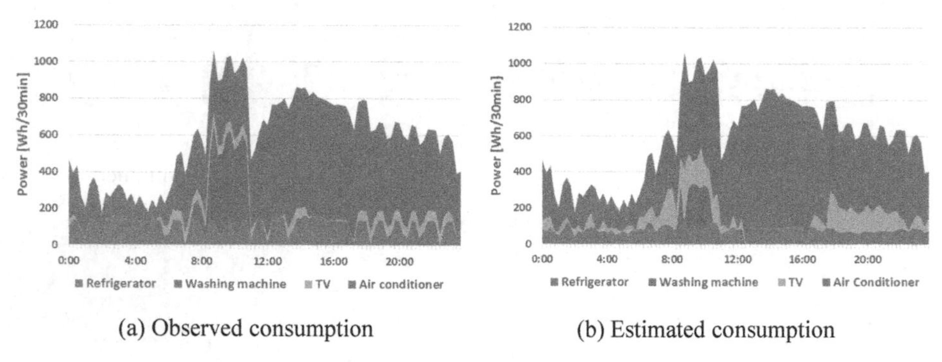

(a) Observed consumption (b) Estimated consumption

Fig. 11. Energy consumption for 15 min

Figures 10 and 11 show the actual appliance-wise energy consumption and the estimated results based on shift-invariant weighted SBNMF under 15 min and 30 min data set, respectively. In this figure, the estimated energy consumption appropriately corresponds with the actual energy consumption. In particular, the constant behavior of the refrigerator is appropriately simulated by introducing a base load assumption. However, the power consumption for the TV was inaccurately estimated because it was the lowest among all appliances.

Figure 12 shows observed and estimated energy consumption of the washing machine. Result of ordinary SBNMF tends to show large energy consumption and long operating time. However, Result of shift-invariant weighted SBNMF well approximates the observed data; especially operating period of estimate result is consistent with that of the observed.

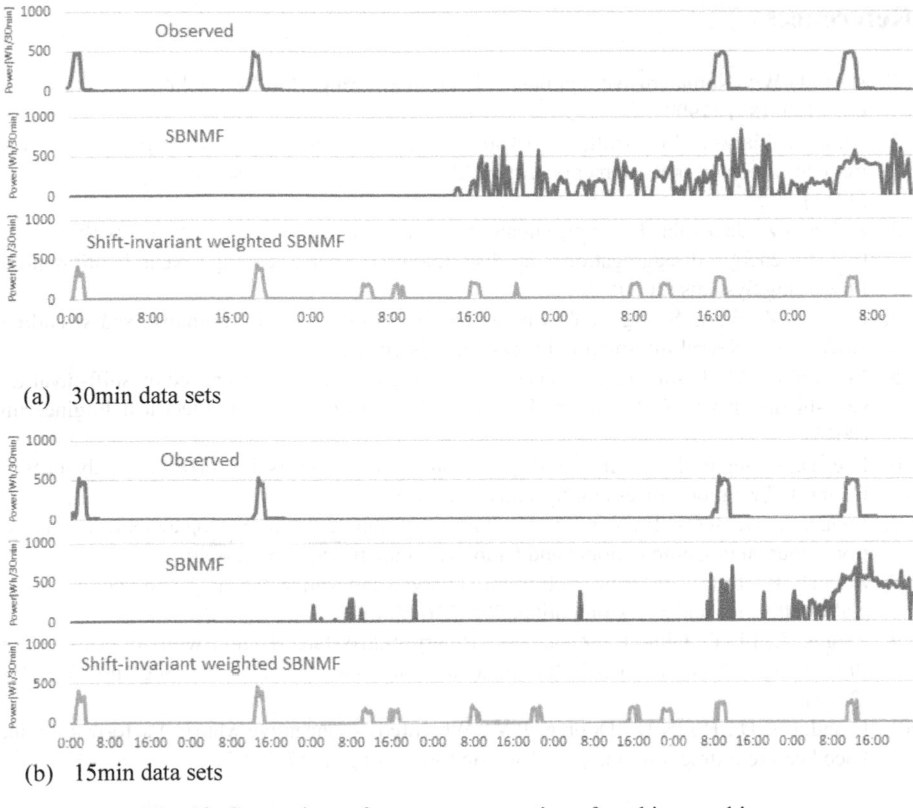

Fig. 12. Comparison of energy consumption of washing machine

5 Conclusion

In this paper, we focus on unsupervised energy disaggregation task and proposed a novel energy disaggregation method based on shift-invariant weighted SBNMF. We introduced the three assumptions for based on SBNMF. The experimental results show that the appliance-wise energy consumption derived by using the proposed framework improves disaggregation accuracy. The proposed method is expected to be a useful tool for disaggregation of various energy consumption patterns.

In this study, we focused only on four typical appliances in the experiments. In future work, we will experimentally evaluate our method for various energy consumption patterns composed of various appliances. Our experimental results suggest difficulty in energy disaggregation of appliances with low power consumption; further discussions are needed for more natural and appropriate constraints in disaggregation of low power consumption appliances.

Acknowledgements. We are deeply grateful to the staff members of Informetis Co., Ltd., and wish to thank them for providing real data and discussion about the evaluation index.

References

1. Hart, G.W.: Noninstrusive appliance load monitoring. In: Proceedings of the IEEE, pp. 1870–1891 (1992)
2. Kim, H., Marwah, M., Arlitt, M., Lyon, G., Han, J.: Unsupervised disaggregation of low frequency power measurements. In: SIAM Conference on Data Mining, pp. 747–758 (2011)
3. Kolter, J.Z., Jaakkola, T.: Approximate inference in additive factorial hmms with application to energy disaggregation via discriminative sparse coding. Neural Information Processing Systems (2010)
4. Kolter, J.Z., Batra, S., Ng, A.Y.: Energy Disaggregation via Discriminative sparse coding. Advances in Neural Information Processing Systems (2010)
5. Matsumoto, M., Fujimoto, Y., Hayashi, Y.: Energy disaggregation based on shift-invariant semi-binary matrix factorization. In: International Conference on Electrical Engineering (2015)
6. Lee, D.D., Seung, H.S.: Algorithms for Non-negative Matrix Factorization. Advances in Neural Information Processing Systems **13** (2000)
7. Lange, K., Hunter, D.R., Yang, I.: Optimization Transfer Using Surrogate Objective Functions. Journal of Computational and Graphical Statistics **9**, 1–59 (2000)
8. Zdunek, R.: Data Clustering with Semi-Binary Nonnegative Matrix Factorization. Artificial Intelligence and Soft Computing, 705–716 (2008)
9. Zhang, Z., Li, T., Ding, C., Zhang, X.: Binary matrix factorization with applications. In: Proceedings of the 7th IEEE International Conference on Data Mining, pp. 391–400 (2007)
10. Blondel, V.D., Ho, N.D., Dooren, P.V.: Weighted Nonnegative Matrix Factorization and Face Feature Extraction. Image and Vision Computing, 1–17 (2007)

A Probabilistic Matrix Factorization Method for Link Sign Prediction in Social Networks

Qiang You(✉), Ou Wu, Guan Luo, and Weiming Hu

CAS Center for Excellence in Brain Science and Intelligence Technology, National Laboratory of Pattern Recognition, Institute of Automation, Chinese Academy of Sciences, Beijing 100190, China
{qyou,wuou,gluo,wmhu}@nlpr.ia.ac.cn

Abstract. In this paper, we consider the link sign prediction in social networks with friend and foe relationships. We view the sign prediction as a user-to-user recommendation problem with trust or distrust information. Not only do we take the topological relationships such as the social structural balance and status theories into consideration, but also the social factors that whether a user is trustworthy and whether the user easily trust others are involved. We propose a probabilistic matrix factorization method with social trust and distrust ensembles and the structural theories from social psychology in order to predict link signs in social networks. The experimental results show that our proposed method outperforms those of the previous studies on this problem.

Keywords: Link sign prediction · Matrix factorization · Social psychology · Signed networks

1 Introduction

The majority of online social networks have only positive (i.e., friend, altruism, or trust) relationship which does not completely express the social interaction in real life. Nowadays, increasing social networks such as *Epinions* and *Slashdot* become to support both positive and negative relationships where people can form links to indicate friendship, approval or to distrust or be "foes" so as to express their disagreement. The authors in [1] investigate the signed networks which have two opposite relationships and connect their analysis to the theories of structural balance and status from social psychology. However, the edge sign prediction in [2] only considers the topological relationship between both the end nodes of an edge and their neighbors manly.

Actually, on one hand, if a user in a social network has been trusted or distrusted (e.g. labeled "friend" or "foe" online) by a large amount of users, then the user should be trustworthy or unreliable; on the other hand, if a user himself is easily believed in or suspicious of others, he may choose to trust or distrust other users online. Inspired by the two facts about trust and distrust in social networks, in this paper, we further study the trust prediction in signed social networks where the sign is defined to be positive or negative depending on whether it expresses trust or distrust from the generator to the recipient.

P. Perner (Ed.): MLDM 2016, LNAI 9729, pp. 415–420, 2016.
DOI: 10.1007/978-3-319-41920-6_32

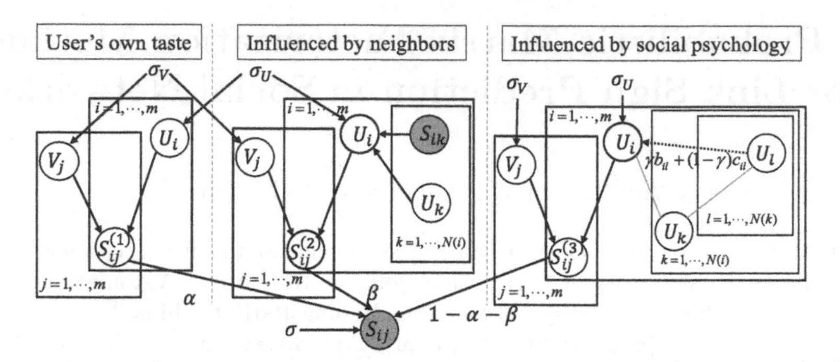

Fig. 1. Graphical model of the proposed method

2 The Proposed Method

Given a social network $G(N, E, S)$ where N stands for the set of web users. Every two web users can label the others as "friend" or "foe" to show their trust or distrust, which creates the set of edges E and the sign set $S \in \{-1, +1\}^{|E|}$. While, only part of the whole edges are effectively labeled and the rest are unknown. How can we get the labels of the rest edges in the social network with the existing partial labeled edges?

Different from the learning method in predicting only one of two states (trust or distrust) in signed social networks [2], we view the link prediction as a user-to-user recommendation problem with trust or distrust information. The process that a user u chooses to label "friend" or "foe" to another user v, is like whether a user recommends an "item" or not in a recommendation system. In order to learn the features of the users, we conduct the matrix factorization to factorize the user-user labeled matrix. The basic idea is to find a latent feature representation of column users (as users) U and a representation of row users (as "items") V according to the analysis of the labeled matrix. Suppose that in a signed social networks, we have m web users, and the label a user marked another can be either "friend" or "foe".

Unlike the previous studies of recommendation that the weight in the recommendation matrix can only be positive, the weight in our paper can be positive or negative, which shows the degree of trust or distrust between users. Given the user-user labeled matrix $S \in \{-1, +1\}^{m \times m}$, the observed label S_{ij} is interpreted by the user u_i's label on user u_j. Each user has his own taste about the others, and at the same time, the user may be influenced by his friends or foes. Meanwhile, the user may be also influenced by the circle of his friends or foes with respect to the social psychology about structural balance and status theories. To carefully tackle the three factors which influence the prediction problem, we define $N(i)(N(k))$ as the set of the neighbors of the user $u_i(u_k)$. S_{ik} represents the relationship the user u_i labels his neighbor u_k. b_{il} and c_{il} represent the balance and status factors from the social structural balance and status theories. γ weights the two factors which influence the trust prediction. The parameters α

and β control how much do users trust themselves, their neighbors or their neighbors' neighbors from the theory of social psychology. Accordingly, the average value of the latent feature is defined as

$$\begin{aligned}\bar{u} = \alpha U_i^T V_j + \beta \sum_{k \in N(i)} S_{ik} U_k^T V_j \\ + (1 - \alpha - \beta) \sum_{k \in N(i)} \sum_{l \in N(k)} (\gamma b_{il} + (1 - \gamma) c_{il}) U_l^T V_j\end{aligned} \tag{1}$$

To describe concisely, let $\pi = (b, c, \sigma_U, \sigma_V, \sigma)$ represent part of the hyper parameters. The conditional distribution over the observed labels is defined as

$$p(S|U, V, \pi) = \prod_{i=1}^{m} \prod_{j=1}^{m} [\mathcal{N}(S_{ij}|\bar{u}, \sigma)]^{I_{ij}^S} \tag{2}$$

where I_{ij}^S is the indicator function that is equal to 1 if u_i labeled u_j and 0 otherwise. The graphical model is shown in Fig. 1. Based on the Bayesian inference, the conditional distribution of the latent features is represented as

$$p(U, V|S, \pi) \propto p(S|U, V, \pi) p(U|b, c, \sigma_U) p(V|b, c, \sigma_V) \tag{3}$$

To make it simply discussed, the social structural parameters b and c are assumed independent to U, V. Thus Eq. 3 can be changed to

$$\begin{aligned}p(U, V|S, \pi) \propto p(S|U, V, \pi) p(U|\sigma_U) p(V|\sigma_V) \\ = \prod_{i=1}^{m} \prod_{j=1}^{m} [\mathcal{N}(S_{ij}|\bar{u}, \sigma)]^{I_{ij}^S} \times \mathcal{N}(U|\mathbf{0}, \sigma_U I) \times \mathcal{N}(V|\mathbf{0}, \sigma_V I)\end{aligned} \tag{4}$$

Maximizing the log-posterior over the latent features with hyper parameters kept fixed is equivalent to minimizing the sum-of-squared-errors objective function with quadratic regularization terms as follows:

$$\mathcal{L}(S, U, V, \bar{u}) = \frac{1}{2} \sum_{i=1}^{m} \sum_{j=1}^{m} I_{ij}^S (S_{ij} - \bar{u}) + \frac{\lambda_U}{2} \|U\|_F + \frac{\lambda_V}{2} \|V\|_F \tag{5}$$

where $\lambda_U = \sigma^2 / \sigma_U^2$ and $\lambda_V = \sigma^2 / \sigma_V^2$. To reduce the model complexity, we set $\lambda_U = \lambda_V$ in our experimental settings. We perform the gradient descent on U and V until the local minimum of the object function is reached using the Lagrange multiplier method so as to inference the latent features U and V.

The probabilistic matrix factorization method described above can be easily extended to a more detailed social networks with different levels of relationships other than the two opposite relationships without many changes.

3 Experimental Results

We collect two large online social networks where each edge is explicitly labeled as positive or negative: *Epinions* and *Slashdot*[1]. The edge signs in the online social

[1] Datasets are available at http://snap.stanford.edu

Table 1. The comparison of the best results on original and balanced data sets

Dataset	Random	16Triads	All23	Ours
Epinions	85.30%	92.02%	92.54%	**94.06%**
Slashdot	77.40%	88.95%	89.95%	**91.28%**
Epinions(balanced)	50.00%	90.26%	91.56%	**91.94%**
Slashdot(balanced)	50.00%	85.78%	89.02%	**89.10%**

networks which we researched are mostly positive. Referencing the methodology in [2], we create two approaches when using the data set. First, we conduct our experiments on the full data set where about 80% edge signs are positive. Second, we create a balanced data set where all the negative edges are used, and the equal number of positive edges are randomly sampled to use. The balance and status factors $b, c \in \{-1, 0, +1\}$ are specified as in [2]. The hyper parameter γ weights the balance and status factors; we set $\gamma = 0.8$ for simplicity, which means the balance factor is much more significant than the status factor in our consideration. The hyper parameter α, β should be learned according to the users' personal interest in real applications. To make it simple, we just set $\alpha = 0.6, \beta = 0.3$ though cross-validation on our data sets which are lack of users' information. The random guess is as the baseline on our balanced data sets. As the Table 1 shows, the result of our method outperforms the work in [2]. Especially on the original unbalanced data sets, our method makes an explicit improvement than the binary logistic classification method in [2] whatever choosing all the 23 features (All23) or the triangle features (16Triads) from the structural theories from social psychology. However, based on the experimental result on balanced data sets, we only make a small progress in the trust and distrust prediction problem. There are two main reasons to explain the small improvement. One is that the data sets are actually lack of the probability of the users' trust or distrust to the others whatever in the parameter learning of probabilistic matrix factorization or testing for the final result. The other is that the criterion cannot effectively measure the performance of our probabilistic matrix factorization method because the output is only right or wrong which doesn't support the probabilities for different grades.

In the following simulation experiments, we add the uncertainty to the original data sets to simulate the belief of the labeling process when users make their decisions. We assume that the uncertainty of users' belief follows the uniform distribution $[0, \tau]$ and add it to the edge sign prediction, where τ controls the threshold of the maximum uncertainty. In our experimental settings, We set $\tau \leq 0.5$ to assure that the added uncertainty doesn't hurt the original decision making. We choose the root mean square error (RMSE) as the final criterion which better measures the different levels of trust or distrust prediction. The performance is better when RMSE is lower. To fairly measure the learning method in [2] and our probabilistic matrix factorization method, we select

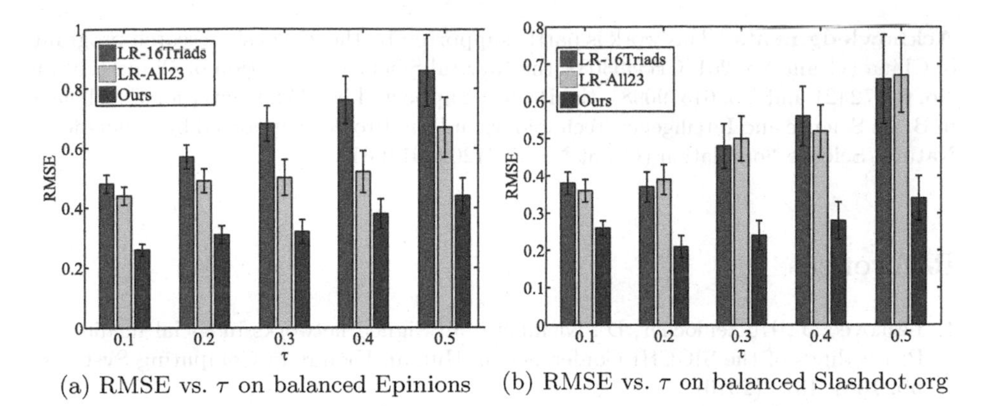

(a) RMSE vs. τ on balanced Epinions (b) RMSE vs. τ on balanced Slashdot.org

Fig. 2. The comparison between our method and learning methods in [2]

logistic regression method as the learning method instead of the binary logistic classification method in [2] with its outputs ranging from -1 to 1. All the hyper parameter settings are equivalent to the previous experiment settings. We get the simulation experimental results as shown in Fig. 2. From the comparison of our method and the logistic regression in Fig. 2, we can get the two basic yet useful points. One is that the probabilistic matrix factorization fusing the user-user relationships performs much better than the linear regression methods with the features studied in previous work about signed social networks. The other is that our method is stabler than the previous work when the uncertainty increases. This means that our method can better handle the trust prediction when the web users are hard to choose to trust or distrust other users in social networks with multiple relationships.

4 Conclusions and Future Work

In this paper, we have studied the link prediction in social networks with friend and foe relationships. We have changed the trust prediction to a user-to-user recommendation problem with trust or distrust information. To solve this problem, We have proposed a matrix factorization method with social trust and distrust ensembles and the structural balance and status theories from social psychology. The experimental results have demonstrated the effectiveness of the proposed method. However, there is a significant issue not involved in our paper. The influencer parameters such as α, β should customized by different users according to their interests, relationships, etc, which will be the key point we should consider in our future work.

Acknowledgments. This work is partly supported by the 973 basic research program of China (Grant No. 2014CB349303), the Natural Science Foundation of China (Grant No. 61472421 and No. 61379098), the Project Supported by CAS Center for Excellence in Brain Science and Intelligence Technology, and the Project Supported by Guangdong Natural Science Foundation (Grant No. S2012020011081).

References

1. Leskovec, J., Huttenlocher, D., Kleinberg, J.: Signed networks in social media. In: Proceedings of the SIGCHI Conference on Human Factors in Computing Systems, pp. 1361–1370 (2010)
2. Leskovec, J., Huttenlocher, D., Kleinberg, J.: Predicting positive and negative links in online social networks. In: Proceedings of the 19th International Conference on World Wide Web, pp. 641–650 (2010)

Metadata-Based Clustered Multi-task Learning for Thread Mining in Web Communities

Qiang You(✉), Ou Wu, Guan Luo, and Weiming Hu

CAS Center for Excellence in Brain Science and Intelligence Technology, National Laboratory of Pattern Recognition, Institute of Automation, Chinese Academy of Sciences, Beijing 100190, China
{qyou,wuou,gluo,wmhu}@nlpr.ia.ac.cn

Abstract. With user-generated content explosively growing, how to find valuable posts from discussion threads in web communities becomes a hot topic. Although many learning algorithms have been proposed for mining the thread contents, there are still two problems that are not effectively considered. First, the learning algorithms are usually complicated so as to deal with various kinds of threads in web communities, which damages the generalization performance of the algorithms and takes the risk of overfitting to the learning models. Second, the small sample size problem exists when the training data for learning is divided into many isolated groups and each group is trained separately in order to avoid overfitting. In this paper, we propose a metadata-based clustered multi-task learning method, which takes full use of the metadata of threads and fuses it in the multi-task learning based on a divide-and-learn strategy. Our method provides an effective solution to the above problems by finding the geometric structure or context of semantics of threads in web communities and constructing the relations among training thread groups and their corresponding learning tasks. In addition, a soft-assigned clustered multi-task learning model is employed. Our experimental results show the effectiveness of our method.

Keywords: Metadata · Thread mining · Divide-and-learn · Clustered multi-task learning · Web community

1 Introduction

With the rapid development of the Internet, more and more people would like to participate in the discussions in web communities. As a result, a large amount of user-generated content (UGC) has been accumulating, which becomes urgent to analyze so as to find useful information for decision making in different kinds of areas such as viral marketing, industry research, etc. Throughout the past decade there have been many researches on how to find valuable posts in discussion threads in web communities. The previous researches are mainly classified into content-based or structure-based. The formal method takes the posts in each

P. Perner (Ed.): MLDM 2016, LNAI 9729, pp. 421–434, 2016.
DOI: 10.1007/978-3-319-41920-6_33

discussion thread as the document set and follows the pattern of text classification [1]. In the area of text classification, probabilistic topic models [2,3] have been proved to be effective in the extraction of semantics and document summarization when the corpus to be analyzed is sufficient. However, the posts in web communities are always short and sparse, which makes the result of text classification unsatisfactory. While the structure-based method goes another way, it ignores the semantics of the content and only concentrate on the structure of a web community. Considering the reply-to graph of the posts in web communities, many random-walk-based algorithms are available for measuring the importance of web pages such as HITS [4], PageRank [5] and their successive approaches are introduced to the valuable post finding. However, the reply-to graph of the posts is not explicit or hard to extract in many web communities. What is more, the posts without link-in or link-out are common in web communities, which is not applicable in most of the random-walk-based algorithms.

In this paper, we combine the content-based method and structure-based method together by the concept of metadata. We introduce the metadata into the mining tasks in web communities. If the reply-to graph of the web community apparently exists (e.g. *Slashdot.org*[1]), it is viewed as a kind of metadata. While the reply-to graph doesn't explicitly appear (e.g. many Q&A discussion forums), we reconstruct it through semantic similarity measure. Some data that shows the quality or characteristics of the data set for learning tasks can be viewed as metadata in our consideration. Different from some previous studies, we do not take the metadata just as a kind of data directly for learning and add to the learning tasks similarly to the other data. The reason is that the distribution of the metadata is different from that of the data for learning tasks and the metadata is also not independent from the data. If we simply add it and combine with the data to the learning tasks, the performance may be degraded seriously. In this paper, we conduct a *divide-and-learn* strategy. Specifically, In a web community, the different discussion threads are not isolated from each other because the users often make discussions around several central topics. Assuming that the discussion threads are clustered according to several topics in a web community, in the *dividing* step, we model the metadata of each thread as an attributed graph, and divide all the metadata attributed graphs into several groups. In the *learning* step, we propose a metadata-based clustered multi-task learning algorithm, which takes full use of the metadata and fuses it to the multi-task learning framework. The aim is that each task may benefit from each other by an appropriate sharing of information across different tasks in the framework of multi-task learning, which may significantly reduce the risk of overfitting if we develop our learning model with respect to the adaptive data.

The remainder of this paper is organized as follows. Section 2 briefly reviews related work. Section 3 presents the characteristics of the metadata in web communities, and Section 4 shows the formation of multiple tasks based on the metadata clustering. Section 5 describes the clustered multi-task learning frame-

[1] http://slashdot.org

work in detail. Experimental results are presented in Section 6, followed by the conclusion in Section 7.

2 Related Work

Many researchers have studied the mining tasks such as finding valuable posts or domain experts in web communities from the perspective of semantic understanding of the discussion threads and posts. As for the semantic models, Cong et al. [6] aimed at ranking answers for given questions in web forums. References such as [7] and [8] reconstructed the relationship among posts and threads based on the similarity of topics and semantics. Lin et al. [9] proposed a combination approach for simultaneously modeling semantics and structure of threaded discussions, which was used for junk detection and expert finding. The researches listed above almost all consider separated learning tasks in the whole feature space. Regardless of semantic reconstruction ([7,8]) or achievement in the optimization algorithm with respect to the relation of posts ([9]), there are still two problems commonly existing in mining tasks in web communities. First, the dimension of the whole feature space is high. An effective strategy is to partition the space into sub-regions and reduce the dimension according to the different characteristics of the data. Second, in spite of the large amount of data in the whole web, the data samples for a mining task in a web community is sparse and insufficient. Finally, we review a few previous studies involved in solving the two problems generally.

Several previous studies have implicated the concept of metadata. Researches in [10,11] extracted a large number of quality measures from the biometric traits. With the help of the quality information derived from the data, a unified framework for biometric expert fusion was constructed. The quality measures in biometric authentication can be treated as a kind of metadata in our consideration. Another kind of metadata describes the geometric characteristic of the original data. The quality measure is also adopted to web data classification [12]. In [13,14], a learning model with mixing linear SVMs was proposed to handle the problem of nonlinear classification, and to promote the efficiency while still providing a classification performance comparable to non-linear SVM. Based on the local linearly separable characteristic of the data set, the feature space can be partitioned into sub-regions. As a result, the learning model with mixing linear SVMs is available for nonlinear classification. However, the strategy that simply partitions the data set into subgroups according to the metadata and learns different models with respect to different groups largely ignores the connectivity of each group. Especially in web communities with discussion groups, the central topics are never completely isolated. What is more, the data samples for semantic analysis are more insufficient if we divide them into pieces.

Providing the sparsity and shortage of data samples in multiple related classification tasks under some circumstances, there is a growing interest in multi-task learning (MTL), where multiple related tasks are learned simultaneously by extracting appropriate shared information across tasks. The effectiveness of

MTL has been verified theoretically in researches such as [15–17]. Several methods have been proposed based on how the relatedness of different tasks is modeled. Mean-regularized MTL [18] was proposed under the assumption that the parameter vectors of all tasks are close to each other, which is simple but not hold in real applications such as mining in different topics of discussion threads in web communities. By sharing a different kind of underlying structure among multiple tasks, the relatedness can also be modeled as clusters [19,20], tree [21] or network [22,23]. In practical applications, the tasks may suggest a more sophisticated group structure where the models of tasks from the same group are closer to each other than those from a different group. There have been many researches involved in this line of research, known as clustered multi-task learning (CMTL). Bakker and Heskes [20] adopted a Bayesian approach by considering a mixture of Gaussians instead of single Gaussian priori to realize the clustered multi-task learning. Xu et al. [24] identified subgroups of related tasks using the Dirichlet process prior. Jacob et al. [25] proposed a clustered MTL framework that simultaneously identified clusters and performed multi-task inference. Given that the formulation is non-convex, they introduced a convex relaxation to the original formulation. Zhou et al. [26] found the equivalence between alternating structure optimization and CMTL formulation. They also relaxed the problem and solved it though alternating optimization method and other two gradient optimization algorithms [26]. While the previous researches of CMTL all assume the tasks are clustered into isolated groups, in this paper, we extend and propose a soft assigned CMTL in order to study the semantic context of different task groups.

3 The Metadata

The metadata has been widely used in search engine techniques where the web crawler can easily get the characteristics of the web page such as *charset*, *encoding*, *key words* and other descriptive information without crawling the whole page. Similarly, it is introduced here to show the schema of the data set to be analyzed. Let us take the popular technology-related news web community *Slashdot.org* as an example. *Slashdot.org* is a typical web community organized with threads constituted by posts which are scored by users where the score can be seen the value of the post.

There are mainly two methods to conduct the mining task. As Fig. 1a shows, with the content of the post and the score as its label, we can learn a model with each thread without much difficulty. However, we may face a small sample size problem because the posts in a thread is insufficient. On the contrary, when we take all the posts into learning without separating the posts according to the thread as shown in Fig. 1b, and use vector space models such as term frequency as feature description, the dimension of the feature space is really high and nearly all the posts are sparse, which damages the performance of the learning model. What is more, the learning process is also time-consuming. Rather than dimension reduction via feature selection by different kinds of rules, we extract the metadata to describe the characteristics of the discussion threads. The metadata of the thread in a web community consists two aspects: one is the structure

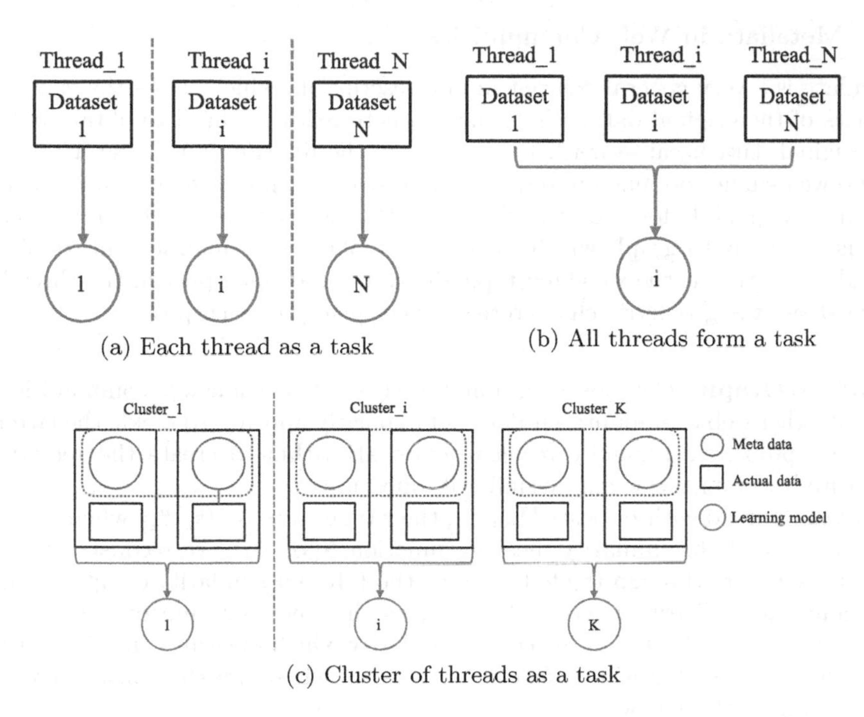

(a) Each thread as a task

(b) All threads form a task

(c) Cluster of threads as a task

Fig. 1. The three methods for learning in a typical web community

of the threads modeled as a reply-to graph, the other is the topic distribution of the posts in the thread. The formal shows the context of the related posts, while the latter suggests the geometric characteristics and quality of themselves. With respect to the two aspects of the metadata of the thread in the web community, we model it as an attributed graph.

We assume that a web community is constituted by N threads, and the i-th thread is represented as a directed graph $G_i(V, E)$. The node $v \in V$ is associated with a post. There is an attribute vector $\mathbf{a}_{iv}$ described the characteristic of the post v in the i-th thread. Inspired by [27,28], we design a clustering algorithm with respect to the metadata as shown in Fig. 1c. Our algorithm is based on the assumption that rather than isolated from each other, the threads are clustered according to several central topics.

3.1 Metadata Modeling

As mentioned above, there are N threads in the web community. The metadata of the i-th thread can be modeled as an attributed graph $G_i(V, E, A)$ where V is the set of nodes, E the set of edges, and A the set of m attributes with nodes in V for describing node properties. The attribute vector is represented as $\mathbf{a}_{iv} = [a_1(v), ..., a_j(v), ..., a_n(v)]$ where $a_j(v)$ is the attribute value of node v on attribute a_j.

3.2 Metadata in Web Communities

Metadata is a very general concept in our description, which shows the characteristics of the original data. While the characteristics are hard to obtain from the original data because neither do we know the distribution of the data set nor do we assume too much in order to avoid decreasing the generalization performance, we mainly focus on two aspects of the metadata in a web community. One is the reply-to graph which we think supplies the semantic context of a thread. The other is the weighted topic distribution of each post in the thread, which shows the geometric characteristics and quality of each post.

Reply-To Graph. The reply-to graph directly exists in some web communities, while in other web communities it does not explicitly appear. To tackle the latter issue, we propose a semantic reconstruction algorithm to create the reply-to structure based on the semantic similarity measure.

Given a thread with m posts $\{\mathbf{L}_i\}_{i=1}^m$, their time stamps $\{ts_i\}_{i=1}^m$ where $ts_i < ts_j$ if $i < j$ and the similarity measure function $S(\mathbf{L}_i, \mathbf{L}_j)$, we reconstruct the reply-to structure through the following method. In our similarity computation, we define the similarity measure function as the weighted sum of two parts: The first part $S_{\cos}(\mathbf{L}_i, \mathbf{L}_j)$ is the cosine similarity which measures the similarity between the directions of two feature vectors, which shows the consistency of the semantics between two posts.

$$S_{\cos}(\mathbf{L}_i, \mathbf{L}_j) = \frac{\mathbf{L}_i \mathbf{L}_j^T}{2\|\mathbf{L}_i\| \, \|\mathbf{L}_j\|} \tag{1}$$

The second part $S_{str}(\mathbf{L}_i, \mathbf{L}_j)$ is the similarity between two posts with respect to the strength of the semantics.

$$S_{str}(\mathbf{L}_i, \mathbf{L}_j) = \frac{\|\mathbf{L}_i\| \, \|\mathbf{L}_j\|}{\|\mathbf{L}_i\|^2 + \|\mathbf{L}_j\|^2} \tag{2}$$

The parameter λ here weights the two parts. Now we get the whole similarity function

$$S(\mathbf{L}_i, \mathbf{L}_j) = \lambda S_{\cos}(\mathbf{L}_i, \mathbf{L}_j) + (1 - \lambda) S_{str}(\mathbf{L}_i, \mathbf{L}_j) \tag{3}$$

As for post $\mathbf{L}_j$, we choose one post as its predecessor from the ahead posts. The predecessor should have the most similarity with $\mathbf{L}_j$. We choose the predecessor of $\mathbf{L}_j$ according to the following maximum problem, where $\mathbf{L}_*$ is the most suitable predecessor.

$$\mathbf{L}_* = \arg_{\mathbf{L}_i} \max_{1 \leq i \leq j-1} S(\mathbf{L}_i, \mathbf{L}_j) \tag{4}$$

Let j decrement from m to 2, then the reply-to structure is reconstructed.

Weighted Topic Distribution. We deem each thread with many posts in a web community as a document with many paragraphs and thus the whole community with many threads can be seen as a document set. After that, we apply the latent Dirichlet allocation [3] algorithm to the document set and extract n hottest topics. In every post of a thread, we can map each word in the post to one of the n hottest topics with a relevance weight. Therefore, the n dimensional weighted topic distribution of every post is extracted, which is the attribute vector in the attributed graph representing the metadata.

4 Formation of Multiple Tasks

In the following, we first define a metric to measure the distance between attributed graphs. Based on the similarity measures of each graph, we can cluster the metadata into different groups. Consequently, the multiple tasks are naturally created with respect to the different groups.

4.1 The Similarity Measure

As aforementioned, the metadata of the i, j-th thread is modeled as an attributed graph $G_i(V, E, A), G_j(V', E', A')$. Their directed product $G_\times$ is a graph with vertex set $V_\times = \{(v_r, v_r') : v_r \in V, v_r' \in V'\}$ and edge set $E_\times = \{((v_r, v_r'), (v_s, v_s')) : (v_r, v_s) \in E \wedge (v_r', v_s') \in E'\}$. The k-th order subgraph of the graph G_i is defined as $G_i^k(V^k, E^k)$ where $V^k \subseteq V, E^k \subseteq E$ and $|V^k| = k \leq |V|$. As we only want to extract the semantic context similarity of the attributed graph without comparing the two whole graphs, unlike the time-consuming calculation of the similarity is conducted from the 1-st order to the full order subgraphs [28], we only calculate the similarity between the second order subgraphs (edges). The similarity measure between G_i and G_j can be defined as the graph kernel:

$$k(G_i, G_j) = \frac{\sum_{e \in E} \sum_{e' \in E'} k(e, e')}{|E_\times|} \tag{5}$$

Assume $e = (v_r, v_s), e' = (v_r', v_s')$ and the node v_r with the attribute $\mathbf{a}_r$, then $k(e, e')$ is defined as the attribute similarity

$$k(e, e') = \phi(\mathbf{a}_r, \mathbf{a}_r') \times \phi(\mathbf{a}_s, \mathbf{a}_s') \tag{6}$$

where in our calculation, we use the Gaussian similarity function defined as follows:

$$\phi(\mathbf{x}, \mathbf{y}) = \exp(-\gamma \|\mathbf{x} - \mathbf{y}\|^2) \tag{7}$$

where γ is a scalar parameter determined the width of the Gaussian kernel. In the following metadata clustering algorithm, we set $\gamma = 1$ for simplicity.

4.2 The Metadata Clustering

Given the metadata set for the whole N threads in a web community, the similarity matrix is $S \in \mathcal{R}^{N \times N}$ where the element $s_{ij} = k(G_i, G_j)$. The diagonal matrix is D where $d_i = \sum_{j=1}^{m} s_{ij}$. Accordingly, the graph Laplacian matrix is represented as $L = D - S$. Once we get the graph Laplacian matrix, referring to the spectral clustering algorithm [29] and the efficient clustering method based on the data fragments [30], we can easily cluster the metadata set into several groups.

5 The Learning Algorithm

The multiple learning tasks are automatically constructed after we cluster the metadata into several groups. Not only the semantic context of the feature space is considered, we also want to study the semantic context of the task graphs. Unlike the previous studies in CMTL, we think the learning tasks are softly clustered in groups instead of independently assigned to each group. We propose the softly clustered multi-task learning (sCMTL) algorithm with Gaussian mixture models.

Given K (clusters of the metadata) learning tasks $\{T_i\}_{i=1}^{K}$, for the i-th task T_i with its feature space $\mathbf{F}_i \subseteq \mathbb{R}^{d_i}$ where d_i is the dictionary dimension of the $i-th$ thread, the training set consists of n_i sample points $\{(\mathbf{x}_j^i, y_j^i)\}_{j=1}^{n_i}$, with $\mathbf{x}_j^i \in \mathbf{F}_i$ and its corresponding output $y_j^i \in \mathbb{R}$ if it is a regression problem. The linear function for T_i is defined as $f_i(\mathbf{x}) = \mathbf{w}_i^T \mathbf{x} + b_i$. The loss function is defined as $l(f(\mathbf{x}), y) = (f(\mathbf{x}) - y)^2$. The basic model is to find an optimal value of $W = \{(\mathbf{w}_i, b_i)\}_{i=1}^{m}$ through minimizing the loss function

$$\mathcal{L}(W) = \frac{1}{K} \sum\nolimits_{i=1}^{K} \sum\nolimits_{j=1}^{n_i} l(f(\mathbf{x}_j^i), y_j^i) \tag{8}$$

As for Eq. (8), there is nothing much different from the single-task learning for K tasks respectively. In order to learn the K tasks simultaneously, we follow the regularized form to minimize the empirical risk where the regularized part $\Omega(W)$ can be designed from priori knowledge to constrain some sharing of information between tasks. The learning framework can be represented as the minimum optimization problem with respect to the learning weight matrix W.

$$\min_{W} \mathcal{L}(W) + \lambda \Omega(W) \tag{9}$$

The whole regularization Ω can be divided into two partial regularization parts, namely, the clustered regularization part $\mathcal{C}$ and the parameter penalty part. Now we have

$$\Omega(W) = \alpha \mathcal{C}(W) + \beta [tr(W^T W)] \tag{10}$$

Suppose that the i-th task with learning weight $\mathbf{w}_i$ can be assigned to the j-th task cluster with the probability p_{ij} and there are $k \leq K$ task clusters, the clustered regularization part can be written as follows

$$\begin{gathered} \mathcal{C}(W) = \sum\nolimits_{j=1}^{k} \sum\nolimits_{i=1}^{m} p_{ij} \left\| \mathbf{w}_i - \overline{\mathbf{w}}_j \right\|_2^2 \\ s.t. \sum\nolimits_{j=1}^{k} p_{ij} = 1; i = 1, ..., K \end{gathered} \tag{11}$$

The softly specified probability matrix is simply written as $\mathcal{P} \in [0,1]^{K \times k}$ as the element p_{ij} is the assigned probability. The clustered regularization part can be simplified as the following form

$$\begin{aligned} \mathcal{C}(W) &= tr(W^T W) - tr(\mathcal{P}^T W^T W \mathcal{P}) \\ s.t. &\sum\nolimits_{j=1}^{k} p_{ij} = 1; i = 1, ..., K \end{aligned} \tag{12}$$

where $tr(.)$ is the trace of a matrix. To learn both the soft assigned matrix $\mathcal{P}$ and learning weight matrix W, the whole regularization Ω can be written as:

$$\Omega(W, \mathcal{P}) = \alpha[tr(W^T W) - tr(\mathcal{P}^T W^T W \mathcal{P})] + \beta[tr(W^T W)] \tag{13}$$

Let $\eta = \beta/\alpha > 0$. Since $tr(W^T W) = tr(W W^T)$,

$$\begin{aligned} \Omega(W, \mathcal{P}) &= \alpha \left((1+\eta) tr(W^T W) - tr(\mathcal{P}^T W^T W \mathcal{P})\right) \\ &= \alpha \left((tr(W^T \left((1+\eta)I - \mathcal{P}\mathcal{P}^T\right) W)\right) \end{aligned} \tag{14}$$

As for $\mathcal{P}$, it is concave, and the formulation in Eq. (14) is non-convex. We conduct an alternating optimization method to inference the parameters. If the softly specified matrix $\mathcal{P}$ is fixed, Eq. (14) is convex with respect to W. It can be solved using gradient methods. After we find the optimal W^* to minimize the loss function with the whole penalty regularization, we simply fix W^*, and recompute the soft assigned matrix $\mathcal{P}$ with the Gaussian mixture models. We repeat the alternating optimization procedure until the constraints (e.g. the constraint steps, or the minimal error rate) achieved.

6 Experimental Results

We collect two data sets over a period of time by a web crawler designed for the threaded discussion communities. One is from the iPad Q&A board in the apple discussion forum; the other is from the technique community *Slashdot.org*. These two data sets are chosen because of the following reasons: (1)The two data sets are from two kinds of typical threaded discussion communities. The first is the Q&A forum, and the second is an open discussion forum where everyone can participate and judge the comments. Both of them have interested hot topics and the reply-to structure can be extracted or reconstructed without much difficulty. (2) Both data sets contain labeled information. The iPad Q&A data set can label the answers "Helpful" by other users or "Solved" by the questioner, which we quantize as 2, 1, 0 respectively, while *Slashdot.org* can give each comment a score ranging from -1 to 5 by all the participators. In the preprocess of text feature extraction, we first remove the stop words, and then collect the terms whose number is no less than 3. For each data set, we select 5 hottest topics and ignore the unqualified threads that have posts fewer than three or without labels or ratings. The basic statistic results are shown in Table 1. The two kinds of threaded discussion communities are quite different in average thread

Table 1. The basic statistics of the data sets

Data set	iPad Q&A	Slashdot.org
Number of threads	1130	664
Number of posts	8489	146569
Number of users	2175	14241
Average thread length	7.51	220.74
Average words per post	63.09	76.33
Average posts per user	3.90	10.29
Number of topics	5	5

lengthes, user active degrees and so on. However, by computing the similarity in content and structure organization, we can obtain valuable answers to the questions or recommend the popular comments in our clustered multi-task learning framework.

Throughout the experiments, we use the root mean square error (RMSE) across the tasks as a criterion. The performance is better when RMSE is lower. In the learning process, the results are evaluated by 5-fold cross validation. In the following, first we conduct the experiments of the performance between the explicit reply-to graph and the reconstructed reply-to graph for *Slashdot.org*. Due to the lack of explicit reply-to graph in iPad Q&A data set, we reconstruct the reply-to graph with semantic similarity measure. Second, we discuss the two significant hyper parameters: one is the number of original multiple tasks K automatically formed by the metadata clustering, and the other is the number of task clusters k used for CMTL. We compare our sCMTL with the classical single-task learning methods such as the linear SVM, the Gaussian kernel SVM, and the hard-assigned CMTL. In the inference of W in multi-task learning algorithms, the logistic loss function is unified chosen for simplicity.

6.1 Evaluation for the Explicit v.s. Reconstructed Reply-to Graph

The experiments are only conducted on the *Slashdot.org* data set because the other data set is lack of explicit reply-to graph. To better measure the performance influenced by the partial metadata which shows the semantic context of each discussion thread, after the metadata clustering procedure, we compare them in the framework of the single-task learning and multi-task learning separately. In single-task learning, we choose the linear SVM for regression; while in MTL, we select our sCMTL with $k = 3$. The number of data clusters $K \geq k$ is changed from 5 to 30.

As shown in Fig. 2, it suggests that whatever in single-task learning or MTL, the performance the metadata modeling based on reconstructed reply-to graph is comparable to that based on the explicit graph when we set the suitable K. Because the explicit reply-to graph truly shows the semantic interaction between the web users, throughout the semantic reconstruction method, the reconstructed graph can basically suggest the realistic semantic interaction.

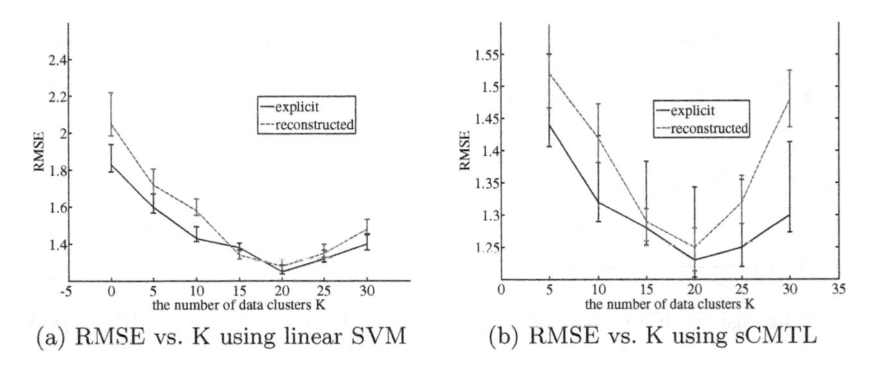

(a) RMSE vs. K using linear SVM (b) RMSE vs. K using sCMTL

Fig. 2. The comparison between explicit and reconstructed reply-to graphs

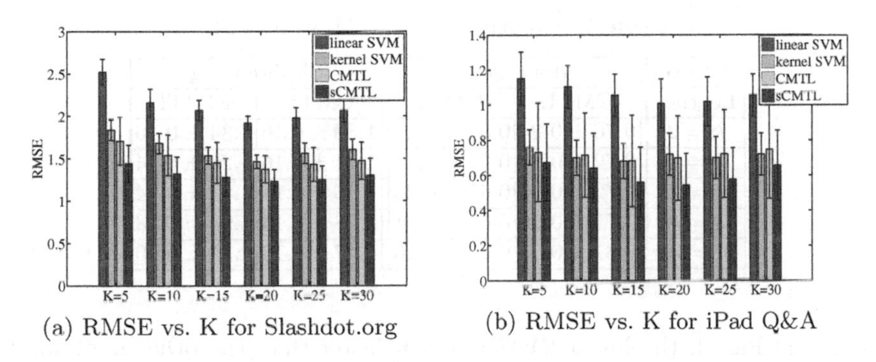

(a) RMSE vs. K for Slashdot.org (b) RMSE vs. K for iPad Q&A

Fig. 3. The learning algorithms comparison on two data sets

Accordingly, in the following experiments, when the explicit graph does not exist, the reconstruction method is available as alternative.

6.2 Evaluation for the Number of Data Clusters K

The hyper parameter K is significant as it shows how many semantic subspaces can be set apart in the whole feature space. On one hand, we can develop the corresponding learning model which needs not to meet all kinds of conditions; on the other hand, the semantics can be shared in each cluster, which alleviates the small sample size problem. Similar to the previous experiment settings, we set $k = 3$ for MTL algorithms, and change K from 5 to 30. We compare the single-task learning with MTL algorithms.

As shown in Fig. 3, both on two data sets, both MTL (CMTL and sCMTL) algorithms are generally more effective than the classical single-task learning algorithms. The proposed sCMTL outperforms the other three methods. Besides the comparison of performance, we also record the time cost for the learning methods on both of the two data sets. Through overall comparing the results in

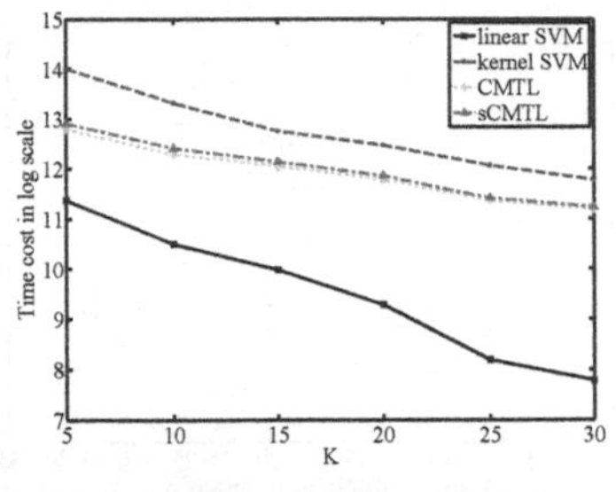

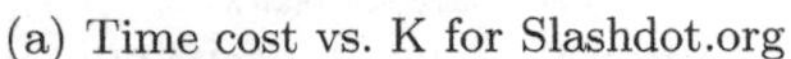
(a) Time cost vs. K for Slashdot.org

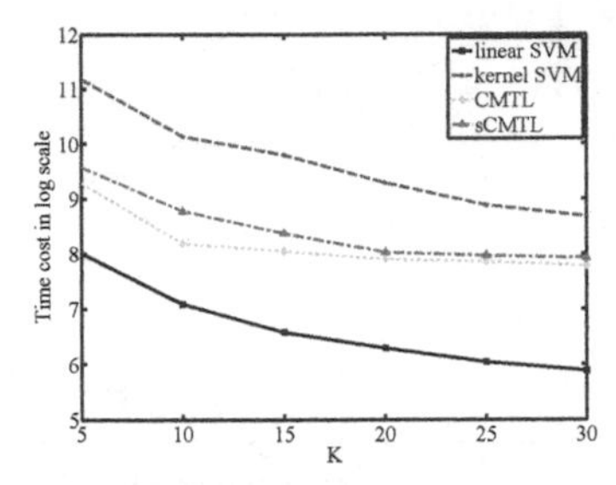

(b) Time cost vs. K for iPad Q&A

Fig. 4. The comparison of time costs among the learning methods on two data sets

Table 2. Evaluation for CMTL vs sCMTL

Data set	iPad Q&A		Slashdot.org	
Learner	CMTL	sCMTL	CMTL	sCMTL
$k = 2$	0.76 ± 0.24	0.62 ± 0.24	1.38 ± 0.16	1.34 ± 0.16
$k = 4$	0.72 ± 0.20	0.58 ± 0.18	1.31 ± 0.16	1.23 ± 0.14
$k = 6$	0.72 ± 0.20	0.61 ± 0.20	1.85 ± 0.14	1.51 ± 0.14
$k = 8$	0.77 ± 0.20	0.68 ± 0.21	2.28 ± 0.14	1.96 ± 0.13
$k = 10$	0.81 ± 0.21	0.75 ± 0.20	2.47 ± 0.13	2.31 ± 0.10

Fig. 3 and Fig. 4, the linear SVM is much faster than the other methods but with worst performance even inapplicable to the thread mining tasks. Although the kernel SVM performs well comparable to the CMTL, it is much more time-consuming than the other algorithms. The proposed sCMTL costs almost equally with the CMTL, but performs much better.

6.3 Evaluation for the Number of Task Clusters k

The hyper parameter k shows the geometric characteristics of the learning tasks. With the CMTL, the learning weight can be shared among tasks. We set $K = 20$, and change k from 2 to 10. We compare the two CMTL algorithms based on the two data sets. As shown in Table 2, the sCMTL is more effective than CMTL on both of the two data sets.

7 Conclusion

In this paper, we have studied the metadata-based clustered multi-task learning for thread mining in web communities, which takes use of the metadata and fuses it in the framework of multi-task learning based on the divide-and-learn strategy. We divide the data set according to the metadata clustering, and learn multiple tasks simultaneously in the framework of softly assigned clustered multi-task learning. We have conducted the experiments on two real data sets from two kinds of web communities. The experimental results show that our method is more effective than many previous learning algorithms, and the moderate time cost makes the propose method acceptable to the thread mining tasks.

Acknowledgments. This work is partly supported by the 973 basic research program of China (Grant No. 2014CB349303), the Natural Science Foundation of China (Grant No. 61472421 and No. 61379098), the Project Supported by CAS Center for Excellence in Brain Science and Intelligence Technology, and the Project Supported by Guangdong Natural Science Foundation (Grant No. S2012020011081).

References

1. Pang, B., Lee, L., Vaithyanathan, S.: Thumbs up?: sentiment classification using machine learning techniques. In: Proceedings of the ACL-02 Conference on Empirical Methods in Natural Language Processing-Volume 10, pp. 79–86. Association for Computational Linguistics (2002)
2. Hofmann, T.: Probabilistic latent semantic indexing. In: Proceedings of the 22nd Annual International ACM SIGIR Conference on Research and Development in Information Retrieval, pp. 50–57. ACM (1999)
3. Blei, D.M., Ng, A.Y., Jordan, M.I.: Latent dirichlet allocation. The Journal of Machine Learning Research **3**, 993–1022 (2003)
4. Kleinberg, J.M.: Authoritative sources in a hyperlinked environment. Journal of the ACM (JACM) **46**(5), 604–632 (1999)
5. Page, L., Brin, S., Motwani, R., Winograd, T.: The pagerank citation ranking: bringing order to the web (1999)
6. Cong, G., Wang, L., Lin, C.Y., Song, Y.I., Sun, Y.: Finding question-answer pairs from online forums. In: Proceedings of the 31st Annual International ACM SIGIR Conference on Research and Development in Information Retrieval, pp. 467–474. ACM (2008)
7. Blei, D.M., Moreno, P.J.: Topic segmentation with an aspect hidden markov model. In: Proceedings of the 24th Annual International ACM SIGIR Conference on Research and Development in Information Retrieval, pp. 343–348. ACM (2001)
8. Shen, D., Yang, Q., Sun, J.T., Chen, Z.: Thread detection in dynamic text message streams. In: Proceedings of the 29th Annual International ACM SIGIR Conference on Research and Development in Information Retrieval, pp. 35–42. ACM (2006)
9. Lin, C., Yang, J.M., Cai, R., Wang, X.J., Wang, W.: Simultaneously modeling semantics and structure of threaded discussions: a sparse coding approach and its applications. In: Proceedings of the 32nd International ACM SIGIR Conference on Research and Development in Information Retrieval, pp. 131–138. ACM (2009)
10. Poh, N., Kittler, J., Bourlai, T.: Quality-based score normalization with device qualitative information for multimodal biometric fusion. IEEE Transactions on Systems, Man and Cybernetics, Part A: Systems and Humans **40**(3), 539–554 (2010)
11. Poh, N., Kittler, J.: A unified framework for biometric expert fusion incorporating quality measures. IEEE Transactions on Pattern Analysis and Machine Intelligence **34**(1), 3–18 (2012)

12. Wu, O., Hu, R., Mao, X., Hu, W.: Quality-based learning for web data classification. In: Twenty-Eighth AAAI Conference on Artificial Intelligence (2014)
13. Fu, Z., Robles-Kelly, A., Zhou, J.: Mixing linear svms for nonlinear classification. IEEE Transactions on Neural Networks **21**(12), 1963–1975 (2010)
14. Gu, Q., Han, J.: Clustered support vector machines. In: Proceedings of the Sixteenth International Conference on Artificial Intelligence and Statistics, pp. 307–315 (2013)
15. Ben-David, S., Schuller, R.: Exploiting task relatedness for multiple task learning. In: Schölkopf, B., Warmuth, M.K. (eds.) COLT/Kernel 2003. LNCS (LNAI), vol. 2777, pp. 567–580. Springer, Heidelberg (2003)
16. Torralba, A., Murphy, K.P., Freeman, W.T.: Sharing features: efficient boosting procedures for multiclass object detection. In: Proceedings of the 2004 IEEE Computer Society Conference on Computer Vision and Pattern Recognition, CVPR 2004, vol. 2, p. II-762. IEEE
17. Ando, R.K., Zhang, T.: A framework for learning predictive structures from multiple tasks and unlabeled data. The Journal of Machine Learning Research **6**, 1817–1853 (2005)
18. Evgeniou, T., Pontil, M.: Regularized multi-task learning. In: Proceedings of the tenth ACM SIGKDD International Conference on Knowledge Discovery and Data Mining, pp. 109–117. ACM (2004)
19. Thrun, S., O'Sullivan, J.: Clustering learning tasks and the selective cross-task transfer of knowledge. Springer (1998)
20. Bakker, B., Heskes, T.: Task clustering and gating for bayesian multitask learning. The Journal of Machine Learning Research **4**, 83–99 (2003)
21. Kim, S., Xing, E.P.: Tree-guided group lasso for multi-task regression with structured sparsity. In: Proceedings of the 27th International Conference on Machine Learning (ICML 2010), pp. 543–550 (2010)
22. Chen, J., Liu, J., Ye, J.: Learning incoherent sparse and low-rank patterns from multiple tasks. ACM Transactions on Knowledge Discovery from Data (TKDD) **5**(4), 22 (2012)
23. Chen, J., Tang, L., Liu, J., Ye, J.: A convex formulation for learning shared structures from multiple tasks. In: Proceedings of the 26th Annual International Conference on Machine Learning, pp. 137–144. ACM (2009)
24. Xue, Y., Liao, X., Carin, L., Krishnapuram, B.: Multi-task learning for classification with dirichlet process priors. The Journal of Machine Learning Research **8**, 35–63 (2007)
25. Jacob, L., Bach, F., Vert, J.P., et al.: Clustered multi-task learning: a convex formulation. In: NIPS, vol. 21, pp. 745–752 (2008)
26. Zhou, J., Chen, J., Ye, J.: Clustered multi-task learning via alternating structure optimization. In: NIPS, pp. 702–710 (2011)
27. Zhou, Y., Cheng, H., Yu, J.X.: Graph clustering based on structural/attribute similarities. Proceedings of the VLDB Endowment **2**(1), 718–729 (2009)
28. Cheng, H., Zhou, Y., Yu, J.X.: Clustering large attributed graphs: A balance between structural and attribute similarities. ACM Transactions on Knowledge Discovery from Data (TKDD) **5**(2), 12 (2011)
29. Von Luxburg, U.: A tutorial on spectral clustering. Statistics and Computing **17**(4), 395–416 (2007)
30. Wu, O., Hu, W., Maybank, S.J., Zhu, M., Li, B.: Efficient clustering aggregation based on data fragments. IEEE Transactions on Systems, Man, and Cybernetics, Part B: Cybernetics **42**(3), 913–926 (2012)

On Genetic Algorithms for Detecting Frequent Item Sets And Large Bite Sets

Roman A. Sizov and Dan A. Simovici(✉)

Computer Science Department, University of Massachusetts Boston, Boston, USA
{rsizov,dsim}@cs.umb.edu

Abstract. This paper introduces the use of genetic algorithms to mine binary datasets for obtaining frequent item sets and large bite item sets, two classes of problems that are important for optimal exposure of item sets to customers and for efficient advertising campaigns. Whereas both problems can be approached in a common framework, we highlight specific features of the fitness functions suitable for each of the problems.

Keywords: Genetic algorithm · Support set · Bite set · Covering

1 Introduction

In this paper we propose the use of a bi-objective optimization approach via genetic algorithms to find solutions for two related problems both of which occur in mining certain datasets. These datasets consist of a set of items I and a multiset of transactions T that is a collection of subsets of I and have natural binary representations. Namely, each transaction is represented by its binary characteristic vector of the corresponding subset of I.

The first problem, which has an extensive literature [3], [2], [18], is to find the large subset of items (frequent item sets) that are present in large sets of transaction. Since each transaction can be regarded as a basket of items, this problem is known as exploring market basket datasets (MBD) and identifying large set of items that are bought together by many customers. Finding frequent item sets allows us to identify association rules that satisfy certain probabilistic criteria [3], [2], [18].

The second related problem, which we identify, is to find the smallest subset of items that intersects as many transactions as possible. This is a significant issue for marketeers that wish to find the smallest subset of items that intersects large subset of transactions (i.e. items that are bought by many customers) in order to make sure that as many customers see an add placed on these items.

Market basket binary datasets often contain millions or even billions of transactions. Thus, finding solutions to the above problems becomes very challenging. The proposed methods allow to find solutions in reasonable amount of time, although it does not guarantee that these solutions are optimal.

P. Perner (Ed.): MLDM 2016, LNAI 9729, pp. 435–445, 2016.
DOI: 10.1007/978-3-319-41920-6_34

Many multi-objective optimization problems have been approached with genetic algorithms. For example, [11] applies genetic algorithms to multi-objective problem in groundwater quality management, [7] uses genetic algorithm to produce fuzzy autopilot controllers for engineering problems, and [19] compares the methods of multi objective genetic algorithm and simulated annealing applied to analogue filter tuning.

Various evolutionary techniques applied in this work are introduced in [9], [14], [4]. A review of evolutionary multi-objective optimization can be found in [8], [12].

In Section 2 we introduce basic concepts of this paper. The proposed genetic algorithm is introduced in Section 3. In Section 4 the results of the algorithm application are discussed. In Section 5 we present future directions of investigation.

2 Support and Bite Set Definitions

Suppose that we have a set of transactions $T = \{t_1, ..., t_n\}$ involving a set of items $I = \{i_1, ..., i_m\}$. A *transaction* on I is a subset t of I. Collections of items are denoted using letters from the middle of the alphabet: J, K, L, etc. The set of transactions is denoted by T. We denote subsets of T using capital letters from the end of the alphabet: U, V, W, etc. Define the functions $\phi, \alpha : \mathcal{P}(I) \longrightarrow \mathcal{P}(T)$ and $\psi, \beta : \mathcal{P}(T) \longrightarrow \mathcal{P}(I)$ as

$$\begin{aligned}
\phi(J) &= \{t \in T \mid J \subseteq t\} \\
&= \{t \in T \mid (\forall j \in J) j \in t\}, \\
\alpha(J) &= \{t \in T \mid J \cap t \neq \emptyset\} \\
&= \{t \in T \mid (\exists j \in J) j \in t\}, \\
\psi(U) &= \{i \in I \mid (\forall t \in U) i \in t\}, \\
\beta(U) &= \{i \in I \mid (\exists t \in U) i \in t\}.
\end{aligned}$$

Note that $J_1 \subseteq J_2$ implies $\phi(J_2) \subseteq \phi(J_1)$ and $\alpha(J_1) \subseteq \alpha(J_2)$. Also, $U_1 \subseteq U_2$ implies $\psi(U_2) \subseteq \psi(U_1)$ and $\beta(U_1) \subseteq \beta(U_2)$. In other words, the mappings ϕ and ψ are antimonotonic, while α and β are monotonic mappings.

Also, note that

$$J \subseteq \psi(\phi(J)), U \subseteq \phi(\psi(U)).$$

for every $J \in \mathcal{P}(I)$ and $U \in \mathcal{P}(T)$.

The above inequalities show that the pair (ϕ, ψ) is a Galois connection between $\mathcal{P}(T)$ and $\mathcal{P}(I)$ (see [6]).

For the mappings α, β we have

$$\beta(\alpha(J)) = \{i \in I \mid (\exists j \in J) j \in t \text{ and } i \in t\}.$$

In other words, $i \in \beta(\alpha(J))$ if and only if i is contained by some transaction t that also contains an item $j \in J$. Thus, it is clear that $J \subseteq \beta(\alpha(J))$ for every set of items J.

Similarly,

$$\begin{aligned}\alpha(\beta(U)) &= \{t \in T \mid \beta(U) \cap t \neq \emptyset\} \\ &= \{t \in T \mid (\exists t' \in U) t' \cap t \neq \emptyset\},\end{aligned}$$

which shows that $U \subseteq \alpha(\beta(U))$. Thus, the pair (α, β) is an *axiality* between the same sets.

Note that

$$\phi(J) = \phi(\psi(\phi(J))), \psi(U) = \psi(\phi(\psi(U))).$$

The *bite set* of J is $\alpha(J)$ and the *bite* of J is $|\alpha(J)|$. The *resistance set* of J is $T - \alpha(J)$ and the *resistance* of J is $|T| - |\alpha(J)|$. The *support set* of J is $\phi(J)$ and the *support* of J is $|\phi(J)|$. Since $\phi(J) \subseteq \alpha(J)$ we have $\text{supp}(J) \leqslant \text{bite}(J)$.

The problem of identifying large items sets having high support has been extensively examined in the data mining literature in connection with the identification of association rules. A fundamental observation is that the size of an item set varies inversely with its support. We propose to identify large item sets with large values of support using a bi-objective optimization approach via genetic algorithms. Both [16] and [13] use genetic algorithms to mine datasets but their aim is to find association rules.

The problem of identifying small items sets covering large numbers of transactions is related to the minimal covering problem, although there is no requirement to cover all of the transactions. For an item set J, its cardinality and its resistance $|T| - |\alpha(J)|$ are in conflict, that is, the resistance increases while the cardinality decreases, and vice versa. Thus, this problem can also be approached using a bi-objective optimization approach via genetic algorithms seeking minimal item sets that intersect as many transactions as possible, that is, have minimal resistance. The latter problem occurs in marketing campaigns where an advertiser is interested to find a small set of items bearing its advertising in as many baskets (transactions) as possible.

3 The Design of the Genetic Algorithm

The algorithm that we propose is identical for both problems that we investigate. However, the choice of the fitness function is specific to the problem that we seek to solve as mentioned above.

Chromosomes are binary strings of size k, where $k = |I|$. Namely, they are characteristic vectors of subsets of I. If $S \subseteq I$, then its corresponding chromosome is $\mathbf{c}_S = (c_1, \ldots, c_k)$, where

$$c_j = \begin{cases} 1 & \text{if } i_j \in S, \\ 0 & \text{otherwise} \end{cases}$$

for $1 \leqslant j \leqslant I$.

Genetic algorithm operations considered here are *p-crossover* and *q-mutations* defined as follows.

Given two chromosomes $\mathbf{c}_S$ and $\mathbf{c}_T$ defined as above and a number p, where $0 < p < 1$, the two children produced by *p-crossover* are the random variables $\mathbf{c}_{S,T}$ and $\mathbf{c}_{T,S}$, where the jth allele of the first child is

$$(\mathbf{c}_{S,T})_j : \begin{pmatrix} (\mathbf{c}_S)_j & (\mathbf{c}_T)_j \\ p & 1-p \end{pmatrix},$$

and the jth allele of the second child is

$$(\mathbf{c}_{T,S})_j = \begin{cases} (\mathbf{c}_T)_j & \text{if } (\mathbf{c}_{S,T})_j = (\mathbf{c}_S)_j \\ (\mathbf{c}_S)_j & \text{if } (\mathbf{c}_{S,T})_j = (\mathbf{c}_T)_j. \end{cases}$$

The *q-mutation* of a chromosome switches the alleles at each location of the chromosome with the probability q, where $0 < q \ll 1$.

Let $\mathbf{C}$ be the population of chromosomes. Fitness functions of the form $F : \mathbf{C} \longrightarrow \mathbb{R}_{\geqslant 0}$ are specific to the the problems mentioned above and are parametrized by $\alpha \in [0, 1]$.

For the first problem (the identification of large item sets with large support) the fitness function F is defined as

$$F(\mathbf{c}) = f_1(\mathbf{c})^{\alpha} f_2(\mathbf{c})^{1-\alpha},$$

where

- $f_1(\mathbf{c})$ is the number of items in the solution (that is, the number of 1s in the chromosome), and
- $f_2(\mathbf{c})$ is the number of transactions in the dataset that are supported by the selected items.

For the second problem (the identification of small item sets with large bite) the fitness function F is obtained as convex combination

$$F(\mathbf{c}) = \alpha f_1(\mathbf{c}) + (1-\alpha) f_2(\mathbf{c}),$$

where

- $f_1(\mathbf{c})$ is the number of items not in the solution (that is, the number of 0s in the chromosome), and
- $f_2(\mathbf{c})$ is the number of transactions in the dataset that are intersected by the selected items.

Note that for the second problem $f_1(\mathbf{c})$ is increasing as solutions improve.

Better solution are obtained with higher values of $F(\mathbf{c})$. The value of the fitness function is highly dependent on a particular dataset.

The pseudo-code of the algorithm is shown at Algorithm 1.

The *java.util.BitSet* class [17] was used to implement chromosomes in the algorithm.

Algorithm 1. Genetic Algorithm

Input : Selection ratio s, Crossover ratio p, Mutation ratio q, Fitness function weight α, Dataset D, Chromosome population size N

begin
  Generate initial population of chromosomes $\mathbf{C}$ of size N
  $i = 0$
  while $i \leq N$ **do**
    Apply fitness function to population $\mathbf{C}_i$ and obtain fitness function values for each chromosome in the population
    Select $s\%$ of the best chromosome from the population $\mathbf{C}_i$
    Split the best chromosomes into two equal groups
    $j = 0$
    while $j <$ *size of group 1* **do**
      Apply p-crossover to the best chromosomes such that parent 1 is jth chromosome from the first group and parent 2 is a random chromosome from the second group
      $j = j + 1$
    end
    $j = 0$
    while $j <$ *size of group 2* **do**
      Apply p-crossover to the best chromosomes such that parent 1 is jth chromosome from the second group and parent 2 is a random chromosome from the first group
      $j = j + 1$
    end
    Save all the obtained children as well as the parents in the new population
    if *the size of the new population* $< N$ **then**
      pick random chromosomes from the population $\mathbf{C}_i$ to get N chromosomes
    end
    if *the size of the new population* $> N$ **then**
      remove random chromosomes from the new population to get N chromosomes
    end
    Apply q-mutation to the new Population
    $i = i + 1$
  end
  Pick the best solutions, i.e. solutions with higher fitness function values
end

---

4 Results

Our experiments involved two data sets: the *Right Heart Catheterization Dataset (RHC)* obtained from [5] and a synthetic data set randomly created based on lognormal distribution [10] and having the same size as the RHC data set.

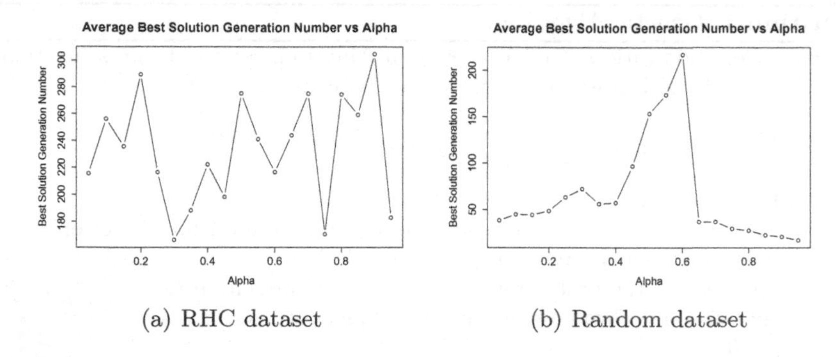

(a) RHC dataset (b) Random dataset

Fig. 1. Support set average best solution number of generation vs alpha

For the first data set only binary fields (1/0 or $Yes = 1/No = 0$) were retained with the exception of the *death* and *ortho* fields; all the 0-only rows were deleted from both data sets.

Both resulting datasets contain 23 columns and 5712 rows.

The algorithm was coded in *Java*. For each experiment the population size was set to 50, the number of generations was 500, and we averaged the results for 10 runs.

Each part of the experiment was run for both datasets with various values of α. For both problems, larger values of α make the item part of the fitness function more important; smaller αs, on the other hand, make the transaction part of the fitness function more important. Hence, by changing the parameter α we can obtain various solutions for the problems and produce the Pareto front. A somewhat similar approach was used in [15].

Figures 1-4 show the experiment results for the support set problem.

Figure 1(a) demonstrates no obvious dependency of the average best solution generation number on α in the case of the RHC dataset. In contrast, Figure 1(b) demonstrates a clear dependency of the average best solution generation number on α in the case of the random dataset. A sudden increase of the number of generations can be observed for α in the range $[0.45, 0.60]$. One possible reason for this is that it is harder to find optimal solutions in the middle range of α (importance of the item and transaction parts in the fitness function is almost equal) for binary datasets with no structure. Furthermore, we noticed that the generation number for the RHC dataset is on average much higher than for the random dataset. This suggests that optimal solutions for structured datasets are trickier to find.

Figures 2(a) and 3(a) show numbers of items and transactions versus α for the RHC dataset, and Figures 2(b) and 3(b) show the same for the random dataset.

Somewhat reciprocal behavior of the number of items and the number of transactions is due to the fact that there is a conflict between the numbers of items and transactions for the support set problem. For all graphs, we can see

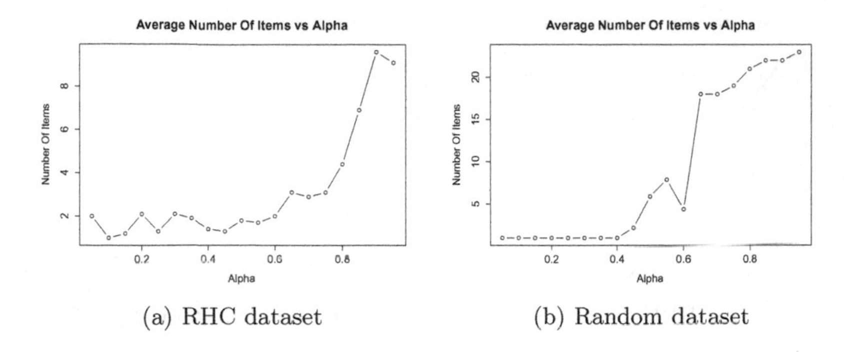

(a) RHC dataset (b) Random dataset

Fig. 2. Support set average number of items vs alpha

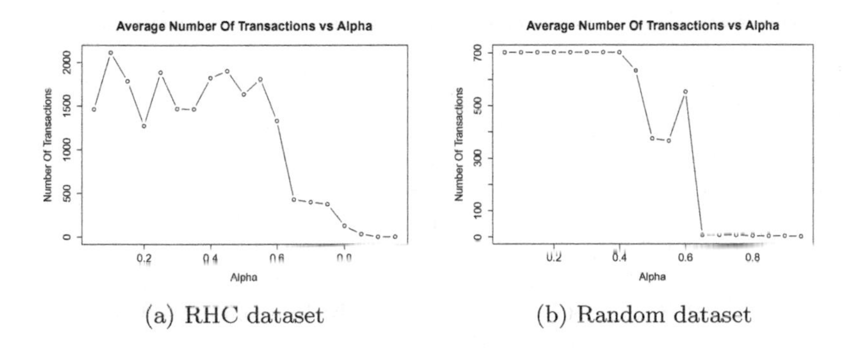

(a) RHC dataset (b) Random dataset

Fig. 3. Support set average number of transactions vs alpha

that at the α value close to 0.6 (with an exception of the graph at Figure 2(a) where it is close to 0.8), the behavior of the graphs resembles something that looks like a phase change. We notice that at low and high values of α there is not much change in the optimality of solutions for both of the datasets.

Nevertheless, one might observe that the solutions for the RHC dataset have, in general, a higher number of transactions than the solutions for the random datasets. The solutions for the random dataset with a high number of items have a very low support.

These observations suggest that the RHC dataset is more structured than the random dataset, which we certainly expect. This difference in optimal solutions for the RHC dataset and the random dataset can be readily observed if Figures 4(a) and 4(b) are compared. These figures show the Pareto fronts of the support set problem for both of the datasets, and are obtained by changing the α parameter.

Figures 5-8 show the experiment results for the bite set problem.

Figures 5(a) and 5(b) demonstrate a clear dependency of the average best solution generation number on α for both the RHC and random datasets.

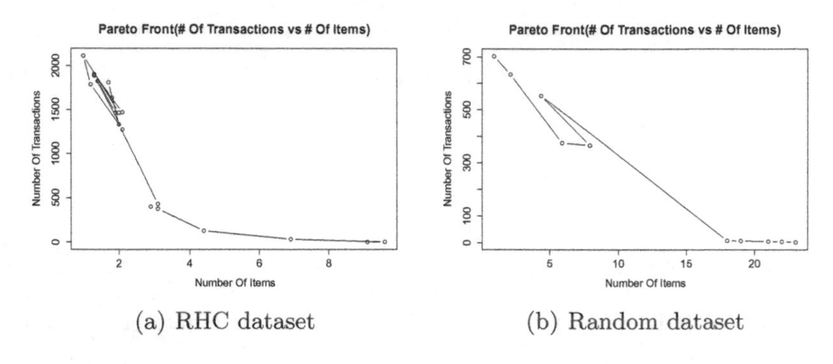

(a) RHC dataset (b) Random dataset

Fig. 4. Support set Pareto front

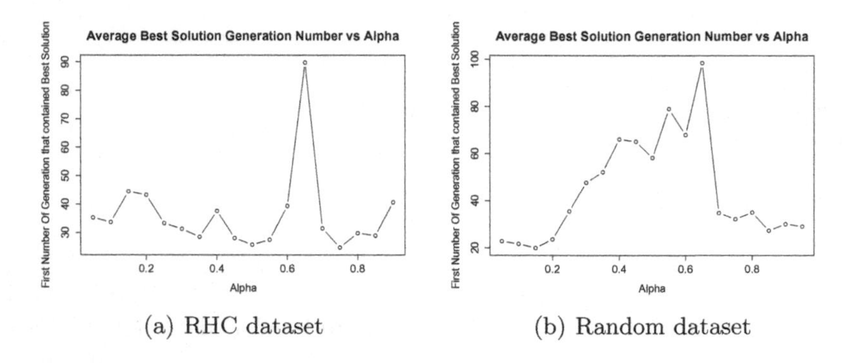

(a) RHC dataset (b) Random dataset

Fig. 5. Bite set average best solution number of generation vs alpha

A spike in the number of generations can be observed for α at the value of 0.65 for the RHC dataset, whereas in case of the random dataset we see a slow increase of the number of generations from α at the value of 0.20 up to the value of 0.65 and then a sharp decrease when the value of α equals 0.70. Although the reason for such behavior is not clear, it is possible that the importance of the number of items and the number of transactions is at an *equilibrium*, and that makes it harder to find a better solution. The equilibrium value of α might depend on the dimensions of a dataset. In our case, the dimensions are the same for both datasets, which could explain a similar behavior in both experiments.

Figures 6(a) and 7(a) show numbers of items and transactions versus α for the RHC dataset, and Figures 6(b) and 7(b) show the same for the random dataset.

The experiment results for the RHC dataset show a drastic difference from the respective results for the random dataset. In case of the random dataset we can see much smoother behavior. This is expected since the random dataset presumably has no structure. Also, we can notice that at the value of α equals to 0.65 there is a sharp decrease in the numbers of items and transactions, and

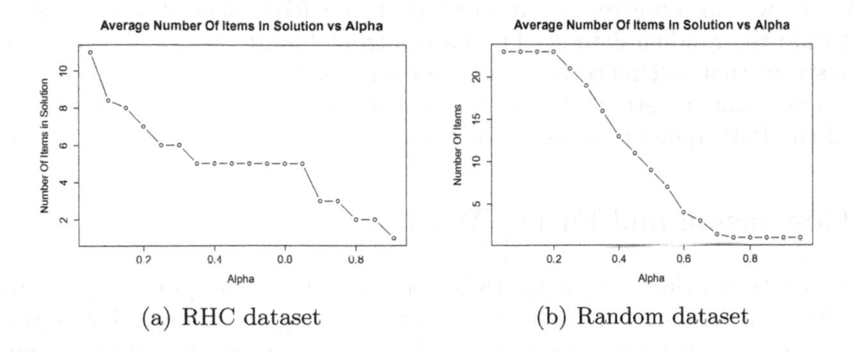

(a) RHC dataset (b) Random dataset

Fig. 6. Bite set average number of items vs alpha

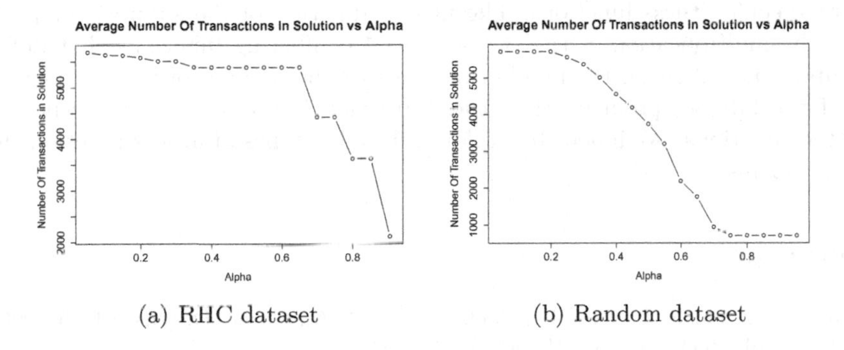

(a) RHC dataset (b) Random dataset

Fig. 7. Bite set average number of transactions vs alpha

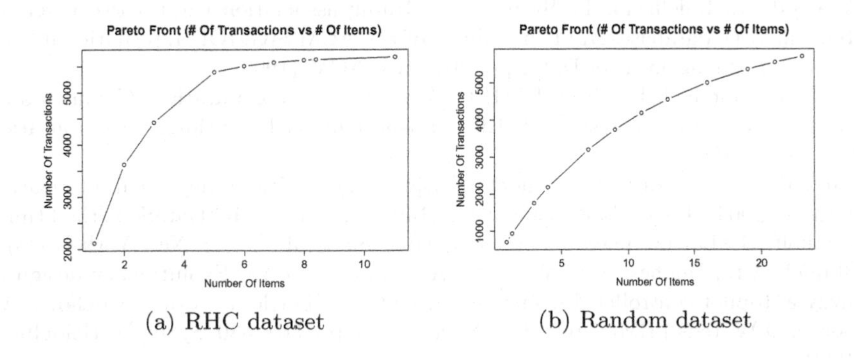

(a) RHC dataset (b) Random dataset

Fig. 8. Bite set Pareto front

at the α range of $[0.35, 0.60]$ there is almost no change in both the number of items and the number of transactions. Recall that in case of the bite set problem, the smaller the number of items and the higher the number of transactions the better the solution.

Again, we can observe, as expected, that the RHC dataset is more structured than the random dataset. The Pareto front Figures 8(a) and 8(b) clearly demonstrate that difference between the datasets. We can readily notice that the Pareto front in case of the random dataset is much smoother, whereas in case of the RHC dataset the behavior is more drastic.

5 Conclusion and Future Work

We demonstrated that genetic algorithms are useful in identifying frequent item sets with high support and minimal item sets with high bite. Even though there is no guarantee that solutions obtained by the proposed algorithm are optimal, we regard these solutions as reasonable considering the complexity of the problems. Furthermore, despite the common framework of both problems they require specific fitness functions. The fitness function of the support set problem is the multiplication of the two weighted conflicting functions that define the importance of either the number of items or the number of transactions. In case of the bite set problem, the fitness function is the sum of the respective weighted functions. We indent to explore related families of fitness functions for these problems.

References

1. Zitzler, E., Thiele, L., Deb, K., Coello Coello, C.A., Corne, D. (eds.): EMO 2001. LNCS, vol. 1993. Springer, Heidelberg (2001)
2. Adamo, J.M.: Data Mining for Association Rules and Sequential Patterns: Sequential and Parallel Algorithms. Springer, New York (2001)
3. Agarwal, R., Imielinski, T., Swami, A.: Mining association rules between sets of items in large databases. In: Proceedings of the ACM SIGMOD International Conference of Management of Data, pp. 207–216. ACM (1993)
4. Back, T., Hammel, U., Schwefel, H.P.: Evolutionary computation: Comments on the history and current state. IEEE Transactions on Evolutionary Computation **1**(1), 3–17 (1997)
5. Vanderbilt University Department of Biostatistics: Right heart catheterization dataset (2004). http://biostat.mc.vanderbilt.edu/wiki/Main/DataSets/rhc.html
6. Birkhoff, G.: Lattice Theory. American Mathematical Society, New York (1948)
7. Blumel, A.L., Hughes, E.J., White, B.A.: Multi-objective Evolutionary design of fuzzy autopilot controller. In: Zitzler, E., Deb, K., Thiele, L., Coello Coello, C.A., Corne, D.W. (eds.) EMO 2001. LNCS, vol. 1993, pp. 668–680. Springer, Heidelberg (2001)
8. Coello Coello, C.A.: A Short tutorial on evolutionary multiobjective optimization. In: Zitzler, E., Deb, K., Thiele, L., Coello Coello, C.A., Corne, D.W. (eds.) EMO 2001. LNCS, vol. 1993, pp. 21–40. Springer, Heidelberg (2001)
9. De Jong, K.A.: Evolutionary Computation. A Unified Approach. MIT Press and A Bradford Book, Cambridge (2006)
10. Devore, J.: Probability and Statistics. Brooks Cole-Thomson Learning, Belton, CA (2004)

11. Erickson, M., Mayer, A., Horn, J.: The niched pareto genetic algorithm 2 applied to the design of groundwater remediation systems. In: Zitzler, E., Deb, K., Thiele, L., Coello Coello, C.A., Corne, D.W. (eds.) EMO 2001. LNCS, vol. 1993, pp. 681–695. Springer, Heidelberg (2001)
12. Fonseca, C.M., Fleming, P.J.: An overview of evolutionary algorithms in multiobjective optimization. Evolutionary Computation **3**(1), 1–16 (1995)
13. Ghosh, S., Biswas, S., Sarkar, D., Sarkar, P.P.: Mining frequent itemsets using genetic algorithms. International Journal of Artificial Intelligence and Applications **1**(4) (October 2010)
14. Goldberg, D.E.: Genetic Algorithms in Search, Optimization, and Machine Learning. Addison-Wesley, Reading (1989)
15. Jin, Y., Okabe, T., Sendhoff, B.: Adapting weighted aggregation for multiobjective evolution strategies. In: Zitzler, E., Deb, K., Thiele, L., Coello Coello, C.A., Corne, D.W. (eds.) EMO 2001. LNCS, vol. 1993, pp. 96–110. Springer, Heidelberg (2001)
16. Prakash, V., Govardhan, D., Sarma, S.: Mining frequent itemsets from large data sets using genetic algorithms. International Journal of Computer Application Special Issue on "Artificial Intellegence Techniques - Novel Approaches and Practical Applications" (2011)
17. Schildt, H.: Java: The Complete Reference, 9th edn. McGraw-Hill Education, New York (2014)
18. Tan, P.N., Steinbach, M., Kumar, V.: Introduction to Data Mining. Pearson Education, Boston (2006)
19. Thompson, M.: Application of multi objective evolutionary algorithms to analogue filter tuning. In: Zitzler, E., Deb, K., Thiele, L., Coello Coello, C.A., Corne, D.W. (eds.) EMO 2001. LNCS, vol. 1993, pp. 546–559. Springer, Heidelberg (2001)

Multi-view Learning for Classification of X-Ray Crystallography Images

B.M. Thamali Lekamge[1(✉)], Arcot Sowmya[1], and Janet Newman[2]

[1] School of Computer Science and Engineering, University of New South Wales, Sydney, NSW 2052, Australia
{thamalil,sowmya}@cse.unsw.edu.au
[2] CSIRO Manufacturing, 343 Royal Parade, Parkville, VIC 3052, Australia
Janet.Newman@csiro.au

Abstract. Multi-view learning is a very useful classification technique when multiple, conditionally independent feature sets are available in a dataset. In this paper multi-view learning is used to classify sequences of protein crystallization images that were obtained over a period of time, varying between a few hours to a few months. We introduce the use of the difference image features, along with the original image features, as a second feature set in classifying x-ray crystallography images, after arranging the images according to the timeline of an experiment. Usage of multi-view learning is proposed after carrying out experiments to determine the features that should be used in each view to increase classification accuracy. Random forests are used as the classifier in each view, as preliminary experiments have suggested that it provides higher classification accuracy in crystallography datasets. Accuracy of 97.2% was obtained using multi-view learning based on original and difference features, which is the highest obtained so far in the classification of protein crystallography images.

Keywords: X-ray crystallography · Protein crystals · Multi-view learning · Co-training · Random forests · Difference images

1 Introduction

Determining the three dimensional structure of a protein crystal is important in understanding the link between the activity of the protein and its shape – more commonly called the structure/function relationship. X-ray crystallography is one of the main techniques used to achieve this task [1, 2]. This process comprises production of a suitable crystalline protein sample and irradiation of the crystal with X-rays. The diffraction patterns obtained from this process are evaluated to produce an illustration of electron density associated with each specific protein [3-5].

Crystal growth is a time dependent process and involves setting up thousands of individual experiments, which are monitored either directly or by the automatic collection of a timecourse of images, where the timecourse can range from a few hours to a couple of months. An individual trial most commonly consists of a sub-microliter liquid droplet on a plastic support. Automatic imagers are used to obtain the protein

P. Perner (Ed.): MLDM 2016, LNAI 9729, pp. 446–458, 2016.
DOI: 10.1007/978-3-319-41920-6_35

crystal images, and the machines play no part in analyzing or classifying the images. The automatic imagers may generate millions of images annually in a single laboratory, leaving it to the crystallographers to study these images manually and identify the important results [6].

Although thousands of experiments are set up and observed over time, not all experiments end in desirable crystals. Different outcomes may eventuate, depending on external environmental conditions and the chemical conditions of each crystallization trial. Even though the main goal of a crystallization experiment is to obtain a protein crystal, identifying other outputs is also important for the crystallographer in order to extract information such as:

(i) conditions under which a particular protein may crystallise
(ii) the phase behaviour of a particular sample
(iii) stability of the protein
(iv) non-crystalline outcomes that provide solubility and other information in the protein crystallisation space rather than for a specific protein.

By studying the information associated with a sample, it is possible to gain significant insights into the general process of crystallization [4]. Therefore, it is important to analyze and classify the whole database of crystallography images.

In the medium-throughput crystallization laboratory within CSIRO (Collaborative Crystallisation Centre, C3, http://crystal.csiro.au/), 10,000 to 20,000 images are generated on a daily basis. Due to the enormous amount of data produced daily and the lack of time to study each image manually, a large amount of the data is left uninspected. Therefore, it is very likely that interesting phenomena that may occur during the crystallization process are missed. Automated analysis and classification can address many of these issues.

In order to become accustomed to a learning setting, the traditional machine learning algorithms combine multiple views in to a single view but multi-view learning presents with a function to create a specific view and to jointly improve all the functions to consider all the available views of the same input data and improve the learning performance [7]. To find the changes that occur during timecourses of protein crystallography experiments, several experiments were carried out using original images and multiple difference images in Mele et al [8]. These experiments provided with multiple sets of features, hence it was suitable to explore the route of multi-view learning to increase the classification accuracy of protein crystallography images.

In this paper we focus on automating the classification of x-ray crystallography images using multi-view learning with random forests. The next section of this paper provides a brief overview of related work. Our approach to solving the problem is then described, followed by the results and analysis. The final section outlines our conclusions and discusses potential future developments.

2 Related Work

Protein crystallization image classification is an interesting research area that is relatively underexplored, compared to other image databases. So far the classification experiments carried out by other research groups have only considered original image frames. All analyses require an initial segmentation step, to identify the region of the image associated with the experimental droplet, as image features outside the droplet are unrelated to the crystallization experiment.

A two-stage system has been used [5] to classify x-ray crystallography images. First the raw image data was processed into a vector of numeric features and then these features were used for classification using multiple random forests. Features used were Radon-Laplacian features, Sobel features, Sobel-edge features and microcrystal features, as well as grey-level co-occurrence matrix (GLCM) features. Kotseruba et al. [9] also used GLCM for classification of crystallography images. Cumbaa and Jurisica [10] used random forests to increase the classification performance, as it is considered to be one of the most accurate learning algorithms available.

A wavelet based method to classify crystallography images has been presented [11]. Several techniques have been used to exploit different sources of information. To extract features from a crystallization drop as a whole, wavelet transforms were used [11, 12]. Support vector machines with both linear (SVM_linear) and radial basis kernels (SVM_RBF), learning vector quantization (LVQ), self-organizing maps (SOMs) and linear discriminant analysis (LDA) were used as the main classification technique or combination of classification techniques that provided the highest accuracy [11]. Through this experiment it was found that the best results were obtained using SVM classifier with an RBF kernel.

Fourier transform is another technique for classification of crystallography images, as the Fourier transform has the ability to extract features to differentiate image texture and classify images. Fourier transform with a mask has also been used to classify crystallization images [13].

Another technique used by the x-ray crystallography community is the Hidden Markov model. They are used for both recognition and classification. In experiments the Hidden Markov model was used to interpret time sequence images [9]. This is one of the few instances where time sequence images have been considered, rather than single frames.

2.1 Contributions

So far, there is no work that has exploited the difference images of the X-ray crystallography sequence, and their properties. Most work has treated the sequence of images as individual frames. Nor has multi-view learning been applied to this problem, as far is known. In this paper, we advocate the use of the difference image sequence along with the original image sequence to provide two sets of independent features, that may then be utilized in multi-view learning. We employ statistical texture features of both the original and difference image sequences, along with random forests and multi-view learning, to classify x-ray crystallography images.

3 Approach

Our experiments were carried out on several x-ray crystallography data sets and the process has a number of steps. Firstly, each image was pre-processed to find the area of the image associated with the droplet, to align timecourse images using that information and to calculate the difference images. Then the feature calculation and basic classification was carried out. After analysing the results from the basic classification experiments, multi-view learning experiments were carried out. The process carried out for this experiment is illustrated in Fig.1.

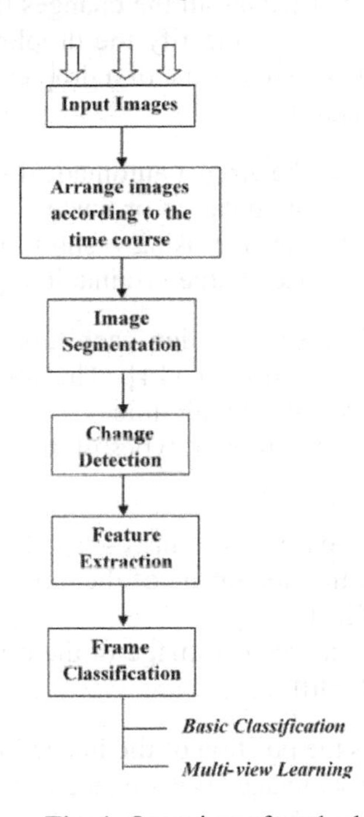

Fig. 1. Overview of method

3.1 Data

C3 facility in Melbourne, Australia provided data for our experiments. The image datasets were obtained by photographing crystallization experiments over a period of time varying from one hour to 10 weeks after start of the experiments. Each set of images belonging to a single crystallization experiment is called a 'time sequence'. Each time sequence consists of 8-16 images, and Fig. 2 shows an example of a single time sequence. Each experiment is carried out on a crystallization plate that contains 96 experimental wells, with each plate producing 96 time sequences. The images are grey scale, each providing a snapshot of changes inside the well at a point in time.

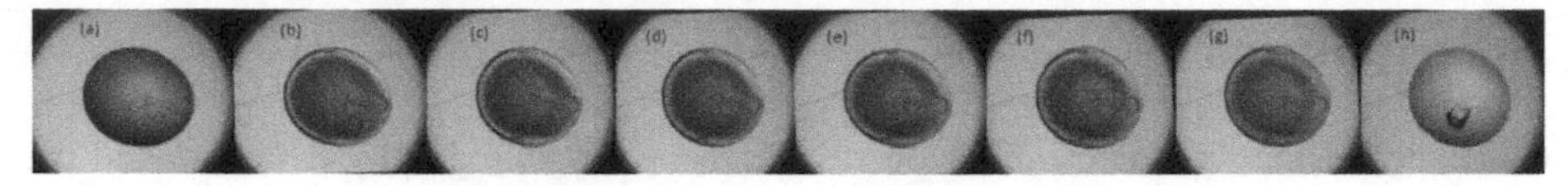

Fig. 2. Time sequence; Time intervals after the experiment is set up (a) 1 hour (b) 5 hours (c) 10 hours (d) 1 day (e) 2 days (f) 5 days, (g) 1 week and (h) 2 weeks

3.2 Pre-processing

The original images in each time sequence are arranged according to the acquisition time, and the droplet area is identified, as all the changes that occur during an experiment do so inside the droplet area. To identify the droplet area (Fig. 3) we used the DroplIt algorithm [14]. It identifies the outline of a droplet on the crystallization plate. This algorithm follows three main steps:

(i) identify a point within the droplet automatically
(ii) transform the domain into polar co-ordinates
(iii) find the outline of the droplet using a shortest path algorithm and transform the shortest path back to the original image.

After identifying the droplet area, we align each time sequence according to the masks obtained by the DroplIt algorithm [14]. The alignment is complicated by changes in the shape of the droplet itself with time.

After aligning the images, three different types of difference images of a sequence were computed.

(i) difference between consecutive images ($I_n - I_{n+1}$) (Diff_consec)
(ii) difference between the first image of the time sequence and the rest of the images ($I_1 - I_n$) (Diff_1)
(iii) difference between the second image of the time sequence and the rest of the images ($I_2 - I_n$) (Diff_2)

where I is the image and n is the position of the image in a time sequence.

Simple difference between two images was obtained first and the median filter was used to remove background noise. The pre-processing step is elaborated elsewhere [8].

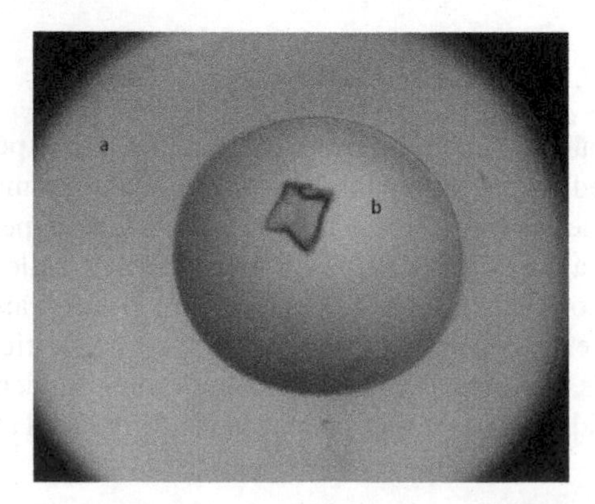

Fig. 3. Droplet in a well (a) Well area (b) Droplet area

3.3 Feature Extraction and Basic Classification

Feature extraction is an important step in any image classification experiment. For this work, we computed 31 statistical texture features for each image [6]. These appear in Table 1.

Table 1. Statistical texture features and their properties

Feature Name	Feature properties
Co-occurrence matrix	Contrast Correlation Energy Homogeneity
Moments of histogram	Mean Standard deviation Entropy Energy
Grey level difference method (Image was divided into 4 blocks)	Mean Standard deviation Entropy Contrast
Grey level run length method	Total number of runs Grey level uniformity
Discrete wavelet transform (Entropy for each matrix was computed)	Approximation coefficients matrix Horizontal coefficients matrix Vertical coefficients matrix Diagonal coefficients matrix

Table 2. Image feature set combinations

Combination no.	Image Feature Set Combinations
1	Features obtained from Diff_consec images
2	Features obtained from Diff_consec images with PCA
3	Features obtained from original images
4	Features obtained from original images with PCA
5	Features obtained from Diff_consec and original images
6	Features obtained from Diff_consec images with PCA and original images with PCA
7	Features obtained from Diff_1 images
8	Features obtained from Diff_1 images with PCA
9	Features obtained from Diff_2 images
10	Features obtained from Diff_2 images with PCA
11	Features obtained from Diff_consec, Diff_1 and original images
12	Features obtained from Diff_consec images with PCA, Diff_1images with PCA and original images with PCA
13	Features obtained from Diff_consec, Diff_1, Diff_2 and original images
14	Features obtained from Diff_consec images with PCA, Diff_1images with PCA, Diff_2 images with PCA and original images with PCA
15	Features obtained from Diff_1 and original images
16	Features obtained from Diff_1images with PCA and original images with PCA

These features were computed for both the original and difference images in each sequence. In preliminary classification experiments, each feature set on its own and different combinations of features for the different image combinations were used for classifying the images using support vector machines [15], J48 decision trees [16] and random forests [17, 18] with and without bagging/boosting [15, 16]. The experiments were carried out on different image feature set combinations of original and difference images, with and without principal component analysis (PCA). The images feature set combinations used are given in Table 2:

Table 3. Basic classification results for different combinations of feature sets, corresponding to the image feature set combinations listed in Table 2)

Combi-nation no	J48			SVM			Random Forests		
		Bag-ging	Boost ing		Bag-ging	Boosti ng		Bag-ging	Boost-ing
1	73.93 (3.97)	76.4 (10.46)	68.14 (4.05)	73.13 (3.58)	73.1 (3.58)	73.13 (3.58)	76.7 (4.88)	79.13 (4.16)	79.71 (3.55)
2	74.03 (4.70)	76 (8.91)	67.64 (4.49)	73.68 (4.17)	73.22 (4.17)	73.66 (3.75)	77.25 (4.73)	79.14 (4.36)	79.7 (3.88)
3	85.14 (3.02)	86.99 (10.98)	80.81 (2.64)	82.83 (2.82)	81.3 (2.82)	82.83 (3.03)	86.54 (2.77)	87.82 (2.83)	88.13 (2.43)
4	84.91 (3.62)	87.44 (12.59)	74.42 (3.08)	81.35 (3.47)	82.16 (3.79)	80.87 (3.38)	86.38 (3.14)	87.46 (3.20)	87.81 (2.64)
5	87.45 (3.49)	89.45 (13.89)	78.23 (2.26)	84.22 (2.93)	83.56 (2.93)	84.22 (2.78)	89.65 (3.07)	90.31 (2.98)	91.2 (2.45)
6	86.79 (3.38)	88.52 (17.0)	77.93 (2.81)	83.33 (3.24)	76.98 (3.24)	83.33 (3.63)	87.61 (3.48)	89.44 (2.88)	90.12 (2.65)
7	90.49 (2.52)	90.54 (8.22)	89.03 (2.76)	87.12 (3.19)	86.98 (3.19)	87.1 (3.06)	88.97 (3.14)	89.89 (2.48)	91.25 (2.30)
8	88.49 (2.74)	88.81 (10.80)	85.64 (2.54)	84.43 (3.54)	84.29 (3.60)	84.52 (3.31)	86.91 (4.41)	88.61 (3.47)	89.4 (2.48)
9	90.12 (2.54)	90.53 (12.61)	79.06 (2.53)	85.69 (3.14)	84.87 (2.98)	85.69 (3.14)	88.72 (3.32)	90.25 (2.50)	91.17 (3.38)
10	88.59 (2.95)	89.87 (2.58)	74.81 (17.08)	85.19 (3.47)	84.27 (3.47)	85.35 (3.83)	86.35 (4.21)	89.21 (2.75)	89.31 (3.99)
11	87 (3.67)	89.56 (2.90)	78.21 (13.85)	85.91 (2.36)	84.87 (2.43)	85.91 (2.36)	89.39 (3.33)	90.33 (2.65)	91.3 (2.72)
12	83.79 (3.38)	88.52 (2.81)	77.93 (17.00)	85.9 (3.24)	84.83 (3.63)	85.9 (3.24)	86.21 (2.65)	88.7 (3.79)	88.88 (2.84)
13	88.39 (3.72)	90.24 (2.89)	80 (11.04)	83.53 (2.80)	81.32 (2.99)	83.53 (2.80)	89.74 (3.20)	90.65 (2.70)	91.55 (3.14)
14	88.84 (3.41)	90.06 (2.97)	78.37 (13.11)	88.04 (2.73)	87.09 (3.22)	88.75 (2.83)	88.98 (4.10)	90.22 (2.53)	91.71 (3.25)
15	87.49 (3.80)	90.41 (2.48)	79.95 (13.60)	88.34 (2.56)	87.34 (2.56)	88.34 (2.44)	89.68 (2.94)	90.72 (2.60)	91.81 (2.65)
16	87.29 (3.44)	90.45 (2.91)	81.63 (9.85)	88.56 (2.65)	88.01 (3.03)	88.54 (2.65)	90.32 (3.48)	91.39 (2.65)	92.05 (2.72)

The classification accuracy and their respective variance results for these experiments are illustrated in Table 3. The images were classified into 5 classes:

1. crystal
2. precipitation
3. skin
4. clear and
5. other.

From the results, the highest classification accuracy of 92.05% with a variance of 2.72 (as highlighted on Table 3) is obtained using a combination of original image features and the features of the difference images obtained by subtracting other images from the first image in a time sequence (I_1-I_n). The classification method that produced this accuracy was random forests. Moreover, after performing a pair-wise Wilcoxon test it also confirms that using random forests provides a higher accuracy compared to the other methods.

3.4 Classification Using Multi-view Learning

As the basic classification results in Table 3 indicate that a combination of feature sets (original image features and difference image features) provides better classification results than single feature sets, multi-view learning [7, 19-24] can be used to exploit the multiple feature sets available. The two views considered for multi-view learning were:

(i) original image features
(ii) image features of the difference images obtained by subtracting other images from the first image in a time sequence (I_1-I_n)

as these provided the best results in basic classification experiments described in section 3.3. Random forests is the classifier of choice for both views for the same reason. To provide feedback to the classification algorithm, co-training was used [24-27]. Co-training is based three assumptions:

(i) sufficiency: each view is sufficient for classification on its own
(ii) compatibility: the target functions in both views predict the same labels for co-occurring features with high probability
(iii) conditional independence: the views are conditionally independent given the class label [7, 22, 24, 28]

The average correlation between the two feature sets was computed and found to be 0.08. As the above assumptions were satisfied by the two feature sets, further experiments were carried out.

Three different data sets were used to carry out the MVL experiments. These datasets contained few labelled images, where the labels were a user-created score that was created manually by the users.

1. XC1000: 1000 image data set, with all images labelled; this is the same set that was used to carry out the basic classification experiments in section 3.3
2. XC5000: 5000 image data set, of which only 150 images are initially labelled
3. XC4000: 4000 image data set with no labelled images.

4 Experiments and Results

Each MVL experiment starts with a 100 image labelled training set and random forests was trained on it. Each test iteration comprises another 100 randomly selected unlabelled images. Testing is repeated 10 times with 10 different randomly selected (without replacement) 100 image set for each iteration. For each iteration:

(i) non-matching image labels by the two view classifiers are identified
(ii) the corresponding image is manually checked and its label corrected by an expert, and provided to the system as feedback
(iii) the training and testing cycle is repeated

For each iteration, testing was also carried out using a completely separate 1000 image test data set (XC_Test1000). This test data set was not used as feedback to the system and solely used for the purpose of testing the system after each iteration. As the XC4000 dataset does not have any labelled images, it was combined with the XC5000 dataset and experiments were carried out on all 9000 images (XC9000).

Each of Figs 4-6 displays the results of one of the datasets. Each figure contains 2 or 3 line graphs which represent:

Average: the average accuracy on each dataset after running the experiment for 10 iterations while providing feedback to the system at each iteration with random test samples.

Baseline: Highest classification obtained for each data set using random forests but without MVL.

Average test set: Average accuracy obtained by testing the classifer at each iteration on XC_Test1000.

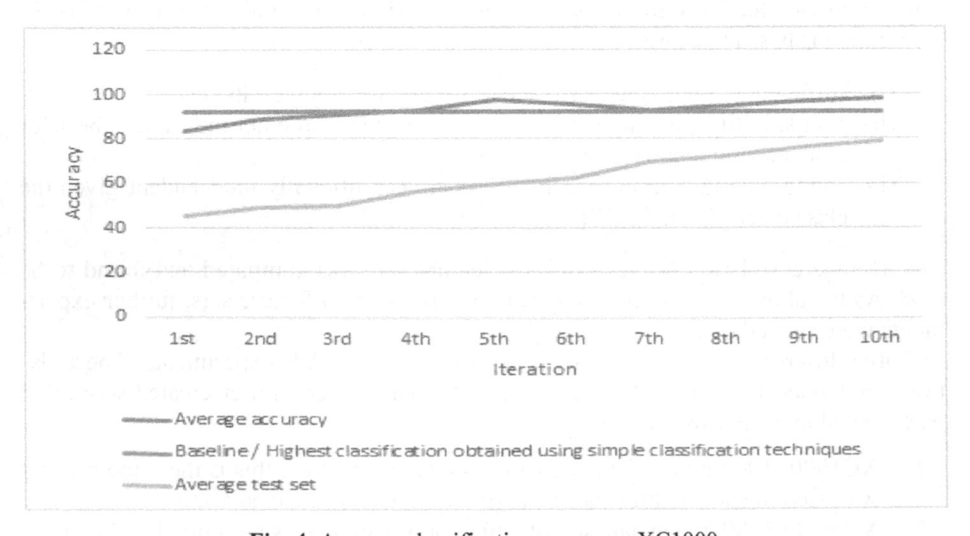

Fig. 4. Average classification accuracy: XC1000

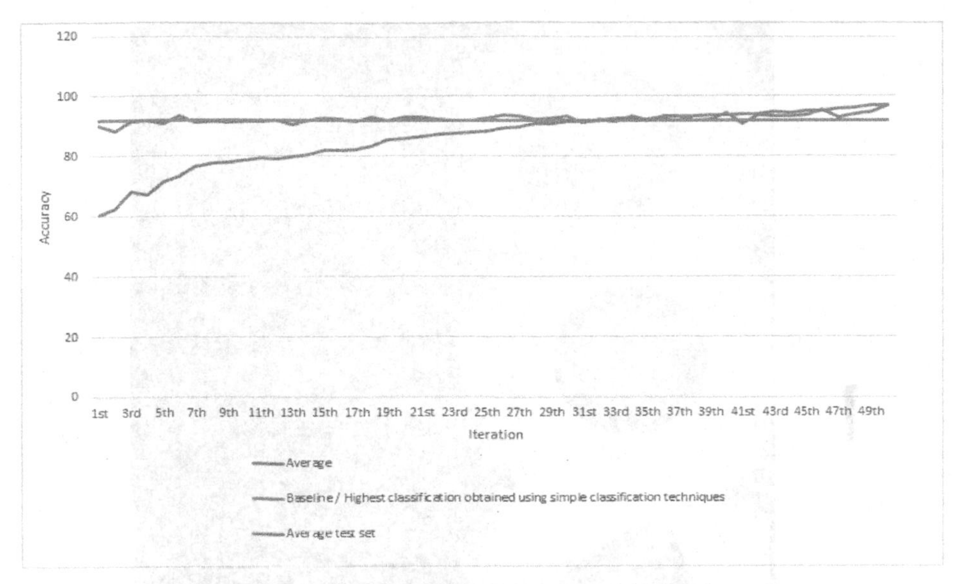

Fig. 5. Average classification accuracy: XC5000

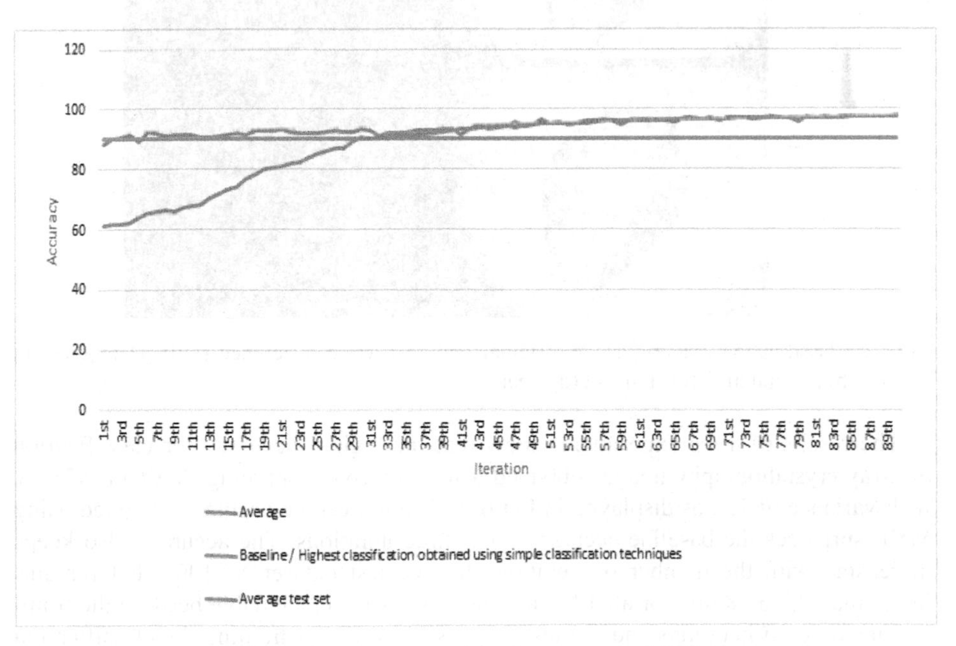

Fig. 6. Average classification accuracy: XC9000

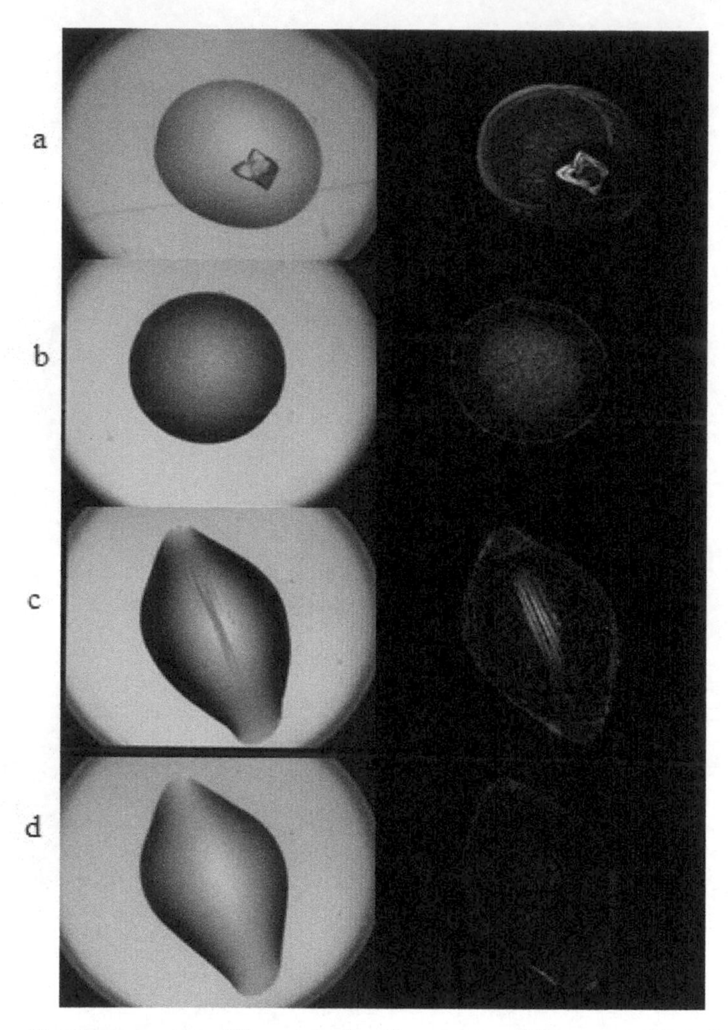

Fig. 7. Classification examples. The original and respective difference (Diff_1) images (a) Crystal (b) precipitation (c) skin and (d) clear

The results of these experiments show that the highest accuracy for classification of x-ray crystallography images obtained using multi-view learning (MVL) is 97.8%, with variance of 7.3, as displayed in Fig 6. In Fig 4, average accuracy obtained using MVL surpasses the baseline accuracy after three iterations. The accuracy also keeps increasing with the number of iterations for the test dataset XC1000, but remains lower than the base line for all 10 iterations. This is understandable because the training has only 10 iterations and the test set was not used for training the classification algorithms. It can be seen that this problem is overcome when the number of iterations increases further, as in Figs 5 (XC5000) and 6 (XC9000). The accuracy obtained in the iterations therein shows that even though there is lot of variation in average accuracy during the training process, accuracy does keep increasing with time. In both cases, the accuracy achieved on the XC_Test1000 test dataset rapidly increases till the 25-27th iterations and then keeps gradually increasing till the end.

An example of each of the first four classes is provided in Fig. 7. If an image does not belong to any of the 4 classes, it is classified as other.

In these experiments we have continued to train the classifier with every iteration. A logical next step is to determine a termination condition for training, which would reduce the training time and produce a stable classifier faster.

5 Conclusion

The highest accuracy obtained from multi-view learning for X-ray crystallography datasets is 97.8% using random forests as the classification technique for both the views. This is significantly higher than results obtained by other basic classification techniques used before and reported in the literature. It is reasonable to conclude that the combination of multi-view learning and the difference image features helps to increase the classification accuracy of protein X-ray crystallography image classification. To further improve the multi-view learning application, we are currently experimenting with obtaining the most informative/ most uncertain samples in each test set and using them to train the next iteration, rather than using the whole test set. Also the other base classifiers may be utilized in a multi-view learning setting to check for improvements over the base-classifier results.

References

1. Publications.nigms.nih.gov. Chapter 2: X-ray Crystallography: Art Marries Science - The Structures of Life - Science Education - National Institute of General Medical Sciences (2011). http://publications.nigms.nih.gov/structlife/chapter2.html (cited 2012)
2. Hofmann, A., et al.: Methods of Molecular Analysis in the Life Sciences. Cambridge University Press (2014)
3. Gray, E.D., et al.: What is x-ray crystallography? n.d. http://www.chem.ed.ac.uk/bunsen_learner/bunsen_xray.html (cited 2012)
4. Newman, J., et al.: On the need for an international effort to capture, share and use crystallization screening data. Acta. Crystallogr. Sect. F Struct. Biol. Cryst. Commun. **68**(Pt 3), 253–258 (2012)
5. Cumbaa, C., Jurisica, I.: Protein crystallization analysis on the World Community Grid. Journal of Structural and Functional Genomics **11**(1), 61–69 (2010)
6. Lekamge, B.M.T., et al.: Classification of protein crystallisation images using texture-based statistical features. In: AIP Conference Proceedings, vol. 1559, no. 1, pp. 270–276 (2013)
7. Xu, C., Tao, D., Xu, C.: A Survey on Multi-view Learning (2013)
8. Mele, K., et al.: Using Time Courses To Enrich the Information Obtained from Images of Crystallization Trials. Crystal Growth & Design **14**(1), 261–269 (2013)
9. Kotseruba, Y., Cumbaa, C.A., Jurisica, I.: High-throughput protein crystallization on the World Community Grid and the GPU. Journal of Physics: Conference Series **341**(1), 012027 (2012)

10. Cumbaa, C., Jurisica, I.: Automatic classification and pattern discovery in high-throughput protein crystallization trials. J. Struct. Funct. Genomics **6**(2–3), 195–202 (2005)
11. Buchala, S., Wilson, J.C.: Improved classification of crystallization images using data fusion and multiple classifiers. Acta Crystallographica Section D **64**(8), 823–833 (2008)
12. Watts, D., Cowtan, K., Wilson, J.: Automated classification of crystallization experiments using wavelets and statistical texture characterization techniques. Journal of Applied Crystallography **41**(1), 8–17 (2008)
13. Walker, C.G., Foadi, J., Wilson, J.: Classification of protein crystallization images using Fourier descriptors. Journal of Applied Crystallography **40**(3), 418–426 (2007)
14. Vallotton, P., et al.: DroplIT, an improved image analysis method for droplet identification in high-throughput crystallization trials. Journal of Applied Crystallography **43**(6), 1548–1552 (2010)
15. Mitchell, T.M.: Machine Learning, p. 45. McGraw Hill, Burr Ridge (1997)
16. Witten, I.H., Frank, E., Hall, M.A.: Data Mining: Practical Machine Learning Tools and Techniques, 3rd edn, p. 629. Elsevier Inc., Burlington (2011)
17. Aiping, W., et al.: An incremental extremely random forest classifier for online learning and tracking. In: 2009 16th IEEE International Conference on Image Processing (ICIP) (2009)
18. Shotton, J., Johnson, M., Cipolla, R.: Semantic texton forests for image categorization and segmentation. In: IEEE Conference on Computer Vision and Pattern Recognition, CVPR 2008, Anchorage, AK, p. 1–8 (2008)
19. Wang, Z., et al.: Feature Extraction via Multi-View Non-Negative Matrix Factorization with Local Graph Regularization
20. Hady, M.F.A., et al.: Multi-view forests based on Dempster-Shafer evidence theory: a new classifier ensemble method. In: Proceedings of the Fifth IASTED International Conference on Signal Processing, Pattern Recognition and Applications. ACTA Press, Innsbruck, pp. 18–23 (2008)
21. Li, S.-Y., Jiang, Y., Zhou, Z.-H.: Partial multi-view clustering. In: Twenty-Eighth AAAI Conference on Artificial Intelligence (2014)
22. Sun, S.: A survey of multi-view machine learning. Neural Computing and Applications **23**(7–8), 2031–2038 (2013)
23. Wang, M., Hua, X.-S.: Active learning in multimedia annotation and retrieval: A survey. ACM Trans. Intell. Syst. Technol. **2**(2), 1–21 (2011)
24. Settles, B.: Active Learning Literature Survey (2010)
25. Wang, W., Zhou, Z.-H.: Analyzing co-training style algorithms. In: Kok, J.N., Koronacki, J., Lopez de Mantaras, R., Matwin, S., Mladenič, D., Skowron, A. (eds.) ECML 2007. LNCS (LNAI), vol. 4701, pp. 454–465. Springer, Heidelberg (2007)
26. Blum, A., Mitchell, T.: Combining labeled and unlabeled data with co-training. In: Proceedings of the Eleventh Annual Conference on Computational Learning Theory, pp. 92–100. ACM, Madison (1998)
27. Hillebrand, M., Kreßel, U., Wöhler, C., Kummert, F.: Traffic Sign classifier adaption by semi-supervised co-training. In: Mana, N., Schwenker, F., Trentin, E. (eds.) ANNPR 2012. LNCS, vol. 7477, pp. 193–200. Springer, Heidelberg (2012)
28. Lazarova, G., Koychev, I.: A semi-supervised multi-view genetic algorithm. In: 2014 2nd International Conference on Artificial Intelligence, Modelling and Simulation (AIMS) (2014)

C-KPCA: Custom Kernel PCA for Cancer Classification

Van-Sang Ha[1(✉)] and Ha-Nam Nguyen[2]

[1] Department of Economic Information System, Academy of Finance, Hanoi, Viet Nam
sanghv@hvtc.edu.vn

[2] Department of Information Technology, VNU-University of Engineering and Technology, Hanoi, Viet Nam
namnh@vnu.edu.vn

Abstract. Principal component analysis (PCA) is an effective and well-known method for reducing high-dimensional data sets. Recently, KPCA (Kernel PCA), a nonlinear form of PCA, has been introduced into many fields. In this paper, we propose a new gene selection, namely Custom Kernel principal component analysis (C-KPCA). The new kernel function for KPCA is created by combining a set of kernel functions. First, Singular Value Decomposition (SVD) is used to reduce the dimension of microarray data. Input space is then mapped to a higher-dimensional feature space using the proposed custom kernel function. The main objective of our method is to extract nonlinear features for classification process. In order to test the accuracy of our method, a number of experiments are carried out on four binary gene datasets: Colon Tumor, Leukemia, Lymphoma, and Prostate. The experimental results show that our proposed method results in a higher prediction rate as comparing with several recently published algorithms.

Keywords: Feature extract · KPCA · SVD · Cancer classification · Dimension reduction

1 Introduction

Traditional cancer diagnosis takes a long time because it is based on clinical evaluation and physical examination. Inexpensive microarray gene expression data, having an adequate number of genes to detect diseases, can help diagnostic research develop a fast diagnostic procedure.

In the field of data mining, processing high dimensional dataset is an important task. Most of existing classification algorithms only can handle limited and low dimension data. Microarray technology has generated terabytes of biological data. Microarray data usually contains a small number of samples which have a large number (thousands to tens of thousands) of gene expression levels as features [1]. It leads to the curse of dimensionality. In addition, irrelevant and redundant gene expression data require a high computational complexity and make it impossible to discover relevant genes. The redundancy and the deficiency in data may result in reducing the classification accuracy and lead to incorrect decision [2] [3]. To solve those problems,

P. Perner (Ed.): MLDM 2016, LNAI 9729, pp. 459–467, 2016.
DOI: 10.1007/978-3-319-41920-6_36

feature selection and feature extraction are two techniques commonly used in dimensionality reduction. Indeed, filter approach and wrapper approach are widely used for feature selection [4]. On one hand, the filter approach works as a pre-processing step that selects features without considering the accuracy of classification algorithm. On the other hand, the wrapper approach relies on the predictive accuracy of an algorithm to evaluate the possible subset of features and select a subset of features that provides the highest accuracy. Feature selection method was introduced in the 1970s in the literature of probability statistics, machine learning and data mining [5][6][7].

Feature extraction has been studied for a long time. Many methods have been proposed, for example principle component analysis (PCA) [8], linear discriminant analysis (LDA) [9], Independent Component Analysis (ICA) [10], etc. The advantages of feature extraction are speeding up the algorithm, improving the data quality and increasing the performance of the mining results. Though being successfully applied for pattern classification, these methods usually miss some discriminant information while extracting relevant features for the classification task.

Principal component analysis (PCA) is a well-known and famous statistical technique that can be used to reduce the dimensionality of microarray data. Recently, a method proposed by Scholkhof et al., namely KPCA, applies kernel technique in PCA in order to solve nonlinear problems [11]. In the field of machine learning, KPCA has been extensively studied and was successful in many applications [12]. Liu et al. developed an algorithm for classification using KPCA [13]. Moreover, a number of optimization strategies for KPCA have been proposed, for example Pochet et al. suggested to use RBF kernel followed by Fisher Discriminant Analysis (FDA) to find the parameters of KPCA [14].

In the last few years, many classification models and algorithms were applied to analyze cancer, for example decision tree [15], nearest neighbor K-NN [16], [17], support vector machine (SVM) [18], [19] [20] [21] and neural network [22][23]. In order to achieve higher classification performance, SVM-recursive feature elimination (SVM-RFE) method was introduced to filter relevant features and remove relatively insignificant feature variables [20][24][25]. Recently, other methods like evolutionary algorithms [22], stochastic optimization technique [23] and neural network have shown promising results in terms of prediction accuracy. Indeed, the main goal of our research is to reduce the dimensionality of microarray datasets in order to design a powerful pre-processing step. Our study proposed a feature extraction method based on the custom kernel method for KPCA to improve performance of cancer classification. The proposed method reduces features via SVD. We applied the custom kernel to several cancer datasets such as Colon Tumor, Leukemia, Lymphoma and Prostate. The C-KPCA method shows a better classification accuracy than KPCA and sometimes higher than some existing algorithms.

The rest of the paper is organized as follows: Section 2 presents the background of KPCA, Custom kernel method. Next, Section 3 describes the details of the proposed model. Then, Experimental results are discussed in Section 4. Finally, concluding remarks and future works are presented in Section 5.

2 Background

2.1 Kernel Method

Kernel methods are important pillars to machine learning system theory. Many researchers have applied kernel method in supervised and unsupervised learning. Kernel methods have received growing attention and have been extensively covered in many machine learning and pattern recognition textbooks [26][27][28]. A kernel function provides a flexible and effective learning mechanism in SVM, and the choice of a kernel function should reflect prior knowledge about the problem at hand. However, it is often difficult for us to exploit the prior knowledge on patterns to choose a kernel function, and it is an open question how to choose the best kernel function for a given dataset. According to no free lunch theorem [7] on machine learning, there is no superior kernel function in general, and the performance of a kernel function rather depends on applications.

Table 1. The types of kernel functions

Kernel function	Formula
Inverse Multi-Quadric	$1/\sqrt{\|x-y\|^2+c^2}$
Radial	$e^{(-\gamma\|x-y\|^2)}$
Neural	$tanh(s \cdot \langle x,y \rangle - c)$

2.2 Kernel PCA

Principal component analysis (PCA) is a common method applied to dimensionality reduction and feature extraction. PCA transforms the original features of a dataset to a reduced number of uncorrelated ones, termed principal components. The projected coordinates can recover the original data quite accurately. However, PCA has some disadvantages: it is an orthogonal linear transformation, and PCA is not optimal for discovering structures for input data with a highly non-Gaussian distribution. In many practical applications, the effectiveness of this method is often limited by the nonlinear data. In order to apply this method to nonlinear datasets, many researcher have applied neural network in transferring data into linear. Kramer et al. [29] proposed nonlinear PCA based on neural network. However, this neural network is quite complex and difficult to find optimal result due to its five layers. The new nonlinear techniques do not outperform the traditional PCA on real- world tasks.

Kernel principal component analysis is the approach of generating the traditional linear PCA in a high-dimensional space using a kernel function. KPCA maps the original space into a very high dimension features space z through some nonlinear u(xt), z = u(xt), in this feature space the PCA is performed similar to applying nonlinear PCA in the original input space.

Given a set of observations $x_i \in R^p, i = 1, \ldots, n$, consider a feature space F related to the input space by a map $\emptyset: R^p \to F$, which may be nonlinear. We assume that we are dealing with centered data$\sum_{i-1}^{n} \emptyset(x_i) = 0$. In F the covariance matrix takes the form

$$C = \frac{1}{n} \sum_{j-1}^{n} \emptyset(x_j) \emptyset^T(x_j) \quad (1)$$

We seek eigenvalues λ ≥0 and nonzero eigenvectors v∈F\{0} satisfying Cv=λv. All solutions v with λ≠0 lie in the span of $\{\emptyset(x_j)\}_{i-1}^{n}$ as shown in the literature. This has two consequences: first we may instead consider the set of equations

$$\langle \emptyset(x_j), \mathbf{C}\mathrm{v} \rangle = \lambda \langle \emptyset(x_j), \mathrm{v} \rangle \quad (2)$$

For all j=1,...,n, where ⟨.,.⟩ denotes the dot product defined in F. Second, there exits coefficients α_i, i=1,…,n, such that

$$v = \sum_{i-1}^{n} \alpha_i \emptyset(x_i) \quad (3)$$

Combining Formulas 2 and 3 we obtain the dual representation of the eigenvalue problem for nonzero eigenvalues:

$$\mathbf{K}\boldsymbol{\alpha} = n\lambda\boldsymbol{\alpha} \quad (4)$$

Where $\mathbf{K}=(K(x_i,x_j))$, i, j=1,…,n, is the kernel matrix and K is a kernel function such that the dot product in F satisfies $\langle \emptyset(x_i), \emptyset(x_j) \rangle = K(x_i, x_j)$ Let $\lambda_1 \geq \lambda_2 \geq \ldots \geq \lambda_n$ be the eigenvalues of **K** and $\alpha^1, \ldots, \alpha^n$ be the corresponding set of normalized eigenvectors, with λ_r being the last nonzero eigenvalue. For the purpose of principal component extraction, we need to compute the projections onto the eigenvectors v^j in F, j=1,…,r. Let x be a test point, with an image$\emptyset$(x)) in F. Then

$$\langle \mathrm{v}^j, \emptyset(x) \rangle = \sum_{i-1}^{n} \propto_i^j K(x_i, x) \quad (5)$$

which is the j-th nonlinear principal component corresponding to $\emptyset$

3 The Proposed Method

3.1 The Proposed Framework

The basic framework of the system consists of three components: data preprocessing, dimensionality reduction and data classification. Our proposed method can be described in the following figure:

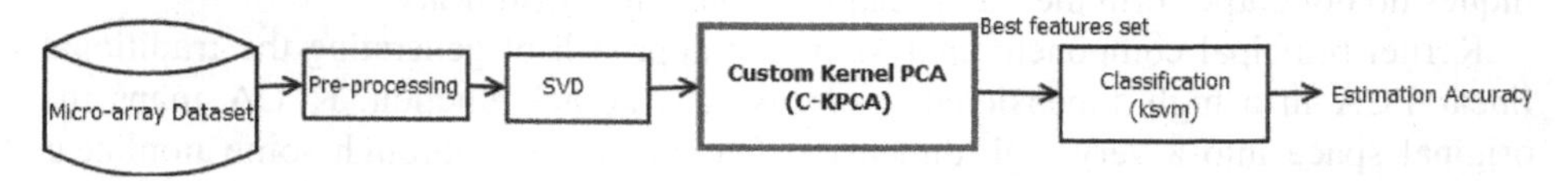

Fig. 1. Framework of the proposed method

In this framework, input microarray data is first preprocessed by a scaling function, a normalization function and a transformation function in preprocessing module. Next, SVD [30] is used to reduce the dimension of microarray data. As a result, input data matrix is decomposed into singular values.

In order to understand the relationships between samples (microarrays) and variables (genes), we extract nonlinear features by using C-KPCA procedure. The variables are mapped into a feature space, then the main nonlinear features of variables are extracted.

Since microarray data are generally nonlinear ksvm classifier from kernlab (an R-library) is used to perform cancer classification. By default, the ksvm() function uses the Gaussian RBF kernel, but our work propose a custom kernel as shown in Formula (7).

3.2 Custom Kernel Function

There are several ways to construct a kernel such as inferring from other kernels or from features [31]. In this paper, we only focus on how-to build a new kernel from other kernels. A symmetric function is a kernel if the matrix defined by restricting the function to any finite set of points is positive semi-definite. The proposition that helps to construct more complicated kernel based on other kernels is restated as following:

Proposition 1. Let K_1 and K_2 be kernels over $X*X$, $X \subseteq R^n$, $a \in R^+$, $f(.)$ is a real valued function on X

$$\Phi: X \rightarrow R^m \tag{6}$$

with K_3 a kernel over $R^m * R^m$ and B is a symmetric positive semi-definite $n*n$ matrix. Then the following functions on X are kernels:

$$
\begin{aligned}
&1.\ K(x,z) = K_1(x,z) + K_2(x,z)\\
&2.\ K(x,z) = a * K_1(x,z)\\
&3.\ K(x,z) = K_1(x,z) * K_2(x,z)\\
&4.\ K(x,z) = f(x) * f(z)\\
&5.\ K(x,z) = K_3(\phi(x), \phi(z))\\
&6.\ K(x,z) = x' Bz
\end{aligned}
$$

The proofs of the proposition can be found in [31]. This proposition is the principle for us to create a new kernel function. A new kernel function is created by combining the set of kernel functions. The custom kernel function has the form of

$$K_c = (K_1)^{e_1} \circ \cdots \circ (K_m)^{e_m} \tag{7}$$

where $\{K_i \mid i = 1, \ldots, m\}$ is the set of kernel functions to be custom, e_i is the exponent of i-th kernel function, and $\circ$ denotes an operator between two kernel functions such as plus, minus, multiply or divide operators. In our case, three types of the kernel functions listed in Table 1 are custom, and multiplication or addition operators are used to combine kernel functions.

4 Experiment and Results

Our proposed algorithm was implemented using R language (http://www.r-project.org). We tested the proposed algorithm with several public datasets including colon tumor, leukemia, lymphoma and prostate to validate our approach. The learning and validation accuracies were determined by means of 10-fold cross validation. In this paper, we used KPCA and ksvm with the original dataset as the base-line method. The proposed method and the base-line method were executed on the same datasets to compare their efficiency.

4.1 Data Set

Colon Tumor Dataset. The colon tumor dataset composes of 2,000 genes in 40 colon tumor samples and 22 normal colon tissue samples. Gene expression levels are analyzed with Affymetrix oligonucleotide array. The dataset is available on the web at http://www.molbio.princeton.edu/colondata. We complete the preprocessing of the gene expression data with a microarray standardization and gene centering.

Leukemia Dataset. The leukemia dataset composes of 3,051 gene expressions in three classes of leukemia: 19 cases of B-cell acute lymphoblastic leukemia (ALL), 8 cases of T-cell ALL and 11 cases of acute myeloid leukemia (AML). Gene expression levels are measured using Affymetrix high-density oligonucleotide arrays. The data can be downloaded from http://www.genome.wi.mit.edu. The data are preprocessed according to the protocol described by Dudoit et al [32].

Lymphoma Dataset. The lymphoma dataset comes from a study of gene expression of three prevalent lymphoid malignancies: B-cell chronic lymphocytic leukemia (B-CLL), follicular lymphoma (FL) and diffuse large B-cell lymphoma (DLCL). Among 96 samples we took 62 samples containing 4,026 genes in three classes: 11 cases of B-CLL, 9 cases of FL and 42 cases of DLCL. Gene expression levels were measured using two-channel cDNA microarrays. The data can be obtained from http://genome-www. stanford.edu/lymphoma. After preprocessing, all gene expression profiles were base-10 log-transformed and standardized to zero mean and unit variance in order to prevent single arrays from dominating the analysis.

Prostate Dataset. The prostate dataset comprises 12,533 genes with 102 samples. Among them, prostate tumor occupies 51% with 52 samples. The normal tissues account for 49% with 50 examples. The data can be downloaded from http://www-genome.wi.mit.edu/mpr/prostate. After preprocessing, all gene expression profiles were base-10 log-transformed and standardized to zero mean and unit variance in order to prevent single arrays from dominating the analysis.

4.2 Results

In the experiments carried out on four aforementioned cancer datasets, the proposed method with the custom kernel function frequently lead to much higher accuracy than the traditional KPCA method with default kernels (see Table 2).

Table 2. The classification accuracy of the custom kernel PCA method and the traditional KPCA method on four microarray datasets.

Number of genes	Colon		Leukemia		Lymphoma		Prostate	
	KPCA	C-KPCA	KPCA	C-KPCA	KPCA	C-KPCA	KPCA	C-KPCA
5	87.5	**98.39**	76.25	**78.25**	98.70	98.70	91.13	79.41
10	91.93	**96.74**	87.70	80.56	98.05	**98.70**	94.11	**97.05**
15	91.93	**100**	88.06	**91.67**	99.74	98.70	94.46	**98.04**
20	92.50	**100**	88.13	**100**	100	100	94.17	**100**
50	87.50	**100**	87.15	**100**	96.30	**100**	97.79	**100**
100	83.71	**100**	88.33	**100**	76.88	**100**	95.63	**100**
200	82.74	**100**	85.42	**100**	84.80	**100**	100	**100**
500	81.29	**100**	81.88	**100**	83.64	**100**	100	**100**

Table 3. The classification accuracy of the custom kernel PCA method and the mRMR+SVM method

Number of genes	Colon		Leukemia		Lymphoma		Prostate	
	mRMR+SVM	C-KPCA	mRMR+SVM	C-KPCA	mRMR+SVM	C-KPCA	mRMR+SVM	C-KPCA
50	91.94	**100**	91.66	**100**	93.33	**100**	97.79	**100**
100	93.55	**100**	97.22	**100**	98.48	**100**	95.63	**100**
200	96.77	**100**	100	100	100	100	100	100
500	100	100	100	100	100	100	100	100

Table 3 shows the classification accuracy of our methods and that of the existing method mRMR+SVM [33]. The accuracy of our method is always better than that of mRMR+SVM. Our framework even can achieve 100 % classification accuracy in most of experimented cases, i.e. number of genes are 50, 100, 200, and 500. These results also indicate that the new kernel and the learning method result in more stable classification accuracies than the existing ones, i.e. traditional KPCA, and mRMR+SVM.

5 Conclusion

In this paper, we focused on studying the kernel method. Features extraction involves in determining the highest classifier accuracy of a subset or seeking the acceptable accuracy of the smallest subset of features. We have introduced a new feature extraction approach based on KPCA. The accuracy of classifier using the selected features is better than several recently published methods. Fewer features allow us to concentrate on collecting relevant and essential variables. The proposed custom kernel leads to a significant decrement in runtime. The experimental results show that our method is effective in cancer classification. It makes not only the evaluation be more quickly but also the accuracy of the classification be considerably increased.

References

1. Wang, Y., Tetko, I.V., Hall, M.A., Frank, E., Facius, A., Mayer, K.F.X., Mewes, H.W.: Gene selection from microarray data for cancer classification - A machine learning approach. Comput. Biol. Chem. **29**(1), 37–46 (2005)
2. Liu, H., Motoda, H.: Feature Selection for Knowledge Discovery and Data Mining (1998)
3. Guyon, I., Elisseeff, A.: An Introduction to Variable and Feature Selection. J. Mach. Learn. Res. **3**, 1157–1182 (2003)
4. Kohavi, R., John, G.H.: Wrappers for feature subset selection. Artif. Intell. **97**(1–2), 273–324 (1997)
5. Blum, A.L., Langley, P.: Selection of relevant features and examples in machine learning. Artif. Intell. **97**(1–2), 245–271 (1997)
6. Tan, P.-N., Steinbach, M., Kumar, V.: Introduction to Data Mining, p. 500. Addison Wesley (2005)
7. Duda, R.O., Hart, P.E., Stork, D.G.: Pattern Classification, p. 680. John Wiley Section, New York (2001)
8. Kirby, M., Sirovich, L.: Application of the Karhunen-Loeve procedure for the characterization of human faces. IEEE Trans. Pattern Anal. Mach. Intell. **12**(1), 103–108 (1990)
9. Swets, D.L.: Using discriminant eigenfeatures for image retrieval. IEEE Trans. Pattern Anal. Mach. Intell. **18**(8), 831–836 (1996)
10. Comon, P.: Independent component analysis, A new concept? Signal Processing **36**(3), 287–314 (1994)
11. Scholkopf, B., Smola, A., Muller, K.: Nonlinear Component Analysis as a Kernel Eigenvalue Problem. Neural Comput. **10**, 1299–1319 (1998)
12. Ng, A.Y., Jordan, M.I., Weiss, Y.: On Spectral Clustering: Analysis and an algorithm. Adv. Neural Inf. Process. Syst., 849–856 (2001)
13. Liu, Z., Chen, D., Bensmail, H.: Gene expression data classification with kernel principal component analysis. J. Biomed. Biotechnol. **2005**(2), 155–159 (2005)
14. Pochet, N., De Smet, F., Suykens, J.A.K., De Moor, B.L.R.: Systematic benchmarking of microarray data classification: Assessing the role of non-linearity and dimensionality reduction. Bioinformatics **20**(17), 3185–3195 (2004)
15. Czajkowski, M., Grześ, M., Kretowski, M.: Multi-test decision tree and its application to microarray data classification. Artif. Intell. Med. **61**(1), 35–44 (2014)
16. Aha, D.W., Kibler, D., Albert, M.K.: Instance-Based Learning Algorithms. Mach. Learn. **6**(1), 37–66 (1991)
17. Li, L., Weinberg, C.R., Darden, T.A., Pedersen, L.G.: Gene selection for sample classification based on gene expression data: study of sensitivity to choice of parameters of the GA/KNN method. Bioinformatics **17**(12), 1131–1142 (2001)
18. Guyon, I., Weston, J., Barnhill, S., Vapnik, V.: A gene selection method for cancer classification using Support Vector Machines. Mach. Learn. **46**, 389–422 (2002)
19. Vapnik, V.: The Nature of Statistical Learning Theory, vol. 8 (1995)

20. Mundra, P.A., Rajapakse, J.C.: SVM-RFE with MRMR filter for gene selection. IEEE Trans. Nanobioscience **9**, 31–37 (2010)
21. Kim, S.: Margin-maximized redundancy-minimized SVM-RFE for diagnostic classification of mammograms. In: 2011 IEEE Int. Conf. Bioinforma. Biomed. Work., pp. 562–569 (2011)
22. Tong, D.L., Schierz, A.C.: Hybrid genetic algorithm-neural network: Feature extraction for unpreprocessed microarray data. Artif. Intell. Med. **53**(1), 47–56 (2011)
23. Vimaladevi, M., Kalaavathi, B.: Cancer Classification using Hybrid Fast Particle Swarm Optimization with Backpropagation Neural Network **3**(11), 8410–8414 (2014)
24. Duan, K.B., Rajapakse, J.C., Wang, H., Azuaje, F.: Multiple SVM-RFE for gene selection in cancer classification with expression data. IEEE Trans. Nanobioscience **4**(3), 228–233 (2005)
25. Yoon, S., Kim, S.: AdaBoost-based multiple SVM-RFE for classification of mammograms in DDSM. BMC Med. Inform. Decis. Mak. **9**(Suppl 1), S1 (2009)
26. Bishop, C.M.C.C.M.: Pattern Recognition and Machine Learning **4**(4) (2006)
27. Williams, C.K.I.: Learning With Kernels: Support Vector Machines, Regularization, Optimization, and Beyond **98**(462) (2003)
28. Theodoridis, S., Koutroumbas, K.: Pattern Recognition, 4th edn. (2009)
29. Kramer, M.A.: Nonlinear principal component analysis using autoassociative neural networks. AIChE J. **37**(2), 233–243 (1991)
30. Alter, O., Brown, P.O., Botstein, D.: Singular value decomposition for genome-wide expression data processing and modeling. Proc. Natl. Acad. Sci. U.S.A. **97**(18), 10101–10106 (2000)
31. Nello Cristianini, J.S.-T.: An Introduction to Support Vector Machines and Other Kernel-based Learning Methods. Cambridge University Press (2000)
32. Dudoit, S., Fridlyand, J., Speed, T.P.: Comparison of Discrimination Methods for the Classification of Tumors Using Gene Expression Data. J. Am. Stat. Assoc. **97**(457), 77–87 (2002)
33. Alshamlan, H.M., Badr, G.H., Alohali, Y.A.: Genetic Bee Colony (GBC) algorithm: A new gene selection method for microarray cancer classification. Comput. Biol. Chem. **56**, 49–60 (2015)

LiveDoc: Showing Contextual Information Using Topic Modeling Techniques

Jayati Deshmukh(✉), K.M. Annervaz, Shubhashis Sengupta, and Neetu Pathak

Accenture Technology Labs, Bangalore, India
{jayati.deshmukh,annervaz.k.m, shubhashis.sengupta,neetu.pathak}@accenture.com

Abstract. We present a solution named LiveDoc, which augments natural language text documents with relevant contextual background information. This background information helps readers to understand the context of the discourse better by fetching relevant information from other sources such as Wikipedia. Often the readers do not possess all background and supplementary information required for comprehending the purport of a narrative such as a news op-ed article. At the same time, it is not possible for authors to provide all contextual information while addressing a particular topic. LiveDoc processes the information in a document; uses extracted entities to fetch relevant background information in the context of the document from various sources (as defined by user) using semantic matching and topic modeling techniques like Latent Dirichlet Allocation and Hierarchical Dirichlet Process; and presents the background information to the user by augmenting the original document with the fetched information. Reader is then equipped better to understand the document with this additional background information. We present the effectiveness of our solution through extensive experimentation and associated results.

Keywords: Information retrieval · Topic modeling · Natural language processing · Data contextualization · Latent Dirichlet Allocation · Hierarchical Dirichlet Process

1 Introduction and Motivation

We are living in the age of online information deluge. An enormous number of news articles, on a variety of topics, get published on the web every day. Apart from news articles, blog posts and other similar content also get added on large counts. The variety of the available content and the speed of generation present certain challenges to the users of the information. It is very common that when we read a news article, we stumble upon names, locations or events that we don't know or we can't recollect. Even if we knew, it would be hard for us to recognize the significance of that person or place in context of the topic or theme of current article. Often we take recourse to web search to get related content

P. Perner (Ed.): MLDM 2016, LNAI 9729, pp. 468–482, 2016.
DOI: 10.1007/978-3-319-41920-6_37

for background check, and try to relate that information to the current article in order to understand the significance of that name or location. However, web search may throw upon a lot of information for us to sift through to get the relevant context. This poses a challenge, and in most cases the article doesn't get completely comprehended by the reader.

As an illustration, suppose a user is reading a news article on the political conflicts in Iraq. In the article the writer has quoted a comment on this issue from Iran's President. Now, a reader who is not adequately informed about the history of political conflicts and tension between Iran and Iraq, will not understand why Iran's president's comment is relevant in that context. However, from the news writer's perspective, it may not be feasible to give all related background information. In another hypothetical example, suppose reader A is reading a financial article about the decrease in the prices of stock of a company where Donald Trump's comment is quoted, and reader B is reading a political article on the policies of Barack Obama where also Donald Trump's comment is quoted. The information that A and B need to know about Donald Trump are very distinct in nature. A needs to know about Donald Trump's business interest in that company, where as B needs to know that Trump is in the race for US presidential candidature. The research question in this work is *How to fetch this contextual background information which is absent in the source document, but required for the comprehensive understanding of it?* The examples chosen here to motivate are simple. The real life scenarios where contextualization of background information is needed may be far more complex.

In this paper we present LiveDoc - a prototypical system that processes the information in natural language documents and fetches additional relevant information from predefined sources (like web and Wikipedia) using various topic modeling and semantic similarity matching techniques. The augmented information is presented to the user along with the document. This helps the user to understand the original document in a more contextual manner.

Rest of the paper is organized as follows. We cover some of the related works in the next section 2 and articulate how our work is different. The approach and the algorithms are explained in detail in section 3. We present and evaluate the experimental results in section 4. In this section, we also provide some interesting examples from corpus that we have come across during experimentation. We provide implementation details in section 5 for building LiveDoc. Finally, we conclude with section 6 by giving some future work directions.

2 Related Work

Substantial work have been reported in the area of Wikification [11,17,20,21] that involve identifying entities in the input document and linking these entities to their corresponding Wikipages. Wikification helps the reader of a document to understand more about the entities referred in the text. Understanding of individual entity sense is important to understand the overall context of the document. Wikification has been done for both large documents as well as for smaller texts like tweets [7,18].

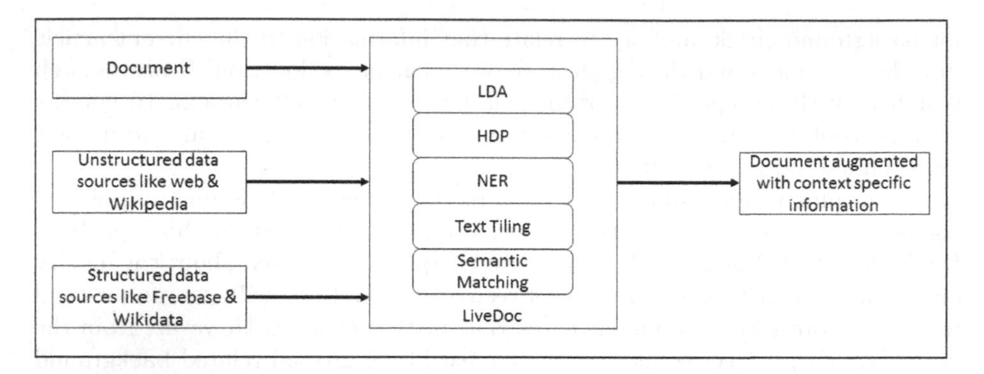

Fig. 1. Flow & Context

An entity with the same surface form may have different meanings in different contexts. For example: Mercury may refer to an element in the periodic table or a planet in the solar system in different contexts. The technique of finding the correct context is known as word or entity sense disambiguation. Wikipedia has been used in this regard [8,19], where techniques are applied to increase the agreement between the input document and corresponding Wikipages. Techniques like matching of tags and annotations of documents to Wikipage tags have also been tried. Wikipages with the highest match are mostly related to the entity mentioned in the document.

In all these cases, most relevant Wikipage of an entity is found. However, these works fail to consider the segregation of information in Wikipages or similar reference sources according to the topics and context of the document at hand. We try to take the next step of finding the relevant information for the referred entity in the topic of the document, using topic modeling and semantic matching techniques. For example, we try to discern whether planet Mercury is being talked in context of its compositional structure and geology or in reference to its astrological observation history. This information is much more specific and of topical interest to the reader. Also such specific background information is much more relevant and useful as compared to generic information details about an entity.

Another important task in information retrieval is to find relationships between to co-occuring entity pairs. Relationship extraction has been studied and experimented using various techniques. Some of the earlier methods [6,28] used kernel functions and shallow parsing to build a dependency tree and extract relations from it. Statistical techniques [30] like Markov logic networks (MLNs) also give good results in entity relationship extraction. Feature based relation extraction [14,27] using SVM and other techniques has also been tried.

However, like entity extraction most of the work on relationship extraction also focuses on finding the conventional relationship between entities. Hence the techniques can derive relationships like 'CapitalOf', 'PresidentOf', 'MotherOf' etc. But it may not be possible to derive topical relationships like if two are firms are competitors in a specific domain or if a person has favorite destinations for summer etc. These kind of relationships are needed to be extracted to give fine-grained relations among entities contextual to the narrative. Typically, such topical relationships are not directly available in structured sources like Freebase.

Generic parameters of entities and relationship between entities can be found using various structured and unstructured sources. However all this information might not be relevant in a specific context - only a few are. While research has been quite extensive for entity and relationship extraction, filtering background content based on topic of the discourse has not been studied much. We try to address this problem here.

3 Our Approach & Algorithm

The central problem LiveDoc addresses is to provide selective but relevant contextual background information from various reference sources that will help the reader to comprehend the discourse better. Although this issue may have unlimited scope, we currently limit the definition of the context to the theme(s) or topic(s)(using topic modeling) of the document and the named entities present in the document. Figure 1 shows the simple schematic for LiveDoc. Any natural language document can be fed into LiveDoc, LiveDoc fetches data from public structured and unstructured sources (which can be configured) that are relevant for understanding the document and presents it to the user in an easily consumable form. We have described the details of current implementation and how a user can see the augmented information along with the document in section 5.

In LiveDoc, context is defined as the theme of the neighborhood in which an entity is mentioned in the article. The neighborhood can be user defined or can be automatically found using text segmentation techniques. The neighborhood becomes relevant when the theme distinguishes itself from the theme of the entire article. It might be very much possible that, all segments of an article have same theme, especially in short articles. However in long articles this may not be the case, and the segment themes varies widely across.

A natural language document is first processed to identify the named entities [22]. Towards this purpose we used the Conditional Random Field(CRF) classifier [25][12]. This classifier is trained with appropriate training data from multiple corpus, so it performs reasonably well across domains. LiveDoc mainly identifies Person, Organization and Location as entities in the document. The remaining algorithm deals with these identified entities. We can also plug-in other classifiers if required, which can extract other entities. LiveDoc provides four kinds of contextual information based on these entities. The first two are based on individual entities and the context in which they are referred. The next two are based on the relationship between a pair of entities and the context in which they are referred to. Let us see each of the approaches in detail in the subsequent subsections.

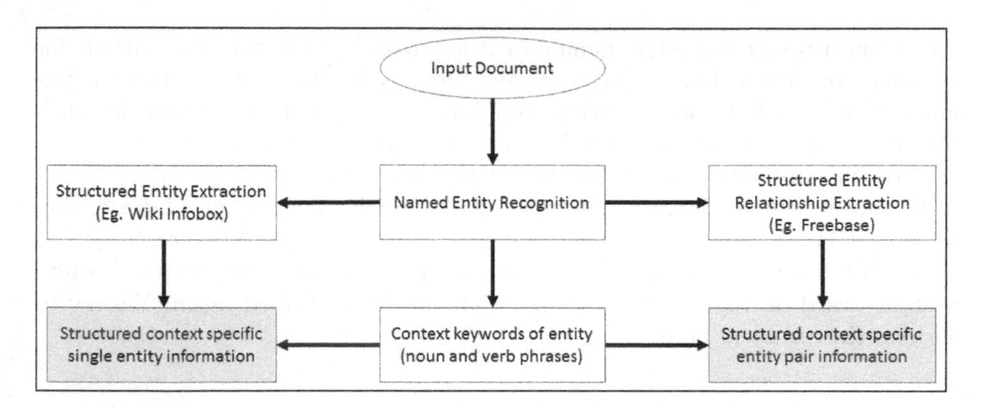

Fig. 2. Information extraction from Structured data

3.1 Contextual Information from Structured Sources

Figure 2 shows contexual information from structured sources for single entity and entity pairs.

Single Entity. The key-value pair information available from structured data sources like infobox[1] of Wikipedia page(s) provide first set of reference information pertaining to an extracted named entity. For a given entity, there will be a large number of infobox entries, covering various aspects / attributes of the entity and only a very small subset of these may be relevant for the discourse at hand. For each entity a neighborhood is defined in the text. This is done by defining a proportional neighborhood, where the entity is being referred, as compared to the length of the document. Alternatively we used Text Tiling [15] for this purpose. Text Tiling finds variance in occurrence of words across the textual blocks and splits them into separate sections if variance is high. Each section or tile where the entity reference happens is taken as its neighborhood.

The text is tokenized and key noun phrases and verb phrases are extracted after performing the steps of stop word removal and cleanup. These are further weighed and filtered according to TF-IDF [10]. For our purpose, very low IDF value words are retained as they might carry information of the entire context of the document. Let us call this set of context words for an entity as $\mathcal{C}$. The Wikipedia page(s) of the entity is retrieved subsequently. The surface form of the entity reference may not match directly to the Wikipedia page entries. Surface form disambiguation and matching entity reference to Wikipages is an area of research [29] discussed in prior work section. However, we have restricted ourselves to simple fuzzy matching techniques on the entity references, $\mathcal{C}$ and Wikipage topics. The key-value pair entries from sources like infobox are retrieved from the corresponding pages.

[1] https://en.wikipedia.org/wiki/Infobox

We then compute the semantic similarity measures, like Jiang Conrath [16] or Lin measure [5], based on open lexical resources like Wordnet[2] for all tokens t in $\mathcal{C}$ and keys for each of the key-value pair entry. Jiang Conrath Similarity as shown in Equation 1 calculates semantic similarity based on information content of input. Here t is a token in $\mathcal{C}$ which is the set of context keywords of an entity and c is the set of tokens of the key of a key-value pair. The distance() function is computed on the basis of information content of lowest common subsumer of t and c. Lin measure also calculates the similarity score on the basis of information content of the input. As both are sets of tokens, Jaccard Set similarity as shown in Equation 2 is applied thereafter to compute the final similarity score. The keys are ordered, according to the final similarity score of the tokens in them with the context words of the entity. Top x percentile of the keys and corresponding values are retained.

$$sim_{JC}(t, c) = 1/distance(t, c) \tag{1}$$

$$J(\mathcal{C}, c) = \frac{|\mathcal{C} \cap c|}{|\mathcal{C} \cup c|} \tag{2}$$

Entity Pair. To find objective topical relationships between entity pairs; for each pair of entities, a set of relationships are retrieved from *structured* sources like Freebase[3] or more recent ones like Wikidata[4]. Please note that the relationship retrieved may not be necessarily single edge relationship. For example, a relationship edge between a country A and city B, may be that B is located in A. Another edge could be that A has a president called X and X was born in B. The context word set $\mathcal{C}$ is extracted as in the earlier case, and for each relation semantic similarity measure is computed for the description tokens in the relation and $\mathcal{C}$ as earlier. The relationships that matches the context with high semantic similarity are used for generating the output.

3.2 Contextual Information from Unstructured Sources

Figure 3 shows contexual information from unstructured sources for single entity and entity pairs.

We leverage topic modeling techniques to generate augmented information from unstructured reference sources. Given a large corpus of documents, topic modeling techniques identify topics or themes present in these documents, and this information can be used to organize and segregate the documents based on topics. Various techniques ranging from non probabilistic models like Latent Semantic Indexing [9] to probabilistic models [3] like Latent Dirichlet Allocation(LDA) [4] have been developed for topic modeling. In probabilistic topic

[2] http://wordnet.princeton.edu
[3] http://www.freebase.com/
[4] https://www.wikidata.org/

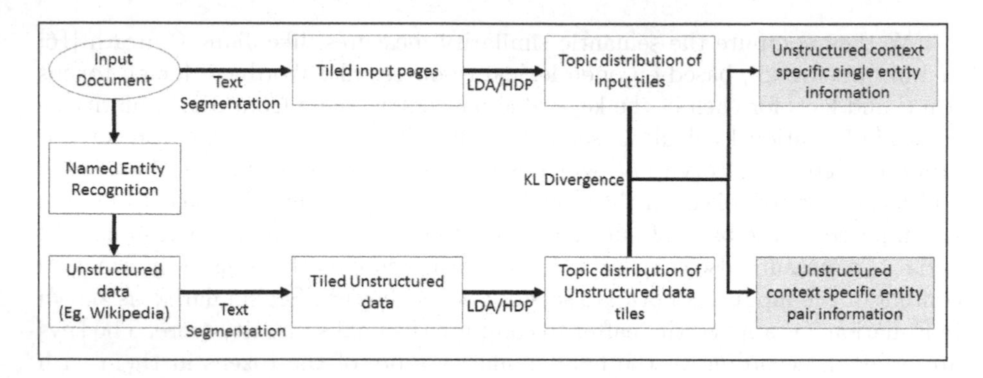

Fig. 3. Information extraction from Unstructured data

modeling techniques, a topic is defined as a probability distribution over words. Recently non parametric Bayesian methods like Hierarchical Dirichlet Process(HDP) [26] have gained prominence for topic modeling. In the current work, we have used classical LDA as well as HDP techniques, trained on appropriate corpus like Wikipedia, to find context or topics of complete input documents and parts thereof. Results for experiments conducted with both the techniques are reported in the sections ahead.

Single Entity. The augmented information about an entity is the related description of the entity in natural language from reference sources. For the entity of reference a neighborhood is drawn as in the earlier cases. The Wikipedia page(s) of the entity is retrieved as earlier, and this page(s) is segmented into smaller snippets based on Text Tiling [15]. Topic distribution of the tile where entity is referred to is computed using the trained LDA and HDP models. Similarly, topic distribution of the tiles of the Wikipedia entry is also computed. A matching score is computed based on Kull-back?Leibler(KL) divergence [2] (Equation 3) between td_{tile} which is the topic distribution of the tile where entity appears and td_{wiki} which is the topic distribution of the tiles from the Wikipedia page(s) of the entity. The tiles from the Wikipedia page(s) with minimum KL divergence of topic distribution from the context tile are retained for use in the final output. If no tile from Wikipedia page(s) meets the minimum matching criteria, matching is repeated with the topic distribution of the entire document against the tiles from Wikipedia page(s). Currently in LiveDoc, tiles only from Wikipedia page(s) of the entities are shown, however scheme can be extended to show information from reference open web pages as well

$$D_{KL}(td_{tile}||td_{wiki}) = \sum_i td_{tile}(i) log \frac{td_{tile}(i)}{td_{wiki}(i)} \tag{3}$$

Entity Pair. For this type of augmentation (through the natural language description in the reference source) a neighborhood is again drawn surrounding the text where both the entities are referenced in the document. Wikipedia page(s) of both the entities are retrieved and tiled. Furthermore, tiles where one entity refers another are filtered. Topic distribution of the context and the tiles from Wikipedia based on trained LDA and HDP models is computed. The tiles with minimal KL diverging topic distribution from the topic distribution of the context are retained for the output. Again as in the single entity case if the topic distribution of the context does not match any of the topic distribution of the tiles from Wikipedia page(s), the matching attempt is repeated with the full topic distribution of the document.

Before proceeding to the result section, a word on the application of the solution in other contexts. Although the approach is presented in context of Wikipedia, Freebase, Wikidata etc, the solution does not depend on the structure of any of these. This solution can be applied to show augmented information for a document based on any set of related past documents or structured information sources.

4 Experimental Results & Evaluation

We performed experiments on LiveDoc using the BBC news dataset [13] and some of the best speeches[5] of all time as narratives. BBC news dataset comprised 2225 documents from the BBC news website for the time period 2004 – 2005. The documents belonged to one of the following five topics: business, entertainment, politics, sports and tech. Around 967 speeches of personalities like Obama, Algore, Kennedy, Churchill etc were also processed as narrative inputs for the experiments. Wikipedia was used as reference set (for both structured and unstructured matching) Here the results are presented with respect to single entity and relation between a pair of entities, structured and unstructured information for both the cases. User verification is independently done by 10 general users to calculate scores on a random subset of 100+ documents from both BBC news and speech datasets and associated outputs of these documents.

The results for structured information, for both single entity and entity pair were almost all relevant for the context of the document. As judgment on topical appropriateness would be very subjective, we are not presenting the numerical results for these, but instead give factual examples. We present numerical results from the experiments for unstructured data in both single entity and entity pair information fetching, as well as illustrative examples from the experiments. Precision is calculated as usual as the ratio of number of correct outputs to the total number of entries in the output. However as a standard comparison dataset is not available for this custom task, we calculated recall as the ratio of number of entities for which relevant information was fetched by our algorithm with a high confidence to the total number of entities present in the text. For the experiments, LDA and HDP models were trained using full Wikipedia dump of size around 10GB. We have experimented with various parameters of the

[5] http://www.americanrhetoric.com/

models, especially topic count for LDA and HDP. The results reported are the best found values in the experiments, for various parameter values.

4.1 Single Entity and Entity Pair: Contextual Information from Structured Sources

In Wikipedia infobox, a 'location' entity like New York can have properties like legislature, admittance date, latitude and longitude, nickname, state anthem, flag, governor, total area and much more. For 'person' entity like Stevie Wonder, an Wiki page infobox will also contain some of the properties are occupation, background, birthplace, instruments he plays, his alias names etc. These infobox properties give the overall generic information about an entity. LiveDoc filters the entries according to the context of the document. For a political document on New York, only relevant entries like legislature will be fetched and entries like latitude and longitude will be skipped. Similar example for Argentina is shown in Figures 7 and 8. As the number of entries in the infobox is huge, this helps the user to see only the context relevant entries. Similarly from multiple relationships possible between a pair of entities, LiveDoc filters the one that is relevant for the context in which it is mentioned in the document. For example for a political document, LiveDoc shows relationship like 'GovernorOf' between a person and a city, rather than 'BornIn' relationships between them. Another interesting example, on an article on fund allocation to states by the federal government of country A, which is a political article, and all states would have 'StateOf' relationship. But LiveDoc identifies a multi-edge relationship for a particular state, S. X 'PresidentOf' A and X 'NativeOf' S. This will be more relevant to understand the document, in terms of percentage fund allocation across states.

4.2 Single Entity: Contextual Information from Unstructured Sources

Table 1 shows precision and recall of LiveDoc using both LDA and HDP models for BBC and Speech datasets. We have high precision and recall numbers, at least using one of the models across all the datasets which illustrates the effectiveness of LiveDoc in picking up relevant contextual information. A couple of illustrative

Table 1. Context Specific Single Entity Results

	BBC Dataset		Speech Dataset	
	LDA	HDP	LDA	HDP
Precision	87.31%	69.57%	85.48%	52.38%
Recall	88.25%	76.36%	87.78%	82.26%

examples from the experiments are as follows. In a BBC business article about a firm investing in Google, the corresponding most suitable Wikipedia page tile of Google is about its financial information: revenue, growth in share price and so on. LiveDoc picks up this tile for augmentation. Thus a person reading the news article can see the financial details of Google rather than seeing general information like date of incorporation and founders etc. Another example as shown in Figure 4 is an entertainment article on the book The Da Vinci Code where its author Dan Brown is mentioned. The corresponding most suitable Wikipedia page tile for Dan Brown is about his writing career and bibliography rather than topics on the author's early life or education. This information can be dynamically found for entities in different contexts.

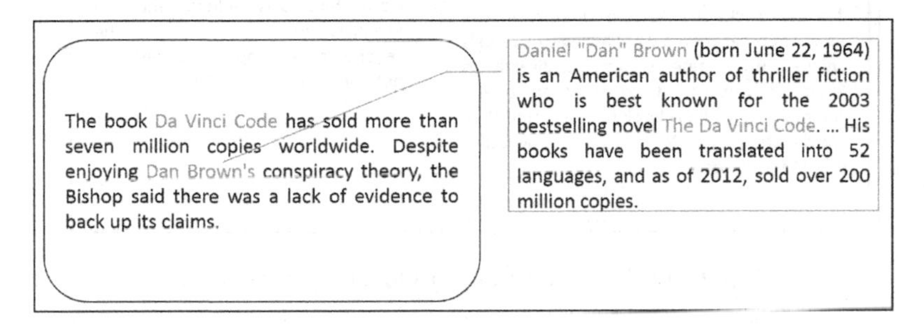

Fig. 4. Information of Dan Brown in an article about Da Vinci Code

4.3 Entity Pair: Contextual Information from Unstructured Sources

Table 2 shows precision and recall for entity pairs information fetched by LiveDoc, using LDA and HDP models separately for BBC and speech dataset. Precision represents the accuracy of correctly identified entities and entity pairs in the documents. Recall represents the number of entities and entity pairs that were identified out of all entities or entity pairs present in the document. The precision numbers are comparatively lower than single entity numbers. This is because multiple possibilities of the interactions between two entities exist in varied contexts. Nonetheless, the numbers are good enough for the practical

Table 2. Context Specific Entity Pair Results

	BBC Dataset		Speech Dataset	
	LDA	HDP	LDA	HDP
Precision	76.36%	68.63%	75.93%	76.47%
Recall	82.75%	50.19%	82.59%	54.91%

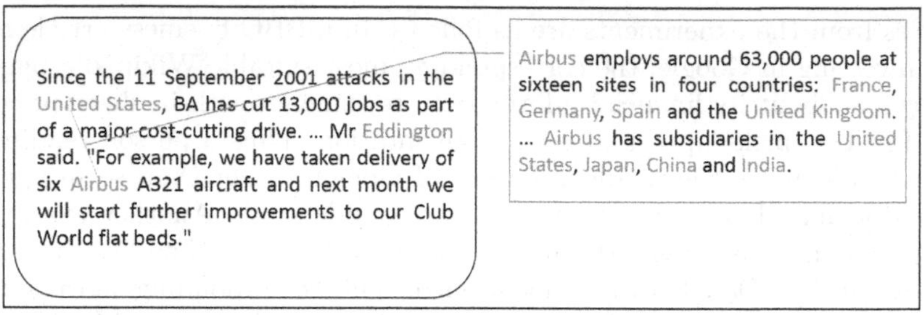

Since the 11 September 2001 attacks in the United States, BA has cut 13,000 jobs as part of a major cost-cutting drive. ... Mr Eddington said. "For example, we have taken delivery of six Airbus A321 aircraft and next month we will start further improvements to our Club World flat beds."

Airbus employs around 63,000 people at sixteen sites in four countries: France, Germany, Spain and the United Kingdom. ... Airbus has subsidiaries in the United States, Japan, China and India.

Fig. 5. Entity Pair details of Airbus and United States

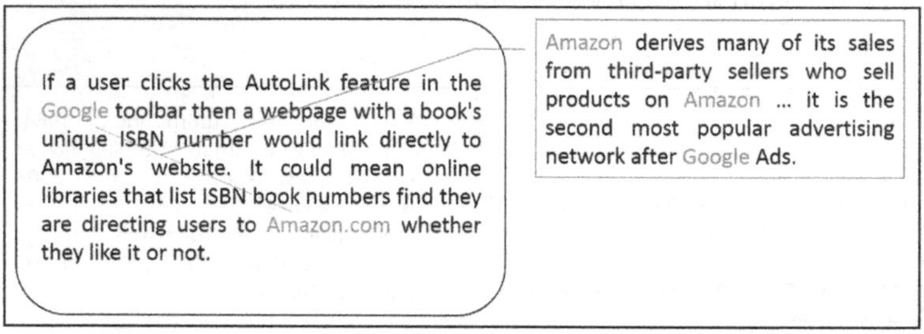

If a user clicks the AutoLink feature in the Google toolbar then a webpage with a book's unique ISBN number would link directly to Amazon's website. It could mean online libraries that list ISBN book numbers find they are directing users to Amazon.com whether they like it or not.

Amazon derives many of its sales from third-party sellers who sell products on Amazon ... it is the second most popular advertising network after Google Ads.

Fig. 6. Entity Pair details of Google and Amazon

purpose. We also observe that results obtained with LDA are better than those with HDP. A couple of illustrative examples from the experiments are as follows. A business article in BBC set on Airbus also mentioned United States as shown in Figure 5. In the most relevant Wikipedia page tile of Airbus containing the entity United States we find that Airbus has subsidiaries in United States. Thus LiveDoc could highlight the US subsidiary of Airbus for reference and this is very context specific relationship. In another example as shown in Figure 6, a technology news article on Google toolbar mentions that the toolbar directly redirects to Amazon if a specific book is searched. The most relevant Wikipedia page tile of Google identified by LiveDoc which also mentions Amazon is about third party sellers where its mentioned that Amazon is the second most popular advertising network after Google Ads. Thus LiveDoc fetches very context specific information concerning both the companies.

Precision and recall values of tables 1 and 2 as discussed above are obtained from user evaluation. The target group of readers independently verified whether the additional context specific information obtained by LiveDoc for single entities and entity pairs is relevant in context to the part of document or tile in which it is mentioned.

5 Implementation Details

LiveDoc is currently implemented as a standalone application. Any input document can be uploaded by user and it is automatically augmented with the

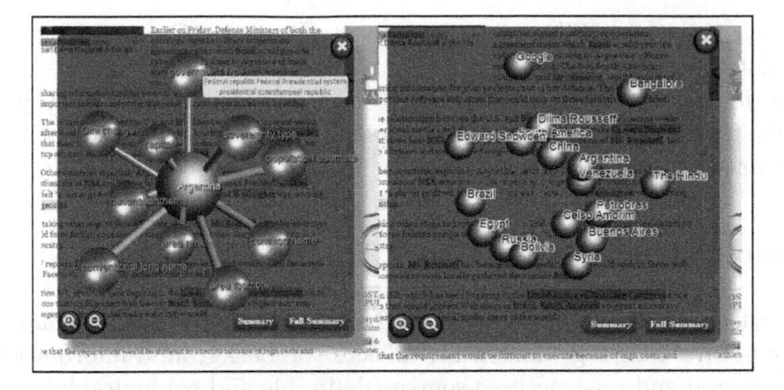

Fig. 7. Screenshots of the prototype application - 1

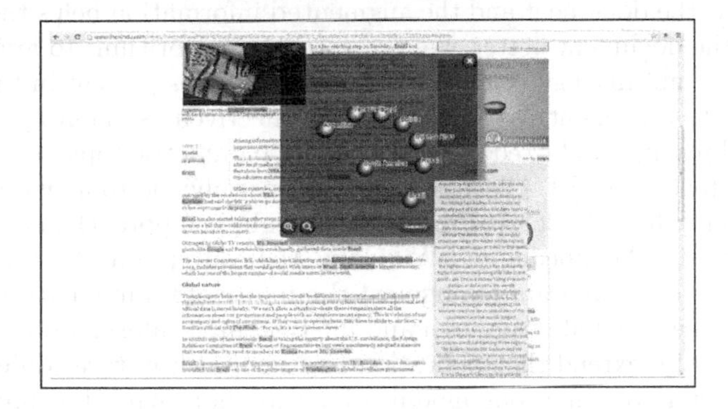

Fig. 8. Screenshots of the prototype application - 2

additional context specific information found by LiveDoc. The augmented information is visually displayed along with the original document. Figures 7 and 8 depict some screen-shots of LiveDoc showing augmented information from structured sources. In the background is the news article, with the named entities mentioned highlighted in various colors. The left part of Figure 7 shows information shown about Argentina when hovered over it in the text. As you can see as the article is political, the shown information is more gravitated towards political features of Argentina. Its right part shows various entity pairs in the document. Figure 8 shows the relationship in context between the hovered over entity pair(geographic/political relationships between South America and Brazil).

The application invokes a back-end engine for processing the document and fetching the relevant information, which is exposed via REST services. The back-end engine is implemented in Java and Python using a number of tools and libraries. For Java modules, we have used OpenNLP[6] for basic natural language processing tasks and Stanford NLP[7] for Named Entity Recognition using CRF.

[6] http://opennlp.apache.org/

[7] http://nlp.stanford.edu/software/

We have used bliki[8] for Wikipedia information processing. For calculating semantic similarity measures we used ws4j[9]. For the Python components, we have used NLTK [1] for basic natural language processing and Gensim [24] for Topic Modeling techniques. Our algorithm was implemented using a combination of all these libraries.

6 Conclusion & Future Work

We presented a novel application "LiveDoc" which processes the information from any natural language document, collects the background information from open structured and unstructured sources(definable and configurable), and augments this information with the document. The user can see this information when he reads the document and the augmented information helps the reader to understand the document better. We presented the algorithms to fetch four different kinds of information, centered around the entities present in the text and the topics of the document. We illustrated the effectiveness of our approach with experimental results conducted separately using parametric topic modeling technique like Latent Dirichlet Allocation and non-parametric topic modeling technique like Hierarchical Dirichlet Process. Although the approach and application is illustrated using Wikipedia and Wikidata, the concept can be generalized to other contexts where we want to augment the information in a natural language article from other related structured and unstructured information sources.

We envisage to extend the application and approach in three different directions. The first is to carry out modifications for improving the precision and recall. In this case various different alternatives of key components of the algorithm like neighborhood definition of entities can be tried. Extensive experiments are required on the parameters of the topic modeling techniques to fine tune and improve the final numbers. The second direction is by the way of including new approaches like Text Summarization [23] and information retrieval techniques which have not been currently considered for enhancing the functional output. The third direction is through the changes required for making the approach effective in domain specific data. For example, We intend to use LiveDoc in enterprise context such as finding related Request for Proposal (RfP) response documents against a particular proposal. We are working in all three directions, with specific focus and priority on the third one.

References

1. Bird, S.: NLTK: the natural language toolkit. In: Proceedings of the COLING/ACL on Interactive Presentation Sessions, pp. 69–72. Association for Computational Linguistics (2006)
2. Bishop, C.M.: Pattern recognition and machine learning. Springer (2006)

[8] https://code.google.com/p/gwtwiki/
[9] https://code.google.com/p/ws4j/

3. Blei, D.M.: Probabilistic topic models. Communications of the ACM **55**(4), 77–84 (2012)
4. Blei, D.M., Ng, A.Y., Jordan, M.I.: Latent dirichlet allocation. The Journal of Machine Learning Research **3**, 993–1022 (2003)
5. Budanitsky, A., Hirst, G.: Semantic distance in wordnet: An experimental, application-oriented evaluation of five measures (2001)
6. Bunescu, R.C., Mooney, R.J.: A shortest path dependency kernel for relation extraction. In: Proceedings of the Conference on Human Language Technology and Empirical Methods in Natural Language Processing, pp. 724–731. Association for Computational Linguistics (2005)
7. Cassidy, T., Ji, H., Ratinov, L.A., Zubiaga, A., Huang, H.: Analysis and enhancement of wikification for microblogs with context expansion. In: COLING, vol. 12, pp. 441–456 (2012)
8. Cucerzan, S.: Large-scale named entity disambiguation based on wikipedia data. In: EMNLP-CoNLL, vol. 7, pp. 708–716 (2007)
9. Dumais, S.T.: Latent semantic analysis. Annual Review of Information Science and Technology **38**(1), 188–230 (2004)
10. Feldman, R., Sanger, J.: The text mining handbook: advanced approaches in analyzing unstructured data. Cambridge University Press (2007)
11. Ferragina, P., Scaiella, U.: Tagme: On-the-fly annotation of short text fragments (by wikipedia entities). In: Proceedings of the 19th ACM International Conference on Information and Knowledge Management, CIKM 2010, pp. 1625–1628 (2010). http://doi.acm.org/10.1145/1871437.1871689
12. Finkel, J.R., Grenager, T., Manning, C.: Incorporating non-local information into information extraction systems by gibbs sampling. In: Proceedings of the 43rd Annual Meeting on Association for Computational Linguistics, pp. 363–370. Association for Computational Linguistics (2005)
13. Greene, D., Cunningham, P.: Practical solutions to the problem of diagonal dominance in kernel document clustering. In: Proc. 23rd International Conference on Machine learning (ICML 2006), pp. 377–384. ACM Press (2006)
14. GuoDong, Z., Jian, S., Jie, Z., Min, Z.: Exploring various knowledge in relation extraction. In: Proceedings of the 43rd Annual Meeting on Association for Computational Linguistics, pp. 427–434. Association for Computational Linguistics (2005)
15. Hearst, M.A.: Texttiling: Segmenting text into multi-paragraph subtopic passages. Computational Linguistics **23**(1), 33–64 (1997)
16. Jiang, J.J., Conrath, D.W.: Semantic similarity based on corpus statistics and lexical taxonomy. CoRR cmp-lg/9709008 (1997)
17. Kulkarni, S., Singh, A., Ramakrishnan, G., Chakrabarti, S.: Collective annotation of wikipedia entities in web text. In: Proceedings of the 15th ACM SIGKDD International Conference on Knowledge Discovery and Data Mining, KDD 2009, pp. 457–466 (2009). http://doi.acm.org/10.1145/1557019.1557073
18. Liu, X., Li, Y., Wu, H., Zhou, M., Wei, F., Lu, Y.: Entity linking for tweets. In: ACL (1), pp. 1304–1311 (2013)
19. Mihalcea, R.: Using wikipedia for automatic word sense disambiguation. In: HLT-NAACL, pp. 196–203 (2007)
20. Mihalcea, R., Csomai, A.: Wikify!: linking documents to encyclopedic knowledge. In: Proceedings of the Sixteenth ACM Conference on Information and Knowledge Management, pp. 233–242. ACM (2007)
21. Milne, D., Witten, I.H.: Learning to link with wikipedia. In: Proceedings of the 17th ACM Conference on Information and Knowledge Management, pp. 509–518. ACM (2008)

22. Nadeau, D., Sekine, S.: A survey of named entity recognition and classification. Lingvisticae Investigationes **30**(1), 3–26 (2007)
23. Nenkova, A., McKeown, K.: A survey of text summarization techniques. In: Mining Text Data, pp. 43–76. Springer (2012)
24. Řehůřek, R., Sojka, P.: Software framework for topic modelling with large corpora. In: Proceedings of the LREC 2010 Workshop on New Challenges for NLP Frameworks, pp. 45–50. ELRA, Valletta. http://is.muni.cz/publication/884893/en
25. Sutton, C., McCallum, A.: An introduction to conditional random fields for relational learning. Introduction to Statistical Relational Learning, 93–128 (2006)
26. Teh, Y.W., Jordan, M.I., Beal, M.J., Blei, D.M.: Hierarchical dirichlet processes. Journal of the American Statistical Association **101**(476) (2006)
27. Wang, T., Li, Y., Bontcheva, K., Cunningham, H., Wang, J.: Automatic extraction of hierarchical relations from text. Springer (2006)
28. Zelenko, D., Aone, C., Richardella, A.: Kernel methods for relation extraction. The Journal of Machine Learning Research **3**, 1083–1106 (2003)
29. Zhou, Y., Nie, L., Rouhani-Kalleh, O., Vasile, F., Gaffney, S.: Resolving surface forms to wikipedia topics. In: Proceedings of the 23rd International Conference on Computational Linguistics, pp. 1335–1343. Association for Computational Linguistics (2010)
30. Zhu, J., Nie, Z., Liu, X., Zhang, B., Wen, J.R.: Statsnowball: a statistical approach to extracting entity relationships. In: Proceedings of the 18th International Conference on World Wide Web, pp. 101–110. ACM (2009)

Genetic-Based Thresholds for Multi Histogram Equalization Image Enhancement

Saeed Sedighi[1(✉)], Mehdi Roopaei[1], and Sos Agaian[2]

[1] Department of Computer, Shiraz Branch, Islamic Azad University, Shiraz, Iran
sedighi.saeed@gmail.com
[2] University of Texas at San Antonio, San Antonio, USA

Abstract. Image Enhancement is one of the important pre-processing part in any image processing system. It attempts to make Image/Video more understandable while keeping original information for the rest of an image-processing system. Histogram-based image enhancements divide histogram of the original image by one or more separating points and apply the conventional histogram equalization techniques on each sub-image. In this paper, a Genetic-Algorithm scheme tries to find the best point for separating the Histogram. The fitness function of the designed GA is chosen by an image quality measurement to preserve the information of the original image. The experimental results show the superiority of the proposed method than traditional histogram based image-enhancement techniques.

Keywords: Image enhancement · Histogram equalization · Thresholds · Genetic algorithm · Image quality measurement

1 Introduction

Histogram is described as the probabilistic distribution of each gray level in a digital image [1]. It provides a general overview of an image such as gray level distribution and its density, the average luminance of an image, image contrast, and so on [1]. Histogram equalization (HE) is the one of the well-known methods for enhancing the contrast of captured images by mapping the gray levels based on the probability distribution of the input gray levels. HE attempts to flat and stretch the dynamics range of the image's histogram and makes overall contrast improvement [2,3].

The mean brightness of enhanced image doesn't preserve by HE [3]. Many techniques has been proposed to overcome the aforementioned problem [4,5,6,7,8,9,10,11,12]. Mean preserving Bi-histogram equalization (BBHE) firstly separates the input image's histogram into two sub-histograms based on its mean and then equalizes the mentioned sub-histograms independently. Later, equal area Dualistic Sub-Image Histogram Equalization (DSIHE) has been proposed and declared to outperform BBHE both in term of brightness and also image entropy preservation. However, there are still cases that are not handled well by both the BBHE and DSIHE. These images require higher degree of brightness preservation to avoid annoying artifacts [3]. Multi- HE (MHE) techniques has been introduced to enhance

P. Perner (Ed.): MLDM 2016, LNAI 9729, pp. 483–490, 2016.
DOI: 10.1007/978-3-319-41920-6_38

contrast, preserve brightness and produce natural looking images. MHE first decomposes the input image into several sub-images, and then applies the classical HE process to each of them. Otsu proposed a dynamic thresholding selection method [13]. Robustness and speed in partitioning background from the object make Otsu as one of the best thresholding techniques. Two discrepancy functions to decompose the image, conceiving two MHE methods for image contrast enhancement, Minimum Within-Class Variance MHE (MWCVMHE) and Minimum Middle Level Squared Error MHE (MMLSEMHE) are presented in [14].

Genetic algorithms (GAs) are search algorithms based on the mechanism of natural selection and natural genetic systems. There are many problems in the area of pattern recognition and image processing where we need to perform efficient search in complex spaces in order to achieve an optimal solution [15]. Various GA approaches have been applied for image contrast enhancement [16,17,18]. The proposed method in [16] is based on a local enhancement technique similar to statistical scaling method. In this method, a transformation function is applied to each pixel of the input image. The parameters of the proposed transformation function are adapted using a GA according to an objective fitness criterion. In another genetic approach, a relation between input and output gray levels is determined based on a curve by a genetic algorithm, which are represented in a lookup table [17]. The parameters of the contrast enhancement function are determined using a genetic algorithm in [18]. Proposed method in [18] employs a transform, GA, and wavelet neural network to enhance contrast for an image.

In this paper, a new GA based image enhancement is addressed. Histogram of the captured image is illustrated at first. The intensity is divided by two separating points and makes three sub-histograms. The separating points are chosen based on a GA equipped with a Peak Signal to Noise Ratio (PSNR) as its fitness function. In another word, a Genetic-Algorithm tries to find the best points for separating the Histogram. The fitness function of the designed GA is chosen as PSNR to keep the information of original image however it could be developed by combination of other Image Quality Measurements (IQMs) [19-21]. The current paper is organized as follows: section 2 presents the proposed method, section 3 expresses the simulation and results and finally conclusion is stated in section 4.

2 Proposed GA-Based Multi Histogram Equalization (GAMHE)

MHE attempts to dived the histogram into some sub-histograms and then apply HE on each of them. Number of separating points and their locations in the intensity range are still challenges.

There are several methods presented to overcome the mentioned problems. Otsu [13] partitions the background from the object using a cost function includes of between or within-class variance. Minimum Within-Class Variance MHE (MWCVMHE) and Minimum Middle Level Squared Error MHE (MMLSEMHE) are presented in [14]. In this paper, new method is proposed to find the separating points of the histogram according to a GA demonstrated in Fig. 1.

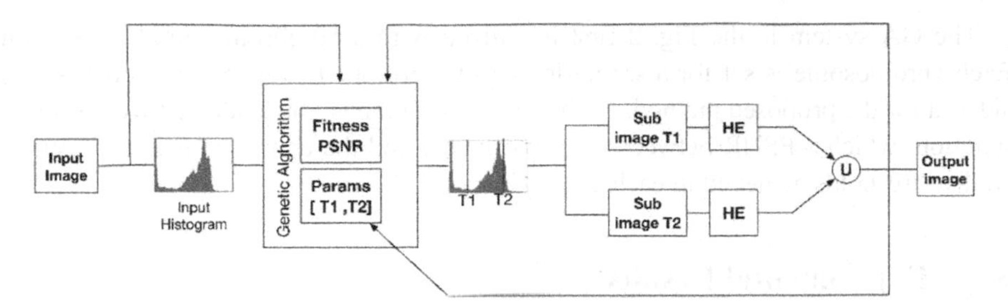

Fig. 1. Propose GAMHE method

According to the proposed structure, a GA system is initialized and then the original and enhanced images are evaluated with a PSNR fitness function in each generation. The enhanced image in each generation made by updated thresholds as output parameters of the GA. At the end of generation, the optimal thresholds are achieved and the structure divides the original image with obtained optimal points, and applies HE separately on each sub-image. The structure of the GA system used in the presented technique is illustrated in the Fig. 2.

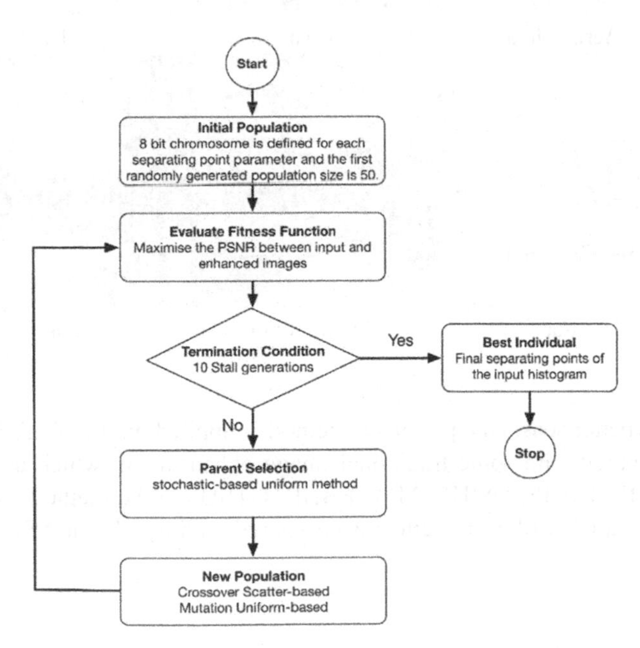

Fig. 2. GA structure

The GA system in the Fig. 2 first initialized with a 50 chromosomes population. Each chromosome is set for a separating point therefore 16 bits chromosome is considered for the proposed method. Each individual in the GA is evaluated with a fitness function, which a PSNR between the input image and the enhanced image constructed, by thresholds achieved in each generation.

3 Experimental Results

The data set addressed in this paper is derived from the [22]. There are six gray-scale images chosen from the mentioned data set which illustrate in the Table 1.

Table 1. Database [22]

To make a better sense, the presented method is applied on the "Car" image the results are compared with some traditional image enhancement which are depicted in the Table 1. HE, BBHE, DSIHE, MMBEBHE, OTSU are the methods selected to be compared. The quality of each method is compared with PSNR and SSIM depicted in the Table 2.

Table 2. Propose method comparison with traditional image enhancement techniques such as HE, BBHE, DSIHE, CLAHE, MMBEBHE, and Otsu

picture			
Algorithm	Original	HE	BBHE
Threshold	-	-	159
PSNR	-	14.9704	21.0583
SSIM	-	0.7389	0.8826
picture			
Algorithm	CLAHE	MMBEBHE	DSIHE
Threshold	-	90	167
PSNR	19.4199	26.4964	18.5645
SSIM	0.8063	0.9518	0.8269
picture			
Algorithm	Genetic 1 point	Otsu 2 Point	Genetic 2 Point
Threshold	113	105 , 171	77 , 129
PSNR	29.6537	30.5217	32.4342
SSIM	0.9635	0.9699	0.9743

The proposed method is applied on the dataset of Table 1 and the results are illustrated in Figs. 3 and 4. These figures demonstrate comparison between the proposed method and other well-known histogram based image enhancements.

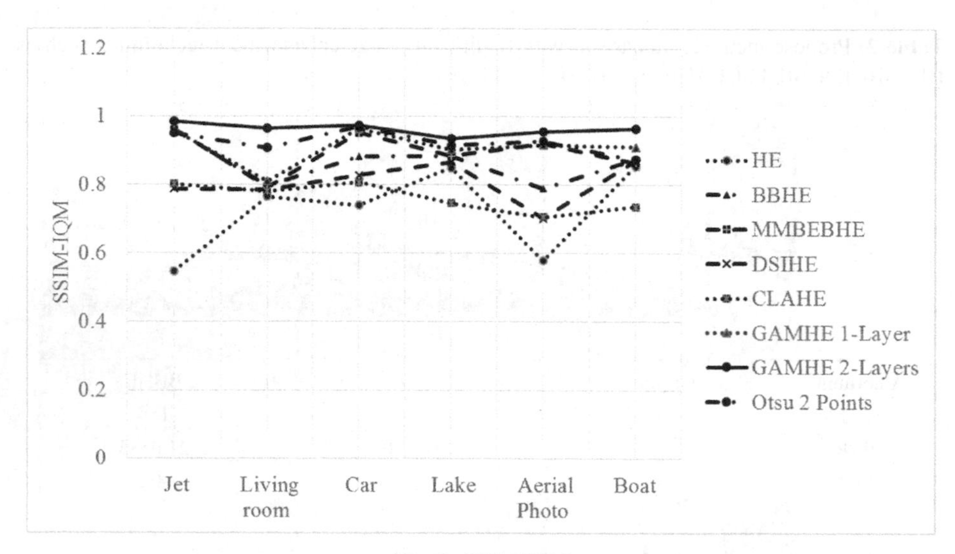

Fig. 3. SSIM IQM

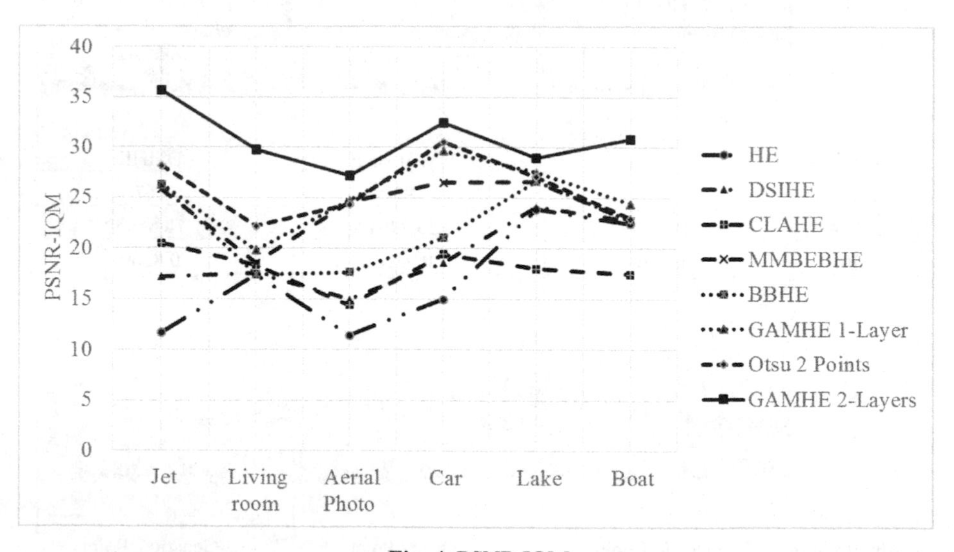

Fig. 4. PSNR IQM

The GAMHE 2-layer shows higher PSNR rather than other enhancement algorithm however the similarity of the proposed method to the original image is more preserved. The achieved results of applying the GAMHE 2-layer are depicted on the Table 3.

Table 3. Proposed method outcome using 2 points GA

The results are more natural rather than other methods while PSNR and SSIM as image quality measurements prove this naturalness.

4 Conclusion

A Genetic-Algorithm addressed in this article tries to find the best points for separating the Histogram. The fitness function of the proposed GA structure is chosen as any proper image quality assessment such as PSNR and SSIM to keep the information of original image and make image enhancement. The experimental results show the superiority of the proposed method than traditional histogram based image-enhancement techniques.

References

1. Wang, Y., Chen, Q., Zhang, B.: Image enhancement based on equal area dualistic sub-image histogram equalization method. IEEE Transactions on Consumer Electronics **45**(1), 68–75 (1999)
2. Kim, Y.-T.: Contrast enhancement using brightness preserving bi-histogram equalization. IEEE Transactions on Consumer Electronics **43**(1), 1–8 (1997)
3. Chen, S.-D., Ramli, A.R.: Contrast enhancement using recursive mean-separate histogram equalization for scalable brightness preservation. IEEE Transactions on Consumer Electronics **49**(4), 1301–1309 (2003)

4. Sheet, D., et al.: Brightness preserving dynamic fuzzy histogram equalization. IEEE Transactions on Consumer Electronics **56**(4), 2475–2480 (2010)
5. Kim, Y.-T., Cho, Y.-H.: Image enhancing method using mean-separate histogram equalization and a circuit therefor. U.S. Patent No. 5,963,665, October 5, 1999
6. Pizer, S.M.: The medical image display and analysis group at the University of North Carolina: Reminiscences and philosophy. IEEE Transactions on Medical Imaging **22**(1), 2–10 (2003)
7. Kim, Y.-T.: Image enhancing method and circuit using mean separate/quantized mean separate histogram equalization and color compensation. U.S. Patent No. 6,049,626, April 11, 2000
8. Huang, L.-L., et al.: Face detection from cluttered images using a polynomial neural network. Elsevier Neurocomputing **51**, 197–211 (2003)
9. Agaian, S., Roopaei, M.: Method And Systems For Thermal Image/Video Measurements And Processing. U.S. Patent No. 20,150,244,946, August 27, 2015
10. Agaian, S., Roopaei, M.: New haze removal scheme and novel measure of enhancement. In: 2013 IEEE International Conference on Cybernetics (CYBCONF). IEEE (2013)
11. Roopaei, M., et al.: Cross-entropy histogram equalization. In: 2014 IEEE International Conference on Systems, Man and Cybernetics (SMC). IEEE (2014)
12. Agaian, S., et al.: Bright and dark distance-based image decomposition and enhancement. In: 2014 IEEE International Conference on Imaging Systems and Techniques (IST). IEEE (2014)
13. Otsu, N.: A threshold selection method from gray-level histograms. Automatica **11**(285-296), 23–27 (1975)
14. Menotti, D., et al.: Multi-histogram equalization methods for contrast enhancement and brightness preserving. IEEE Transactions on Consumer Electronics **53**(3), 1186–1194 (2007)
15. Pal, S.K., Bhandari, D., Kundu, M.K.: Genetic algorithms for optimal image enhancement. Pattern Recognition Letters **15**(3), 261–271 (1994)
16. Munteanu, C., Rosa, A.: Towards automatic image enhancement using genetic algorithms. In: Proceedings of the 2000 Congress on Evolutionary Computation, 2000, vol. 2. IEEE (2000)
17. Saitoh, F.: Image contrast enhancement using genetic algorithm. In: 1999 IEEE International Conference on Systems, Man, and Cybernetics, IEEE SMC 1999 Conference Proceedings, vol. 4. IEEE (1999)
18. Carbonaro, A., Zingaretti, P.: A comprehensive approach to image-contrast enhancement. In: Proceedings. International Conference on Image Analysis and Processing, 1999. IEEE (1999)
19. Agaian, S., Roopaei, M., Akopian, D.: Thermal-image quality measurements. In: 2014 IEEE International Conference on Acoustics, Speech and Signal Processing (ICASSP). IEEE (2014)
20. Roopaei, M., Agaian, S., Jamshidi, M.: Thermal imaging in fuzzy condition monitoring. In: World Automation Congress (WAC), 2014. IEEE (2014)
21. Roopaei, M., et al.: Noise-Free Rule-Based Fuzzy Image Enhancement. Electronic Imaging **2016**(13), 1–5 (2016)
22. Miscelaneous gray level images. http://decsai.ugr.es/cvg/dbimagenes/g512.php

A Clustering Approach for Discovering Intrinsic Clusters in Multivariate Geostatistical Data

Francky Fouedjio(✉)

CSIRO Mineral Resources, Perth, WA, Australia
francky.fouedjiokameni@csiro.au

Abstract. Multivariate georeferenced data have become omnipresent in the many scientific fields and pose substantial analysis challenges. One of them is the grouping of data locations into spatially contiguous clusters so that data locations within the same cluster are more similar while clusters are different from each other, in terms of a concept of dissimilarity. In this work, we develop an agglomerative hierarchical clustering approach that takes into account the spatial dependency between observations. It relies on a dissimilarity matrix built from a non-parametric kernel estimator of the multivariate spatial dependence structure of data. It integrates existing methods to find the optimal cluster number. The capability of the proposed approach to provide spatially compact, connected and meaningful clusters is illustrated to the National Geochemical Survey of Australia data.

Keywords: Clustering · Geostatistics · Multivariate data · Non-parametric

1 Introduction

Multivariate data indexed by geographical coordinates have become increasingly frequent in scientific disciplines and pose real analysis challenges. A classical problem is the clustering of observations into spatially contiguous groups so that observations in the same group are similar to each other and different from those in other groups, in some sense. Some typical examples in the geosciences are [16]: (i) defining climate zones; (ii) determining zones of similar land use; (iii) identifying archaeological sites; (iv) delineation of agricultural management areas; (v) establishment of ore typologies.

In the non-spatial framework, the problem of clustering observations is well-known and described in many textbooks from descriptive to theoretical viewpoint. There are two principal clustering approaches namely, hierarchical and partitioning. In the hierarchical approach, a hierarchy of a tree-like structure is constructed using agglomerative or divisive procedures. In the partitioning approach, observations are divided into clusters once the number of clusters to be formed is specified. Very often, applying on geostatistical data, these non-spatial clustering algorithms have a tendency to produce significant spatial scattered

P. Perner (Ed.): MLDM 2016, LNAI 9729, pp. 491–500, 2016.
DOI: 10.1007/978-3-319-41920-6_39

clusters. However, this characteristic is undesirable for many applications (e. g., delineation of agricultural management zones).

In the geostatistical framework, a more specific approach is needed. Geostatistical data often show properties of spatial dependency and heterogeneity, over the region under study. Observations located close to one another in the geographical space might have similar characteristics. In addition, the mean, variance and/ or spatial dependence structure can be different from one subregion to another. Hence, the necessity to obtain a close related or contiguous clusters of data locations with similar attribute values. The clustering can be achieved in different ways, depending mainly on the measure used to quantify proximity among the observations. It is important to point out that proximity in attribute space does not ensure proximity in geographical space. Thus, in addition to the proximity in the attribute space, proximity in the geographical space needs to be taken into account and data locations belonging to the same cluster should usually be close to one another in the geographical space. To take into account all these constraints, conventional non-spatial clustering approaches have been adapted to the geostatistical context. Existing approaches can be distinguished into four different categories: (i) non-spatial clustering with geographical coordinates as additional variables, (ii) non-spatial clustering based on a spatial dissimilarity measure, (iii) spatially constrained clustering and (iv) model-based clustering.

The first category incorporates the spatial information by just considering geographical coordinates as additional variables. Each observation is seen as a point in a dimensional space, including both geographical space and attribute space. Thereby existing non-spatial clustering methods like K-means or agglomerative hierarchical can be applied to this new space. In practice, the resulting clusters provided by this approach look too scattered because it does not distinguish between geographical space and attribute space.

The second category uses existing non-spatial clustering methods by modifying the dissimilarity measure between two observations to take explicitly into account the spatial dependency. Olivier and Webster [12] were the first to propose an approach in this context. In a univariate case, they suggested using a stationary variogram to weight the original dissimilarities between data locations, while in the multivariate case, a stationary variogram of the first principal component is used. Bourgault et al. [5] used rather a stationary multivariate variogram as a weighting function to decrease similarities of distant data locations. Romary et al. [14] pointed out that these methods have a tendency to produce a smooth dissimilarity matrix without however reinforce the contiguity between resulting clusters.

The third category is different from the second one in that it considers spatial contiguity constraints rather than spatial dissimilarities. Specifically, data locations are grouped together through a non-spatial clustering technique and according to a set of spatial contiguity constraints. Pawitan and Huang [13] developed two spatially constrained clustering algorithms (hierarchical and nonhierarchical) in the univariate case. Spatial connexity of resulting clusters is imposed by a graph structuring the data locations in the geographical space such as Delaunay triangulation. However, the lengths of the edges of the graph are not accounted; this might produce spurious results. In the multivriate case, Romary et al. [14] proposed two spatially constrained clustering algorithms which are adaptations of agglomerative hierarchical and spectral algorithms. The Delaunay triangulation is used to define the neighbourhood structure of data, and edges lengths are taken into account. The main shortcoming of these algorithms is the sensitivity to different hyper-parameters used, no selection method of these latter being proposed.

Contrarily to the three previous ones, the fourth category is not model-free. It relies on the assumption that observations are drawn from a specific distribution like a mixture of Gaussian random fields or Markov random fields. Generally, parameter inference and assignment of each data location to a cluster is carried out through an expectation-maximization algorithm [9]. Ambroise et al. [4] proposed a Markov random field based clustering algorithm applicable to geostatistical data. Data locations in the geographical space are structured via a Delaunay graph. Although this approach is proven to be effective for lattice data, Allard and Monestiez [3] highlighted that it is not sure that it performs well on geostatistical data, since several approximations have been done. Allard and Guillot [2] and Romary et al. [14] also pointed out that the underlying structure of data can not be well-represented by the neighborhood structure used. In contrast, this latter reflects the structure in the data sampling scheme. In the univariate case, Allard [1], Allard and Monestiez [3] and Allard and Guillot [2] proposed a mixture of Gaussian random fields based clustering algorithms. The clustering approach proposed by Allard [1] and Allard and Monestiez [3] is based on the minimization of ratio within variance on between variance. The clustering approach proposed by Allard and Guillot [2] is a based on an approximation of the expectation-maximization algorithm. This latter has been further generalized to the multivariate case by Guillot et al. [10]. As mentioned by Romary et al. [14], the main drawbacks related to these clustering algorithms based on a mixture of Gaussian random fields are the Gaussianity assumption, the independence assumption, and the computational burden. Indeed, observations are supposed to be Gaussian and observations assigned to distinct groups are supposed to be independent. These assumptions might be doubtful in practice. Concerning the large datasets, the estimation procedure is time and resource consuming, since the covariance matrix inverse is required to compute the maximum likelihood at each iteration of the expectation-maximization algorithm.

In this work, we develop an agglomerative hierarchical clustering approach that takes into account the spatial dependency between data. It is based on a non-parametric kernel estimator of the multivariate spatial dependence structure. The idea is to include the spatial information in the clustering procedure by taking the direct and cross variogram to built a measure of dissimilarity between two locations, emphasizing the spatial dependence among data locations. The developed approach is model-free, adapted to irregularly spaced data, and can produce spatially contiguous clusters without including any geometrical constraints.

It incorporates existing methods to determine the optimal cluster number. The developed clustering approach is applied to the National Geochemical Survey of Australia data. The results are compared with those obtained by using: (i) K-means clustering with geographical coordinates as additional variables; (ii) Ward's agglomerative hierarchical clustering with geographical coordinates as additional variables; (iii) Oliver's geostatistical clustering [12]; (iv) Bourgault's geostatistical clustering [5].

The remainder of the paper is organized as follows. Section 2 describes the proposed clustering method through its basic ingredients. Section 3 illustrates on the National Geochemical Survey of Australia data, the capability of the developed clustering approach to providing spatially compact, connected, and meaningful clusters. Section 4 outlines concluding remarks.

2 Methodology

The purpose of a geostatistical clustering technique is the grouping of data locations into spatially compact and connected clusters so that data locations belonging to the same cluster are more similar than those in different clusters. To achieve that the proposed clustering approach starts by specifying a dissimilarity measure between two locations through a non-parametric kernel estimator of the multivariate spatial dependence structure. Then, the resulting dissimilarity matrix at data locations is supplied to a classical agglomerative clustering algorithm. The optimal number cluster is determined through an internal cluster validity index. The resulting clusters are obtained by cutting the dendogram at the corresponding level or height. Different basic ingredients required to implement the proposed clustering approach are described in this section.

2.1 Dissimilarity Measure

Consider a set of p standardized variables of interest $\{Z_1, \ldots, Z_p\}$ defined on a continuous domain of interest $G \subset \mathbb{R}^d, d \geq 1$ and all measured at a set of distinct locations $\{\mathbf{s}_1, \ldots, \mathbf{s}_n\}$. A non-parametric kernel estimator of the multivariate spatial dependence structure described by the direct and cross variogram between two variables Z_i and Z_j $(i, j = 1, \ldots, p)$ at locations $\mathbf{x}$ and $\mathbf{y}$ respectively, is given as follows:

$$\widehat{\gamma}_{ij}(\mathbf{x}, \mathbf{y}; \lambda) = \frac{\sum_{k,l=1}^{n} K_\lambda\left((\mathbf{x}, \mathbf{y}), (\mathbf{s}_k, \mathbf{s}_l)\right) \left(Z_i(\mathbf{s}_k) - Z_i(\mathbf{s}_l)\right) \left(Z_j(\mathbf{s}_k) - Z_j(\mathbf{s}_l)\right)}{2 \sum_{k,l=1}^{n} K_\lambda\left((\mathbf{x}, \mathbf{y}), (\mathbf{s}_k, \mathbf{s}_l)\right)} \mathbb{1}_{\{\mathbf{x} \neq \mathbf{y}\}}, \tag{1}$$

where $K_\lambda\left((\mathbf{x}, \mathbf{y}), (\mathbf{s}_k, \mathbf{s}_l)\right) = K_\lambda(\|\mathbf{x} - \mathbf{s}_k\|) K_\lambda(\|\mathbf{y} - \mathbf{s}_l\|)$, with $K_\lambda(\cdot)$ a non-negative kernel function with bandwidth parameter $\lambda > 0$.

Given the set of estimated direct and cross variograms $\{\widehat{\gamma}_{ij}(\cdot,\cdot;\lambda)\}_{i,j=1}^{p}$, the dissimilarity between two sample locations $\mathbf{s}_k$ and $\mathbf{s}_l$ $(k,l=1,\ldots,n)$ is defined as follows:

$$d_\lambda(\mathbf{s}_k,\mathbf{s}_l)=\sum_{i,j=1}^{p}|\widehat{\gamma}_{ij}(\mathbf{s}_k,\mathbf{s}_l;\lambda)|. \tag{2}$$

Thus, the dissimilarity between two observed locations is defined as the sum of absolute values of all direct and cross variograms at these two observed locations. Equation (2) well defines a measure of dissimilarity [17]. The normalized dissimilarity between two sample locations $\mathbf{s}_k$ and $\mathbf{s}_l$ $(k,l=1,\ldots,n)$ is given by:

$$\tilde{d}_\lambda(\mathbf{s}_k,\mathbf{s}_l)=\frac{1}{D}d_\lambda(\mathbf{s}_k,\mathbf{s}_l),\quad \text{with} \quad D=\max_{(k,l)\in\{1,\ldots,n\}^2} d_\lambda(\mathbf{s}_k,\mathbf{s}_l). \tag{3}$$

2.2 Agglomerative Hierarchical Clustering

Given the dissimilarity measure specified in Equation (3), a dissimilarity matrix at sample locations is built.Then, an agglomerative hierarchical clustering algorithm can operate directly on this dissimilarity matrix to find clusters of sample locations. An agglomerative method begins by the trivial partition of n singletons, letting each sample location be its own cluster. Next, the method forms $n-1$ clusters by grouping the two sample locations that are most similar with respect to a prespecified criterion. The method proceeds in this manner until all sample locations are combined into a single cluster or a stopping rule is activated. The agglomerative hierarchical clustering can be represented by a binary tree called dendrogram.

At each step of the clustering process, it is necessary to update the matrix of dissimilarities. After each grouping of two data locations or two clusters or one data location to a cluster, the dissimilarities between the newly formed cluster and the others are calculated and are replaced, in the matrix of dissimilarities coming be aggregated. Different approaches are possible at this level corresponding to different agglomerative hierarchical clustering. The distinction between these is relative by the way they specify the dissimilarity from a newly formed cluster to one data location, or to other existing clusters. We advocate the use of the complete linkage criterion. Indeed, under the complete linkage clustering, observations belonging to the same cluster are more similar comparatively to other linkage procedures and thus are more likely to be geographically close. This procedure tends to produce tight, compact clusters.

2.3 Hyper-parameters Selection

The proposed clustering approach relies on the kernel function $K_\lambda(\cdot)$ used in the computation of the non-parametric kernel estimator of the spatial dependence structure defined in Equation (1). The choice of the kernel function $K_\lambda(\cdot)$ is less important than the choice of its bandwidth parameter λ. We opt for the Epanechnikov kernel whose support is compact, showing optimality properties in density estimation [18]. Indeed, the use of compactly supported kernel functions considerably reduces the computational burden.

For λ very small, there will not be enough data locations inside the support of the kernel function $K_\lambda(\cdot)$ to estimate reliably the spatial dependence structure. Thus, the resulting dissimilarity matrix can be too rough and can contain spurious features that are artefacts of the sampling process. Furthermore, very small values of the bandwidth parameter λ will tend to produce clusters that are spatially non-contiguous. For λ very large, the non-parametric kernel estimator will over-smooth the underlying multivariate spatial dependence structure. Thus, important features of the underlying structure are smoothed away; this can lead to having clusters that do not reflect the underlying clusters albeit spatially contiguous. The bandwidth parameter λ is chosen by using a common heuristic approach in geostatistics: λ is chosen so that the support of the kernel function $K_\lambda(\cdot)$ centered at each data location contains, at least, 35 observations. Thus, the bandwidth parameter λ corresponds to the maximum distance of the 35th neighbour.

The optimal number of clusters is chosen so that it corresponds to the best clustering identified in terms of an internal clustering performance index. A variety of internal clustering performance indices have been proposed in the literature (see [8], [17] for a review). We choose the silhouette index [11,15] which relies on the pairwise difference of between and within cluster dissimilarities. Given various number of clusters $q = 2, 3, \ldots$, the optimal cluster number is one that maximizes the silhouette index:

$$S(q) = \frac{1}{q}\sum_{i=1}^{q}\left(\frac{1}{n_i}\sum_{\mathbf{x}\in C_i} S^i(\mathbf{x})\right) \quad \text{with} \quad S^i(\mathbf{x}) = \frac{b^i(\mathbf{x}) - a^i(\mathbf{x})}{\max(b^i(\mathbf{x}), a^i(\mathbf{x}))}. \tag{4}$$

where $a^i(\mathbf{x}) = \frac{1}{n_i-1}\sum_{\mathbf{y}\in C_i, \mathbf{y}\neq\mathbf{x}} \tilde{d}_\lambda(\mathbf{x},\mathbf{y})$, $b^i(\mathbf{x}) = \min_{j, j\neq i}\left[\frac{1}{n_j}\sum_{\mathbf{y}\in C_j}\tilde{d}_\lambda(\mathbf{x},\mathbf{y})\right]$; $\{C_i\}_{i=1}^{q}$ denotes the ith cluster of cardinality n_i.

The quantity $a^i(\mathbf{x})$ represents the average dissimilarity of location $\mathbf{x}$ to all other locations belonging to the same cluster i. More $a^i(\mathbf{x})$ is little better is the assignment of $\mathbf{x}$ to his class; $a^i(\mathbf{x})$ is the average dissimilarity of location $\mathbf{x}$ in this class. $b^i(\mathbf{x})$ is the lowest average dissimilarities of location $\mathbf{x}$ to every other cluster that $\mathbf{x}$ does not belong. The cluster with this lowest average dissimilarity is the called nearest cluster of $\mathbf{x}$ because it is the second best choice for location $\mathbf{x}$. $S^i(\mathbf{x})$ denotes the silhouette width of $\mathbf{x}$. By definition, $S^i(\mathbf{x})$ is between -1 and 1. When the value of $S^i(\mathbf{x})$ is close to 1, then dissimilarity of $\mathbf{x}$ from the cluster where it belongs is much less than the dissimilarity between $\mathbf{x}$ and its nearest cluster; this indicates that $\mathbf{x}$ is probably correctly classified. When the values of $S^i(\mathbf{x})$ is close to -1, then the dissimilarity between $\mathbf{x}$ and the cluster where it belongs is larger than the dissimilarity between $\mathbf{x}$ and its nearest cluster; this indicates that $\mathbf{x}$ is probably misclassified. When the value of $S^i(\mathbf{x})$ is close to 0, then $\mathbf{x}$ falls close to the boundaries between the two clusters. The average of $S^i(\mathbf{x})$ for all locations in a cluster i denotes the average silhouette width of that cluster. Given the optimal number of clusters, the resulting clusters are obtained by cutting the dendogram at the corresponding level or height.

3 Application

The proposed clustering method is applied to the National Geochemical Survey of Australia data [7]. The National Geochemical Survey of Australia is a low-density geochemical survey that collected catchment sediment samples covering most of the Australia. The data correspond to 50 concentration elements (Al, As, Au, Ba, Be, Bi,Ca, Ce, Co, Cr, Cs, Cu, Dy, Er, Eu, F, FeT, Ga, Gd, Ge, Hf, Ho, K, La, Lu, Mg, Mn, Na, Nb, Nd, Ni, P, Pb, Pr, Rb, Sc, Se, Si, Sm, Sn, Sr, Tb, Th, Ti, U, V, Y, Yb, Zn, and Zr) from topsoil ($0 - 10$cm depth) and coarse grain-size fraction (< 2mm). The dataset contains 1315 georeferenced observations. The results produced by the proposed clustering method (M1) are compared with those given by: (i) K-means clustering with geographical coordinates as additional variables (M2); (ii) Ward's agglomerative hierarchical clustering with geographical coordinates as additional variables (M3); (iii) Oliver's geostatistical clustering [12] (M4); (iv) Bourgault's geostatistical clustering [5] (M5). For every clustering method, all the variables are logit-transformed and standardized. For clustering methods M2 and M3, geographical coordinates are also standardized.

In the proposed clustering method (M1), the optimal cluster number through the silhouette index corresponds to two, as shown in Figure 1a. The corresponding clusters are shown in Figure 1b. As we can see, the proposed clustering method succeeds in providing spatially contiguous clusters. Figures 1c, 1d, 1e, and 1f give respectively, the clustering results obtained by methods M2, M3, M4, and M5. As we can see, non-spatial clustering with geographical coordinates as additional variables (M2 and M3) as well as non-spatial clustering based on a spatial dissimilarity measure (M4 and M5) fail to produce spatially contiguous clusters.

Given the resulting clusters produced by the proposed clustering method, Figure 2 indicates the relative contribution of variables in the clustering through the random forest classifier [6]. Thus, we can note that the ten most important variables are respectively, Mn, Yb, Na, Y, Mg, Er, P, Cs, Dy, and U. Table 1 shows the main descriptive statistics (respect to raw data) of the two optimal clusters according to these variables. The contrast between the two groups is substantial. Figure 2 and Table 1 reveal that we can distinguish two spatially contiguous clusters with low (blue points) and high (green points) overall concentrations. The group of lower values contains 679 data locations while the group of high values contains 636 data locations.

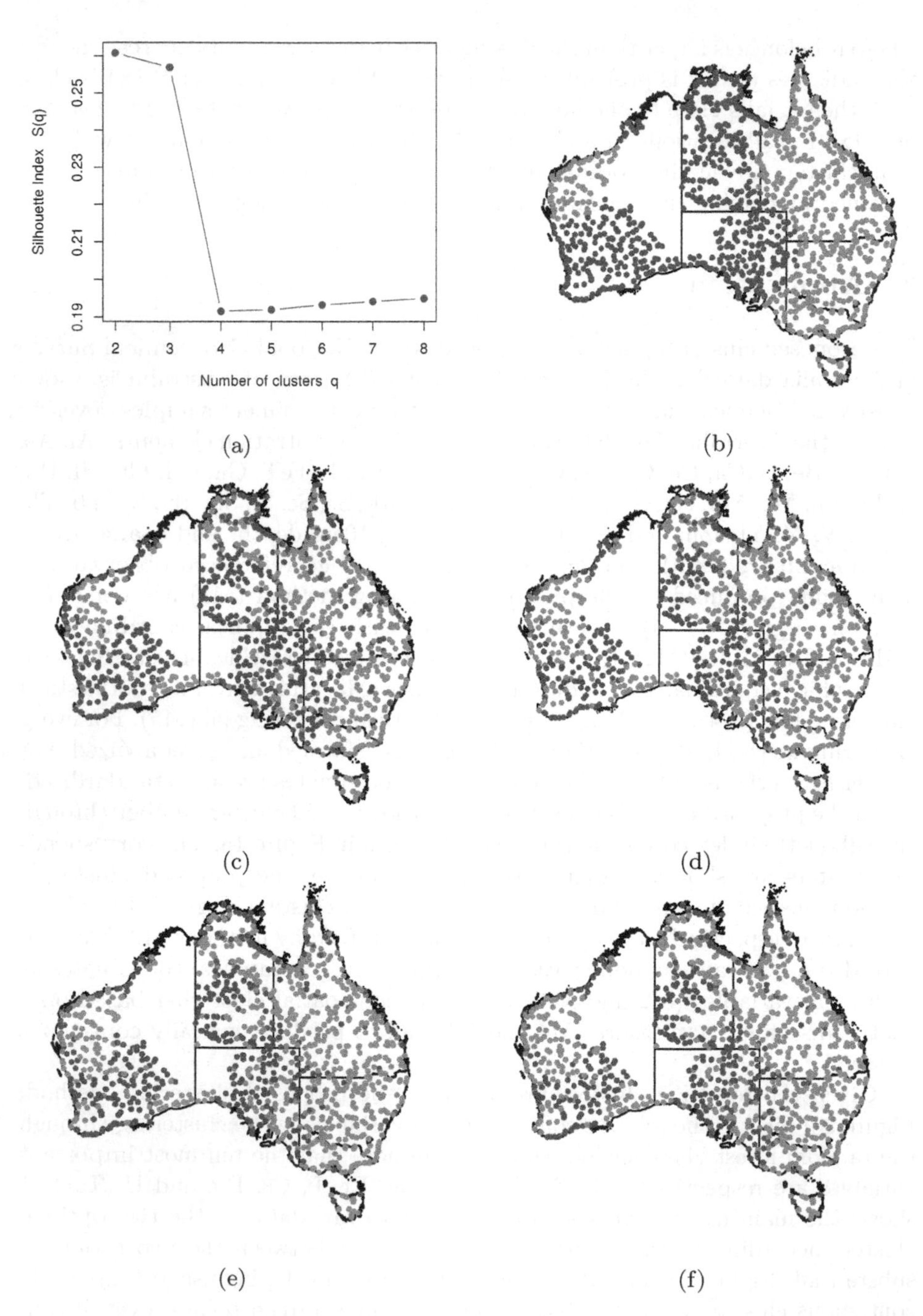

Fig. 1. (a,b) Proposed clustering method (M1); (c) K-means clustering with geographical coordinates as additional variables (M2); (d) Ward's agglomerative hierarchical clustering with geographical coordinates as additional variables (M3); (e) Oliver's geostatistical clustering (M4); (f) Bourgault's geostatistical clustering (M5).

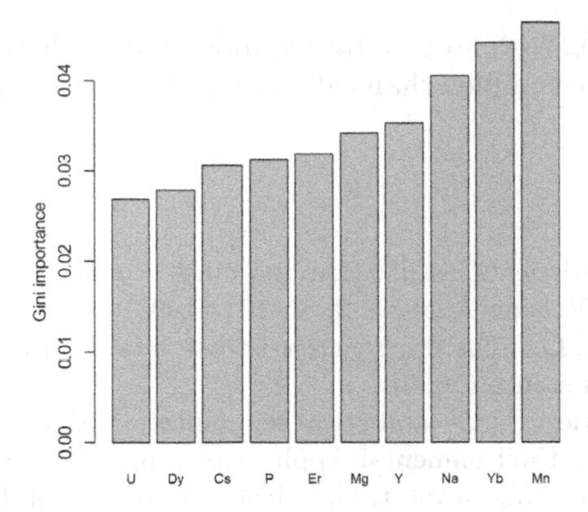

Fig. 2. Proposed clustering method: relative contribution of the ten most important variables in the clustering through the random forest classifier.

Table 1. Proposed clustering method: descriptive statistics (respect to raw data) of the two optimal clusters according to the ten most important variables.

	Cluster 1 ($n_1 = 679$)		Cluster 2 ($n_2 = 636$)	
	Mean	Std.	Mean	Std.
Mn	5.55e+02	4.05e+02	2.62e+02	2.05e+02
Yb	2.54e+00	9.06e-01	1.59e+00	9.29e-01
Na	4.87e+03	4.31e+03	2.61e+03	3.80e+03
Y	2.57e+01	9.51e+00	1.59e+01	9.84e+00
Mg	4.33e+03	3.41e+03	3.92e+03	5.13e+03
Er	2.41e+00	8.82e-01	1.49e+00	9.20e-01
P	4.15e+02	2.68e+02	2.27e+02	1.60e+02
Cs	3.39e+00	2.02e+00	1.88e+00	1.60e+00
Dy	3.99e+00	1.48e+00	2.48e+00	1.54e+00
U	2.35e+00	1.82e+00	1.78e+00	1.58e+00

4 Conclusion

In this work, we developed a new clustering approach aimed to discover spatially compact, connected and meaningful clusters in multivariate point referenced data, in which spatial dependence plays an important role. By taking a non-parametric kernel estimator of the multivariate spatial dependence structure to built a dissimilarity measure, the proposed clustering method reinforces the spatial coherency of the resulting clusters. The developed clustering approach is model-free, adapted to irregularly sampled data, and can produce spatially contiguous clusters without including any geometrical constraints. It incorporates existing methods to find the optimal cluster number. Applied to the National Geochemical Survey of Australia data, the proposed clustering approach highlights two spatially contiguous clusters with significant meaning. The developed clustering approach is computationally intensive when dealing with large datasets. Indeed, the calculation of the dissimilarity measure is more complex than calculating the sum of squared deviations.

References

1. Allard, D.: Geostatistical classification and class kriging. Journal of Geographical Information and Decision Analysis **2**, 87–101 (1998)
2. Allard, D., Guillot, G.: Clustering geostatistical data. In: Proceedings of the Sixth Geostatistical Conference (2000)
3. Allard, D., Monestiez, P.: Geostatistical segmentation of rainfall data. In: geoENV II: Geostatistics for Environmental Applications, pp. 139–150 (1999)
4. Ambroise, C., Dang, M., Govaert, G.: Clustering of spatial data by the EM algorithm. In: geoENV I: Geostatistics for Environmental Applications, pp. 493–504 (1995)
5. Bourgault, G., Marcotte, D., Legendre, P.: The multivariate (co)variogram as a spatial weighting function in classification methods. Mathematical Geology **24**(5), 463–478 (1992)
6. Breiman, L.: Random forests. Machine Learning **45**(1), 5–32 (2001)
7. Caritat, P., Cooper, M.: National geochemical survey of australia: The geochemical atlas of australia. Geoscience Australia Record 2011/020 (2011)
8. Charu, C., Chandan, K.: Data Clustering: Algorithms and Applications. Chapman and Hall/CRC (2013)
9. Dempster, A.P., Laird, N.M., Rubin, D.B.: Maximum likelihood from incomplete data via EM algorithm (with discussion). Journal of the Royal Statistical Society Ser. **39**, 1–38 (1977)
10. Guillot, G., Kan-King-Yu, D., Michelin, J., Huet, P.: Inference of a hidden spatial tessellation from multivariate data: application to the delineation of homogeneous regions in an agricultural field. Journal of the Royal Statistical Society: Series C (Applied Statistics) **55**(3) (2006)
11. Kaufman, L., Rousseeuw, P.: Finding Groups in Data: An Introduction to Cluster Analysis. John Wiley & Sons, New York (1990)
12. Olivier, M., Webster, R.: A geostatistical basis for spatial weighting in multivariate classification. Mathematical Geology **21**, 15–35 (1989)
13. Pawitan, Y., Huang, J.: Constrained clustering of irregularly sampled spatial data. Journal of Statistical Computation and Simulation **73**(12), 853–865 (2003)
14. Romary, T., Ors, F., Rivoirard, J., Deraisme, J.: Unsupervised classification of multivariate geostatistical data: Two algorithms. Computers & Geosciences (2015)
15. Rousseeuw, P.: Silhouettes: A graphical aid to the interpretation and validation of cluster analysis. Journal of Computational and Applied Mathematics **20**, 53–65 (1987)
16. Schuenemeyer, J., Drew, L.: Statistics for Earth and Environmental Scientists. Wiley (2011)
17. Theodoridis, S., Koutroumbas, K.: Pattern Recognition, 4th edn. Academic Press (2009)
18. Wand, M., Jones, C.: Kernel Smoothing. Monographs on Statistics and Applied-Probability. Chapman and Hall (1995)

On Learning and Exploiting Time Domain Traffic Patterns in Cellular Radio Access Networks

Jordi Pérez-Romero(✉), Juan Sánchez-González, Oriol Sallent, and Ramon Agustí

Universitat Politècnica de Catalunya (UPC), Barcelona, Spain
{jorperez,juansanchez,sallent,ramon}@tsc.upc.edu

Abstract. This paper presents a vision of how the different management procedures of future Fifth Generation (5G) wireless networks can be built upon the pillar of artificial intelligence concepts. After a general description of a cellular network and its management functionalities, highlighting the trends towards automatization, the paper focuses on the particular case of extracting knowledge about the time domain traffic pattern of the cells deployed by an operator. A general methodology for supervised classification of this traffic pattern is presented and it is particularized in two applicability use cases. The first use case addresses the reduction of energy consumption in the cellular network by automatically identifying cells that are candidates to be switched-off when they serve low traffic. The second use case focuses on the spectrum planning and identifies the cells whose capacity can be boosted through additional unlicensed spectrum. In both cases the outcomes of different classification tools are assessed. This capability to automatically classify cells according to some expert guidance is fundamental in future networks, where an operator deploys tenths of thousands of cells, so manual intervention of the expert is unfeasible.

Keywords: Classification · Cellular networks · 5G · Radio access network management

1 Introduction

Our interconnected world is increasingly marked by fluid boundaries, tighter interlinkages and globally coordinated actions. Among these complexities, one of the most influential factors shaping our global society are networks. Networks serve as a central metaphor for describing the complexities of modern life. But they are also an undeniable technological foundation for unlocking tremendous social and economic benefits. Understanding the dynamics of networks – and their potential for positive change - can help us collectively meet our greatest social, economic and environmental challenges [1]. In this context, cellular networks have become pivotal: currently, there are as many mobile subscriptions as people in the world, and every second, 20 new mobile broadband subscriptions are activated. In addition to the increase in subscribers, data consumption also continues to rise.

Then, as a next step in the evolution of cellular communication systems, research carried out by industry and academia is nowadays focused on the development of the

P. Perner (Ed.): MLDM 2016, LNAI 9729, pp. 501–515, 2016.
DOI: 10.1007/978-3-319-41920-6_40

new generation of mobile and wireless systems, known as 5th Generation (5G) that targets a time horizon beyond 2020. 5G intends to provide solutions to the continuously increasing demand for mobile broadband services associated with the massive penetration of wireless equipment such as smartphones, tablets, the tremendous expected increase in the demand for wireless Machine To Machine communications [2], and the proliferation of bandwidth-intensive applications including high definition video, 3D, virtual reality, etc.

To cope with the abovementioned demands, requirements of future 5G system have been already identified and discussed at different fora [3][4]. Examples of these requirements include1000 times higher mobile data volume per area, 10 times to 100 times higher number of connected devices, 10 times to 100 times higher typical user data rate, 10 times longer battery life for low power devices and 5 times reduced End-to-End latency.

Furthermore, 5G networks will be fueled by the advent of big data and big data analytics [5]. The volume, variety and velocity of big data are simply overwhelming. Nowadays, there are already tools and platforms readily available to efficiently handle this big amount of data and turn it into value by gaining insight and understanding data structures and relationships, extracting exploitable knowledge and deriving successful decision-making. Applications of big data and big data analytics are already present in different sectors (e.g. entertainment, financial services industry, automotive industry, logistics, etc.). Therefore, with the huge amount of data generated by mobile networks, it can be envisaged that big data technologies will play a key role in 5G to extract the most of possible value of the available data exploiting it for enhancing the efficiency in mobile service provisioning.

In this context, this paper supports the idea that Artificial Intelligence (AI) mechanisms, which intend to develop intelligent systems able to perceive and analyze the environment and take the appropriate actions, will fully fertilize in the 5G ecosystem. While many seeds can be found in the literature both from an academic/theoretical perspective (e.g., connected to the so-called Cognitive Networks [6]) and from a practical perspective in current Third Generation (3G) and Fourth Generation (4G) networks (e.g., connected to the so-called Self-Organizing Networks [7]), more ambitious objectives can be targeted and the 5G era is the proper time for AI-based networks to happen.

Based on the above considerations, this paper intends to present a vision of how the different Radio Access Network (RAN) management procedures of future 5G networks can be built upon the pillar of AI concepts. For that purpose, the paper provides in Section 2 a general description of a cellular network and corresponding RAN management functionalities and highlights the trends towards automatization. This is mainly addressed to the non-specialized reader. In turn, Section 3 deepens on the framework to support RAN management in the context of 5G networks. Section 4 focuses on the particular case of extracting knowledge about the time domain traffic pattern of the cells and provides a number of potential applications. With this, two applicability use cases are presented in Section 5 and Section 6. Conclusions close the paper in Section 7.

2 Radio Access Network Management

A generic view of a cellular network is depicted in Fig. 1. Its main components are briefly described in the following: The User Equipment (UE) is the device that enables the Mobile Network Operator (MNO)'s customer to gain access to the network services (e.g., voice, data). The UE connects to the Radio Access Network (RAN) through the so-called radio interface (i.e., a wireless interface). Typically, the UE is multi-technology (e.g., 3G, 4G, WiFi) and can operate at different frequency bands.

The RAN is the network subsystem responsible to provide the connectivity between the UE and the Core Network (CN), which manages the provision of final services to the users. The RAN is composed of multiple base stations (BS), also known in general as "cells", and, for some technologies such as Second Generation (2G) and 3G, it also includes additional network controller nodes. The RAN includes a number of management functionalities in order to provide the wireless connectivity in an efficient way. In turn, the CN takes care of aspects such as the interconnection with other networks.

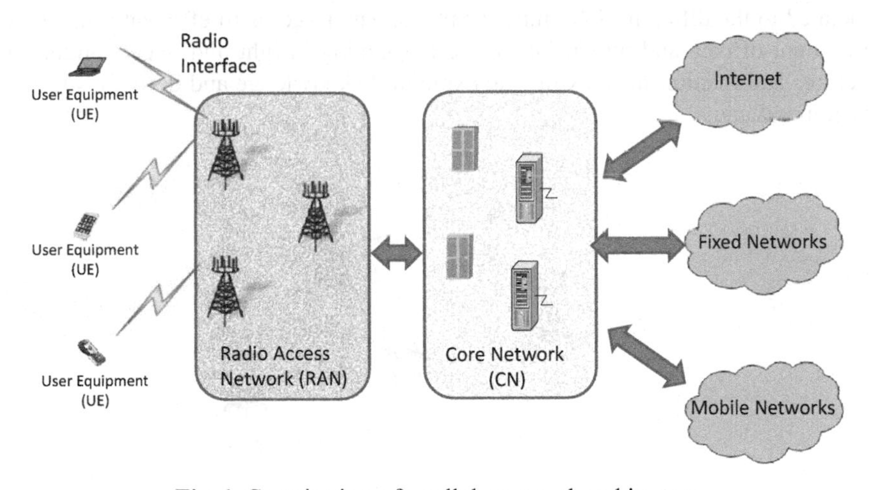

Fig. 1. Generic view of a cellular network architecture

Given that the RAN is composed of multiple cells and there will be multiple UEs moving around, as illustrated in Fig. 2, the set of RAN management functionalities takes care of deciding aspects such as to which cell a specific UE is attached to, the time at which a moving UE needs to switch the connectivity from one cell to another neighboring cell, how a given cell splits its capacity (i.e., data rate) among the different UEs that are attached to it, etc.

The vision of the future 5G RAN corresponds to a highly heterogeneous network at different levels, including multiple technologies, multiple cell layers, multiple spectrum bands, multiple types of devices and services, etc., with unprecedented requirements in terms of capacity, latency or data rates. The resulting network easily comprises 10.000-20.000 cells for a wide coverage service area (e.g., medium size

European country). Consequently, the overall RAN management processes that constitute a key point for the success of the 5G concept will exhibit tremendous complexity. In this direction, legacy systems such as 2G/3G/4G already started the path towards a higher degree of automation in the planning and optimization processes through the introduction of Self-Organizing Network (SON) functionalities, in order to carry out these processes in a more efficient way.

SON refers to a set of features and capabilities designed to reduce or remove the need for manual activities in the lifecycle of the network, so that operating costs can be reduced as well as revenue can be protected by minimizing human errors. As such, with the introduction of SON features, classical manual planning, deployment, optimization and maintenance activities of the network can be replaced and/or supported by more autonomous and automated processes.

SON can greatly benefit from AI-based tools able to smartly process input data from the environment and come up with exploitable knowledge (i.e., knowledge that can be formalized in terms of models and/or structured metrics that represent the network behavior in a way that can be directly used to make smart network planning and optimization decisions). The obtained knowledge will drive the appropriate actions associated to the different SON functionalities. The target is to efficiently handle this big amount of data and turn it into value by gaining insight and understanding data structures and relationships, extracting exploitable knowledge and deriving successful decision-making.

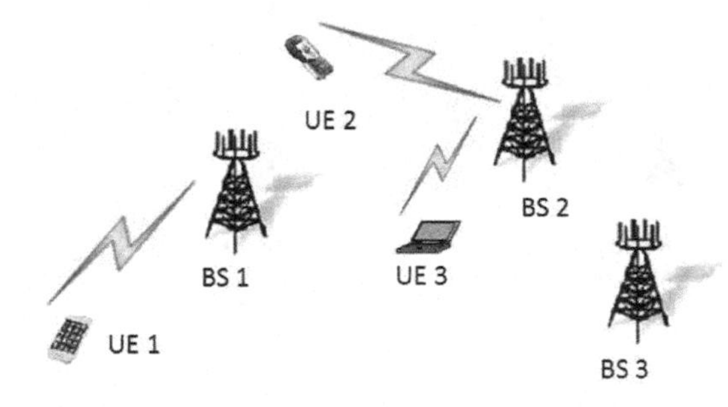

Fig. 2. Closer view of the RAN subsystem

3 Knowledge Discovery to Support 5G RAN Management

3.1 Data Acquisition and Pre-processing

MNOs have traditionally operated with complex, disparate sets of data, with useful information residing in multiple systems such as customer relationship management systems, network management systems, billing, inventories, network elements, service management systems, deep packet inspection devices, application-specific databases, etc., [8]. In addition, MNOs have to deal with the concurrent operation of network elements belonging to multiple network generations (2G/3G/4G, etc.) and/or to multiple vendors, each one holding different types of data, in various formats.

This huge heterogeneity of data and the associated difficulties in carrying out an efficient processing has led to perform the network management processes relying on a limited amount of data both in terms of variety of data considered (i.e., many counters and measurements that can be captured are not exploited at their possible extent) and time spam that are stored in the management systems (i.e., many counters and measurements are just retained for a short period of time in support of certain functionalities and then are either deleted or forgotten in back-up systems, while their applicability to keep the memory of the system is disregarded).

While this limited approach has been the rule in legacy 3G/4G systems, with SON deployment still at its infancy, a substantial evolution is necessary to deal with the increased complexity and stringent efficiency constraints of 5G. Therefore, the challenge for an efficient 5G RAN Management is to build this complete network vision by smartly analyzing and correlating all the different data sources in order to extract the most relevant information contained in them.

In general, gathered input data can belong to different categories (e.g., network data, user data, content data, external data). Network data characterizes the behavior of the network in terms of different measurements collected and recorded by the network nodes. Measurements include network traffic levels (e.g. traffic load at the radio interface, signaling traffic, active UEs per cell, etc.), resource access measurements, Quality of Service (QoS) measurements (e.g. throughput, latency, etc.) or cell availability measurements. Measurements can be performed by the network nodes (e.g. the cells) and also by the UEs that report them to the network.

The time span of the data collection will tightly depend on the targeted applicability of each type of data (e.g., planning actions will consider input data recorded over longer periods of time that can spam over several days or weeks while optimization actions will usually consider input data collected over much shorter time frames).

The proposed framework relies on the application of data mining techniques over the collected input data in order to distil all the available information and identify meaningful models and patterns that will drive the subsequent decisions. In this respect, the collected data coming from multiple heterogeneous sources needs to be preprocessed in order to prepare it for mining. This includes different tasks such as [9]: data cleaning to remove noise and inconsistent data (e.g. discard network counters that exhibit errors); data integration to combine network data collected at different nodes and exhibiting different time stamps or different periodicities; data selection to choose the relevant data for each specific analysis; and data transformation where data are consolidated through e.g. summary or aggregation operations (e.g. aggregating measurements collected with a periodicity of 15min to derive the equivalent measurement with 1h periodicity).

3.2 Knowledge Discovery

The Knowledge Discovery stage performs inference on the pre-processed data in order to build models that reflect the relevant knowledge that will drive the optimization and planning decisions. The core functionality of the knowledge discovery consists in learning from the users and the network in order to extract models that reveal their behavior. It is worth emphasizing here that, given the ultra-high level of efficiency that is associated to the design of future 5G systems, the target is to gain in-depth and detailed knowledge about the whole ecosystem, which in turn will enable ultra-efficient management and optimization. In this respect, the higher level of knowledge about the network and its users constitutes a key differential factor between 5G and legacy systems.

This stage will be based on machine learning tools used to carry out the mining of the pre-processed data to extract relevant knowledge at different levels: cell level (contains the characterization of the conditions on a per cell basis), cell cluster level (characterization of groups of cells built according to their similarities) and user level (contains the characterization of the existing conditions at the user equipment level).

The general goal of machine learning is to build computer systems that can adapt and learn from their experience [10]. Machine learning techniques are usually subdivided into three big categories, namely supervised learning, unsupervised learning and reinforcement learning. From the perspective of the knowledge discovery stage considered here, both supervised and unsupervised learning techniques are the ones that exhibit more applicability, while reinforcement learning tools will normally be more relevant for the decision making processes associated to the management functionalities.

Specific machine learning functions that are relevant in the framework considered here for RAN management are classification, prediction and clustering [9]. Among them, the focus of this paper is on the classification applied for knowledge discovery related to the time domain traffic variations of the cells deployed in a network. Classification is the process of finding a model or function that describes and distinguishes data classes or concepts. The obtained model (i.e. the classifier) is then used to determine the class to which an object belongs. The object to be classified is represented by a tuple that includes a set of attribute values. Classification process assumes that the possible classes are predefined in advance. Then, the classification model is usually obtained from a supervised learning algorithm that analyses a set of training tuples associated with known classes.

4 Classification of the Cell-Level Time Domain Traffic

The cell-level time domain traffic defines how the traffic of a cell varies as a function of time. Traffic can be measured in different ways, such as the load factor, the total number of users connected to the cell, the total data rate, etc., and it can be aggregated or split among QoS classes. The traffic in a cell will be tightly related with the environment where the cell is deployed and with the characteristics and profiles of the users served by the cell. This will lead to time correlations in the traffic evolution of a given cell at different levels (e.g. intra-day variations in which the traffic can substantially differ between mornings or nights, variations during the week between working days and weekend, etc.). The detailed analysis of these correlations will allow extracting valuable knowledge that can be used for making management decisions regarding the configuration of a cell. In this respect, this paper focuses on the application of classification techniques to extract this knowledge. In particular, the cells will be classified based on their historical traffic samples. The possible classes will indicate certain behaviors of the cells that are relevant for different RAN management processes. In the following we start by providing the general classification methodology and then we particularize it according to its applicability in some selected use cases, providing some results obtained using data extracted from a real mobile network.

4.1 General Classification Methodology

The input data for each cell i is a time series $\mathbf{X}_i=(x_i(t),, x_i(t-1), ..., x_i(t-(N-1)))$ composed of N samples of the measured traffic in the cell i at different times t. The objective of the classifier is to make an association between the input time series $\mathbf{X}_i$ and a class $C(\mathbf{X}_i)$ that characterizes the behavior of the cell's traffic in the time domain. The number and the type of classes will depend on the specific applicability of the classification outcomes, as it will be detailed in the use cases that will be presented later on.

Since the number of time samples N will typically be a very large value (e.g. reflecting the traffic measured in a cell in periods of some minutes and collected during several weeks, months, etc.), it will not be feasible to use the time series $\mathbf{X}_i$ directly as input of a classifier tool. Therefore, an initial processing is carried out to come up with a vector $\mathbf{F}(\mathbf{X}_i)$ of shorter dimension M that preserves the relevant characteristics of the traffic pattern. This vector will be the input of the classifier. Following the usual terminology in classification [9], vector $\mathbf{F}(\mathbf{X}_i)$ represents the tuple to be classified and each of its components represents a feature or attribute. Again, the definition of the mapping between $\mathbf{X}_i$ and $\mathbf{F}(\mathbf{X}_i)$ will be dependent on the specific applicability of the classification, so it will be detailed later on when analyzing the different use cases.

The classifier will perform the association between the input $\mathbf{F}(\mathbf{X}_i)$ and the class $C(\mathbf{X}_i)$, as illustrated in Fig. 3. The internal structure of the classifier will be given by the specific classification tool being used and its settings will be automatically configured through a supervised learning process executed during an initial training stage. This training will use as input S time series $\mathbf{X}_j$ j=1,...,S of some cells whose associated classes $C(\mathbf{X}_j)$ are pre-defined by an expert. In this way, the training set will be composed by the S tuples $\mathbf{F}(\mathbf{X}_j)$, j=1,...,S and their associated classes $C(\mathbf{X}_j)$. The supervised learning process will analyze this training set to determine the appropriate configuration of the classification tool. The overall process is illustrated in Fig. 3.

4.2 Classification Tools

Regarding the classification tool, the following alternatives are considered [9]:

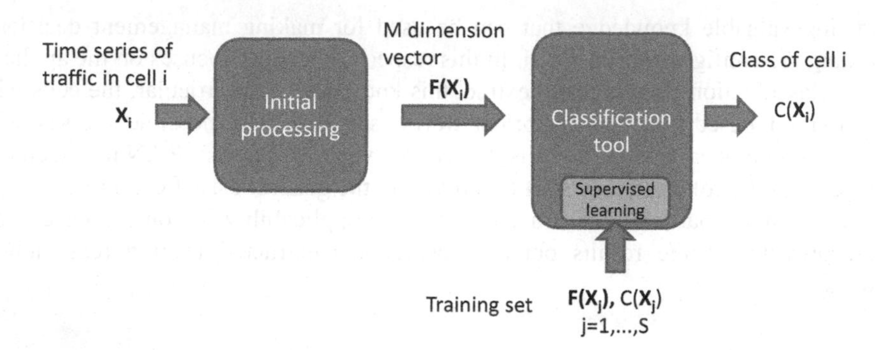

Fig. 3. General classification methodology

- Decision tree induction: The classification is done by means of a decision tree, which is a flow-chart structure where each node denotes a test on a feature value, i.e. a component of vector $\mathbf{F}(\mathbf{X}_i)$, each branch represents an outcome of the test, and tree leaves represent the classes. The tree structure is built during the supervised learning stage through a top-down recursive divide-and-conquer manner, starting from the training set which is recursively partitioned into smaller subsets.
- Naive Bayes classifier: In this case the classifier evaluates the probability $\text{Prob}(C(\mathbf{X}_i)|\mathbf{F}(\mathbf{X}_i))$ that a given cell $\mathbf{X}_i$ belongs to a class $C(\mathbf{X}_i)$ based on the values of the features $\mathbf{F}(\mathbf{X}_i)$. The resulting class is the one with the highest probability. The computation of this probability is done using Bayes' theorem under the "naive" assumption of class conditional independence, which presumes that the effect of a feature value on a given class is independent of the values of the other features. In turn, the different terms in the computation of the Bayes' theorem are obtained from the analysis of the training set.
- Support Vector Machine (SVM): A SVM is a classification algorithm based on obtaining, during the training stage, the optimal boundary that separates the vectors $\mathbf{F}(\mathbf{X}_j)$ of the training set in their corresponding classes $C(\mathbf{X}_j)$. The obtained boundary is then used to perform the classification of any other input vector $\mathbf{F}(\mathbf{X}_i)$. To find this optimal boundary, it uses a nonlinear mapping to transform the original training data into a higher dimension so that the optimal boundary becomes an hyperplane. Although SVM classifier is originally intended to do a binary classification, a multi-class SVM classifier can easily built by hierarchically combining multiple binary SVM classifiers. Each of these binary classifiers specifies whether the cell belongs or not to a given class.
- Neural Network: The classification is done by means of a feed-forward neural network that consists of an input layer, one or more hidden layers and an output layer. Each layer is made up of processing units called neurons. The inputs to the classifier, i.e. each of the components of vector $\mathbf{F}(\mathbf{X}_i)$, are fed simultaneously into the neurons making up the input layer. These inputs pass through the input layer and are then weighted and fed simultaneously to a second layer. The process is repeated until reaching the output layer, whose neurons provide the selected class $C(\mathbf{X}_i)$. The weights of the connections between neurons are learnt during the training phase using a back propagation algorithm.

The abovementioned general classification methodology presents applicability in different management process, such as network planning, optimization of radio resource management algorithms, energy saving, spectrum planning or load balancing. In the following sections, the methodology is particularized for two of these use cases.

5 Use Case 1: Energy Saving

This use case aims at reducing the energy consumption in the deployed cellular network. According to the Mobile's Green Manifesto report [11], approximately 80% of the energy consumption and Green House Gas emissions of mobile operators is caused from their networks. From an economical perspective, if all networks with above-average energy consumption were improved to the industry average, there is a potential energy cost saving for mobile operators of $1 billion annually at 2010 prices. In case of improving to levels of the top quartile the cost saving could be more than $2 billion a year [11]. Therefore, techniques intended to reduce the energy consumption are relevant for operators of current and future networks.

In this use case, the energy reduction is done by switching off the cells that carry very little traffic at certain periods of the day (e.g. at night) and making the necessary adjustments in the neighbor cells so that the existing traffic can be served through some other cell. In this context, the classification methodology of section 4 can be used to identify candidate cells to be switched-off based on their traffic patterns. The automation of this procedure based on expert criteria captured in the training set becomes particularly useful considering that networks in the envisaged ultra-dense scenarios for future 5G systems can comprise several tens of thousands of cells. Therefore, it is not practical that a human expert can make this classification manually. It is worth mentioning, however, that the final decision on whether or not to switch off a cell would make use of this classification as well as other possible inputs which are out of the scope of this paper (e.g. the neighbor cell lists to ensure that a call that is generated in a cell that has been switched-off can be served through another cell).

In this use case a cell can be classified in two different classes:

- Class A: Candidate cell to be switched off
- Class B: Cell that cannot be switched off.

5.1 Data Acquisition and Pre-processing

In this case, the components of vector $\mathbf{F}(\mathbf{X}_i)$ correspond to the average normalized traffic of the cell during the nights (i.e. from 0h to 8h), the mornings (i.e. from 8h to 16h) and the afternoons (i.e. from 16h to 24h) for each day of the week (Monday to Sunday). This leads to a total of M = 21 components that can be easily obtained by normalizing the time series $\mathbf{X}_i$ so that the traffic ranges from 0 to 1 and by averaging the time series in each of the abovementioned periods.

To assess the behavior of the classification methodology in this use case, a set of real traffic measurements for a total of 419 cells deployed by an operator in a certain geographical region has been used. For each cell i, the time series $\mathbf{X}_i$ is composed by the data traffic measurements done every 15 min, and collected during a whole week. Therefore, each time series is composed by N = 672 traffic samples. The traffic in a period of 15 min is given by the average number of users in the cell with an active data session.

5.2 Knowledge Discovery

The different classification tools discussed in section 4.1 have been implemented by means of RapidMiner Studio Basic [12]. The different parameters have been manually adjusted to obtain good accuracy levels of the different classification tools. In particular, the SVM is configured with radial kernel type, complexity constant which sets the tolerance for misclassification C = 30, kernel cache 200 MB, convergence precision 0.001, a maximum of 10^5 iterations and the loss function is defined with complexity constants equal to 1 for both positive and negative examples and insensitivity constant equal to 0. The neural network classifier is configured with one hidden layer, 500 training cycles, learning rate 0.3, momentum 0.6 and the optimization is stopped if the training error gets below 10^{-5}. The decision tree is configured with maximal depth 20, minimal leaf size 2, confidence level 0.25, minimal size for split 4, minimal gain 0.1 and applying pruning and prepruning with 3 alternatives. Finally the Naive Bayes classifier is configured with Laplace correction, greedy estimation mode and 10 kernels.

5.3 Results

To illustrate the expert criteria to be learnt by the classification tool, Fig. 4 plots the time series $\mathbf{X}_i$ of 4 example cells included in the training set. Two of them are classified by the expert as A and two of them are classified as B. Then, different training sets have been built including these cells together with other examples in order to train the classification tools.

First, several tests have been done to derive the accuracy of the considered classification tools as a function of the training set size S. For a given S, the accuracy is measured by executing the classification over the cells of the training set and calculating the percentage of cells that are classified in the same category that was declared by the expert in the training. The test has been applied for all 4 classification tools and training set sizes ranging from S = 10 to S = 200. The best accuracy is obtained by the SVM, which provides 100% accuracy in all the cases, followed by the Neural Network and Decision Tree, which exhibit accuracy above 98.5%. The worst behavior is obtained with the Naive Bayes classifier with a minimum accuracy of 96.4%.

After completing the training process, the classification of the 419 available cells is performed. Then, as a first result that illustrates the operation of the classification process, Fig. 5 depicts the time series $\mathbf{X}_i$ of two example cells that didn't belong to the training set: Cell 260, which is classified as Class A by all 4 classification tools considered, and Cell 240, which all 4 classification tools categorize as Class B. From visual inspection, and by comparing these cells with the examples given by the expert in Fig. 4, it appears an adequate decision given that Cell 260 exhibits relatively long periods at night serving no traffic at all and Cell 240 has traffic during all the time periods in the week.

Fig. 6 presents the total number of cells that are classified as A by each classification tool as a function of the training set size S. It is observed that, for low values of S (e.g. S = 10) roughly half of the cells are classified as A and half are classified as B by all the tools. This indicates that, due to the low number of examples in the training set, the classification tools are not able to clearly distinguish the traffic patterns and the classification exhibits high randomness. Instead, when increasing the training set size S, the number of cells belonging to class A is substantially reduced for all the classifiers (e.g. for the case of the largest training set size S = 200 the number of cells classified as A ranges from 46 with SVM up to 90 for the Naive Bayes case). It is worth emphasizing that the SVM exhibits a more efficient operation compared to the rest of classification tools since it is less sensitive to the value of S: as soon as the training set is S ≥ 20, the result of the classification is very similar (i.e., there are around 50 cells classified as A).

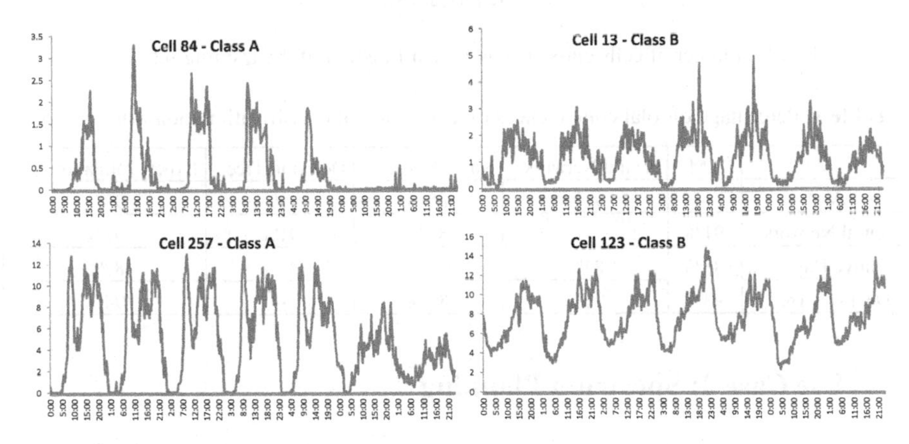

Fig. 4. Examples of cells of the training set belonging to classes A and B.

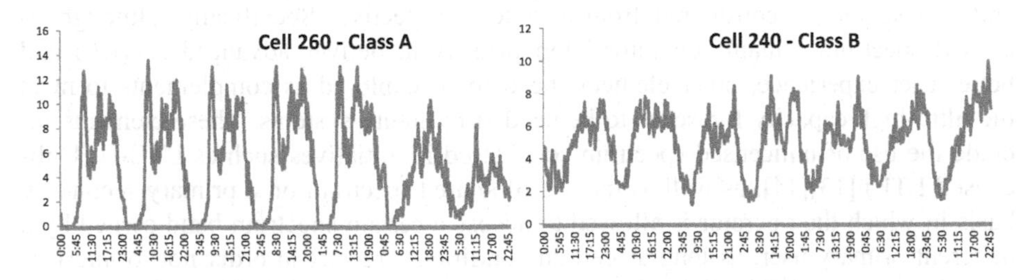

Fig. 5. Examples of two cells classified as A (Cell 260) and B (Cell 240).

Table 1 compares the outcomes of the different classification tools by presenting the percentage of coincidences between every pair of tools for the case S = 200. For example, the table shows that 91% of the cells (i.e. 381 out of 419 cells) have been classified equally by the SVM and the Neural Network. The table also presents the

"Expert validation", which measures the percentage of coincidences with respect to the classification made by the expert. It can be observed that the largest percentages of coincidences are obtained with SVM.

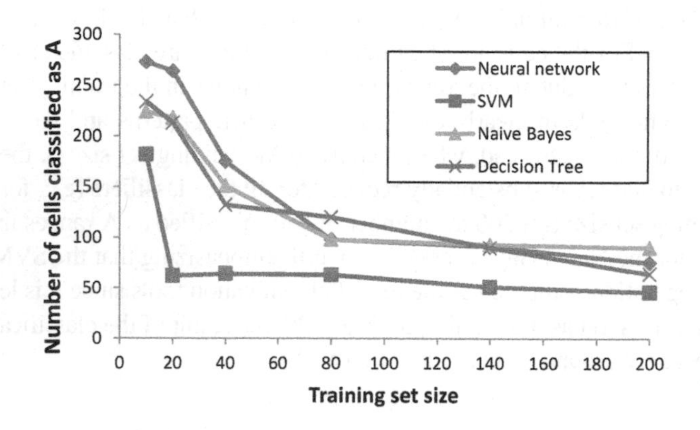

Fig. 6. Number of cells classified as A as a function of the training set size.

Table 1. Percentage of total coincidences by every pair of classification tools with S=200

	SVM	Neural Network	Naive Bayes	Decision Tree	Expert validation
SVM	--	91%	88%	93%	98%
Neural Network	91%	--	87%	91%	91%
Naive Bayes	88%	87%	--	88%	87%
Decision Tree	93%	91%	88%	--	94%

6 Use Case 2: Spectrum Planning

In light of the more advanced spectrum management models envisioned for future 5G systems, the provisioning of the spectrum resources to be exploited at a given time and cell should be considered from a wider perspective. Specifically, although licensed spectrum remains operators' top priority to deliver advanced services and better user experience, other elements need to be explored as complements to meet the ultra-high capacity foreseen to be needed by future systems. These elements include the use of unlicensed spectrum considered in initiatives such as LTE-U (Unlicensed LTE) [13][14], as well as the use of shared spectrum on a primary/secondary basis in which the operator is allowed to access a certain spectrum band owned by a different primary user, as long as certain conditions are met in order not to interfere the primary users. With all these considerations, the use case considered here intends to decide whether it is possible or not to boost the capacity of a cell by exploiting unlicensed spectrum bands. This decision will exploit the knowledge about the time evolution of the cell's traffic, in the sense that typically unlicensed spectrum could be adequate to cope with sporadic traffic increases. Then, this use case intends to classify the cells according to the following classes:

- Class A: Candidate cell to boost capacity through additional unlicensed spectrum.
- Class B: Cell that does not need capacity boost through unlicensed spectrum.

6.1 Data Acquisition and Pre-processing

This use case has been assessed considering a total of 300 cells from a real cellular network deployed in an urban area, under the rationality that this type of scenario is where capacity boosting will be more likely needed. Besides, assuming that spectrum demands will be mainly associated to the periods of the day when there is more traffic, in this use case the components of vector $\mathbf{F}(\mathbf{X}_i)$ correspond to the average traffic of a cell on a per hour basis, between 6h and 22h. This leads to a total of M=16 components. As a difference from the previous use case, here the traffic is not normalized, since the absolute value of the traffic is also relevant to decide whether additional unlicensed spectrum may be needed.

6.2 Knowledge Discovery

The same classification tools as in section 5.2 are considered here.

6.3 Results

Fig. 7 plots the components of vector of $\mathbf{F}(\mathbf{X}_i)$ for 2 cells of the training set categorized as A and B by the expert. Class A cells use to exhibit peaks of high traffic levels while class B cells exhibit lower traffic values and more homogeneity. Like in the previous use case, different training set sizes have been used to train the considered classification tools. After the training process, the 300 cells have been classified. Fig. 8 depicts two example cells that were not included in the training set and that are classified as A and B by all the considered classifiers. It is observed that both cells present similar characteristics like the cells of the training set shown in Fig. 7, meaning that the classification tools have been able to identify also the relevant characteristics of the time evolution in this use case.

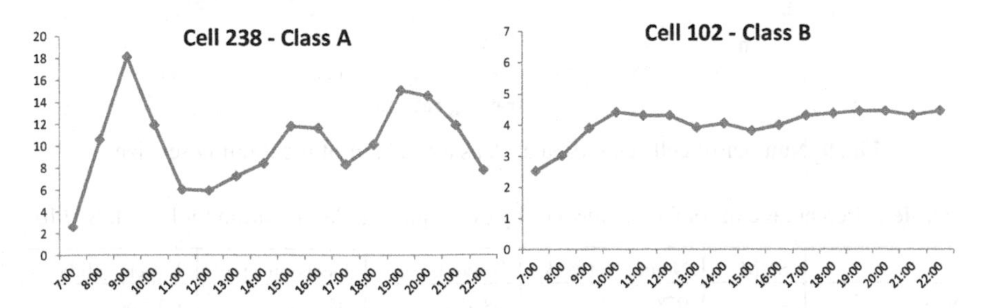

Fig. 7. Examples of cells of the training set belonging to classes A and B.

Fig. 9 presents the number of cells classified as A by each classifier as a function of the training set size with the different classifiers. Like in the previous use case it is observed that the SVM is able to converge more quickly than the other classifiers when the training set size is small. It is also noticed that for the case of S=140 very small differences are observed between the classifiers. This can also be corroborated in Table 2 that presents the percentage of coincidences between every pair of classifiers and with the expert validation. It can be observed that the percentages of coincidence with the expert in this use case are higher than in the previous one. This reflects that the characteristics that make a cell to be classified as A (e.g. sporadic traffic peaks) are more easily distinguishable than in the previous use case. Table 2 also shows that the best performance in terms of coincidences with the expert validation is achieved by both SVM and Neural Network classifiers.

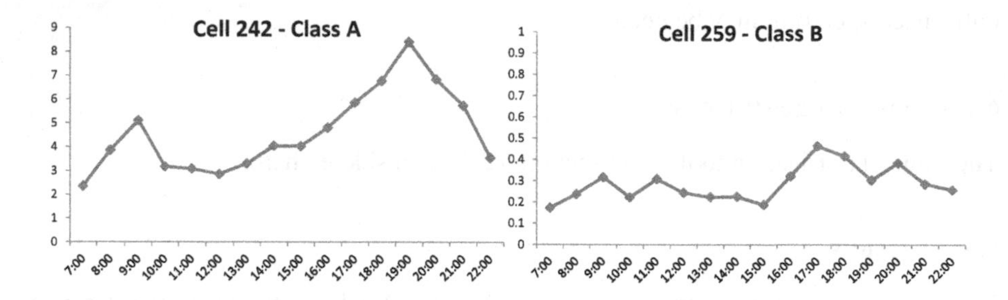

Fig. 8. Examples of two cells classified as A and B.

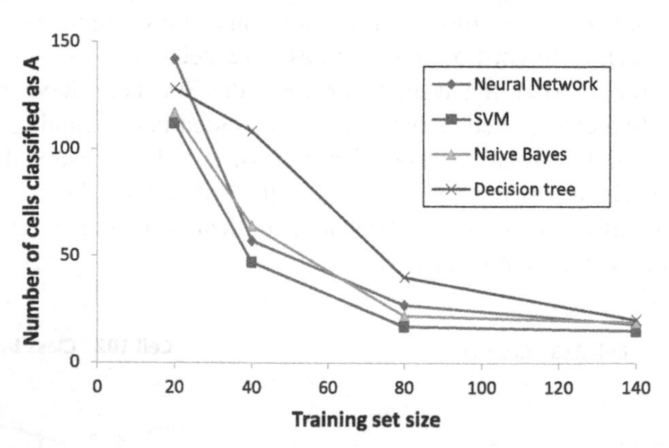

Fig. 9. Number of cells classified as A as a function of the training set size.

Table 2. Percentage of total coincidences by every pair of classification tools with S=140

	SVM	Neural Network	Naive Bayes	Decision Tree	Expert validation
SVM	--	97%	98.7%	97.7%	99.7%
Neural Network	97%	--	98.4%	99.4%	99.7%
Naive Bayes	98.7%	98.4%	--	99%	98.4%
Decision Tree	97.7%	99.4%	99%	--	97.4%

7 Conclusions

This paper has focused on the application of artificial intelligence and data mining concepts to support the radio access management in future cellular networks, where automatization is fundamental to cope with the huge number of cells that an operator can deploy, so manual intervention from a human expert becomes impractical. In particular, the paper has focused on extracting knowledge about the time domain traffic pattern of the cells. A general methodology for supervised classification of this traffic pattern has been presented and particularized in two applicability use cases, addressing energy saving and spectrum planning processes. In both cases the outcomes of different classification tools are assessed, concluding that the SVM technique is in general the one that best captures in the classification process the expert knowledge provided in the examples of the training set.

Acknowledgements. This work has been supported by the EU funded H2020 5G-PPP project SESAME under the grant agreement no 671596 and by the Spanish Research Council and FEDER funds under RAMSES grant (ref. TEC2013-41698-R).

References

1. World Economic Forum. Enabling Transformation: Information and Communications Technologies and the Networked Society (2009)
2. Ericsson White Paper. More than 50 billion connected devices, February 2011. http://www.ericsson.com/res/docs/whitepapers/wp-50-billions.pdf
3. Fallgren, M., Timus, B. (eds.): Scenarios, requirements and KPIs for 5G mobile and wireless system, Deliverable D1.1. of the METIS project, May 2013
4. El Hattachi, R., Erfanian, J. (eds.): NGMN 5G White Paper, NGMN Alliance, February 2015
5. Ericsson, Big Data Analytics. White paper, August 2013
6. Thomas, R.W., DaSilva, L.A., MacKenzie, A.B.: Cognitive networks. In: 1st IEEE International Symposium on New Frontiers in Dynamic Spectrum Access Networks, (DySPAN), pp. 352–360 (2005)
7. Ramiro, J., Hamied, K.: Self-Organizing Networks. Self-planning, self-optimization and self-healing for GSM, UMTS and LTE. John Wiley & Sons (2012)
8. Banerjee, A.: Advanced Predictive Network Analytics: Optimize your Network Investments and Transform Customer Experience. White Paper, Heavy Reading (2014)
9. Han, J., Kamber, M.: Data Mining Concepts and Techniques, 2nd edn. Elsevier (2006)
10. Wilson, R.A., Keil, F.C.: The MIT Encyclopedia of the Cognitive Sciences. MIT Press (1999)
11. GSMA, Mobile's Green Manifesto, 2nd edn., June 2012
12. RapidMiner Studio. http://www.rapidminer.com
13. GPP workshop on LTE in unlicensed spectrum, Sophia Antipolis, France, June 13, 2014. http://www.3gpp.org/ftp/workshop/2014-06-13_LTE-U/
14. Al-Dulaimi, A., Al-Rubaye, S., Quiang, N., Sousa, E.: 5G Communications Race: Pursuit of More Capacity Triggers LTE in Unlincensed Band. IEEE Vehicular Technology Magazine **10**(1), 43–51 (2015). doi:10.1109/MVT.2014.2380631

Pruning a Random Forest by Learning a Learning Algorithm

Kumar Dheenadayalan(✉), G. Srinivasaraghavan, and V.N. Muralidhara

International Institute of Information Technology, Bangalore, India
d.kumar@iiitb.org, {gsr,murali}@iiitb.ac.in

Abstract. Ensemble Learning is a popular learning paradigm and finds its application in many diverse fields. Random Forest, a decision tree based ensemble learning algorithm has received constant attention in the research community due to its ability to learn complex rules and generalize well for unknown data. Identifying the number of base classifiers (trees) required for a particular dataset is one of the key questions addressed in this paper. Statistical analyses of individual base classifiers are carried out to prune the ensemble model without compromising the classification accuracy of the model. Learning the learned model, i.e., learning the statistics of the forest in its entirety along with the information available in the dataset can reveal the optimal thresholds that should be used to prune an ensemble model. Experimental results reveal that, on an average 78% of the trees were pruned on 26 different datasets obtained from the UCI repository. The impact of pruning was positive with 22 out of 26 datasets showing equal or better classification accuracy in comparison with the Classical Random Forest algorithm.

Keywords: Ensemble learning · Random forest · Pruning · Matthews correlation coefficient · Meta-learning

1 Introduction

Ensemble Learning is a class of supervised learning algorithms where a single base learning algorithm is used to train multiple hypotheses (classifiers) for learning the same task [5,14]. Given a dataset $\mathcal{D}$ belonging to the input space $\mathcal{X}$ and output space $\mathcal{Y}$, a supervised learning algorithm will try to approximate an unknown target function f from a possible hypothesis set $\mathcal{H}$ to output $h : \mathcal{X} \rightarrow \mathcal{Y}$. Typically, an ensemble learning algorithm consists of growing and combining phases. Growing phase involves generation of k different classifiers. Techniques like Bagging or Boosting are using to grow individual trees as these techniques have shown to improve the accuracy of a learning algorithm in both theoretic and empirical sense [3,7]. Both these techniques re-sample the given dataset $\mathcal{D}$ to generate different training instances for individual classifiers. In the combining phase the outputs of the individual classifiers are combined using a function $e : (h_1(X), \ldots h_k(X))$. Majority prediction or weighted averaging is popular function used in the combining phase in machine learning literature.

P. Perner (Ed.): MLDM 2016, LNAI 9729, pp. 516–529, 2016.
DOI: 10.1007/978-3-319-41920-6_41

Though there are many popular ensemble learning algorithms, the two algorithms that have set the benchmark in ensemble learning are Adaboost(adaptive boosting) and Random Forest algorithms. Both work on the principle of achieving a strong classifier by combining the outputs of many weak classifiers [4,6]. There have been several attempts in the past to improve upon the classic version of Random Forest. Attempts to improve the accuracy of classification can be broadly divided in to two categories:

- Pruning of individual trees in the forest [11,12,20,21]
- Weighing individual trees [18]

We attempt to improve the accuracy of a Random Forest through selective pruning based on the statistical measures of individual trees.

The motivation for pruning of Random Forest arises from our past work on classifying the response state of a storage system where close to 100,000 features are available. Observations on the prediction accuracy of individual trees in the forest have revealed that a number of trees did not generalize well in the live testing environment. Statistical measure of Matthews Correlation Coefficient (MCC) was evaluated for individual trees. As MCC considers the True Positives and True Negatives of the classifier in evaluating the coefficient, it can be used effectively to identify the trees with maximum accuracy.

A pth percentile of MCC was used as the threshold to prune the forest. Trees not satisfying the threshold were pruned to come up with a subset of base classifiers. The selective pruning has shown marked improvement in the classification accuracy on relatively large test sets. The extent of pruning ranges from 33% to 95% on various standard datasets that have been traditionally used in machine learning literature. There are significant portions of the ensemble learner, which can be eliminated to obtain similar or better classification accuracy without compromising with the intra-correlation or strength of individual trees.

The value of p was initially calculated iteratively. To explore the possibility of guiding the user towards near optimal value of p, we collected statistical data related to the un-pruned Random Forest. Along with these, statistics related to the dataset were also collected. A total of 39 statistics were collected for each of the 9 different percentile measures ranging from 55 to 95 to form a dataset $\mathcal{D}'$ for learning a learning algorithm.

We built a new Random Forest ($\mathcal{RF}^g$), which we will refer to as the guiding Random Forest with 10 trees by considering the percentile value p as the class attribute. The guiding Random Forest $\mathcal{RF}^g$ built using $\mathcal{D}'$, was able to suggest the near optimal percentile value of MCC that should be used for pruning a Random Forest built on any new dataset. Hence, we learn the statistics of a number of Random Forests to understand the behavior of the forest for a particular dataset and try to identify the percentile value of MCC that would provide the best pruned forest.

In section 2 we cover the past literature on pruning followed by an overview and analysis of the proposed pruning and learning of learning algorithm in Section 3. Evaluation of the proposed method along with implementation details are discussed in Section 4 followed by the conclusion in Section 5.

2 Literature Survey

A Random Forest classifier uses bagging or bootstrap aggregation to construct k different tree based classifiers $\{h_1(x, \Theta_1), ..., h_k(x, \Theta_k)\}$. Bagging involves sampling with replacement of $|\Theta_i|$ random vectors that are independently and identically distributed from the given dataset $\mathcal{D}$. Each tree in the Random Forest casts a vote to one of the classes with some probability, which is aggregated to predict the class. The final ensemble model generated can be more accurate than its individual components if the necessary and sufficient conditions, namely high strength and diversity of individual classifiers are guaranteed [5,8].

Strength of the classifiers is defined as the expected value of the margin function,

$$s = E_{X,Y}(mr(X, Y)) \tag{1}$$

where the margin function is estimated as shown in Equation 2

$$mr(X, Y) = P_{\Theta}(h(X, \Theta) = Y) - max_{j \neq Y} P_{\Theta}(h(X, \Theta) = j) \tag{2}$$

The margin measures the difference between votes acquired for the right class and the maximum vote acquired for any other class. The expected value of for all the classifiers indicates the strength of the classifier. Large values of strength indicates higher confidence in the predicted class.

The diversity among trees within the forest is essential for generalization as each individual tree will be able to model diverse areas of the input space. Classifiers are diverse if their predictions (or errors) vary on individual data points. The diversity is measured by evaluating the correlation (ρ) of predictions by individual trees in the forest. Lower the correlation, higher is the diversity. Breiman suggests that the ratio of correlation to squared strength gives a good estimation of the generalization error. Hence, each individual tree evaluated should have low correlation and high strength.

Reordering of classifiers and aggregating a sub-ensemble was proposed in [11] with 15% to 30% of the initial ensemble size being used in final aggregation. Similar approach is proposed in [12] with boosting technique being used for ordering the classifier. Both use a fixed percentage of trees in the sub-ensemble. It seems unreasonable however to assume a fixed fraction of the original ensemble to form a pruned forest. The percentages used in the works referred to above seem arbitrary and their use for all datasets does not seem justified. Ideally the best sub-ensemble size should be a function of the dataset. Our work focuses on identifying the best set of trees that can be aggregated to generalize well to the dataset irrespective of the initial size of the ensemble.

An attempt to prune the forest based on the margin function has been suggested in [19]. The pruning strategy is based on the margin function, which is part of the internal estimates evaluated during the construction of the Random Forest. The list of trees is ranked based on different margin metrics after which, the least important tree is eliminated iteratively till the number of trees reduces to 20. Drawback with this approach is that the pruning target is predecided, which may or may not be the ideal size of sub-ensemble of trees.

Our evaluation on multiple datasets showed that, the number of trees ideal for each dataset depends on multiple factors related to data and the forest constructed. The dataset has direct influence on the statistics of the forest and this varies with different datasets. This in turn influences the choice of the best subset of trees required for optimal class accuracy. We effectively address this issue in this paper by automatically determining the forest size and the collection of trees in the pruned forest based on the dataset.

Authors in [20] explore a correlation based pruning by considering both similarity and prediction accuracy of trees. They suggest that pruning of the forest based on the accuracy of individual trees performs slightly better in identifying the sub-ensemble forest compared to similarity of trees. Other approaches, for example in [16], include mixing of different metrics in the construction of the Random Forest namely, Gain ratio, Gini index, minimum description length and many more. The results for forest with several metrics shows marginal increase in the prediction accuracy but utilize the complete forest by incorporating different types of diversity.

The experiments in past literature do not consider the two key aspects that need to be preserved for effective classification, namely the strength and the correlation. Another important aspect that has to be considered is the imbalance of classes in data. Imbalanced datasets have a tendency to mis-classify instances that belong to class with low class probability in the training set. Pruning should not aggravate this tendency of misclassification of imbalanced data.

Collecting statistics related to the model are termed as meta-data of the model. There is research related to meta-learning that can help choose the best algorithm suited for a given problem. Meta-learning involves learning of performance of a base learning algorithm for an application [2,17], which is in contrast to the learning of meta-data to choose the best parameters for a base learner. The later is being considered in this paper with the meta-data being the statistics of Random Forest for different applications. The near optimal percentile value of MCC to be used for pruning the forest has to be learned by using the meta-data. We achieve this by learning the meta-data using the base learning algorithm itself.

2.1 Matthews Correlation Coefficient

Commonly used evaluation measures including Recall, Precision, F-Measure are biased in terms of label and population prevalence as argued in [9]. Experiments reveal that using such evaluation measures can appear to perform better for certain datasets but performs worse in the objective sense of Informedness or Markedness.

Informedness is a measure to quantify how informed a predictor is for the specified condition and specifies the probability that a prediction is informed in relation to the condition (versus chance) [9]. Markedness quantifies how marked a condition is for the specified predictor and specifies the probability that a condition is marked by the predictor (versus chance) [9].

The dependence of Matthews Correlation Coefficient with Informedness and Markedness that forms an unbiased accuracy measure was established in [15]. For our analysis in a Classification setting, Matthews correlation Coefficient can be evaluated from a contingency matrix using the following formula

$$MCC = \frac{TP \times TN - FP \times FN}{\sqrt{(TP+FP)(TP+FN)(TN+FP)(TN+FN)}} \tag{3}$$

where TP = True Positive, TN = True Negative, FP = False Positive, FN = False Negative

MCC [13] provides an unbiased measure of accuracy that is widely used in bioinformatics [1]. The same measure is used in our analysis to prune out trees of the Classical Random Forest. Systematical pruning of the weak classifiers based on MCC measure can help in enhancing the classification accuracy without compromising with the generalization error.

WEKA (Waikato Environment for Knowledge Analysis) is a popular open source machine learning workbench targeted towards domain specialists who can directly apply the existing Machine Learning techniques to real world problems in various domains. The workbench though a couple of decades old, is actively being updated with the latest state-of-the-art Machine Learning techniques. We use the WEKA's implementation of Random Forest with bagging as the base implementation for Classical Random Forest algorithm and compare it with the proposed statistically pruned Random Forest.

3 Pruning a Random Forest

3.1 Modeling

Given a dataset $\mathcal{D}$, Random Forest is built in the classical way by choosing a random subset of the training set to build a tree. This process is repeated k times to build a Random Forest of k trees. As part of the internal estimates to guide the generalization measures, a part of the training set is kept aside for each tree to estimate the Out Of Bag (OOB) error rate.

$$\vartheta_k = \Theta \setminus \Theta_k \tag{4}$$

This OOB data instances (ϑ_k) are utilized to evaluate the weighted MCC for each tree(k) individually.

The prediction or vote for OOB instances by each tree are evaluated and compared against the true class value.

$$\forall_{i \leq k} \forall_{x_j \in \vartheta_k} \{T_i(x_j)\} \tag{5}$$

The statistics for each individual tree like TP, TN, FP, FN are calculated and the weighted MCC is noted. $classWeight_c$ in (6) is the class probability of each class c in the Test set.

$$\forall_{i \leq k} wMCC_i = \sum_{c \in C} MCC_c \times classWeight_c \tag{6}$$

Iterative pruning of the Random Forest is carried out based on various percentiles of weighted MCC of trees in the Classical Random Forest (CRF). All the trees with weighted MCC below the reference percentile MCC score will be pruned off.

$$PRF = \begin{cases} PRF \cup T_i & \text{if } wMCC_i > wMCC(p) \\ \text{prune } T_i & \text{otherwise} \end{cases} \tag{7}$$

The Pruned Random Forest (PRF) is tested with a separate test set and its test accuracy along with the Area Under the Curve (AUC) for Receiver Operating Characteristic (ROC) is noted. The value of percentile is iterated to see its impact on the number of trees and the test accuracy until optimal test set accuracy is obtained.

Identifying the optimal number of trees for a dataset is possible with the above mentioned process. We now examine the two key conditions of maintaining higher strength for the classifiers and lower correlation among the individual classifiers. Strength of the classifier defined by (1) measures the expected maximum margin (true class predictions) of the classifier over a dataset. wMCC is used in the proposed pruning model and this can be used as a weaker replacement of strength. This is true because wMCC is directly proportional to the difference in true predictions and false predictions. The numerator of MCC measure is similar but a weaker representation of the margin function shown in (2). Hence, the trees with higher strength will be the once that will be retained. This is experimentally verified and a high correlation value of +0.62 was observed between Optimal wMCC and strength of classifiers. MCC being an unbiased measure of estimating the accuracy of a model gives another advantage to shortlist trees that are best in the forest in an unbiased probabilistic sense. We experimentally show that this pruning actually identifies trees, which have lower correlation for a number of different datasets especially binary classification problems.

Figure 1 shows the process of varying the percentile and the corresponding number of trees retained after pruning for autos dataset. As the percentile varies, the number of trees pruned also varies. The accuracy of the test set during this variation is shown in Figure 2. The best accuracy of 83.61 was obtained for a 55th percentile score of wMCC with 45 trees. Classical Random Forest with 100 trees could achieve only 78% classification accuracy giving an indication of the effect of pruning based on a wMCC.

3.2 Learning a Learning Algorithm

There are at least 5 iterations that can be carried out to identify the percentile score for wMCC, which facilitates the retention of optimal number of trees. We try to eliminate this by learning the statistics of a learning algorithm, i.e. Classical Random Forest in the current scenario. If we can learn the properties of the dataset and statistics of the Classical Random Forest that is constructed from the dataset, identifying the percentile for which wMCC value can generate optimal number of trees can be determined.

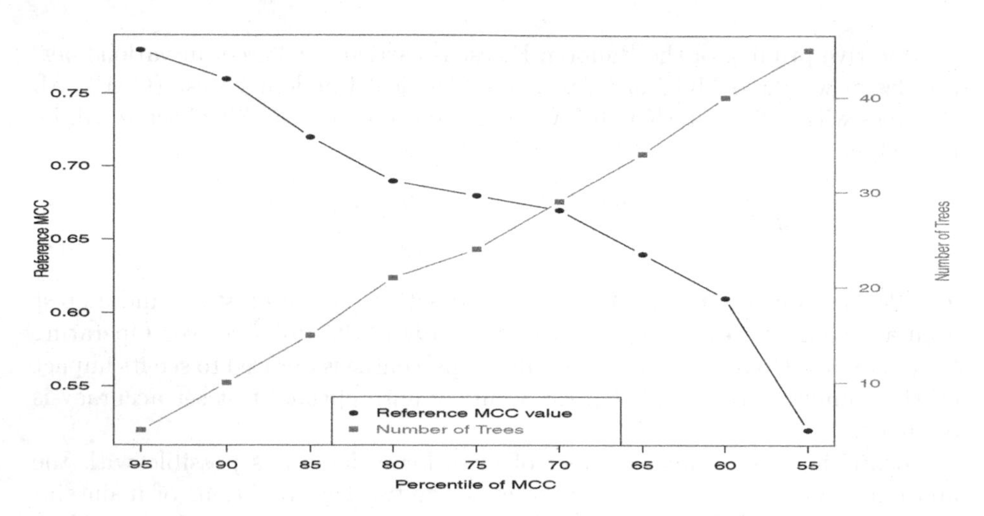

Fig. 1. Variation in the number of trees with varying Matthews Correlation Coefficient.

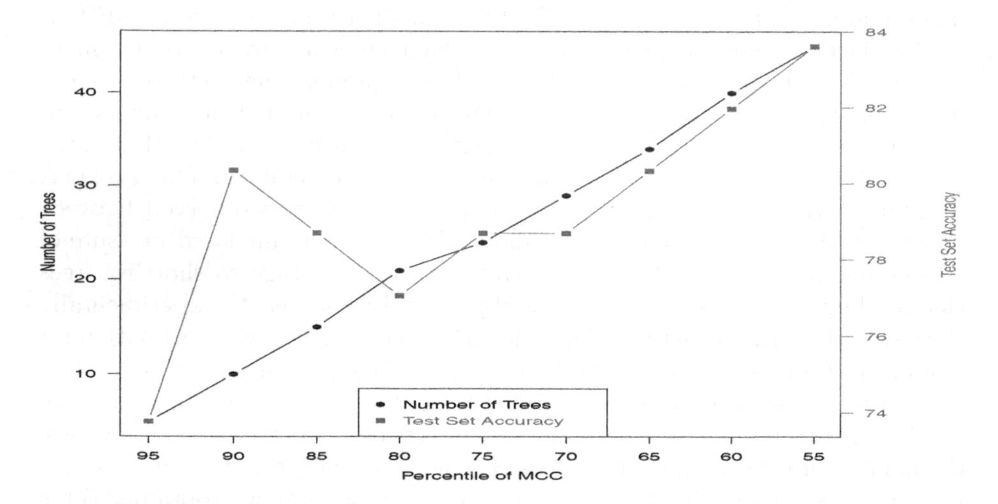

Fig. 2. Variance in the Test Set Classification Accuracy with varying number of trees in the Pruned Forest.

During the iterative process of varying the percentiles of wMCC, a number of statistics related to the CRF and dataset were collected. The list of statistics collected are listed in Table 1. WEKA's Random Forest has a convenient way of extracting these statistics in each iteration through the Evaluation class implementation. Statistics listed in Table 1 are collected for every dataset during each iteration of percentile value. The optimal percentile score for each dataset is used as the class variable in the dataset to learn a learning algorithm.

Some key statistics that are of interest are class complexity stats, Kononenko & Bratko Information [10], Kappa Statistics . . . , which gives an insight about the

Table 1. Statistics collected for each Classical Random Forest

CorrectClassification_Count	Correct Classification_Percentage	IncorrectClassification_Count
IncorrectClassification_Percent	Kappa_statistic	Total_Cost
Average_Cost	Relative_Info_Score	Info_Score_bits
Info_Score_bits_per_instance	Correlation_coefficient	Class_complexity_order_0_bits
Class_complexity_order_0_bits_per_instance	Class_complexity_scheme_bits	Class_complexity_scheme_bits_per_instance
Complexity_improvement_Sf_bits	Complexity_improvement_Sf_bits_per_instance	Mean_absolute_error
Root_mean_squared_error	Relative_absolute_error	Root_relative_squared_error
Coverage_0.95level	Mean_rel_region_size_0.95level	UnClassified_Instances
Total_Number_of_Instances	Ignored_Class_Unknown_Instances	Total_Number_Of_Classes
Total_Number_Of_Attributes	Training_AUCROC	Training_Accuracy
Weighted_Precision	Weighted_Recall	Weighted_TruePositive
Weighted_FalsePositive	Weighted_FMeasure	Weighted_MCC
Weighted_PRCArea	OOB_Error_Rate	Optimal_Percentile(Class Variable)

class distribution, the information available in the dataset and to what extent were they used in building the Random Forest. These statistics by itself have a lot of information that can help in analyzing the number of trees required for good classification accuracy. We use the data collected in Table 1 as our dataset $\mathcal{D}'$ and learn the learning algorithm through another Random Forest ($\mathcal{RF}^g$). This can turn out to be a useful way of guiding the pruning algorithm by predicting the optimal percentile for wMCC score to be used as a thresholding measure. It can eliminate multiple iterations of pruning and evaluating the test set. UCI repository has a number of datasets for which statistics can be collected and $\mathcal{RF}^g$ can be built. The entire iterative process of building $\mathcal{D}'$ from a dataset repository is explained in Algorithm 1.

4 Results

WEKA workbench was used to implement the proposed pruning method. Bagging class was modified and the pruning methodology was embedded during the bagging process. WEKA's current implementation of Random Forest, which we call the Classical Random Forest (CRF) is used as the reference that will be compared with our Pruned Random Forest (PRF). WEKA package from sourceforge contains datasets obtained from the UCI repository. These datasets have been used widely in the original paper on Random Forest [4] and other subsequent work on pruning.

The experimental setup for PRF consists of 26 datasets. For a particular percentile, PRF was built for each dataset for 10 iterations. The percentile values were varied from 95 to 55 giving a total of 9 (percentile values) $\times$ 10 (iterations) $=$ 90 instances for each dataset. The results reported here are the average values over ten iterations for each dataset. Results for CRF are obtained for 26 datasets, over 10 iterations and the results are averaged over these ten iterations. Both CRF and PRF are executed with a 70:30 split of training and testing samples from the dataset. If the dataset has an explicit test set (example: spect), then the test set is used instead of splitting the dataset into training and testing set. During the process of data collection, statistics from Table 1 is collected for each dataset while running PRF to build $\mathcal{D}'$.

Algorithm 1. Learning a learning algorithm

```
1: procedure PRUNE(trainIndex, testIndex)
2:     for all D ∈ repository do
3:         trainData ← D[trainIndex, ]
4:         testData ← D[testIndex, ]
5:         for all p ∈ [55, 60, 65, 70, 75, 80, 85, 90, 95] do
6:             i ← 10
7:             PRF_best ← NULL
8:             while i > 0 do
9:                 Build RF
10:                for all T_i ∈ RF do
11:                    Evaluate wMCC(p)_i
12:                    Update PRF using Equation (7)
13:                end for
14:                PRF_err ← test(PRF_err, testData)
15:                extract D'
16:                if PRF_best > PRF_err then
17:                    PRF_best = PRF
18:                end if
19:                i ← i − 1
20:            end while
21:        end for
22:    end for
23:    Build RF^G
24: end procedure
```

Table 2 shows the results for PRF and CRF along with the details of the dataset. The dataset size, number of classes and number of attributes are extracted from the dataset input file. Test accuracy for PRF is the average test classification accuracy (of ten iterations for a single percentile) for the best of 9 different percentile values. Test accuracy for CRF was the average test classification accuracy out of the 10 iterations. Optimal wMCC indicates the value of wMCC that was used to prune and obtain the test accuracy reported in PRF Test (%). PRF Size represents the size of pruned forest. The row number in the table is used as the dataset index in the subsequent figures.

Results for 26 datasets presented in Table 2 shows that PRF has equal or better classification accuracy when compared to CRF for 22 out of the 26 dataset. CRF was better than PRF by more than 1% for only two datasets. Rest of the two datasets had better accuracy with very low statistical significance. The number of trees required to achieve these results ranges from a minimum of 2 trees to a maximum of 67 trees and 21 trees on an average. The optimal percentile varies from 95 for a majority of datasets to 55 for very few datasets. Optimal wMCC values that gave the best test accuracy was also high with an average optimal wMCC value of 0.81 for 26 datasets. Pearson's Correlation Coefficient of optimal wMCC with strength of the PRF was observed at +0.62, which indicates that pruning based wMCC actually retains trees with high strength.

Table 2. Success Rate Comparison of Pruning with Classical Random Forest

Dataset Index	Dataset Name	Classes	Attributes	Total Instances	Optimal Percentile	Optimal wMCC	PRF Test(%)	CRF Test(%)	PRF Size
1	anneal.ORIG	6	39	898	95	0.94	**94.42**	93.68	5
2	autos	7	26	205	55	0.52	**83.61**	78.68	45
3	balance-scale	3	5	625	90	0.82	77.01	**79.14**	9
4	breast-cancer	2	10	286	70	0.68	**71.00**	69.76	29
5	breast-w	2	10	699	85	0.96	**96.19**	95.71	23
6	colic	2	23	368	95	0.88	**85.45**	83.63	4
7	contact-lenses	3	5	24	95	0.83	**42.86**	42.85	2
8	credit-a	2	16	690	90	0.79	85.80	**85.99**	8
9	diabetes	2	9	768	70	0.75	**77.48**	76.52	24
10	glass	7	10	214	60	0.88	77.19	**78.12**	4
11	hayes-roth	4	5	160	85	0.87	85.71	85.71	20
12	heart-c	5	14	303	80	0.74	**85.71**	84.61	27
13	heart-h	5	14	294	90	0.75	78.41	**79.54**	7
14	heart-statlog	2	14	270	70	0.86	**80.62**	79.01	30
15	hepatitis	2	20	155	75	0.85	**86.96**	86.95	24
16	hypothyroid	4	30	3772	95	0.99	**99.20**	99.11	5
17	ionosphere	2	35	351	60	0.89	**91.43**	90.47	44
18	iris	3	5	150	95	0.81	**95.56**	95.55	5
19	labor	2	17	57	90	0.83	**94.12**	94.11	6
20	lymph	2	19	148	90	0.7	**92.95**	88.63	10
21	mushroom	2	23	8124	95	1	100.00	100.00	67
22	sonar	2	61	208	70	0.9	**84.84**	82.25	27
23	spect	2	23	267	85	0.65	**71.12**	68.44	15
24	splice	3	62	3190	60	0.73	**94.25**	91.22	42
25	vowel	11	14	990	55	0.8	**94.28**	93.60	45
26	zoo	7	18	101	75	0.81	**93.33**	80.00	25

To see if PRF actually generalizes as well if not better than CRF, we evaluated the strength and mean correlation for each dataset. Any good Random Forest model requires the forest to have high strength and low mean correlation. Plot of mean correlation and strength for CRF is shown in Figure 3. We observe that 18 out of 26 datasets have strength higher than correlation but for a number of datasets, the strength and correlation are very close to each other. Figure 4 shows strength and correlation for PRF measured on all the 26 datasets. Even for PRF, 18 out of 26 have higher strength than correlation with wider gap compared to CRF, which is a desirable property for effective generalization.

Finally, we compare the $\frac{c}{s^2}$ ratio of CRF and PRF for each dataset in Figure 5. c is the mean correlation of trees in the forest and s^2 is the squared strength of the forest. The desired property of lower $\frac{c}{s^2}$ is better achieved through PRF than CRF for 23 out of 26 datasets. Average improvement in $\frac{c}{s^2}$ ratio for 26 datasets is 1.82. The upper bound on the generalization error as derived in [4] is given by (8) where ρ is the mean correlation of the trees within the forest.

$$\frac{\rho \times (1 - s^2)}{s^2} \tag{8}$$

The plot for the upper bound on the generalization error of 26 datasets for PRF and CRF is shown in Figure 6 which is a scaled version Figure 5.

The value of optimal wMCC obtained by the optimal percentile has a big effect on the classification accuracy for a number of datasets. Many datasets are sensitive to the value of optimal wMCC and the number of trees in PRF, but some datasets have shown minimum to no variation in the classification accuracy with change in the percentile value. This can be observed in Figure 7 where high variance in test accuracy is observed for 9 different percentile values.

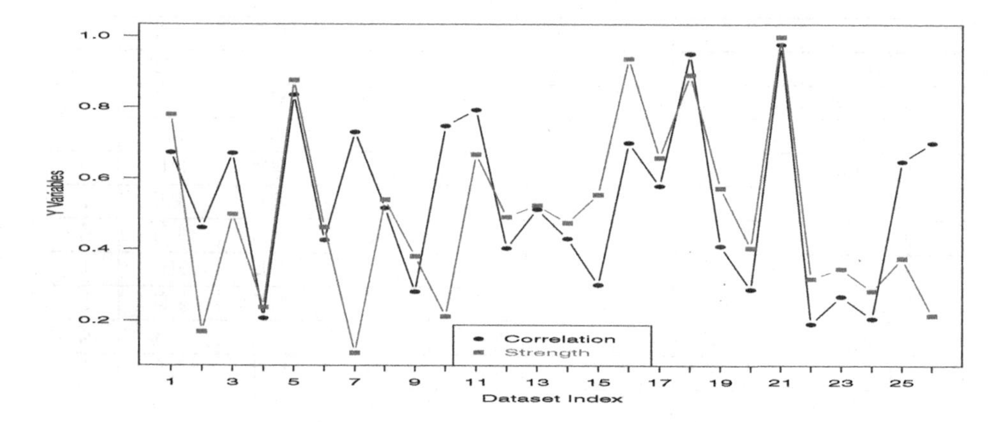

Fig. 3. Correlation and Strength of the Classical Random Forest.

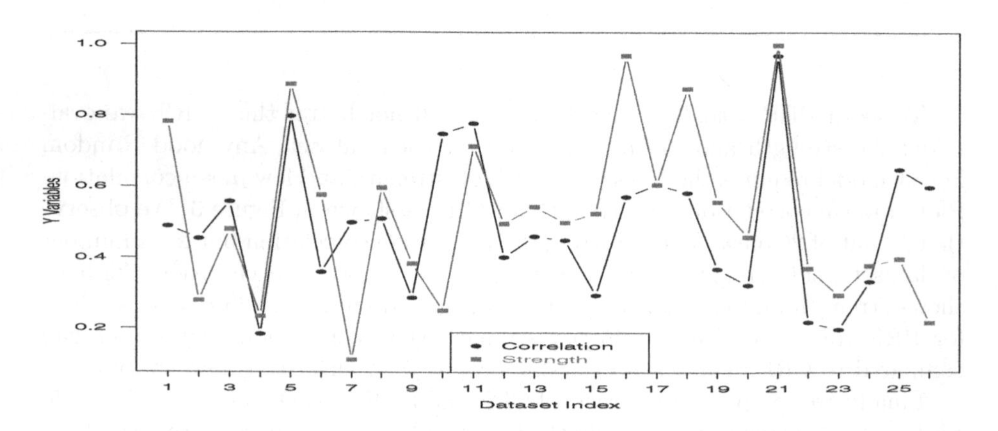

Fig. 4. Correlation and Strength of the Pruned Random Forest.

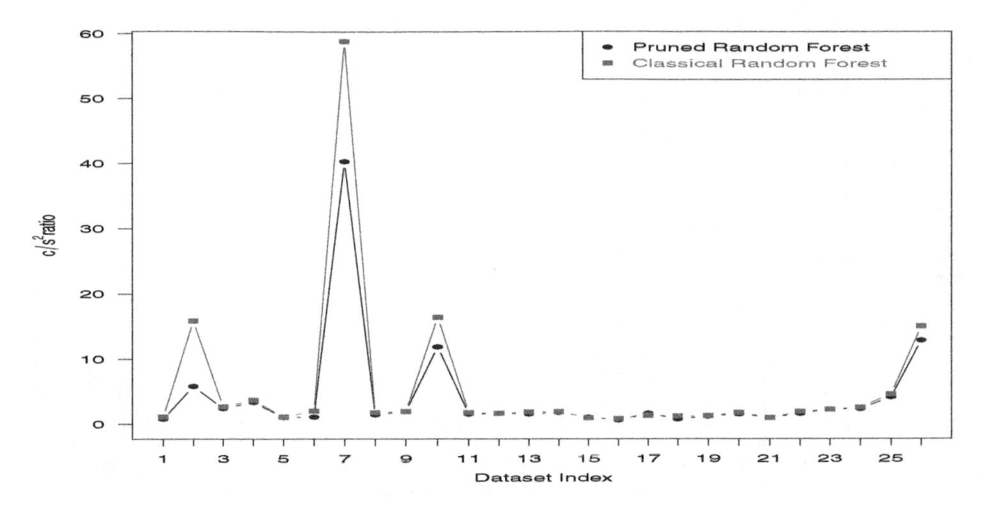

Fig. 5. $\frac{c}{s^2}$ ratio for Classical Random Forest and Pruned Random Forest.

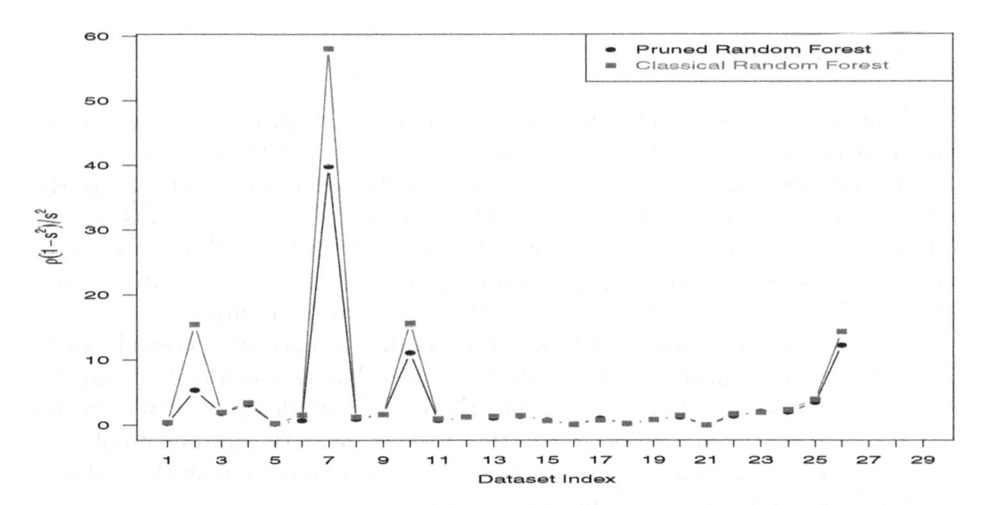

Fig. 6. $\frac{\bar{\rho}(1-s^2)}{s^2}$ ratio for Classical Random Forest and Pruned Random Forest.

Hence, it becomes increasingly important to identify the optimal percentile as achieved by learning a learning algorithm. Dataset $\mathcal{D}'$ containing 1800 instances obtained from 20 datasets with the optimal percentile value as the class variable is used to train a guiding Random Forest. $\mathcal{RF}^g$ was used to predict the optimal percentile value for the remaining 6 datasets. $\mathcal{RF}^g$ was able to guide the optimal percentile value for 4 datasets, while percentile value for 2 datasets were off by 5 percentile value. The effect of this was the classification accuracy of the test set was 0% and 1.2%.

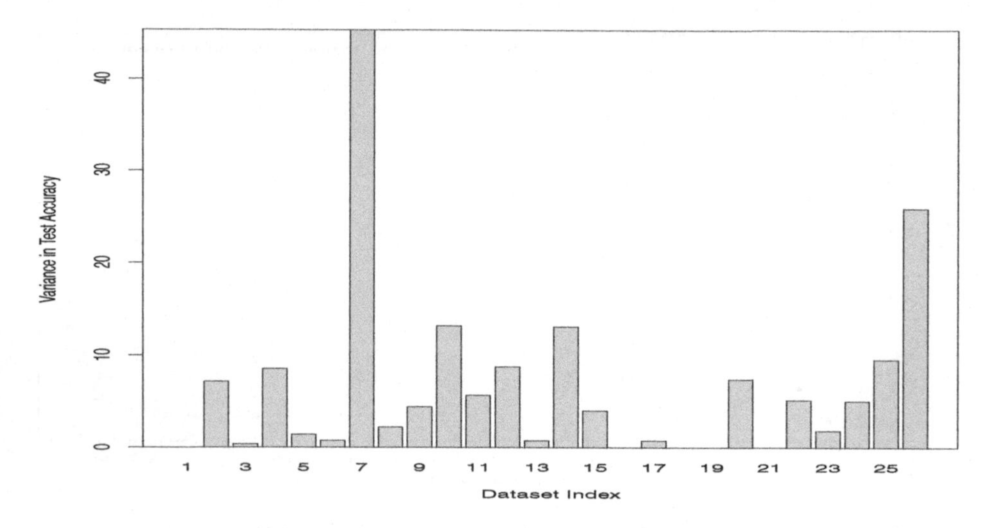

Fig. 7. Variance in Testset accuracy for Pruned Random Forest.

5 Conclusion

Pruning of an ensemble classifier like Random Forest without compromising on the generalization goals is presented in this paper. MCC was used to have an unbiased measure of accuracy for individual classifiers, which has effectively retained trees with relatively high strength. High percentage of trees was pruned off indicating that a lot of redundant trees were part of CRF. When used on a large scale datasets or in image processing, pruned forests will help in reducing the test time with a possible increase in the classification accuracy.

By learning a learning algorithm, we were able to uncover a possible guiding mechanism for pruning. The guiding Random Forest was able to learn the statistics very accurately indicating that there are statistics present within the dataset and un-pruned Random Forest that can decide on the optimal number of trees. A more theoretical process of uncovering the relation among the features of $\mathcal{D}'$ will be the next step of our work.

References

1. Baldi, P., Brunak, S., Chauvin, Y., Andersen, C.A.F., Nielsen, H.: Assessing the accuracy of prediction algorithms for classification: an overview. Bioinformatics **16**(5), 412–424 (2000)
2. Brazdil, P., Giraud-Carrier, C., Soares, C., Vilalta, R.: Metalearning: Applications to Data Mining, 1st edn. Springer Publishing Company, Incorporated (2008)
3. Breiman, L.: Bagging predictors. Machine Learning **24**(2), 123–140
4. Breiman, L.: Random forests. Mach. Learn. **45**(1), 5–32 (2001)
5. Dietterich, T.G.: Ensemble methods in machine learning. In: Kittler, J., Roli, F. (eds.) MCS 2000. LNCS, vol. 1857, pp. 1–15. Springer, Heidelberg (2000)

6. Freund, Y., Schapire, R.E.: A decision-theoretic generalization of on-line learning and an application to boosting. J. Comput. Syst. Sci. **55**(1), 119–139 (1997)
7. Freund, Y., Schapire, R.E.: A Short Introduction to Boosting. Journal of Japanese Society for Artificial Intelligence **14**(5), 771–780 (1999)
8. Hansen, L., Salamon, P.: Neural network ensembles. IEEE Transactions on Pattern Analysis and Machine Intelligence **12**(10), 993–1001 (1990)
9. Hu, B.G., He, R., Yuan, X.T.: Information-theoretic measures for objective evaluation of classifications. Acta Automatica Sinica **38**(7), 1169–1182 (2012)
10. Kononenko, I., Bratko, I.: Information-based evaluation criterion for classifier's performance. Mach. Learn. **6**(1), 67–80 (1991)
11. Martínez-Muñoz, G., Suárez, A.: Pruning in ordered bagging ensembles. In: Proceedings of the 23rd International Conference on Machine Learning, ICML 2006, pp. 609–616. ACM, New York (2006)
12. Martínez-Muñoz, G., Suárez, A.: Using boosting to prune bagging ensembles. Pattern Recogn. Lett. **28**(1), 156–165 (2007). http://dx.org/10.1016/j.patrec.2006.06.018
13. Matthews, B.W.: Comparison of the predicted and observed secondary structure of T4 phage lysozyme. Biochim. Biophys. Acta **405**, 442–451 (1975)
14. Opitz, D., Maclin, R.: Popular ensemble methods: An empirical study. Journal of Artificial Intelligence Research **11**, 169–198 (1999)
15. Powers, D.M.W.: Evaluation: From precision, recall and f-measure to roc., informedness, markedness & correlation. Journal of Machine Learning Technologies **2**(1), 37–63 (2011)
16. Robnik-Šikonja, M.: Improving random forests. In: Boulicaut, J.-F., Esposito, F., Giannotti, F., Pedreschi, D. (eds.) ECML 2004. LNCS (LNAI), vol. 3201, pp. 359–370. Springer, Heidelberg (2004)
17. Vilalta, R., Giraud-Carrier, C., Brazdil, P.: Meta-learning - concepts and techniques. In: Data Mining and Knowledge Discovery Handbook, pp. 717–731. Springer, Boston (2010). http://dx.org/10.1007/978-0-387-09823-4_36
18. Winham, S.J., Freimuth, R.R., Biernacka, J.M.: A weighted random forests approach to improve predictive performance. Statistical Analysis and Data Mining **6**(6), 496–505 (2013)
19. Yang, F., Hang Lu, W., Kai Luo, L., Li, T.: Margin optimization based pruningfor random forest. Neurocomputing **94**, 54–63 (2012)
20. Zhang, H., Wang, W.: Search for the Smallest Random Forest, pp. 381–388 (2009)
21. Zhou, Z.H., Tang, W.: Selective ensemble of decision trees. In: Wang, G., Liu, Q., Yao, Y., Skowron, A. (eds.) Rough Sets, Fuzzy Sets, Data Mining, and Granular Computing. LNCS, vol. 2639, pp. 476–483. Springer, Heidelberg (2003)

Interestingness Hotspot Discovery in Spatial Datasets Using a Graph-Based Approach

Fatih Akdag(✉) and Christoph F. Eick

Department of Computer Science, University of Houston, Houston, USA
{fatihak,ceick}@cs.uh.edu

Abstract. This paper proposes a novel methodology for discovering interestingness hotspots in spatial datasets using a graph-based algorithm. We define interestingness hotspots as contiguous regions in space which are interesting based on a domain expert's notion of interestingness captured by an interestingness function. In our recent work, we proposed a computational framework which discovers interestingness hotspots in *gridded* datasets using a 3-step approach which consists of seeding, hotspot growing and post-processing steps. In this work, we extend our framework to discover hotspots in any given spatial dataset. We propose a methodology which firstly creates a neighborhood graph for the given dataset and then identifies seed regions in the graph using the interestingness measure. Next, we grow interestingness hotspots from seed regions by adding neighboring nodes, maximizing the given interestingness function. Finally after all interestingness hotspots are identified, we create a polygon model for each hotspot using an approach that uses Voronoi tessellations and the convex hull of the objects belonging to the hotspot. The proposed methodology is evaluated in a case study for a 2-dimensional earthquake dataset in which we find interestingness hotspots based on variance and correlation interestingness functions.

Keywords: Spatial data mining · Interestingness hotspot · Interestingness function · Hotspot discovery · Graph-based hotspot growing algorithm · Spatial polygon models

1 Introduction

Interestingness hotspots are contiguous regions in space which are interesting based on a domain expert's notion of interestingness which is captured in an interestingness function. Typically, spatial clustering algorithms have been used to find interestingness hotspot in spatial datasets using interestingness functions; however, in our previous works [1,2] we proposed an alternative, non-clustering approach to obtain interestingness hotspots in gridded datasets, which grows interestingness hotspots from seed hotspots. In this paper, we extend our framework to discover hotspots in any given spatial dataset, relying on a graph-based framework. In gridded datasets, determining the contiguity of region is trivial: grid cells are neighboring if they share an edge. However,

P. Perner (Ed.): MLDM 2016, LNAI 9729, pp. 530–544, 2016.
DOI: 10.1007/978-3-319-41920-6_42

contiguity or neighborhood relation is not well-defined for point based datasets, and an approach is needed to define neighborhood relation between points. In this paper, we propose an approach which employs Gabriel graphs [3] as neighborhood graphs of the spatial objects in the dataset. Next, we introduce a graph-based hotspot discovery algorithm that identifies contiguous, interestingness hotspots in Gabriel graphs, maximizing a plug-in interesting function. Furthermore, we generate polygon models that describe the scope of each hotspot based on Voronoi diagram.

Finding hotspots in a dataset maximizing an interestingness function that is based on a domain expert's notion of interestingness allows domain experts to discover regions with interesting patterns in the data. For example, Miller et al. [4] identifies regions with strong association of internet ad performance and demographic data to be used for geo-targeted advertising. In our previous work [1], we identify regions with high correlation of air pollutants and Ozone levels in an air pollution dataset which can be used to find associations between air pollutants and Ozone levels.

The proposed graph-based hotspot discovery framework can be summarized as follows:

1. We propose a methodology for finding hotspots in spatial datasets that consists of 4 steps: 1) building neighborhood graph 2) finding hotspot seeds 3) growing hotspot seeds 4) generating polygon models.
2. We propose methods for creating a neighborhood relation between spatial objects.
3. A heap-based hotspot growing algorithm is proposed to find interestingness hotspots using the neighborhood graph in spatial datasets.
4. We propose an approach to generate a polygon model for two-dimensional hotspots based on Voronoi diagram.
5. The proposed interestingness hotspot discovery framework is evaluated in a case study involving a two-dimensional earthquake dataset.

The rest of the paper is organized as follows. In Section 2, we compare the existing methods for discovering hotspots in spatial datasets. Section 3 introduces our framework and Section 4 provides a detailed discussion of our methodology. We present the experimental evaluation in Section 5, and Section 6 concludes the paper.

2 Related Work

Spatial scan statistics introduced by Kulldorff [5] is the most widely used hotspot discovery tool. It searches spatial circular regions occurring within a certain time interval and can obtain circular hotspots by growing circles from a point of origin by increasing the radius of the circle. However, our framework is quite different as we compute hotspots based on a given interestingness measure rather than using distance-based features of the regions.

There are also spatial clustering algorithms which can be used for computing spatial hotspots. Varlaro et al. [6] describe the goal of spatial clustering "*to group nearby sites and form clusters of homogeneous regions...Only similar sites (transitively) connected in the discrete spatial structure may be clustered together*". However, in a clustering approach all hotspots are obtained in a single run of the clustering and the obtained clusters are always disjoint in contrast to hotspot growing approaches. DBSCAN [7] has been used and extended by many for performing spatial clustering. Another popular clustering algorithm SNN [8] (Shared Nearest Neighbor) uses the sharing of objects in k-nearest neighbor lists to assess the similarity of spatial object which enables the algorithm to identify clusters of varying densities. Wang et al. compare different spatial clustering approaches [9].

Most clustering algorithms compute clusters based on only the distance information. A new group of clustering algorithms has been introduced in the literature that find contiguous clusters by maximizing plug-in interestingness functions similar to the approach used in this paper. These algorithms are capable of considering non-spatial attributes in objective functions that drive the clustering process. Clusters are computed maximizing the sum of the rewards for each cluster based on a cluster interestingness function. CLEVER [10] is a k-medoids-style clustering algorithm which exchanges cluster representatives as long as the overall reward grows, whereas MOSAIC [11] is an agglomerative clustering algorithm which starts with a large number of small clusters, and then merges neighboring clusters as long as merging increases the overall interestingness. We use an algorithm similar to MOSAIC in our framework for reducing the number of hotspot seeds by merging seed regions if merging them increases the interestingness.

3 Interestingness Hotspot Discovery Framework

In this section, we describe the framework for discovering hotspots in spatial datasets using interestingness functions.

Interestingness hotspots are contiguous areas in space for which an interestingness function i assigns a reward $w \geq 0$, indicating "news-worthy" regional associations. Our goal is to mine spatial patterns for performance attributes in a predefined space. The scope of an interestingness hotspot is a contiguous spatial region for which the association is valid and validity is assessed using interestingness functions. Formally, we assume a spatial dataset O is given in which objects $o \in O$ are characterized by a set of performance attributes P, a set of spatial attributes S, a set of continuous attributes M, which provide meta data under which the performance attributes P are analyzed in the spatial space. Moreover, we assume that we have an interestingness measure $i{:}2^{O} \rightarrow \{0\} \cup \Re^{+}$ that assesses the interestingness of subsets of the objects in O by assigning rewards to a particular cluster H. Moreover, we assume a spatial neighboring relationship $N \subseteq O \times O$ is given that describes which objects belonging to O are neighbors. N is usually computed using spatial attributes S of objects in O. Finally, we assume an interestingness threshold θ is given that defines which patterns are interesting.

In this research we find interestingness hotspots $H \subseteq O$; where H is an interestingness hotspot with respect to *i*, if the following 2 conditions are met:

1. $i(H) \geq \theta$

2. H is contiguous with respect to a neighborhood relation N; that is, for each pair of objects (o,v) with $o,v \in H$, there has to be a path from o to v that traverses

neighboring objects (w.r.t. N) belonging to H. In summary, interestingness hotspots H are contiguous regions in space that are interesting ($i(H) \geq \theta$).

The most simplistic interestingness i_p measure we can think of is one that directly uses the value of a single performance attribute p, which is defined as follows:

$$i_p(\mathrm{H}) = \frac{\sum_{h \in H} h.p}{|H|} \tag{1}$$

where $H \subseteq O$ is an interestingness hotspot, |H| denotes cardinality of H and h.p denotes the value for attribute p for cell h in H.

Another interestingness function considers the correlation of two performance attributes p_1 and p_2; the corresponding interestingness function $i_{corr(p1,p2)}$ is defined as follows:

$$i_{corr(p1,p2)}(H) = \begin{cases} 0, & if\ |correl(H, p_1, p_2)| < \theta \\ |correl(H, p_1, p_2)| - \theta, & otherwise \end{cases} \tag{2}$$

where $0 < \theta < 1$ is the interestingness threshold, and $correl(H,p_1,p_2)$ is the correlation of attributes p_1 and p_2 with respect to the objects belonging to hotspot H. This interestingness function is used to find regions in a dataset where the performance attributes p_1 and p_2 are correlated and therefore allows for the identification of regional correlation patterns in spatial datasets which is needed for commercial applications, such as geo-targeting.

Finally, variance interestingness function considers the variance of a performance attribute p and we define the corresponding interestingness function $i_{var\,(p)}$ as follows:

$$i_{var(p)}(H) = \begin{cases} \theta - variance(H, \mathrm{p}), & if\ variance(H, p) < \theta \\ 0, & otherwise \end{cases} \tag{3}$$

where $\theta > 0$ is the variance threshold, and $variance(H,p)$ is the variance of an attribute p with respect to the objects that form hotspot H. This interestingness function is used to find regions in a dataset where the performance attribute p does not change significantly. The obtained hotspots can be used to generate maps for the performance attribute and for generating prediction models for the performance attribute—similar to regression trees.

The proposed hotspot discovery framework firstly identifies some small regions with high interestingness as seed regions, and then grow these seed regions by adding neighboring objects which increase the reward most when added.

4 Methodology

In this section, we describe the graph-based interestingness hotspot discovery algorithm which works in 4 phases:

1) Building a neighborhood graph
2) Finding "small" hotspot seed regions with high interestingness
3) Finding hotspots by growing the hotspot from the seed regions
4) Generating a polygon model for the hotspots found in Phase 3

4.1 Building a Neighborhood Graph

As interestingness hotspots are defined as contiguous regions in space, a neighborhood relation has to be defined for the spatial objects in the dataset that indicates which objects are neighboring. Various neighborhood graphs have been proposed in the literature. The most popular graphs include: Delaunay triangulation, Gabriel graphs, relative neighborhood graphs, Euclidian minimum spanning trees and Beta skeletons. Fig. 1 shows comparison of 4 popular graph types for a dog shaped point set. As seen in the figure, the Delaunay triangulation (DT), in general, contains many edges between distant points in the dataset which are non-intuitive as they capture irrelevant relationships; therefore it is not a good choice for a neighborhood graph. On the other hand, minimum spanning tree and relative neighborhood graph contain only a small amount of connections between points and many close points are not connected, losing important relationships. In contrast, Gabriel graphs strike a good balance: many edges between distant points in the DT are eliminated, yet edges between close points are preserved. Thus, we use Gabriel graphs to identify neighboring objects in spatial datasets. For a more detailed discussion of various neighborhood graphs we refer to [12, 13].

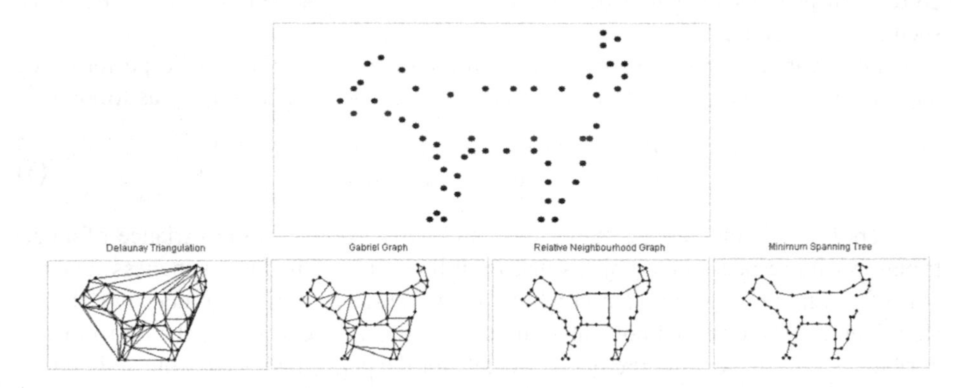

Fig. 1. Popular proximity graph types for a dog shaped dataset

Building Gabriel Graphs: Gabriel graphs are defined by its authors [3] as "Any two localities [points in the plane] A,B, are said to be contiguous iff all other localities are outside the A-B circle, that is, the circle on whose circumference A and B are at opposite points. In other words, two localities A and B are contiguous, unless there exists some other locality C such that in the triangle ABC the angle subtended at C is of 90 degree or more." Any points P and Q in a dataset are adjacent in the Gabriel graph if P and Q are distinct and the closed disc *D*, of which the line segment PQ is a diameter, contains no other points. Unlike Delaunay graph, Gabriel graphs generalize to higher dimensions, with the empty disks replaced by empty closed balls. For 2-dimensional data, the Gabriel graph can be computed from the Delaunay graph in O(n) time [12], in a total of O(*n*log*n*) complexity. For higher dimensional data, Gabriel graph can be computed in $O(n^3)$ time by brute force.

4.2 Finding Small Hotspot Seeds with High Interestingness

Once we create the Gabriel graph for the spatial dataset, in this phase, we identify the small, contiguous subgraphs with high interestingness, which we call "hotspot seed regions" and grow these seed regions in Phase 3 to obtain larger hotspots. In order to identify seed regions, we visit each vertex in the graph, and create a region consisting of the vertex and all of its neighbors. The interestingness value for each region is calculated by applying the plugin interestingness function on the set of objects in the region. A "seed interestingness threshold" is used to determine which of these regions can be used to grow larger hotspots.

4.2.1 Merging Seed Regions

The seeding phase finds many regions with high interestingness which can be used to grow hotspots in in the next phase. However, the case study reported in [1] shows that many seed regions grow to the same hotspot. This is not surprising as large hotspots usually have smaller sub-regions with high interestingness which were identified as seeds. Thus, we try to eliminate some of these seeds before growing by merging them. We use the following method to reduce the number of seed regions grown:

1) Find neighboring seed regions
2) Create a neighborhood graph of seed regions where each seed region is a node. Seed regions are neighbors if they share at least 1 node. Create an edge between nodes in this graph if the corresponding seed regions are neighbors *and* the union of these regions yields a region with an acceptable reward value. A merge threshold is used to assess if the union of two seed regions is acceptable. Weight of the edge is the reward gain when the two regions connected to the edge is merged.
3) Merge the seed regions connected to the edge with the highest weight.
4) Update seed neighborhood graph after the merge operation: Create an edge between the new node and neighbors of the merged nodes using the same procedure.
5) Continue merging seed regions as long as there are nodes to be merged in the graph.

We use a merge threshold μ to define if the union of seed regions is acceptable. If the reward of the new region is higher than the total reward of merged regions multiplied by μ, then the merge is acceptable:

$$\text{merge}(s_1, s_2) \text{ if } R(\cup(s_1,s_2)) > (R(s_1) + R(s_2)) * \mu \quad (4)$$

where R(si) represents the reward of seed region s_i. Fig. 2 gives a shortened version of seed processing algorithm explained above and Fig. 3 depicts pseudocode for the seed merge procedure. We assign the reward gain which is calculated by $R(\cup(s_1,s_2)) - (R(s_1) + R(s_2))$ as the weight of an edge. We keep all edges to be processed in a max-heap where the edge with the highest weight (reward gain) is the root of the heap tree and processed first.

```
1:  Create an neighborhood graph G of all seed regions
2:  Create a max-Heap H of all edges using edge weights as priority
3:  while H has elements
4:       nextEdge = H.dequeue()
5:       if graph contains both nodes connected by nextEdge then
6:            Merge(nextEdge)
7:     end if
8:  end while
```

Fig. 2. Pseudocode for seed processing algorithm

```
 1: Procedure Merge (Edge e)
 2:    Set s1 = e.source, s2 = e.target
 3:    Merge s1 and s2 by adding all elements in s1 and s2
       in a newregion snew
 4:    Add snew into G as a new vertex
 5:    Remove e from G
 6:    foreach neighbor si of s1 connected by edge ei
 7:        if R(U(si,snew)) >  (R(si) + R(snew)) * μ then
 8:             Create an edge enew connecting nodes si and snew
 9:             enew.weight = R(U(si,snew))  -  (R(si) + R(snew))
10:             G.AddEdge(enew)
11:            G.RemoveEdge(ei)
12:       end if
13:    end foreach
14:    foreach neighbor sj of s2 connected by edge ej
15:        if R(U(sj,snew)) >  (R(sj) + R(snew)) * μ then
16:            Create an edge enew connecting nodes sj and snew
17:            enew.weight = R(U(sj,snew))  -  (R(sj) + R(snew))
18:            G.AddEdge(enew)
19:            G.RemoveEdge(ej)
20:       end if
21:    end foreach
22:    Remove s1 and s2 from the graph G
23:  end procedure
```

Fig. 3. Pseudocode for Seed Merge procedure

4.3 Hotspot Growing Phase

In this section, we describe the heap-based hotspot growing algorithm. In hotspot growing phase, we search the best neighbor among all neighbors in each step, and after each time we add a new neighbor we do this search again. Searching for the best neighbor each time increases the complexity of hotspot growing algorithm. Instead, when new neighbors are encountered as a result of growing the hotspot, we assign each new neighbor a fitness value by evaluating the reward gain in case the neighbor is added to the region. We use a max-heap data structure to keep the list of neighbors where the neighbor with the highest fitness value is the root of the heap tree. We add each neighbor into the heap using the fitness value as the priority. Fig. 4 gives the algorithm for each step of hotspot growing algorithm. We continue growing the region as long as there are more neighbors to be added and the interestingness of the region is higher than the interestingness threshold.

```
1:  Procedure AddNextNeighbor(region)
2:     set bestNeighbor = Heap.dequeue()
3:     add bestNeighbor to region
4:     set newReward = CalculateReward(region)
5:     foreach neighbor n of bestNeighbor
6:        if n is not in region and n is not in the neighbors list
          then
7:           set fitness = CalculateFitness(region, n)
8:           Heap.enqueue(n,fitness)
9:        end if
10:    end foreach
11:    if newReward > region.alltimeBestReward then
12:       set region.alltimeBestReward = newReward
13:       set region's overallBestGridCells =region.currentGridCells
14:    end if
15: end procedure
```

Fig. 4. Pseudocode for heap-based hotspot growing algorithm

The runtime complexity of the heap-based hotspot growing algorithm is O(nlogn) as a total of O(logn) time is spent in each step where n is the number of objects in the hotspot. We refer to [2] for more details about the runtime complexity calculation. We also implemented incremental calculation of correlation and variance interestingness function to calculate the new reward in O(1) time. Details of incremental calculation is also available in [2]. Moreover, we grow seeds in parallel using a parallel processing framework which is also discussed in [2].

4.4 Generating Polygon Models for Hotspots

In this phase, we present a method to create polygon models for 2-dimensional hotspots. Polygons serve an important role in the analysis of spatial data as they can be used as higher order representations for spatial clusters. Computationally it is much cheaper to perform certain calculations on polygons than on sets of objects. Moreover, relationships and change analysis between spatial clusters can be studied more efficiently and quantitatively by representing each spatial cluster as a polygon. Furthermore, many database systems support operations on polygons, increasing the importance of polygons as models for spatial data.

We use the Voronoi diagram for the spatial dataset as the basis for creating a polygon model for hotspots. Each point in the hotspot will either be in a Voronoi polygon, or if the point is on the convex hull of the dataset, it will not be enclosed by a Voronoi polygon (in which case it will be in an unbounded Voronoi region). In this case, we propose enclosing such points in a polygon by intersecting the convex hull of the dataset with the Voronoi edges. Moreover, some points will not lie on the convex hull, but they will be enclosed by a polygon which crosses the convex hull. Such points and their Voronoi cells usually lie on the edges of the dataset and their Voronoi polygons are quite large, beyond the convex hull. In this case, we intersect such Voronoi polygons with the convex hull to obtain more compact hotspots. Once we create the Voronoi diagram and the convex hull of the dataset, we propose the following algorithm for creating a polygon model for a spatial hotspot:

1. Find the Voronoi polygons or edges for each point in the hotspot.
2. For each point P in the hotspot:

   a. If P is in a closed Voronoi cell (Voronoi polygon), check if it crosses with the convex hull:

      i. If the convex hull does not cross Voronoi polygon, then add this polygon into the polygon model for the hotspot.

      ii. Else if the convex hull crosses the Voronoi polygon, then the convex hull splits this polygon into 2 polygons. In this case, the point will be inside one of these polygons. Add the polygon with the point into the polygon model.

   b. If the point is not in a Voronoi polygon: find the intersection of the Voronoi edges around the point and the convex hull. The intersection will create a polygon; add this polygon into the polygon model.

This method will create polygons for all points in the hotspot and merge them. Since all points in the hotspot are connected, the union of all polygons will create one large polygon model for the hotspot.

5 Experimental Evaluation

We tested our methodology in a case study involving an earthquake dataset containing all earthquakes of magnitude 6.0 or higher in Japan and Korea region from January 1st 2000 to January 1st 2016. The dataset contains latitude, longitude, depth and magnitude of 236 earthquakes in the region. The data was downloaded from USGS (United States Geological Survey) website [14] for latitudes from 29.091 to 45.841 and longitudes from -234.756 to -210.41. Fig. 5 shows the earthquakes on a map. In the first experiment, we find hotspots with very high correlations of earthquake depth and magnitude. Next, we find hotspots where variance of earthquake depth is very low.

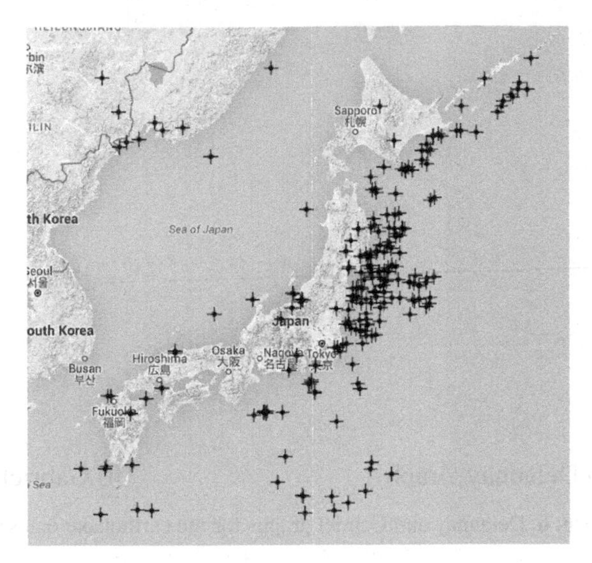

Fig. 5. Earthquake dataset visualized on a map

Following reward function is used for evaluating the quality of a region Ri:

$$\varphi(R_i) = \mathit{interestingness}(R_i) \times \mathit{size}(R_i)^{\beta} \tag{5}$$

where β is a real number determining the preference for larger regions. In the case study, we set β to 1.01 as we prefer smaller regions with high interestingness to larger regions with low interestingness.

5.1 Finding High Correlation Hotspots

In this case study, we find hotspots in the area in which depth and magnitude of the earthquake is highly correlated. We use the correlation interestingness function defined in (2) and set 0.75 as the interestingness threshold.

Step 1: Building the Neighborhood Graph: Figure 6 shows the Delaunay and Gabriel graphs for this dataset. As seen on Figure 6a, Delaunay graph contains too many long edges which connect very distant objects in the dataset. Many of those edges are removed in the Gabriel graph. This provides additional evidence for our choice to use Gabriel graph to determine the neighborhood of the objects. In case it is desired, our framework allows using Delaunay graphs too. However, Delaunay graphs cannot be used for higher dimensions.

Step 2: Identifying Hotspot Seeds: We used 0.95 as the seed threshold to identify regions with very high correlation of depth and magnitude. The correlation of these variables in the dataset is 0.029, which is very low. Out of 235 regions evaluated in the dataset (1 region around each object), 33 regions had an absolute correlation value greater than 0.95. After applying seed merge operation, we obtained 30 seed regions which were grown in the next phase.

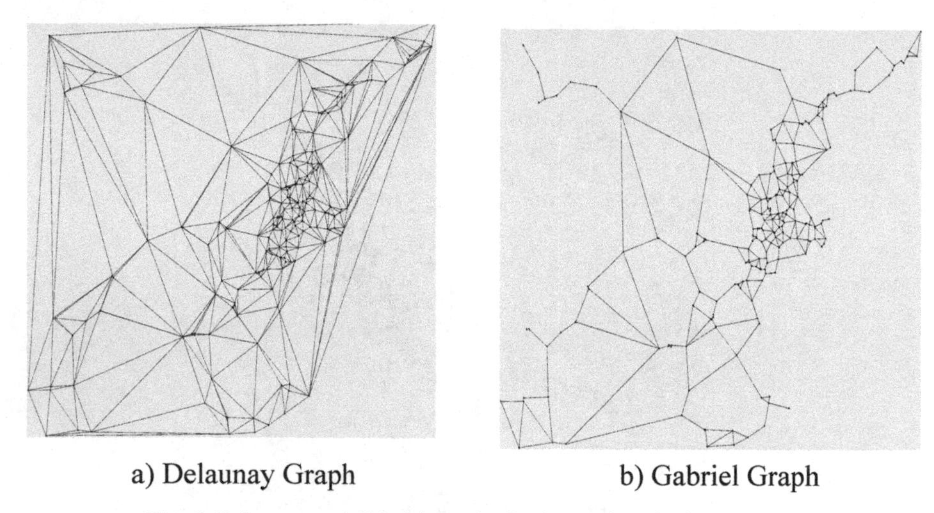

a) Delaunay Graph b) Gabriel Graph

Fig. 6. Delaunay and Gabriel graphs for the earthquake dataset

Step 3: Growing Seed Regions: Out of 30 seed regions grown, 29 of them had high positive correlation (greater than 0.75) and only 1 region had very high negative correlation (-0.93). Average positive correlation was 0.86. 3 seeds grew to the hotspots which were already discovered so they were deleted.

Step 4: Creating Polygon Models for Hotspots: In this step, we create polygon models for hotspots by merging the Voronoi cells of each vertex in the hotspot. As many hotspots share objects, it is not feasible to visualize all hotspots, thus we will only visualize two sample hotspots.

Figure 7 shows two hotspots discovered. As seen in the figure, the hotspots share an object in the middle. This makes sense, as that object's attributes are positively correlated with objects in the orange region and negatively correlated with objects in the green region.

When creating the polygon model for the green colored hotspot, there was one Voronoi polygon crossing the convex hull (which is the cell on the bottom right corner in Fig 7a and that polygon was split into two parts, and the polygon with the point was added to the polygon model.

5.2 Finding Low Variation Hotspots

In this case study, we find low variation hotspots in the same geographic area in which the variance of the depth of the earthquakes is lower than 5. We used the variance interestingness function defined in (3) and set 5 as the interestingness threshold. We used 3 as the seed interestingness threshold to find small regions with variance less than 3. There were 10 seed regions in the dataset. 2 of the seed regions were merged and the resulting 9 seed regions were grown. While growing these seed regions, 2 of them grew to the already discovered hotspots so they were deleted. Table I lists the remaining 7 hotspots. Average hotspot size (number of earthquakes in the region) is 5.57. Figure 8 depicts the 3 non-overlapping low variation hotspots (hotspot 2, 6 and 7) generated.

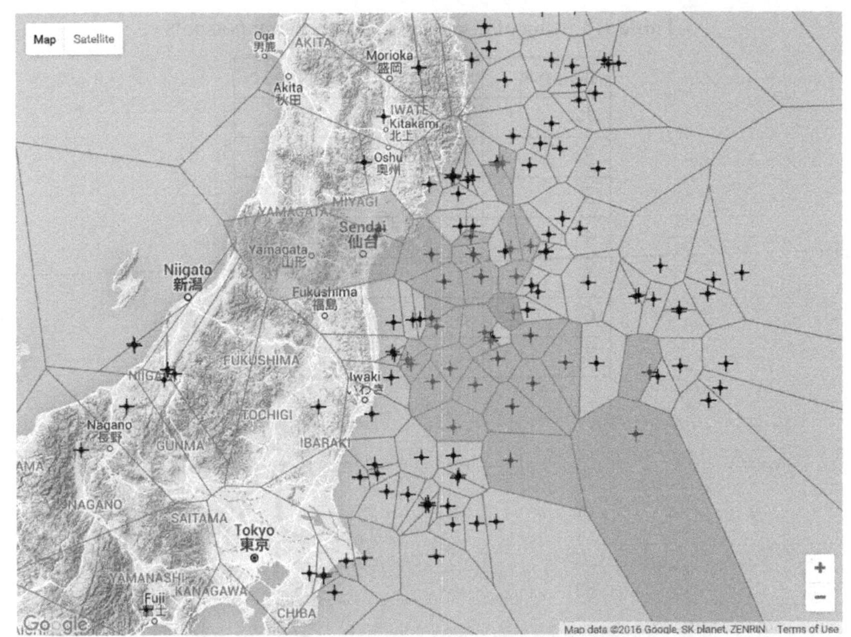

a) Green hotspot: Negative correlation, Orange hotspot: Positive correlation

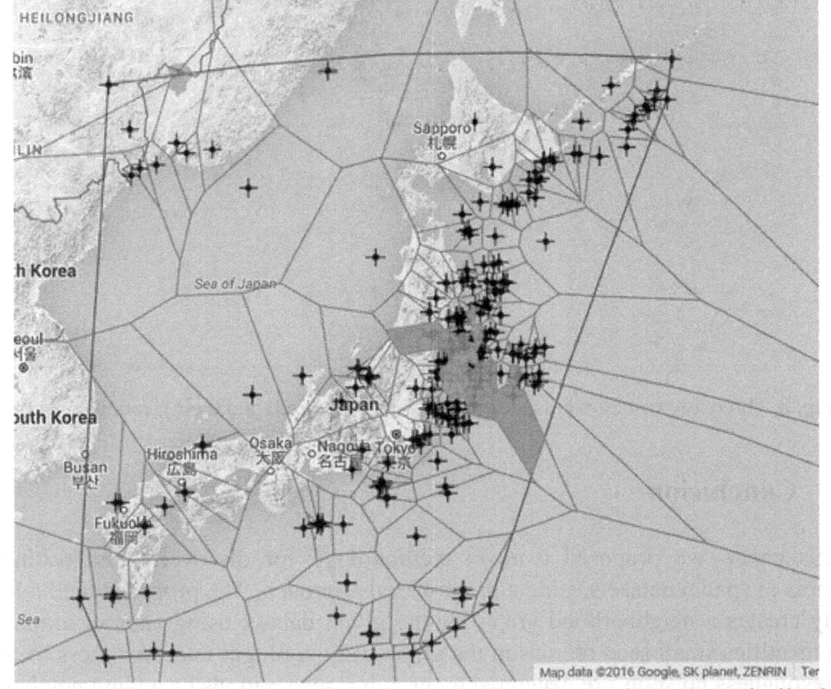

b) Hotspots on a map with smaller scale (black lines are the convex hull edges)

Fig. 7. Two hotspots and their location on maps with different scales

Table 1. Listing of discovered low variation hotspots

hotspot	size	variance
1	4	4.25
2	4	3.9
3	7	2.41
4	4	2.21
5	8	1.79
6	8	1.48
7	4	0.32

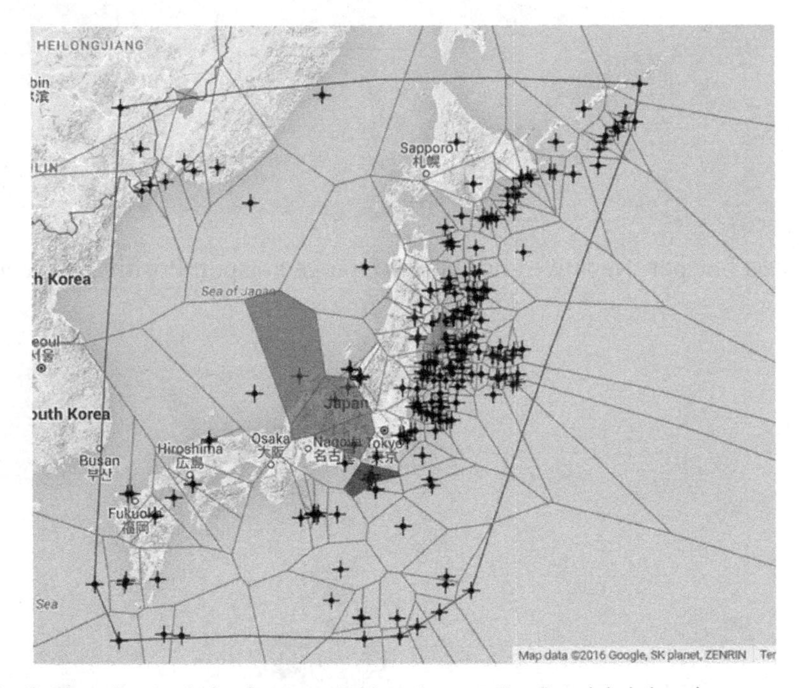

Fig. 8. Three low variation hotspots (2:blue, 6:green, 7:red) and their locations on a map

6 Conclusion

In this paper, we proposed a novel methodology for discovering interestingness hotspots in spatial datasets using a graph-based approach. The proposed methodology firstly creates a neighborhood graph for the given dataset using Gabriel graphs and then identifies small seed regions in the graph using a plugin interestingness measure. Interestingness hotspots are generated by growing seed regions by adding neighboring nodes, maximizing the given reward function. To the best of our knowledge, this is the only algorithm that grows hotspots from seed regions using a reward function. Furthermore, we create a polygon model for each hotspot by merging Voronoi cells for each spatial object in the hotspot, identifying the scope of the identified also allowing for the quantitative assessment of the size of the hotspots and relationships to other objects in the spatial dataset.

The proposed methodology is evaluated in a case study for a 2-dimensional earthquake dataset in which we find interestingness hotspots based on variance and correlation interestingness functions. The methodology proved to be successful in finding hotspots based on plugin interestingness and reward functions. We plan to extend our framework for higher dimensional datasets in which we create higher dimensional Gabriel graphs and polygonal models.

Compared to clustering approaches we believe that our approach has more potential to compute "better", more interesting hotspots, as the clustering approach searches for all hotspots in parallel, being forced to make compromises, as switching one sub region from one to another cluster might increase the reward of one cluster but decrease the reward of the other cluster. We plan to compare our approach to clustering approaches in a future work.

References

1. Akdag, F., Davis, J.U., Eick, C.F.: A computational framework for finding interestingness hotspots in large spatio-temporal grids. In: Proceedings of the 3rd ACM SIGSPATIAL International Workshop on Analytics for Big Geospatial Data, pp. 21–29. ACM, November 2014
2. Akdag, F., Eick, C.F.: An optimized interestingness hotspot discovery framework for large gridded spatio-temporal datasets. In: 2015 IEEE International Conference on Big Data (Big Data), pp. 2010–2019. IEEE, October 2015
3. Gabriel, K.R., Sokal, R.R.: A New Statistical Approach to Geographic Variation Analysis. Systematic Zoology **18**, 259–278 (1969)
4. Miller, R., Chen, C., Eick, C.F., Bagherjeiran, A.: A framework for spatial feature selection and scoping and its application to geo-targeting. In: 2011 IEEE International Conference on Spatial Data Mining and Geographical Knowledge Services (ICSDM), pp. 26–31. IEEE (2011)
5. Kulldorff, M.: A spatial scan statistic. Communications in Statistics-Theory and methods **26**(6), 1481–1496 (1997)
6. Varlaro, A., Appice, A., Lanza, A., Malerba, D.: An ILP Approach to Spatial Clustering. Convegno Italiano di Logica Computazionale, Roma (2005)
7. Ester, M., Kriegel, H.P., Sander, J., Xu, X.: A density-based algorithm for discovering clusters in large spatial databases with noise. Kdd **96**(34), 226–231 (1996)
8. Ertöz, L., Steinbach, M., Kumar, V.: Finding clusters of different sizes, shapes, and densities in noisy, high dimensional data. SDM, pp. 47–58 (2003)
9. Wang, X., Hamilton, H.J.: A comparative study of two density-based spatial clustering algorithms for very large datasets. In: Kégl, B., Lee, H.-H. (eds.) Canadian AI 2005. LNCS (LNAI), vol. 3501, pp. 120–132. Springer, Heidelberg (2005)
10. Cao, Z., Wang, S., Forestier, G., Puissant, A., Eick, C.F.: Analyzing the composition of cities using spatial clustering. In: Proceedings of the 2nd ACM SIGKDD International Workshop on Urban Computing, p. 14. ACM (2013)

11. Choo, J., Jiamthapthaksin, R., Chen, C.-S., Celepcikay, O.U., Giusti, C., Eick, C.F.: MOSAIC: a proximity graph approach for agglomerative clustering. In: Song, I.-Y., Eder, J., Nguyen, T.M. (eds.) DaWaK 2007. LNCS, vol. 4654, pp. 231–240. Springer, Heidelberg (2007)
12. Matula, D.W., Sokal, R.R.: Properties of Gabriel graphs relevant to geographic variation research and the clustering of points in the plane. Geographical analysis **12**(3), 205–222 (1980)
13. Jaromczyk, J.W., Toussaint, G.T.: Relative neighborhood graphs and their relatives. Proceedings of the IEEE **80**(9), 1502–1517 (1992)
14. United States Geological Survey (USGS). http://earthquake.usgs.gov/earthquakes/search/

MapReduce-Based Growing Neural Gas for Scalable Cluster Environments

Johannes Fliege(✉) and Wolfgang Benn

Department of Computer Science, Technische Universität Chemnitz, Straße der Nationen 62, 09107 Chemnitz, Germany
flj@cs.tu-chemnitz.de

Abstract. Growing Neural Gas (GNG) constitutes a neural network algorithm to create topology preserving representations of data, thus, being applicable in cluster analysis. With fast growing amounts of data, cluster analysis tasks face distributed data sets managed by cluster environments requiring scalable, parallel computation methods. In this paper we present a MapReduce-based version of the GNG training method. The algorithm is able to process large data sets on scalable cluster systems. We discuss its complexity and consider communication costs that arise from its structure. We conduct experiments on artificial data in different cluster environments to evaluate the algorithms scalability. Finally, we show that the algorithm is applicable for cluster analysis of large data sets in scalable cluster systems.

1 Introduction

Todays fast growing data amounts require scalable cluster systems to store, manage, and analyze data. Cluster environments like Apache Hadoop [14] offer implementations of distributed file systems that address the first two requirements. The requirement to analyze distributed data sets on an arbitrary number of machines in parallel requires parallel computation models, e. g. the MapReduce [4] model. Beside implementations in cluster storage environments, more specific systems like Apache Spark [15] address data analysis and provide more performant solutions to these computation models.

Such parallel computation models may be applied in Knowledge Discovery in Databases (KDD), the process of preparing and processing data in order to gain information from it. In this, KDD is used for data mining which often includes cluster analysis on data sets. If neural networks are used for cluster analysis, a large amount of time is spent to train these networks; training time increases with growing data sizes and is proportional to network quality requirements. Thus, if distributed data management is a requirement, e. g, due to a data sets size, distributed knowledge discovery becomes necessary as well.

We present a MapReduce-based variant of the GNG algorithm [6], the MapReduce GNG algorithm, discuss its runtime behavior and consider the algorithms scalability. The GNG creates a neural net to topologically represent a data set by "competetive Hebbian learning" [11]. The number of neurons is not

P. Perner (Ed.): MLDM 2016, LNAI 9729, pp. 545–559, 2016.
DOI: 10.1007/978-3-319-41920-6_43

fixed in this model but only limited by an upper bound. After the learning phase, clusters are formed by connected components in the graph representation of the neural network. This clustering property has already been used in ICIx [8], a database indexing structure. A MapReduce-based training method will make ICIx applicable to distributed data sets.

The paper is organized as follows. Related work in the field of parallel and distributed approaches of the GNG algorithm and work that discusses MapReduce approaches concerning neural networks is outlined in the following section. Section 3 gives a basic overview to the GNG algorithm and its batch variant. In section 4, the MapReduce GNG algorithm is presented and its complexity as well as its communication costs concerning its application in scalable cluster environments are discussed. Finally, the algorithm's scalability in cluster environments of different sizes with data sets of different sizes is evaluated in section 5.

2 Related Work

The GNG algorithm was first presented by Fritzke in [6]. Since then, several approaches to speedup or parallelize this learning algorithm have been proposed. There are approaches that focus on parallelizing the complex find winner phase by using GPUs [12] or dedicated cluster environments [13]. Other approaches address parallelization of the learning algorithm by introducing batch computation versions of the GNG algorithm [1,2]. Nevertheless, all these approaches do not utilize the MapReduce [4] paradigm, which was found by other authors to be generically adaptable to machine learning algorithms [3,7,10].

In order to obtain a better overview, we split this section into two parts: part 1 introduces work that is related to GNG parallelization in particular, part 2 introduces work that discusses MapReduce in the machine learning domain in general.

2.1 GNG Parallelization

In [1] Adam et al. present two different parallelization approaches to the GNG: they introduce data parallel approaches that make use of partitioning data into smaller fractions in order to train several neural networks in parallel with smaller input signal set sizes. First, an approach is presented that trains two neural networks independently and merges their results to obtain a global representation of the trained data. Second, a batch approach is presented, that determines position and error gradients of the neurons based on an initial neural network for each data partition. These partial results are then synchronized into a global result that updates a central neural net. In addition to data parallel processing, they introduce an approach that also partitions the neural network and synchronizes these parts after computation. They show in [2] that the batch approach is most promising in terms of clustering quality and performance.

In [12], Parigi et al. focus GPU-based parallelization for growing neural networks in general. They propose a multi-signal variant that considers more than one signal in each iteration of the neural network algorithm. The parallelization approach described in their paper is limited to the find winners phase of the GNG algorithm. It is realized in a concurrent manner for multiple input signals assigning each signal to a single thread. This is done to better adapt to GPU specific characteristics. They show that the multi-signal approach converges faster than a single signal approach in a GPU-based implementation.

In [13], Vojácek and Dvorsky present an approach that parallelizes the find winner phase of the GNG algorithm for application on an HPC cluster. They partition the GNG graph according to the number of cluster nodes in order to determine local winner neurons. Multiple input signals are regarded in one step, each on one cluster node. Next, their algorithm determines the global winner out of the local winners and updates the neural network. Their experiments show a speedup compared to a non-parallel standard GNG without a loss of clustering quality.

A MapReduce-based GNG variant that is designed to be applicable in scalable, general cluster environments requires a parallelization approach that is able to process data partitions independently and has low, ideally constant communication costs. The GPU based approaches in [12] and [13] require communication for every input signal which introduces further algorithmic complexity that depends on the number of input signals. Since the batch computation approach in [1] only requires communication after processing the whole data set, the communication costs are constant concerning the number of input signals. Hence, we base the MapReduce GNG algorithm on batch processing.

2.2 MapReduce-Based Approaches

Since its publication in 2004, MapReduce [4] has been referred to as the large scale data processing paradigm, even for machine learning algorithms. Thus, different machine learning algorithms and generic models using this paradigm have been discussed in literature. However, to the best of our knowledge, there is no evidence that shows the GNG algorithm operating on MapReduce.

Gillick et al. address single pass, iterative, and query-based learning paradigms within the MapReduce model in [7]. They show benefits and limitations of these paradigms on a Hadoop-based implementation concluding, that a larger set of algorithms may also be realized with the MapReduce approach by following the same pattern. This illustrates that the MapReduce paradigm may offer substantial benefits to machine learning algorithms in general and especially in scalable cluster environments.

In [10], Liu et al. propose a generic MapReduce-based model that is intended for upscaling similarity-based algorithms and iterative learning algorithms. In particular, they adapt non-negative matrix factorization, support vector machines, and page rank to the MapReduce model by applying a partitioning and summation process to the general matrix multiplication realized through a MapReduce scheme. Their results show good speedup rates for varying numbers of machines, although the speedup is behind the ideal linear speedup which is caused by additional communication overhead in growing cluster environments.

In [3] several machine learning algorithms are implemented on MapReduce following the Statistical Query Model [9]. According to this model, statistical properties of the data are used for learning. When these statistics are combined with an objective function, e. g. the error function of a learning algorithm, learning algorithms may be regarded as optimization algorithms which can be written in summation form. The summation form follows the MapReduce model, since partial results, i. e. map results, may be aggregated into a global sum. Algorithms that follow this model first compute the statistics and gradients from a statistical query oracle before these are aggregated over all data points. Therefore, this model may be applied in a data parallel approach, where a Map function determines partial gradients while a Reduce function aggregates these partial results to form a global result.

Although we do not use statistical queries, we rely on the summation form by directly presenting input signals to the training algorithm. Since batch processing already requires the summation form, this generic method forms the foundation of our proposed MapReduce GNG algorithm as well.

3 Growing Neural Gas

The GNG algorithm was first described in [6] by Fritzke as a data topology preserving neural network. Its aim is to adapt a set of neurons A connected by edges in order to represent the distribution of a given data set D. Neurons or units, respectively, are described by n-dimensional weight vectors $w = (w_1, ..., w_n)$ which represent the neurons positions in an n-dimensional space. All input signals ξ provided by D are propagated through the network leading to an adaption process in which the weight vectors of the neurons are adapted to those of the input signals. After a certain number of input signals, the insertions threshold λ is reached. Upon this threshold, a new neuron is inserted and neurons may be removed from the neural network. The removal is only executed for neurons that are no longer connected to other neurons whereas the existence of a connection is indicated by an edge age value a. The steps described previously are repeated until a stop criterion is reached and run in $O\left(|D| \cdot \lambda \cdot |A|^2 \cdot d\right)$ [8]. Algorithm 1 shows the detailed version of the GNG learning phase. For an extensive discussion of the algorithm, see [6].

3.1 Data Parallel Batch Processing

The batch variant of the GNG, discussed by [1], introduces a modified learning phase. It does not move neurons during its learning phase but accumulates the movement values into gradient variables Δi for each neuron. Therefore, the learning phase may be executed in parallel without interdependencies on dedicated partitions D_i of a data set D. The neural network is updated, after all input signals have been processed. The central updated process is equal to the

Table 1. Symbol Definitions

Symbol	Description
A	set of neurons
D	training data set
w	weight vector representing a neuron
ϵ_b	winner unit adaption rate
ϵ_n	neighbor adaption rate
α	error adaption rate
a_{max}	maximum edge age

Algorithm 1. The Growing Neural Gas Algorithm

1: Start with two units a and b at random positions w_a and w_b in R^n.
2: Generate an input signal ξ according to $P(\xi)$.
3: Find the nearest unit s_1 and the second nearest unit s_2.
4: Increment the age of all edges emanating from s_1.
5: Add the squared distance between the input signal and the nearest unit in input space to a local counter variable:

$$\Delta\text{error}(s_1) = \|w_{s_1} - \xi\|^2$$

6: Move s_1 and its direct topological neighbors towards ξ by fractions of ϵ_b and ϵ_n, respectively, of the total distance:

$$\Delta w_{s_1} = \epsilon_b (\xi - w_{s_1})$$
$$\Delta w_n = \epsilon_n (\xi - w_n) \text{ for all direct neighbors } n \text{ of } s_1$$

7: If s_1 and s_2 are connected by an edge, set the age of this edge to zero. If such an edge does not exist, create it.
8: Remove edges with age larger than a_{max}. If this results in points having no emanating edges, remove them as well.
9: If the number of input signals created so far is an integer multiple of a parameter λ, insert a new unit as follows:
  - Determine the unit q with the maximum accumulated error.
  - Insert a new unit r halfway between q and its neighbor f with the largest error variable:
$$w_r = 0.5\,(w_q + w_f)\,.$$
  - Insert edges connecting the unit r with units q and f, and remove the original edge between q and f.
  - Decrease the error variables of q and f by multiplying them with a constant α. Initialize the error variable of r with the new value of the error variable of q.

10: Decrease all error variables by multiplying them with a constant α.
11: If a stopping criterion (e. g. network size or some performance measure) is not yet fulfilled go to step 2.

updated phase of the original GNG algorithm. Consequently, the batch approach leads to an algorithm consisting of two parts: part 1 initializes the batch computation and finally updates the neural network with the partial batch results; part 2 computes the learning phase in a batch manner. The algorithm shows a complexity of $O\left(|D| \cdot \lambda \cdot |A|^2 \cdot d/p\right)$; p denotes the level of parallelism. If communication costs are regarded, a term indicating the synchronization overhead should be added. For example, logarithmic complexity is required if a tree-like synchronization scheme is applied.

4 The MapReduce GNG Algorithm

We present a MapReduce-based version of the GNG algorithm in this section. The goal of this algorithm is to provide a method to process distributed data sets in scalable computation cluster environments. The foundation of this algorithm is formed by the statistical query model and the batch approach described in [1].

In a MapReduce environment, partitions of a data set D are propagated through Map processes. Map results are then aggregated into a single result by reducers. Since these Map processes need to be mutually independent, communication amongst them should be avoided. As already discussed, this requires batch computation. If a MapReduce approach is applied to the GNG, fragmentation of the algorithm into independent Map and Reduce needs to be ensured. The Map function is applied to each input signal of D and returns partial gradient values of the neurons' weight vectors, error values and edge ages. Subsequently, the Reduce function performs the summation of these partial gradients and returns global gradient values that are then applied to the neural network by a central process.

For this MapReduce-based approach, the insertion factor λ gains special interest. In Fritzke's GNG algorithm, λ expresses an insertion threshold, i. e. "If the number of input signals created so far is an integer multiple of a parameter λ, insert a new unit..." [6]. Due to the nature of batch processing, all input signals are processed in one portion before update operations are applied. Thus, λ is inherently equal to the cardinality of a given data set D. Hence, insertion is only performed once in each batch and causes the batch variant to be independent of λ. Consequently, we do not consider an insertion threshold for the formulation of our MapReduce GNG algorithm in contrast to the GNG variants introduced before. We present the MapReduce GNG algorithm in the following section.

4.1 The Algorithm

The proposed MapReduce GNG algorithm is built of three parts:

1. a central *Driver algorithm* that performs central update operations on the neural network and initializes the Map and the Reduce processes,

2. a *Map algorithm* that is executed in parallel on a set of mapper processes $M = \{m_1, ..., m_p\}$, and
3. a *Reduce algorithm* that aggregates the mapper results.

Assuming a distributed data set D, partitioned into p fragments, the three parts of the algorithm work as follows:

The Driver Algorithm initializes the neural network with two neurons according to the original GNG algorithm. In each iteration i, it initializes the Map process by distributing the state of the neural network n_i to the machines hosting the Map processes and starts their computation. After the Map processes have finished, the driver program initializes the execution of the Reduce phase by propagating the Map algorithms results to the Reduce processes. Note that the data set D is not transferred between any of these parts. Finally, the driver part updates the neural network by inserting or removing a neuron, again according to the original GNG algorithm. This process is repeated until the algorithm converges to a certain criterion, e. g. a minimum error or a maximum number of iterations i_{max}. Algorithm 2 shows a detailed description of the Driver routine.

Algorithm 2. MapReduce GNG: Driver Algorithm

1: Start with two units a and b at random positions w_a and w_b in R^n.
2: Initialize p parallel mapper processes with the Map Algorithm (see algorithm 3) and transmit the neural networks state to those.
3: Execute all mapper processes in parallel.
4: Synchronize all mapper results with the Reduce Algorithm (see algorithm 4).
5: Update the neural network by accumulating the Reduce results into it.
  - Synchronize the units positions and errors.
  - For all pairs of units s_k and s_l that were marked as nearest and second nearest unit to any input signal ξ update the edge ages as follows:
    - If these units s_k and s_l are connected by an edge, set the age of this edge to zero. If such an edge does not exist, create it.
  - Remove edges with age larger than a_{max}. If this results in points having no emanating edges, remove them as well.
6: If $|A| < |A_{max}|$ insert a new unit r. (cp. algorithm 1)
7: Decrease all error variables by multiplying them with a constant α.
8: If a stopping criterion (e. g. network size or some performance measure) is not yet fulfilled go to step 2.

The Map Algorithm determines the local position, error and edge age gradients for a partition D_k of the data set D and, thus, forms the training phase of the algorithm. Each instance m_k of the Map algorithm is assigned to a data partition. In contrast to standard MapReduce processing, the Map algorithm processes the whole data partition before a results is returned instead of returning one result for each input signal. This reduces communication costs. For each input signal $\xi \in D_k$, the find winner phase is evaluated returning the closest and second closest neurons u_1 and u_2 to ξ. Next, the position, error and edge age gradients are determined without moving the neurons. For each D_k these values are accumulated into the appropriate gradient variables. Algorithm 3 gives a detailed description of the Map algorithm.

Algorithm 3. MapReduce GNG: Map Algorithm

1: Select an input vector ξ from the assigned data partition D_i.
2: Find the nearest unit s_1 and the second nearest unit s_2.
3: Add the squared distance between the input signal and the nearest unit in input space to a local error gradient variable:

$$\Delta \text{error}(s_1) = \|w_{s_1} - \xi\|^2$$

4: Add the indicated movement of s_1 and its direct topological neighbors towards ξ by fractions of ϵ_b and ϵ_n, respectively, of the total distance to a local weight vector gradient variable:

$$\Delta w_{s_1} = \epsilon_b\,(\xi - w_{s_1})$$
$$\Delta w_n = \epsilon_n\,(\xi - w_n) \text{ for all direct neighbors n of } s_1$$

5: Increment a local counter variable for the edge e_{12} connecting s_1 and s_2.

The Reduce Algorithm aggregates the Map results by computing an average of the gradients determined by the Map algorithms instances. Since one Reduce instance is capable of aggregating two Map results, this aggregation process is performed in a tree like structure to determine a global result. Algorithm 4 gives a detailed description of the Reduce algorithm.

Algorithm 4. MapReduce GNG: Reduce Algorithm

1: Input: mapper results m_k and m_l
2: Aggregate the weight vector and error gradient variables of m_k and m_l and aggregate the edge counter variables.

4.2 Complexity

In the following sections, the computational complexity of the parts of the proposed algorithm are discussed. We also consider communication costs which may be crucial to a MapReduce-based algorithm with each part potentially running on different machines in a cluster.

Map Algorithm. The map algorithm determines the closest neuron and the second closest neuron amongst its neighbors foreach $\xi \in D$. It performs update operations on the local gradient variables of the determined neurons weight vectors and their local error gradients. Its runtime depends on $|D|, v$, and d (see (1)).

$$\begin{aligned} t_{map_{single}} &= |D| \cdot (2 \cdot v \cdot d + v + 2) \\ &= O(|D| \cdot v \cdot d) \end{aligned} \tag{1}$$

If D is processed in parallel by p Map processes, the total complexity of a single Map process iteration will be the pth fraction of $t_{map_{single}}$:

$$t_{map_{||}} = O\left(\frac{|D| \cdot v \cdot d}{p}\right). \tag{2}$$

Accumulated over all iterations, the mapping function runs in

$$\begin{aligned} t_{map} &= \sum_{v=2}^{|A|} \frac{|D| \cdot (v\,(2d + 1) + 2)}{p} \\ &= O\left(\frac{|D| \cdot |A|^2 \cdot d}{p}\right). \end{aligned} \tag{3}$$

Reduce Algorithm. Each instance of the Reduce algorithm merges two mapping results in a pairwise manner. That is, it aggregates all gradient variables, i. e. the gradient variables of neurons $(O\,(v \cdot d))$, edges $(O\,(v^2))$, and errors $(O\,(v))$. The complexity of a single invocation of the Reduce Algorithm is given in (4).

$$t_{reduce\_single} = O\left(v^2 + v \cdot d\right) \tag{4}$$

In a tree-like Reduce scheme the function is invoked $\log p$ times to fully aggregate all Map processes results. Thus, the computational complexity stated in (4) is increased by factor $\log p$ for each iteration of the algorithm. The total runtime of the Reduce scheme accumulated over all iterations of the algorithm is shown in (5). Please note that the cubic term is negligible for $d \gg |A|$.

$$\begin{aligned} t_{reduce} &= \log p \cdot \sum_{v=2}^{|A|} \left(v^2 + v \cdot d + v\right) \\ &= O\left(\left(|A|^3 + |A|^2 \cdot d\right) \log p\right) \end{aligned} \tag{5}$$

Overall Complexity. Following the Map algorithms learning phase and the synchronization in the Reduce algorithm, the driver updates the state of the neural network performing neuron insertion $O(v + d)$ and removal $O(v)$. Hence, the Map algorithm introduces the major part of the algorithms complexity. Accumulated over all steps, the total runtime of the presented approach is linearly depending on $|D|$ and d, and may be cubic in $|A|$ for small D (see (6)).

$$t_{all} = t_{map} + t_{reduce}$$
$$= O\left(\frac{|D| \cdot |A|^2 \cdot d}{p} + \log p \cdot \left(|A|^3 + |A|^2 \cdot d\right)\right) \quad (6)$$

Note: if $\log p \cdot \left(|A|^3 + |A|^2 \cdot d\right) \ll |D|$, the logarithmic term of the Reduce algorithm becomes negligible.

Communication Costs. In a general cluster environment the different parts of the algorithm may reside on different machines. Communication between these parts may considerably increase its total runtime depending on the data amount to be transferred and the bandwidth of the utilized communication medium. To express these costs in complexity considerations, we introduce a factor $k_c, 1 \leq k_c < \infty$. For an ideal communication medium, k_c is set to 1. In our algorithm, this factor is to be considered whenever its three parts communicate their results. Thus, communication complexity is regarded when the Driver transmits the initial state to the Mappers and when the latter transmit the gradient results to the Reduce components. The synchronization costs are increased as well. In every transmission, the neuron gradient values have to be transmitted. Accumulated over all iterations, one transmission introduces costs of

$$t_{trans} = O\left(k_c \cdot \left(|A|^3 + |A|^2 \cdot d\right)\right). \quad (7)$$

Under consideration of the transmission costs, the overall complexity of the MapReduce GNG is in

$$O\left(\frac{|D| \cdot |A|^2 \cdot d}{p} + k_c \left(|A|^3 + |A|^2 \cdot d\right) \cdot \log(p)\right). \quad (8)$$

5 Experiments

The aim of our experiments was to evaluate the scalability of the MapReduce GNG algorithm. We conducted a series of experiments with artificial data sets of different sizes in a cluster environment. Comparisons were performed with 1, 2, 4, 8, and 16 cluster nodes. These nodes were set up with one dual core AMD Opteron 2.6 GHz CPUs and 4 GB physical memory. Scientific Linux 6.3 was used as operating system.

The tested MapReduce GNG algorithm was implemented on top of the Apache Spark [15] framework (version 1.5.1). We integrated an Apache Spark cluster into the described cluster system. It was used in standalone mode.

5.1 Test Data

The scalability tests were performed using five synthetic 2D data sets of different sizes. Data set 1 contained 0.5 million instances, set 2 contained one million

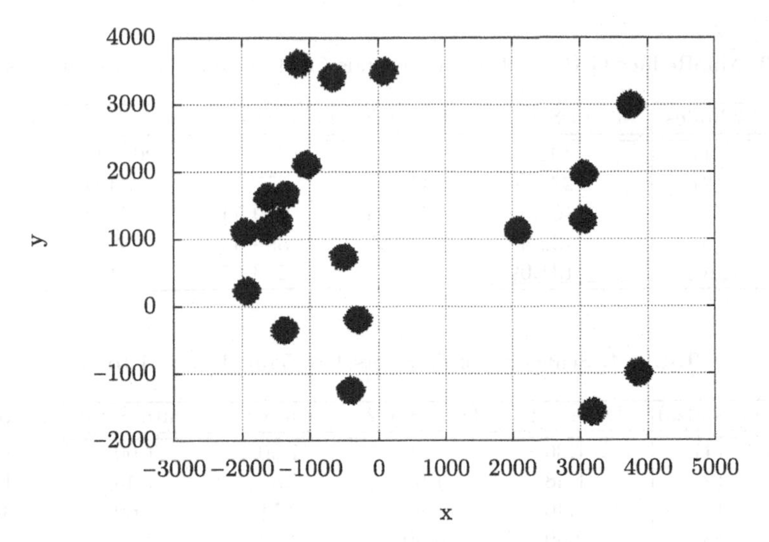

Fig. 1. Synthetic data set containing 500,000 signals uniformly distributed over 20 clusters.

Table 2. MapReduce GNG Algorithmic Parameters

Description	Symbol	Value
winner adaption rate	ϵ_b	0.1
neighbor adaption rate	ϵ_n	0.002
error normalization value	α	0.001
dimensionality of the data records	d	2
maximum number of neurons	$\|A\|_{max}$	40
maximum edge age	a_{max}	88

instances, set 3 two million instances, etc. up to eight million instances in data set 5. Each of the 20 clusters contained the same number of instances that were arranged using Gaussian distribution; the maximum extent of a cluster was limited by a fixed radius value. Again, the clusters were positioned by a Gaussian distribution in the data space. For all data sets, the cluster positions of data set 1 were used as a template to provide equal conditions. Figure 1 illustrates data set 1 as an example.

5.2 Results and Discussion

We performed the evaluation with the before named data sets using equal parameter settings for the MapReduce GNG algorithm throughout all test runs. The parameters that were used are shown in Table 2. In order to provide a comparable convergence criterion, we set the MapReduce GNGs number of iterations to a fixed size of i_{max} = 1000. Each data set was run 10 times on each cluster setting. The data sets were partitioned according to the number of machines in the appropriate cluster setting.

Table 3. MapReduce GNG runtimes in seconds for data sets 1 to 5 on clusters 1 to 5.

Cluster	(#Nodes)	Data Set 1	Data Set 2	Data Set 3	Data Set 4	Data Set 5
1	(1)	402.75	772.27	1356.12	2661.09	6165.35
2	(2)	272.16	475.64	877.62	1834.86	3652.98
3	(4)	167.56	293.63	539.54	989.19	1777.48
4	(8)	134.49	220.35	399.53	838.65	1306.32
5	(16)	131.09	164.11	263.17	464.69	826.84

Table 4. Scale out for data sets 1 to 5 on clusters 1 to 5.

Cluster	(#Nodes)	Data Set 1	Data Set 2	Data Set 3	Data Set 4	Data Set 5
1	(1)	1.00	1.00	1.00	1.00	1.00
2	(2)	1.48	1.62	1.55	1.45	1.69
3	(4)	2.40	2.63	2.51	2.69	3.47
4	(8)	2.99	3.50	3.39	3.17	4.72
5	(16)	3.07	4.71	5.15	5.73	7.46

Table 3 provides the average runtime of all data sets on all clusters. The resulting scale out values are provided in Table 4. As can bee seen, the algorithms scale out increases with growing level of parallelization and with growing data set sizes. The scale out values are sublinear in reference to the number of nodes in the cluster which is mainly caused by synchronization and communication costs (cp. (7)). These costs increase logarithmically in reference to the level of parallelism. Consequently, the smaller the data set and the higher the level of parallelism, the synchronization component in the algorithm's complexity increasingly outweighs the speedup that is gained through parallelization. In addition to the algorithmic complexity, the resulting scalability is also influenced by general runtime effects that occur in a distributed system. That is, startup costs, interference effects, and skew [5] reduce the final scale out. Since parallel instances of the Map algorithm operate mutually independent, interference effects that are caused by concurrent access to shared resources will be of minor importance in this phase. Nevertheless, interference occurs in a tree-like reduce scheme, increasing the described communication overhead. Furthermore, the skew that describes increasing variance in execution times proportional to the level of parallelism also influences our results since partition sizes were decreased with increasing p.

In figure 2, the scale out values mentioned before are visualized. It can be observed that the scale out of each data set is basically logarithmic. Only data set 5 shows a more linear behavior. In the results, especially data set 1 and 2 show the increasing effect of the logarithmic factor caused by the synchronization. The scale out of data set one is almost equal for 8 and 16 nodes. Thus, its slope almost converges to zero for a cluster of 16 nodes. Furthermore, the increasing communication overhead can be observed by a decreasing slope throughout all data sets when increasing the level of parallelism. Finally, it is to be expected,

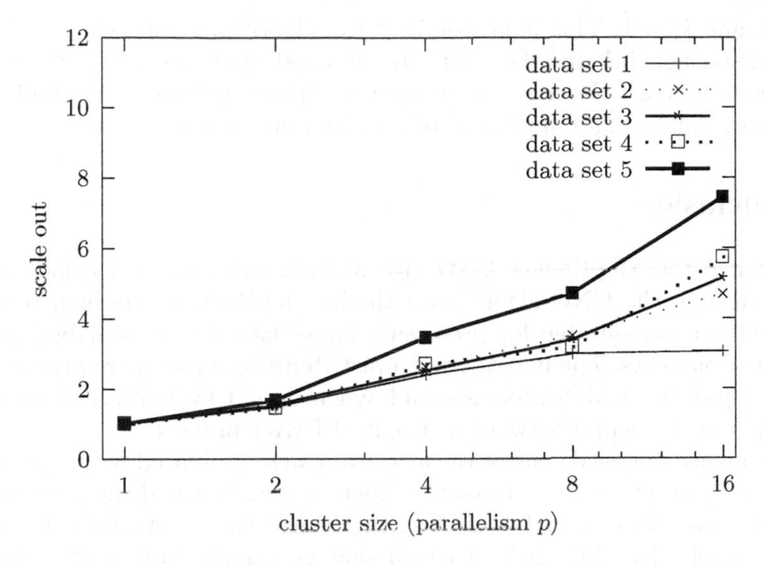

Fig. 2. Scale out values for data sets 1 to 5 depending on the level of parallelism. The scale out was determined in comparison to cluster 1 (1 node).

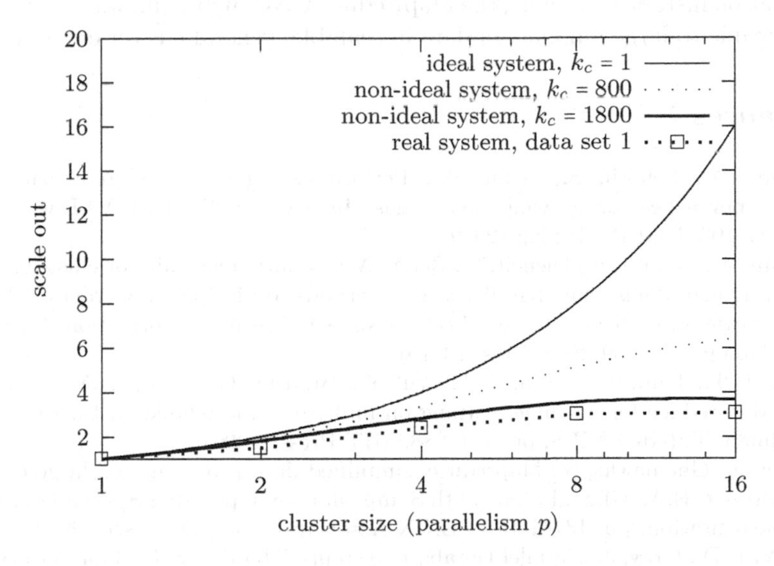

Fig. 3. Scale out curves for different factors k_c. The figure shows the scale out of the algorithm in cases of different network speeds indicated by k_c. $k_c = 1$ denotes an ideal system with no communication delay.

that the scale out of all data sets will converge to a constant value at a certain level of parallelism.

The reduction of the slope is also increased by k_c in (8). Figure 3 shows the scale out curve of data set 1 compared to three curves of (8) in case of $k_c = 1800$,

$k_c = 800$, and $k_c = 1$. This indicates, that the algorithms scalability is directly dependent on network speeds due to its intended application in a distributed cluster environment. Nevertheless, as network speeds increase, the MapReduce GNG algorithm becomes applicable to a wider range of data set sizes.

6 Conclusion

We presented the MapReduce GNG algorithm in this paper. It combined the batch variant of the GNG algorithm with the MapReduce paradigm resulting in a GNG variant suitable for processing large data sets in scalable, general cluster environments. The partitions of a distributed data set were processed by a batch approach in Map processes and synchronized by Reduce processes to finally update the neural network in a central Driver process.

Performance and scale out of the algorithm were evaluated with a synthetic data set of different sizes on clusters of different size. It was demonstrated that the performance gain through increasing the cluster size is potentially higher for larger data sets. In addition, it emerged that communication effort and other side effects are of significance in a real cluster system. These effects introduced reasonable damping effects to the scale out upon enlarging the number of parallel computation instances. Finally, the MapReduce GNG algorithm was evaluated to be applicable to large amounts of data in scalable, general cluster environments.

References

1. Adam, A., Leuoth, S., Benn, W.: Performance gain of different parallelization approaches for growing neural gas. In: Perner, P. (ed.) MLDM Posters, pp. 94–103. IBaI Publishing (2009)
2. Adam, A., Leuoth, S., Dienelt, S., Benn, W.: Performance gain for clustering with growing neural gas using parallelization methods. In: Filipe, J., Cordeiro, J. (eds.) Enterprise Information Systems. Lecture Notes in Business Information Processing, vol. 73, pp. 264–269. SciTePress (2010)
3. Chu, C.T., Kim, S.K., Lin, Y.A., Yu, Y., Bradski, G.R., Ng, A.Y., Olukotun, K.: Map-reduce for machine learning on multicore. In: Schölkopf, B., Platt, J.C., Hoffman, T. (eds.) NIPS, pp. 281–288. MIT Press (2006)
4. Dean, J., Ghemawat, S.: Mapreduce: simplified data processing on large clusters. In: Brewer, E.A., Chen, P. (eds.) 6th Symposium on Operating System Design and Implementation, pp. 137–150. USENIX Association, San Francisco (2004)
5. DeWitt, D., Gray, J.: Parallel Database Systems: The Future of High Performance Database Systems. Commun. ACM **35**(6), 85–98 (1992)
6. Fritzke, B.: A growing neural gas network learns topologies. In: Advances in Neural Information Processing Systems, vol. 7, pp. 625–632. MIT Press (1995)
7. Gillick, D., Gillick, D., Faria, A., Denero, J.: MapReduce: Distributed Computing for Machine Learning. Commun. ACM **51**, 107–113 (2006)
8. Görlitz, O.: Inhaltsorientierte Indexierung auf Basis künstlicher neuronaler Netze. Ph.D. thesis, Chemnitz University of Technology (2005)
9. Kearns, M.: Efficient noise-tolerant learning from statistical queries. J. ACM **45**(6), 983–1006 (1998)

10. Liu, S., Flach, P., Cristianini, N.: Generic Multiplicative Methods for Implementing Machine Learning Algorithms on MapReduce. CoRR (2011)
11. Martinetz, T.M.: Competitive Hebbian learning rule forms perfectly topology preserving maps. In: International Conference on Artificial Neural Networks, ICANN 1993, pp. 427–434. Springer, Amsterdam (1993)
12. Parigi, G., Stramieri, A., Pau, D., Piastra, M.: A multi-signal variant for the GPU-based parallelization of growing self-organizing networks. In: Ferrier, J.L., Bernard, A., Gusikhin, O., Madani, K. (eds.) Informatics in Control, Automation and Robotics. LNEE, vol. 283, pp. 83–100. Springer International Publishing, Switzerland (2014)
13. Vojáček, L., Dvorský, J.: Growing neural gas – a parallel approach. In: Saeed, K., Chaki, R., Cortesi, A., Wierzchoń, S. (eds.) CISIM 2013. LNCS, vol. 8104, pp. 408–419. Springer, Heidelberg (2013)
14. White, T.: Hadoop: The Definitive Guide, 4th edn. O'Reilly, Beijing (2015)
15. Zaharia, M., Chowdhury, M., Franklin, M.J., Shenker, S., Stoica, I.: Spark: cluster computing with working sets. In: 2nd USENIX Conference on Hot Topics in Cloud Computing. USENIX Association, Berkeley (2010)

Comprehensible Enzyme Function Classification Using Reactive Motifs with Negative Patterns

Thanapat Kangkachit(✉) and Kitsana Waiyamai

Department of Computer Engineering, Kasetsart University, Bangkok, Thailand
{g5185041,fengknw}@ku.ac.th

Abstract. The main objective of this research work is to build a comprehensible enzyme function classification by using high-coverage and high-precision reactive motifs. Reactive motifs are directly extracted from the protein active and binding sites. Main advantage of the reactive motifs is their high-coverage, however their generalization make them over-generalized with low-precision quality. In this paper, a method for generating reactive motifs with negative patterns is proposed. Reactive motifs with negative patterns are able to control the level of motif generalization. As result, non over-generalized reactive motifs with high-precision are generated. Each of the reactive motifs is associated with a specific enzyme function. They can directly predict enzyme function of an unknown protein sequence. Without use of a complex classification model, set of voting methods are proposed and used to construct a comprehensible enzyme function classification. Essential clues of the enzyme mechanism are provided to the biologist end-users.

Keywords: Comprehensible enzyme function classification · Reactive motifs with negative patterns · Generalized reactive motifs

1 Introduction

Enzymes are known as important catalysts in living organs. They serve a variety of biological processes e.g. digestive process inside human body. Thus, knowing and understanding enzymes function is able to facilitate the development of new drugs.This is a reason why classification of enzyme function is one of the important bioinformatics tasks.

To automatically classify enzyme function, several methods have been proposed [1]. Most of them obtained high classification accuracy but lack of comprehensibility. Among those methods, motif-based enzyme function classification are able to provide both accuracy and comprehensibility to biologists end-users. An enzyme function is predicted in terms of motifs or their combination which give essential clues to the biologists. Different types of motifs have been proposed [2–6], high-coverage and high-precision motifs have been proved [7,8] to give the best performance in terms of accuracy and comprehensibility. High-coverage motifs have very high possibility to appear in the unseen sequences. Meanwhile, high-precision motifs are able to generate high classification accuracy.

P. Perner (Ed.): MLDM 2016, LNAI 9729, pp. 560–568, 2016.
DOI: 10.1007/978-3-319-41920-6_44

Expert-based motifs such as PROSITE [9] have been widely used. Since, they offer high-precision and often correspond to the functional regions of an enzyme (e.g. active sites, ligand and DNA binding sites). Those functional regions are essential clues for enzymes to function [5,10]. However, PROSITE has one main drawback i.e. their low-coverage in enzyme sequences database due to the manual process to discover them by the experts.

Reactive motifs have been proposed in [7,8]. They are short conserved regions discovered from active and biding sites information of enzyme sequences which are directly involved in the chemical reaction mechanism. Thus, reactive motifs provide biological evidences, i.e., potentially contain sites information. However their generalization make them over-generalized with low-precision quality. Due to their low precision, directly use reactive motifs to classify enzyme function may result in unsatisfied classification accuracy. Thus, complex machine learning algorithms (e.g. SVM [11], AdaBoost [12]) are needed.

To increase precision of reactive motifs, this paper proposes a method to discover the negative-patterns of reactive motifs. The main idea is to prevent the appearance of reactive motifs in misclassified enzyme sequences. Negative patterns of the reactive motifs are then discovered by extracting their superset-patterns in each of the functional group of the misclassified enzyme sequences. Without use of a complex classification model, set of voting methods are proposed and used to construct a comprehensible enzyme function classification. Experimental results show that the negative patterns yield higher precision than reactive motifs but also retain satisfied coverage. To summarize, the reactive motifs with negative patterns are capable to provide satisfied classification accuracy, high-coverage and comprehensibility to the biologists end-users.

2 Preliminaries and Related Works

In this section, reactive motifs discovery process as described in [8] is briefly explained.

2.1 Extraction of the Initial Reactive Motifs

Since, enzymes perform their functions in relatively small regions i.e. active and binding sites. Directly extracting motifs from these regions may mis-classify the function of unknown enzyme sequences. Thus, an environment around these regions is considered. Designating the active or binding site position as the center, sub-sequences of length 15 amino acids are retrieved following the recommended substrate size of 7 amino acids on each position side [13]. These sub-sequences form set of initial reactive motifs. Figure 1(a) shows an *initial motifs* `EMGGCDSCHVSDKGG` with active site `C` as center position.

2.2 Generalization of the Initial Reactive Motifs

Due to the lack of sites information, the coverage of *initial motifs* is generally low in protein sequences databases. To increase coverage, motif generalization is performed. For each initial reactive motif, the dataset is scanned to retrieve all the similar sub-sequences having the same function. Sub-sequence similarity score is computed using BLOSUM62 table [14]. Result of this step is a block of ranked motifs. In Figure 1(a), the sub-sequences DMGSCQSCHAKPIKV, ..., to HVLSSNTCTNVGTEG constitute the block associated to the initial reactive motif EMGGCDSCHVSDKGG.

Following the recommendation of [15], a block is considered as high quality when all sub-sequences in the block have at least 3 positions presenting the same type of amino acids. First sub-sequence which does not preserve at least the 3 positions is a filter-out criteria. Sub-sequences having lower similarity score than the filter-out criteria are then filtered out from the block. Figure 1(a) the block contains high quality candidate motifs {DMGSCQSCHAKPIKV, HVLSCNTCHNVGTGG, GFISCNSCHNLSMGG} in which the sub-sequence HFGSTNSCHNKSPGV and other low-similarity sub-sequences are filtered out.

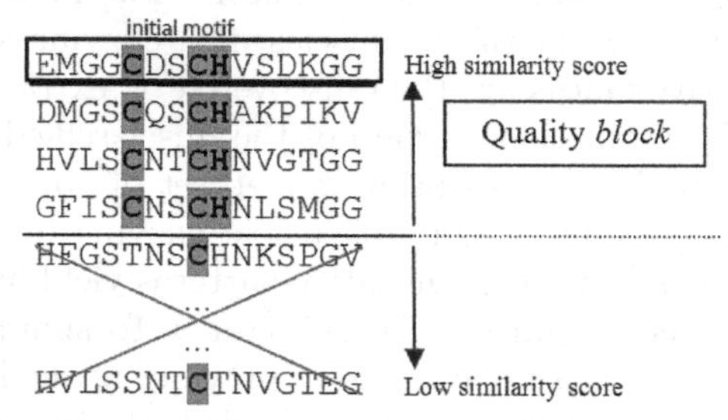

(a) High quality block extraction with an initial motif EMGGCDSCHVSDKGG

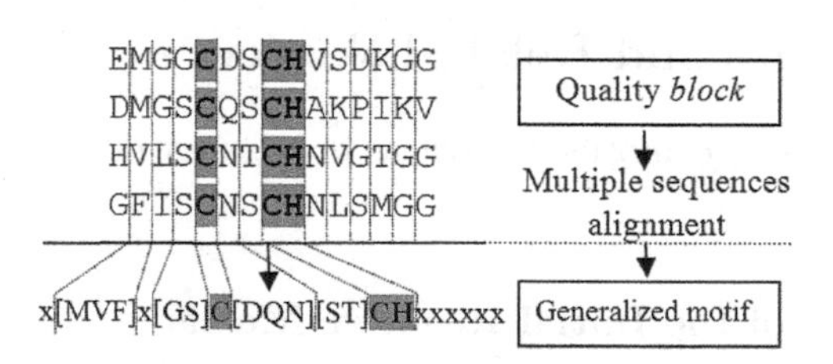

(b) *Generalized motif* obtained from multiple sequences alignment operation on a quality *block*

Fig. 1. Generalization of initial motifs

To discard the positions that may not be involved in the protein functional mechanism, all the candidate motifs are pruned with a multiple sequence alignment [16]. The output of this step is a representative block pattern containing *generalized motifs* in the form of regular expressions. Figure 1(b) shows a reactive motif x[FMV]x[GS]C[DQN][ST]CHxxxxx, where x be any amino acid, and [] be a set of possible amino acids that can be substituted at a given position, *called substitution group.*

Substitution groups in reactive motifs may not be complete groups since they may not cover the diversity of unseen patterns in unknown protein sequences. Thus, a complete substitution groups operation is needed to obtain more general reactive motifs . Various background knowledge are used and integrated into a unique fuzzy-concept lattice structure. This structure facilitates the process of determining all possible amino acids which share the similar properties i.e. complete substitution group. Finally, an initial reactive motif `x[FMV]x[GS]C[DQN][ST]CHxxxxx` is generalized to `x[FMV]x[GS]C[DEKNQR][ST]CHxxxxx`. Explanation and details of complete substitution group determination process can be found in [8].

3 Enzyme Classification Using Reactive Motifs with Negative Patterns

3.1 Negative Pattern

Output of the generalization process is the set of reactive motifs with high-coverage, however they have very low precision accuracy due to their over-generalization. To increase precision of reactive motifs, their level of generalization are constrained by their counterpart negative patterns. The definition of a negative pattern is defined as follow.

Given a reactive motif rm having function EC, a negative pattern Neg of rm is its superset pattern having a specific function that is not EC. All the negative patterns associated with rm are able to detect its superset patterns that are misclassified. By using them as constraints to rm, its precision is increased since misclassified sequences can be discarded.

Fig. 2 illustrates an example of a reactive motif and its associated negative patterns. Given a reactive motif rm, $x[FMV]x[GS]C$ $[DEKNQR][ST]$ $CHxxxxx$ having function EC 1.1.1.1. rm appears not only in enzyme sequences having function EC 1.1.1.1 but its superset patterns also appear in enzyme sequences having specific functions EC 1.1.2.1 and EC 2.1.3.1. Thus, the set of negative patterns associated with rm are $x[FV][DN][GS]C[DEK][ST]$ $CHxCxxx$ and $x[FM]x[GS]C[DNQR][ST]CHxxxxx$. The two negative patterns are used as constraints to make rm more precise.

3.2 Reactive Motifs with Negative Patterns Discovery

A direct extraction of all the negative patterns associated with each of the reactive motifs is a time-consuming process. This is due to the fact that each reactive motif can appear in many groups of enzyme functions. To reduce the number of negative patterns to discover, reactive motifs with higher precision but appear in the misclassified sequences are considered.

The detail of a process to extract negative patterns is shown in Table 1. In line 1-2, we first calculate the precision of every single reactive motifs Rm with EC function. For each Rm, (in line 5-8) we search for the different-function

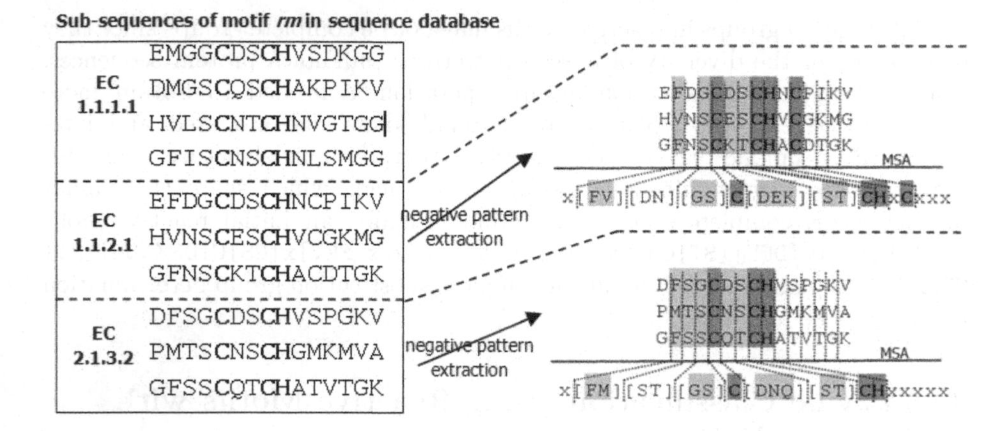

Fig. 2. An example of motif *rm* with EC 1.1.1.1 and its negative patterns

motifs *Ro* obtained higher precision than *Rm* but appear in the *EC* enzyme sequences. These *Ro* motifs will lead to misclassification. In line 9, a set of *Ro* is then ranked according to their precision in descending order. In line 10, a set of enzyme sequences having *EC* function, D_{ro}, is prepared. In line 11-17, negative patterns extracted from each *Ro* that increase the precision of *Rm* are discovered. In line 12, a set of sub-sequences S_{ro} that matches *ro* in D_{ro} is extracted. In line 13, multiple alignment is applied to S_{ro} to generate negative pattern Neg_{ro}. In line 14-17, Neg_{ro} is added to *rm* if precision P_{rmN} of *rm* with Neg_{ro} is higher than P_{rm}. Finally, *Rm* with negative patterns that gain more precision is obtained.

3.3 Enzyme Function Classifier

After the process of extracting reactive motifs with negative patterns is ended, each enzyme sequences is then represented by those motifs and their weights (i.e. their precision). To build an effective classifier while retains comprehensibility, simple voting methods are used instead of complex machine learning algorithms. The reasons are that a reactive motif itself is able to identify function (i.e. it comes with EC function) and the precision of a motif is quite high (by adding negative patterns). Here, 4 different types of voting methods are used. The details of voting methods are as follows.

Majority votes (Mv): By simply count the EC function of motifs in such enzyme sequence, the predicted result is the most frequent EC function.

Weighted Majority votes (WMv): Instead of counting the most-frequent EC function in such enzyme sequence, the predicted result is the EC function with highest sum of weight (i.e. precision of each motif).

Average-Weighted Majority votes (AWMv): Unlike the weighted majority votes, the average weight of each EC is calculated by dividing sum

Table 1. Algorithm for discovering reactive motifs with negative patterns

Function NegativeReactiveMotifs(D, RM, Ln)	
Parameter D : training enzyme sequences with their EC class RM : set of reactive motifs with their EC class	
Output $NegRM$: set of reactive motifs with negative patterns	
1	find the occurrence of RM in D
2	calculate **precision** P_{rm} of each reactive motif rm in RM
3	for each reactive motif rm in RM
4	/* find other motifs obtained higher precision but having different EC*/
5	for each enzyme sequence D_i in D which contained rm
6	for each reactive motif ro occurs in D_i
7	if ($P_{rm} < P_{ro}$)
8	$++Count[ro]$
9	sort $Count$ in descending order
10	find a set of enzyme sequences D_{rm} having the same EC as rm
11	for each ro in $Count$
12	extract a set of sub-sequences of length 15-amino acids S_{ro} that match ro in D_{rm}
13	generate negative reactive motif Neg_{ro} by applying multiple alignment to S_{ro}
14	calculate precision P_{rmN} of rm regarding to negative reactive motif Neg_{ro}
15	if ($P_{rmN} > P_{rm}$) {
16	add Neg_{ro} into negative patterns of rm
17	set $P_{rm} = P_{rmN}$ }
18	return RM and their precision

of weight by the number of motifs. The predicted result is the EC class with highest average weight.

Max-Precision votes (MPv): The predicted result is the EC function of the motif having the highest precision in such enzyme sequence.

4 Experimental Results

4.1 Experimental Setup

As in [8], only the UNiProtKB/Swiss-Prot release 2011-10 [17] is used as a benchmark dataset. This is the recent release protein sequence databases, with sites information significantly increased and spanned into large number of EC functions. Only the 4th-level EC functions having active and binding sites annotated by the experts are selected. Finally, a dataset composed of 407 EC functions and 67,915 sequences is obtained and used as a benchmark.

4.2 Reactive Motifs with Negative Patterns vs. Reactive Motifs

In this section, we compare the performance of the reactive motifs with using negative patterns and without using the negative patterns as constraints. The performance is given in terms of precision and recall.

Table 2 shows performance of the reactive motifs using various voting methods. Unsurprisingly, for all the voting methods, the most generalized reactive motifs (i.e. with lowest FCL- = 0.8) give the smallest performance. In contrast, the most specific reactive motifs (i.e. without use of background knowledge (Baseline)) are able to generate the best performance. We notice that, with the max-precision voting method (Mpv), it is able to generate 0.946 precision which is very close to 1.

Table 3 shows performance of the reactive motifs with negative patterns using various voting methods. For all the voting methods, the reactive motifs with negative patterns yield better performance than those motifs without constraints of their negative patterns. With the Max-precision voting method (MPv), the best performance is obtained. This is due to the ability to control the generalization process of the reactive motifs via their negative patterns.

With the use of machine learning algorithms, table 4 shows performance of the reactive motifs with negative patterns against reactive motifs. We notice that the use of machine learning algorithms with reactive motifs is able to generate comparable precision but higher recall, compared to the methods that use reactive motifs with negative patterns. However, more than the high performance obtained, the reactive motifs with negative patterns are capable of providing evidences (i.e. explanation ability) while retaining high classification accuracy.

Table 2. Performance of **reactive motifs** using various voting methods in terms of precision and recall

Background Knowledge	δ	Mv		WMv		AWMv		MPv	
		Precision	Recall	Precision	Recall	Precision	Recall	Precision	Recall
Integrated knowledge	1.00	0.421	0.213	0.575	0.479	0.499	0.495	0.702	0.603
	0.95	0.409	0.204	0.567	0.461	0.485	0.482	0.689	0.591
	0.90	0.375	0.197	0.550	0.442	0.469	0.459	0.671	0.572
	0.85	0.359	0.189	0.538	0.428	0.455	0.444	0.663	0.562
	0.80	0.295	0.168	0.497	0.389	0.442	0.402	0.617	0.529
Without use of knowledge	-	**0.634**	**0.368**	**0.791**	**0.680**	**0.788**	**0.724**	**0.946**	**0.745**

Table 3. Performance of **reactive motifs with negative patterns** using various voting methods in terms of precision and recall

Background Knowledge	δ	Mv		WMv		AWMv		MPv	
		Precision	Recall	Precision	Recall	Precision	Recall	Precision	Recall
Integrated knowledge	1.00	0.421	0.213	0.575	0.479	0.499	0.495	0.702	0.603
	0.95	0.434	0.214	0.602	0.470	0.499	0.501	0.709	0.626
	0.90	0.390	0.206	0.583	0.459	0.492	0.468	0.684	0.583
	0.85	0.367	0.201	0.553	0.462	0.473	0.452	0.696	0.601
	0.80	0.303	0.182	0.527	0.408	0.451	0.433	0.666	0.544
Without use of knowledge	-	**0.641**	**0.373**	**0.781**	**0.711**	**0.793**	**0.741**	**0.947**	**0.784**

Table 4. Performance of reactive motifs with negative patterns vs reactive motifs on a benchmark dataset in terms of precision and recall

Background Knowledge	δ	MPv[1]		SVM[2]		Ensemble[2,3]	
		Precision	Recall	Precision	Recall	Precision	Recall
Integrated knowledge	1.00	0.702	0.603	0.917	0.913	**0.893**	**0.893**
	0.80	0.666	0.544	**0.952**	**0.947**	0.868	0.869
Without use of knowledge	-	**0.947**	**0.784**	0.917	0.913	0.893	0.893

[1] Reactive motifs with negative patterns
[2] Reactive motifs
[3] Using AdaBoost(M1) with 10 iterations of C4.5

5 Conclusions

Reactive motifs provide essential clues to the biologists since they are extracted from the functional regions of enzyme sequences. However, the generalization of the reactive motifs make them cover most of the enzyme sequences. Thus, this process generates high-coverage reactive motifs, but very low-precision. To control the reactive motifs generalization process, reactive motifs with negative patterns are introduced. Negative patterns of the reactive motifs are then discovered by extracting their superset-patterns in each of the functional group of the misclassified enzyme sequences. The main idea is to prevent the appearance of reactive motifs in misclassified enzyme sequences. As result, reactive motifs with higher precision are obtained. Thus, interpretable classification models such as voting methods can be used to classify unseen sequence. Experimental results show, among these methods, the max-precision voting method offers the highest classification accuracy.

Acknowledgements. The authors gratefully acknowledge the Thailand Research Fund (TRF) and Kasetsart University for financial support through the Royal Golden Jubilee Ph.D. scholarship program (1.0.KU/49/A.1).

References

1. Freitas, A.A., Wieser, D.C., Apweiler, R.: On the importance of comprehensible classification models for protein function prediction. IEEE/ACM Trans. Comput. Biol. Bioinformatics **7**(1), 172–182 (2010)
2. Kunik, V., Solan, Z., Edelman, S., Ruppin, E., Horn, D.: Motif extraction and protein classification. In: CSB, pp. 80–85. IEEE Computer Society (2005)
3. Sander, C., Schneider, R.: Database of homology-derived protein structures and the structural meaning of sequence alignment. Proteins: Structure, Function, and Genetics **9**(1), 56–68 (1991)
4. Eidhammer, I., Jonassen, I., Taylor, W.R.: Structure comparison and structure patterns. Journal of Computational Biology **7**, 685–716 (1999)

5. Huang, J.Y., Brutlag, D.L.: The emotif database. Nucleic Acids Research **29**(1), 202–204 (2001)
6. Bennett, S.P., Lu, L., Brutlag, D.L.: 3matrix and 3motif: a protein structure visualization system for conserved sequence motifs. Nucleic Acids Research **31**(13), 3328–3332 (2003)
7. Waiyamai, K., Liewlom, P., Kangkachit, T., Rakthanmanon, T.: Concept lattice–based mutation control for reactive motifs discovery. In: Washio, T., Suzuki, E., Ting, K.M., Inokuchi, A. (eds.) PAKDD 2008. LNCS (LNAI), vol. 5012, pp. 767–776. Springer, Heidelberg (2008)
8. Kangkachit, T., Waiyamai, K., Lenca, P.: Enzyme classification using reactive motifs. I. J. Functional Informatics and Personalised Medicine **4**(3/4), 243–258 (2014)
9. Bairoch, A.: The prosite dictionary of sites and patterns in proteins, its current status. Nucleic Acids Research **21**(13), 3097–3103 (1993)
10. Bork, P., Koonin, E.: Protein sequence motifs. Curr. Opin. Struct. Biol. **6**(3), 366–376 (1996)
11. Boser, B.E., Guyon, I., Vapnik, V.: A training algorithm for optimal margin classifiers. In: COLT, pp. 144–152 (1992)
12. Freund, Y., Schapire, R.E.: A decision-theoretic generalization of on-line learning and an application to boosting. Journal of Computer and System Sciences **55**(1), 119–139 (1997)
13. Schomburg, I., Chang, A., Ebeling, C., Gremse, M., Heldt, C., Huhn, G., Schomburg, D.: BRENDA, the enzyme database: updates and major new developments. Nucleic Acids Research **32**(Database issue), D431–D433 (2004)
14. Henikoff, S., Henikoff, J.G.: Amino acid substitution matrices from protein blocks. Proceedings of the National Academy of Sciences **89**(22), 10915–10919 (1992)
15. Smith, H.O., Annau, T.M., Chandrasegaran, S.: Finding sequence motifs in groups of functionally related proteins. Proceedings of the National Academy of Sciences **87**(2), 826–830 (1990)
16. Chenna, R., Sugawara, H., Koike, T., Lopez, R., Gibson, T.J., Higgins, D.G., Thompson, J.D.: Multiple sequence alignment with the clustal series of programs. Nucleic Acids Res. **31**(13), 3497–3500 (2003)
17. Apweiler, R., Bairoch, A., Wu, C.H., Barker, W.C., Boeckmann, B., Ferro, S., Gasteiger, E., Huang, H., Lopez, R., Magrane, M., Martin, M.J., Natale, D.A., O'Donovan, C., Redaschi, N., Yeh, L.S.L.: UniProt: the Universal Protein knowledgebase. Nucleic Acids Research **32**(Database–Issue), 115–119 (2004)

Enumerating Maximal Isolated Cliques Based on Vertex-Dependent Connection Lower Bound

Yoshiaki Okubo[1], Makoto Haraguchi[1(✉)], and Etsuji Tomita[2]

[1] Graduate School of Information Scienece and Technology, Hokkaido University, N-14 W-9, Sapporo 060-0814, Japan
mh@ist.hokudai.ac.jp

[2] The Advanced Algorithms Research Laboratory, The University of Electro-Communications, Chofugaoka 1-5-1, Chofu, Tokyo 182-8585, Japan

Abstract. In this paper, we are concerned with a problem of enumerating *maximal isolated cliques* (*MIC*) in a given undirected graph. Each target to be extracted is defined as a maximal subset X of vertices which is a clique and satisfies an isolatedness. Our isolation concept is based on the notion of *variable* j*-coreness* which imposes a connection lower bound depending on each vertex in X. Based on a standard clique enumerator, we design a depth-first algorithm for the problem. In our algorithm, we can prune many hopeless cliques from which we can never obtain our solution. It is noted that our solution MICs can be divided into two classes, *fixpoint solutions* and *non-fixpoint solutions*, where the latter ones are not trivial to detect. Based on a degree descending order of vertices, we show a theoretical property of non-fixpoint MICs and present an effective method of detecting them. Our experimental results for real world (benchmark) networks show that the proposed algorithm can work very well even for a large network with over a million vertices.

Keywords: Isolated cliques · Maximal cliques · Variable j-core · Vertex-dependent connection lower bound

1 Introduction

In Social Network Analysis, detecting communities in a given network or graph has been an important central task [12]. As a standard approach to community detection, methods of *graph clustering* or *graph partitioning* are widely used and well known to be useful (e.g. [7,8]). Since they usually suppose small numbers of clusters, they would be quite helpful when we prefer to detect relatively larger communities. On the other hand, it is easily imagined that there also exist small or moderate-sized communities which are possibly valuable for us. However, those communities are invisibly merged and absorbed into larger clusters.

The notion of *clique* is a typical vertex set understood as such a potential community [10]. With the help of anti-monotonicity property of cliques, many

P. Perner (Ed.): MLDM 2016, LNAI 9729, pp. 569–583, 2016.
DOI: 10.1007/978-3-319-41920-6_45

algorithms for efficiently enumerating (maximal) cliques have been proposed (e.g. [15]). However, the clique model is too restrictive in the sense that it is rare for actual communities to appear as cliques in a real world network. It is, therefore, reasonable to consider *clique relaxation models* as practical communities.

Several classes of pseudo-cliques have already been proposed for community detection in social network analysis [10], including e.g., k-cliques, k-clans and k-plexes [14]. Although they have been found to be useful in, e.g., a link-prediction problem [6] and a detection of structural change of graphs [9], there in general exist a large number of pseudo-cliques. Particularly, many of them often overlap and hence it is quite difficult to recognize boundaries of communities. This motivates us to investigate a model of *isolated pseudo-cliques* each of which consists of densely-connected vertices (cohesiveness) and has a small number of outgoing edges (isolatedness) [16]. As a step toward further investigating isolated pseudo-cliques, we formalize in this paper a class of *τ-isolated cliques* and discuss its enumeration problem.

A notion of *c-isolated cliques* has already been proposed in [3]. In the framework, a target vertex set is defined as a maximal clique which is c-isolated. In other words, we unfortunately miss every non-maximal clique even if it is actually isolated. Moreover, c-isolatedness requires the total number of outgoing edges from the community to be less than a threshold determined by the parameter c. As the result, we often obtain a community in which just a few vertices have relatively many outgoing edges.

The notion of *maximal τ-isolated cliques* we propose in this paper can be regarded as a revised model resolving the above shortcomings. A maximal τ-isolated clique is a maximal set of vertices which is a clique and satisfies τ-isolatedness. We can, therefore, obtain many isolated cliques unfortunately missed in the framework of [3]. Furthermore, our τ-isolation concept is based on the notion of *variable j-coreness* which imposes a connection lower bound depending on an individual member vertex in the community. With an adequate function for the lower bound, we can control the degree of isolation depending on each member vertex.

It is noted here that generalizing the framework in [3], a notion of maximal c-isolated cliques has already been investigated in [4], where as is similar to ours, a maximal c-isolated clique is defined as a maximal one among all c-isolated cliques. Although the class certainly covers c-isolated maximal cliques [3], the isolation concept is still based on c-isolatedness. Therefore, we can never control the isolation degree depending on each member vertex.

We design an algorithm for enumerating maximal τ-isolated cliques in a given graph G. As will be discussed later, a maximal τ-isolated clique can be identified as j_τ-core of some maximal clique in G. As a naive approach to our enumeration task, we can extract every maximal clique C and then compute j_τ-core of C. Such a naive method seems to work well because several efficient maximal clique enumerators such as `CLIQUES` in [15] have already been proposed and the computation of j_τ-core can be performed with the order $O(|C| \cdot D)$, where D is the maximum degree in G. In general, however, several maximal cliques can results in the same τ-isolated clique. That is, we have to cope with the problem of excluding duplicated solutions which is not trivial.

Instead, we present an algorithm based on a standard depth-first clique search in which we try to expand a clique by adding a candidate vertex step-by-step. We show some theoretical properties of maximal τ-isolated cliques. Based on a necessary condition to detect our solutions by expanding a clique, we can prune many hopeless cliques which can never result in any solution during our search process. It is also shown that we can divide our solutions into two classes, *fixpoint solutions* and *solutions with false candidates*. Although we are forced to pay some additional cost to detect the latter solutions, we show a useful property to reduce the cost and present an efficient detection procedure. In our experimentation, we observe computational performance of our algorithm and the numbers of solutions for several real world benchmark networks. Our experimental results show that the proposed algorithm can work well even for a large scale network with over a million vertices.

2 Preliminaries

An undirected graph G is denoted by $G = (V, \Gamma)$, where $V = \{v_1, ..., v_{|V|}\}$ is a set of vertices and $\Gamma(v)$ the set of vertices adjacent to $v \in V$. $\Gamma(v)$ is assumed to not include v itself, that is, $v \notin \Gamma(v)$. For a vertex $v \in V$, $|\Gamma(v)|$ is called the degree of v and is referred to as $deg(v)$. For a set of vertices $X \subseteq V$ and a vertex $v \in X$, $\Gamma(v) \cap X$ is denoted by $\Gamma_X(v)$ and $|\Gamma_X(v)|$ is referred to as $deg_X(v)$.

For a set of vertices $X \subseteq V$, the subgraph of G induced by X is denoted by $G[X]$. Particularly, if any pair of vertices are adjacent, the $G[X]$ is called a *clique* in G. A clique $G[X]$ is often referred to as simply X.

For a clique $X \subseteq V$ and a vertex $v \in (V \setminus X)$, if $X \cup \{v\}$ forms a clique, then v is called a *candidate* for X and the set of candidates for X is denoted by $Cand(X)$. It is noted that for any pair of cliques $X, Y \subseteq V$ such that $X \subseteq Y$, we observe $Cand(X) \supseteq Cand(Y)$.

3 Isolated Vertex Set Based on Variable j-Core

3.1 Constant j-Core

Let $G = (V, \Gamma)$ be a graph. Given a set of vertices $X \subseteq V$, if the degree of any vertex $v \in X$ in $G[X]$ is no less than j, that is, $deg_X(v) \geq j$ for any $v \in X$, then X is said to have *j-core property* [13].

From the definition, for any vertex sets X and Y with j-core property, $X \cup Y$ always has j-core property. This means that for any set of vertices $W \subset V$, there exists the *maximum* subset of W with the property. The maximum subset is called the *j-core* of $G[W]$ [1].

Intuitively speaking, a set of vertices with j-core property gives a cohesive subgraph of G. Identifying cohesive subgraphs is one of the important tasks in social network analysis.

3.2 Variable j-Core

In the definition of j-core property, the parameter j is supposed to be *independent of vertices.* In this paper, we try to extend such a constant j-core property to *variable j-core property.*

More precisely speaking, we consider a function $j : V \to \mathbb{R}_{\geq 0}$ which assigns each vertex a non-negative real value. Then, for a set of vertices $X \subseteq V$, if $deg_X(x) \geq j(x)$ for any $v \in X$, then we say X has a *variable j-core property.* As opposed to constant j-core property, variable j-core property imposes a lower bound of required degree in the subgraph *depending on each vertex.* Thus, we can consider cohesive subgraphs more flexibly.

As is similar to constant j-core property, it is easy to see that for any vertex sets X and Y with variable j-core property, $X \cup Y$ has also the property. For a set of vertices $W \subseteq V$, therefore, there exists the maximum subset of W which has variable j-core property. The maximum subset is called the *variable j-core* of $G[W]$ and referred to as $core_j(W)$.

We can observe a monotonicity for variable j-core operation.

Observation 1. (Monotonicity of Variable j-Core Operation). *Let $X, Y \subseteq V$ be a pair of vertex sets such that $X \subseteq Y$. Then, $core_j(X) \subseteq core_j(Y)$.* ■

Verification 1. It is easy to see $core_j(X) \subseteq X \subseteq Y$. Since for each $v \in core_j(X)$, $deg_X(v) \geq j(v)$ and $deg_Y(v) \geq deg_X(v)$, we have $v \in core_j(Y)$ for each $v \in core_j(X)$. ■

3.3 Isolated Vertex Set as Variable j-Core

In a set of vertices X with variable j-core property, each vertex $v \in X$ is required to have a certain number of adjacent vertices in X defined as $j(v)$. In other words, the requirement gives an upper bound of the number of adjacent vertices outside X, namely $deg(v) - j(v)$. Based on this simple fact, we can consider an *isolatedness* of vertex set.

More precisely speaking, we introduce a parameter τ with the range $0 < \tau \leq 1.0$, called an *isolation factor*. Then, for a vertex $x \in V$, we define a function j_τ as

$$j_\tau(x) = \tau \cdot deg(x).$$

In a vertex set X with variable j_τ-core property and any vertex $v \in X$, the number of vertices adjacent to v and outside X is at most $(1.0 - \tau) \cdot deg(v)$. If, therefore, τ is large enough, X can be considered almost isolated in the sense that it has few outgoing edge from X. Particularly, X is completely isolated with $\tau = 1.0$. Thus we can control isolatedness of vertex set with the parameter τ. Needless to say, we can also observe a monotonicity for variable j_τ-core operation.

Given a set of vertices $X \subseteq V$, the procedure for computing the variable j_τ-core of X, $core_{j_\tau}(X)$, is almost the same as that for constant j-cores [1]. Concretely speaking, we iteratively remove a vertex v from X such

```
procedure JCore(G, X, τ):
  begin
    Create vert with |X|-elements;
    head = 0; tail = |X| − 1;
    for each v ∈ X do
      if τ ≤ deg_X(v)/deg(v) then // v is a core candidiate
        vert[tail] = v; d(v) = deg_X(v); index(v) = tail;
        tail--;
      else // v must be removed
        vert[head] = v; d(v) = deg_X(v); index(v) = head;
        head++;
      endif
    endfor
    head = 0;
    while head ≤ tail do
      v = vert[head];
      for each w ∈ Γ_X(v) do
        if tail < index(w) then // w is a core candidate
          d(w)--;
          if τ < d(w)/deg(w) then // w must be removed
            tail++;
            if tail ≠ index(w) then
              vert[index(w)] = vert[tail]; vert[tail] = w;
            endif
          endif
        endif
      endfor
      head++;
    endwhile
    while head ≤ |X| − 1 do
      print vert[head]; head++;
    endwhile
  end
```

Fig. 1. Algorithm for Computing Variable j_τ-Core

that $deg_X(v) < j_\tau(v)$ until no vertex remains to be removed. A pseudo-code of the algorithm is shown in Figure 1. In the algorithm, vertices are stored in an array `vert`, where for a vertex v, the index of v in `vert` is referred to as $index(v)$.

The most dominant part of the algorithm is `while` loop beginning at the middle. In the loop, since for each vertex in X, all of its adjacent vertices are checked, the worst case time complexity is given as $O(|X| \cdot D)$, where D is the maximum degree in G.

4 Maximal τ-Isolated Clique Problem

Based on variable j_τ-cores, in this section, we introduce a notion of *maximal τ-isolated cliques* and define our problem of enumerating them.

Definition 1 (Maximal τ-Isolated Cliques). Let $G = (V, \Gamma)$ be a graph and τ an isolation factor with the range $0 < \tau \leq 1.0$. A set of vertices $X \subseteq V$ is called a *τ-isolated clique*, if X is a clique in G and has variable j_τ-core property, that is, for each $v \in X$, $|X| > \tau \cdot deg(v)$. Particularly, if there exists no τ-isolated clique Y such that $X \subset Y$, X is called a *maximal τ-isolated clique*. ∎

Definition 2 (Maximal τ-Isolated Clique Problem). Given a graph $G = (V, \Gamma)$ and an isolation factor τ, *Maximal τ-Isolated Clique Problem* is to enumerate every maximal τ-isolated clique in G. ■

A maximal clique in G is not always j_τ-cored. In general, a maximal τ-isolated clique is obtained as the variable j_τ-core of some maximal clique. As a naive strategy, therefore, we can enumerate maximal τ-isolated cliques by extracting every maximal clique X in G and then computing $core_{j_\tau}(X)$. Several efficient algorithms for enumerating maximal cliques have already been proposed (e.g., [15]) and $core_{j_\tau}(X)$ can be efficiently computed. One might, therefore, expect that even such a naive strategy would work well. However, there usually exist several maximal cliques each of which is a superset of the same maximal τ-isolated clique. This means that we must have some mechanism for extracting our targets *without any duplication*. Since it is unfortunately not trivial, we take in this paper another approach.

5 Enumerating Maximal τ-Isolated Cliques

In this section, we present an algorithm for enumerating maximal τ-isolated cliques. Our algorithm is based on a standard method for extracting cliques in depth-first manner. During our search, a clique is examined whether it is our target or not. Based on a theoretical property of variable j_τ-cores (discussed later), we can exclude many useless cliques from which we can never obtain our targets.

5.1 Depth-First Clique Search

Given a graph $G = (V, \Gamma)$, every clique in G can be completely explored along a *set enumeration tree* [11].

Let us here assume a total ordering $\prec$ on V and any set of vertices is sorted in the order, where the last element of a vertex set X is referred to as $tail(X)$. A clique $X \subseteq V$ can be expanded into a larger clique $Xv = X \cup \{v\}$ by adding a candidate vertex $v \in Cand(X)$. Since we can observe anti-monotonicity of cliques, that is, any subset of a clique is also a clique, we can completely detect every clique starting with the initial clique $X = \emptyset$ and $Cand(X) = V$ and iterating such a clique expansion process. It is noted here that we can find every clique without duplication by expanding a clique X with a candidate $v \in Cand(X)$ such that $tail(X) \prec v$.

This is a standard procedure for clique enumeration widely adopted in many search algorithms for cliques. In actual computation process, a depth-first search strategy is usually preferred from the view point of space complexity.

5.2 Hopeful Cliques

Before going into details of our algorithm, we present some important properties for variable j_τ-cores. In what follows, for a clique X and its candidate vertex set $Cand(X)$, $core_{j_\tau}(X \cup Cand(X))$ is denoted by $ext(X)$.

Observation 2. Let X be a clique and Y a τ-isolated clique such that $X \subseteq Y$. Then,

$$X \subseteq Y \subseteq ext(X) = core_{j_\tau}(X \cup Cand(X)).$$

■

Verification 2. Since Y is a clique such that $X \subseteq Y$, it is easy to see $X \subseteq Y \subseteq X \cup Cand(X)$. From the monotonicity of variable j_τ-core operation, we have $core_{j_\tau}(Y) \subseteq core_{j_\tau}(X \cup Cand(X))$. Note here that $Y = core_{j_\tau}(Y)$ because Y is j_τ-cored. Therefore, we have $X \subseteq Y \subseteq ext(X) = core_{j_\tau}(X \cup Cand(X))$. ■

The above property provides us a necessary condition to detect our solution, a maximal τ-isolated clique, by expanding a clique X. In case of $X \subseteq ext(X)$, we have a chance to get a solution by expanding X. On the other hand, in case of $X \not\subseteq ext(X)$, any superset of X can never be a solution. Therefore, we can immediately stop expanding X without missing any solution. We call a clique X in the former case a *hopeful clique*.

5.3 Maximal τ-Isolated Clique as Fixpoint

Observation 2 tells us that if for a (hopeful) clique X, there exists a τ-isolated clique Y such that $X \subsetneq Y$, X is a *proper* subset of $ext(X)$. This implies that if $X = ext(X) = core_{j_\tau}(X \cup Cand(X))$, that is, X is a *fixpoint* w.r.t. $ext(\cdot)$, then X is definitely a maximal τ-isolated clique because X is a j_τ-cored clique and there exists no τ-isolated clique properly subsuming X.

Observation 3. A clique X such that $X = ext(X)$ is a maximal τ-isolated clique, that is, X is a *fixpoint solution.* ■

From Observation 2, any maximal τ-isolated clique Y can be identified as a subset of $ext(X)$ for some hopeful clique X. It is, therefore, sufficient to expand X with only vertices in $ext(X) \setminus X$. In the following discussion, $K(X)$ stands for $ext(X) \setminus X$ and is called the *K-candidate set* for X.

It is easy to see the following.

Observation 4. A hopeful clique X such that $K(X) = \emptyset$ is a maximal τ-isolated clique. ■

Conversely, for a maximal τ-isolated clique X, we cannot necessarily observe $K(X) = \emptyset$. That is, in order to completely enumerate our solutions, we have to take a careful consideration for a solution X such that $K(X) \neq \emptyset$.

5.4 Maximal τ-Isolated Clique with False K-Candidates

Let X be a (hopeful) clique such that $K(X) \neq \emptyset$. Since $K(X) \neq \emptyset$, there seems to exist a maximal τ-isolated clique which subsumes X. Even in such a case, however, X could be a solution to be extracted. In other words, a maximal τ-isolated clique possibly has some K-candidates which are quite useless for obtaining solutions. We call those needless candidates *false K-candidates.*

Figure 2 illustrates a maximal τ-isolated clique with false K-candidates. For $\tau = 2/5$, the clique X consisting of 3 black vertices forms a τ-isolated clique because for each $x \in X$, $deg(x) = 5$ and $deg_X(x) = 2$, thus, $deg_X(x) \geq \tau$. It is noted that we cannot obtain any 2/5-isolated clique by expanding X with any of 3 white vertices in $K(X)$. We therefore know X is a maximal 2/5-isolated clique and the white vertices are false K-candidates.

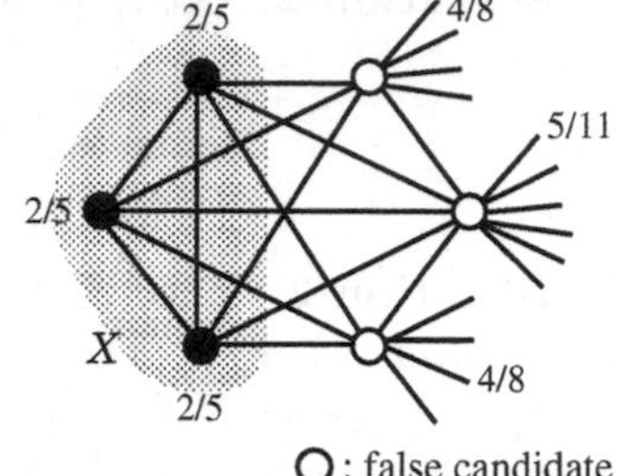

Fig. 2. Example of Maximal 2/5-Isolated Clique with False K-Candidates

In order to detect a maximal τ-isolated clique X with false K-candidates, it is sufficient to check if for every clique C in $G[K(X)]$, $X \cup C$ is not j_τ-cored. However, it is easy to imagine that the cost is not negligible.

Assuming a degree descending order on vertices, we present a theoretical property of solutions with false K-candidates. Based on the property, we can reduce the cost for identifying those solutions.

For the graph $G = (V, \Gamma)$, let $\prec$ be a degree descending ordering on V, that is,

$$\text{for any pair of vertices } u, v \in V, u \prec v \Leftrightarrow deg(u) \geq deg(v),$$

where for any vertices with the same degree, we can assume an arbitrary order on them. In the following discussion, for any set of vertices $X \subseteq V$, we suppose the vertices in X are ordered based on $\prec$.

As has already been stated, a hopeful clique X can be expanded with K-candidates in $K(X)$. Particularly, in order to examine cliques without duplication, we actually try to expand X with $x \in K(X)$ such that $tail(X) \prec x$. The set of those vertices is denoted by $NL(X)$, that is,

$$NL(X) = \{x \in K(X) \mid tail(X) \prec x\}.$$

For a maximal τ-isolated clique with false K-candidates, we can observe the following property.

Observation 5. For a maximal τ-isolated clique X such that $K(X) \neq \emptyset$, $NL(X) = \emptyset$. ■

Verification 3. Suppose the converse, that is, $NL(X) \neq \emptyset$. Since X is a j_τ-cored clique, $deg_X(x) = |X| - 1 \geq \tau \cdot deg(x)$ for each $x \in X$. From the vertex ordering $\prec$, we see $deg(x) \geq deg(v)$ for each $v \in NL(X)$. Moreover, since $NL(X) \subseteq Cand(X)$, each $v \in NL(X)$ is adjacent to all vertices in X. Then we have $deg_{Xv}(v) = |X| > |X| - 1 \geq \tau \cdot deg(v)$. This means that Xv is a j_τ-cored clique, contradicting X is maximal. We therefore have $NL(X) = \emptyset$. ■

From Observation 5, we have to check if a hopeful clique X is a solution with false K-candidates only when $K(X) \neq \emptyset$ and $NL(X) = \emptyset$. As a naive way, such a check can be done by verifying $X \cup C$ is not j_τ-cored for each (non-empty) clique C in $G[K(X)]$. For an efficient enumeration of our solutions, however, it would be desired to have another checking procedure with less computation cost.

5.5 Detecting Solutions with False K-Candidates

Let X be a hopeful clique such that $K(X) \neq \emptyset$ and $NL(X) = \emptyset$. We now try to check if there exists a (non-empty) clique C such that $C \subseteq K(X)$ and $X \cup C$ is j_τ-cored. Then, a theoretical property can be observed as follows.

Observation 6. If $X \cup C$ such that $C \subseteq K(X)$ is j_τ-cored, then for each vertex $x \in C$, the degree of x in $G[K(X)]$ is at least $\max\{0, \tau \cdot deg(x) - |X|\}$, that is,

$$deg_{K(X)}(x) \geq \max\{0, \tau \cdot deg(x) - |X|\}.$$ ■

Verification 4. If a set of vertices $X \cup C$ is j_τ-cored, then for each $x \in C$, the degree of x in $G[X \cup C]$, $deg_{X \cup C}(x)$, is at least $\tau \cdot deg(x)$. Moreover, since Xx is a clique for any $x \in C$, $deg_C(x) \geq \tau \cdot deg(x) - |X|$. From $C \subseteq K(X)$, hence, we have $deg_{K(X)}(x) \geq \tau \cdot deg(x) - |X|$. Note here that although $\tau \cdot deg(x) - |X|$ could be negative, any vertex degree is actually non-negative. Therefore, we have $deg_{K(X)}(x) \geq \max\{0, \tau \cdot deg(x) - |X|\}$ for any $x \in C$. ■

We now introduce a function $j_{K(X)} : K(X) \to \mathbb{R}_{\geq 0}$ such that for any $x \in K(X)$,

$$j_{K(X)}(x) = max\{0, \tau \cdot deg(x) - |X|\}. \tag{1}$$

For a vertex $x \in K(X)$, $j_{K(X)}(x)$ gives the minimum degree of x in $G[K(X)]$ required in order for $X \cup C$ to be j_τ-cored. Therefore, if there exists such a $X \cup C$, C must be a subset of some maximal clique Q_i in $G[K(X)]$. Actually, C appears as a subset of the variable $j_{K(X)}$-core of Q_i.

Recall that the main purpose here is to check whether X is a maximal τ-isolated clique with false K-candidates or not. Then, we need to examine only X that is j_τ-cored. For the $j_{K(X)}$-core of Q_i, $core_{j_{K(X)}}(Q_i)$, if we have $core_{j_{K(X)}}(Q_i) \neq \emptyset$, then $X \cup core_{j_{K(X)}}(Q_i)$ is j_τ-cored and therefore X is not a maximal τ-isolated clique (solution). On the other hand, for any maximal clique Q_i in $G[K(X)]$, if we have $core_{j_{K(X)}}(Q_i) = \emptyset$, then X is a maximal τ-isolated clique to be enumerated. As the result, it becomes clear that X is a solution with false K-candidates.

In the above computation process, although we need to enumerate maximal cliques in $G[K(X)]$, we anticipate that the enumeration cost would not be high. Because since $K(X) \subseteq Cand(X)$, each vertex $vinK(X)$ must be adjacent to every $x \in X$ and the number of such v is at most $min_{x \in X}\{deg(x)\} - |X| + 1$. Furthermore, since we probably have $K(X) \neq \emptyset$ and $NL(X) = \emptyset$ for a larger X, the size of $K(X)$, $|K(X)|$, would be small enough. An efficient enumerator for maximal cliques (e.g. [15]) can work well particularly for such a small graph $G[K(X)]$.

In practice, whenever we have $core_{j_{K(X)}}(C) \neq \emptyset$ even for some non-maximal clique C, it is certainly true X is not a solution. Incorporating such a practical judgment, we can further improve our procedure for detecting solutions with false K-candidates because we can find X is not a solution before we arrive at a maximal clique.

5.6 Pruning Useless K-Suffix Candidates

Let X be a hopeful clique. For efficient enumeration of our solutions, it is strongly desired to prune useless expansions of X which can never yield our solutions. Based on the idea discussed in [15], we can identify several useless candidates in $K(X)$.

Recall that X is tried to expand with $x \in K(X)$ such that $tail(X) \prec x$, that is, $x \in NL(X)$. For a vertex $v \in NL(X)$ and a suffix $S \subseteq NL(X)$ such that $v \prec s$ for any $s \in S$, let us assume $S \subseteq \Gamma(v)$, that is, v is adjacent to each $s \in S$. Then we observe the following property.

Observation 7. For each $Z \subseteq S$, whenever $X \cup Z$ is a τ-isolated clique, $X \cup Z$ is never a solution. ■

Verification 5. Since $X \cup Z$ is a τ-isolated clique, Z is a clique. From $Z \subseteq S \subseteq \Gamma(v)$, $Zv = Z \cup \{v\}$ is a clique. Moreover, $Xv = X \cup \{v\}$ is also a clique because of $v \in NL(X)$. Thus, $XZv = X \cup Z \cup \{v\}$ is a clique.

Since $X \cup Z$ is τ-isolated, for each $xinX$, we have $|X| + |Z| - 1 \geq \tau \cdot deg(x)$, hence $deg_{X \cup Z}(x) \geq \tau \cdot deg(x)$.

Since v is adjacent to every vertex in $X \cup Z$, it is easy to see $deg_{X \cup Z}(v) = |X| + |Z|$. On the other hand, for each $xinX$, $deg_{X \cup Z}(x) = |X| + |Z| - 1$. Then we have $deg_{X \cup Z}(v) > deg_{X \cup Z}(x)$ for each $xinX$.

From the ordering $\prec$, since $deg(x) \geq deg(v)$ for each $x \in X$, we have $deg_{X \cup Z}(v) > deg_{X \cup Z}(x) \geq \tau \cdot deg(x) \geq \tau \cdot deg(v)$, then, $deg_{X \cup Z}(v) > \tau \cdot deg(v)$. Thus XZv is a τ-isolated clique which subsumes $X \cup Z$. ■

The property tells us that we need not to expand X with any candidate z in the suffix S of $NL(X)$ ($\subseteq K(X)$) because expanding X with such a z yields only subsets Z of S and $X \cup Z$ can never be a solution. Therefore, we can safely prune expanding X with any suffix candidate in S of $K(X)$.

6 Algorithm for Enumerating Maximal Isolated Cliques

Summarizing the discussion above, we present in this section our algorithm for enumerating maximal isolated cliques. A pseudo-code of the algorithm is shown in 3. In the algorithm, the procedure MaxCliqueEnum($\cdot$) is to enumerate maximal cliques for a given graph.

It is noted here that the pseudo-code is presented as an overview of our whole enumeration process. We can customize several parts in our actual implementation. In the procedure Expand, for example, the computation of $NewExt$ can be replaced with $NewExt \leftarrow core_{j_\tau}(Xv \cup (K \cap NewCand))$. By first taking intersection of $NewCand$ and K, we can obtain a smaller $NewExt$ as the j_τ-core of $Xv \cup (K \cap NewCand)$. Since such a smaller $NewExt$ gives a less number of search branches from Xv, it provides us a certain improvement in efficiency of our enumeration process. As has been just mentioned above, moreover, the procedure SolWithFalseCand can be improved. Excluding useless suffix candidates in our expansion process is also effective in pruning search branches resulting in non-maximal isolated cliques.

```
procedure MAIN(G, τ):
  [Input] G = (V, Γ): an undirected graph.
          τ: an isolation parameter (0 < τ ≤ 1.0).
  [Output] All maximal j_τ-cored cliques.
  [Global Variables] G, τ
  begin
    Sort V in degree decsending order;
    EXPAND(∅, V, V);
  end
```

```
procedure EXPAND(X, Ext, Cand):
  begin
    K ← Ext \ X;
    NL ← {v ∈ K | tail(X) ≺ v};
    if K = ∅ then
      Output X; // as a fixpoint solution
      return;
    endif
    if NL = ∅ then
      if X is j(x)-cored then
        if SOLWITHFALSECAND(X, K) == true then
          Output X; // as a solution with false candidates
        endif
      endif
      return;
    endif
    for each v ∈ NL
      NewCand ← Cand ∩ Γ(v);
      NewExt ← core_{j_τ}(Xv ∪ NewCand);
      if Xv ⊆ NewExt then
        EXPAND(Xv, NewExt, NewCand);
      endif
    endfor
  end
```

```
procedure SOLWITHFALSECAND(X, K):
  begin
    Let j_K be a function defined as
          j_K(x) = max{0, τ · deg_G(x) − |X|} for each x ∈ K;
    for each Q presented by MAXCLIQUEENUM(G[K]) then
      if core_{j_K}(Q) ≠ ∅ then
        return false; // X is not a solution
      endif
    endfor
    return true; // X is a solution with false candidates
  end
```

Fig. 3. Algorithm for Enumerating Maximal τ-Isolated Cliques

7 Experiments

In this section, we present our experimental results. The proposed system, referred to as `MIC`, has been coded in C and executed on a PC with Intel® Core™-i7 (3.1GHz) processor and 16GB memory. For several real world networks, we observe computation times and the number of solutions.

As far as we know, since no algorithm for enumerating maximal τ-isolated cliques has so far been investigated, we have no system with which we can farely compare `MIC`. We, therefore, observe computation times for enumerating maximal cliques and the numbers of those cliques, because the naive method, extracting every maximal clique X and computing $core_{j_\tau}(X)$, is only one alternative

Table 1. Scale of Networks

Name	# of Nodes	# of Edges	Density	Max Deg.	Ave. Deg.
Gowalla	196,591	950,327	0.000049	14,730	9.67
Amazon	334,863	925,872	0.000017	549	5.53
DBLP	317,080	1,049,866	0.000021	343	6.62
Stanford	281,903	1,992,636	0.000050	38,625	14.14
Google	875,713	4,322,051	0.000011	6,332	9.87
Youtube	1,134,890	2,987,624	0.000005	28,754	5.27

and behavior of the clique enumeration process deeply affects the whole computational performance.

7.1 Datasets

For our experimentation, we have prepared a collection of real world benchmark networks, `Gowalla`, `Stanford`, `Youtube`, `Amazon`, `DBLP` and `Google`[5].

`DBLP` is a collaboration network constructed from the DBLP Computer Science Bibliography. Authors as nodes are connected if they have published a paper together. `Google` is a Web graph consisting of web pages and their hyperlinks. It has been released in 2002 by Google as a part of Google Programming Contest. `Stanford` is also a Web graph consisting of web pages in Stanford University (stanford.edu) and their hyperlinks. `Youtube` is a social network representing users of Youtube, a video-sharing web site, and their friendship relation. `Gowalla` is also a social network consisting of users of Gowalla, a location-based social networking website, and their friendship relation. `Amazon` is a network in which vertices represent products and edges connect commonly co-purchased products. Scale of those networks are given in Table 1.

7.2 Computational Performance

For various values of isolation factor τ, we present here computation times and numbers of solutions for each network in Figure 4. In each figure, the x-axis is for τ-value, the first (left) y-axis for computation times in seconds and the second (right) y-axis for numbers of maximal τ-isolated cliques. Computation times are given by solid lines and numbers of solutions by dashed lines. We have imposed a time limit on computation in each trial and aborted every trial with over 6-hours. We have varied τ-values from 0.3 to 1.0. Missing points in a larger range of τ means there exists no solution, that is, no clique cannot be regarded as isolated one. On the other hand, missing points in a smaller range are due to time limitation.

Computation Times: In most of the cases, we have been successful in enumerating our solutions with reasonable computation times. Particularly, in larger ranges of τ-values, `MIC`can detect our solutions faster than the maximal clique

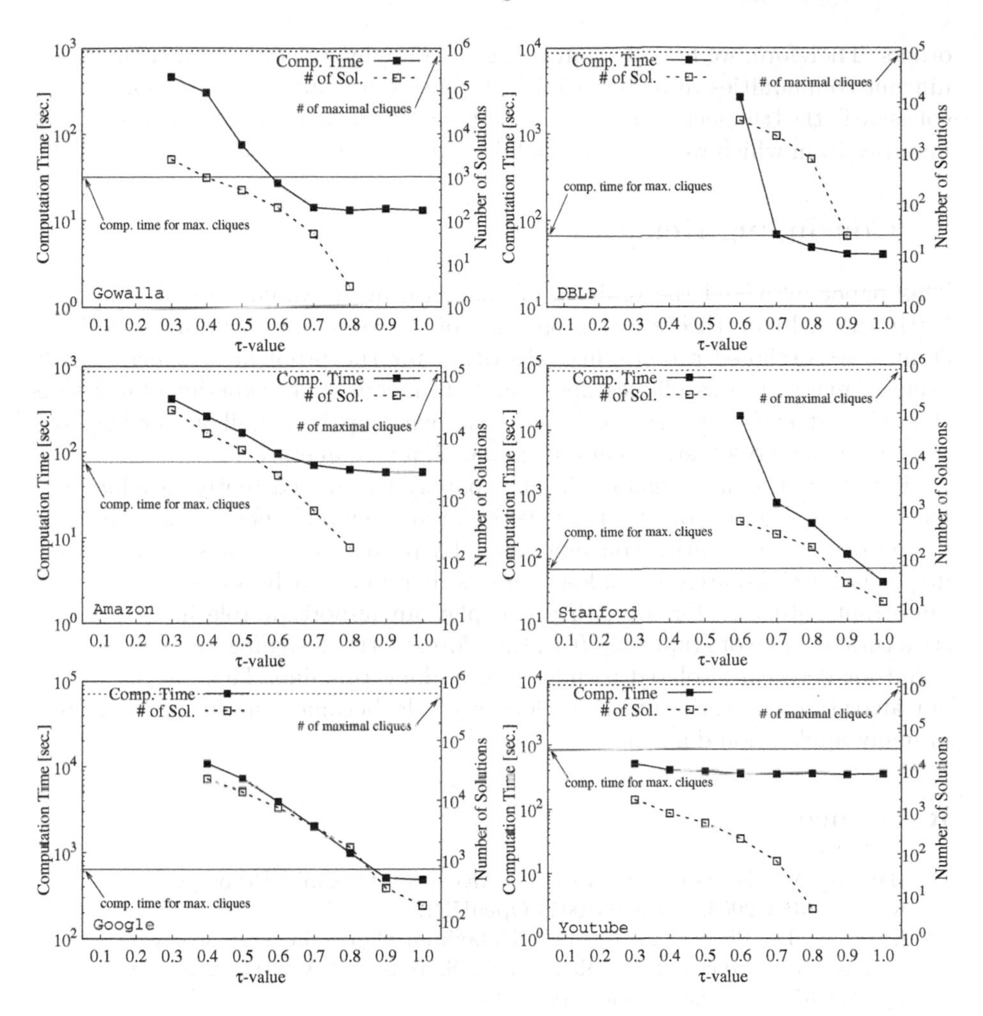

Fig. 4. Computation Times and Numbers of Solutions

enumeration which is just a part of the naive method. On the other hand, for DBLP and Stanford, we have failed to complete our enumeration tasks within the time limit in smaller ranges of τ. Actually, however, any maximal τ-isolated clique for such a small τ would not be valuable for us because its degree of isolation seems to be not sufficient. The authors, therefore, consider that those failures are not critical in our algorithm. A remarkable point of the proposed method is that it can efficiently detect maximal vertex sets which are *actually isolated.*

Numbers of Solutions: For each network, the numbers of maximal τ-isolated cliques are several orders of magnitude smaller than those of maximal cliques. Conversely speaking, most of the maximal cliques are *frequently* connected to others. Therefore, we have no convincing reason why we can regard them as significant communities to be distinguished. Since our solution isolated cliques are cohesive in the true sense, they can possibly provide us some useful aspects of communities from which we can newly obtain valuable information and knowledge.

8 Concluding Remarks

This paper discussed the problem of enumerating maximal τ-isolated cliques. With the help of theoretical properties of j_τ-cores and maximal τ-isolated cliques, we designed a depth-first algorithm for the problem in which we can exclude many useless cliques and search branches. Our experimental results showed that even for large scale networks with up to a million vertices, our algorithm can work faster than a maximal clique enumerator.

Based on the investigation in this paper, we are currently developing a method for extracting our final targets, isolated pseudo-cliques, by extending the framework of enumerating constant j-cored k-plexes [16]. As has been reported in [4], isolated k-plexes (pseudo-cliques) still remain to be investigated as an important subject. Moreover, they can play an important role in detecting a structural change of graphs [9]. It is, therefore, worth designing an efficient algorithm for detecting isolated pseudo-cliques along this line. The authors expect our algorithm currently under developing would become a useful practical tool in many application domains.

References

1. Batagelj, V., Zaversnik, M.: An $O(m)$ Algorithm for Cores Decomposition of Networks. CoRR 2003, cs.DS/0310049 OpenURL
2. Eppstein, D., Strash, D.: Listing all maximal cliques in large sparse real-world graphs. In: Pardalos, P.M., Rebennack, S. (eds.) SEA 2011. LNCS, vol. 6630, pp. 364–375. Springer, Heidelberg (2011)
3. Ito, H., Iwama, K.: Enumeration of Isolated Cliques and Pseudo-Cliques. ACM Transactions on Algorithms **5**(4), Article 40 (2009)
4. Komusiewicz, C., Hüffner, F., Moser, H., Niedermeier, R.: Isolation Concepts for Efficiently Enuerating Dense Subgraphs. Theoretical Computer Science **410**(38–40), 3640–3654 (2009)
5. Leskovec, J., Krevl, A.: SNAP Datasets: Stanford Large Network Dataset Collection (2014). http://snap.stanford.edu/data
6. Liben-Nowell, D., Kleinberg, J.M.: The Link-Prediction Problem for Social Networks. Journal of the Association for Information Science and Technology **58**(7), 1019–1031 (2007)
7. Luxburg, U.: A Tutorial on Spectral Clustering. Statistics and Computing **17**(4), 395–416 (2007)
8. Newman, M.E.J.: Finding Community Structure in Networks Using the Eigenvectors of Matrices. Physical Review E. **74**(3), 036104 (2006)
9. Okubo, Y., Haraguchi, M., Tomita, E.: Structural change pattern mining based on constrained maximal k-plex search. In: Ganascia, J.-G., Lenca, P., Petit, J.-M. (eds.) DS 2012. LNCS, vol. 7569, pp. 284–298. Springer, Heidelberg (2012)

10. Pattillo, J., Youssef, N., Butenko, S.: Clique Relaxation Models in Social Network Analysis. Handbook of Optimization in Complex Networks: Communication and Social Networks, Springer Optimization and Its Applications **58**, 143–162 (2012)
11. Rymon, R.: Search through systematic set enumeration. In: Proceedings of International Conference on Principles of Knowledge Representation Reasoning - KR 1992, pp. 539–550 (1992)
12. Scott, J.P., Carrington, P.J. (eds.): The SAGE Handbook of Social Network Analysis (2011)
13. Seidman, S.B.: Network Structure and Minimum Degree. Social Networks **5**(3), 269–287 (1983)
14. Seidman, S.B., Foster, B.L.: A Graph-Theoretic Generalization of the Clique Concept. Journal of Mathematical Sociology **6**(1), 139–154 (1978)
15. Tomita, E., Tanaka, A., Takahashi, H.: The Worst-Case Time Complexity for Generating All Maximal Cliques and Computational Experiments. Theoretical Computer Science **363**(1), 28–42 (2006)
16. Zhai, H., Haraguchi, M., Okubo, Y., Tomita, E.: A fast and complete enumeration of pseudo-cliques for large graphs. In: Khan, L., et al. (eds.) PAKDD 2016. LNCS, vol. 9651, pp. 423–435. Springer, Heidelberg (2016). doi:10.1007/978-3-319-31753-3_34

Simulator Prototyping Through Graphical Dependency Modeling

Kumar Dheenadayalan(✉), V.N. Muralidhara, and G. Srinivasaraghavan

International Institute of Information Technology, Bangalore, India
d.kumar@iiitb.org, {murali,gsr}@iiitb.ac.in

Abstract. Given a complex system, simulating the behavior of the system under various conditions and inputs is a common requirement in different domains. Designing a simulator by non-domain expert is not an easy task. However, the vast amount of performance related parameters that can be monitored on any physical system can help in probabilistic modeling of the system. Accurate modeling of a complex system requires the identification of dependency among the individual components. These dependencies can be deduced by probabilistic analysis of the performance parameters generated by the physical system. We propose constrained Hill Climbing for structure learning of a Bayesian Network to learn the dependency among internal components of the system. Dependency graph, thus generated will act as a basis for developing a prototype simulator. Multivariate Adaptive Regression Splines are compared with the proposed constrained hill climbing to deduce the mathematical relation between the interdependent components. Two prototype simulators are designed, one for a complex large-scale storage system and another for a group of computer servers. Behavior of both the physical systems is tested on workload traces. The normalized mean absolute error for simulation of storage systems and server was 4.6% & 3.75% respectively. Results indicate that the proposed simulation prototyping can be a useful and unique way of understanding complex systems without the help of a domain expert.

Keywords: Probabilistic graphical model · Bayesian network · Regression · Multivariate adaptive regression splines · Simulator

1 Introduction

Simulation and modeling of a physical system is a discipline in itself and widely used to analyze the behavior of a physical system. Modeling defines the interaction between components of a system and simulation provides insights on the implications of interaction between the components. Models help in simplifying the reality and represent interactions into a form that users can comprehend. Research in simulation and modeling has spread across various domains and the methodologies can vary in each research domain. There is no standard mechanism for users to extract a model and build a quick prototype simulator for a

P. Perner (Ed.): MLDM 2016, LNAI 9729, pp. 584–598, 2016.
DOI: 10.1007/978-3-319-41920-6_46

physical system. Consider a novice user, trying to understand the behavior of a complex physical system for different inputs and configurations. Running a physical system for different type of inputs and combination of configurations can be taxing on the physical system and a costly, tedious and time consuming process for the users. Users of a physical system typically depend on a simulator to test its performance. In certain cases where a simulator does not exist, choices are limited and users are forced to go through multiple runs on the system. To facilitate the understanding of a system and its behavior in various scenarios, learning algorithms can be effectively used to build a quick prototype simulator.

Let us assume that the physical system, $\mathcal{S}$ has a management/monitoring interface that monitors all the components within a system. For example, a computer system, where performances of various components like hard disk, network, CPU etc. can be monitored (disk_read_rate, disk_latency, network_packets_sent etc.). Collecting performance parameters $\mathcal{P}$, of all the components over a period can act as a valuable dataset $\mathcal{D}$ for non-domain experts to extract a high level model of $\mathcal{S}$. In this paper we propose a generic scheme to develop a prototype simulator. The proposed scheme is evaluated on two different datasets, i.e., performance parameters for two different physical systems. Firstly, a group of computer servers are considered for modeling and simulation. Secondly, a multi-component storage system (NetApp filer) is used. Both the physical systems have their own way of extracting performance parameters relevant to the embedded components. Number of hardware, software and virtual components in the systems enables complex interactions, which need to be decoded for developing a simulator. The performance parameters of various components are collected over multiple days to build $\mathcal{D}$. Process of extracting a model from $\mathcal{D}$ is called the *modeling phase*. It builds the dependency between various performance parameters of $\mathcal{S}$. We propose the learning of dependency structure of a Bayesian Network using a score based algorithms, constrained by parameters of input traces. Constraining the structure based on input traces helps in optimizing the same for use in simulation. Constrained Hill Climbing is used as the heuristics to learn the Bayesian Network $\mathcal{B}$ to produce a Directed Acyclic Graph (DAG).

Acyclic property of the DAG helps in deducing the values of all the performance parameters through probabilistic analysis or through regression schemes during the *Simulation phase*. Topological ordering of nodes are extracted to identify the best path for simulation. Using the topological order of nodes, both Bayesian framework and a Multivariate Adaptive Regression Splines (MARS) based multi-model regression scheme are developed and their simulating capabilities are compared. It must be noted that all the performance parameters are treated as response variable with very few predictor variables (workload parameters) making this a complex error prone task. We propose Normalized Mean Absolute Error (NMAE) as a measure to estimate regression error. This is because, the systems being modeled and simulated tend to have parameters with large range of values. The NMAE rates for the two case studies discussed turn out to be 4.6% and 3.75%.

In the next section we discuss the past literature followed by an overview and analysis of the proposed scheme for prototype development in Section 3. Evaluation of the proposed method along with implementation details are discussed in Section 4 followed by the conclusion in Section 5.

2 Literature Survey

Modeling and Simulation exists in almost all the fields like aeronautics, medicine, physics, transportation, environmental studies, astronautics, climatology to name a few [7,16,17,19]. Past literature on applying learning algorithms for simulation appear in two diverse fields. One of the literature talks about applying clustering algorithm to guide architecture simulation [8]. Structured sequences of a few recurring behaviors are identified during program execution using clustering algorithms. These are uniquely sampled to create a complete representation of program's execution. This enables evaluation of computer architecture without having to evaluate the entire program. Another application of using learning algorithm for simulation is in the field of manufacturing [21]. A generic simulation system is developed for the manufacturing industry using Neural network based adaptive simulation system to independently model the interdependencies of production processes. Modeling and Simulation phases proposed in our work are also generic and simulation phase can be adapted to any domain.

2.1 Bayesian Network

A Bayesian network is a graphical model representing probabilistic relationships among a set of variables [10]. Bayesian Networks are used to model uncertain domain knowledge and find probable configurations of variables [11]. Features of the dataset are a set of random variables $X = \{X_1, ...X_n\}$, represented as nodes in a network. Learning the structure identifies the dependency among random variables and this can handle missing data in a probabilistic sense. Direct dependencies between variables is represented by a set of directed arcs between pairs of nodes, $\{X_i \rightarrow X_j\}$. Learning the structure of Bayesian Network from $\mathcal{D}$ is a way of modeling the domain that might have many inherent uncertainties. Identifying the best Bayesian Network structure from the Network space, $\mathcal{B} \in \mathcal{B}_n$, is quantified by a score function. Scoring function is a measure of goodness-of-fit of the network [11]. Best network structure is the one, which maximizes the scoring function for a given $\mathcal{D}$.

Application of Bayesian Network was initially limited to the field of medicine [4,9]. Later, Bayesian Networks have been used to model diverse systems and environments and the areas of application have continuously been growing. Diagnosis of physical systems like printers [18] and taking actions based on the diagnosis [1] have also been achieved using Bayesian Network. The level of uncertainties in the ecosystems and other components within the earths' environment is high. The risks of oil spill and the environmental impacts of human activities is modeled using Bayesian Network [12]. Applications as diverse as Software Fault

Prediction to dataset compression has been achieved through this learning algorithm [5]. Bayesian Network learns the dependencies of features from the data that makes it one of the most popular tool for modeling a physical environment.

3 Simulation Scheme

Simulation scheme proposed in this paper broadly consists of two steps.

- Construct the Constrained Bayesian Network from data and fit a Conditional Probability Table[CPT].
- Use the dependencies to build a multi-model regression.

A good model represents the dynamics of a system and allows exploring its behavior systematically. The key for simulation is the amount of information that is available to model a real world physical system. The information extracted from a physical system that is available in the form of a dataset is described here. A single data instance of performance parameters $\mathcal{D}_t$ consists of all the monitorable components and related parameters. If $\mathcal{C}_i$ is the ith component of $\mathcal{S}$, then $\mathcal{P}_{\mathcal{C}_i}$ is a row vector of performance parameters related to the component $\mathcal{C}_i$.

$$\mathcal{D}_t = (\mathcal{P}_{t_{\mathcal{C}_1}}, \ldots\ldots, \mathcal{P}_{t_{\mathcal{C}_m}})$$

Let $\mathcal{T}$ be the features of workload trace, i.e. the trace of inputs to the system. *Spec* is a vector of specifications of the physical system being modeled. The system to be simulated will be monitored to record two sets of information. $\mathcal{T}_t$ is the input trace recorded on the system at time t and the response of the system to input trace are recorded in $\mathcal{D}_t$. The workload traces here are the actual inputs to the live system. We build the dataset for our simulation as a combination of workload traces, system specifications and response of the system.

$$\mathcal{D} = \begin{bmatrix} \mathcal{T}_1 & Spec & \mathcal{P}_{1_{\mathcal{C}_1}} & \cdots & \mathcal{P}_{t_{\mathcal{C}_m}} \\ \mathcal{T}_2 & Spec & \mathcal{P}_{2_{\mathcal{C}_1}} & \cdots & \mathcal{P}_{t_{\mathcal{C}_m}} \\ \vdots & \vdots & & \ddots & \vdots \\ \mathcal{T}_n & Spec & \mathcal{P}_{n_{\mathcal{C}_1}} & \cdots & \mathcal{P}_{t_{\mathcal{C}_m}} \end{bmatrix}$$

System specifications are also recorded, as certain environments might contain multiple physical systems with various specifications. Model should be robust enough to simulate the inputs on physical systems with various specifications. The need for recording and combining the workload traces with performance parameters is explained in the next section.

3.1 Dependency Modeling

In this section, we describe Structure learning of a Bayesian Network constrained by traces for a given $\mathcal{D}$. Learning the most probable posterior probability of a Bayesian Network from data is known to be NP-hard [3]. Optimization strategy is applied through search-and-score approach that consists of a scoring function and a search strategy. Search Strategy involves heuristics to efficiently search the network structure space and identify the best network for the given data. The notion of best network is measured by the scoring function, which is a measure of goodness-of-fit for the data. One such scoring function is the Bayesian Information Criterion (BIC) defined as

$$\mathcal{BIC}score(\mathcal{B}) = loglikelihood(\mathcal{B}) - (log(n)/2) \times log(N)$$

n $\leftarrow$ Number of rows in $\mathcal{D}$
N $\leftarrow$ Number of free parameters
The number of free parameters is defined as the sum of logically independent parameters of each node given its parents [2].

Hill climbing is a popular heuristic to search the network space locally until a local maxima is found [20]. Strategies like random perturbation of the network structure for a fixed set of iterations are used to prevent getting stuck at local maxima. Behavior of a physical system is a consequence of the input provided to the system. Hence, we want to induce this logic as a constraint in our structure learning algorithm. The reason for combining the workload traces with performance parameters was to make sure the structure is built with the above logic. All variables from $\mathcal{T}$ and *Spec* will be applied as constraints forcing them to be treated as a root node in the Bayesian Network. This constraint is imposed in the Hill climbing algorithm described in Algorithm 1.

Consider an example with two features related to input traces (T_1, T_2), one feature related to specification (S_1) and 3 performance parameters (P_1, P_2, P_3). The effect of constructing constrained and unconstrained networks using hill climbing is shown in Figure 1. An Unconstrained network has unwanted dependencies deriving input trace from parameters $(\{P_3 \rightarrow T_2\}, \{P_1 \rightarrow T_2\})$ whereas constrained network fixes these situations. Inputs affect the performance parameters and not the opposite. Although input traces can sometime be derived from performance parameters, they are not of much use for the simulation task considered here. As a result of these constraints, this workload traces to be simulated along with the *Spec* would be enough to simulate an entire system.

The hill climbing algorithm is initialized with an empty structure to induce sparsity. Similar to standard hill climbing, addition, deletion and reversal of node operations are applied to the graph $\mathcal{G}$. During this exploratory process, constraints are evaluated to ensure that addition of a node or reversal of an edge does not create a dependency from a performance parameter to a workload parameter. As a result, parameters of workload traces will most likely end up having no parents. $updateNetwork()$ function updates the network $\mathcal{B}$ every time a better structure is identified. Once the model of the physical system is available, we process the network by utilizing the identified dependencies in the simulation process.

Algorithm 1. Learn Bayesian Network Structure

1: **procedure** LEARNSTRUCTURE($\mathcal{D}, \mathcal{G}, \mathcal{B}, trainIndex, constraintFeatures$)
2: $trainData \leftarrow \mathcal{D}[trainIndex,]$
3: $\mathcal{BIC}score \leftarrow -\infty$
4: **while** $\mathcal{BIC}score \geq max\mathcal{BIC}score$ **do**
5: **for all** $\{node_i, node_j\} \in features(\mathcal{D})$ **do**
6: **if** $acyclic(\mathcal{G} \cup \{node_i \rightarrow node_j\})\&\&\{node_j \cap constraintFeatures\} == \emptyset$ **then**
7: $\mathcal{G} \leftarrow \{\mathcal{G} \cup \{node_i \rightarrow node_j\}\}$
8: $update(\mathcal{D}, \mathcal{G}, \mathcal{B}, trainData)$
9: **end if**
10: **if** $acyclic(\mathcal{G} - \{node_i \rightarrow node_j\})$ **then**
11: $\mathcal{G} \leftarrow \{\mathcal{G} - \{node_i \rightarrow node_j\}\}$
12: $update(\mathcal{D}, \mathcal{G}, \mathcal{B}, trainData)$
13: **end if**
14: **if** $acyclic(\{\mathcal{G} - \{node_i \rightarrow node_j\}\} \cup \{node_j \rightarrow node_i\})\&\&\{node_i \cap constraintFeatures\} == \emptyset$ **then**
15: $\mathcal{G} \leftarrow \{\{\mathcal{G} - \{node_i \rightarrow node_j\}\} \cup \{node_j \rightarrow node_i\}\}$
16: $update(\mathcal{D}, \mathcal{G}, \mathcal{B}, trainData)$
17: **end if**
18: **end for**
19: **end while**
20: **end procedure**

Algorithm 2. Update Bayesian Network Structure

1: **procedure** UPDATE($\mathcal{D}, \mathcal{G}, \mathcal{B}, trainData, \mathcal{BIC}score$)
2: **if** $\mathcal{BIC}score \leq \mathcal{BIC}(\mathcal{G}, trainData)$ **then**
3: $\mathcal{BIC}score \leftarrow \mathcal{BIC}(\mathcal{G}, trainData)$
4: $\mathcal{B} \leftarrow updateNetwork()$
5: **end if**
6: **end procedure**

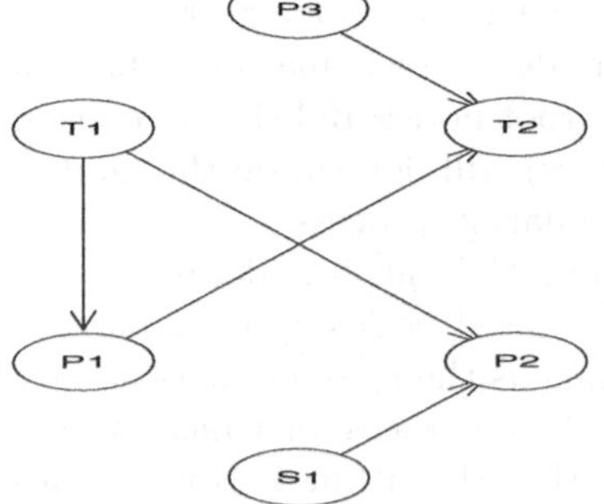

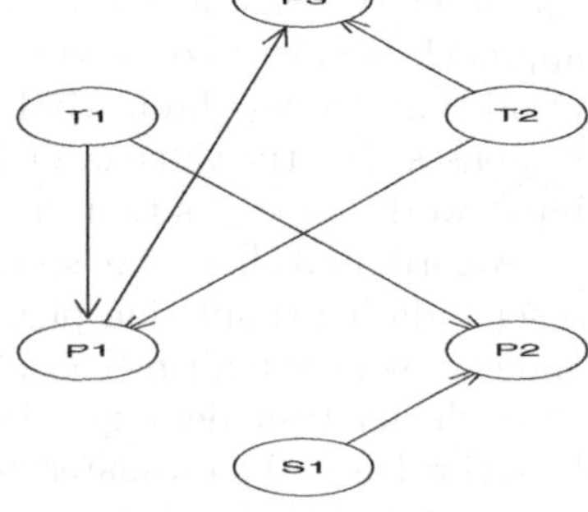

Fig. 1. Variation in Network Structure with input traces and specification used as constraints.

3.2 Node Value Prediction

Simulation phase in the proposed scheme consists of two steps:

- Topologically order the network structure and predict the responses of node values to input trace.
- Use Bayesian inference or a multi-model regression for predicting node values during simulation.

Lack of an edge is true in the physical system if the Bayesian network follows the Markov property. Verifying the existence of Markov property depends on domain knowledge, which we assume is unavailable. Experts can verify and nullify any unwanted dependency among nodes by parameterization of conditional probability table. Using constrained hill climbing, we use the basic knowledge of always treating performance parameters as dependent variables.

As there is lack of any other domain knowledge, we try an alternative approach by using MARS [6] to build a multi-model regression scheme. MARS algorithm uses variable importance estimation that can identify the best subset of features required in building the regression model. This may compensate for the lack of domain knowledge and indirectly eliminate any unrealistic dependency. Each non-root node in the modeled network is treated as a response variable and their parents as predictor variables. Using this strategy, at most set of $n-1$ different regression datasets for non-root nodes are extracted from $\mathcal{D}$. Each of the $n-1$ datasets will generate unique models for prediction of $n-1$ different nodes.

Another possible approach for extracting $n-1$ datasets would be to use non-root node as a response variable and the Markov Blanket as predictor variables. Markov blanket of a node consists of the node's parents, its children, and its children's parents. It is useful in finding optimal feature subsets [15]. Low prediction errors for nodes with high Markov Blanket is necessary. If prediction errors for such nodes are high, then the rate of propagation of these errors across network might also be high.

Inputs to physical systems are not random in a lot of domains. Fixed patterns or a time series are observed in the inputs to a physical system (e.g., Inputs to computer systems do show such behavior as shown in [8]). Hence, regression approach is suited to capture the response of nodes, which may also have a trend. The network built will have input traces as root nodes and the simulation responses (i.e., the value of performance parameters) will depend on the kind of input workload traces that are fed during the simulation process.

Algorithm 3 shows the steps followed in building the multi-model regression scheme during simulation phase. Regression for isolated nodes with no parents and children are built by considering the constraints as their predictor variables. Once the multi-model regression is built, past workload traces for which system behavior has to be evaluated is used as input to the Algorithm 4. Node values are predicted in topological order to avoid encountering missing data during the simulation process. Alternate approach of predicting the node values using classical Bayesian Network involves building of Condition Probability Table that

Algorithm 3. Extract Rules

```
1: procedure EXTRACTSIMULATIONRULES(B, D, trainIndex, constraintFeatures)
2:     trainData ← D[trainIndex, ]
3:     for all node in B do
4:         node.parents ← getParents(node)
5:         if N(node.parents) == 0 then
6:             node.parents ← constraintFeatures
7:         end if
8:         responseVar ← node
9:         rules[node] ← fitModel(trainData, node.Parents, responseVar)
10:    end for
11:    return rules
12: end procedure
```

best fits the network structure. We compare the prediction capabilities of using Constrained Bayesian Network(CBN) and Constrained Bayesian Network with MARS(CBNM) on real world data and the results are discussed in Section 4

Algorithm 4. Simulator Response

```
1: procedure SIMULATERESPONSES(B, orderedRules, testworkloadtrace, cpt)
2:     response_MARS[] ← zeroes()
3:     response_BN[] ← zeroes()
4:     orderedRules ← topologicalSort(B)
5:     for all rule in orderedRules do
6:         response_MARS ← predict(rule, node, testworkloadtrace, response)
7:         response_BN ← predict(cpt, node, testworkloadtrace, response)
8:     end for
9:     return response
10: end procedure
```

3.3 Simulation Efficiency

The simulation efficiency is measured by comparing the predicted performance parameters with actual values of the system for an input trace. Three measures are used to evaluate the simulation efficiency as shown in Table 1.

In all the three error measures e_t is the prediction error at time t. As the range of values for each parameter within the dataset can vary, we introduce the normalized mean absolute error. RMSE and MAE may be dominated by error instances, which are too far from true values. Such scenarios are common in cases where range of values for a regression algorithm is high. Hence, NMAE is used to give better measure the overall simulation ability in our experiments. We report the percentage NMAE in this paper, as NMAE $\times$ 100

Table 1. Error Measure for Simulation

Error Measure	Formula
Mean Absolute Error (MAE)	$\frac{1}{n}\sum_{t=1}^{n}\lvert e_t\rvert$
Normalized Mean Absolute Error (NMAE)	$\frac{1}{n}\sum_{t=1}^{n}\frac{\lvert e_t\rvert}{\max(Y)-\min(Y)}$
Root Mean Square Error (RMSE)	$\sqrt{\frac{1}{n}\sum_{t=1}^{n}e_t^2}$

4 Results

4.1 Data Description

Two environments are simulated to evaluate the CBN and CBNM. First environment consists of four servers with different specifications (*Spec*). A set of applications are executed on this clustered setup and their performance parameters ($\mathcal{P}_{tc_m}$) related to the components of each server are recorded. Real world applications with traces ($\mathcal{T}$) are also recorded. Performance parameters and the workload traces are measured at specified time interval on each server. The monitored components ($\mathcal{C}_m$) in this experiment include processor, RAM, memory manager, network and I/O devices. Each component has variable number of performance parameters recorded in $\mathcal{P}_{tc_m}$ for the m_{th} component at time t. *Spec* vector consists of specifications related to processor, network card, RAM, graphics card and hard drive. Test application traces $\mathcal{T}$ are provided to simulate the performance parameters and error measures are evaluated in the test phase. Simulation for this dataset will be referred to as SIM1.

The second environment considered for simulation is a NetApp storage system also called a filer Each component within the storage architecture can produce useful information about its performance through multiple performance parameters called counters. More details about the components and their related parameters are available in [13,14]. The workload trace contains information about the type of I/O operation, and its destination for performing the operation. $\mathcal{D}$ consists of performance parameters, features of workload trace and specification of the filer. Simulation related to this dataset will be referred to as SIM2. Bayesian Networks constructed might turn out to have a different structure in each iteration. Hence, the average error rates for each node over 10 iterations are presented.

4.2 Analysis

We first discuss the results of SIM1 as it involves variation in the physical system (servers) and the *Spec* vector has a role in the modeling phase. There are 30 performance parameters, 8 specification parameters and 5 workload trace parameters. Given a dataset with 90,000 instances and 43 features, 13 non-performance related parameters are treated as constraints to build the Bayesian Network. 80,000 instances are used as input for building the network. Application traces available in rest of the 10,000 instances are used as input to Algorithm 4. The built network has 30 non-root nodes and 13 root nodes resulting in 30 different regression schemes have to be built using MARS.

Algorithm 3 is applied to the Bayesian Network to build CPT and multi-model regression for SIM1. Any new application trace that needs to be simulated will be provided as a matrix with each column representing the workload trace features. The entire trace of application execution is fed to Algorithm 4, which predicts the responses of each performance parameter for both CBN and CBNM. Figures 2 shows the plot for three different simulation efficiency measures defined in the previous section. Lower the values in all the three subplots, better is the simulation efficiency. Range of values in the y-axis should be observed carefully as it explains the need for normalization. An example of how NMAE is able to estimate errors more accurately is seen in the plots. Though predictions for node index 4 have very high MAE and RMSE, the true estimate of prediction is seen in NMAE. As the range of values for the parameter indexed at 4 is very high, RMSE and MAE will be dominated by few bad predictions. In certain cases, even though the error might be perceived to be high, the relative error with respect to the range of values might be low. Hence, RMSE and MAE can over estimate the prediction errors, which can lead to wrong conclusions. In a majority of parameter predictions, CBN has better simulation results compared to CBNM.

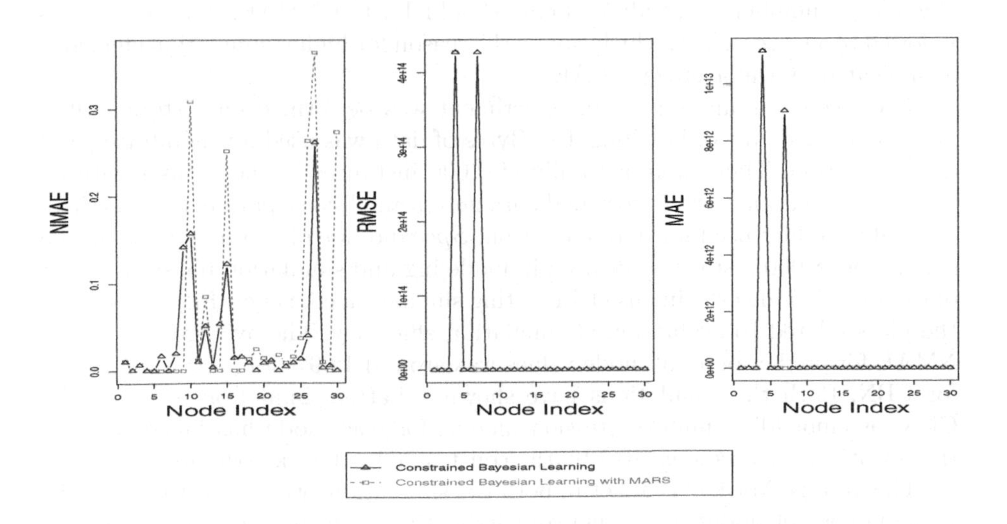

Fig. 2. Normalized MAE, RMSE and MAE for estimating simulation efficiency of server.

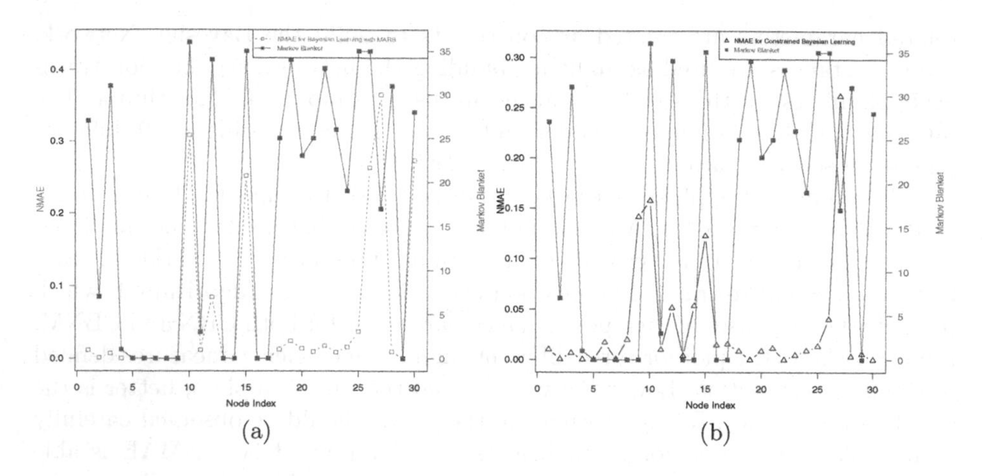

Fig. 3. Variation NMAE with the size of Markov Blanket for Server simulation.

The average percentage NMAE over all the 30 predictions was observed at 6% for CBNM and 3.75 CBN%. A plot comparing NMAE and Markov Blanket for each node is shown in Figure 3. Nodes with larger Markov Blanket should ideally have higher prediction accuracy (lower NMAE) as they influence large number of nodes. CBN has better values of NMAE compared to CBNM for most of the nodes in general and nodes with higher Markov Blanket in particular. The same idea can be extended to the number of children of a node as shown in Figure 4. Higher the number of children, lower should be the NMAE. This pattern is followed to a large extent, which can be the reason for higher simulation efficiency as indicated by the average NMAE.

The second simulation in our experiment was performed on a storage filer. Storage filer capable of handling TeraBytes of data was used for monitoring and data collection. There were a totally of 9,000 instances in the dataset with 88 performance parameters, 5 workload parameters and 6 *Spec* parameters. As there was only one filer used for data collection, *Spec* did not play a major role in the simulation. 8,000 instances are used in modeling and simulation phase with rest of the 1,000 instances being used for testing simulation efficiency. Figures 5 shows the plots related to evaluation of simulation efficiency. The average percentage NMAE for prediction of all node values was around 12.9% for CBNM and 4.6% for CBN. Both the simulations have shown a better simulation efficiency for CBN, as embedding multi-regression models for each node has failed to learn the set of parameters identified by the constrained network structure.

The average Markov Blanket in both the simulations were close to half of the total number of nodes in the network indicating high interdependence among components. As seen in SIM1, the nodes with higher number of children or with a larger Markov Blanket should be analyzed to see if their NMAE values are low. Figure 6 & 7 show certain nodes with higher children or larger Markov Blanket have higher values of NMAE, which can lead to error propagation.

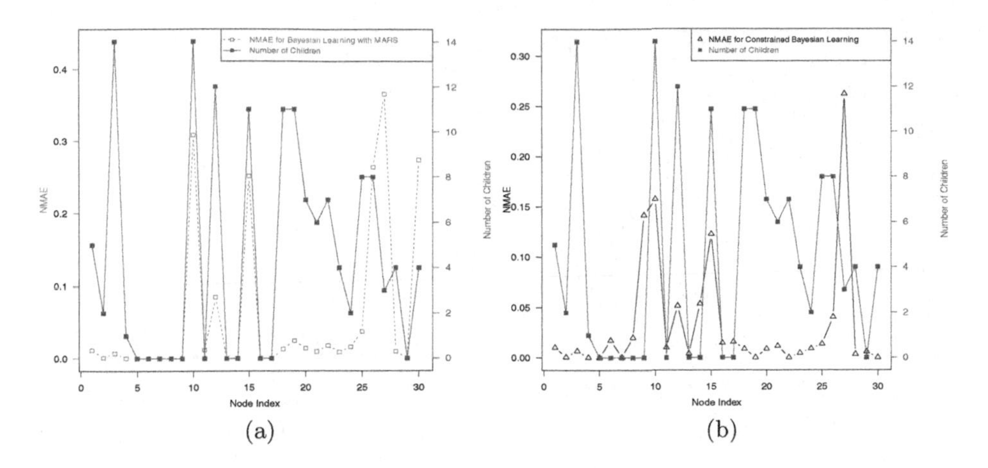

Fig. 4. Variation NMAE with the size of Children for Server simulation.

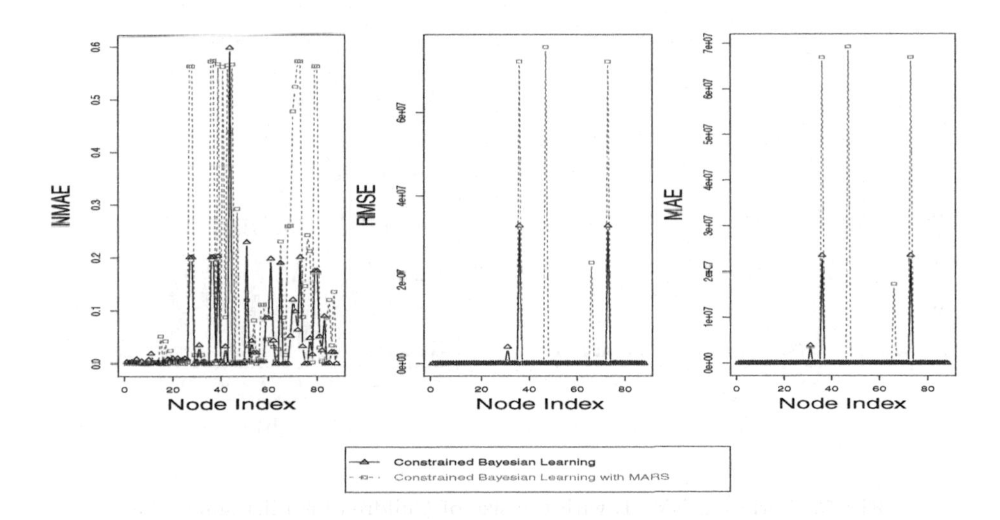

Fig. 5. Normalized MAE, RMSE and MAE for estimating simulation efficiency of Filer.

Also, node index obtained from topological ordering are a function of the depth of a node in the DAG. Lower the node index, higher is the probability of error propagation. Hence, nodes with lower index should be carefully analyzed during model construction to see how well the structure truly represents reality. Fig 6(a) & 6(b) clearly shows that majority of nodes with large Markov Blanket have lower NMAE values for CBN when compared to CBNM leading lesser error propagation.

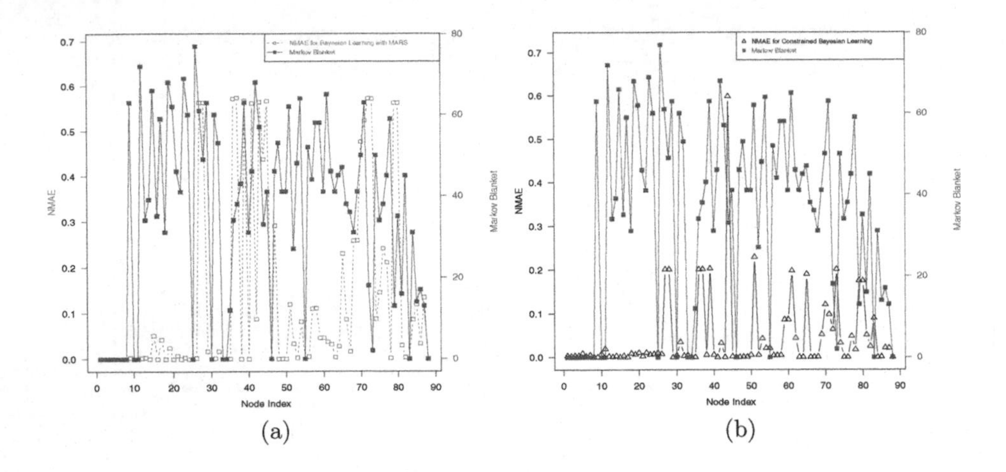

Fig. 6. Variation NMAE with the size of Markov Blanket for Filer simulation.

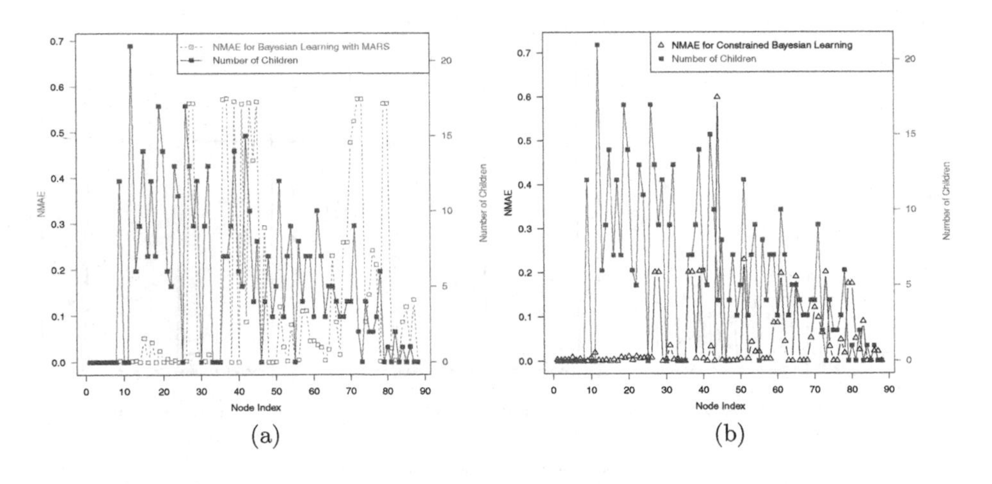

Fig. 7. Variation NMAE with the size of Children for Filer simulation.

5 Conclusion

Simulation provides major advantage in designing, developing, analyzing and verifying the performance of a physical system. A generic framework for non-domain experts to simulate system behavior was presented in this paper. Learning structure of Bayesian Network using Constrained Hill climbing algorithm has shown promising results in simulating the system performance. Workload traces can be directly provided to the Simulation process that can simulate multiple component behavior. Only assumption in the proposed scheme is the presence of a management interface to collect parameters. Lack of any other assumption helps in extending this scheme to other domains for simulation. Topological ordering of the nodes ensures that there are no missing information while predicting the node values. Exploring regression scheme during simulation phase can help in not just simulating but also scaling the simulation process. Future work can focus on increasing simulation accuracy, especially nodes with higher number of children or large Markov Blanket. Integration of prototype simulator with a discrete event system will be attempted in our future work.

References

1. Breese, J.S., Heckerman, D.: Decision-Theoretic Troubleshooting: A Framework for Repair and Experiment. CoRR abs/1302.3563 (2013)
2. Chickering, D.M., Geiger, D., Heckerman, D.: Learning Bayesian networks: search methods and experimental results. In: Proceedings of the Fifth International Workshop on Artificial Intelligence and Statistics, pp. 112–128, January 1995
3. Chickering, D.M.: Learning Bayesian networks is NP-complete. In: Learning from Data: Artificial Intelligence and Statistics V, pp. 121–130. Springer-Verlag (1996)
4. Daly, R., Shen, Q., Aitken, S.: Review: Learning bayesian networks: Approaches and issues. Knowl. Eng. Rev. **26**(2), 99–157 (2011)
5. Davies, S., Moore, A.: Bayesian networks for lossless dataset compression. In: Proceedings of the Fifth ACM SIGKDD International Conference on Knowledge Discovery and Data Mining, KDD 1999, pp. 387–391. ACM, New York (1999)
6. Friedman, J.H.: Multivariate adaptive regression splines. The Annals of Statistics **19**(1), 1–67 (1991)
7. Gerathewohl, S.J.: United States: Fidelity of simulation and transfer of training [electronic resource]: a review of the problem / by Gerathewohl, S.J., Dept. of Transportation, Federal Aviation Administration, Office of Aviation Medicine [Washington, D.C.] (1969)
8. Hamerly, G., Perelman, E., Lau, J., Calder, B., Sherwood, T.: Using machine learning to guide architecture simulation. J. Mach. Learn. Res. **7**, 343–378 (2006). http://dl.acm.org/citation.cfm?id=1248547.1248559
9. Heckerman, D.E., Horvitz, E.J., Nathwani, B.N.: Toward normative expert systems: Part I. The Pathfinder project. Methods of Information in Medicine **31**(2), 90–105 (1992)
10. Heckerman, D.: A tutorial on learning with bayesian networks. Tech. Rep. MSR-TR-95-06, Microsoft Research, March 1995
11. Heckerman, D., Geiger, D., Chickering, D.M.: Learning Bayesian Networks: The Combination of Knowledge and Statistical Data. CoRR abs/1302.6815 (2013)
12. Helle, I.: Assessing oil spill risks in the northern baltic sea with bayesian network applications. abs/1302.3563 (2015)
13. Inc, N.: Report fields and performance counters (2013). https://library.netapp.com/ecmdocs/ECMP1222473/html/GUID-33405732-F0C7-4FD3-8AB8-39A5B76ACC77.html
14. Inc, N.A.: Netapp Unified Storage Performance Management Using Open Interfaces, March 2010
15. Koller, D., Sahami, M.: Toward optimal feature selection. In: 13th International Conference on Machine Learning, pp. 284–292 (1995)
16. Lozano, A.C., Li, H., Niculescu-Mizil, A., Liu, Y., Perlich, C., Hosking, J., Abe, N.: Spatial-temporal causal modeling for climate change attribution. In: Proceedings of the 15th ACM SIGKDD International Conference on Knowledge Discovery and Data Mining, KDD 2009, pp. 587–596. ACM, New York (2009)

17. Olstam, J.J., Lundgren, J., Adlers, M., Matstoms, P.: A framework for simulation of surrounding vehicles in driving simulators. ACM Trans. Model. Comput. Simul. **18**(3), 9:1–9:24 (2008)
18. Skaanning, C., Jensen, F.V., Kjærulff, U.B.: Printer troubleshooting using Bayesian networks. In: Logananthara, R., Palm, G., Ali, M. (eds.) IEA/AIE 2000. LNCS (LNAI), vol. 1821, pp. 367–380. Springer, Heidelberg (2000)
19. Sterbenz, J., Cetinkaya, E., Hameed, M., Jabbar, A., Rohrer, J.: Modelling and analysis of network resilience. In: 2011 Third International Conference on Communication Systems and Networks (COMSNETS), pp. 1–10, January 2011
20. Tsamardinos, I., Brown, L.E., Aliferis, C.F.: The max-min hill-climbing bayesian network structure learning algorithm. Machine Learning **65**(1), 31–78 (2006)
21. Westkamper, E.: Simulation based on learning methods. Journal of Intelligent Manufacturing **9**(4), 331–338

Applying Data Mining to Healthcare: A Study of Social Network of Physicians and Patient Journeys

Shruti Kaushik(✉), Abhinav Choudhury, Kaustubh Mallik, Anzer Moid, and Varun Dutt

Applied Cognitive Science Laboratory, Indian Institute of Technology Mandi, Mandi 175005, Himachal Pradesh, India
{shruti_kaushik,abhinav_choudhury, kaustubh_priya,md_anzer}@students.iitmandi.ac.in, varun@iitmandi.ac.in
http://www.acslab.org/index.html

Abstract. In 2004, the US President launched an initiative to make healthcare medical records available electronically [27]. This initiative gives researchers an opportunity to study and mine healthcare data across hospitals, pharmacies, and physicians in order to improve the quality of care. Physicians can make better informed decisions regarding care of patients if physicians have proper understanding of patient journeys. In addition, physician healthcare decisions are influenced by their social networks. In this paper, we find patterns among patient journeys for pain medications from sickness to recovery or death. Next, we combine social network analysis and diffusion of innovation theory to analyze the diffusion patterns among physicians prescribing pain medications. Finally, we suggest an interactive visualization interface for visualizing demographic distribution of patients. The main implication of this research is a better understanding of patient journeys via data-mining and visualizations; and, improved decision-making by physicians in treating patients.

Keywords: Diffusion of innovation · Patient journey · Social network analysis · Physicians · Visualization · Pain medications

1 Introduction

Modern healthcare has started using a patient-centered approach by building and evaluating patient journeys through their sickness and recovery. The study of patient journeys is relatively recent innovation in the healthcare quality improvement process [1]. A patient's journey involves a sequence of events that a patient proceeds through from the point of entry into the healthcare system (triggered by sickness) until the complete recovery or death. Thus, patient journeys include filling prescriptions at a pharmacy, visiting a doctor, being admitted to the hospital, undergoing lab tests, getting treatment, and recovering from sicknesses. Understanding the whole journey from patient's point of view is important as the patient is the only person who experiences the whole journey [2]. Patient journeys highlight bottlenecks in the healthcare

P. Perner (Ed.): MLDM 2016, LNAI 9729, pp. 599–613, 2016.
DOI: 10.1007/978-3-319-41920-6_47

system, which helps providers improve the system. Hence, a comprehensive approach to patient journeys can help impart provocative vision that may lead physicians to revise their treatment plan.

During a patient's journey, the decisions of the physicians may be influenced by their social network. Most importantly, the medication that a physician prescribes or any new innovation that he adopts is highly influenced by the interpersonal communication of the physician with the members of his personal network [5]. Here, social network analysis investigates various direct and indirect interpersonal communications and interaction patterns between the members of a social system.

Since patients are the end-users of medications, the understanding of patient journeys will help us specify how and when patients consume pharmaceutical products. Proper evaluation of patient journey will help a physician know the unspoken needs of patients. Thus, it is important to investigate factors involved in creating good patient journeys. Up to now, researchers have created these journeys using process mapping and visualization tools by conducting surveys for a few hundred patients [1]. In this paper, however, we have adopted a bottom-up approach to understand patient journeys. We focus on pain medications being used by patients and prescribed by physicians in the United States of America. We build patient journeys by mining through billions of patient records and creating a social network of the physicians using their prescribing histories, specialty, and diffusion of innovation theory [7]. Furthermore, using social network analysis, we pinpoint key-opinion leaders in the physician's social network. These opinion leaders have high influencing power and can bring about behavioral change in ways that medications are prescribed [7]. Moreover, in order to visualize multidimensional patients' data we have implemented visualization techniques for exploratory data analyses. Starting in the next section, we discuss background work in the area of patient journeys, social network analysis and visualization. Then, we divide our analysis into three sections: The first section deals with patient journeys; second section deals with diffusion of innovation of the medication and social network analysis of the physicians' network; and, the last section shows the visualization of multivariate dataset. We close this paper by discussing the significance of our results for improving existing healthcare system.

2 Background

Patient journey motivates us to observe and examine the various paths a patient goes through and the decisions that are taken by various stakeholders through the amelioration of a disease and its treatment. Process mapping is a framework used for building patient journeys through patient's perspective. It allows us to figure out a series of sequential-events linked with the patient's experience [3]. It aims to boost the quality of clinical role, eliminate unnecessary activities from the care, and finally focus on more valuable activities [4]. Though process-mapping technique successfully maps patient's experiences; however, in order to visualize each stakeholder's role we need graphical communication tools. These graphical communication tools act as communication medium which encourages communication between various participants and helps to improve the journey experiences of the patient. Furthermore, the communication tool encourages stakeholder involvement and highlights the importance of relevant variables that contribute to the whole experience of the patient. However, less attention has been paid to the analysis of patient journeys using bottom-up approach, i.e., from low-level data. In this paper, we use data from US for pain medications and build patient journeys bottom-up. We believe that building such journeys will provide insights into improving patient experience across their journeys.

Furthermore, the rate of adoption of a new medication/innovation varies from physician-to-physician [5]. While some physicians adopt a new medication early, most of them tend to test the waters first. Thus, most physicians wait for others in his social system to have tried the innovation first. The diffusion of innovation is the process in which an initial few people adopt an innovation first and, through their social network, the innovation diffuses to others in the network. As time goes on the rate of adoption increases and all or most members of the social system start adopting the innovation [6-8]. A social network/system is the pattern of friendship, advice, communication or support which exists among the members of a social system [9-12]. Such networks can be used to find key-opinion leaders inside a social system. Thus, a major contribution of social networks to diffusion research has been the categorization of adopters based on innovativeness as measured by the time-of-adoption [13]. Here, innovativeness is the degree to which an individual is relatively early in adopting new ideas compared to other members of a social system [15]. As per theory of innovation of diffusion, there are five adopter categories among members of a social system on the basis of their innovativeness: 1) Innovators, 2) Early adopters, 3) Early majority, 4) Late majority, and 4) Laggards [6, 14, 7]. According to theory of diffusion of innovation given by Everett Roger [7, 17] adopter distribution takes the form of a bell-shaped curve. Using two basic statistical parameters of the normal adopter distribution-mean time of adoption (t) and its standard deviation (σ) we obtain the five adopter categories [17] (Table 1). In this paper, we use this categorization and apply it to physicians prescribing pain medications in the US. As suggested by theory of innovation-diffusion, physicians' relative location in the social network with other physicians affect their decisions concerning the adoption of new innovation [16]. Thus, categorization based upon theory of innovation-diffusion helps us understand how pain medications diffuse over a social network of physicians and allows us to measure the rate of diffusion of innovation over time.

Table 1. Adopter categories based on Innovativeness [17]

Adopter Categories	% adopters	Area covered under curve
Innovators	2.5	Between t - 2σ
Early adopters	13.5	Between $t - \sigma$ *and* t - 2σ
Early majority	34	Between t *and* $t - \sigma$
Late Majority	34	Between t *and* $t + \sigma$
Laggards	16	Between $t + \sigma$

Furthermore, huge amounts of multivariate data are being generated on a daily basis nowadays about patient journeys and physician prescribing histories. In order to find patterns in the growing data, we need to be able to visualize it efficiently. Conventional visualization methods like 2-d plots, scatter grams, histograms are limited in the sense that they can only depict 2-dimensions at one time. In contrast, Parallel Coordinates system [24] has been recently proposed and it allows end-users to visualize entire data together at one time. Furthermore, various functionalities like brushing, scattering and distribution can be attained using Parallel Coordinates. Currently, healthcare industry lacks a generalized, interactive and easy-to-use visualization interface which incorporates features like brushing, distribution, visualization of selective dimensions, and correlation among dimensions. Parallel Coordinates helps us overcome this necessity. In this paper, we build a tool based upon Parallel Coordinates that accepts a CSV dataset of any size and generates the visualization plot, correlation matrix, and distribution lists from this data.

3 Method

The patient journeys, social networks, and visualizations were created for patients and physicians residing in the US. We used a large medical-prescriptions dataset[1] in order to build patient journeys, social network of the physicians, and visualizations. The dataset, containing patients and physicians, was provided by a pharmaceutical company.

3.1 Patient Journey

We have focused our analyses on outpatient refill data and inpatient hospital-visitation data. The data is Big in nature as it has more than 100 million records between years 2008 and 2014. Inpatients are those who consume pain medications and are admitted to hospitals. Outpatients are those patients who consume pain medications but were never admitted to hospitals. We used a Big-Data architecture consisting of q-programming language to query a kdb+ database (from Kx systems) in order to find patterns among inpatients and outpatients [18]. In our patient journeys, each activity of patients' is coded using a letter code. For example, *H* represents that a patient has been admitted to the hospital, *D* represents discharge from hospital. Furthermore, *px* and *dy* represent procedure and diagnostic-test codes corresponding to procedure *x* and diagnostic-test *y* performed on patients, respectively. Medicine consumption is coded as amount of potency consumed by the patient (e.g., 5 mcg/hr, 10 mcg/hr, 15 mcg/hr, and 20 mcg/hr). After building the long chain of sequences for each patient who consumed the pain medications, we applied Apriori algorithm [19] to find out strong association rules across journeys. The Apriori algorithm helps us to find the frequently appearing item sets in a large database. The frequent item sets are used to determine the association rules that highlight patterns in data. We have performed a demographic analysis based on sex and age-group of patients to see how

[1] Data shown has been altered to protect privacy.

these characteristics affect the refill behavior among patients. Furthermore, we analyzed the switch behavior to see how patients switched from one medicine level to another during their journeys.

3.2 Social Network Analysis

The dataset consisted of billions of physicians prescribing medicines between years 2011 and 2015. The nature of data was prescription records of each physician during this period. Using social-network analysis, we created a social network for 50,000+ physicians, who prescribed pain medications and satisfied the constraints discussed below.

The following assumptions were used to create a social network:

1. Physicians living in close vicinity are likely to be in contact with each other and influence each other
2. Physicians having the same specialty are likely to be in contact with each other and influence each other

The distance d (i, j) between the physician i and j was computed using the haversine distance formula [20]

$$d(i,j) = 2\text{rarcsin}\left(\sqrt{sin^2\left(\frac{\varphi_j - \varphi_i}{2}\right) + \cos(\varphi_i)\cos(\varphi_j)sin^2\left(\frac{\lambda_j - \lambda_i}{2}\right)}\right) \quad (1)$$

where φ_i, φ_j and λ_i, λ_j are the latitude and longitude respectively, of prescribers i and j. The distances calculated are in miles. Thus, all physicians prescribing pain medications, having the same specialty, and within a certain threshold distance of each other were in each other's social network. In this paper, the threshold distances selected were the 12.5^{th}, 25th and 50th percentiles of all distances calculated between each physician (12.5^{th} percentile: 323 miles; 25th percentile: 527 miles; 50th percentile: 919 miles). Once the social network was created, each physician was given a diffuser level based on when he/she first prescribed pain medications with respect to his/her personal network. The algorithm for assigning a diffuser level is as follows:

1. Prescribing physicians are given an integer diffuser level in the set {1, 2…n}
2. A physician *i* is given diffuser level (n) = 1 if he/she prescribed in the first month or is the first person in his/her personal network to prescribe the pain medications, i.e., all members of his/her personal network had a time of adoption later than *i*
3. The diffuser level assigned to a physician is n+1 where n is the diffuser level of physician i who is in the personal network of j but has prescribed the pain medications earlier than j.

3.2.1 Algorithm for Social Network Analysis

The algorithm used for social network analysis can be summarized as follows:

1. Calculate distances between all prescribers of pain medications using *haversine* [20] distance formula
2. Set a distance threshold for creating the social network

3. Create a social network by connecting physicians satisfying the vicinity and specialty assumptions
4. Find the key-opinion leaders i.e., prescribers with the highest number of connections
5. Assign diffuser level to each prescriber
6. Apply the diffusion of innovation theory to the social network to categorize the physicians into the five adopter categories

Based upon the algorithm above, key-opinion leaders were those physicians who were categorized as early adopters and whose personal network had the highest number of prescribers of the medication.

3.3 Visualization

Parallel Coordinates are becoming popular methods for data visualization, especially for multivariate data. The technique was proposed by Inselberg for analysis of hyper-dimensional geometry [21]. To show a set of points in an n-dimensional space, a backdrop is drawn consisting of parallel lines, typically vertically and equally spaced. A point in an n-dimensional space is represented as a polyline with vertices on the parallel axes; the position of the vertex on the i-th axis corresponds to the i-th coordinate of the point. In order to make the interface more interactive various functionalities were incorporated by making use of the d3 parallel coordinates library of JavaScript and enabling brushing experience for the user [22, 23]. This library follows an object-oriented design and consists of core functionalities implemented in JavaScript. It provides APIs that are used for further development of new features [23]. For testing our implementation, we used Parallel Coordinates on a large medication refill dataset (5000+ patients).

4 Results

4.1 Patient Journey

The Apriori algorithm [19] measures the quality of association rules using confidence of the rule. The confidence of a rule is the number of cases in which the rule is correct related to the number of cases in which it is applicable. Based upon the Apriori algorithm, we found the following three rules with 100% confidence among patients:

1. *If patients suffer from morbid obesity, then they go for gastric bypass and gastric restrictive procedures and consume pain medications*
2. *If patients are females and they experience infections related to giving birth, then they go for spine-related surgeries and consume pain medications*
3. *If patients have severe knee conditions (e.g. osteoarthritis), then they go for total knee replacements and consume pain medications*

As the above rules had 100% confidence, whenever the "if" condition occurred in data, then the "then" part of the rule occurred with 100% probability. These rules indicated that the same pain medications were used to treat a number of post-operative patient conditions.

Furthermore, results showed that more females went for refills as compared to males among both inpatients and outpatients (Fig. 1a and 1b). Total number of males and females going for refill were 4,800 and 11,057, respectively, for inpatients (Fig. 1a) and 8,543 and 14,762, respectively, for outpatients (Fig. 1b). On average both males and females went for 4 refills among inpatients (Fig. 1a) and outpatients (Fig. 1b) (shown as the vertical dotted line). In addition, the average number of refills and the distribution of refills were the same among both inpatients and outpatients.

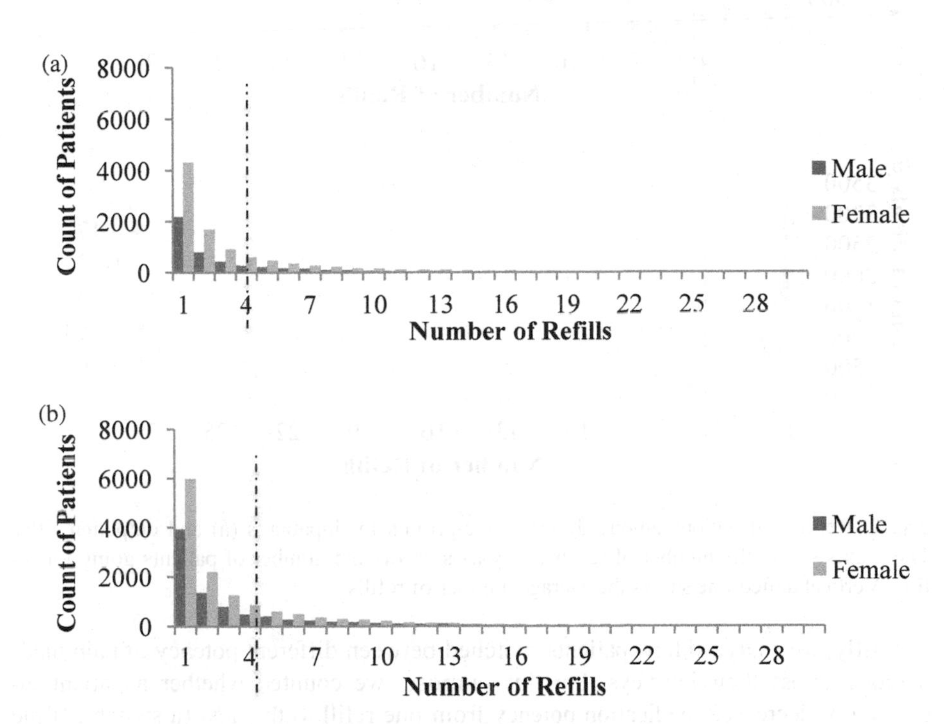

Fig. 1. Refill distributions among males and females for inpatients (a) and outpatients (b). The x-axis shows the number of refills and y-axis shows the number of patients going for refills. Vertical dotted line shows the average number of refills.

Next, we analyzed the distribution of refills with respect to different age groups among inpatients and outpatients (Fig. 2a and 2b). As shown in Fig. 2a, the total number of inpatients belonging to the age-group 0-17, 18-34, 35-44, 45-54, and 55-64 were 35, 2,468, 3,164, 4,705, and 4,664, respectively. Similarly, the total number of outpatients belonging to the age-group 0-17, 18-34, 35-44, 45-54, and 55-64 were 44, 3,964, 5,643, 7,247, and 5,449, respectively. Overall, there was more number of patients belonging to 45-54 and 55-64 age-groups refilling pain medications compared to other age groups. Interestingly, the average number of refills were same (~ 4) for both inpatients and outpatients.

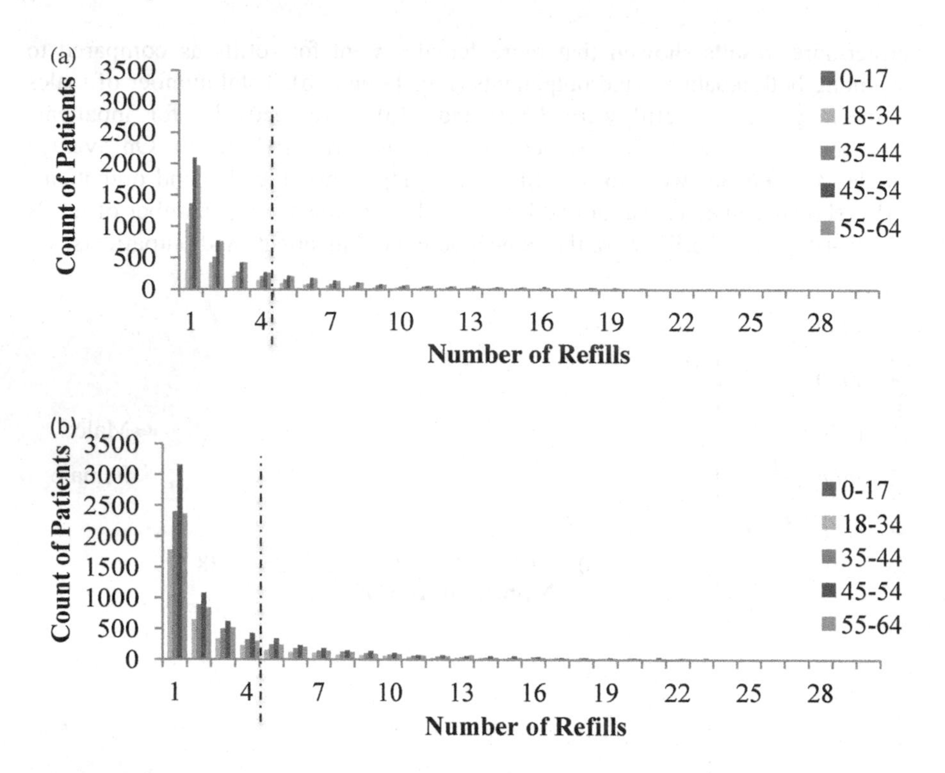

Fig. 2. Refill distributions among different age-groups for inpatients (a) and outpatients (b). The x-axis shows the number of refills and y-axis shows the number of patients going for refills. Vertical dotted line shows the average number of refills.

Lastly, we analyzed how patients switched between different potency of pain medications across their journeys. For this purpose, we counted whether a patient increased or decreased medication potency from one refill to the next (a switch). If the patient consumed the same potency across two consecutive refills, then she did not switch. Next, using this switch data, we computed the switch ratio as:

$$\text{Switch Ratio} = \frac{m}{n} \tag{2}$$

Where, m is the total number of times a patient switched medication potency across her journey and n is the total number of possible switches (= number of refills – 1). Fig. 3 shows the distribution of potency switching among patients using the switch ratios. As shown in Fig. 3, the distribution of switch ratios had a bimodal distribution with two peaks at 0.0-0.1 switch ratio and 0.9-1.0 switch ratio. Thus, there existed two kinds of switch-ratio behaviors: Patients who switched very little between different potencies and patients who switched a lot between different potencies.

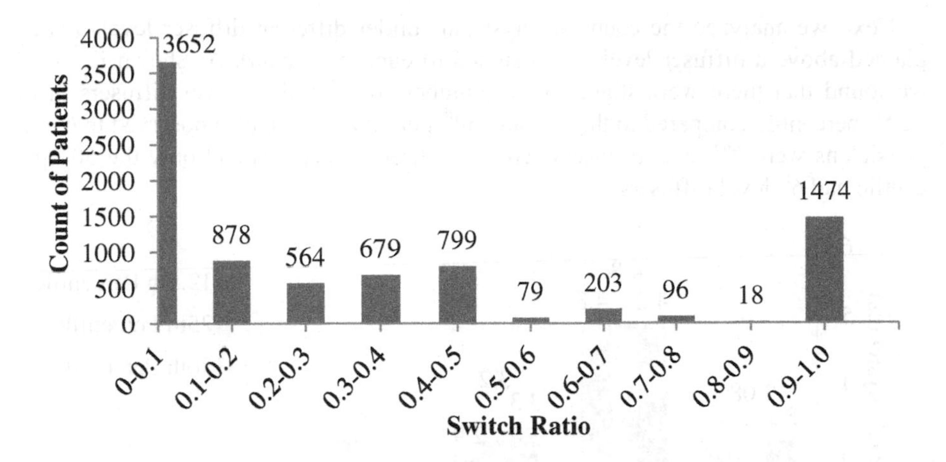

Fig. 3. Distribution of potency switching among patients. The x-axis shows the switch ratio and the y-axis shows the number of patients showing a particular switch ratio.

4.2 Social Network Analysis

In this section, we look into the results of how physicians are connected with other physicians in their network. Using the algorithm described above, first, we constructed a list of top-500 key-opinion leaders. Among this list, we found that most opinion leaders were from mid-western and southern regions of USA[2] for different percentile distances (Fig. 4). Among physicians, the specialties with the highest number of prescribers were the following (in decreasing order): Family Medicine, Internal Medicine, and Nurse practitioners.

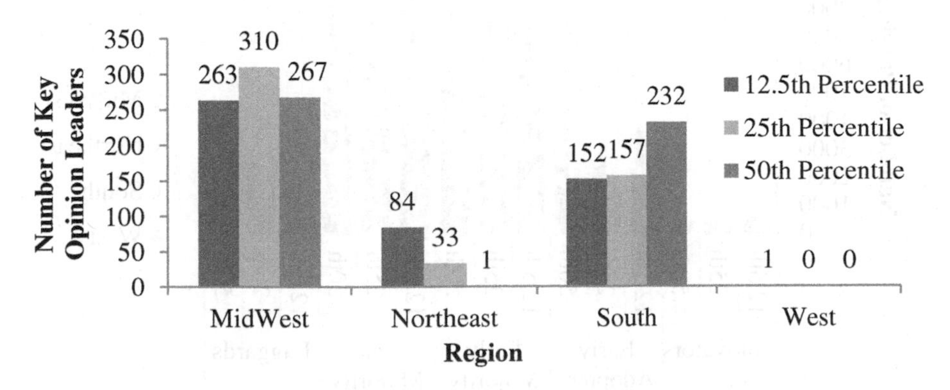

Fig. 4. Region-wise key-opinion leaders for 12.5^{th}, 25th and 50th percentile.

[2] The regions were divided as per U.S. Census Bureau Regions and Divisions.

Next, we analyzed the count of physicians under different diffuser levels (as explained above a diffuser level was assigned to each physician). As shown in Fig. 5, we found that there were slightly more number of 3rd and 4th level diffusers in the 12.5th percentile compared to the 25th and 50th percentiles. We also observed that most physicians were 2nd level diffusers across all three percentiles and only the 50th percentile had 6th level diffusers.

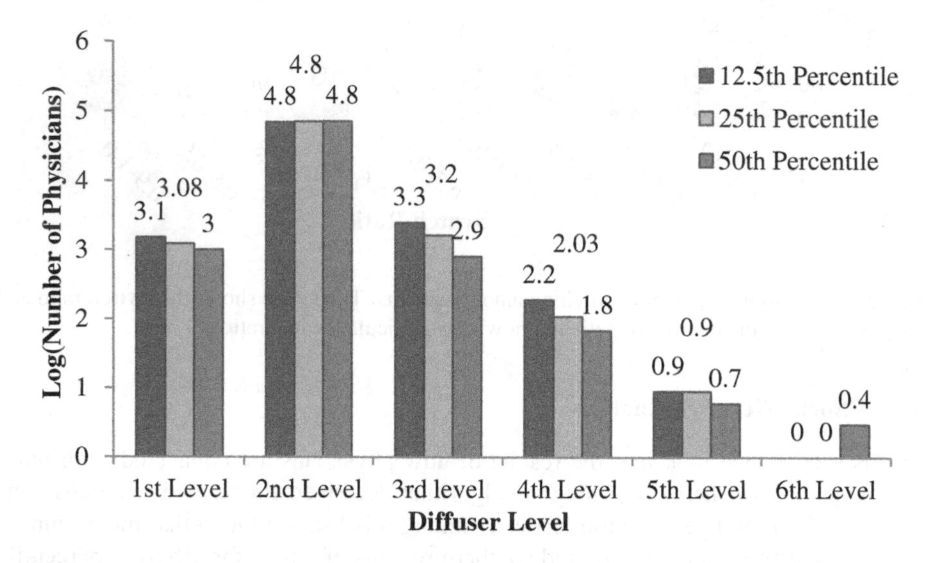

Fig. 5. Log (Number of Physicians) of Diffuser Level at 12.5th, 25th, and 50th percentile.

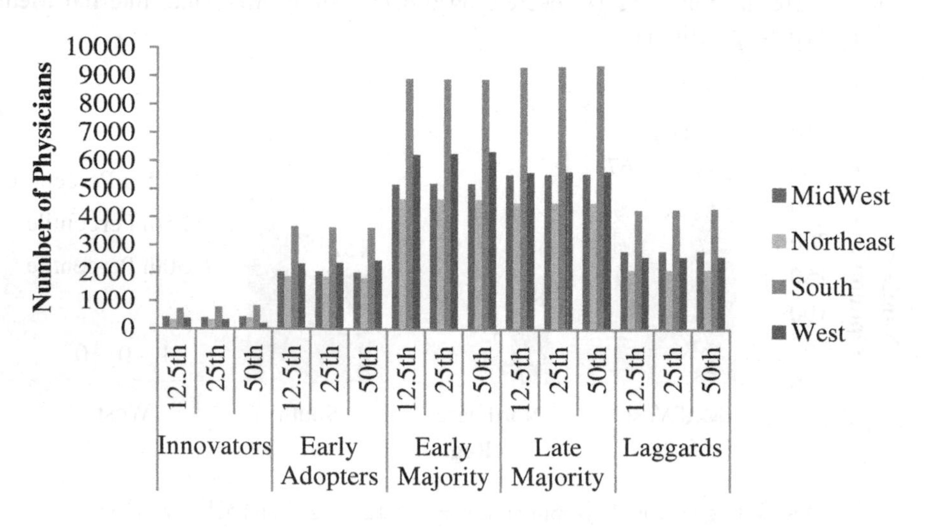

Fig. 6. Region-wise distribution of adopter categories across 12.5th, 25th and 50th percentile.

Next, using the theory of diffusion of innovation [7], we calculated the number of physicians in the different diffuser categories as well as distance percentiles (Fig. 6). As shown in Fig. 6, majority of physicians were from the southern region across all diffuser categories and different percentiles. The second highest numbers of physicians were from mid-west for innovator and laggards categories and from west for early-adopters, early-majority, and late-majority categories, respectively.

4.3 Visualization

In this section, we describe results of visualizing data of patients using parallel coordinates.

4.3.1 Distribution of Data in Different Dimensions by Brushing a Particular Dimension

It is important to visualize how a selected (brushed) dimension is distributed among other dimensions. For example, how a patient age-group 35 to 44 years is distributed among their healthcare plan type (Fig. 7). For this functionality, a user first needs to select a subset of data via brushing on a dimension (e.g., AGEGRP in Fig. 7). Then, the user needs to click on a different dimension's header (PLANTYP) to display a list showing the count and percentages of various values within the clicked dimension (shown in the box on the right in Fig. 7).

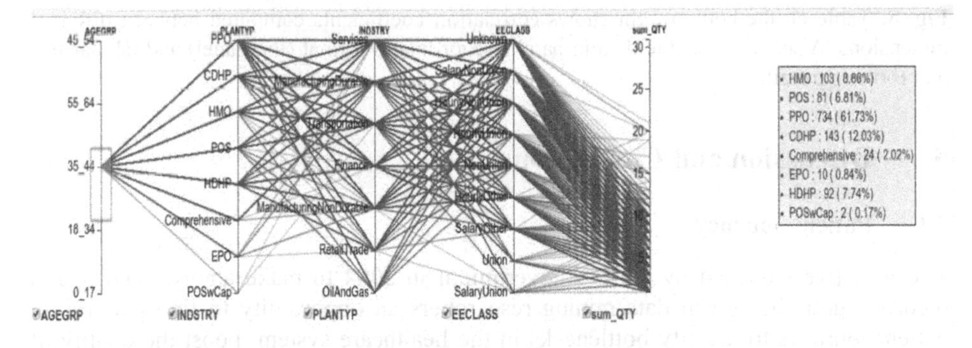

Fig. 7. Figure shows different dimensions (AGEGRP, PLANTYP, INDSTRY etc.) in parallel coordinate system. AGEGRP is brushed at the point 35_44 and PLANTYP header is clicked. The table on the right shows the distribution of different PLANTYP with values and percentages.

4.3.2 Correlation Between Dimensions

It is important to visualize if there is interdependency (correlation) among the different dimensions of a multivariate dataset. In order to find correlation among categorical dimensions (e.g., gender, region, etc.) integer codes (starting from 1 and above) were assigned in lexicographical order of the category. For numerical categories values were directly used to calculate correlation coefficients. Fig. 8 shows an example where correlation coefficients have been computed between different dimensions (some numerical and some categorical). The correlation matrix is dynamic in nature and changes whenever a user brushes a subset of points in a certain dimension (e.g., AGEGRP is brushed between points 18_34 and 35_44 and correlations are computed only for this data range).

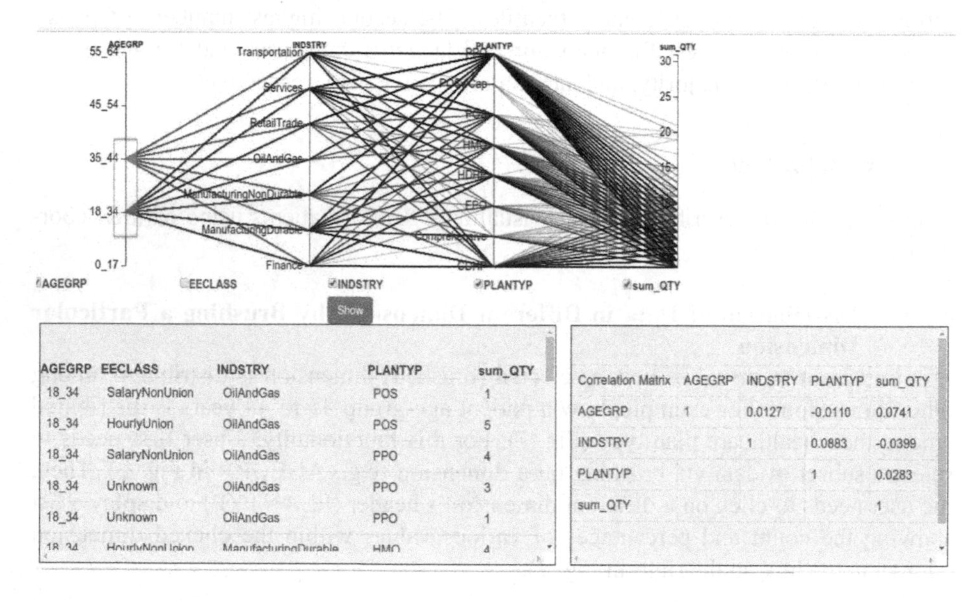

Fig. 8. Table on the bottom-right shows correlation coefficients computed between different dimensions. Also, shown is the data in parallel-coordinates format (top panel) and tabular format (bottom panel).

5 Discussion and Conclusions

5.1 Patient Journey

The initiative launched by the US government in 2004 to make anonymized health records public has given data-mining researchers an opportunity to find patterns in patient journeys to identify bottlenecks in the healthcare system, boost the quality of clinical role, and eliminate unnecessary activities.

First, we found that certain pain medications were used to treat a number of post-operative patient conditions. This result shows that pain medications are being used by pharmaceutical industry across a wide spectrum of ailments among patients. Thus, it would be best for start-up companies to align their drug production strategies to those medications that are robust against a number of illnesses; rather, being focused on only a few ailments. Second, results showed that more females went for refills as compared to males among both inpatients and outpatients. One reason for this result could be the fact that females go through a number of medical procedures (like child-birth and spinal-cord injuries), which make them consume more pain medications compared to males. Third, results showed that there were more number of patients belonging to 45-54 and 55-64 age-groups refilling pain medications compared to other age groups. As per evidence from literature [26], knee problems are one of the most common health problems among middle-aged patients. Lastly, we found there existed two kinds of switch-ratio behaviors among patients' potency-consumption patterns. Thus, it seems that physicians are good at prescribing the right quantity of medication to majority of people. However, there also exists a large population of patients, where the pain medications require trial-and-error adjustments before the right potency is prescribed.

5.2 Social Network Analysis

As per theory of diffusion of innovation [7], innovators and early-adopters are known for their innovativeness and ability to take risks. Innovators are global leaders while early adopters are local leaders. As such, early-adopters have a higher influence as compared to innovators. In this paper, we classify physicians as key-opinion leader who are early-adopters, and have the highest number of prescribers in their personal network that they have influenced in diffusing pain medications.

First, we found that even though the innovation was pain medications, the highest number of prescribers were from the specialty family medicine rather than pain medicine. This finding seems counter-intuitive; however, it could simply be explained by the fact that family-medicine doctors are general physicians available in larger numbers to whom patients go to at the onset of sickness. In contrast, pain medicine is a specialized field where doctors would be fewer in number and only referred to by family-medicine doctors as a second step in patient journeys. Second, results show that most physicians are second level diffusers i.e. most physicians wait for other physicians in their personal network to prescribe before they themselves prescribe. This is also consistent with the diffusion of innovation theory [7], which shows fence-sitting effects, as bulk percentage of adopters are early-majority (34%) and late-majority (34%) rather than innovators or early-adopters. Third, our results show that there were slightly more number of 3rd and 4th level diffusers in the 12.5th percentile compared to the 25th and 50th percentiles. This is because in a smaller geographical area (12.5th percentile), people are likely to know each other and diffuse innovation across their small community; however, in a larger geographical area, people are less likely to know each other due to distance and would be less likely to diffuse innovation. Lastly, we found that southern region of US shows the highest concentration of physician prescribing pain medications across all the adopter categories. This peak for southern region was followed by western region; with a slight deviation in the case of innovators and laggards, where mid-western region was higher. This may be due to the fact that southern region has warmer temperatures and denser population followed by west and mid-western region [25].

5.3 Visualization

It became quite evident that scatter-grams and conventional methods are not sufficient enough for visualization of Big-Data in healthcare and other domains. The tool created on parallel coordinates technologies is very interactive and helps data analysts to mine data more efficiently and find meaningful patterns.

6 Future Scope

Patient journeys can provide a roadmap to create a better healthcare plan. In the future we will enhance patient journeys to highlight reasons that made patients or physicians to change their medication/treatment. We will also interleave diagnosis and procedures in the patient journeys which will explain reasons behind prescribing a particular medication.

Social network analysis can be an efficient and powerful tool to find key opinion leaders. In the future we will consider referral patterns and nomination studies to strengthen the reliability of the connections between the physicians and also use the centrality measures (e.g., eigen-vector centrality) as a metric for finding key-opinion leaders.

The visualization tools could be extended to incorporate features like pie-chart for distribution within a dimension, manual assignment of codes for non-numeric attributes for calculation of correlation, and showing only significant correlation among dimensions using t-distributions and *p*-values. The tools can also include linear-regression models between dimensions using correlations found between dimensions.

These and other ideas are some of the immediate next steps that we plan to take in this ongoing project.

Acknowledgement. The project was supported from grants (awards: #IITM/CONS/PPLP/VD/03 and #IITM/CONS/PPLP/VD/05) to Varun Dutt.

References

1. Curry, J., McGregor, C., Tracy, S.: A communication tool to improve the patient journey modeling process. In: Conference Proceedings of the International Conference of IEEE Engineering in Medicine and Biology Society, pp. 4726–4730 (2006)
2. Ben-Tovim, D.I., et al.: Patient journeys: the process of clinical redesign. Medical Journal of Australia **188**(6), S14 (2008)
3. Trebble, T.M., et al.: Process mapping the patient journey through health care: an introduction. BMJ **341**(7769), 394–397 (2010)
4. NHS Modernisation Agency. Process mapping, analysis and redesign. Department of Health, London, pp. 1–40 (2005)
5. Valente, T.W.: Social network thresholds in the diffusion of innovations. Social networks **18**(1), 69–89 (1996)
6. Ryan, B., Gross, N.C.: The diffusion of hybrid seed corn in two Iowa communities. Rural Sociology **8**, 15–24 (1943)
7. Rogers, E.M.: Diffusion of Innovation, 3rd edn. Free Press, New York (1983)
8. Valente, T.W.: Diffusion of Innovation and policy decision-making. Journal of Communications **43**, 30–45 (1993)
9. Knoke, D., Kuklinski, J.H.: Network analysis. Sage, Newbury Park (1982)
10. Burt, R.S., Minor, M.J.: Applied Network Analysis. Sage, Newbury Park (1983)
11. Wellman, B.: Structural analysis: From method and metaphor to theory and substance. Contemporary Studies in Sociology **15**, 19–61 (1997)
12. Scott, J.: Network Analysis: A Handbook. Sage, Newbury Park (1991)

13. Rogers, E.M.: Categorizing the adopters of agricultural practices. Rural Sociology **23**, 345–354 (1958)
14. Beal, G.M., Bohlen, J.M.: How Farm People Accept New Ideas. Cooperative Extension Service Report 15, Ames, IA (1955)
15. Valente, T.W., Rogers, E.M.: The origins and development of the diffusion of innovations paradigm as an example of scientific growth. Science communication **16**(3), 242–273 (1995)
16. Anderson, J.G., Jay, S.J.: The diffusion of medical technology: Social network analysis and policy research. The Sociological Quarterly **26**(1), 49–64 (1985)
17. Mahajan, V., et al.: Determination of adopter categories by using innovation diffusion models. Journal of Marketing Research, 37–50 (1990)
18. Big Data Analytics, Time-Series Database, Kx Systems. http://www.kx.com
19. Agrawal, R., Srikant, R.: Fast algorithms for mining association rules. In: Proceedings of the 20th International Conference on Very Large Data Bases, VLDB, vol. 1215 (1994)
20. Don Josef de Mendoza y Rios, F.R.S.: Recherches sur les principaux problemes de l'astronomie nautique. In: Proceedings of the Royal Society (1796)
21. Johansson, J.: On the Usability of 3D Display in Parallel Coordinates: Evaluating the Efficiency of Identifying 2D Relationships (2013)
22. Bostock, M.: Parallel Coordinates, November 5, 2011
23. Parallel coordinates (0.7.0) A visual toolkit for multidimensional detectives
24. Nutrient Contents - Parallel Coordinates. http://www.exposedata.com
25. U.S. Population By Region (1990–2010). http://www.infoplease.com
26. Heidari, B.: Knee osteoarthritis prevalence, risk factors, pathogenesis and features: Part I. Caspian Journal of Internal Medicine **2**(2), 205–212 (2011)
27. The White House, authors. Transforming Health Care: The President's Health Information Technology Plan, December 16, 2004. http://www.whitehouse.gov/infocus/technology/economic_policy200404/chap3.html

Feature Selection for Handling Concept Drift in the Data Stream Classification

Pavel Turkov[1,3], Olga Krasotkina[2,3](✉), Vadim Mottl[2,3], and Alexey Sychugov[1,3]

[1] Tula State University, 92 Lenin Ave., Tula 300600, Russia
ko180177@yandex.ru
[2] Moscow State University, Leninskie Gory, Moscow 119991, Russia
[3] Computing Center of the Russian Academy of Sciences, 40 Vavilov St., Moscow 119333, Russia

Abstract. With the advance in both hardware and software technologies, streaming data is ubiquitous today, and it is often a challenging task to store, analyze and visualize such rapid large volumes of data. One of difficult problems in the data stream domain is the data streams classification problem. The traditional classification algorithms have to be adapted to run in a streaming environment because of the underlying resource constraints in terms of memory and running time. There are at least three hard aspects in the data streams classification: large length, concept drift and feature selection. Concept drift is a common attribute of data streams that occurs as a result of changes in the underlying concepts. Feature selection has been extensively studied from a conventional mining perspective, but it is a much more challenging problem in the data stream domain. The concept drift and large length make impossible applying classical feature selection methods in the learning procedure. This paper proposes a new Bayesian framework to feature selection in data streams pattern recognition problem. We suggest a hierarchical probabilistic model with sparse regularization for estimation of decision rule parameters. The proposed approach gives a strong Bayesian formulation of the shrinkage criterion for predictor selection. Experimental results show that the proposed framework outperforms other methods of concept drift analysis.

Keywords: Feature selection · Concept drift · Changing environment · Data streams classification · Supervised learning · Pattern recognition

1 Introduction

Most of the previous and current research in data stream mining is carried out in static environments, wherein a complete dataset is presented to the learning algorithm. Over these years, many solutions to the static classification have been developed, and several quite accurate classifiers are now available in large scale. However, in some of the newest and latest applications, learning algorithms work

P. Perner (Ed.): MLDM 2016, LNAI 9729, pp. 614–629, 2016.
DOI: 10.1007/978-3-319-41920-6_48

in dynamic environments. Traffic management, sensor networks, monitoring, web log analysis or telecommunication are examples of such applications. Concept drift occurs when the concept about which data are being collected changes in time. Such changes are reflected in incoming instances and decrease the accuracy of classifiers learned from past training tuples (input data examples). In this kind of tasks, both the nature of changes in the environment and the fact of changes itself are often hidden from direct observation, and this makes learning even more difficult [1] . Examples of real life concept drifts include monitoring systems, financial fraud detection, spam categorization, weather predictions, and customer preferences [2]. The issue of how to develop a lightweight learning algorithm on data streams has become a research hotspot. Different approaches to mining data streams with concept drift include instance selection methods, drift detection, ensemble classifiers, option trees and using Hoeffding boundaries to estimate classifier performance [3–7].

Unfortunately, most of existing data stream classification techniques address only the infinite length and concept-drift problems. At the same time, the sheer amount of data that must be collected and processed is one of the main causes of the slow speed and heavy overhead. For this reason, data processing algorithms must reduce the amount of features being processed [8]. There are three main approaches to feature selection: filtering, wrapper and embedded approaches. Filtering approach consists in removing irrelevant features from the original feature set before sending it to the learning algorithm. Typically, the initial dataset is analyzed for identifying a subset of relevant features that appears to be sufficient for classification. In wrapper approach the relevance feature subset is meant to be chosen using the classifier's power. Wrappers use an appropriate search algorithm applied to the space of possible features, and evaluate each candidate subset by running the goal classifier [9,10]. The goal of the method is finding a feature subset that provides the best classification quality. In embedded techniques, the search for an optimal subset of features is built into the classifier construction, and can be seen as a search in the combined space of feature subsets and hypotheses [11–16]. Embedded methods have the advantage that they include the interaction with the classification model. Actually, these techniques are working with classifying entities and searching features at the same time.

Detecting an informative feature subset in a large volume of data stream is a difficult problem. First, the data stream could be infinite, so any off-line feature selection algorithm that store the entire data stream for analysis will run out of memory. Second, the feature importances change dynamically over time due to the concept drift, important features may become insignificant, and vice-versa. Third, for online applications, it is important to obtain the feature subset in close to real-time. In general, we can mark that all the existing algorithms for feature selection, both in the batch [17] and online setting [18,20], can't handle large data streams effectively due to the limited memory and CPU time.

In this paper, we suggest a hierarchical probabilistic model for estimation of changing decision rule parameters with sparse regularization. The proposed approach gives a strong Bayesian formulation of the shrinkage criterion for predictor selection. In our previous papers for rank [21] and Cox regression [22], we have shown that such type of shrinkage is unbiased, has grouping and oracle properties, its maximum risk converges to a finite value. Experimental results show that the proposed framework outperforms the existing methods of concept drift analysis on both simulated data and publicly available real data sets.

2 A Hierarchical Probabilistic Model with Feature Selection for the Pattern Recognition Problem Under Concept Drift

2.1 The Bayesian Approach to Feature Selection in the Pattern Recognition Problem

Let us start with the statement of the feature selection problem in pattern recognition without concept drift. Let every instance of the environment $\omega \in \Omega$ be represented by a point in the linear feature space $\mathbf{x}(\omega) = \big(x^1(\omega), \ldots, x^n(\omega)\big) \in \mathbb{X} = \mathbb{R}^n$, and its hidden membership in one of two classes be determined by the class index $y(\omega) = \mathbb{Y} \in \{1, -1\}$. The function

$$\big(\mathbf{x}(\omega), y(\omega)\big) : \Omega \rightarrow \mathbb{X} \times \mathbb{Y} \tag{1}$$

is known within the bounds of a finite training set $\big\{\big(\mathbf{x}(\omega_j), y(\omega_j)\big) = (\mathbf{x}_j, y_j), j = 1, ..., N\big\}$ where N is the number of observations. It is required to continue the function (1) onto the entire set Ω, so that it would be possible to estimate the class membership of new objects $\omega \in \Omega$ not represented in the training set. Assume that there exists a probability space $\mathbb{X} \times \mathbb{Y}$ with some joint distribution density $\varphi(\mathbf{x}, y)$. As the probabilistic model of the data source, we shall consider two parametric families of distribution densities $\varphi_1(\mathbf{x}|\mathbf{a}, b)$ and $\varphi_{-1}(\mathbf{x}|\mathbf{a}, b)$, $\mathbf{a} \in \mathbb{R}^n$, $b \in \mathbb{R}$, associated with two class indexes $y = \pm 1$. These two class-conditional probability distributions are concentrated primarily at the different sides of an unknown discriminant hyperplane $\mathbf{a}^T\mathbf{x} + b \gtrless 0$:

$$\varphi(\mathbf{x}|y, \mathbf{a}, b, c) \propto \exp\Big\{-c \max\big[0,\ 1 - y(\mathbf{a}^T\mathbf{x} + b)\big]\Big\}. \tag{2}$$

In the strict sense, these functions are no probability densities, because their integrals over $\mathbb{R}^n$ do not equal any finite number. But we use the terminology by Morris De Groot, who calls such density functions improper distribution densities [23]. This pair of improper densities completely defines a probabilistic model of the data source. Both distributions (2) are "uniform" along the infinite axes parallel to the hyperplane, as well as in the infinite areas $\mathbf{a}^T\mathbf{x} + b > 1$ for the entities of class $y = 1$, and $\mathbf{a}^T\mathbf{x} + b < -1$ for the opposite class $y = -1$.

As to the additional structural parameter $c > 0$, it controls the assumed intersection degree of the areas of two clases in the feature space $\mathbf{x} \in \mathbb{R}^n$. If $c \rightarrow \infty$, the points of different classes are completely separable by a discriminant hyperplane. The less the norm of the direction vector $||\mathbf{a}|| = (\mathbf{a}^T\mathbf{a})^{1/2}$, the greater is the guaranted Euclidean margin $\pm\varepsilon$ between the two aries in $\mathbb{R}^n$. But in the case $c < \infty$, the feature vectors of the respective cla1ss $y = \pm 1$ may violate the inequality $\mathbf{a}^T\mathbf{x} + b \gtrless \pm 1$ and even $\mathbf{a}^T\mathbf{x} + b \gtrless 0$.

Actually, the conditional densities (2) express, in the probabilistic form, the essence of the famous Support Vector Machine (SVM) [24–26], whose main idea consists just in finding the hyperplane being a compromise between the mutually contradictory desires to separate the points of two classes with the greatest margin and minimize the violations of this requirement.

We suppose the random feature vectors of single entities in the training set to be conditionally independent, thus, the joint distribution density for fixed class indexes will be the product:

$$\Phi(\mathbf{x}_1, ..., \mathbf{x}_N | y_1, ..., , y_N, \mathbf{a}, b, c) = \prod_{j=1}^{N} \varphi_{y_j}(\mathbf{x}_j | \mathbf{a}, b, c). \tag{3}$$

Let, further, the direction vector $\mathbf{a} = (a_1, ..., a_n) \in \mathbb{R}^n$ of the discriminant hyperplane $\sum_{i=1}^{n} a_i x_i + b \gtrless 0$ be considered as vector of independent normal random variables with zero mean values and, in the general case, different variances $\mathbf{r} = (r_1, ..., r_n)$:

$$\psi(a_i | r_i) = \frac{1}{\sqrt{2\pi r_i}} \exp(-(1/2r_i)a_i^2). \tag{4}$$

No prior information is assumed concering the hyperplane's bias b, hence,

$$\Psi(\mathbf{a}, b | \mathbf{r}) \propto \Psi(\mathbf{a} | \mathbf{r}) \propto \prod_{i=1}^{n} (1/r_i)^{-1/2} \exp\left(-\frac{1}{2} \sum_{i=1}^{n} \frac{1}{r_i} a_i^2\right). \tag{5}$$

Finally, we shall consider independent a priori gamma distributions of inverse variances $\gamma(1/r_i | \alpha, \beta) \propto (1/r_i)^{\alpha - 1} \exp[(-\beta(1/r_i)]$ with identical mathematical expectations $E(1/r_i) = \alpha/\beta$ and variances $Var(1/r_i) = \alpha/\beta^2$. To get rid of the double parametrization, we set $\alpha = 1 + 1/(2\mu)$ and $\beta = 1/(2\mu)$. We have now a parametric family of gamma distributions defined by only one parameter $\mu \geq 0$, such that $E(1/r_i) = 1 + 2\mu$ and $Var(1/r_i) = 2\mu(1 + 2\mu)$:

$$\gamma(1/r_i | \mu) \propto (1/r_i)^{1/(2\mu)} \exp[(-(1/2\mu)(1/r_i)]. \tag{6}$$

The a priori joint density of the variance vector $\mathbf{r} = (r_1, ..., r_n)$ is the product:

$$G(\mathbf{r} | \mu) \propto \prod_{i=1}^{n} \left\{(1/r_i)^{1/(2\mu)} \exp[(-(1/2\mu)(1/r_i)]\right\}. \tag{7}$$

The joint a posteriori distribution density of hyperplane parameters $(\mathbf{a} = (a_1, ..., a_n), b)$ and variances $\mathbf{r} = (r_1, ..., r_n)$ (5) will be defined by the Bayes formula

$$\begin{aligned} &p((\mathbf{a}, b), \mathbf{r} \,|\, \mathbf{x}_j, y_j,\ j = 1, ..., N, c, \mu) \propto \\ &\quad \propto \Psi(\mathbf{a} \,|\, \mathbf{r}) G(\mathbf{r} \,|\, \mu) \Phi(\mathbf{x}_j,\ j = 1, ..., N \,|\, y_j,\ j = 1, ..., N, \mathbf{a}, b, c). \end{aligned} \tag{8}$$

The Bayesian estimation of hyperplane parameters $(\mathbf{a}, b)$ jointly with variances $\mathbf{r}$ according to (8) and (7) results in the adaptive training criterion

$$\begin{aligned} ((\hat{\mathbf{a}}, \hat{b}), \hat{\mathbf{r}} | c, \mu) &= \arg\max p(\mathbf{a}, b | \mathbf{x}_j, y_j, j = 1, ..., N, c) = \arg\max[\ln \Psi(\mathbf{a} | \mathbf{r}) + \\ &\ln G(\mathbf{r} | \mu) + \ln \Phi(\mathbf{x}_j,\ j = 1, ..., N | y_j,\ j = 1, ..., N, \mathbf{a}, b, c)]. \end{aligned} \tag{9}$$

Density $\Psi(\mathbf{a}|\mathbf{r})$ (5) is the normal conditional prior distribution of the direction vector with respect to fixed variances, and $G(\mathbf{r}|\mu)$ (7) is the hyperprior gamma distribution of the conditions. We consider the product $\Psi(\mathbf{a}\,|\,\mathbf{r})G(\mathbf{r}|\mu)$ and its logarithm $\ln\Psi(\mathbf{a}\,|\,\mathbf{r})+\ln G(\mathbf{r}|\mu)$ as a hierarchical prior model of the hidden direction vector $\mathbf{a}$, in this case, with two levels of hierarchy.

It is just this hierarchical prior model, which provides the pronounced property of the Bayesian estimate (9) to emphasize the elements of the direction vector that are adequate to the trainers data $0<\hat{r}_i\ll\infty$, i.e., $\hat{a}_i^2\gg 0$ (5), and to suppress $\hat{r}_i\to\infty$ up to negligibly small values $\hat{a}_i^2\to 0$ the redundant ones.

Indeed, if $\mu\to 0$ in (6), the random values $1/r_i$ approach the nonrandom identity $1/r_i\cong ...\cong 1/r_n\cong 1$ because $\big[E(1/r_i)\to 1, Var(1/r_i)\to 0\big]$, and all the squared elements of the direction vector a_i^2 are equally penalized in (5) by the requirement $\ln\Psi(\mathbf{a}|\mathbf{r})\to\max$. But the growing parameter $\mu\to\infty$ allows the independent nonnegative values $1/r_i$ to arbitrarily differ from each other $\big[E(1/r_i)\to\infty, Var(1/r_i)\to\infty\big]$, and the requirements $\big[\ln G(\mathbf{r}|\mu)\to\max,\ \ln\Psi(\mathbf{a}|\mathbf{r})\to\max\big]$ enforce their growth $1/r_i\to\infty$.

As a result, the estimated discriminant hyperplane $\sum_{i=1}^{n}\hat{a}_i x_i+\hat{b}\gtrless 0$ will take into account only a subset of most informative featutes x_i and practically ignore the others [26]. The parameter $0<\mu<\infty$ serves as a selectivity parameter. If $\mu=0$ the training criterion (9) is the classical SVM without feature selection ability, and if $\mu\to\infty$ it becomes extremely selective.

Thus, such a technique is a probabilistic generalisation of the traditional SVM, which endows it with the ability to suppress redundant features at the preset selectivity level.

2.2 A Probabilistic Concept Drift Model Based on the Markov Property of the Drifting Discriminant Hyperplane

We shall use the combined notaion $\mathbf{w}=(\mathbf{a},b)\in\mathbb{R}^{n+1}$ for the parameters of the discriminant hyperplane, namely, its direction vector $\mathbf{a}\in\mathbb{R}^n$ and bias $b\in\mathbb{R}$, and, respectively, the extended notation $\mathbf{x}=(\mathbf{x},1)\in\mathbb{R}^{n+1}$ for the feature vector. Thus, the hyperplane equation will obtain the form $\mathbf{w}^T\mathbf{x}\gtrless 0$.

The key element of our Bayesian approach to the concept drift problem is treating the time-varying vector parameter of the drifting discriminant hyperplane

$$\mathbf{w}_t=(\mathbf{a}_t,b_t)=(w_{1,t},...,w_{n+1,t})\in\mathbb{R}^{n+1},\ w_{i,t}=a_{i,t}, i=1,...,n,\ w_{n+1,t}=b_t, \quad (10)$$

as a vector random process whose independent components possess the Markov property [27]:

$$\begin{aligned}&\mathbf{w}_t=q\mathbf{w}_{t-1}+\boldsymbol{\xi}_t\in\mathbb{R}^{n+1},\ E(\boldsymbol{\xi}_t)=\mathbf{0},\ E(\boldsymbol{\xi}_t\boldsymbol{\xi}_t^T)=\mathbf{Diag}(d_1,...,d_{n+1}),\\ &d_i=(1-q^2)r_i,\ i=1,...,n,\ d_{n+1}=1-q^2.\end{aligned} \quad (11)$$

Here $\boldsymbol{\xi}_t=\big(\xi_{i,t}, i=1,...,n+1\big)$ is the vector white noise $\big(E(\boldsymbol{\xi}_t\boldsymbol{\xi}_s^T)=\mathbf{0}, s\neq t\big)$ with zero mean value of independent components, each of which has its individual variance $d_i > 0$, $i = 1, ..., n+1$. We shall use the notations

$$\mathbf{r} = (r_1, ..., r_n) \text{ and } E(\boldsymbol{\xi}_t \boldsymbol{\xi}_t^T) = \mathbf{Diag}(\mathbf{r}, 1) = \mathbf{D_r} \left[(n+1)\times(n+1)\right], \tag{12}$$

assuming that coefficient q in (11) remains constant.

If $|q| < 1$, each elementary random process $w_{i,t}$ is stationary and ergodic on the infinite axis of discrete time $t = ..., 1, 2, 3, ...$. The mathematical expectation of each random process equals zero, and its asymptotic stationary variance is completely defined by the parameters of the Markov equation (11):

$$\begin{aligned} E(w_{1,t}) = E(w_{n+1,t}) = 0, \;\; Var(w_{i,t}) = Var(a_{i,t}) = \frac{d_i}{1-q^2} = r_i, \; i = 1, ..., n, \\ Var(w_{n+1,t}) = Var(b_t) = \frac{d_{n+1}}{1-q^2} = 1. \end{aligned} \tag{13}$$

We additionaly assume that the coefficient q obeys the constraints $0 \le q < 1$. This coefficient, which remauns the same for the entire vector process $\mathbf{w}_t = (\mathbf{a}_t, b_t) \in \mathbb{R}^{n+1}$, determines the assumed hidden dynamics of the discriminant hyperplane. The less the difference $1-q > 0$, the slower is the assumed drift, but all the independent random processes $w_{i,t}$ remain ergodic, and their stationary variances $Var(w_{i,t}) = r_i$, $Var(w_{n+1,t}) = 1$ (13). The equality $q = 0$ is out of interest, because in this case the direction vector will chaotically change in time instead of gradually drifting. But if $q = 1$, the Markov equation (11) turns into a strong equality $\mathbf{w}_t = \mathbf{w}_{t-1}$, and retains its probabilistic sense only in combination with some assumption on the distribution of one of the values, for instance, $\mathbf{w}_0$.

In what follows, we consider the stationary variances $\mathbf{r} = \left(r_1 > 0, ..., r_n > 0\right)$ of the first n random processes $\left(w_{1,t}, ..., w_{n,t}\right)$ (13), i.e., of the elements of the direction vector $\mathbf{a}_t = \left(a_{1,t}, ..., a_{n,t}\right)$, to be unknown. Estimation of them will be the instrument of feature selection in our technique of data stream processing. The less $Var(w_{i,t}) = r_i$, the less will be all the estimated values of the i-th element of the drifting direction vector $\left(\hat{a}_{i,t}, t = ..., 1, 2, 3, ...\right)$, and the less the weight of the i-th feature $\left(x_{i,t}, t = ..., 1, 2, 3, ...\right)$ in classification of entities $\hat{\mathbf{a}}_t^T \mathbf{x}_t + b_t = \sum_{i=1}^{n} \hat{a}_{i,t} x_{i,t} + b_t \gtrless 0$. If $r_i \to 0$, the respective feature is almost completely suppressed.

On the contrary, the stationary variance of the last random process $w_{n+1,t}$, i.e., of the hyperplane's bias b_t, is assumed to be predefined $Var(w_{n+1}) = 1$. This means that the drifting bias remains ever active.

The training set is a finite sequence of finite training batches $\left\{(\mathbf{X}_t, \mathbf{Y}_t), t = 1, ...T\right\}$, each of which consists of a finite number of training samples $(\mathbf{X}_t, \mathbf{Y}_t) = \left\{(\mathbf{x}_{j,t}, y_{j,t}), j = 1, ..., N_t\right\}$ and is associated with the respective moment of discrete time $t = 1, ..., T$. If all the direction vectors $(\mathbf{w}_t, t = 1, ..., T)$ are fixed, the feature vectors $\mathbf{X}_t = (\mathbf{x}_{j,t}, j = 1, ..., N_t)$ within and across the batches will be conditionally independent and distributed in accirdance with (3):

$$\Phi(\mathbf{X}_t, t = 1, ..., T | \mathbf{Y}_t, \mathbf{w}_t, t = 1, ..., T, c) = \prod_{t=1}^{T} \prod_{j=1}^{N_t} \varphi_{y_{j,t}}(\mathbf{x}_{j,t} | y_{j,t}, \mathbf{w}_t, c). \tag{14}$$

Under the assumption that the white noise in the Markov equation (11) is assumed to be normally distributed, the conditional probability density of each hyperplane parameter vector $\mathbf{w}_t$ with respect to its immediately previous value $\mathbf{w}_{t-1}$ will be normal, too:

$$\psi(\mathbf{w}_t|\mathbf{w}_{t-1},\mathbf{r}) \propto \mathcal{N}(\mathbf{w}_t|q\mathbf{w}_{t-1},\mathbf{D}_\mathbf{r}) = \frac{1}{|\mathbf{D}_\mathbf{r}|^{1/2}(2\pi)^{n/2}} \exp\left(-\frac{1}{2}(\mathbf{w}_t - q\mathbf{w}_{t-1})^T\mathbf{D}_\mathbf{r}^{-1}(\mathbf{w}_t - q\mathbf{w}_{t-1})\right). \tag{15}$$

If we assume, in addition, that there is no a priori information on the first value of the parameter vector $\mathbf{w}_1$, the a priori distribution density of the hidden sequence of hyperplane parameters over the entire observation interval $(\mathbf{w}_t, t = 1, ..., T)$ will the product of these conditional densities:

$$\Psi(\mathbf{w}_t, t = 1, ..., T|\mathbf{r}) = \prod\nolimits_{t=2}^{T} \psi(\mathbf{w}_t|\mathbf{w}_{t-1},\mathbf{r}). \tag{16}$$

The a priori joint density of the variance vector $G(\mathbf{r}|\mu)$ remains the same as in the case of drift absence (7).

2.3 The Training Criterion

The joint a posteriori distribution density of the entire hidden sequence of the hyperplane parameters and the variance vector will be proportional to the product of (7), (14) and (16):

$$p\Big(\mathbf{w}_t, t = 1, ..., T, \mathbf{r}|(\mathbf{X}_t, \mathbf{Y}_t), t = 1, ..., T|c, \mu\Big) \propto \Psi(\mathbf{w}_t, t = 1, ..., T|\mathbf{r})G(\mathbf{r}|\mu)\Phi\big(\mathbf{X}_t, t = 1, ..., T|\mathbf{Y}_t, \mathbf{w}_t, t = 1, ..., T, c\big). \tag{17}$$

Maximization of this joint a posteriori probability density results in a generalization of the Bayesian training criterion (9) onto the case of drift in hyperplane parameters with respect to (14) and (16):

$$\begin{aligned}(\hat{\mathbf{w}}_t, t = 1, ..., T, \hat{\mathbf{r}}|c, \mu) = \\ \operatorname*{arg\,max}_{\mathbf{w}_t, t=1,...,T,\mathbf{r}} p\Big(\mathbf{w}_t, t = 1, ..., T, \mathbf{r}|(\mathbf{X}_t, \mathbf{Y}_t), t = 1, ..., T, c, \mu\Big) = \\ \operatorname*{arg\,max}_{\mathbf{w}_t, t=1,...,T,\mathbf{r}} \Big[\ln\Psi(\mathbf{w}_t, t = 1, ..., T|\mathbf{r}) + \ln G(\mathbf{r}|\mu) + \\ \ln\Phi\big(\mathbf{X}_t, t{=}1, ..., T|\mathbf{Y}_t, \mathbf{w}_t, t{=}1, ..., T, c\big)\Big].\end{aligned} \tag{18}$$

The pair $\Psi(\mathbf{w}_t, t = 1, ..., T|\mathbf{r})G(\mathbf{r}|\mu)$ in (17) and $\ln\Psi(\mathbf{w}_t, t = 1, ..., T|\mathbf{r}) + \ln G(\mathbf{r}|\mu)$ in (18) is a straightforward generalization of the two-level hierarchical prior model of the static hidden hyperplane $\mathbf{w} = (\mathbf{a}, b)$ (Section 2.1) onto the case of a drifting one $\mathbf{w}_t = (\mathbf{a}_t, b_t)$. The joint density $\Psi(\mathbf{w}_t, t{=}1, ..., T|\mathbf{r})$ (16) is the prior conditional normal model of the drifting discriminant hyperplane with fixed stationary variances $\big[Var(a_{i,t}) = r_i, (r_1, ..., r_n) = \mathbf{r}\big]$ (13) of the drifting direction vector elements, and the hyperprior gamma model of these variances $G(\mathbf{r}|\mu)$ remains the same (7).

By substitution of distribution densities (2), (15) and (16) into (18), we obtain the following learning criterion:

$$\begin{aligned}
&(\hat{\mathbf{w}}_t, t=1,...,T, \hat{\mathbf{r}}|c,\mu) = \underset{\mathbf{w}_t, t=1,...,T,\mathbf{r}}{\arg\min}\ J(\mathbf{w}_t, t=1,...,T,\mathbf{r}|c,\mu),\\
&J(\mathbf{w}_t, t=1,...,T,\mathbf{r}|c,\mu) = (T-1)\ln|\mathbf{D}_\mathbf{r}|+\\
&\quad\sum\nolimits_{t=2}^{T}(\mathbf{w}_t - q\mathbf{w}_{t-1})^T\mathbf{D}_\mathbf{r}^{-1}(\mathbf{w}_t - q\mathbf{w}_{t-1})-\\
&\qquad 2\ln G(\mathbf{r}|\mu) + 2c\sum\nolimits_{t=1}^{T}\sum\nolimits_{j=1}^{N_t}\max\big(0,\ 1-y_{j,t}\mathbf{w}_t^T\mathbf{x}_{j,t}\big).
\end{aligned} \tag{19}$$

We use the group coordinate descent method for two groups of variables, namely, hyperplane parameters $(\mathbf{w}_t, t=1,...,T)$ and variances $\mathbf{r}=(1,...,n)$, Section 3. When one of these groups is fixed, the following two partial optimization problems form an iterative procedure:

$$\left.\begin{aligned}
&J(\mathbf{w}_t, t=1,...,T|\mathbf{r},c) =\\
&\sum\nolimits_{t=2}^{T}(\mathbf{w}_t - q\mathbf{w}_{t-1})^T\mathbf{D}_\mathbf{r}^{-1}(\mathbf{w}_t - q\mathbf{w}_{t-1})+\\
&2c\sum\nolimits_{t=1}^{T}\sum\nolimits_{j=1}^{N_t}\max\big(0,\ 1-y_{j,t}\mathbf{w}_t^T\mathbf{x}_{j,t}\big)\to\min,
\end{aligned}\right\}\quad \text{Section 3.1,} \tag{20}$$

$$\left.\begin{aligned}
&J(\mathbf{r}|\mathbf{w}_t, t=1,...,T,\mu) =\\
&(T-1)\ln|\mathbf{D}_\mathbf{r}^{-1}|+\sum\nolimits_{t=2}^{T}(\mathbf{w}_t - q\mathbf{w}_{t-1})^T\mathbf{D}_\mathbf{r}^{-1}(\mathbf{w}_t - q\mathbf{w}_{t-1})-\\
&2\ln G(\mathbf{r}|\mu)\to\min,
\end{aligned}\right\}\quad \text{Section 3.2.} \tag{21}$$

Both of of these interleaved optimization problems are convex.

3 Parameter Estimation in the Hierarchical Probablility Model

3.1 An Approximate Dynamic Programming Procedure for Estimation of the Drifting Hyperplane with Fixed Feature Weights

The Idea of Dynamic Programming. The objective function (20) depends on T variables $(\mathbf{w}_1,...,\mathbf{w}_T)$ ordered along the time axis, each of which is an $(n+1)$-dimensional vector $\mathbf{w}_t\in\mathbb{R}^{n+1}$ (10). Despite the fact that the entire number of variables equals to $T(n+1)$, the structure of the criterion possesses a specific property that enables its minimization, at least, theoretically, in only T steps.

The criterion (20) is pairwise separable, i.e., may be considered as a sum of elementary functions each of which is that of one vector variable $\mathbf{w}_t$ or two variables $(\mathbf{w}_{t-1},\mathbf{w}_t)$ immediately adjacent in discrete time. The most appropriate conputational way of solving optimization problems of such a kind is the well-known principle of dynamic programming [28,29].

The main notion of dynamic programming is that of the sequence of Bellman functions. Let us consider the criterion (20) with respect only to the initial part of the entire time interval $s=1,...,t$:

$$J_t(\mathbf{w}_s, s=1,...,t|\mathbf{r},c) = \sum\nolimits_{s=2}^{t}(\mathbf{w}_s-q\mathbf{w}_{s-1})^T\mathbf{D}_\mathbf{r}^{-1}(\mathbf{w}_s-q\mathbf{w}_{s-1})+ \\ 2c\sum\nolimits_{s=1}^{t}\sum\nolimits_{j=1}^{N_s}\max(0,\, 1-y_{j,t}\mathbf{w}_s^T\mathbf{x}_{j,s})\to\min, \tag{22}$$

If we mentally fix the last argument $\mathbf{w}_t \in \mathbb{R}^{n+1}$ and mentally minimize this criterion (20) by all the precedent variables $(\mathbf{w}_1,...,\mathbf{w}_{t-1})$, the result will be function of $\mathbf{w}_t$:

$$\tilde{J}_t(\mathbf{w}_t|\mathbf{r},c) = \min_{\mathbf{w}_s,s=1,...,t-1} J_t(\mathbf{w}_s, s=1,...,t|\mathbf{r},c) = \\ \min_{\mathbf{w}_1,...,\mathbf{w}_{t-1}} J_t(\mathbf{w}_1,...,\mathbf{w}_{t-1},\mathbf{w}_t|\mathbf{r},c). \tag{23}$$

These are just the Bellman functions $\tilde{J}_t(\mathbf{w}_t|\mathbf{r},c)$, $\mathbb{R}^{n+1}\to\mathbb{R}$, $t=0,...,T$, $\tilde{J}_0(\mathbf{w}_0|\mathbf{r},c)\equiv 0$, completely defined by the given training set $\{(\mathbf{X}_t,\mathbf{Y}_t), t = 1,...T\}$.

The fundamental property of Bellman functions is the recurrence relation that almost evidently follows from the definition (22)-(23):

$$\tilde{J}_0(\mathbf{w}_0|\mathbf{r},c)\equiv 0,\ \mathbf{w}_0\in\mathbb{R}^{n+1},\ t=0, \\ \tilde{J}_t(\mathbf{w}_t|\mathbf{r},c) = 2c\sum\nolimits_{j=1}^{N_t}\max(0,\, 1-y_{j,t}\mathbf{w}_t^T\mathbf{x}_{j,t})+ \\ \min_{\mathbf{w}_{t-1}\in\mathbb{R}^{n+1}}\Big[(\mathbf{w}_t-q\mathbf{w}_{t-1})^T\mathbf{D}_\mathbf{r}^{-1}(\mathbf{w}_t-q\mathbf{w}_{t-1}) + \tilde{J}_{t-1}(\mathbf{w}_{t-1}|\mathbf{r},c)\Big], \\ \mathbf{w}_t\in\mathbb{R}^{n+1},\ t=1,...,T. \tag{24}$$

If our computer had a "sufficiently huge" amount of memory and "sufficiently huge" computational speed, and, in addition, we had available a "sufficiently effective" method of solving the optimization problem in (24), we could recurrently compute all the Bellman functions and store them in memory. Then, the last of them would completely define the optimal value of the hyperplane parameters at the last time moment $(\hat{\mathbf{w}}_T|\mathbf{r},q,c)$:

$$(\hat{\mathbf{w}}_T|\mathbf{r},q,c) = \min_{\mathbf{w}_T\in\mathbb{R}^{n+1}} \tilde{J}_T(\mathbf{w}_T|\mathbf{r},q,c). \tag{25}$$

All the remaining elements of the sought-for solution $\big[(\hat{\mathbf{w}}_1|\mathbf{r},q,c),...,(\hat{\mathbf{w}}_T|\mathbf{r},q,c)\big]$ of the problem (20) could be found by the backward recurrence relation that almost evidently follows from (24):

$$\hat{\mathbf{w}}_{t-1} = \arg\min_{\mathbf{w}_{t-1}\in\mathbb{R}^{n+1}}\Big[(\hat{\mathbf{w}}_t-q\mathbf{w}_{t-1})^T\mathbf{D}_\mathbf{r}^{-1}(\hat{\mathbf{w}}_t-q\mathbf{w}_{t-1}) + \tilde{J}_{t-1}(\mathbf{w}_{t-1}|\mathbf{r},c)\Big], \\ t=T,\,T-1,...,2. \tag{26}$$

Given the training set $\{(\mathbf{X}_t,\mathbf{Y}_t), t=1,...T\}$, the first summand in the recurrence representation of the sequence of Bellman functions (24) is continuous and piecewise linear in $\mathbb{R}^{n+1}$. The second summand of (24)

$$F_t(\mathbf{w}_t|\mathbf{r},c) = \min_{\mathbf{w}_{t-1}\in\mathbb{R}^{n+1}}\Big[(\mathbf{w}_t-q\mathbf{w}_{t-1})^T\mathbf{D}_\mathbf{r}^{-1}(\mathbf{w}_t-q\mathbf{w}_{t-1})+\tilde{J}_{t-1}(\mathbf{w}_{t-1}|\mathbf{r},c)\Big], \tag{27}$$

is continuous and piecewise quadratic if the preceding Bellman function, namely, $\tilde{J}_{t-1}(\mathbf{w}_{t-1}|\mathbf{r},c)$, is continuous and piecewise quadratic. The initial vacuous function $\tilde{J}_0(\mathbf{w}_0|\mathbf{r},c) \equiv 0$ may be considered as quadratic, hence, all the Bellman functions will be continuous and piecewise quadratic in $\mathbb{R}^{n+1}$.

But the continuous and piecewise quadratic Bellman functions are no quadratic ones, and there exists no parametric way to effectively compute and compactly store them in memory.

In order to save the computational advantages of the dynamic programming procedure, we resort here to the following trick: we heuristically replace the genuine piecewise quadratic Bellman functions $\tilde{J}_t(\mathbf{w}_t|\mathbf{r},c)$ by some appropriate quadratic approximations to them $\tilde{J}'_t(\mathbf{w}_t|\mathbf{r},c)$:

$$\tilde{J}'_t(\mathbf{w}_t|\mathbf{r},c) = (\mathbf{w}_t - \tilde{\mathbf{w}}'_t)^T \tilde{\mathbf{Q}}'_t (\mathbf{w}_t - \tilde{\mathbf{w}}'_t) + const_t \cong \tilde{J}_t(\mathbf{w}_t|\mathbf{r},c). \tag{28}$$

Thus, each Bellman function $\tilde{J}_t(\mathbf{w}_t|\mathbf{r},c)$ will be approximately represented by its Hessian matrix $\tilde{\mathbf{Q}}'_t$ $\big((n+1)\times(n+1)\big)$ and minimum point $\tilde{\mathbf{w}}'_t \in \mathbb{R}^{n+1}$. As to the constant terms $const_t$, we shall ignore them.

Such an idea suggests the quadratic representation of the initial vacuous Bellman function $\tilde{J}'_0(\mathbf{w}_0)$ as $\tilde{\mathbf{Q}}'_0 = \mathbf{0}$ and, for instance, $\tilde{\mathbf{w}}'_0 = \mathbf{0}$.

Let the preceding Bellman function be considered as approximately quadratic $\tilde{J}'_{t-1}(\mathbf{w}_{t-1}|\mathbf{r},c) = (\mathbf{w}_{t-1} - \tilde{\mathbf{w}}'_{t-1})^T \tilde{\mathbf{Q}}'_{t-1} (\mathbf{w}_{t-1} - \tilde{\mathbf{w}}'_{t-1})$ without the constant term, then the function $F_t(\mathbf{w}_t|\mathbf{r},c)$ (27) in (24) can be shown to be quadratic, too:

$$F_t(\mathbf{w}_t|\mathbf{r},c) = (\mathbf{w}_t - q\tilde{\mathbf{w}}'_{t-1})^T \big(\mathbf{D}_{\mathbf{r}} + q^2 \tilde{\mathbf{Q}}'^{-1}_{t-1}\big)^{-1} (\mathbf{w}_t - q\tilde{\mathbf{w}}'_{t-1}). \tag{29}$$

In accordance with (24), the next Bellman function $\tilde{J}_t(\mathbf{w}_t)$ should be approximately defined as the sum

$$\begin{aligned} \tilde{J}_t(\mathbf{w}_t|\mathbf{r},c) = 2c \sum\nolimits_{j=1}^{N_t} & \max\big(0,\, 1 - y_{j,t}\mathbf{w}_t^T \mathbf{x}_{j,t}\big) + \\ & (\mathbf{w}_t - q\tilde{\mathbf{w}}'_{t-1})^T \big(\mathbf{D}_{\mathbf{r}} + q^2 \tilde{\mathbf{Q}}'^{-1}_{t-1}\big)^{-1} (\mathbf{w}_t - q\tilde{\mathbf{w}}'_{t-1}). \end{aligned} \tag{30}$$

But the non-quadratic hinge function as the first summand would turn the entire Bellman function to a non-quadratic form.

Our heuristic idea consists in replacing the functions $\tilde{J}_t(\mathbf{w}_t|\mathbf{r},c)$ by appropriate quadratic ones

$$\tilde{J}'_t(\mathbf{w}_t|\mathbf{r},c) = (\mathbf{w}_t - \tilde{\mathbf{w}}'_t)^T \tilde{\mathbf{Q}}'_t (\mathbf{w}_t - \tilde{\mathbf{w}}'_t) \cong \tilde{J}_t(\mathbf{w}_t|\mathbf{r},c) \tag{31}$$

from the conditions:

$$\begin{cases} \tilde{\mathbf{w}}'_t = \arg\min \tilde{J}'_t(\mathbf{w}_t|\mathbf{r},c), \\ \tilde{\mathbf{Q}}'_t = \nabla^2_{\mathbf{w}_t} \tilde{J}'_t(\mathbf{w}_t|\mathbf{r},c) \text{ at } \mathbf{w}_t = \tilde{\mathbf{w}}'_t. \end{cases} \tag{32}$$

To find vector $\tilde{\mathbf{w}}'_t$, it is enough to solve the convex optimization problem $\tilde{J}'_t(\mathbf{w}_t|\mathbf{r},c) \to \min(\mathbf{w}_t \in \mathbb{R}^{n+1})$ (30) with respect to the next batch $(\mathbf{X}_t, \mathbf{Y}_t)$:

$$\begin{aligned} \tilde{\mathbf{w}}'_t = \underset{\mathbf{w}_t \in \mathbb{R}^{n+1}}{\arg\min} \Bigg[2c \sum\nolimits_{j=1}^{N_t} & \max\big(0,\, 1 - y_{j,t}\mathbf{w}_t^T \mathbf{x}_{j,t}\big) + \\ & (\mathbf{w}_t - q\tilde{\mathbf{w}}'_{t-1})^T \big(\mathbf{D}_{\mathbf{r}} + q^2 \tilde{\mathbf{Q}}'^{-1}_{t-1}\big)^{-1} (\mathbf{w}_t - q\tilde{\mathbf{w}}'_{t-1}) \Bigg]. \end{aligned} \tag{33}$$

This is "almost" the usual SVM problem [24,25] solvable by a standard SVM tool. The only specificity is a more complicated quadratic regularization function $(\mathbf{w}_t - q\tilde{\mathbf{w}}'_{t-1})^T\big(\mathbf{D}_\mathbf{r} + q^2\tilde{\mathbf{Q}}'^{-1}_{t-1}\big)^{-1}(\mathbf{w}_t - q\tilde{\mathbf{w}}'_{t-1})$ instead of the usual $\mathbf{w}_t^T\mathbf{w}_t$. This quadratic function carries crucially important information on the prehistory of the training sequence $\big\{(\mathbf{X}_1,\mathbf{Y}_1),...,(\mathbf{X}_{t-1},\mathbf{Y}_{t-1})\big\}$. Just as in the usual SVM, an integral part of the solution, in addition to $\tilde{\mathbf{w}}'_t$, will be the set of support objects within the bounds of the respective batch $\tilde{\mathbb{J}}_t = \big\{j : y_{j,t}\tilde{\mathbf{w}}'^T_t\mathbf{x}_{j,t} \le 1\big\} \subset \mathbb{J} = \big\{1,...,N_t\big\}$, which naturally fractionizes into the subset of proper support objects $\tilde{\mathbb{J}}'_t = \big\{j : y_{j,t}\tilde{\mathbf{w}}'^T_t\mathbf{x}_{j,t} = 1\big\}$ and that of improper ones $\tilde{\mathbb{J}}''_t = \big\{j : y_{j,t}\tilde{\mathbf{w}}'^T_t\mathbf{x}_{j,t} < 1\big\}$.

Whence both subsets of support objects are found $\tilde{\mathbb{J}}_t = \tilde{\mathbb{J}}'_t \cup \tilde{\mathbb{J}}''_t \subset \{1,...,N_t\}$, the SVM problem (33) can be put in the equivalent form:

$$\tilde{\mathbf{w}}'_t = \underset{\substack{\mathbf{w}_t\in\mathbb{R}^{n+1} \\ y_{j,t}\mathbf{w}_t^T\mathbf{x}_{j,t}=1,\, j\in\tilde{\mathbb{J}}'_t}}{\arg\min} \left[2c\sum\nolimits_{j\in\tilde{\mathbb{J}}''_t}(1-y_{j,t}\mathbf{w}_t^T\mathbf{x}_{j,t}) + (\mathbf{w}_t - q\tilde{\mathbf{w}}'_{t-1})^T\big(\mathbf{D}_\mathbf{r}+q^2\tilde{\mathbf{Q}}'^{-1}_{t-1}\big)^{-1}(\mathbf{w}_t - q\tilde{\mathbf{w}}'_{t-1})\right]. \tag{34}$$

This is a purely quadratic linearly constrained optimization problem. Therefore, it appears reasonable to accept matrix $\tilde{\mathbf{Q}}'_t$ in (32) as the Hessian of the objective function $\nabla^2_{\mathbf{w}_t}[...]$ under the equality constraints $\big\{y_{j,t}\tilde{\mathbf{w}}'^T_t\mathbf{x}_{j,t} = 1,\ j\in\tilde{\mathbb{J}}'_t\big\}$. This Hessian can be found in the subspace there $\nabla_{\tilde{\mathbf{w}}'_t}\big[y_{j,t}\tilde{\mathbf{w}}'^T_t\mathbf{x}_{j,t} - 1\big]^T\mathbf{w}_t = 0$ for each $j\in\tilde{\mathbb{J}}'_t$ or $\mathbf{G}_t\mathbf{w}_t = 0$, there $\mathbf{G}_t = [y_{j,t}\mathbf{x}_{j,t}]_{j\in\tilde{\mathbb{J}}'}$ is the matrix of the left parts of the equality constraints.

Let $\mathbf{Z}_t$ be the matrix constructed from a basis in the null-space $\mathbf{G}_t\mathbf{w}_t = 0$. It can be found for example with QR decomposition of the matrix $\mathbf{G}_t$. It can be shown that the Hessian satisfies the equality $\tilde{\mathbf{Q}}'_t = \mathbf{Z}_t^T\big(\mathbf{D}_\mathbf{r}+q^2\tilde{\mathbf{Q}}'^{-1}_{t-1}\big)^{-1}\mathbf{Z}_t$.

Quadratic Dynamic Programming. Thus, we have heuristically replaced all the Bellman functions in (24) by their quadratic approximations $\tilde{J}'_t(\mathbf{w}_t|\mathbf{r},c) = (\mathbf{w}_t - \tilde{\mathbf{w}}'_t)^T\tilde{\mathbf{Q}}'_t(\mathbf{w}_t - \tilde{\mathbf{w}}'_t)$ (28), each of which is parametrically represented by the minimum point $\hat{\mathbf{w}}'_t$ (33) and Hessian $\tilde{\mathbf{Q}}'_t$.

Let $\big\{(\mathbf{X}_1,\mathbf{Y}_1),...,(\mathbf{X}_T,\mathbf{Y}_T)\big\}$ be the available sequence of training data batches, $\mathbf{r} = (r_1,...,r_n)$ be the fixed vector if feature weights (11), and $c > 0$ be the preset SVM parameter (33). The quadratic version of the dynamic programming procedure (24) consists in recurrently recomputing the parameters $\tilde{\mathbf{w}}'_t\in\mathbb{R}^{n+1}$ and $\tilde{\mathbf{Q}}'_t$ $[(n+1)\times(n+1)]$ of quadratic Bellman functions:

$$\tilde{\mathbf{w}}'_0 = \mathbf{0},\ \tilde{\mathbf{Q}}'_0 = \mathbf{0},\ t = 0;$$

$$\left.\begin{aligned} \tilde{\mathbf{w}}'_t &= \big[\text{solution of the SVM problem (33) for } (\mathbf{X}_t,\mathbf{Y}_t)\big], \\ \tilde{\mathbf{Q}}'_t &= \mathbf{Z}_t^T\big(\mathbf{D}_\mathbf{r}+q^2\tilde{\mathbf{Q}}'^{-1}_{t-1}\big)^{-1}\mathbf{Z}_t \end{aligned}\right\} t = 1,...,T. \tag{35}$$

In accordance with (25), the last result $\tilde{\mathbf{w}}'_T =$ is just the approximate estimate of the hyperplane parameter vector $(\hat{\mathbf{w}}_T|\mathbf{r},q,c)$ for the last batch in the training sequence. The quadratic form of the backward recurrence relation (26) defines the remaining estimates $\left[(\hat{\mathbf{w}}_1|\mathbf{r},q,c),...,(\hat{\mathbf{w}}_{T-1}|\mathbf{r},q,c)\right]$:

$$\hat{\mathbf{w}}_{t-1} = \left(q^2\mathbf{D}_{\mathbf{r}}^{-1} + \tilde{\mathbf{Q}}'_{t-1}\right)^{-1}\left(q\mathbf{D}_{\mathbf{r}}^{-1}\hat{\mathbf{w}}_t + \tilde{\mathbf{Q}}'_{t-1}\tilde{\mathbf{w}}'_{t-1}\right),\; t=T,T-1,...,1. \quad (36)$$

3.2 Re-estimation of Feature Weights for the Fixed Hyperplane Drift

Let the problem (20) be solved, and $\left((\hat{\mathbf{w}}_t|\mathbf{r},c), t=1,...,T\right)$ be the solution for the fixed vector of variances $\mathbf{r}=(r_1,...,r_n)$. In the previous Section we have shown how this solution can be approximately found by a simple dynamic programming algorithm.

As to the criterion (21), it can be easily proved to be the sum of items each of which depends on only one variance r_i:

$$J(\mathbf{r}|\mathbf{w}_t, t=1,...,T,\mu) = \sum\nolimits_{i=1}^{n}\left[\left(T-1+\frac{1}{\mu}\right)\ln\frac{1}{r_i} + \frac{1}{(1-q^2)r_i}\left(\sum\nolimits_{t=2}^{T}(w_{i,t}-qw_{i,t-1})^2 + \frac{1}{\mu}\right)\right]. \quad (37)$$

The summands are convex functions, and their differentiation $\partial/\partial(1/r_i)\left[...\right]=0$ yields simple formulas for the solution $\left(\hat{\mathbf{r}}|\mathbf{w}_1,...,\mathbf{w}_T,\mu\right)$ of (21):

$$\left(\hat{r}_i|\mathbf{w}_1,...,\mathbf{w}_T,\mu\right) = \frac{\frac{1}{1-q^2}\sum\nolimits_{t=2}^{T}(w_{i,t}-qw_{i,t-1})^2 + (1/\mu)}{T-1+(1/\mu)},\; i=1,...,n. \quad (38)$$

3.3 The Iterative Procedure of Parameter Estimation and the Effect of Feature Selection

As accepted in Section 2.2, the elements of the vector $\mathbf{r}=(r_1,...,r_n)$ define the a priori stationary variances of the drifting direction vector coordinates $\mathbf{a}_t = (a_{i,t}, i=1,...,n)$ (13) and are subject to estimation in the course of training. Let the iterative procedure start with some initial values $\mathbf{r}^0 = (r_1^0,...,r_n^0)$, for instance, $r_1^0 = ... = r_n^0 = 1$.

On the k-th step vector $\mathbf{r}^k = (r_1^k,...,r_n^k)$ completely defines the quadratic dynamic programming procedure (35) that results in the estimated drift of the discriminant hyperplane $\left[(\hat{\mathbf{w}}_1^k|\mathbf{r}^k,q,c),...,(\hat{\mathbf{w}}_T^k|\mathbf{r}^k,q,c)\right]$, $\hat{\mathbf{w}}_t^k = (\hat{\mathbf{a}}_t^k,\hat{b}_t^k)$. In its turn, the estimated hyperplane drift yields a new estimate of the vector $\mathbf{r}^{k+1}=(r_1^{k+1},...,r_n^{k+1})$ (38), and so on. The experience shows that this iterative procedure converges in 10 – 15 steps.

4 Experimental Study

To verify the proposed method, we used a synthetic data set having the form of two normal distributions. The class labels in this data set are equi-distributed and take values from the set $\{-1, 1\}$. Two informative features had been generated by two class-dependent normal distributions. Later, 98 synthetic "redundant" features were added to this set. So, each entity was characterized by $n = 100$ features (10), but only two of them were relevant to its class-membership. In the zero time moment $t = 0$, these distributions had the equal variation 0.5 and mathematical expectations -1 and 1 respectively. With time, the centers of distributions were changing, namely, rotated around the origin of the two-dimensional feature space. We generated 100 consecutive data batches $(\mathbf{X}_t, \mathbf{Y}_t)$, $T = 100$ each containing 50 instances $N_t = 50$.

To compare the obtained results, we used three concept drift algorithms realized in the software environment Massive Online Analysis (MOA) [30]:

- OzaBagAdwin - bagging underlied by the ADWIN techniqie [31], which is a change detector and estimator that solves the problem of tracking the average of a stream of bits or real-valued numbers in a well-specified way. The model in the bag is a decision tree for streaming data with adaptive Naive Bayes classification at its leaves.
- SingleClassifierDrift - a single classifier based on the drift detection method EDDM [32]. The decision tree for streaming data with adaptive Naive Bayes classification at its leaves was selected as the classifier for training.
- AdaHoeffdingOptionTree - the adaptive decision option tree for streaming data with adaptive Naive Bayes classification at the leaves.
- DriftFeatureSelection - our method proposed in this paper.

Table 1 demonstrates the elements of the estimated variance vector $\hat{\mathbf{r}} = (\hat{r}_1, ..., \hat{r}_n)$. As we can see, the two first vector components have the biggest values, what means that the corresponding features are significant in contrast to the redundant ones. The final results presented in Table 2 show that the proposed algorithm totally outperforms other algorithms.

Table 1. The experimental results: variance vector

Feature number	1	2	3	...	100
Variation	47,6407	48,1681	0,7355	...	0,0052

4.1 Real-world Data

The KDDCup'99 dataset [33] has been used as real-world data. This dataset is a collection of TCP dumps taken over nine weeks in the framework of DARPA Intrusion Detection Evaluation Program in 1998. In this program local-area network (LAN) simulated a typical U.S. Air Force LAN and logged connections. A connection is a sequence of TCP packets starting and ending at some well defined times, between which data flow to and from a source IP address to a

Table 2. The experimental results: Artificial data

Algorithms	Classification error, %
OzaBagAdwin	14,62
SingleClassifierDrift	15,28
AdaHoeffdingOptionTree	14,76
DriftFeatureSelection	4.52

target IP address under some well defined protocol. Each connection has 41 features and is labelled either as normal, or as an attack, with exactly one specific attack type. As we solve the classification problem concrete attack type isn't important. This data set exists in two variants: full with about 5 millions records and its 10 percents subset. In current work we used 10-percentage set which was normalized and divided into 49 batches by 10000 dumps in each.

As in the previous section we compared the results received by means of our method with the results of three algorithms from MOA programming package. Quality has been calculated by "test-then-train" procedure, which consists in using data batch for testing on current value of decision rule parameters and after the algorithm uses this batch for training. The averaged results are presented in Table 3.

Table 3. The experimental results: KDD Cup Dataset

Algorithms	Classification error, %
OzaBagAdwin	6,144
SingleClassifierDrift	7,12
AdaHoeffdingOptionTree	1,056
DriftFeatureSelection	0,782

5 Conclusion

The main claim of this paper is a new variable selection technique for concept drift problems in data stream analysis. The focus is the hierarchical Bayesian model with supervised selectivity which allows to obtain the asymptotic unbiased parameters estimates, possessing the grouping and oracle properties. The methodological power of the proposed concept drift algorithm with supervised selectivity is demonstrated via carefully designed simulation studies. Comparing to three newest concept drift algorithms, it is far more specific in selecting informative features. As a result, it has much smaller absolute error error of class recognition.

Acknowledgment. This work is supported by grants of the Russian Foundation for Basic Research No. 14-07-00964, 14-07-00661 and 16-07-01008.

References

1. Widmer, G., Kubat, M.: Learning in the presence of concept drift and hidden contexts. Machine Learning **23**(1), 69–101 (1996)
2. Dongre, P., Malik, L.: Stream Data Classification and Adapting to Gradual Concept Drift. International Journal of Advance Research in Computer Science and Management Studies **2**(3), 125–129 (2014)
3. Chen, S., Wang, H., Zhou, S., Yu, P.: Stop chasing trends: discovering high order models in evolving data. In: Proceedings of the ICDE 2008, pp. 923–932 (2008)
4. Hulten, G., Spencer, L., Domingos, P.: Mining time-changing data streams. In: SIGKDD, San Francisco, CA, USA, pp. 97–106, August 2001
5. Yang, Y., Wu, X., Zhu, X.: Combining proactive and reactive predictions for data streams. In: Proceedings of the SIGKDD, pp. 710–715 (2005)
6. Kolter, J., Maloof, M.: Using additive expert ensembles to cope with concept drift. In: ICML, Bonn, Germany, pp. 449–456, August 2005
7. Wang, H., Fan, W., Yu, P.S., Han, J.: Mining concept-drifting data streams using ensemble classifiers. In: KDD 2003, pp. 226–235 (2003)
8. Zhou, X., Li, S., Chang, C., Wu, J., Liu, K.: Information-value-based feature selection algorithm for anomaly detection over data streams. Tehnicki Vjesnik **21**(2), 223–232 (2014)
9. Sauerbrei, W.: The use of resampling methods to simplify regression models in medical statistics. Apply Statistics **48**, 313–339 (1999)
10. Sauerbrei, W., Schumacher, M.: A bootstrap resampling procedure for model building: Application to the cox regression model. Statistics in Medicine (1992)
11. Zou, H., Hastie, T.: Regularization and variable selection via the elastic net. Journal of the Royal Statistical Society (2005)
12. Zou, H.: The adaptive lasso and its oracle properties. Journal of the American Statistical Association (2006)
13. Zou, H., Li, R.: One-step sparse estimates in nonconcave penalized likelihood models (with discussion). Annals of Statistics (2008)
14. Seredin, O., Kopylov, A., Mottl, V.: Selection of subsets of ordered features in machine learning. In: Perner, P. (ed.) MLDM 2009. LNCS, vol. 5632, pp. 16–28. Springer, Heidelberg (2009)
15. Seredin, O., Mottl, V., Tatarchuk, A., Razin, N., Windridge, D.: Convex support and relevance vector machines for selective multimodal pattern recognition. In: 21st International Conference on Pattern Recognition (ICPR 2012), Tsukuba, Japan, November 11–15, 2012, pp. 1647–1650 (2012)
16. Fan, J., Samworth, R., Wu, Y.: Ultrahigh Dimensional Feature Selection: Beyond The Linear Model. J. Mach. Learn. Res. **10**, 2013–2038 (2009)
17. Cai, D., Zhang, C., He, X.: Unsupervised Feature Selection for Multi-cluster Data. SIGKDD (2010)
18. Yang, H., Lyu, M.R., King, I.: Efficient Online Learning for Multitask Feature Selection. TKDD **7**(2), 6 (2013)
19. Song, Q., Ni, J., Wang, G.: A Fast Clustering-based Feature Subset Selection Algorithm for High-dimensional Data. TKDE **25**(1) 2013

20. Maung, C., Schweitzer, H.: Pass-efficient Unsupervised Feature Selection. NIPS (2013)
21. Krasotkina, O., Mottl, V.: A Bayesian approach to sparse learning-to-rank for search engine optimization. In: Perner, P. (ed.) MLDM 2015. LNCS, vol. 9166, pp. 382–394. Springer, Heidelberg (2015)
22. Krasotkina, O., Mottl, V.: A Bayesian approach to sparse Cox regression in high-dimentional survival analysis. In: Perner, P. (ed.) MLDM 2015. LNCS, vol. 9166, pp. 425–437. Springer, Heidelberg (2015)
23. De Groot, M.: Optimal Statistical Decisions. McGraw-Hill Book Company (1970)
24. Cortes, C., Vapnik, V.: Support-Vector Networks. Machine Learning **20**, 273–297 (1995)
25. Vapnik, V.: Statistical Learning Theory. J. Wiley, NY (1998)
26. Tatarchuk, A., Mottl, V., Eliseyev, A., Windridge, D.: Selectivity supervision in com-bining pattern-recognition modalities by feature-and kernel-selective Support Vector Machines. In: Proceedings of the 19th International Conference on Pattern Recognition, vol. 1–6, pp. 2336–2339. IEEE (2008). ISBN 978-1-4244-2174-9
27. Markov, M., Krasotkina, O., Mottl, V., Muchnik, I.: Time-varying regression model with unknown time-volatility for nonstationary signal analysis. In: Proceedings of the 8th IASTED International Conference on Signal and Image Processing, Honolulu, Hawaii, USA, August 14–16, 2006, pp. 534–196 (2006)
28. Bellman, R.: Dynamic Programming. Princeton University Press, Princeton (1957)
29. Sniedovich, M.: Dynamic Programming. Marcel Dekker, NY (1991)
30. Bifet, A., Holmes, G., Kirkby, R., Pfahringer, B.: MOA: Massive Online Analysis. Journal of Machine Learning Research (JMLR) (2010). http://sourceforge.net/projects/moa-datastream/
31. Bifet, A., Holmes, G., Pfahringer, B., Kirkby, R., Gavalda, R.: New ensemble methods for evolving data streams. In: 15th ACM SIGKDD International Conference on Knowledge Discovery and Data Mining (2009)
32. Gama, J., Medas, P., Castillo, G., Rodrigues, P.: Learning with drift detection. In: Bazzan, A.L.C., Labidi, S. (eds.) SBIA 2004. LNCS (LNAI), vol. 3171, pp. 286–295. Springer, Heidelberg (2004)
33. KDD Cup 1999 Data. http://kdd.ics.uci.edu/databases/kddcup99/

Feature Reduction for Multi Label Classification of Discrete Data

V. Susheela Devi(✉) and Bhupesh Akhand

Department of Computer Science and Automation, Indian Institute of Science, Bangalore 560012, India
{susheela,akhand123}@csa.iisc.ernet.in

Abstract. We describe a novel multi-label classification algorithm which works for discrete data. A matrix which gives the membership value of each discrete value of each attribute for every class. For a test pattern, looking at the values taken by each attribute, we find the subset of classes to which the pattern belongs. If the number of classes are large or the number of features are large, the space and time complexity of this algorithm will go up. To mitigate this problems, we have carried out feature selection before carrying out classification. We have compared two feature reduction techniques for getting good results. The results have been compared with the algorithm multi-label KNN or ML-KNN and found to give good results. Using feature reduction our classification accuracy and running time for algorithm is improved. The performance of the above algorithm is evaluated using some benchmark datasets and the results have been compared with the algorithm multi-label KNN or ML-KNN and found to give good results.

Traditionally, single label classification is assumed where each instance is associated with a single label. If X is the domain of instances and $Y = \{l_1, l_2, ..., l_q\}$ is the set of labels or classes, then each instance $x \in$ X is mapped to a label learnt by a function h : X → Y. Multi-label classification is the prediction of class labels of instances where each instance can belong to more than one class. In multi-label learning, a function h : X → 2^Y is learnt which maps each instance x ∈ X to a set of labels h(x)⊆ Y, that is, $h(x) \subseteq Y$.

There are a number of applications where multi-label classification is used. For example, if the news items in a newspaper are classified into politics, sports, entertainment etc., a newsclip can be about a movie star who is a politician so that it belongs to both politics and entertainment simultaneously. Other applications include annotation of images and video, determining the emotions from music or sentiment analysis of a document, scene classification etc.

The performance metrics used for single label classification cannot be used for multi-label classification. A number of metrics such as Hamming loss [16], one-error, coverage, ranking loss and average precision [10] have been discussed in the literature. [25] discusses these performance metrics.

In this study , we have carried out multi-label classification as proposed by the authors in [1] for discrete data. The feature values are discretized if they

P. Perner (Ed.): MLDM 2016, LNAI 9729, pp. 630–642, 2016.
DOI: 10.1007/978-3-319-41920-6_49

are continuous. A membership matrix is formed which gives the membership value of each value of each attribute to each class. If the number of features are very large, the fuzzy matrix which is formed from the training data set will be very large. To mitigate this problem, we have first carried out dimensionality reduction. Two methods have been used for dimensionality reduction namely Multi-label Dimensionality Reduction via Dependence Maximization(MDDM) [24] and simulated annealing [8]. The results obtained for classification after carrying out dimensionality reduction using the above two methods have been compared with the results obtained without dimensionality reduction using our proposed method and also mL-kNN [22]. A number of performance metrics have been used to carry out this comparison.

1 Related Work

Several methods have been proposed for multi label classification although most of them are for continuous data. The literature on discrete data is very sparse as not much work has been done in this regard. The approaches for multi-label problems either transform the problem to a single label problem and then solve it or use an algorithm which work with multiple labels. Some of the methods are described below.

ML-KNN [22] is a multi-label lazy learning approach which is derived from the traditional k-nearest neighbour algorithm. In this method prior probability and the posterior probability of a label using k nearest algorithm [21] is calculated. Using these probabilities we find the labels of a given instance.

Label power-set(LP) [5] multi-label learning uses subsets of the labelset $Y = \{l_1, l_2, ..., l_q\}$ and considers each subset as a label. This may lead to large number of classes with few instances of each class. An improvement on this is the algorithm RAkEL(RAndom k labELsets) [19] where the label-sets formed are divided into smaller random subsets and LP is applied on it. Here k is the size of the subsets. Another method is the random-walk KNN method [20] in which each instance is associated with a link graph using its k nearest neighbours. To classify a new pattern, a random walk is performed in the link graph.

Another approach is the Binary Relevance(BR) [18]. Some approaches use label correlations [7,12] or use a hierarchy in the labels [3].

The multi-label classification approach [1] proposed by the authors uses discrete data and a fuzzy matrix M is formed. If d features are present and each feature falls into p discrete ranges, then the matrix will be of size q X (d*m) where q is the number of classes. If the number of features d is large, since we need to consider p discrete ranges, the matrix M can become very large. To take care of this problem, we have carried out multi-label dimensionality reduction before performing multi-label classification. This has reduced the space and time complexity of our multi-label classification algorithm.

Dimensionality reduction can be categorized into feature selection and feature extraction. In feature selection subset of original features is selected and the original representation of the features is not changed but in transformation based methods input features are modified to a new feature space. Feature selection methods can be broadly divided into three main families: filter, wrapper and embedded approaches. In filter methods, features are independent of learning algorithm and features are evaluated by looking only at the intrinsic properties of the data. Wrapper methods select features by the search algorithm with learning algorithm and evaluate feature subsets based on feature relevance score for each feature subset. Unlike feature extraction, in feature selection physical meaning of feature will remain same.

Embedded methods feature search and the learning algorithm is combined into a single optimization problem. Wrapper and embedded methods perform better than filter methods on a particular classifier, but the features chosen for one classifier may not be appropriate for other classifiers because they select features specific to the classifier. Wrapper methods are computationally expensive because they need to train and test the classifier for each feature subset, which will not be possible when working with high-dimensional data.

Feature extraction methods can be broadly divided into two categories , i.e., unsupervised and supervised. One unsupervised dimensionality reduction method is PCA [15], in which linear mapping of the data to a lower dimensional space is done in such a way that the variance of the data is maximized in the low-dimensional representation. Another supervised dimensionality reduction method is LDA [9], which identify a lower-dimensional space minimizing the inter-class similarity while maximizing the intra-class similarity simultaneously. Another popular supervised dimensionality reduction method is CCA [11], which aims at extracting the representation of the object by correlating linear relationships between two views of the object.

In [13], dimensionality reduction is carried out by projection using compressed sensing. [17] have used principal component analysis. In [23], canonical correlation analysis has been used and in [6], singular value decomposition has been used.

When the labels are very large and multi-label classification has to be carried out, other methods besides dimensionality reduction have been used. In [2] only a subset of the labels are used and the labels are selected using a group-sparse learning method. In [4], label selection is carried out using randomized sampling.

2 Methodology

Our approach uses a fuzzy matrix M which is updated using the training data. In this matrix M, each row stands for a class label. The entries in a row give the fuzzy membership value for each discrete value of each feature.

2.1 Proposed Algorithm for Discrete data

This algorithm has been explained in detail in [1]. We partition attributes of continuous data into intervals for converting it to discrete data. For preprocessing step of modified algorithm we used a discretization method which is based on k- means algorithm [14]. We assume that the number of cluster is already given to us. In this algorithm we used K=8 for discretization.

Discretization of Data. Discretization mainly consists of converting the continuous data into intervals and the data values will fall into one of the intervals. We have used the k-means algorithm for discretization [14]. The initial centers of the clusters are chosen in increasing order. The attribute values are assigned to the closest cluster center. Here the number of clusters(or the number of discrete intervals for an attribute) is fixed at k. We have taken k to be 8. To reduce the effect of discretization on the performance of the classifier, the value of k can be increased.

Proposed Methodology. Consider data set $\{a_1, ...a_n\}$ which has n continuous attributes. After carrying out discretization, each attribute a_i has p discrete values $\{a_{i1}, ...a_{ip}\}$. The dataset has q classes $\{c_1, ...c_q\}$. Now consider a matrix M of size q×np where n is the no. of attributes. Each row of M corresponds to one class label. The values in a row r give the fuzzy membership values of each attribute value to the class c_r. The total number of columns is therefore n×p.

In the training phase, the matrix M is updated for each training data as given below. After the training phase, matrix M shows the membership value of each discrete value of each attribute to the class labels. Using M, a test pattern is classified to the set of class labels as detailed below.

Pseudo Code for Updation of M Matrix. Step 1: For every training instance, Note the class labels to which this pattern belongs. Update the rows of M which pertain to these classes. In each of these rows, the elements which represent the values of the attribute, are incremented by one.

Step 2: For a row i, the values are divided by the number of patterns in the training data which belong to the class i.

for i=1 to c_q do
{
for j=1 to n×p do
{ $M[i][j] = M[i][j]/N_i$
}
}

After training, matrix M gives the membership value of each value of each attribute to every class.

Classification of a Test Pattern

Step 1: For a test pattern, form vector N of size np×1. Discretize the values of the attribute in the test pattern. According to the discretized values, update the corresponding entries in N to 1, keeping the other values as zero.

Step 2: Multiply M and N to get P of size q×1 i.e. P = M×N

Step 3: Let *max* be the maximum value in P. Divide every value in P by *max*. The ith value in P is the membership value of the test pattern to $class_i$.

Step 4: For every class i, if $P_i > threshold$, then pattern belongs to class c_i.

Selecting the Threshold. The threshold can be selected in two ways.

1. For a test pattern, find P and put the values in P in increasing order. Find the difference between successive values. Put the threshold in the position where the difference is the maximum.
2. Use a validation set. Obtain M using the training patterns. Carry out classification on the validation set for various threshold values. Choose the threshold value giving the best performance on the validation set.

2.2 Dimensionality Reduction

Since this approach uses discrete data and each attribute has a number of discrete values, the matrix M will be very large if the number of features is very high. To mitigate this problem, we carried out dimensionality reduction on the training data.

The proposed algorithm increases the dimension of the matrix M because of the discretization process. This leads to both increase in space and time complexity. For example, if we have a dataset with k classes and 300 features, ff the discretization process gives 8 values per feature, then the number of columns of matrix M will become $300 \times 8 = 2400$. It can be seen that if the number of features is very high, this will lead to M being of very large size.

Using feature selection, we choose only those features which are important for classification. This leads to reduction in the running time of the algorithm. For example, if the number of features is reduced to 200, the number of columns in matrix M comes down from 2400 to 1600.

There are some key benefits of feature reduction techniques when constructing predictive models, because of which we have used these techniques:

1. Reduce storage requirement and training time: In proposed algorithm we are discretizing data set before applying classification algorithm, As a result our data size is increasing and classification time is also increasing. For the solution of this problem we are using feature reduction technique. We reduce the feature set size so that storage requirement and training time also reduce.
2. Enhance generalization by reducing overfitting: Feature reduction is effective for critical analysis, because data set frequently contains more information than required. Some features are noisy or redundant which makes it more difficult for classification of pattern. In our case also after feature reduction, classification accuracy is improved which directly shows overfitting problem is reduced by feature reduction. Experiments confirm that removing noisy or redundant features helps to reduce the generalization error and to avoid overfitting on both synthetic and real world data sets.

3. Improve model interpretability: To make any prediction model more "interpretable", it is necessary to reduce the number of input variables that are used in the model. We can make a simpler model using feature reduction. We have selected only those features which are useful in building a model and excluded other features that are statistically insignificant.

We have used two methods for dimensionality reduction, which are described in the next two subsections.

Multi-label Dimensionality Reduction via Dependence Maximization. One of the techniques we have used for dimensionality reduction is Multi- label Dimensionality reduction via Dependence Maximization(MDDM) [24] in which original features are reduced to a lower dimensional space. The other method we have used is the Simulated Annealing [8] algorithm in which features are selected and original features are not changed.

MDDM aims at identifying a lower-dimensional space maximizing the dependence between the original feature description and class labels associated with the same object.

MDDM mainly considers relation between the feature description and the labels associated with the same instance. In this method lower-dimensional feature space is generated in which dependence between the input and output is maximized. We consider the linear case of uncorrelated projection dimensionality reduction.

Suppose projection vector is p. An instance x after projection will be $\phi(x)$ = $\mathrm{p^T x}$ in the new space and the kernel function is $k(x_i, x_j) \triangleq \langle\phi(x_i), \phi(x_j)\rangle =$ $\langle p^T x_i, p^T x_j\rangle$ We consider linear kernel for output $y \in Y$ which is $l(y_i, y_j) \triangleq$ $\langle y_i, y_j\rangle$ For $\{(x_1, y_1) \ldots (x_N, y_N)\}$ with joint distribution P_{xy}, the kernel matrix for feature space will be K = $[K_{ij}]_{N\times N}$, K_{ij} $=k(x_i, x_j)$ and the kernel matrix for label space will be L = $[L_{ij}]_{N\times N}$, L_{ij} $= l(y_i, y_j)$. We then maximize the dependence between the feature description and the class labels. We calculate XHLHX^T where H = $\frac{1}{N}ee^T$, where e is a column vector with all ones and X is $D\times N$ feature matrix. We apply eigen-decomposition on XHLHX^T and then construct $D\times d$ matrix P^*. Here d is the input parameter which shows how much dimension we want to reduced. In P^*, columns are composed of the eigenvectors corresponding to largest d eigenvalues.

Simulated Annealing. The other technique we have used for feature reduction is the Simulated Annealing [8] which is a global search technique derived by Statistical Mechanics. Simulated Annealing is based upon metropolis algorithm. Metropolis algorithm has been proposed to simulate small fluctuations and behaviour of a system of atoms starting from an initial configuration. In the Metropolis algorithm sequence of iterations is generated. Each iteration is composed by random permutation of the actual configuration and computation of the corresponding energy variation($\triangle$E). If $\triangle$E < 0 the transition is

Algorithm 1. MDDM

Input: X : D×N feature matrix
Y : M×N label matrix
d : the dimension to be reduced
th : threshold

Output: P* : the projection from R^D to R^d

1: Construct the label matrix L.
2: Calculate XHLHX^T.
3: **if** d is given
4: Do eigen-decomposition on XHLHX^T, then construct D×d matrix P* whose columns are composed by the eigenvectors corresponding to the largest d eigenvalues.
5: **else** (i.e., th is given)
6: Construct D×r matrix Q* in a way similar to Step 4 where r is the rank of L, then choose the first d eigenvectors that enable $\sum_{i=1}^{d} \lambda_i \geq th \times \langle \sum_{i=1}^{D} \lambda_i \rangle$ to compose P*
7: **end if**

unconditionally accepted, otherwise the transition of system is accepted with probability given by the Boltzmann distribution:

$$P(\Delta E) = \exp\langle \frac{-\Delta E}{KT} \rangle$$

where T is Temperature and K is the Boltzmann constant. Temperature T is the control parameter of search space and K is usually set to 1. Perturbation is done by switching 2×r bits of g,by flipping the values of randomly selected r bits set to 0 and r bits set to 1.

Simulated Annealing works as a probabilistic hill-climbing procedure and search for the global optimum of cost function. The temperature T works as the control parameter of the search space and temperature is gradually lowered until there is no further improvement in cost function.

3 Evaluation metrics

The evaluation metrics used for single label classification is not suitable for multi label classification. Since an instance is associated with more than one label simultaneously, finding the classification accuracy is more difficult. The criteria used differs from the criteria used for single label classification.

In this paper we use five criteria often used for multi-label learning [25]. In a given multi-label data set S = { (X_1,Y_1), (X_2,Y_2),......,(X_p,Y_p) }, these criteria are defined as below. Here, h(X_i) returns a set of proper labels of X_i ; h(X_i,y) returns a real-value indicating the confidence for y to be a proper label of X_i ; rank h(X_i,y) returns the rank of y derived from h(X_i,y).

Algorithm 2. Simulated Annealing

Input: s(number of features to be selected)
r(number of bits to be selected)
T(initial temperature of the system)
α(cooling parameter)
h_{min}(minimum number of success for each T)
f_{max}(maximum number of success for each T)
Output: Best features to be selected
1: Initialize g at random(binary mask).
2: Perform classification using proposed algorithm and evaluate the E(error) using performance metrics.
3: **do**
4: Initialize f=0(number of iterations)
h=0(number of success)
5: **do**
6: Increment the number of iterations f
7: Perturb mask g
8: Perform classification using proposed algorithm and evaluate the E(error) using performance metrics.
9: Generate a random number rand in interval[0,1]
10: **if** rand < P(ΔE) **then**
11: Accept the new g mask
12: Increment number of successes h
13: **endif**
14: **loop until** h$\leqslant$ h_{min} and f$\leqslant$ f_{max}
15: update T = αT
16: **loop until** h>0

Hamming loss:
$hloss_S(h) =$

$$\frac{1}{p}\sum_{i=1}^{p}\frac{1}{|y|}|h(X_i)\triangle Y_i| \tag{1}$$

Here $\triangle$ denotes the symmetric difference between two sets. The hamming loss evaluates how many times an example-label pair is misclassified,i.e., a proper label is missed or a wrong label is predicted.

One error:
$one - error_S(h) =$

$$\frac{1}{p}\sum_{i=1}^{p}||[argmax_{y\in Y}h(X_i, y) \notin Y_i]|| \tag{2}$$

The one-error evaluates how many times the top-ranked label is not a proper label of the object.

Table 1. Data sets

DATA SET	DOMAIN	INSTANCES	ATTRIBUTES	LABELS	CARDINALITY	DENSITY
image	images	2000	294	5	1.236	0.247
scene	images	2407	294	6	1.074	0.179

URL 1: http://mulan.sourceforge.net/datasets.html
URL 2: http://cse.seu.edu.cn/people/zhangml/Resources.htm#data

Coverage:
$coverate_S(h) =$

$$\frac{1}{p}\sum_{i=1}^{p} max_{y \in Y_i} rank^h(X_i, y) - 1 \tag{3}$$

The coverage evaluates how far it is needed, on the average, to go down the list of labels in order to cover all the proper labels of the object.

Ranking Loss:
$rloss_S(h) =$

$$\sum_{n=1}^{p} \frac{1}{p|Y_i||\bar{Y_i}|\{(y_1, y_2)|h(X_i, y_1) \leq h(X_i, y_2), (y_1, y_2) \in Y_i \times Y_i\}|} \tag{4}$$

where $\bar{Y_i}$ denotes the complementary set of Y_i in Y. The ranking loss evaluates the average fraction of label pairs that are misordered for the object.

Average precision:
$avgprec_S(h) =$

$$\frac{1}{p}\sum_{n=1}^{p} \frac{1}{|Y_i|} \sum_{y \in Y_i} \frac{|\{y'|rank^h(X_i, y') \leq rank^h(X_i, y), y' \in Y_i\}|}{rank^h(X_i, y)} \tag{5}$$

The average precision evaluates the average fraction of proper labels ranked above a particular label $y \in Y_i$.

The performance of algorithm will be perfect when $hloss_S(h) = 0$ and the smaller the value of $hloss_S(h)$ the better the performance of h on S. Similarly, performance is perfect when $one - error_S(h) = 0$ and the smaller the value of $one - error_S(h)$ the better the performance of h. For $coverages_S(h)$ also, the smaller the value of $coverages_S(h)$, the better the performance of h. The performance is perfect when $rloss_S(h) = 0$ and the smaller the value of $rloss_S(h)$, the better the performance of h. If $avgprec_S(h) = 1$ performance will be perfect and the larger the value of $avgprec_S(h)$, the better the performance of h.

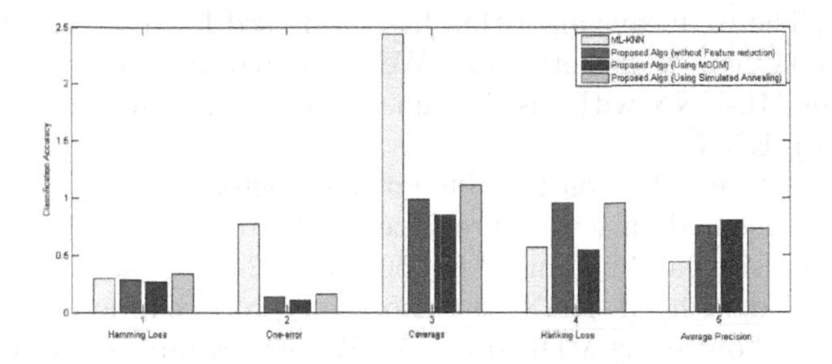

Fig. 1. Comparison of ML-KNN and Proposed Algorithm on image data set

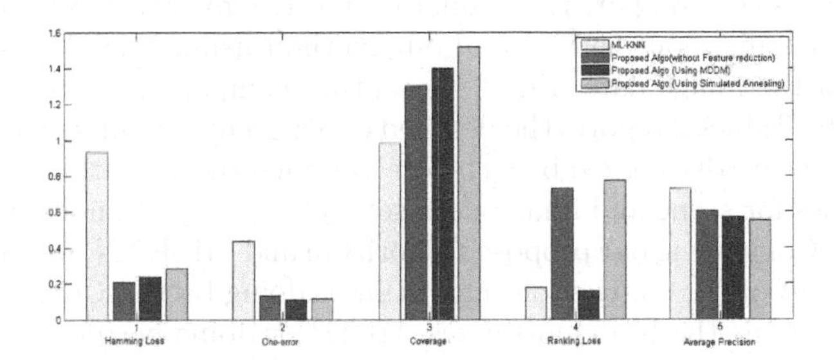

Fig. 2. Comparison of ML-KNN and Proposed Algorithm on scene data set

Table 2. Experimental results of the compared algorithms on image and scene data set. For each evaluation criterion, ↓ indicates the smaller the better while ↑ indicates the bigger the better.

Evaluation criteria	Data set	ML-KNN	Proposed Algo	Simulated Ann	MDDM
Hamming Loss ↓	image	0.2960	0.2820	0.3373	0.2680
	scene	0.9331	0.2094	0.2837	0.2392
One-error ↓	image	0.7700	0.1317	0.1600	0.1067
	scene	0.4391	0.1302	0.1177	0.1107
Coverage ↓	image	2.4417	0.9850	1.1100	0.8483
	scene	0.9834	1.3033	1.5166	1.3989
Ranking Loss ↓	image	0.5717	0.9565	0.9503	0.5460
	scene	0.1772	0.7304	0.7723	0.1603
Average Precision ↑	image	0.4401	0.7525	0.7284	0.8041
	scene	0.7262	0.6063	0.5533	0.5709

4 Implementation and Results

We implemented the proposed algorithm and fixed the threshold using validation set. We used multi-label data sets image and scene. The description of data sets is given in Table 1. These data sets have continuous attributes. These data sets were discretized by us before applying the algorithm. We used eight discrete values for scene and image data sets for every continuous value to cover the entire range. The K- means algorithm has been used for discretization which is a supervised method of discretization. We compared our algorithm with multi-label KNN or ML-KNN with discrete image and scene data sets. The value of K is 10 for ML-KNN.

We tried feature reduction for different parameter and compared them. We found that for image data set features size can be reduced to 100 and for scene data set feature size can be reduced to 150. Table 2 gives the comparison of the proposed method with ML-KNN and also using proposed method with feature reduction carried out using MDDM & SA . We also compared feature reduction algorithm MDDM and simulated annealing for reduced feature size. We observe that MDDM works better as compared to simulated annealing.

Fig. 1 shows the comparitive evaluation metrics for ML-KNN, our proposed multi-label classification method and our method using MDDM and simulated annealing for the image data. Fig. 2 shows that comparative evaluation metrics for scene data. Tables 2 reports the detailed experimental results on two data sets. These data sets are discretized by us before applying the algorithm. We used eight discrete values for scene and image data sets for every continuous value to cover entire range. Comparing our proposed algorithm and ML-KNN, we can see that in the case of most of the metrics, our algorithm is doing better. Comparing the two feature selection methods, in most cases, MDDM is doing better.

5 Conclusions

In this paper, we have given a new approach for multi-label classification where a fuzzy matrix is computed which can be used for classifying any test pattern. Feature reduction has been carried out on the multi-label data so as to reduce the space and time complexity of using the fuzzy matrix. This is especially useful when the data is discretized. Feature reduction not only reduced the time complexity of algorithm but it also saves on storage requirement for large data sets.

In future, we need to experiment with larger data sets with more number of classes. If the number of classes are high, the complexity of multilabel classification will go up.

References

1. Akhand, B., Devi, V.S.: Multi label classification of discrete data. In: IEEE-FUZZ, pp. 1–5 (2013)
2. Balasubramanian, K., Lebanon, G.: The landmark selection method for multiple output prediction. In: Proceedings of the 29th International Conference on Machine Learning, pp. 983–990 (2012)
3. Bi, K.: Multi-label classification on tree- and DAG- structured hierarchies. In: 28th International Conference on Machine Learning, pp. 17–24 (2011)

4. Bi, W., Kwok, J.T.: Efficient multi-label classification with many labels. In: Proceedings of the 30th International Conference on Machine Learning, pp. 405–413 (2013)
5. Boutell, M.R., Luo, J., Shen, X., Brown, C.M.: Learning multi-label scene classification. Pattern Recognition **37**(9), 1757–1771 (2004)
6. Chen, Y.N., Lin, H.T.: Feature-aware label space dimension reduction for multi-label classification. Advances in Neural Information Processing Systems **25**, 1538–1546 (2012)
7. Dembczynski, K., Cheng, W., Hullermeier, E.: Bayes optimal multilabel classification via probabilistic classifier chains. In: 27th International Conference on Machine Learning, pp. 279–286 (2010)
8. Filippone, M., Masulli, F., Rovetta, S.: Unsupervised gene selection and clustering using simulated annealing. In: Bloch, I., Petrosino, A., Tettamanzi, A.G.B. (eds.) WILF 2005. LNCS (LNAI), vol. 3849, pp. 229–235. Springer, Heidelberg (2006)
9. Fisher, R.A.: The use of multiple measurements in taxonomic problems. Annals of Eugenics **7**(2), 179–188 (1936)
10. Godbole, S., Sarawagi, S.: Discriminative methods for multi-labeled classification. In: Dai, H., Srikant, R., Zhang, C. (eds.) PAKDD 2004. LNCS (LNAI), vol. 3056, pp. 22–30. Springer, Heidelberg (2004)
11. Hardoon, D.R., Szedmak, S., Shawe-Taylor, J.: Canonical correlation analysis: An overview with application to learning methods. Neural Computation **16**(12), 2639–2664 (2004)
12. Hariharan, B., Zelnik-Manor, L., Vishwanathan, S.V.N., Varma, M.: Large scale max-margin multi-labelclassification with priors. In: 27th International Conference on Machine Learning, pp. 423–430 (2010)
13. Hsu, D., Kakade, S.M., Langford, J., Zhang, T.: Multi-label prediction via compressed sensing. Advances in Neural Information Processing Systems **22**, 772–780 (2009)
14. Joiţa, D.: Unsupervised Static Discretization Methods in Data Mining. Revista Mega, Bytes, vol. 9 (2010)
15. Jolliffe, I.T. (ed.): Principal Component Analysis. Springer, New York (1986)
16. Schapire, S.: Boostexter: a boosting-based system for text categorization. Machine Learning **39**(2/3), 135–168 (2000)
17. Tai, F., Lin, H.T.: Multilabel classification with principal label space transformation. Neural Computation **24**(9), 2508–2542 (2012)
18. Tsoumakas, G., Katakis, I., Vlahavas, I.: Mining multi-label data. In: Maimon, O., Rokach, L. (eds.) Data Mining and Knowledge Discovery Handbook, pp. 667–685 (2008)
19. Tsoumakas, G., Katakis, I., Vlahavas, I.: Random k-labelsets for multilabel classification. IEEE Transactions on Knowledge and Data Engineering **23**(7), 1079–1089 (2011)
20. Xia, X., Yang, X., Li, S., Wu, C., Zhou, L.: Rw.knn: a proposed random walk knn algorithm for multi-label classification. In: Proceedings of the 4th Workshop on Workshop for Ph.D. Students in Information and Knowledge Management, PIKM 2011, New York, USA, pp. 87–90 (2011)
21. Zhang, M.L., Zhou, Z.H.: A k-nearest neighbor based algorithm for multi-label classification. In: 2005 IEEE International Conference on Granular Computing, vol. 2, pp. 718–721, July 2005

22. Zhang, M.L., Zhou, Z.H.: Ml-knn: A lazy learning approach to multi-label learning. Pattern Recognition **40**(7), 2038–2048 (2007). http://www.sciencedirect.com/science/article/pii/S0031320307000027
23. Zhang, Y., Schneider, J.: Multi-label output codes using canonical correlation analysis. In: 14th International Conference on Artificial Intelligence and Statistics, pp. 873–882 (2012)
24. Zhang, Y., Zhou, Z.H.: Multilabel dimensionality reduction via dependence maximization. ACM Trans. Knowl. Discov. Data **4**(3), 14:1–14:21 (2010). http://doi.acm.org/10.1145/1839490.1839495
25. Zhou, Z.H., Zhang, M.L., Huang, S.J., Li, Y.F.: Multi instance multi-label learning. Artificial Intelligence **176**(1), 2291–2320 (2012)

Accurate On-Road Vehicle Detection with Deep Fully Convolutional Networks

Zequn Jie(✉), Wen Feng Lu, and Eng Hock Francis Tay

Department of Mechanical Engineering,
National University of Singapore, Singapore, Singapore
jiezequn@u.nus.edu

Abstract. Vision-based on-road vehicle detection is one of the key problems for autonomous vehicles. Conventional vision-based on-road vehicle detection methods mainly rely on hand-crafted features, such as SIFT and HOG. These hand-crafted features normally require expensive human labor and expert knowledge. Also, they suffer from poor generalization and slow running speed. Therefore, they are difficult to be applied in realistic application which demands accurate and fast detection in all kinds of unpredictable complex environmental conditions. This paper presents a framework utilizing fully convolutional networks (FCN) to produce bounding boxes with high confidence to contain a vehicle, and bounding box location refinement with SVM to further improve localization accuracy. Experiments on the PASCAL VOC 2007 and LISA-Q benchmarks show that using high-level semantic vehicle confidence obtained by FCN, higher precision and recall are achieved. Additionally, FCN enables whole image inference, which makes the proposed method much faster than the object proposal or hand-crafted feature based detectors.

1 Introduction

Nowadays, intelligent autonomous driving is becoming increasingly popular in the field of artificial intelligence dues to the bright prospects for commercial application in daily lives. Passenger safety is the main concern that determines the feasibility of the intelligent vehicles in real use. Therefore, a large amount of research are dedicated to the development of intelligent traffic monitoring system which aims to detect and track pedestrians, on-road vehicles and other obstacles from the view of ego-vehicles. Different techniques are available for on-road vehicle detection including radar, LIDAR and computer vision. Computer vision techniques provide richer information of the surrounding environments than the other sensors, because they try to model what humans can see in real-life scenarios. Thanks to the great progress of both image processing techniques and computing power of hardware, real-time implementation of on-road vehicle detection using computer vision techniques is becoming achievable.

However, vision-based on-road vehicle detection is very challenging due to the large within class variance in vehicle appearance which may come from the

P. Perner (Ed.): MLDM 2016, LNAI 9729, pp. 643–658, 2016.
DOI: 10.1007/978-3-319-41920-6_50

following reasons: great diversity of the inherent visual appearance (e.g., shape, color and size) of the vehicles; complex outdoor environmental conditions (e.g., illumination, weather and view angle); unpredictable interactions and occlusions between multiple vehicles or background.

Conventional vision-based methods for on-road vehicle detection primarily rely on hand-crafted image features. These proposed features are usually inspired by the statistical analysis over the pixels within the local image patches. For example, the popular Scale-Invariant Feature Transform (SIFT) describes a patch with the local accumulation of the magnitude of pixel gradients in each orientation, and finally generates a histogram vector with 128 dimensions (16 sub-regions multiplied by 8 orientations) [1]. Such hand-crafted local features require much efforts to design and may not generalize well in some complex detection problems. In other words, it is difficult to design a local feature descriptor that has strong discrimination power for all the detection scenarios. Therefore, it is difficult to use hand-crafted features to differentiate vehicles when many categories are available.

Another problem brought by the hand-crafted features is the feature computation time cost. Generally, due to the limitation that one single hand-crafted feature descriptor can only characterize the image in a certain aspect well, e.g., texture, shape or color, a combination of multiple hand-crafted feature descriptors is usually utilized to improve the recognition accuracy. Such combinations will lead to high time cost in feature computation (usually several seconds per image), thus is not feasible for real-time on-road vehicle detection.

In this paper, a data-driven learning framework is proposed to directly process the original raw data from the images captured by cameras and produce a high-level semantic confidence score which shows to what extent a specific region may contain a vehicle. Briefly, we train a vehicle/non-vehicle binary classifier using a fully convolutional network (FCN) [2] on patches from images with annotated vehicles. The fully convolutional network can take an input image of arbitrary size and output a dense "confidence map" showing the probability of containing a vehicle for each corresponding box region in the original image. To predict the confidences for boxes of different scales, we rescale the original image into multiscales and feed them into the network to obtain the confidence maps of different scales correspondingly. Then, non-maximal suppression (NMS) is performed to remove redundant boxes. Finally, we train a support vector machine (SVM) on the image gradients to refine the boxes by finding the box with the highest confidence score among the neighboring boxes of each rough box obtained by FCN.

2 Related Works

In this section, we first introduce the previous works on vehicle detection which are based on conventional low-level hand-crafted features. Then some representative works based on modern deep convolutional neural networks (CNN) are introduced.

2.1 Low-Level Feature Based Vehicle Detection

Previous approaches are either based on the low-level image cues including edge, shadow and corners or robust hand-crafted feature descriptors like Histogram of Oriented Gradient (HOG) [3], Haar-like features and Scale Invariant Feature Transform (SIFT). These features are usually then fed into various learning models to perform a vehicle/non-vehicle classification.

Matthews et al. [4] used edge detection to find vertical edges to localize left and right boundaries of vehicles. They summed the pixels in each column of the edge map and utilized a triangular filter to smooth th results.

Bertozzi et al. [5] proposed a corner-based method to hypothesize vehicle locations. The approach follows template matching strategy. Four templates, each of them corresponding to one of the four corners, were used to match all the corners in an image.

Carrafi et al. [6] proposed a cascaded detection framework to reject easy identified negative choices in the early stage, thus speeding up the detection process. Histogram of Oriented Gradients (HOG) is used as input features to pass a latent support vector machine (LSVM) to obtain the final detection results. Due to the slow computation of HOG features, they can only achieve 1fps processing speed.

Jazayeri et al. [7] utilized low-level features including edges and corners combined with a Hidden Markov Model (HMM) to segment the vehicles out of the background in each frame of a video.

Sovaraman et al. [8] relied on active learning and obtained good results on vehicle detection. A simple classifier is trained by supervised learning. Later, informative samples are queried and archived by performing selective sampling and then can be used for retraining the classifier. HOG and Haar-like features are employed as the initial input features.

[9] proposed a method based on active learning and used an SVM to determine different parts of vehicles through split and merge mechanism. The part-based cues were verified to be able to provide more robust cues implying the presence of vehicles.

Sun el al. [10] investigated and evaluated different feature extraction methods including Principal Component Analysis (PCA), Gabor filters and Wavelets in pre-crach vehicle detection. They also experimented with two popular classifiers – SVM and neural networks.

Optical flow, as an informative motion-based feature, contains the information about the moving direction and speed of objects in the videos, thus is widely used in vehicle detection in videos. Choi [11] used optical flow and Haar-like feature with AdaBoost to achieve real-time on-road vehicle detection.

2.2 Deep Convolutional Neural Networks

Hand-crafted features usually involve lots of tricks to design thus are difficult to ensure good performance in various complex vision tasks. Recently, learned features by deep learning methods have shown great potentials in computer vision tasks [12–14]. Deep learning tries to model high-level abstractions of visual data by using architectures containing multiple layers of non-linear transformations.

Specifically, deep Convolutional Neural Network (CNN) has shown outstanding performance in large-scale image classification datasets, such as ImageNet [15] and CIFAR-10 [16]. Later lots of works [17–19] consider to transferring CNN features pre-trained on ImageNet to small-scale computer vision tasks in which only limited amount of task-specific training samples are available. Off-line CNN features extracted by the model pre-trained on ImageNet are successfully applied to various vision tasks, including object detection [19], image retrieval [20]. To further improve the adaptation and representation power of CNN features in the specific tasks, the fine-tuned CNN features based on the pre-trained ImageNet CNN features are also used and achieved better performance in these specific tasks [17,18].

3 Multiscale Fully Convolutional Networks

3.1 Fully Convolutional Networks

Convolutional Neural Network (CNN) can be regarded as an automatic hierarchical feature extractor. Such a learning-based deep feature extraction pipeline avoids hand-crafted feature designing which may not be suitable for a specific task. Recently, as an extension of the classic CNN for classification problems [12–14], fully convolutional networks can take an input of arbitrary size and output a map whose size corresponds to the input, which can be used for dense prediction problems (e.g., semantic segmentation [2,21] and image restoration [22]).

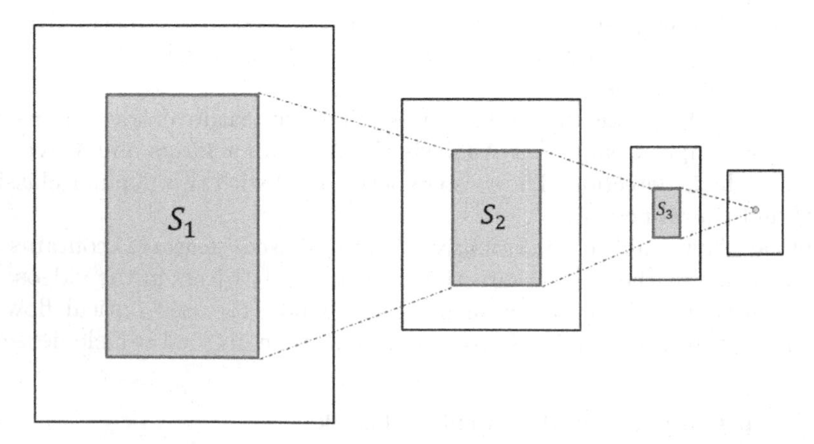

Fig. 1. Illustration of fully convolutional network. The yellow pixel in the output map shows the classification confidence of the yellow window region S_1 in the input image and will not be affected by other regions in the input image.

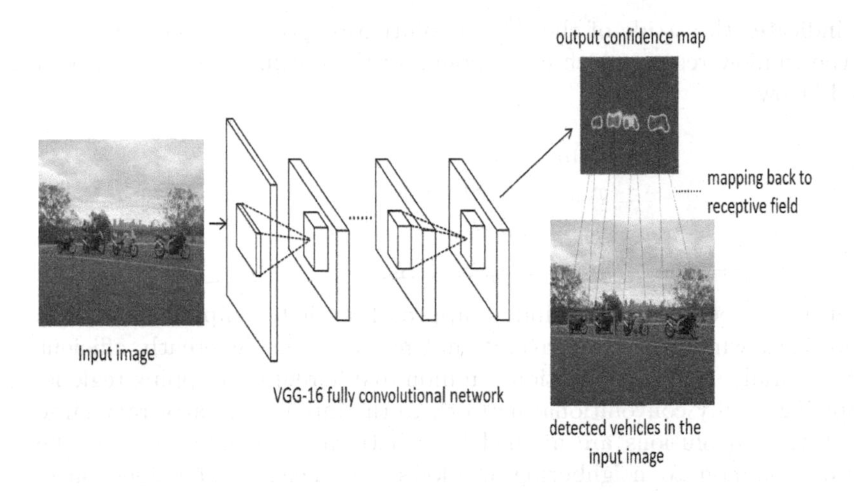

Fig. 2. The illustration of detection pipeline using FCN. An input image is fed into a fully convolutional network to generate the output vehicle confidence map. Then based on the receptive field computation, the vehicle bounding boxes can be obtained by mapping the pixels with high vehicle confidences back to the corresponding bounding boxes (receptive fields) in the input image.

A whole input image is fed into the fully convolutional network to obtain a pixel-wise vehicle confidence map. This feed-forward process can be seen as vehicle/non-vehicle binary classification for the densely sampled sliding windows in the input image. Each output pixel in the confidence map shows the classification confidence of one specific sliding window in the input image, as illustrated in Figure 1. To map back to the input vehicle detection bounding boxes from the output vehicle confidence map, there is a need to decide how big of an area the output pixel can correspond to in the input image (receptive field size). Assuming the receptive field size of each layer is S_i (i=1, 2, ..., n) and S_1 is the receptive field size in the input image, the receptive field size of each layer can be computed using the recursive formula below:

$$S_{i-1} = up(S_i) = s_i(S_i - 1) + k_i \tag{1}$$

where s_i and k_i represent the stride and the convolution kernel size of the ith convolutional or pooling layer. S_{i-1} and S_i denote the receptive field size of the $(i-1)^{th}$ and the ith layer respectively. To accurately map an output pixel back to the window region it covers in the input image, sliding window sampling stride *Str* is also indispensable. Fully convolutional network has its inherent sampling stride *Str*, which is the product of the strides of all the layers, i.e.,

$$Str = \prod_{i=1}^{n} s_i \tag{2}$$

where s_i indicates the stride of the ith convolution or pooling layer. Given S_1 and *Str*, the window region which corresponds to the output pixel (x_o, y_o) can be decided below:

$$\begin{aligned} x_{min} &= x_o Str \\ x_{max} &= x_o Str + S_1 \\ y_{min} &= y_o Str \\ y_{max} &= y_o Str + S_1. \end{aligned} \tag{3}$$

In contrast to other sliding window approaches that compute the entire pipeline for each window, fully convolutional networks are inherently efficient since they naturally share computation common to different overlapping regions. When applying a fully convolutional network to the input of an arbitrary large size in testing, convolutions are applied in a bottom-up manner so that the computation common to neighboring windows only needs to be done once. Therefore, we now are able to map the pixels with high vehicle confidences in the output map back to the corresponding bounding boxes in the input image which are highly possible to contain a vehicle. The whole pipeline of bounding box inference with FCN is shown in Figure 2.

3.2 Network Training

As VGG-16 layer network [13] shows outstanding discrimination power in the ILSVRC classification task, we adopt the publicly released VGG-16 layer model as our pre-trained model for further fine-tuning on the vehicle detection datasets. The two fully-connected layers in VGG-16 network are replaced by two convolution layers with 1×1 convolution kernels to meet the requirements of fully convolutional networks. The last fully-connected layer in VGG-16 has 1,000 neurons as there are 1,000 categories in ILSVRC classification task. Therefore, we replace the last layer of VGG-16 with a 1×1 convolution layer with 2 output neurons as we only have 2 categories (vehicle and non-vehicle). The loss function to be optimized during fine-tuning is the cross-entropy loss, i.e.,

$$E = t_k \ln(y_k) + (1 - t_k) \ln(1 - y_k) \tag{4}$$

where t_k denotes the kth target value and y_k represents the kth prediction value.

In terms of fine-tuning, we treat the network as an vehicle/non-vehicle binary classification network and use a patch-wise training strategy instead of using the whole images to train a dense structured prediction network. To this end, we crop the patches from the images with annotated objects and resize them to 256×256, the same as the input size of VGG-16 model. Among the cropped patches, those having intersection over union (IoU) ≥ 0.5 with a ground-truth box are treated as positive samples and the rest as negatives. To balance the number between the positives and the negatives, we crop multiple patches around each ground-truth box while sparsely sampling the patches in the background regions.

For the stochastic gradient descent (SGD) training process, the weights of all the pre-trained layers are initialized with the weights of the publicly released VGG-16 model. The last layer is initialized with a zero-mean Gaussian distribution with the standard deviation 0.01. The learning rate is set as 0.01 and 0.001 for the last layer and all the pre-trained layers, respectively. The main reasons for this setting are as follows: the first few convolution layers mainly extract low-level invariant features, such as texture, shape, thus the weights are more consistent from the pre-trained dataset to target dataset, whose learning rate are set as a relative low value (0.001); while for the last fully-connected layer, which is specifically adapted to the new target dataset, a higher learning rate is required to guarantee its fast convergence to the new optimum. The learning rates of all the layers are reduced by a scale of 10 after every 20 epochs.

3.3 Multiscale Inputs Inference

Using the aforementioned fully convolutional network, each pixel in the output map only covers a window region with a fixed size S_1. To enable the network to predict the vehicle bounding boxes with different sizes and aspect-ratios, we rescale the original image to different scales. By doing this, a window with the size equaling to the receptive field size S_1 in the rescaled inputs will correspond to windows of different scales and aspect-ratios in the original image.

Subsequently, the multiscale inputs after rescaling are fed into the network individually to obtain the multiscale vehicle confidence maps. Here we present the multiscale setting in detail to specify the scales needed in our approach. Generally, the more and the denser the scales are, the more a concentrated set of bounding boxes near the areas is likely to contain a vehicle. However, the downside is that noisy bounding boxes which may lower the precision will be produced as well. This issue introduces a trade-off in parameter selection for the multiscale setting.

Specifically, we define α as the stepsize indicating the IoU for neighboring boxes. In other words, the step sizes in scale and aspect ratio are determined such that one step results in neighboring boxes having an IoU of α. The scale values range from a minimum box area of 1,000 pixels to the full image. The aspect ratio changes from 1/3 to 3. The exact values of scales and aspect ratio can be computed with Eqn. (5) and Eqn. (6).

$$\text{scale} = \sqrt{1000}(\sqrt{1/\alpha})^{\text{s}}, \tag{5}$$

$$\text{aspect ratio} = (\frac{1+\alpha}{2\alpha})^{\text{r}}. \tag{6}$$

Here the index s can be any integer from 0 to $\lfloor \log(\text{image size}/\sqrt{1000})/\log(\sqrt{1/\alpha}) \rfloor$, and the index r can be any integer from $-\lfloor \log(3)/\log(\frac{1+\alpha}{2\alpha})^2 \rfloor$ to $\lfloor \log(3)/\log(\frac{1+\alpha}{2\alpha})^2 \rfloor$. For the multiscale detection bounding boxes, we first remove those with vehicle confidences lower than 0.2, reducing the total number of detection bounding boxes from several tens of thousand to less than 10,000. Next, we sort all the remained bounding boxes based on their vehicle confidences in a descending order. Finally, non-maximal suppression (NMS) is performed on the sorted bounding boxes. Specifically, we find the bounding boxes with the maximum vehicle confidences and remove all the bounding boxes with an IoU larger than an overlap threshold (we use 0.8 in all our experiments).

4 Bounding Box Refinement

Due to the fixed multiscale setting and box sampling strategy, the above obtained raw bounding boxes have their inherent weakness of being pre-defined both in scales and locations which may cause misdetection of ground-truth boxes. To overcome this, we adopt a greedy iterative search method to refine each raw bounding box.

Previous works show that objects of interest are stand-alone things with well-defined closed boundaries [23–25]. Based on this observation, gradient and edge information are widely used for implying the presence of objects in early works, e.g., BING [26] and Edge Boxes [27]. Considering this, we also rely on the gradient cues instead of 3-channel color information for the efficient implementation of our method. Specifically, we train a linear SVM vehicle/non-vehicle classifier on the gradient maps of the patches from the images with annotated vehicles. We use the ground-truth boxes of the annotated vehicles as positive samples, and crop the patches in the images and treat those with IoU < 0.3 for all the ground-truth boxes as negative samples. For all the chosen samples, we resize them to 16×16 before training the SVM. After training the

Algorithm 1. Refine the bounding boxes $[B_1, B_2, ..., B_n]$

Require: : A set of raw bounding boxes $[B_1, B_2, ..., B_n]$
  for $B = [B_1, ...B_n]$ **do**
    $Confidence \Leftarrow SVM(B)$
    $S_c \Leftarrow 0.2B_w$ (B_w is the bounding box width)
    $S_r \Leftarrow 0.2B_h$ (B_h is the bounding box height)
    while $S_c > 2$ *and* $S_r > 2$ **do**
      $[B_{c1}, B_{c2}, ..., B_{cn}] \Leftarrow ColomnNeighbors(B, S_c)$
      $B_{cmax} \Leftarrow argmax(SVM(B_{ci})),\ i = [1, 2, ..., n]$
      if $svm(B_{cmax}) > Confidence$ **then**
        $Confidence \Leftarrow SVM(B_{cmax})$
        $B \Leftarrow B_{cmax}$
      end if
      $[B_{r1}, B_{r2}, ..., B_{rn}] \Leftarrow RowNeighbors(B, S_r)$
      $B_{rmax} \Leftarrow argmax(SVM(B_{ri})),\ i = [1, 2, ..., n]$
      if $SVM(B_{rmax}) > Confidence$ **then**
        $Confidence \Leftarrow SVM(B_{rmax})$
        $B \Leftarrow B_{rmax}$
      end if
      $S_c \Leftarrow S_c/2$
      $S_r \Leftarrow S_r/2$
    end while
  end for

256-d SVM, to refine the bounding boxes, we maximize the SVM score of each box over the neighboring positions, scales and aspect ratios. After each iteration, the search step is reduced in half. The search is stopped once the translational step size is less than 2 pixels. The procedure is summarized in pseudo-code in Algorithm 1.

5 Experiments

5.1 Experimental Datasets

To verify the effectiveness of the proposed approach for on-road vehicle detection, extensive experiments are conducted on two benchmark datasets, i.e., PASCAL VOC 2007 dataset [28] and LISA-Q Front FOV dataset [8].

PASCAL VOC 2007 dataset is a popular standard benchmark which contains 20 pre-defined categories for general object classification, detection and segmentation tasks in computer vision area. We manually select the images containing vehicles (including car, bus, motorbike and bicycle four categories) for both training and testing in the experiments. In this way, we obtain 1,292 images for training the model and 1,309 images for testing. For the four categories of vehicles in PASCAL VOC, various view-angles of the vehicles may be included as the images are obtained by the ways not only restricted in the front-view cameras in the cars. Thus the visual appearance may be even more diverse and brings higher difficulty for vehicle detection.

In LISA dataset, on-road data are captured daily by LISA-Q testbed [29], which has synchronized the capture of vehicle controller area network data, Global Positioning System, and video from six cameras [30]. The videos from the front-facing camera comprises the LISA-Q Front FOV data sets. It consists

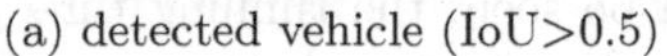
(a) detected vehicle (IoU>0.5) (b) missed vehicle (IoU<0.5)

Fig. 3. Examples of detected ground-truth vehicle (left) and missed ground-truth vehicle (right). In both images, red bounding boxes are the ground-truth annotation bounding boxes, and blue bounding boxes are the detections. White regions are the intersections between the ground-truth vehicles and the detections.

of three video sequences, consisting of 1,600, 300, and 300 consecutive frames, respectively. We randomly sample 70% frames from each of the three video sequences as training images and use the rest for testing.

5.2 Performance Metrics

We consider recall, precision and efficiency as the performance metrics when evaluating our detection method. In specific, recall is defined as the proportion of the truly detected vehicles in all the annotated ground-truth vehicles, see Eqn. (7); precision is defined as the proportion of detection that are true vehicles, see Eqn. (8). In all our experiments, a ground-truth vehicle is regarded as detected only when there is a detection bounding box has an intersection over union (IoU) larger than 0.5 with this ground-truth vehicle bounding box, see Figure 3. This criterion is widely used in the evaluations of several computer vision detection benchmarks, e.g., ImageNet Large Scale Visual Challenge Competition (ILSVRC) [31], PASCAL VOC challenge and Microsoft COCO [32]. For efficiency evaluation, we report the running speed using per image using our method.

$$\text{Recall} = \frac{\#\ \text{detected vehicles}}{\#\ \text{all ground-truth vehicles}} \tag{7}$$

$$\text{Precision} = \frac{\#\ \text{detected vehicles}}{\#\ \text{detected vehicles} + \#\ \text{false positives}} \tag{8}$$

5.3 Experimental Results

Variant Analysis. We begin the experiments by analyzing the effects of the granularity of the multi-scale search on the PASCAL VOC 2007 testing set. Specifically, we vary the stepsize parameter α and plot precision-recall curves for different α. From Figure 4, it is found that when α is between 0.45 to 0.65, as α increases, both recall and precision can be improved generally. This is natural as more scales provide more chances to have a detection close to the groundtruth bounding box. However, when α exceeds 0.65, as α increases, recall is enhanced while precision drops. The reason probably lies in that too many detections concentrated on a small area are introduced , resulting in a higher possibility of false positives and loss of the precision. From Figure 4, α should be set as 0.65 or 0.75.

We also conduct the running time comparison experiment for each search stepsize α on the PASCAL VOC 2007 testing set. Table 1 presents the detailed running time for various values of α. As can be seen, the running time grows exponentially with the increasing of α. Since the balance between recall and precision is similar good when α equals to 0.65 and 0.75, we thus fix α as 0.65 in all the later experiments considering that α as 0.75 has a quite high time cost.

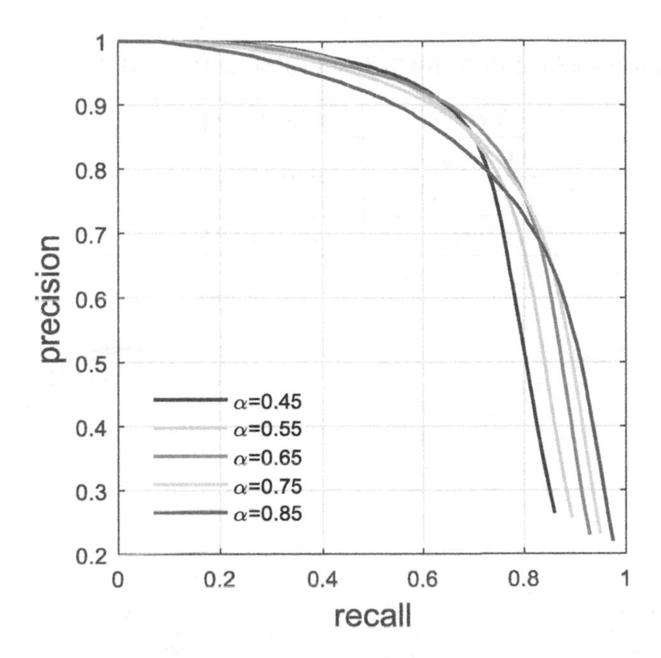

Fig. 4. Precision-recall curves of our FCN method with different multiscale stepsize α on PASCAL VOC 2007 testing set.

Table 1. Runing speed of FCN with different α on PASCAL VOC 2007 testing set.

α	Runing time per image
0.45	0.55s
0.55	0.67s
0.65	0.87s
0.75	1.93s
0.85	4.52s

Comparison to the State-of-the-Art. To verify the effectiveness of the proposed method, we conduct the comparisons between our FCN method and two state-of-the-art object detection algorithms in computer vision, i.e., Regions with CNN features (R-CNN) [17] and Deformable Part Models (DPM) [33]. The comparisons are also focused on recall, precision and running speed.

We plot the precision-recall curves for our FCN method, R-CNN and DPM in Figure 5. It shows the comparisons on both PASCAL VOC 2007 testing set and LISA-Q testing set. As can be seen, on both benchmark datasets, the proposed FCN method outperforms the two state-of-the-art detectors significantly. In specific, the proposed FCN method can achieve recall and precision both higher

Table 2. Runing speed of FCN, R-CNN and DPM on PASCAL VOC 2007 testing set.

Method	Runing time per image
FCN (ours)	0.87s
R-CNN	9.03s
DPM	12.23s

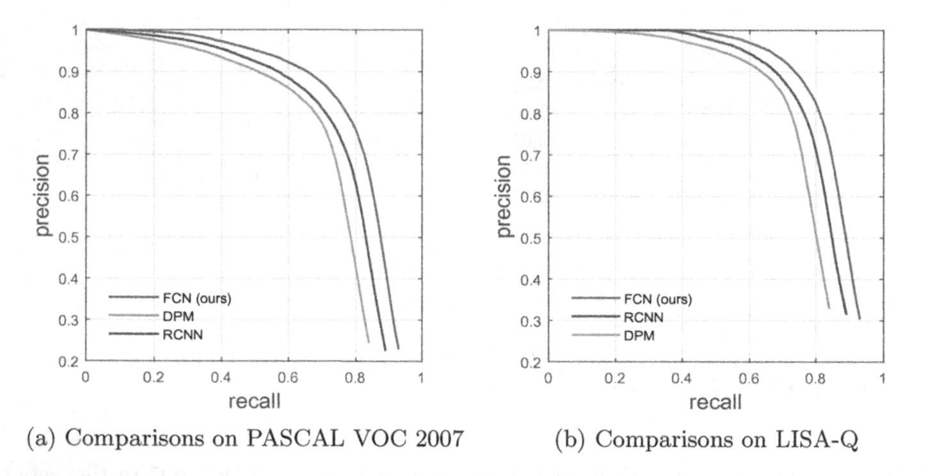

(a) Comparisons on PASCAL VOC 2007 (b) Comparisons on LISA-Q

Fig. 5. Precision-recall curves of our FCN method and other state-of-the-art detectors on PASCAL VOC 2007 testing set (left) and LISA-Q testing set (right).

than 80% at the same time in LISA-Q dataset, which shows promising in realistic applications.

The running speed comparison is presented in Table 2. Please note that for the two CNN-based methods (i.e., FCN (ours) and R-CNN), the implementation is on GPU and based on the popular deep learning open source platform Caffe [34]. We adopt the standard setting of R-CNN and DPM. In specific, 2,000 object proposals generated by Selective Search [35] are used for post-classification in R-CNN. HOG [3] feature is utilized in DPM. From Table 2, it is found that the proposed FCN method is much faster than the two state-of-the-art detectors. Considering that both FCN and R-CNN are CNN-based and require the use of GPU, the big difference in speed comes from the difference in the number of feed-forward computation of CNN. In R-CNN, the number of feed-forward computation of CNN equals to the number of object proposals (2,000 by default). By contrast, our FCN method only require N times of feed-forward computation of CNN, which is the number of scales in the multi-scale inference (less than 100). As for DPM, the computation of hand-crafted feature–HOG, takes most of the time.

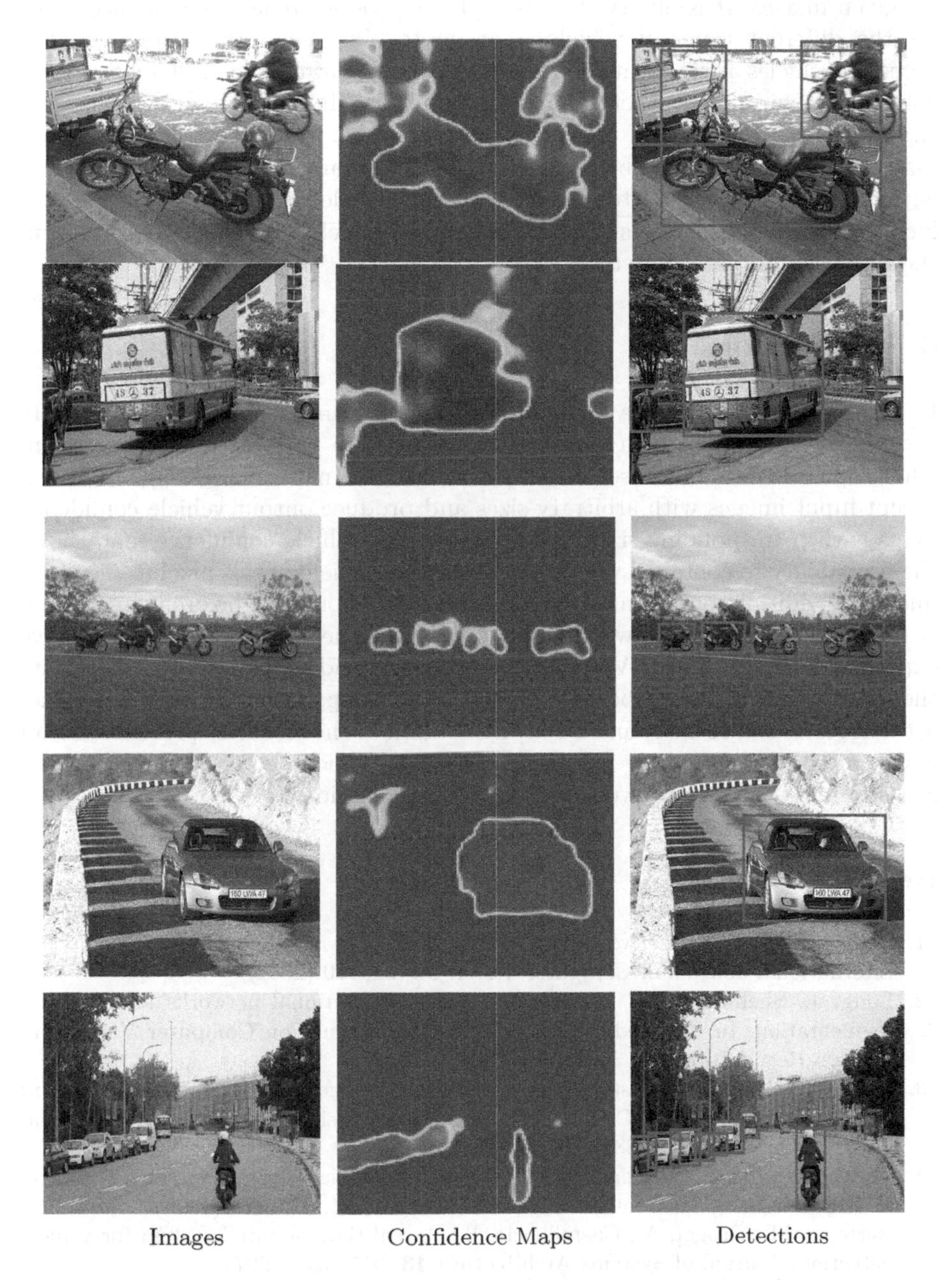

Fig. 6. Examples of predicted vehicle confidence maps (second column) and detection bounding boxes (third column) for the input images (first column). In the confidence maps, red color indicates high confidence to be a vehicle while blue color represents low confidence.

Visualizations. Figure 6 shows examples of confidence maps and detections for given images. It is observed that the FCN produce reliable confidence maps for the different types of vehicles (car, motorbike, bus, etc.). Moreover, the detected vehicles are in great diversity of visual appearances (e.g., shape, size and color) and environmental conditions (e.g., weather, illumination and view angle), showing the robustness of the FCN method. After producing the confidence maps, detection bounding boxes can be directly obtained by mapping the pixels with high confidences back to their CNN receptive fields in the input images. This localization manner enables effective detection of occluded vehicles, as shown in the third column of Figure 6.

6 Conclusions

In this paper, we propose a data-driven learning framework which can directly process the raw visual data captured by cameras to perform vehicle detection. The proposed method is based on fully convolutional networks (FCN), which can accept input images with arbitrary sizes and produce output vehicle confidence maps with corresponding sizes. Each pixel in the vehicle confidence map shows the probability to contain a vehicle for the receptive field of this pixel in the input image. In addition, based on the image gradients, a bounding box refinement step is utilized to refine the raw bounding boxes obtained by FCN. The extensive experiments on PASCAL VOC 2007 testing set and LISA-Q dataset validate the effectiveness of the proposed FCN method. Compared with two state-of-the-art detectors, i.e., R-CNN and DPM, FCN method shows better precision, recall and computation efficiency in vehicle detection, which shows promising in the realistic real-time on-road vehicle detection application.

References

1. Lowe, D.G.: Distinctive image features from scale-invariant keypoints. International Journal of Computer Vision **60**, 91–110 (2004)
2. Long, J., Shelhamer, E., Darrell, T.: Fully convolutional networks for semantic segmentation. In: Proceedings of the IEEE Conference on Computer Vision and Pattern Recognition, pp. 3431–3440 (2015)
3. Dalal, N., Triggs, B.: Histograms of oriented gradients for human detection. In: IEEE Computer Society Conference on Computer Vision and Pattern Recognition, CVPR 2005, vol. 1, pp. 886–893. IEEE (2005)
4. Matthews, N., An, P., Charnley, D., Harris, C.: Vehicle detection and recognition in greyscale imagery. Control Engineering Practice **4**, 473–479 (1996)
5. Bertozzi, M., Broggi, A., Castelluccio, S.: A real-time oriented system for vehicle detection. Journal of Systems Architecture **43**, 317–325 (1997)
6. Caraffi, C., Vojíř, T., Trefný, J., Šochman, J., Matas, J.: A system for real-time detection and tracking of vehicles from a single car-mounted camera. In: 2012 15th International IEEE Conference on Intelligent Transportation Systems (ITSC), pp. 975–982. IEEE (2012)

7. Jazayeri, A., Cai, H., Zheng, J.Y., Tuceryan, M.: Vehicle detection and tracking in car video based on motion model. IEEE Transactions on Intelligent Transportation Systems **12**, 583–595 (2011)
8. Sivaraman, S., Trivedi, M.M.: A general active-learning framework for on-road vehicle recognition and tracking. IEEE Transactions on Intelligent Transportation Systems **11**, 267–276 (2010)
9. Sivaraman, S., Trivedi, M.M.: Real-time vehicle detection using parts at intersections. In: 2012 15th International IEEE Conference on Intelligent Transportation Systems (ITSC), pp. 1519–1524. IEEE (2012)
10. Sun, Z., Bebis, G., Miller, R.: Monocular precrash vehicle detection: features and classifiers. IEEE Transactions on Image Processing **15**, 2019–2034 (2006)
11. Choi, J.: Realtime on-Road Vehicle Detection with Optical Flows and Haar-Like Feature Detectors (2012)
12. Krizhevsky, A., Sutskever, I., Hinton, G.E.: Imagenet classification with deep convolutional neural networks. In: NIPS, pp. 1097–1105 (2012)
13. Simonyan, K., Zisserman, A.: Very deep convolutional networks for large-scale image recognition. arXiv preprint (2014). arXiv:1409.1556
14. Szegedy, C., Liu, W., Jia, Y., Sermanet, P., Reed, S., Anguelov, D., Erhan, D., Vanhoucke, V., Rabinovich, A.: Going deeper with convolutions. arXiv preprint (2014). arXiv:1409.4842
15. Deng, J., Dong, W., Socher, R., Li, L.J., Li, K., Fei-Fei, L.: Imagenet: a large-scale hierarchical image database. In: IEEE CVPR, pp. 248–255 (2009)
16. Krizhevsky, A., Hinton, G.: Learning multiple layers of features from tiny images. Computer Science Department, University of Toronto, Tech. Rep. (2009)
17. Girshick, R., Donahue, J., Darrell, T., Malik, J.: Rich feature hierarchies for accurate object detection and semantic segmentation. arXiv preprint (2013). arXiv:1311.2524
18. Oquab, M., Bottou, L., Laptev, I., Sivic, J., et al.: Learning and transferring mid-level image representations using convolutional neural networks. arXiv preprint (2013)
19. Sermanet, P., Eigen, D., Zhang, X., Mathieu, M., Fergus, R., LeCun, Y.: Overfeat: Integrated recognition, localization and detection using convolutional networks. arXiv preprint (2013). arXiv:1312.6229
20. Gong, Y., Wang, L., Guo, R., Lazebnik, S.: Multi-scale orderless pooling of deep convolutional activation features. arXiv preprint (2014). arXiv:1403.1840
21. Pinheiro, P.H., Collobert, R.: Recurrent convolutional neural networks for scene parsing. arXiv preprint (2013). arXiv:1306.2795
22. Eigen, D., Krishnan, D., Fergus, R.: Restoring an image taken through a window covered with dirt or rain. In: 2013 IEEE International Conference on Computer Vision (ICCV), pp. 633–640. IEEE (2013)
23. Alexe, B., Deselaers, T., Ferrari, V.: Measuring the objectness of image windows. IEEE Transactions on Pattern Analysis and Machine Intelligence **34**, 2189–2202 (2012)
24. Forsyth, D.A., Malik, J., Fleck, M.M., Greenspan, H., Leung, T., Belongie, S., Carson, C., Bregler, C.: Finding Pictures of Objects in Large Collections of Images. Springer (1996)
25. Heitz, G., Koller, D.: Learning spatial context: using stuff to find things. In: Forsyth, D., Torr, P., Zisserman, A. (eds.) ECCV 2008, Part I. LNCS, vol. 5302, pp. 30–43. Springer, Heidelberg (2008)

26. Cheng, M.M., Zhang, Z., Lin, W.Y., Torr, P.: Bing: binarized normed gradients for objectness estimation at 300fps. In: 2014 IEEE Conference on Computer Vision and Pattern Recognition (CVPR), pp. 3286–3293. IEEE (2014)
27. Zitnick, C.L., Dollár, P.: Edge boxes: locating object proposals from edges. In: Fleet, D., Pajdla, T., Schiele, B., Tuytelaars, T. (eds.) ECCV 2014, Part V. LNCS, vol. 8693, pp. 391–405. Springer, Heidelberg (2014)
28. Everingham, M., Van Gool, L., Williams, C.K., Winn, J., Zisserman, A.: The pascal visual object classes (voc) challenge. International Journal of Computer Vision **88**, 303–338 (2010)
29. McCall, J.C., Achler, O., Trivedi, M.M.: Design of an instrumented vehicle test bed for developing a human centered driver support system. In: 2004 IEEE Intelligent Vehicles Symposium, pp. 483–488. IEEE (2004)
30. Trivedi, M.M., Gandhi, T., McCall, J.: Looking-in and looking-out of a vehicle: Computer-vision-based enhanced vehicle safety. IEEE Transactions on Intelligent Transportation Systems **8**, 108–120 (2007)
31. Russakovsky, O., Deng, J., Su, H., Krause, J., Satheesh, S., Ma, S., Huang, Z., Karpathy, A., Khosla, A., Bernstein, M., et al.: Imagenet large scale visual recognition challenge. International Journal of Computer Vision, 1–42 (2014)
32. Lin, T.Y., Maire, M., Belongie, S., Bourdev, L., Girshick, R., Hays, J., Perona, P., Ramanan, D., Zitnick, C.L., Dollr, P.: Microsoft coco: Common objects in context. arXiv preprint (2015). arXiv:1506.06204
33. Felzenszwalb, P.F., Girshick, R.B., McAllester, D., Ramanan, D.: Discriminatively Trained Deformable Part Models, Release 4 (2010)
34. Jia, Y., Shelhamer, E., Donahue, J., Karayev, S., Long, J., Girshick, R., Guadarrama, S., Darrell, T.: Caffe: Convolutional architecture for fast feature embedding. arXiv preprint (2014). arXiv:1408.5093
35. Uijlings, J.R., van de Sande, K.E., Gevers, T., Smeulders, A.W.: Selective search for object recognition. International Journal of Computer Vision **104**, 154–171 (2013)

On the Abstraction of a Categorical Clustering Algorithm

Mina Sheikhalishahi(✉), Mohamed Mejri, and Nadia Tawbi

Department of Computer Science, Université Laval, Québec City, Canada
mina.sheikh-alishahi.1@ulaval.ca,
{mohamed.mejri,nadia.tawbi}@ift.ulaval.ca

Abstract. Despite being one of the most common approach in unsupervised data analysis, a very small literature exists on the formalization of clustering algorithms. This paper proposes a semiring-based methodology, named Feature-Cluster Algebra, which is applied to abstract the representation of a labeled tree structure representing a hierarchical categorical clustering algorithm, named CCTree. The elements of the feature-cluster algebra are called terms. We prove that a specific kind of a term, under some conditions, fully abstracts a labeled tree structure. The abstraction methodology maps the original problem to a new representation by removing unwanted details, which makes it simpler to handle. Moreover, we present a set of relations and functions on the algebraic structure to shape the requirements of a term to represent a CCTree structure. The proposed formal approach can be generalized to other categorical clustering (classification) algorithms in which features play key roles in specifying the clusters (classes).

Keywords: Formal methods · Categorical clustering · Algebraic formalization · Clustering algorithm · Semiring · Abstraction

1 Introduction

Clustering is a very well-known tool in unsupervised data analysis, which has been the focus of significant researches in different studies, spanning from information retrieval, text mining, data exploration, to medical diagnosis [2]. Clustering refers to the process of partitioning a set of data points into groups, in a way that the elements in the same group are more similar to each other rather than to the ones in other groups. The problem of clustering becomes more challenging when data are described in terms of *categorical attributes*, for which, differently from numerical attributes, it is hard to establish an ordering relationship. Although being vastly used in the literature, a very few works exist to express categorical clustering algorithms and the related issues in terms of formal methods.

This research has been supported by Natural Sciences and Engineering Research Council of Canada (NSERC).

P. Perner (Ed.): MLDM 2016, LNAI 9729, pp. 659–675, 2016.
DOI: 10.1007/978-3-319-41920-6_51

In present work, we provide an abstract representation of the clusters through the formal definition of a categorical clustering algorithm, named CCTree. This abstract representation facilitates the analysis of cluster properties, while getting rid of confronting a large amount of data in each cluster. The proposed formal scheme can also be used to formalize challenging tasks in categorical clustering algorithms, e.g. parallel clustering, feature selection.

CCTree (Categorical Clustering Tree) [11] has a decision tree-like structure, which iteratively divides the data of a node on the base of an attribute, or domain of features, yielding the greatest entropy. The division of data is represented with edges coming out from a parent node to its children, where the edges are labeled with the associated features. A node which respects the specified stop conditions is considered as a leaf. The leaves of the tree are the desired clusters. Features play a key role in the CCTree algorithm. In fact, the feature structure is used to uniquely identify each cluster.

Abstraction is a mathematical concept which maps a representation of a problem, which is called the ground (semantic), onto a new representation, which is called the abstract (syntax) representation. Through abstraction, it is possible to deal with the problem in the new space by preserving certain desirable properties and in a simpler way to handle, since it is constructed from ground representation by removing unwanted details [4].

To abstract the CCTree representation, we propose a semiring-based algebraic structure, named "*Feature-Cluster Algebra*". The elements of the proposed algebraic structure are called *terms*. We show that the proposed approach, under some conditions, is able to fully abstract the labeled tree structure. The full abstraction guarantees that the semantic and syntax forms of a problem can be used alternatively. Furthermore, we present a set of functions and relations on the feature-cluster algebra, which is used to present the conditions for a term to represent a CCTree structure.
The contributions of the present work can be summarized as follows:

- A semiring-based formal approach, named *Feature-Cluster Algebra*, is proposed to *abstract* the representation of feature-based clusters. We call the elements of the proposed algebra as *term*.
- We show that a specific term in the feature-cluster algebra, under some conditions, fully abstracts a labeled tree structure, and specifically CCTree. The full abstraction allows to apply the desired calculation on simple abstract form, and ensure that the result is also satisfied in the main semantic representation.
- We present the conditions and requirements for a term in the feature-cluster algebra to represent a CCTree structure.

The paper is organized as follows. In Section 2, we present a review of the literature on formalization methods applied in feature-based problems. In Section 3, the preliminary background notions are provided. We construct the feature-cluster algebra, the semiring based methodology for abstracting CCTree in Section 4. The process of transforming a tree structure to a term and vice versa is presented in Section 5. The relations on feature-cluster algebra are introduced in Section 6 which are used to identify the CCTree term in proposed algebraic system. We conclude and point to future directions in Section 7.

2 Related Work

In the following we present some work on feature models and the associated formal approaches.

Feature models, in computer science, were first defined as information models where a set of products, e.g. software products or DVD player products, are represented as hierarchically arrangement of features, with different relationships among features [1]. Feature models are used in many applications as the result of being able to model complex systems, being interpretable, and ability to handle both ordered and unordered features [10]. Benavides et. al. [1] argue that designing a family of software system in terms of features, makes it easy to be understood by all stakeholders, rather than the time they are expressed in terms of objects or classes. Representing feature models as a tree of features, was first introduced by Kang et. al. [9] to be used in software product line. Some studies [3] show that tree models combined with ensemble techniques, lead to an accurate performance on variety of domains. In feature model tree, differently from CCTree, the root is the desired product, the nodes are the features, and different representation of edges demonstrates the mandatory or optional presence of features. Hofner et. al. [8] [7], were the first who applied idempotent semiring as the basis for the formalization of tree models of products, called it *feature algebra*. The concept of semiring is used to answer the needs of product family abstract form of expression, refinements, multi-view reconciliation, and product development and classification. The elements of semiring in proposed methodology are the sets of products, or product families. We import the idea of exploited in the work to abstract the clustering algorithm.

To the best of our knowledge, we are the first to apply an algebraic structure to abstract a clustering algorithm and formalize the associated issues.

3 Background

In this section we present the preliminary notions and definitions exploited by our proposed methodology.

Categorical Clustering Tree (CCTree). The CCTree [11] is constructed iteratively through a decision tree-like structure, where the leaves of the tree are the desired clusters. The root of the CCTree contains all the elements to be clustered. Each element is described through a set of *categorical* attributes. Being categorical, each attribute may assume a finite set of discrete values, constituting its domain. At each step, a new level of the tree is generated by splitting the nodes of the previous levels, when they are not homogeneous enough. *Shannon Entropy* is used both to define a homogeneity measure called *node purity*, and to select the attribute used to split a node. In particular non-leaf nodes are divided

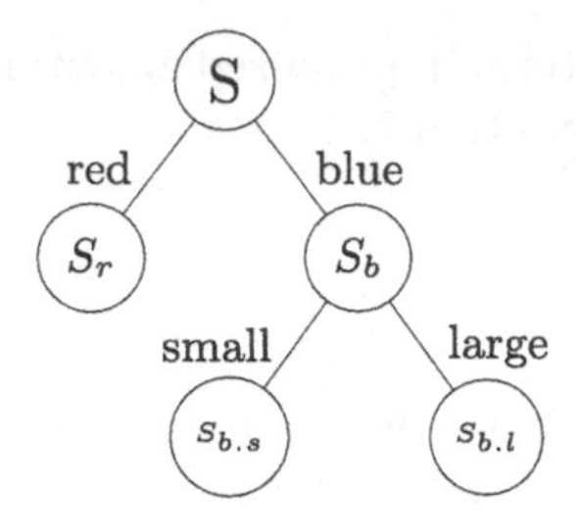

Fig. 1. A small CCTree: the root contains all the data desired to be clustered and the leaves are the desired clusters

on the base of the attribute yielding the maximum value for Shannon entropy. The separation is represented through a branch for each possible outcome of the specific attribute. Each branch or edge extracted from parent node is labeled with the selected feature which directs data to the child node [12].
Figure 1 depicts a simple CCTree.

Graph Theory Preliminaries. In graph theory [5], a *tree* is an undirected graph in which any two vertices are connected by exactly one path. A *leaf* is a vertex of degree 1. A tree is called a *rooted tree* if one vertex has been designated the root, which means that the edges have a natural orientation, towards or away from the root [5]. A tree is a labeled tree if the edges of the tree are labeled. A *branch* of a tree, refers to the path between the root and a leaf in a rooted tree [5]. A *tree structure*, in our context, is a triple (Σ, Q, δ) where Σ represents the set of edge labels; Q is the set of states or nodes; δ is the set of state (node) transition function as $\delta : Q \times \Sigma \rightarrow Q$, such that there is no cycle in transitions. The transitions connect each parent node to its children moving from root to leaves. This means that there exists a state where no transition enters it, which is the root, and there are states for them no transition exits, which are the leaves of the tree. For a transition as $\delta(s_1, f) = s_2$, we show it as the triple (s_1, f, s_2). *Graph homomorphism* ζ from a graph $G = (V, E)$ to a graph $G' = (V', E')$, written as $\zeta : G \rightarrow G'$, is a mapping $\zeta : V \rightarrow V'$ from the vertex set of G to the vertex set of G' such that $\{u, v\} \in E$ implies $\{\zeta(u), \zeta(v)\} \in E'$ [6]. If the homomorphism $\zeta : G \rightarrow G'$ is bijection whose inverse function is also a graph homomorphism, then ζ is a *graph isomorphism*. In our context, except having the same graphical structure, it is important that both $\{u, v\} \in E$ and $\{\zeta(u), \zeta(v)\} \in E'$ have the same edge label. Under this condition, we say that two graphs $G = (V, E)$ and $G' = (V', E')$ are *isomorphic*, we denote it as $G \approx G'$.

Abstraction Theory Preliminaries. The process of *abstraction* is related to the process of extracting from a representation of an object or subject an "*abstract*" representation that consists of a brief sketch of the original representation [4]. More precisely, the abstraction is the process of mapping a representation of a problem, called the "ground"(semantic) representation, onto a new representation, called the "abstract"(syntax) representation, such that it helps to deal with the problem in the original space by preserving certain desirable properties and is simpler to handle as it is constructed from the ground representation by " not considering the details" [4]. An abstraction can formally be written as a function $[[.]] : \mathcal{X} \rightarrow \mathcal{Y}$ from the ground representation of an object to its abstract form. We say $[[.]]$ *adequately abstracts* $\mathcal{X}$ if from the equivalence of two elements of semantic forms, we get the equivalence of their counterpart syntax forms. Formally, let show the equivalence of elements in $\mathcal{X}$ with $\simeq$ and the equivalence of elements in $\mathcal{Y}$ with $\cong$, then for adequate abstraction we have $[[X_1]] \cong [[X_2]] \Rightarrow X_1 \simeq X_2$. We say $[[.]]$ *abstracts* $\mathcal{X}$ if we have $X_1 \simeq X_2 \Rightarrow [[X_1]] \cong [[X_2]]$. In the case that the above relations respect simultaneously, we say $[[.]]$ *fully abstracts* $\mathcal{X}$, i.e. we have $[[X_1]] \cong [[X_2]] \Leftrightarrow X_1 \simeq X_2$.

4 Feature-Cluster Algebra

A cluster in a CCTree can uniquely be identified with the set of elements respecting a set of features. To construct a semiring which also contains the clusters, we firstly propose two semirings on the set of features and the set of elements, respectively. The proofs of some theorems are provided in Appendix.

(I) Semiring of Features. Let a *set of disjoint sorts*, denoted as $\mathcal{A}$, is given, where the *carrier set* of each sort $A_i \in \mathcal{A}$ is denoted by $\mathcal{V}_{A_i}$. In our context, we call the given set of sorts as the set of *attributes*, and we call the *union* of sorts, denoted as $\mathbb{V} = \bigcup_{A_i \in \mathcal{A}} \mathcal{V}_{A_i}$ as the set of *values* or *features*. For example, we may consider the set of attributes as $\mathcal{A} = \{color, size\}$, where the carrier set of each attribute can be considered as $\mathcal{V}_{color} = \{red, blue\}$ and $\mathcal{V}_{size} = \{small, large\}$. In this case, we have $\mathbb{V} = \{red, blue, small, large\}$.

Definition 1 (Sort). *We define the* sort *function which gets a set of features and returns a set of the associated sorts of received feature as follows:*

$$sort : \mathcal{P}(\mathbb{V}) \rightarrow \mathcal{P}(\mathbb{V})$$
$$sort(\{f\}) = \mathcal{V}_A \qquad for \quad f \in \mathcal{V}_A$$
$$sort(V_1 \cup V_2) = sort(\{V_1\}) \cup sort(\{V_2\})$$

Consider $\mathbb{F} = \mathcal{P}(\mathcal{P}(\mathbb{V}))$ be the power set of the power set of $\mathbb{V}$, whilst we denote $1 = \{\emptyset\}$ and $0 = \emptyset$. We define the operations "+" and "·", as *choice* and *composition* operators on $\mathbb{F}$ as the following:

$$\cdot : \mathbb{F} \times \mathbb{F} \rightarrow \mathbb{F} \qquad\qquad + : \mathbb{F} \times \mathbb{F} \rightarrow \mathbb{F}$$
$$F_i \cdot F_j = \{X_s \cup Y_t \; : X_s \in F_i \, , \, Y_t \in F_j\} \qquad\qquad F_i + F_j = F_i \cup F_j$$

We say that F belongs to $\mathbb{F}$, if it respects one of the following syntax forms $F := 0 \,|\, \{\{f\}\} \,|\, F \cdot F \,|\, F + F \,|\, 1$, where $f \in \mathbb{V}$.

Example 1. Some elements of $\mathbb{F}$ on $\mathbb{V} = \{red, blue, small, large\}$ are $F_1 = \{\{red, large\}, \{blue\}\}$, $F_2 = \{\{small\}\}$, $F_1 \cdot F_2 = \{\{red, large, small\}, \{blue, small\}\}$, and $F_1 + F_2 = \{\{red, large\}, \{blue\}, \{small\}\}$.

Definition 2. *Lets consider* $|.|$ *returns the number of elements in a set. Then, we say* $F \in \mathbb{F}$ *belongs to the set* $\mathbb{F}_n$*, if* $|F| = n$*. Under this definition,* $\mathbb{F}_1$*, i.e. the subset of* $\mathbb{F}$*, where each element contains just one dataset of features, is the desired one according to our problem. In this case, for* $F \in \mathbb{F}_1$*, we remove the brackets and separate the features belonging to the same set by multiplication. Hence, we consider* $F \in \mathbb{F}_1$ *if it can be written as one of the syntax forms as:* $0 \,|\, f \,|\, F_1 \cdot F_2 \,|\, 1$*, for* $f \in \mathbb{V}$*.*

It is noticeable when two elements of $\mathbb{F}_1$ are added or multiplied, they follow the same properties following the main semiring defined on $\mathbb{F}$.

Example 2. We simplify the elements of Example 1 according to Definition 2, as $F_1 = \{\{red, large\}, \{blue\}\} = \{\{red, large\}\} + \{\{blue\}\} = red \cdot large + blue$, $F_2 = \{\{small\}\} = small$, $F_1 \cdot F_2 = \{\{red, large, small\}, \{blue, small\}\} = \{\{red, large, small\}\} + \{\{blue, small\}\} = red \cdot large \cdot small + blue \cdot small$, $F_1 + F_2 = \{\{red, large\}, \{blue\}, \{small\}\} = \{\{red, large\}\} + \{\{blue\}\} + \{\{small\}\} = red \cdot large + blue + small$.

(II) Semiring of Elements. Let us consider that the set of the sorts, or the set of attributes $\mathcal{A}$ with an order among attributes, is given. Suppose $|A| = k$, and without loss of generality $A_1, A_2, \ldots, A_k$ are the ordered sorts which range over $\mathcal{A}$. We say s belongs to *the set of elements* $\mathbb{S}$, if $s \in \mathcal{V}_{A_1} \times \mathcal{V}_{A_2} \times \ldots \times \mathcal{V}_{A_k} \times \mathbb{N}$, where $\mathbb{N}$ is the set of *natural* numbers. Hence, $\mathbb{S} \subseteq \mathcal{V}_{A_1} \times \mathcal{V}_{A_2} \times \ldots \times \mathcal{V}_{A_k} \times \mathbb{N}$, i.e. $s \in \mathbb{S}$ can be written as $s = (x_1, x_2, \cdots, x_k, n)$, where $x_i \in \mathcal{V}_{A_i}$ for $1 \leq i \leq k$, and $n \in \mathbb{N}$ represents the *ID* of an element. For the sake of simplicity, we may use the alternative representation $x_i \in A_i$ instead of $x_i \in \mathcal{V}_{A_i}$. We formally define two operations "+" and "·" as *union* and *intersection* of elements of $\mathcal{P}(\mathbb{S})$ (the power set of $\mathbb{S}$) as follows:

$$\cdot : \mathcal{P}(\mathbb{S}) \times \mathcal{P}(\mathbb{S}) \rightarrow \mathcal{P}(\mathbb{S}) \qquad\qquad + : \mathcal{P}(\mathbb{S}) \times \mathcal{P}(\mathbb{S}) \rightarrow \mathcal{P}(\mathbb{S})$$
$$S_i \cdot S_j = S_i \cap S_j \qquad\qquad S_i + S_j = S_i \cup S_j$$

Formally, we say S belongs to the *power set of the elements* $\mathcal{P}(\mathbb{S})$, if it respects one of the form as $S := \emptyset \,|\, S' \,|\, S + S \,|\, S \cdot S \,|\, \mathbb{S}$, where $S' \subseteq \mathbb{S}$. The quintuple $(\mathcal{P}(\mathbb{S}), +, \cdot, \emptyset, \mathbb{S})$ is an idempotent commutative semiring.

Note: It should be noted that the operations "+" and "·" are overloaded to the kind of elements that they are applied on. This means that if the operation "+" is used between two sets of elements, it refers to the addition operation in semiring of elements, and when the operation "+" is applied between two sets of features, it refers to the addition operation in semiring of features. The same property satisfies for multiplication operation "·".

(III) Semiring of Terms. In what follows, we construct the semiring of terms with the use of previous semrirings, which will be used to abstract the tree structure. In the rest of the chapter, we use the same notions and symbols introduced in previous sections. Recalling that a cluster in CCTree can uniquely be identified by a set of elements respecting a set of features, we define the *satisfaction* relation to formally express the concept of cluster.

Definition 3 (Satisfaction Relation $\diamond$). *Recalling that when the elements of $\mathbb{F}$ contain just one dataset of features we remove the brackets (Definition 2), we define* satisfaction *relation, denoted with $\diamond$, as the following:*

$$\begin{array}{ll} \diamond : \mathbb{F} \times \mathcal{P}(\mathbb{S}) \rightarrow \mathcal{P}(\mathbb{S}) & \\ \diamond(f, \{(x_1, x_2, \cdots, x_k)\}) = \{(x_1, x_2, \cdots, x_k)\} & if \quad \exists i, 1 \leq i \leq k, \ s.t\, x_i = f \\ \diamond(f, \{(x_1, x_2, \cdots, x_k)\}) = \emptyset & if \quad \nexists i, 1 \leq i \leq k, \ s.t\ x_i = f \\ \diamond(f, S_1 \cup S_2) \ = \ \diamond(f, S_1) \ \cup \ \diamond(f, S_2) & \\ \diamond(F_1 \cdot F_2, S) \ = \ \diamond(F_1, S) \ \cap \ \diamond(F_2, S) & \end{array}$$

In the case that $\diamond(F, S) \neq \emptyset$, we say that S satisfies *F, and we apply the alternative representation $F \diamond S$ instead of $\diamond(F, S)$.*
We define that the multiplication "$\cdot$" and "+" over $\diamond$ to respect the following properties:

$$(F_1 \diamond S_1) \cdot (F_2 \diamond S_2) = (F_1 \cdot F_2) \diamond S_2 \qquad if \quad S_1 \cdot S_2 = S_2 \tag{1}$$
$$(F_1 \diamond S_1) + (F_2 \diamond S_2) = (F_1 + F_2) \diamond S_2 \qquad if \quad S_1 + S_2 = S_2 \tag{2}$$

where $S_1 \cdot S_2 = S_2$ means $S_2 \subseteq S_1$, and $S_1 + S_2 = S_2$ means $S_1 \subseteq S_2$. In the case neither set is a subset of the other, the multiplication and addition return the received elements unchanged. It should be noted that "$\cdot$" and "+" are overloaded to their own definition for the semiring of features and the semiring of elements when they are applied between two sets of features and two sets of elements, respectively. In our context, the property 1 is applied to address the concept of division of a cluster to new smaller clusters, where each new cluster satisfies the features of the main cluster, plus more restricted features. Moreover, the property 2 is used to get the simpler form of clusters according to Definition 2.

Proposition 1. *For $F_1, F_2 \in \mathbb{F}$ and $S \in \mathcal{P}(\mathbb{S})$, the symbol "$\diamond$" satisfies the following properties with respect to "+" and "$\cdot$":*

$$(F_1 \cdot F_2) \diamond S = (F_1 \diamond S) \cdot (F_2 \diamond S)$$
$$(F_1 + F_2) \diamond S = (F_1 \diamond S) + (F_2 \diamond S)$$

Proof. The proof is straightforward from the properties 1 and 2, since we have $S \cdot S = S$ and $S + S = S$. The above equations express how we can transform the different forms of $F \in \mathbb{F}$ to the form of $F \in \mathbb{F}_1$.

Example 3. The following equation shows the transformation of relations of Proposition 1 to a set of features as $F \in \mathbb{F}_1$ as Definition 2.

$$\{\{f_1, f_2\}, \{f_3\}\} \diamond S = \{\{f_1, f_2\}\} \diamond S + \{\{f_3\}\} \diamond S = f_1 \cdot f_2 \diamond S + f_3 \diamond S.$$

The form of $F \in \mathbb{F}_1$ is a particular desired representation of the set of features which will be used in our context. Hence, we attribute a specific name to it.

Definition 4 (Feature-Cluster (Family) Term). *The set of* feature-cluster family terms *on $\mathbb{V}$ and $\mathbb{S}$ denoted as $\mathbb{FC}_{\mathbb{V},\mathbb{S}}$ (or simply $\mathbb{FC}$) is the smallest set containing elements satisfying the following conditions:*

$$\begin{array}{llll} if & S \subseteq \mathbb{S} & then & S \in \mathbb{FC} \\ if & F \in \mathbb{F}_1, S \subseteq \mathbb{S} & then & F \diamond S \in \mathbb{FC} \\ if & \tau_1 \in \mathbb{FC}, \tau_2 \in \mathbb{FC} & then & \tau_1 + \tau_2 \in \mathbb{FC} \end{array}$$

In this case, we call S and $F \diamond S$ a "feature-cluster term" *and the addition of one or more feature-cluster terms is called* "feature-cluster family term"*. We may simply use "$\mathbb{FC}$-term" to refer to a feature-cluster (family) term.*
We define the block *function, which receives a feature-cluster family term and returns the set of its blocks. Formally, block* : $\mathbb{FC} \to \mathcal{P}(\mathbb{FC})$, *such that:*
$block(S) = \{S\}$, $block(F \diamond S) = \{F \diamond S\}$, $block(\tau_1 + \tau_2) = block(\tau_1) \cup block(\tau_2)$

In the case that no feature specifies S directly, it is called an *atomic* term. The set of all atomic terms is denoted as $\mathscr{A}$.

Definition 5 ($\mathbb{FC}$-Term Comparison). *We say two $\mathbb{FC}$-terms τ_1 and τ_2 are* equal, *denoted by $\tau_1 \equiv \tau_2$, if for different representations of $\mathbb{FC}$-terms, it satisfies the following relations:*

$$\begin{array}{ll} S_1 \equiv S_2 & \Leftrightarrow S_1 = S_2 \\ F_1 \diamond S_1 \equiv F_2 \diamond S_2 & \Leftrightarrow S_1 = S_2\,,\, F_1 = F_2 \\ \tau \equiv \tau' & \Leftrightarrow block(\tau) = block(\tau'). \end{array}$$

Definition 6 (Term). *We call τ a* term, *if it respects one of the syntax forms as "$\tau := S \mid F \diamond S \mid \tau + \tau \mid \tau \cdot \tau$", where S and F come from $\mathcal{P}(\mathbb{S})$ and $\mathbb{F}$, respectively. Furthermore, we consider the operations "+" and "·" be commutative and associative among terms, whilst "·" left and right distributes over "+".* The set of terms *on $\mathcal{P}(\mathbb{S})$ and $\mathbb{F}$ is shown with $\mathbb{C}_{\mathcal{P}(\mathbb{S}),\mathbb{F}}$, or abbreviated as $\mathbb{C}$ where it is known from the context.*

Example 4. Some examples of terms on $\mathbb{V} = \{red, blue, small, large\}$ and dataset S, can be considered as "$red \cdot small \diamond S$", "$red \cdot small \diamond S + blue \diamond S'$", "$(red \cdot small \diamond S) \cdot (blue \diamond S')$", "$\{\{red, large\}, \{blue\}\} \diamond S$".

Theorem 1. *Two identity elements of $\mathbb{C}$ with respect to "+" and "·" are $0 \diamond \emptyset$ and $1 \diamond \mathbb{S}$, respectively.*

Theorem 2. *The quantiple $(\mathbb{C}, \text{"}+\text{"}, \text{"}\cdot\text{"}, 0 \diamond \emptyset, 1 \diamond \mathbb{S})$ is an idempotent commutative semiring.*

Proof. The proof is straightforward from the semrirng definition , semiring of features, semiring of elements, and the properties mentioned in 1 and 2.

Definition 7 (Feature-Cluster Algebra). *The semiring $(\mathbb{C}, \text{"}+\text{"}, \text{"}\cdot\text{"}, 0 \diamond \emptyset, 1 \diamond \mathbb{S})$ is called a* feature-cluster algebra.

It is noticeable that in present work our terms in the following sections, mostly, belong to the set of feature-cluster family terms $\mathbb{FC} \subseteq \mathbb{C}$. This means that as the elements of the semiring $\mathbb{C}$, they follow the same operation and properties among the elements of the proposed feature-cluster algebra.

5 Feature-Cluster (Family) Term Abstraction

In this section, we explain how tree structure and feature-cluster family term can be transformed to each other. To this end, we first present the "*meaning*" relation to transform a feature-cluster family term to a labeled graph structure. Afterwards, we present a function to get a feature-cluster family term from a labeled tree structure. Then, we prove in a theorem that if two labeled trees are equivalent, they return equal terms. However, we show that the two equal feature-cluster family terms do not necessarily return two equivalent graph structures. We prove that under the condition of considering a fixed order among the features, the latter requirement will also be respected. In the provided examples, attributes $Color = \{r(ed), b(lue)\}$, $Size = \{s(mall), l(arge)\}$, and $Shape = \{c(ircle), t(riangle)\}$ are used to describe the terms. To avoid the confusion of different representations of an $\mathbb{FC}$-term, in what follows we present the definitions of *factorized* and *non factorized* terms.

Definition 8 (Factorized Term). *We define the* factorization *rewriting rule through an attribute $A \in \mathcal{A}$, denoted as $\xrightarrow{A}$, from an $\mathbb{FC}$-term to its factorized form as "$f \cdot \tau_1 + f \cdot \tau_2 \xrightarrow{A} f \cdot (\tau_1 + \tau_2)$" for $f \in A$.*

We denote the normal form of applying the factorization rewriting rule on term τ through attribute A as $\tau \downarrow_A$, and the set of factorized forms of the terms of $\mathbb{FC}$ is denoted by $\mathbb{FC} \downarrow$. A term after factorization is called a factorized term.

Definition 9 (Defactorization). *We define the* defactorized *rewriting rule on an $\mathbb{FC}$-term as "$f \cdot (\tau_1 + \tau_2) \rightarrow_d f \cdot \tau_1 + f \cdot \tau_2$".*

A normal term resulted from defactorized rewriting rule is called a non factorized term. *A non factorized form of the term τ is denoted as $\tau \uparrow$. The set of non factorized terms of $\mathbb{FC}$ are denoted by $\mathbb{FC} \uparrow$.*

Example 5. For factorization we have $(r \cdot s \diamond S + r \cdot c \diamond S + b \cdot s \diamond S) \downarrow_{color} = r \cdot (s \diamond S + c \diamond S) + b \cdot s \diamond S$ and for defactorizatio $r \cdot (s \diamond S + c \diamond S) + b \cdot s \diamond S \rightarrow_d r \cdot s \diamond S + r \cdot c \diamond S + b \cdot s \diamond S$.

From Feature-Cluster Family Term to Forest Structure. Applying the same notions presented in previous sections, in what follows we define a relation to get a forest structure from $\mathbb{FC}$-terms.

Definition 10 (Meaning Relation). *Considering $\mathcal{G}_{\mathbb{V},\mathbb{FC}}$ as the set of all possible forest structures on the set of edge labels $\mathbb{V}$ and nodes $\mathbb{FC}$, the* meaning *relation, denoted as $[[.]]$, receives a feature-cluster family term and returns a triple as following:*

$$
\begin{aligned}
&[[\cdot]] : \mathbb{FC} \uparrow \rightarrow \mathcal{G}_{\mathbb{V},\mathbb{FC}} \\
&[[S]] = (\emptyset, \{S\}, \emptyset) \\
&[[f \diamond S]] = (\{f\}, \{f \diamond S, S\}, \{(S, f, f \diamond S)\}) \\
&[[f \cdot F \diamond S]] = (\{f\}, \{f \cdot F \diamond S\}, \{(F \diamond S, f, f \cdot F \diamond S)\}) + [[F \diamond S]] \\
&[[\tau_1 + \tau_2]] = [[\tau_1]] + [[\tau_2]]
\end{aligned}
$$

Example 6. In what follows, we show how a feature-cluster family term is transformed to its equivalent tree structure according to above rules:

$$[[r \diamond S + b \cdot l \diamond S + b \cdot s \diamond S]] = (\{b,\ r,\ l,\ s\}, \{S,\ r \diamond S,\ b \diamond S,\ b.l \diamond S,\ b.s \diamond S\}, \\ \{(S,\ r,\ r \diamond S),\ (b \diamond S,\ l,\ b \cdot l \diamond S),\ (S,\ b,\ b \diamond S),\ (b \diamond S,\ s,\ b \cdot s \diamond S)\})$$

From Tree Structure to a Feature-Cluster Family Term. We define the function *root*, denoted as $r : \mathcal{T}_{\mathbb{V},\mathbb{FC}} \rightarrow Q$, which gets a tree and returns the root of the tree, as $r(T) = \{s \mid \cup_{s_i \in Q} \{(s_i, f, s)\} = \emptyset\}$, where $\mathcal{T}_{\mathbb{V},\mathbb{FC}}$ is the set of rooted trees on $\mathbb{V}$ and $\mathbb{FC}$. Moreover, we define the set of edge labels of the children of $r(T)$ as $\delta(T) = \{f \mid \exists\, s' \in Q \ s.t.\ (r(T), f, s') \in \omega\}$. Furthermore, in a tree T, the *descendant tree* directly after edge f, as the *derivative tree* of T following edge f, is denoted by $\partial_f(T)$.

With the use of above notations, we define the Ψ function which gets a tree structure T, and returns the features as $\Psi(T) = \sum_{f \in \delta(T)} f \cdot \Psi(\partial_f(T))$, where $\Psi(T) = 1$ when $\delta(T) = 1$, and $f \cdot 1$ is denoted as f.

We define the *transform* function, denoted by ψ, which gets a set of k labeled trees (forest) and returns an $\mathbb{FC}$-term as follows:

$$\psi : \mathcal{G}_{\mathbb{V},\mathbb{FC}} \rightarrow \mathbb{FC}$$
$$\psi(\emptyset) = 0 \quad , \quad \psi(T_1 \cup T_2) = \Psi(T_1) \diamond r(T_1) + \Psi(T_2) \diamond r(T_2)$$

Example 7. Suppose the following tree $M = (\{f_1, f_2\}, \{s, s_1, s_2\}, \{(s, f_1, s_1), (s, f_2, s_2)\})$ is given. Then, the only state to which there is no transition (the root) is the node s. Hence, we have: $\Psi(M) = f_1 \cdot \Psi(\emptyset, \{s_1\}, \emptyset) + f_2 \cdot \Psi(\emptyset, \{s_2\}, \emptyset) = f_1 \cdot 1 + f_2 \cdot 1 = f_1 + f_2$. Hence, the resulting term is equal to $\psi(M) = \Psi(M) \diamond s$.

Definition 11. *A term resulting from a CCTree structure, or equivalently transformable to a tree structure representing a CCTree, is called a* CCTree term.

Example 8. Suppose the CCTree of Figure 1 is given. The tree structure of this CCTree can be written as $(\{red, blue, small, large\}, \{S, S_r, S_b, S_{b\cdot s}, S_{b\cdot l}\}, \{(S, red, S_r), (S, blue, S_b), (S_b, small, S_{b.s}), (S_b, large, S_{b.l})\})$. Hence, the CCTree term resulting from this CCTree equals to :

$$red \diamond S + blue \cdot small \diamond S + blue \cdot large \diamond S$$

Theorem 3. *The* meaning *relation* $[[.]]$ adequately abstracts *a graph structure resulting from a feature-cluster (family) term on* $\mathbb{V}$ *and the same fixed dataset of elements* $S \subseteq \mathbb{S}$. *This means that for two non factorized* $\mathbb{FC}$*-terms* τ *and* τ' *we have* $[[\tau]] \approx [[\tau']] \Rightarrow \tau \equiv \tau'$.

Intuitively, the above relation expresses that if two forest structures resulting from two $\mathbb{FC}$-terms are equal, by certain the original terms were equal as well. In other words, if $\tau \not\equiv \tau'$ then we can conclude that $[[\tau]] \not\approx [[\tau']]$. However, the following example contradicts the satisfiability of the relation of Theorem 3 from right to left.

Example 9. The two following feature-cluster family terms are equivalent in terms of term comparison (Definition 5), i.e. we have:

$$f_1 \cdot f_2 \diamond S + f_1 \cdot f_3 \diamond S \quad \equiv \quad f_2 \cdot f_1 \diamond S + f_3 \cdot f_1 \diamond S$$

but their equivalent tree representation are not equivalent, since we have:

$$[[f_1 \cdot f_2 \diamond S + f_1 \cdot f_3 \diamond S]] = (\{f_1, f_2, f_3\}, \{S, f_1 \diamond S, f_1 \cdot f_2 \diamond S, f_1 \cdot f_3 \diamond S\},$$
$$\{(f_1 \diamond S, f_2, f_1.f_2 \diamond S), (f_1 \diamond S, f_3, f_1.f_3 \diamond S), (S, f_1, f_1 \diamond S)\})$$
$$[[f_2 \cdot f_1 \diamond S + f_3 \cdot f_1 \diamond S]] = (\{f_1, f_2, f_3\}, \{S, f_2 \diamond S, f_3 \diamond S, f_2.f_1 \diamond S, f_3.f_2 \diamond S\},$$
$$\{(f_2 \diamond S, f_1, f_2.f_1 \diamond S), (S, f_2, f_2 \diamond S), (f_1 \diamond S, f_3, f_1.f_3 \diamond S), (S, f_1, f_1 \diamond S)\})$$

where the first one contains five nodes, whilst the second one contains four nodes.

This example shows that commutativity of "$\cdot$" is not an appropriate property for *full abstraction.* In what follows, we will show that the reverse of this relation is satisfied if an order of features is identified on the set of features, which solves the problem of multiplication ("$\cdot$") commutativity.

Definition 12 (Ordered Features). *We say that the set of features $\mathbb{V}$ is an* "ordered set of features" *if there is an order relation "$<$" on $\mathbb{V}$, such that $(\mathbb{V}, <)$ is a total ordered set. This means that for any $f_1, f_2 \in \mathbb{V}$ we either have $f_1 < f_2$ or $f_2 < f_1$. We say $F_1 \in \mathbb{F}_1$ is* exactly equal to *$F_2 \in \mathbb{F}_1$, denoted by $F_1 \cong F_2$, if considering the order of features in multiplication representation, they are equal.*

Definition 13 (Order Rewriting Rule). *Let an ordered set of features $(\mathbb{V}, <)$ be given. We say an $\mathbb{FC}$-term is an* ordered $\mathbb{FC}$-term *on $(\mathbb{V}, <)$, if it is the normal form of applying the following rewriting rule:*

$$f_1 \cdot f_2 \diamond S \rightarrow_O f_2 \cdot f_1 \diamond S \qquad if \qquad f_1 < f_2 \quad \forall\, f_1, f_2 \in \mathbb{V}$$

Moreover, we define a rewriting rule which orders the features of an $\mathbb{FC}$-term based on an attribute $A \in \mathfrak{A}$ as $f_2 \cdot f_1 \diamond S \xrightarrow{A}_O f_1 \cdot f_2 \diamond S$ if $f_1 \in A$.
We represent the normal for of a term τ applying above rewriting rule, based on attribute A, as $\tau \Downarrow_A$.

Example 10. Suppose that the set of features $V_1 = \{red, blue, small, large\}$ is given. Without loss of generality, fixing a strict order "$<$" among them as "$red < blue < small < large$" results in having $(V_1, <)$ as a total ordered set. The following terms show how ordered $\mathbb{FC}$-terms on $\mathbb{V}_1$ are obtained by applying the order rewriting rule:

$$red \cdot small \diamond S + blue \cdot large \diamond S \rightarrow small \cdot red \diamond S + large \cdot blue \diamond S$$

Moreover "$red \cdot small \ncong small \cdot red$", whilst "$red \cdot small \cong red \cdot small$".

Definition 14 (Ordered $\mathbb{FC}$-term Comparison). *We say two ordered $\mathbb{FC}$-terms on $(\mathbb{V}, <)$ are* exactly equal, *denoted by $\sim$, as the smallest relation for which the terms respect one of the following relations:*

$$\begin{array}{lll} if & S_1 = S_2 & then \quad S_1 \sim S_2 \\ if & S_1 = S_2 \wedge F_1 \cong F_2 & then \quad F_1 \diamond S_1 \sim F_2 \diamond S_2 \\ if & \forall \tau_i \in block(\tau)\ \exists \tau_j \in block(\tau')\ s.t\ \tau_i \sim \tau_j, & \\ & \forall \tau_j \in block(\tau')\ \exists \tau_i \in block(\tau)\ s.t\ \tau_j \sim \tau_i & then \quad \tau \sim \tau' \end{array}$$

Theorem 4. *Let $(\mathbb{V}, <)$ be a total ordered set of features and $S \subseteq \mathbb{S}$. The* meaning *relation* $[[.]]$ abstracts *the forest (tree) structure resulted from the ordered non factorized $\mathbb{FC}$-terms on $\mathbb{V}$ and S. This means considering τ and τ' be two arbitrary ordered non factorized $\mathbb{FC}$-terms on $(\mathbb{V}, <)$ and $S \subseteq \mathbb{S}$, we have:*

$$\tau \sim \tau' \Rightarrow [[\tau]] \approx [[\tau']]$$

Theorem 5 (Main Theorem). *Let the ordered set of features* $(\mathbb{V}, <)$, *the set of elements* $S \subseteq \mathbb{S}$ *are given. The* meaning *function* $[[.]]$ fully abstracts *the ordered feature-cluster family terms on* $(\mathbb{V}, <)$ *and* S. *This means that for two arbitrary ordered feature-cluster family terms* τ *and* τ' *on* $\mathbb{V}$ *and* S, *we have:*

$$[[\tau]] \approx [[\tau']] \Leftrightarrow \tau \sim \tau'$$

Proof. The proof is straightforward from the proofs of Theorems 3 and 4.

6 Relations on Feature-Cluster Algebra

In this section, we define several relations on feature-cluster algebra and discuss the properties of the proposed relations.

Definition 15 (Attribute Division). Attribute division *($\mathscr{D}_{\mathcal{A}}$) is a function from* $\mathcal{A} \times \mathbb{FC}$ *to* $\{True, False\}$, *which gets an attribute and a non factorized* $\mathbb{FC}$*-term as input; it returns "True" or "False" as follows:*

$$\begin{aligned}
&\mathscr{D}_{\mathcal{A}} : \mathcal{A} \times \mathbb{FC} \uparrow \rightarrow \{True, False\}\\
&\mathscr{D}_{\mathcal{A}}(A, S) = False\\
&\mathscr{D}_{\mathcal{A}}(A, f \diamond S) = True \qquad\qquad if \quad f \in A\\
&\mathscr{D}_{\mathcal{A}}(A, f \diamond S) = False \qquad\qquad if \quad f \notin A\\
&\mathscr{D}_{\mathcal{A}}(A, f \cdot F \diamond S) = \mathscr{D}_{\mathcal{A}}(A, f \diamond S) \vee \mathscr{D}_{\mathcal{A}}(A, F \diamond S)\\
&\mathscr{D}_{\mathcal{A}}(A, \tau_1 + \tau_2) = \mathscr{D}_{\mathcal{A}}(A, \tau_1) \wedge \mathscr{D}_{\mathcal{A}}(A, \tau_2)
\end{aligned}$$

The concept of attribute division is used to order the attributes presented in a term, which will be discussed later.

Example 11. In the following we show how attribute division performs:

$$\begin{aligned}
&\mathscr{D}_{\mathcal{A}}(color, r \cdot s \diamond S + r \cdot c \diamond S + b \cdot s \diamond S) =\\
&\mathscr{D}_{\mathcal{A}}(color, r \cdot s \diamond S) \wedge \mathscr{D}_{\mathcal{A}}(color, r \cdot c \diamond S) \wedge \mathscr{D}_{\mathcal{A}}(color, b \cdot s \diamond S) = True
\end{aligned}$$

Definition 16 (Initial). *We define the* initial *(δ) function from* $\mathcal{P}(\mathbb{FC} \uparrow)$ *to* $\mathcal{P}(\mathbb{F})$, *i.e.* $\delta : \mathcal{P}(\mathbb{FC} \uparrow) \rightarrow \mathcal{P}(\mathbb{F})$, *which gets a set of ordered non factorized terms on* $(\mathbb{V}, <)$ *and returns a set of the first features of each term as follows:*

$$\delta(\emptyset) = \{0\}, \qquad \delta(\{S\}) = \{1\}, \qquad \delta(\{f \cdot F \diamond S\}) = \{f\}$$
$$\delta(\{\tau_1 + \tau_2\}) = \delta(\{\tau_1\}) \cup \delta(\{\tau_2\}), \qquad \delta(\{\tau_1, \tau_2\}) = \delta(\{\tau_1\}) \cup \delta(\{\tau_2\})$$

with the property $\delta(\{X, Y\}) = \delta(X) \cup \delta(Y)$, *where* $X, Y \in \mathcal{P}(\mathbb{FC} \uparrow)$.
In the case that the input set contains just one term, we remove the brackets, i.e. $\delta(\{\tau\}) = \delta(\tau)$ *when* $|\{\tau\}| = 1$. *Moreover, when the output set also contains just one element, for the sake of simplicity we remove the brackets, i.e.* $\delta(X) = \{f\} = f$ *for* $X \in \mathcal{P}(\mathbb{FC} \uparrow)$.

Definition 17 (Derivative). *We define the* derivative, *denoted by* ∂, *as a function which gets an ordered non factorized* $\mathbb{FC}$*-term on* $(\mathbb{V}, <)$ *and returns the term (set of terms) by cutting off the first features as follows:*

$$\begin{aligned}
&\partial : \mathbb{FC} \uparrow \rightarrow \mathcal{P}(\mathbb{FC})\\
&\partial(S) = \emptyset, \qquad\qquad \partial(f \diamond S) = \{S\}
\end{aligned}$$

$$\partial(f \cdot F \diamond S) = \{F \diamond S\}, \qquad \partial(\tau_1 + \tau_2) = \partial(\tau_1) \cup \partial(\tau_2)$$

Note that the functions initial *(δ) and* derivative *(∂) are overloaded to the input, depending to the input that if it is a tree or a term.*

Definition 18 (Order of Attributes). *We say attribute B is smaller or equal to attribute A on the non factorized term $\tau \in \mathbb{FC} \uparrow$, denoted as $B \preceq_\tau A$, if the number of blocks of τ that B divides, is less than (equal to) the number of blocks that A divides. Formally, $B \preceq_\tau A$ implies that:*

$$|\{\tau_i \in block(\tau) \mid \mathscr{D}_{\mathcal{A}}(B, \tau_i) = True\}| \leq |\{\tau_i \in block(\tau) \mid \mathscr{D}_{\mathcal{A}}(A, \tau_i) = True\}|$$

Given a set of attributes $\mathcal{A}$ and a term τ, the set $(\mathcal{A}, \preceq_\tau)$ is a lattice. We denote the upper bound *of this set as $\sqcap_{\mathcal{A},\tau}$. This mean we have $\forall A \in \mathcal{A} \Rightarrow A \preceq_\tau \sqcap_{\mathcal{A},\tau}$.*

Example 12. In the following we show how the order of attributes of a term is identified. Suppose that the term $\tau = r \cdot s \diamond S + r \cdot c \diamond S + b \cdot s \diamond S$ is given. We have $block(\tau) = \{r \cdot s \diamond S, r \cdot c \diamond S, b \cdot s \diamond S\}$. Consequently,

$$\begin{aligned} &|\{\tau_i \in block(\tau) \mid \mathscr{D}_{\mathcal{A}}(shape, \tau_i) = True\}| = 1 \\ \leq\ &|\{\tau_i \in block(\tau) \mid \mathscr{D}_{\mathcal{A}}(size, \tau_i) = True\}| = 2 \\ \leq\ &|\{\tau_i \in block(\tau) \mid \mathscr{D}_{\mathcal{A}}(color, \tau_i) = True\}| = 3 \end{aligned}$$

which means that we have $shape \preceq_\tau size \preceq_\tau color$.

Recalling that not having the predefined order among features creates a problem in full abstraction of terms. To this end, here we propose a way to order the set of features which is appropriate to our problem. First of all, given a feature-cluster family term τ, we find the order of attributes according to definition 18, whilst if for two arbitrary attributes A and A', we have $A = A'$, without loss of generality, we choose a strict order among them, say $A \prec A'$. Then in each attribute we arbitrarily order the features. It is important that the features of smaller attribute be always smaller than the features of greater attribute. For example, if $size \prec color$, we consider the order of features as $small < large < blue < red$, whilst all the features of *color* are greater than all the features of *size*.

Definition 19 (Ordered Unification). Ordered unification *($\mathcal{F}$) is a partial function from $\mathcal{P}(\mathcal{A}) \times \mathbb{FC} \uparrow$ to $\mathbb{FC} \downarrow$, which gets a set of attributes and a non factorized term; it returns the normal form of applying rewriting rule $\xrightarrow{A}_O$ introduced in Definition 13, iteratively, based on the order of attributes on received term as follows:*

$$\begin{aligned} &\mathcal{F}(\emptyset, \tau \uparrow) = \tau \\ &\mathcal{F}(\{A\}, \tau \uparrow) = \tau \Downarrow_A \\ &\mathcal{F}(\mathcal{A}, \tau) = \mathcal{F}(\sqcap_{\mathcal{A},\tau}, \mathcal{F}(\mathcal{A} - \{\sqcap_{\mathcal{A},\tau}\}, \tau \uparrow)) \end{aligned}$$

The normal form of ordered unification is called a unified term. *By $\mathcal{F}^*(\tau)$ we mean that $\mathcal{F}$ is performed iteratively on the set of ordered attributes on τ to get the unified term.*

Example 13. To find the unified form of $\tau_1 = r \cdot s \diamond S + r \cdot c \diamond S + b \cdot s \diamond S$, we have "$\mathcal{F}^*(\tau_1) = \mathcal{F}(\{shape, color, size\}, \tau_1 \uparrow) = \mathcal{F}(color, \mathcal{F}(size, \mathcal{F}(shape, \tau_1))) = r \cdot s \diamond S + r \cdot c \diamond + b \cdot s \diamond S$".

Definition 20 (Component relation). *Given two ordered non factorized* $\mathbb{FC}$*-terms* τ_1 *and* τ_2 *on* $(\mathbb{V}, <)$*, we define the* component relation*, denoted by* $\sim_1$*, as the first level comparison of the terms as* $\tau_1 \sim_1 \tau_2 \Leftrightarrow \delta(\tau_1) = \delta(\tau_2)$.

Proposition 2. *The component relation is an equivalence relation on the set of ordered non factorized* $\mathbb{FC}$*-terms.*

Definition 21 (Component). *Let consider that the ordered term* $\tau \in \mathbb{FC}\uparrow$ *on* $(\mathbb{V}, <)$ *is given. The equivalence class of* $\tau' \in block(\tau)$ *is called a* component *of* τ*, and it is formally defined as* $[\tau']_\tau = \{\tau_i \in block(\tau) \,|\, \tau' \sim_1 \tau_i\}$*. The set of all components of the term* τ *through the equivalence relation* $\sim_1$*, is denoted by* $block(\tau)/\sim_1$ *or simply* $\tau/\sim_1$*, i.e. we have "*$\tau/\sim_1 = \{[\tau_i]_\tau \mid \tau_i \in block(\tau)\}$*".*

Definition 22 (Component Order). *Let* X *and* Y *be two sets of of ordered non factorized* $\mathbb{FC}$*-terms on* $(\mathbb{V}, <)$*. We say* X *is smaller than* Y*, denoted as* $X < Y$*, if* $\forall f' \in \delta(X)$ *and* $\forall f'' \in \delta(Y)$ *we have* $f' < f''$*. Specifically, let* τ *be an ordered non factorized* $\mathbb{FC}$*-term on* $(\mathbb{V}, <)$*. We order the components of* τ *according to the order of features in* $\mathbb{V}$ *as what follows:*

$$[\tau']_\tau < [\tau'']_\tau \quad \Leftrightarrow \quad (\forall f' \in \delta([\tau']) \;,\; \forall f'' \in \delta([\tau'']) \Rightarrow f' < f'')$$

It is noticeable that $|\delta([\tau'])| = |\delta([\tau''])| = 1$*, for all* $\tau', \tau'' \in block(\tau)$*, since the first features of all elements in a component are equal.*

Definition 23 (Well formed term). Well formed function*, denoted as* W*, is a binary function from* $\mathbb{FC}\uparrow$ *to* $\{True, False\}$*, which gets a unified non factorized* $\mathbb{FC}$*-term; it returns* $True$ *if the set of first features of its components is equal to a sort of* $\mathcal{A}$ *to which these features belong; it returns* $False$ *otherwise. Formally:*

$$W(\tau) = \begin{cases} True \;\; if & \delta(\tau/\sim_1) = sort(\delta([\tau_i]_\tau)) \quad \forall \tau_i \in block(\tau) \\ False & \qquad\qquad\qquad\qquad\qquad\qquad otherwise \end{cases}$$

where $\delta(\tau/\sim_1) = sort(\delta([\tau_1]_\tau))$ *means that the the set of the first features of the components of the term* τ *is equal to the attribute that the first feature belongs to. A unified term* τ *is called a* well formed term*, if* $W(\tau) = True$*. An atomic term is a* well formed term.

Example 14. The unified term of Example 13, $\tau = r \cdot s \diamond S + r \cdot c \diamond S + b \cdot s \diamond S$ is a well formed term, since we have "$\delta(\tau/\sim_1) = \delta(\{\{r \cdot s \diamond S \,,\, r \cdot c \diamond S\}, \{b \cdot s \diamond S\}\}) = \{r, b\}$" and "$sort(\delta([r \cdot s \diamond S])) = sort(\{r \cdot s \diamond S \,,\, r \cdot c \diamond S\}) = \{r, b\}$", i.e. $W(\tau) = True$.

It is noticeable that in an ordered CCTree term all first features belong to the same attribute. Hence, in what follows we exploit the concept of well formed term to identify whether a term represents a CCTree term or not. The knowledge of knowing which feature-cluster family term represents a CCTree term, provides us with the opportunity to iteratively use the rules on CCTree terms.

Theorem 6. *A unified term represents a CCTree term, or it is transformable to a CCTree structure, if and only if, it can be written in the form* $\mathcal{F}^*(\tau) = \sum_i f_i \cdot \tau_i$*, such that "*$W(\mathcal{F}^*(\tau)) = True$*", i.e. the unified form of the received term is a well formed term; and the unified form of each* τ_i *is a well formed term as well (*$W(\tau_i) = True$ *) and it also respects the above formula by itself.*

7 Discussion and Future Directions

In this paper, a semiring-based formal method, named Feature-Cluster Algebra, is proposed to abstract the representation of a tree structure representing a categorical clustering algorithm in general, and specifically CCTree. We proved that the proposed approach, under some conditions, fully abstracts the tree structure. Furthermore, we presented a set of functions and relations on feature-cluster algebra, which is used to shape the requirements for a term to represent a CCTree. The abstraction concept proposed in this work, as a novel methodology, establishes an algebraic structure on clusters, which can be considered as a formal technique exploited to get conclusion on a data mining algorithm.

As future work, we plan to apply the proposed technique to abstract a broad category of feature-based data mining algorithms, e.g. additional categorical clustering algorithms, classification, and association rule mining. Moreover, we expect to use the result of abstraction to address the data mining associated issues, e.g. parallel clustering, feature selection.

References

1. Benavides, D., Segura, S., Ruiz-Cortés, A.: Automated analysis of feature models 20 years later: A literature review. Inf. Syst. **35**(6), 615–636 (2010)
2. Berkhin, P.: A survey of clustering data mining techniques. In: Kogan, J., Nicholas, C., Teboulle, M. (eds.) Grouping Multldimensional Data, pp. 25–71. Springer, Heidelberg (2006)
3. Caruana, R., Niculescu-Mizil, A.: An empirical comparison of supervised learning algorithms. In: Proceedings of the 23rd International Conference on Machine Learning, ICML 2006, pp. 161–168. ACM, NY (2006)
4. Giunchiglia, F., Walsh, T.: A theory of abstraction. Artif. Intell. **57**(2–3), 323–389 (1992)
5. Gross, J.L., Yellen, J.: Graph Theory and Its Applications. Discrete Mathematics and Its Applications, 2nd edn. Chapman & Hall/CRC (2005)
6. Hell, P., Nesetil, J.: Graphs and homomorphisms. Oxford lecture series in mathematics and its applications. Oxford University Press, Oxford, New York (2004)
7. Höfner, P., Khedri, R., Möller, B.: Feature algebra. In: Misra, J., Nipkow, T., Sekerinski, E. (eds.) FM 2006. LNCS, vol. 4085, pp. 300–315. Springer, Heidelberg (2006)
8. Höfner, P., Khédri, R., Möller, B.: An algebra of product families. Software and System Modeling **10**(2), 161–182 (2011)
9. Kang, K.C., Kim, S., Lee, J., Kim, K., Shin, E., Huh, M.: Form: A feature-oriented reuse method with domain-specific reference architectures. Ann. Softw. Eng. **5**, 143–168 (1998)
10. Panda, B., Herbach, J.S., Basu, S., Bayardo, R.J.: Planet: Massively parallel learning of tree ensembles with mapreduce. Proc. VLDB Endow. **2**(2), 1426–1437 (2009)
11. Sheikhalishahi, M., Mejri, M., Tawbi, N.: Clustering spam emails into campaigns. In: Library, S.D. (ed.) 1st International Conference on Information Systems Security and Privacy (2015)
12. Sheikhalishahi, M., Saracino, A., Mejri, M., Tawbi, N., Martinelli, F.: Fast and effective clustering of spam emails based on structural similarity. In: Garcia-Alfaro, J., et al. (eds.) FPS 2015. LNCS, vol. 9482, pp. 195–211. Springer, Heidelberg (2016). doi:10.1007/978-3-319-30303-1_12

Appendix

Proof of Theorem 1

Proof. The proof is resulted from the properties 1 and 2 as the following:

$$(1 \diamond \mathbb{S}) \cdot (F \diamond S) = (1 \cdot F) \diamond S = F \diamond S$$
$$(0 \diamond \emptyset) \cdot (F \diamond S) = (0 \cdot F) \diamond \emptyset = 0 \diamond \emptyset$$
$$(0 \diamond \emptyset) + (F \diamond S) = (0 + F) \diamond S = F \diamond S$$

For the other elements of $\mathbb{C}$, the proof is straightforward from above equations, and the properties 1 and 2.

Proof of Theorem 3

Proof. Suppose that the left hand side of the relation is satisfied. Hence, we have:

$$[[\tau]] \approx [[\tau']] \Rightarrow \Theta(\tau) = \Theta(\tau') \,,\, \Phi(\tau) = \Phi(\tau') \,,\, \Delta(\tau) = \Delta(\tau') \tag{3}$$
$$\Rightarrow block(\tau) = block(\tau') \Rightarrow \tau \equiv \tau' \tag{4}$$

where 3 is resulted from the equivalent graph structures of τ and τ', and 4 is satisfied from $\Phi(\tau) = \Phi(\tau')$ and the fact that two main terms were originated from the same dataset.

Proof of Theorem 4

Proof. Suppose the left side of the relation satisfies. This means that for each feature-cluster term $\tau_i \in \tau$ there exists a feature-cluster term $\tau_j \in \tau'$ such that τ_i and τ_j are exactly equal. This property causes that the set of transitions of $[[\tau_i]]$ to be equal to the set of transitions of $[[\tau_j]]$. Hence, we have:

$$\tau \sim \tau' \Rightarrow \forall \tau_i \in block(\tau) \exists \tau_j \in block(\tau') \; s.t \; \tau_i \sim \tau_j (\Rightarrow [[\tau_i]] \approx [[\tau_j]]), \tag{5}$$
$$\forall \tau_j \in block(\tau') \exists \tau_i \in block(\tau) \; s.t \; \tau_j \sim \tau_i (\Rightarrow [[\tau_i]] \approx [[\tau_j]])$$
$$\Rightarrow [[\tau]] \approx [[\tau']] \tag{6}$$

Proof of Proposition 2

Proof. For ordered non factorized $\mathbb{FC}$-terms τ_1, τ_2 and τ_3, we have:

$$\begin{array}{ll} if \;\; \tau_1 \sim_1 \tau_1 & iff \;\; \delta(\tau_1) = \delta(\tau_1) \\ if \;\; \tau_1 \sim_1 \tau_2 \wedge \tau_2 \sim_1 \tau_1 & iff \;\; \delta(\tau_1) = \delta(\tau_2) \wedge \delta(\tau_2) = \delta(\tau_1) \\ if \;\; \tau_1 \sim_1 \tau_2 \,,\, \tau_2 \sim_1 \tau_3 \; then \; \tau_1 \sim_1 \tau_3 & iff \;\; \delta(\tau_1) = \delta(\tau_2), \delta(\tau_2) = \delta(\tau_3) \; then \; \delta(\tau_1) = \delta(\tau_3) \end{array}$$

this means that $\sim_1$ is an equivalence relation.

Proof of Theorem 6

Proof. First we show that a unified term obtained from a CCTree structure satisfies the equation 6. In a CCTree, the attribute used for division in the root, has the greatest number of occurrence in non factorized CCTree term (all blocks of CCTree term contain one of the features of this attribute). According to 5, for transforming the tree to a term, the first features of components are specified from $\delta(T) = \{f \mid \exists\, s' \in Q \ s.t. \ (s_T, f, s') \in \omega\}$, where in CCTree all belong to the same sort, i.e. we have:

$$\delta(T) = \{f \mid \exists\, s' \in Q \ s.t. \ (s_T, f, s') \in \omega\} = sort(\{f\}) \Rightarrow W(\psi(T)) = True$$

we call the tree following a child of the root as a *new tree*. It is noticeable each new tree is a CCTree by itself; hence, it respects 6. By considering the tree following the new tree as new trees themselves, the aforementioned process is iteratively repeated for all new trees, due to the iterative structure of CCTree, i.e. from 5, we have:

$$W(\psi(\partial_f(T))) = True \qquad \forall\, f \in \delta(T)$$

this means that if the input tree structure is a CCTree, then the obtained term respects the above formula.

On the other hand, a unified term that respects equation 6 can be converted to a CCTree structure. To this end, τ_i's are the components of τ separating their first features (f_i's). The set of the first features of components of the term, constitute the transitions of the first division from the root of CCTree, i.e.:

$$\Omega\Big(\sum_{[\tau_i]\in\tau/\sim_1} \delta([\tau_i]) \cdot \sum_{\tau_k\in[\tau_i]} \partial(\tau_k)\Big) = \bigcup_{[\tau_i]\in\tau/\sim_1} \{(S, \delta([\tau_i]), \sum_{\tau_k\in[\tau_i]} \partial(\tau_k))\}$$

where S is the main dataset the term is originated from.
Since the term is well formed, it guarantees that the label of children belong to the same sort, as required by CCTree. Due to the iterative rule for successive components, iteratively the structure of CCTree is constructed. Note that the condition of equivalence of the first features of components to a sort, guarantees that in the process of transforming the term to its equivalent tree structure, all the features of a selected attribute exist.

Gossip-Based Behavioral Group Identification in Decentralized OSNs

Naeimeh Laleh[1,2](✉), Barbara Carminati[1,2], Elena Ferrari[1,2], and Sarunas Girdzijauskas[1,2]

[1] DiSTA, University of Insubria, Varese, Italy
{nlaleh,barbara.carminati,elena.ferrari}@uninsubria.it
[2] Royal Institute of Technology KTH, Stockholm, Sweden
sarunasg@kth.se

Abstract. DOSNs are distributed systems providing social networking services that become extremely popular in recent years. In DOSNs, the aim is to give the users control over their data and keeping data locally to enhance privacy. Therefore, identifying behavioral groups of users that share the same behavioral patterns in decentralized OSNs is challenging. In the fully distributed social graph, each user has only one feature vector and these vectors can not move to any central storage or other users in a raw form duo to privacy issues. We use a gossip learning approach where all users are involved with their local estimation of the clustering model and improve their estimations and finally converge to a final clustering model available for all users. In order to evaluate our approach, we implement our algorithm and test it in a real Facebook graph.

Keywords: Decentralized Online Social Network (DOSN) · Gossip learning · Newscast EM · Behavioral group identification

1 Introduction

Recently, discovery of meaningful groups of users that share the same behavioral patterns in social networks has become an active research area. Behavioral group identification has many valuable applications. For instance, it can help in improving recommendation systems, it can be used for advertisement purposes, direct marketing, and for risk assessment in online social networks. The key idea in risk assessment is that the more the target user's behavior diverges from those of other similar users, the more the target user is risky [2]. Therefore, risk assessment approaches require to identify similar users that share the same behavioral patterns.

By considering a social network as a graph, each node is depicted as a user, and an edge connecting two users denotes the relationship between them. Here, the main problem is that all users are connected in friend to friend graph, but users that share the same behavioral patterns not necessarily have friendships in the graph. In grouping users we consider the profile and activity information

P. Perner (Ed.): MLDM 2016, LNAI 9729, pp. 676–691, 2016.
DOI: 10.1007/978-3-319-41920-6_52

such as age, gender, education, nationality, number of friends, activity level, etc. Furthermore, in investigating the discovery of behavioral groups, we have cast our attention to decentralized online social networks (DOSNs) [5]. In DOSN, there is no central infrastructure and discovery of behavioral groups is more challenging than in the centralized setup. In the fully distributed social graph, each user can only communicate with his/her direct friends without sending all the private group identification feature values to his/her direct friends in a raw form. In more details, behavioral patterns can be classified into social and individual behavioral patterns. Social behavioral patterns rely on user interactions, while individual behavioral patterns are related to profile information [20]. In this paper, we propose a methodology for identifying both social and individual patterns to group users in DOSNs.

The problem of finding similar users in social networks has been widely studied in the context of community detection. Community detection approaches that are pure link-based, relying on topological structures [6], [7], fail to group users with the same behavioral patterns in that such users might belong to different communities based on their friendship links. Moreover, some of the community detection approaches are content-based that they rely on the analysis of the content generated by each user [21], [18]. The major problem of these approaches is the overhead for building the graph, based on similarity measures, that is not suitable for real-time applications. On the other hand, when each user feature vector includes both discrete and continuous features, the various behavioral patterns may not be obvious by similarity measures and then, this identification can not be made correctly. However, there are some stream-based community detection methods suitable for real-time applications [16], [22]. But, most of these approaches are link-based [17], and they do not consider the personal feature vector of users.

To alleviate the limitations of existing approaches, we propose a fully decentralized clustering algorithm which is capable of clustering distributed information without requiring central control. The selected clustering model requires specific aspects to be considered such as: the final clustering model should maintain a reasonable performance compared to a centralized clustering model and should be robust in that it should not easily fail when some of the users leave the network or do not answer to messages. Also, all users should be able to have the final clustering model at any time after convergence to assign a group for themselves and their direct friends in a local way. Finally, we need to minimize the communication cost by decreasing the number of messages and the size of them as well. These requirements bring us to exploit Newscast EM [13], a probabilistic gossip-based randomized communication clustering approach, originally developed for clustering users in peer-to-peer networks. In Newscast EM, each user initializes a local estimation of the parameters of the clustering model (mean, standard deviation, etc.) and then, contacts a random user from all users in the network, to exchange their parameters estimation and aggregate them by weighted averaging. The choice of random selection is crucial to the wide dissemination of the gossip [13], since, the probability of a user being sampled is proportional to his/her degree [19]. Gossip-based peer-sampling service [10] provides a user with a uniform random subset of all users in the peer-to-peer network. But, the main difference between peer-to-peer and social networks is that in peer-to-peer networks each user can directly communicate with any other user in the network to exchange information. On the contrary, in social networks each user can just communicate directly with his/her direct friends. Therefore, we use the random-sampling implementation for social networks proposed in [11].

The main contribution of this paper is making Newscast EM to be applicable on top of DOSN and apply it to identify behavioral groups of users. Our goal is to achieve an accuracy comparable to a centralized scheme. The advantages of this distributed behavioral group identification are: 1. the usage of both social and individual patterns of users, 2. feature values of users are never send over the network in a raw form and 3. it has low computation and communication cost. The remainder of this paper is organized as follows. We first explain Newscast EM in Section 2. Then, we propose our gossip-based implementation for behavioral group identification in Section 3. In Section 4, we show the result of the clustering model. Section 5 introduces the related work. Finally, Section 6 concludes the paper.

2 Background

In selecting the clustering technique, we focused on soft clustering (i.e., probabilistic-based clustering). Hard clustering techniques, (e.g., k-means) are not proposing a solution to the problem of clustering discrete or categorical data [4] since they are based on distance metrics. We use EM (Expectation Maximization) algorithm, that is, a probabilistic based clustering method. In particular, EM defines K probability distributions to identify K clusters for all users based on their feature vectors, where each distribution represents the likelihood of those feature vectors to belong to a given cluster. In this way, EM first assigns a set of K *membership probabilities* to each user u based on his/her own feature vector. Then, EM maximizes these likelihoods by learning the parameters of the clustering model in order to assign to each user the cluster with the highest probability.

The main idea of distributed EM algorithm is that each user starts the Expectation-step with a local estimation of the parameters of the clustering model. Then, in the Maximization-step, the algorithm employs a gossip-based protocol to learn a final clustering model from these local estimations. Each user exchanges his/her own estimations with several other users by using a randomized communication protocol. By gathering these estimations from random users, the target user can update and re-estimate his/her own estimation.

In the following, we present a summary of Newscast EM. Interested readers are referred to [13] for more details. Let N be the set of users in the OSN, the probability of membership or weight of a target user $\boldsymbol{u}$, $\boldsymbol{u} \in N$, in cluster l is defined as [4]:

$$w_l(\boldsymbol{u}) = \frac{w_l.p_l(\boldsymbol{u}|\theta_l)}{\sum_{i=1}^{K} w_i.p_i(\boldsymbol{u}|\theta_i)} \tag{1}$$

where, w_l is a weight computed as $w_l = \frac{|N_l|}{|N|}$, with N_l denotes the set of users belonging to the lth cluster, where $\sum_{l=1}^{K} w_l = 1$; function $p_l(\boldsymbol{u}|\theta_l)$ is the component density function modeling the feature vector of the lth cluster, where $\theta_l = \{\boldsymbol{\mu}_l, \Sigma_l\}$ represents the parameters for l^{th} distribution, that is, the mean and the covariance.

Newscast EM uses a fully distributed averaging process for estimating the parameters $\Theta = \{w_l, \boldsymbol{\mu}_l, \Sigma_l\}$, $l = 1, \ldots, K$. Assuming that each user has just one feature vector, then the Expectation-step implies that each user locally estimates the parameters based on his/her own feature vector. In this manner, each user u_i, $i = 1, \ldots, N$ starts with a local estimation of $\Theta_i = \{w_{li}, \boldsymbol{\mu}_{li}, \Sigma_{li}\}$ for the parameters of the cluster l. However, in the Maximization-step of the algorithm user u_i needs information from all users in the network to recompute his/her parameters estimation. Therefore, this step is implemented as a sequence of gossip-based cycles. The details of the algorithm, which each user will run in parallel is as follows:

Initialization Phase: We assume that all users agree on the number of clusters K and start the exchanging protocol. Each user $\boldsymbol{u_i}$, sets the membership probability for each cluster as $w_l(\boldsymbol{u_i})$ to some random positive value and then normalizes all to sum to 1 over all l. This phase is completely local for each user.

Maximization-Step: In this step, user u_i initializes the local parameters estimation for each cluster l as follows: $w_{li} = w_l(\boldsymbol{u_i})$, $\boldsymbol{\mu}_{li} = \boldsymbol{u_i}$ and $\tilde{\Sigma}_{li} = \boldsymbol{u_i}.\boldsymbol{u_i}^T$, where T is the transpose of the feature vector of user u_i. Then, user u_i for $\Re$ cycles repeatedly initiates the information exchange with random users, i.e., u_j. Then, users u_i and u_j update their local parameters estimation for each cluster l as follows:

$$w_{li}^{'} = w_{lj}^{'} = \frac{w_{li} + w_{lj}}{2} \tag{2}$$

$$\boldsymbol{\mu}_{li}^{'} = \boldsymbol{\mu}_{lj}^{'} = \frac{w_{li}.\boldsymbol{\mu}_{li} + w_{lj}.\boldsymbol{\mu}_{lj}}{w_{li} + w_{lj}} \tag{3}$$

$$\tilde{\Sigma}_{li}^{'} = \tilde{\Sigma}_{lj}^{'} = \frac{w_{li}.\tilde{\Sigma}_{li} + w_{lj}.\tilde{\Sigma}_{lj}}{w_{li} + w_{lj}} \tag{4}$$

Expectation-Step: User u_i, after waiting for $\Re$ cycles for the convergence of his/her local parameters estimation, computes new membership probabilities for each cluster l using the Maximization-step estimations w_{li}, $\boldsymbol{\mu}_{li}$ and $\Sigma_{li} = \tilde{\Sigma}_{li} - \boldsymbol{\mu}_{li}.\boldsymbol{\mu}_{li}^T$. We denote with Θ^t the parameter values set at iteration t and then $\Theta^{t+1} = \{(\boldsymbol{\mu}_{li}^{t+1}, \Sigma_{li}^{t+1}, w_{li}^{t+1}), l = 1, \ldots, K\}$. The sequence of Θ-values which is then the likelihood of Θ, $L(\Theta)$, is non-decreasing at each iteration. Then, user u_i checks the stopping tolerance by using the estimations from the previous EM-iteration to see if it is satisfied or not, until $|L(\Theta^t) - L(\Theta^{t+1})| \leq \varepsilon$, where $\varepsilon > 0$.

If it is not satisfied, the Maximization-step is repeated, unless a stopping tolerance is satisfied. In the following, we will explain how to run newscast EM in decentralized social networks.

3 Newscast EM in DOSNs

In social networks, each user can just communicate directly with his/her direct friends. Therefore, we propose our gossip-based clustering framework on DOSNs that contains two main components: UserSelection and ClusteringModelUpdate. The same algorithm is run by each user in the network in parallel, as shown in Figure 1.

3.1 User Selection

In randomized user selection, the problem is that if users can be selected randomly with equal probability, the estimation will be unbiased. But, it is well known that the probability of a user being sampled is proportional to his/her degree [19]. Therefore, more popular users have a higher degree and tend to have a higher probability of being sampled. This will lead to overestimate the average value during the gossip process. There are plenty of approaches to select a uniform and unbiased random sampling of users in DOSNs such as: graph traversal techniques [15] and Random Walk [14]. But, most of these approaches are biased towards high degree users [19]. To have a uniform random sample of all users, each user needs to know every other user in the network. Since, accessing each user in the network to gossip with, is unrealistic in a large-scale dynamic networks, we apply a technique for DOSNs proposed [11] for randomized communication, to define a dynamically changing random graph topology over the network. This technique includes two methods *Initialization* and *SelectUser*.

The initialization procedure initializes the service for the new user when he/she joins the social network. First, each user u_i maintains a list of its direct friends and two hops friends in a small fixed size cache, called *Random Neighbors Cache* (*RNC*), including e entries. The set e of entries in the *RNC* contains the list of random users ID, their *longevity* field, and the path to reach them. The field *longevity* is the age of the entry since the moment it was created by the user. Then, in the *SelectRandomEntries()* procedure, a user selects S subset of neighbors from *RNC* and puts them in a cache, called Exchange Cache (*ExC*). After that, the target user u_i continuously selects one of his/her neighbors with the highest *longevity* from *RNC*, i.e., u_j, in *SelectRandomUser()* procedure, to exchange entries of *ExC*. Then, u_i increases by one the *longevity* of all the other entries in u_i's *RNC*.

In social networks, as shown in Figure 2, if user A wants to communicate with user D, he/she needs to reach first B and then C. Therefore, each user needs to maintain both a set of random users ID and also the path to reach them in order to exchange the entries of *ExC*. For instance, let us assume that in Figure 2.(a) user A has six entries in the *RNC* that include C, I, K, D, N and B. User A

selects user I with the highest *longevity* among all neighbors in the RNC and wants to exchange S (i.e., 3 in this example) number of his/her neighbors in the ExC, such as C, D, N, with user I. On the other hand, user A needs to pass through (M, L) to reach user I. Therefore, user I, in order to reach all of the entries in the ExC of user A, needs to first reach user A and then he/she will be able to reach all entries inside A 's ExC. The problem is that this path could be long. For instance, user I, in order to reach user D, first needs to reach user A by passing through (L, M) and then from A to user D via (B, C), i.e., (L, M, A, B, C). This path is long and it is not the shortest path from user I to D. But, user I can reach user D by passing through his/her mutual friends with D, like (E ,D) or (L, D).

More precisely, to decrease the communication cost, the protocol in [11] builds a new path for user I to reach all entries in user A's ExC during the exchanging process, illustrated in procedure *UpdateExC()* in Figure 1. In this way, the source user A asks all users within the direct path from A to I, i.e., (M, L, I) to build a new path for all the entries to be accessible for user I. This process in summary is as followings. First the source user (A) only adds his/her ID to the first part of the address of each entry in ExC and sends the ExC to the next user (M) on the path towards user I, as shown in Figure 2.(1) in red color. Then, every next user (for instance user M) within the path towards user I and also user I him/herself, after receiving the ExC, first reverses the path of each entry in ExC, as shown in Figure 2.(2).a and starts traversing all users inside the reverse path to check, if he/she has any direct friendship or mutual friends with those users. If yes, he/she removes the remaining part of the path and adds his/her friends or mutual friends ID to the path and adds the ID of his/herself to the first part of the path (except user I that does not need to add his/her ID to the first part of the path), as shows in Figure 2.(2).b (the first and second row of ExC). If not, he/she keeps the path and just adds his/her ID to the first part of the path as shown in Figure 2.(2).b (the third row of ExC).

Then, all other users within the path towards user I, i.e., user L, do the same and send the ExC to the next user towards user I. When user I receives the final ExC, checks all entries in the ExC and updates them in his/her own RNC. Then, user I replies by selecting a subset of his/her ExC entries, updates the entries path and, then, sends them to user A from the path (L, M, A).

After exchanging neighbors for a number of cycles, the service will converge to a random overlay where each user connects to a uniform random subset of all users currently in the network. But, in our framework, users in addition of exchanging neighbors, also exchange their parameters estimation of the clustering model. In this way, when the service converges to the random overlay, also converges to a final parameters estimation available for all users and then, users are able to update their local parameters estimation. After some iterations of the EM algorithm, they will converge to a final clustering model. In the next section, we will explain in more details the update of the clustering model.

Algorithm. Gossip-based Clustering Protocol

```
Input: The local Θi for each user ui, i=1,...,N
Output: The global Θi for each user ui.
Initialization()
Loop:
  if Push then//if ui has to push information
    Wait ΔT
    ExC ← RNC.SelectRandomEntries()
    ExC ← UpdateExC(ExC, ui)
    uj ← RNC.SelectRandomUser()
    Sends ExC to ui 's neighbor toward uj
    Sends Θi to ui 's neighbor toward uj
    Θj ← Receive(uj)
    Θi ← UpdateModel(Θi, Θj)
  else if Pull then//If uj has to reply ui
    ExC ← RNC.SelectRandomEntries()
    Sends ExC to uj 's neighbor toward ui
    Sends Θj to uj 's neighbor toward ui
    Receives ExC from uj 's neighbor
    ExC ← UpdateExC(ExC, uj)
    Receives Θi from uj 's neighbor
    Θj ← UpdateModel(Θi, Θj)
  else//Pull user uk within the path
    ExC ← UpdateExC(ExC, uk)
    Sends ExC to uk 's neighbor toward uj
End Loop
procedure INITIALIZATION()
  InitModel()
  InitRNC()
  return Θ, RNC
End procedure

procedure UPDATEMODEL(Θi, Θj)
  for each cluster l do
    w'l = (wli + wlj)/2
    μ'l = (wli.μli + wlj.μlj)/(wli + wlj)
    Σ'l = (wli.Σli + wlj.Σlj)/(wli + wlj)
  return Θ'
End procedure
procedure UPDATEEXC(RxC, u)
  UpdatedRxC= RxC
  if u = ui then
    for all entries path ∈ UpdatedRxC do
      Path.AddFirstID()
  else//(u = uj) or (u = uk within the path to uj)
    for all entries path ∈ UpdatedRxC do
      ReversedPath=path.Reverse()
      for all users ID ∈ ReversedPath do
        if ID ∈ Direct-Friends (uk) then
          Path.AddFirstID(ID)
          Break
        else if ID ∈ Two-Hop-Friends (uk) then
          Path.AddFirstID(ID)
          Path.AddFirstID(GetDirectFriend(ID))
          Break
        else
          Path.AddFirstID(ID)
  if u = uk then
    for all entries ∈ UpdatedRxC do
      Path.AddFirstID(u)
  return UpdatedRxC
End procedure
```

Fig. 1. Gossip-based clustering protocol

3.2 Clutering Model Update

The second main component of our framework is the online clustering algorithm that updates the clustering model based on the local parameters estimation of each user. In our setting, in the set e entries of the user u_i's RNC, in addition to the list of random users ID, their *longevity* field, and the path to reach them, we maintain their parameters estimation of the clustering model. Therefore, the gossip-based clustering algorithm shown in Figure 1 performs the following steps. In the initialization phase, each user u_i, in addition to filling RNC with direct friends and two hops friends, initializes the local parameters estimation for each cluster l. After the initialization phase, users initiate exchanging neighbors and the parameters estimation of the clustering model simultaneously and periodically at a fixed period $\triangle T$. We do assume that the length of the period $\triangle T$ is the same for all users. During a period $\triangle T$, each user initiates one exchanging cycle. There are two types of communication models for exchanging the information. In the Push based model, a target user u_i sends ExC and parameters estimation (Θ_i) to the selected user. In the Push-Pull based model, both the target user u_i and selected user u_j exchange their ExC and parameters estimation. Our communication model is based on Push-Pull, since the Push approach can easily lead to partitioning the set of users in the network [10], [11].

After initialization, the initiating user u_i increases by one the *longevity* of all neighbors in his/her RNC. After that, user u_i selects neighbor u_j with the highest *longevity* among all neighbors in RNC, and set the *longevity* of u_j to zero in his/her RNC. If the information has to be pushed, user u_i replaces u_j's entry in *RNC* with a new entry of *longevity* 0 and with u_i's ID and path to reach user u_i. Then, user u_i selects S subset of neighbors from u_i's own *RNC*, and save them in the temporary *ExC*. Next, user u_i updates the entries path in the *ExC* by building a new path as mentioned in procedure *UpdateExC()* in Figure 1, and sends it to the next user in the path towards user u_j. After that, user u_i sends his/her local parameters estimation to the next user u_k in the path towards user u_j. Later, all users in the path towards user u_j update the entries path in the *ExC* and send the updated *ExC* and parameters estimation received from u_i to the next user in the path towards user u_j.

When user u_j receives from one of his/her direct friends the *ExC* coming from user u_i, user u_j replies by selecting a random subset of S neighbors of his/her own *RNC* and save them in his/her *ExC*. Next, user u_j updates the entries path in *ExC* by adding his/her ID to the first part of the entries path and sends *ExC* to the next user on the path towards user u_i. After that, user u_j sends the local parameters estimation to the next user in the path towards user u_i. Then, user u_j updates the entries path in the received *ExC*. Next, u_j discards entries pointing at u_j and entries already contained in u_j's *RNC* and updates his/her *RNC* to include all remaining entries, by firstly using empty cache slots, and secondly replacing entries among the ones sent to u_i. User u_j set *longevity* to zero for all entries of received *ExC* in the *RNC* and does not increase, though, any entry's *longevity* in the *RNC* until he/she initiates the exchanging process. After that, user u_j updates his/her own parameters estimation by calculating the weighted average with the parameters estimation coming from u_i. Finally, user u_j updates the parameters estimation of u_i and u_i's ID and the path to reach u_i inside his/her *RNC*.

Under this algorithm, after some cycles, the local parameters estimation of users converge exponentially fast to the global parameters estimation in each Maximization-step of the EM algorithm. Therefore, each user is able to compute new membership probabilities and check the stopping tolerance. Each user maintains the newly updated parameters estimation from the previous EM-iteration in a small cache, called *Estimation Cache* (*EsC*), of fixed size. When the cache is full, the parameters estimation stored for the longest time is replaced by the newly added parameters estimation. Therefore, after some iterations of clustering, all users will have a final clustering model and compute a final membership probability to belong to their most fitting cluster.

3.3 User Behaviour-Based Group Identification in DOSNs

Our goal is to have similar users in each group based on their social and individual features. Feature vectors of each user are given as input to the algorithm in Figure 1. Therefore, each user can assign a cluster number (behavior group) to him/herself based on the feature vector by considering the maximum membership probability among them.

User's features vector includes two types of features: individual and social features. Individual features are those, like age, gender, but also those that impact

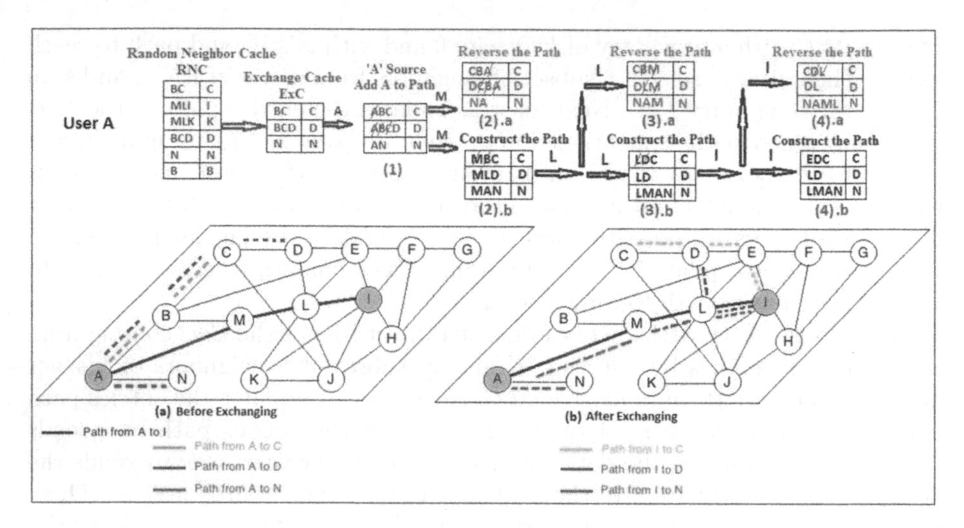

Fig. 2. (a) before and (b) after the neighbours exchanged between A and I

the possible users' behaviors, like, education and nationality. In addition to individual information, in order to measure users' attitude in online socialization, we consider the following social features:

Number of Friends: social users with a lot of friends have different patterns than isolated users with few friends;

Activity Level: unlike active users that write a lot of posts, passive users do not send any information to others. We calculate this feature as the sum of: a) number of posts that a user sends to others from the first day of joining the community, b) number of likes that a user performs on posted items, and c) number of comments a user writes for posted items;

Percentage of Public Profile Items: the assumption is that users with all profile information (100%) public are more social.

4 Experiments

To perform the experiments on a real graph, we used the Facebook dataset crawled and used in [1]. The author crawled the profiles information and friendship links of 75 users that launched the application as seed. Then, the application crawled the information related to these 75 users' friendships. We removed those profiles that have many missing features and obtained a graph by considering the largest connected component which includes 13,000 user profiles, plus the 75 seed users, with a total of about 461,700 friend links, 6,150,892 likes and 1,742,709 comments. Totally, around 7,000 users have more than 75% profile information as public.

4.1 Results for Convergence of the Clustering Model

The experiments are ran with $S = 50$ [10]. During the exchanging process the mean of the local parameters estimation, $\mu_{i,c}$ of each user u_i, $i = 1, \ldots, N$ for each cluster in cycle c, $c = 1, \ldots, \Re$ is always the global correct mean μ_c, but, the variance measure σ_{c^2} that expresses the deviation of the local estimations from the correct mean in the given cycle c decreases over the set of all local estimations on the average by factor γ, with $\gamma < \frac{1}{2\sqrt{e}}$ [10]. In general, when the variance tends to zero, then all users hold the global correct mean μ_c.

$$\mu_c = \frac{1}{N}\sum_{i=1}^{N}\mu_{i,c} \qquad \text{and} \qquad \sigma_c^2 = \frac{1}{N-1}\sum_{i=1}^{N}(\mu_{i,c} - \mu_c)^2$$

The convergence factor between cycle c and $c+1$ is given by $\sigma_{c+1}^2/\sigma_c^2$. We plot the convergence factor (values are averages for 20 independent runs) as a function of the number of cycles, as shown in Figure 3a. It is clear to see that the speed of the convergence of the variance is fast and it decreases exponentially after few cycles. Thus, means that, after a small number of cycles, all users, including the isolated users with low friendship links, will have accurate estimations of the global correct mean μ in each M-step, when no failures occurred. From this experimental result, choosing the number of cycles $\Re$ equal to 120[1] is sufficient to reach a convergence.

4.2 Coping with User Failure

In a dynamic network, users continuously join and leave the network and they fail in some situations. In this section, we consider the performance of the clustering model when some percentage of users fail in each cycle of the exchanging parameters estimation.

As we mention before, each user maintains the parameters estimation he/she receives from other users in the network in his/her *ENC*. If a user u_i sends his/her parameters estimation to a user u_j to exchange and performs averaging and, waiting for $\triangle T$ time, he/she does not receive any answer, u_i checks the *ENC* to verify if he/she has the parameters estimation of user u_j from previous cycles or not. If yes, u_i performs the average with those previous values and updates his/her parameters estimation. Otherwise, he/she skips the exchanging step. We need to mention that the user selection method in [11] takes care of the failure of those users within the path between user u_i and u_j. We consider the effect of these missing exchanges on the final value of the global μ of the clustering model. Towards this goal, we remove 50% of users in each cycle of the exchanging protocol and run independently the Newscast EM for 20 times. We show the result in Figure 3b. The y-axis shows the variance of the 120th cycle to the first cycle. The figure shows that if the failure happens in the first cycles, the result of the global μ of the clustering model is so far from the real

[1] We assume that all users agree on the number $\Re$ of peer sampling cycles, and $\Re$ is large enough to guarantee convergence to the final parameters estimation.

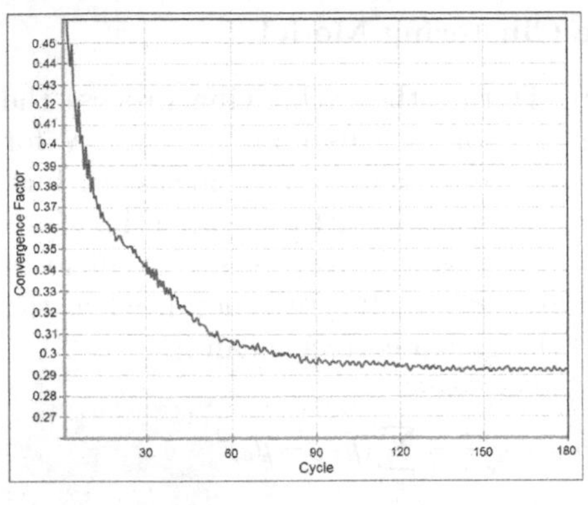

(a) Convergence factor

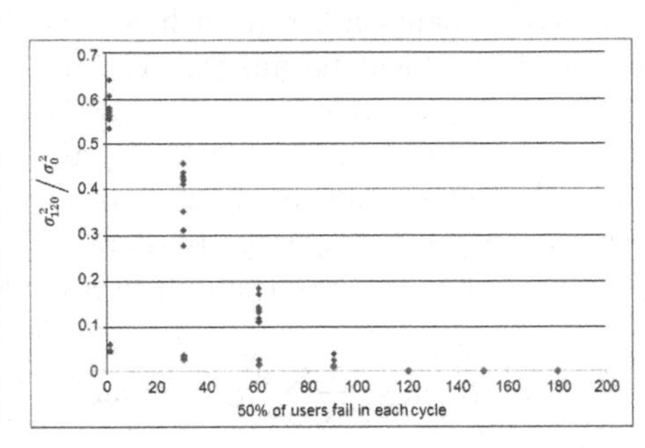

(b) The variance of μ at cycle 120 (for 20 independent run) after the failure of 50% of users in each cycle

Fig. 3. The convergence factor and the variance after users failure

correct global μ. But, if this failure happens in the next cycles (especially after the 100th cycle), the variance tends to zero.

4.3 Clustering Results

In Table 1, we can see the performance of randomly initialized Newscast EM and centralized EM, by setting different number of clusters. The result in this table shows the average number of EM iterations for each user to achieve convergence in Newscast EM (around 42 ± 3) and centralized EM (around 38 ± 2) that are almost near. After convergence, the value of μ_i for each cluster at each user will converge to the true global μ. For example, Table 2 shows this value for feature Activity level in centralized EM and Newscast EM that are almost equal. Also, Table 2 shows the values of μ for each cluster for the feature Number of friends.

Table 1. EM iterations and the ratio of users for different number of clusters

(a) EM iterations

EM Iteration		
NoClusters	CentralizedEM	NewscastEM
3	38	41.25
5	38	42.53
7	39	41.38
10	38	43.19

(b) The ratio of users (RU) in each cluster

ClusterNo	10	7	5	3
Cluster1	20.27%	32.43%	47.29%	41.89%
Cluster2	20.27%	21.62%	22.97%	32.43%
Cluster3	14.86%	18.91%	10.81%	25.67%
Cluster4	10.81%	9.45%	12.16%	
Cluster5	9.45%	6.75%	6.75%	
Cluster6	6.75%	5.4%		
Cluster7	1.35%	5.4%		
Cluster8	5.4%			
Cluster9	5.4%			
Cluster10	5.4%			

Table 2. The value of μ for two features

(a) Activity level

ClusterNo	*CentralEM*	*NewscastEM*
C1	27.5	27.35
C2	52.53	52.25
C3	48.31	48.05
C4	11301.68	11299.987
C5	6869.46	6868.299
C6	335.83	334.062
C7	18338.18	18336.647

(b) Number of friends

ClusterNo	*CentralEM*	*NewscastEM*
C1	536.39	535.56
C2	272.91	271.47
C3	194.42	193.39
C4	438.35	436.04
C5	276.87	275.41
C6	963	961.93
C7	953.6	951.58

4.4 Dominant User Behaviors

Identifying the number of clusters or user behaviors is related to the nature of the dataset. Therefore, there is no correct or incorrect numbers of groups to find. Each increment in the number l of clusters yields to a new group and similarly a new behavior. On the other hand, if we consider a small number of groups by aggregating some behaviors, we will have the most dominant behaviors, but, we may miss some relevant behaviors. In this paper, we assume that we know the best number of clusters. However, we analyze the quality of the clustering for various values of l. In Table 1, we report the ratio of users that belong to each cluster, by considering different numbers of clusters. When we have 10 clusters, the percentage of users that belong to the '7th cluster' is 1%, that is, very small in size. By setting the number of cluster to 5, we will have around 50% of users concentrated in one group (1sh cluster) that will be a huge cluster, and we are not able to precisely identify their behavioral pattern. For the number of clusters equal to 3, we will have the most dominant behaviors, but, we will miss a lot of behavioral patterns. Therefore, we consider the number of clusters equal to 7 in the rest of the experiments.

In the next experiment, we analyze the influence of each feature value on the quality of the discovered groups. Then, among all features, we remove those features that have the same distribution in all clusters. More precisely, we discard those features whose existence in the clustering will not propose a new behavioral group. In order to have a measure for the distribution of each feature value for all clusters, we use the inter-group relative feature value [20], denoted by $Related_{fl}$, which measures how a feature f of cluster l is related to the same feature of the other clusters. It is computed as follows:

$$Related_{fl} = \frac{feature_{fl}}{\sum_{l=1}^{K} feature_{fk}} \quad (5)$$

where $feature_{fl}$ is the value of feature f in cluster l. This allows us to see if the value of a feature is evenly distributed in all clusters or concentrated in a single cluster. For example, we can see in Figure 4a the distribution of all categorical features such as age, number of friends and activity level. This figure shows that the distribution of the feature 'age' in all clusters is nearly the same. Then, we

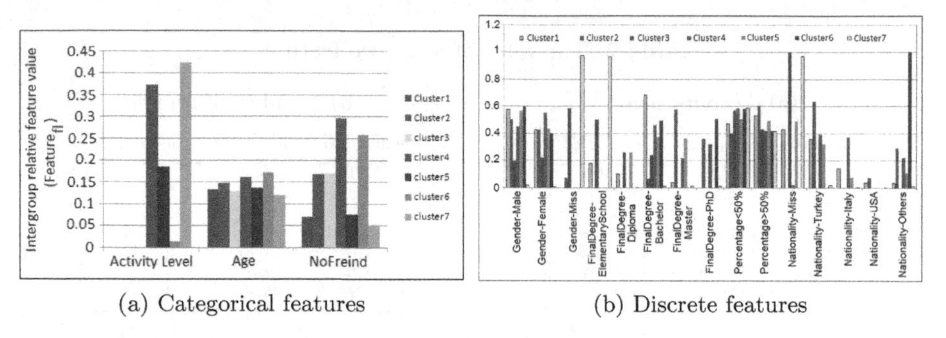

(a) Categorical features (b) Discrete features

Fig. 4. The distribution of features in each cluster

remove this feature since it can not help us to define any behavioral pattern. For discrete features, such as: gender, nationality and percentage of public profile items, we consider the fraction of users that have the same value for that feature in each cluster as shown in Figure 4b. We removed 'gender' and 'percentage of public profile items' and we consider all the remaining features. Based on the experimental results, we are able to find the most dominant behaviors, that are shown in Figure 5. These dominant behaviors are categorized as follows: 1)*most active users with high number of friends:* users in cluster 7 have the highest activity level and the highest number of friends. These users have a bachelor or master degree and all of them are from Italy. 2)*very active users with medium number of friends:* these users are concentrated in cluster 4, have a medium activity level and a medium number of friends. Their education level is either elementary school, bachelor or PhD and they are from all countries except Italy and USA. 3)*active users with low number of friends:* these users are concentrated in cluster 5, and their education level is either diploma, bachelor or master and they are from all countries except USA. 4)*passive users with high number of friends:* these users are concentrated in cluster 6. These users have either a bachelor or PhD degree. They are from all countries except Italy. 5)*very passive users with medium number of friends:* these users are in cluster 1. These users are from all education levels. They are from all countries except Italy. 6)*most passive users with lowest number of friends:* these users are concentrated in cluster 3. Their education level is either bachelor or master and they are from all countries except USA. 7)*most passive users with low number of friends:* these users are in cluster 2 and most of them have bachelor. They are from all countries.

5 Related Work

In decentralized setting, there are several distributed clustering algorithms like building and using a single clustering model for each user individually based on his/her local dataset [9]. Other approaches are sharing the data between

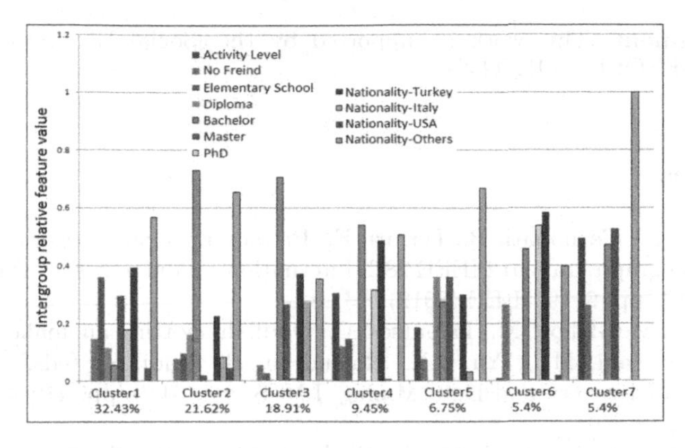

Fig. 5. The most dominant behaviors with percentage of users in each group

users [24]. But, the communication cost is high to share all the raw data between users. Another approach is to organize the clustering model in a hierarchical fashion [23], [9] by which local clustering models are computed first for each user individually, and sent to a logically higher-level user that aggregates local models. But, users need to share all their private information that is a big issue in our application as well as in some other applications.

Other related work are in the area of distributed computation in large distributed systems. In deterministic averaging techniques, each user repeatedly selects all his/her immediate neighbours to update the local parameters estimation. For example, authors in [8] proposed an EM algorithm based on this approach. However, this technique is not suitable for social networks, because of the existence of some high degree users. Another kind of technique is probabilistic gossip-based approaches [3], [12] where, at each iteration, each user repeatedly selects a uniform random user and both users compute the average of their parameters estimation. For instance, the newscast model in [12], the gossip-based protocols in [3] and Newscast EM are all based on gossip learning. Since for applying Newscast EM in social networks, the network needs to be fully connected, we use gossip-based peer sampling service on social networks.

6 Conclusion

We identify behavioral groups of users by applying the Newscast EM algorithm on top of DOSNs. We combine the ability to access random users on social networks with distributed clustering model to identify group of behavioral patterns. For future directions we are planning to find a solution for defining the best number of clusters in a distributed manner and also for exchanging information in a secure way for avoiding malicious users manipulating the local parameters estimation of the clustering model. Furthermore, we will use the identified behavioral patterns for risk assessment in DOSNs.

Acknowledgment. This work is supported by the iSocial EU Marie Curie ITN project (FP7-PEOPLE-2012-ITN).

References

1. Akcora, C.G., Carminati, B., Ferrari, E.: Privacy in social networks how risky is your social graph? In: 2012 IEEE 28th International Conference on Data Engineering (ICDE), pp. 9–19. IEEE (2012)
2. Akoglu, L., McGlohon, M., Faloutsos, C.: Oddball-spotting anomalies in weighted graphs. In: Zaki, M.J., Yu, J.X., Ravindran, B., Pudi, V. (eds.) Advances in Knowledge Discovery and Data Mining. LNCS, vol. 6119, pp. 410–421. Springer, Heidelberg (2010)
3. Boyd, S., Ghosh, A., Prabhakar, B., Shah, D.: Gossip algorithms: design, analysis and applications. In: Proceedings IEEE 24th Annual Joint Conference of the IEEE Computer and Communications Societies, INFOCOM 2005, vol. 3, pp. 1653–1664. IEEE (2005)
4. Bradley, P.S., Fayyad, U., Reina, C.: Scaling em (expectation-maximization) clustering to large databases. Technical report, Technical Report MSR-TR-98-35, Microsoft Research Redmond (1998)
5. Datta, A., Buchegger, S., Vu, L.-H., Strufe, T., Rzadca, K.: Decentralized online social networks. In: Handbook of Social Network Technologies and Applications, pp. 349–378. Springer (2010)
6. De Meo, P., Ferrara, E., Fiumara, G., Provetti, A.: Generalized louvain method for community detection in large networks. In: 2011 11th International Conference on Intelligent Systems Design and Applications (ISDA), pp. 88–93. IEEE (2011)
7. Eliassi-Rad, T., Henderson, K., Papadimitriou, S., Faloutsos, C.: A hybrid community discovery framework for complex networks. In: SIAM Conference on Data Mining (2010)
8. Dongbing, G.: Distributed em algorithm for gaussian mixtures in sensor networks. IEEE Transactions on Neural Networks **19**(7), 1154–1166 (2008)
9. Hido, S., Tokui, S., Oda, S.: Jubatus: an open source platform for distributed online machine learning. In: NIPS 2013 Workshop on Big Learning, Lake Tahoe
10. Jelasity, M., Voulgaris, S., Guerraoui, R., Kermarrec, A.-M., Van Steen, M.: Gossip-based peer sampling. ACM Transactions on Computer Systems (TOCS) **25**(3), 8 (2007)
11. Khelghatdoust, M., Girdzijauskas, S.: Short: Gossip-based sampling in social overlays. In: Noubir, G., Raynal, M. (eds.) Networked Systems. LNCS, vol. 8593, pp. 335–340. Springer, Switzerland (2014)
12. Kowalczyk, W., Jelasity, M., Eiben, A.: Towards data mining in large and fully distributed peer-to-peer overlay networks. In: Proceedings of BNAIC 2003, pp. 203–210 (2003)
13. Kowalczyk, W., Vlassis, N.A.:Newscast EM. In: Advances in Neural Information Processing Systems, pp. 713–720 (2004)
14. Kurant, M., Gjoka, M., Butts, C.T., Markopoulou, A.: Walking on a graph with a magnifying glass: stratified sampling via weighted random walks. In: Proceedings of the ACM SIGMETRICS Joint International Conference on Measurement and Modeling of Computer Systems, pp. 281–292. ACM (2011)
15. Kurant, M., Markopoulou, A., Thiran, P.: Towards unbiased bfs sampling. IEEE Journal on Selected Areas in Communications **29**(9), 1799–1809 (2011)

16. Li, Z., Wang, C., Yang, S., Jiang, C., Li, X.: Lass: Local-activity and social-similarity based data forwarding in mobile social networks. IEEE Transactions on Parallel and Distributed Systems **26**(1), 174–184 (2015)
17. Liu, Y., Liu, Q., Qin, Z.: Community detecting and feature analysis in real directed weighted social networks. Journal of Networks **8**(6), 1432–1439 (2013)
18. Dongyuan, L., Li, Q., Liao, S.S.: A graph-based action network framework to identify prestigious members through member's prestige evolution. Decision Support Systems **53**(1), 44–54 (2012)
19. Lu, J., Li, D.: Sampling online social networks by random walk. In: Proceedings of the First ACM International Workshop on Hot Topics on Interdisciplinary Social Networks Research, pp. 33–40. ACM (2012)
20. Maia, M., Almeida, J., Almeida, V.: Identifying user behavior in online social networks. In: Proceedings of the 1st Workshop on Social Network Systems, pp. 1–6. ACM (2008)
21. Moosavi, S.A., Jalali, M.: Community detection in online social networks using actions of users. In: 2014 Iranian Conference on Intelligent Systems (ICIS), pp. 1–7. IEEE (2014)
22. Rahimian, F., Payberah, A.H., Girdzijauskas, S., Jelasity, M., Haridi, S.: Ja-be-ja: a distributed algorithm for balanced graph partitioning. In: 2013 IEEE 7th International Conference on Self-Adaptive and Self-Organizing Systems (SASO), pp. 51–60. IEEE (2013)
23. Rajasegarar, S., Leckie, C., Palaniswami, M.: Hyperspherical cluster based distributed anomaly detection in wireless sensor networks. Journal of Parallel and Distributed Computing **74**(1), 1833–1847 (2014)
24. Xia, Y., Zhao, Z., Zhang, H.: Distributed anomaly event detection in wireless networks using compressed sensing. In: 2011 11th International Symposium on Communications and Information Technologies (ISCIT), pp. 250–255. IEEE (2011)

A Generalized Framework for Quantifying Trust of Social Media Text Documents

Tuhin Sharma(✉) and Durga Toshniwal

Department of Computer Science and Engineering,
Indian Institute of Technology Roorkee, Roorkee 247667, Uttarakhand, India
tuhinsharma121@gmail.com, durgatoshniwal@gmail.com

Abstract. Social media has become a very popular place for users seeking knowledge about a wide variety of topics. While it contains many helpful documents, it also contains many useless and malicious documents or spams. For a casual observer it is very hard to identify high quality or trustworthy documents. As the volume of such data increases, the task for identifying the trustworthy documents becomes more and more difficult. A huge number of research works have focused on quantifying trust in certain specific social network domains. Some have quantified trust based on social graph. In this work, we use such social graph named Reduced node Social Graph with Relationships (RSGR) and we develop a three-step syntax and semantic based trust mining framework. Here we generalize the concept of trust mining for all structured as well as unstructured unsupervised text documents from all social network domains. We calculate trust based on metadata, trust based on relationships with other documents and finally we propagate the trust calculated so far along various relationship edges to calculate the final trust. Finally we show that our method calculates the trust of social media text documents with more than 80% accuracy.

1 Introduction

Social Media is growing day by day at an increasing rate of growth of millions of documents per day. A document could be a facebook post, a tweet, a blog, a review or even a video. A substantial portion of these documents is useful and is an excellent source for performing various social media analysis like sentiment analysis, segmentation analysis, etc. to obtain knowledge. But due to the availability of this huge amount of information, there is an important need for differentiating between good and bad documents, since all these documents are not useful unless manually read. In the current scenario of social media analysis, documents are fed individually [25][22]. But, as there is no relationship between the documents, we have knowledge of every document individually but not in presence of other documents. So, we cannot say anything about the reliability of the individual documents. Let us call the reliability as Trust. Hence the trust of the data is unknown and there is absolutely no quick method to diagnose if the document could be actually trusted. The need for Trust Mining comes from

P. Perner (Ed.): MLDM 2016, LNAI 9729, pp. 692–713, 2016.
DOI: 10.1007/978-3-319-41920-6_53

the fact that the better understanding of the data we have, the better will be the analysis of that data. So, before feeding the documents into social media analysis, some trust distribution should be assigned to the dataset so that low trustworthy documents can be filtered out for the improvement of the analysis.

In social media we come across various spams and as the documents are processed independent of each other, it results in lack of context, and an inability to remove *noise* from the incoming data. For example if we do not know the context we cannot say if it is worthy to look at a particular review document. Still even if we know the context it is difficult to differentiate between two tweets having same text fields but associated with different urls. So, we need to find a way to deal with these problems. By trust mining we can remove malicious documents which will result in the increase of *SNR*. Thus we shall have much more reliable analysis and visualization.

In this paper, we use Reduced node Social Graph with Relationships (RSGR) [14]. On the basis of the RSGR and various information about the document, such as information about Author, Domain etc. we develop a three-step syntax and semantic based Trust Mining approach. The novelty of our work is as follows. First we develop a Metadata Trust Score Algorithm (MTSA) to calculate trust based on metadata. In the second step we develop a Relative Trust Score Algorithm (RTSA) to calculate trust in the presence of other documents. Finally we develop a Propagated Trust Score Algorithm (PTSA) to propagate trust through various relationship edges.

2 Related Work

Earlier research work on trust mining from social media data normally focus on trust network having explicit trust. Golbeck et al. propose a method for the application of social media analysis on multi-dimensional networks evolved from ontological trust specifications [1]. Guha et al. develop a trust as well as a distrust propagation model [2]. These approaches rely only on the structure of the web of trust. So, the accuracy cannot be guaranteed as there is lack of context and content such as, documents' topics, users' activities etc. To overcome this problem Christian Bizer et al. propose the usage of context and content based trust mechanisms to develop a trust architecture which uses a combination of reputation, context and content-based trust mechanisms [3].

Agichtein et al. [4] propose a classification framework for combining evidence from different information sources. Blumenstock propose a method that simply uses the word count of the articles as a measure of quality of Wikipedia [5]. Zolfaghar et al. develop a system consisting of various classification models such as support vector machine, decision tree etc. to predict trust and distrust relations [6]. But they do not pay any attention to mine information from the unstructured data. McGuinness et al. measure trust based on the frequency of occurrences of the Encyclopedia terms in Wikipedia articles [7].

[27,28] propose a method for trust evaluation by mining feedback comments. [29] proposes a method for user trust modelling in Twitter. But a generalized

trust mining method for social media text documents is still not there irrespective of source domains.

Lim et al. propose a trust antecedent framework which considers ability, integrity and benevolence as the three key factors for the formulation of trust [8][9]. But the problem is that it is very difficult to acquire all the three key factors in online communities. Yu Zhang and Tong Yu uses semantic-based trust mining mechanism to build domain ontology [11]. But they only focus on structured data. Sai T. Moturu and Huan Liu develop an unsupervised approach to quantify trustworthiness [13]. They first identify the features and then quantify the trust. But they do not calculate trust based on relationships in social network. Matsuo and Yamamoto use features extracted from product reviews, user profiles and existing trust relations to predict trust between users [10].

[17]proposes a trust model that leverages the implicit human factor to help quantify the trustworthiness of candidate services. Over 400 publications have reported various types of statistical studies over the DBLP dataset, including coauthorship analysis [18,19], community analysis [20], field analysis [21,23], clustering [17,24], and accessibility analysis [26]. But they do not discuss how to deal with missing data fields.

In our previous work [14] we proposed a syntax and semantic based relationship mining approach for establishing relationships between social media text documents irrespective of the type of document and the source domain. As a result we finally get a Reduced node Social Graph with relationships (RSGR). In the current work we use this RSGR for quantifying trust of social media documents. The next sections describe the work we have done. In Sec. 3 the proposed work is elaborately discussed. In Sec. 4 experimental results are shown.

3 Proposed Work

The framework for the proposed work is shown in Fig. 1. Feature extraction and finding different types of relationships are briefly depicted in Sec. 3.1 and Sec. 3.2 respectively as per [14]. Additionally we include the concept of LOCATION node in Sec. 3.1 and Sec. 3.2. We mainly concentrate on 3.3 which focuses on the trust mining approach.

3.1 Feature Extraction

Each of the incoming documents has 5 types of information regarding its Author, Timestamp, Domain, Location and Text associated with it. So for individual documents we create nodes corresponding to these 5 types and we segregate corresponding information and store them accordingly in graph structure as shown in Fig. 2. We additionally create CONCEPT nodes to store different concepts or keywords of text documents and to store interest of Authors. The edges shown in Fig. 2 are not assigned any weight.

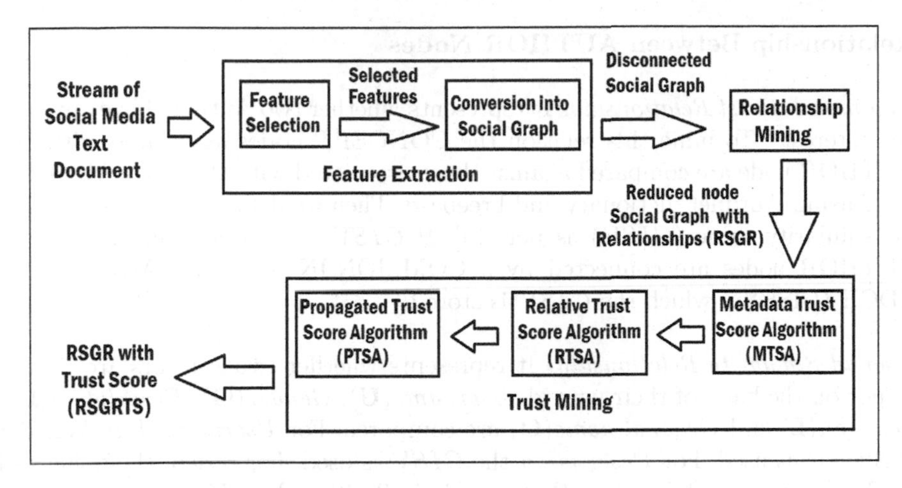

Fig. 1. Proposed framework for Trust mining

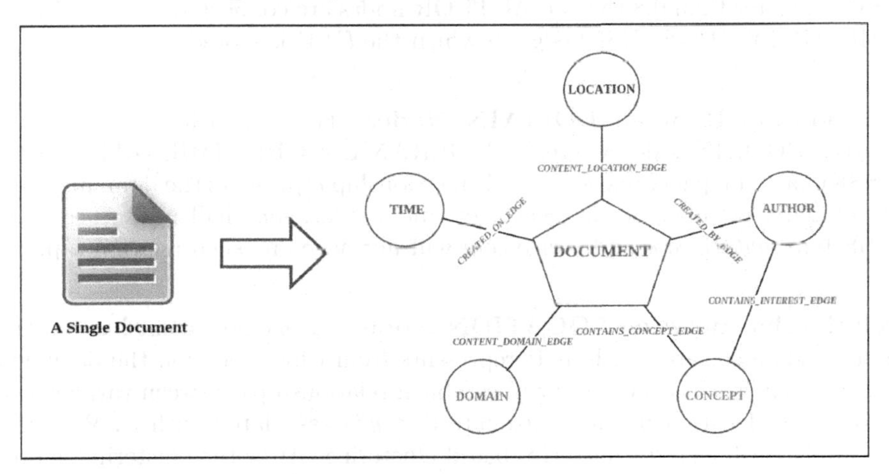

Fig. 2. Documents converted into Graph structure

In this step we just read and store the attribute values for respective documents. We create DOCUMENT, AUTHOR, LOCATION, DOMAIN and TIME nodes. The significance of these nodes is as per [14]. Additional to this we add LOCATION node.

LOCATION Node: It contains the information about the place from where the document is created like, *Location, Country, Co-ordinates,* etc.The LOCATION node is connected with the DOCUMENT node via CONTENT_LOCATION_EDGE.

3.2 Relationship Mining

Depending on the type of nodes different approach is followed to remove duplicates and to establish relationships as per [14].

Relationship Between AUTHOR Nodes

Common Interest Relationship. It represents whether two Authors have any similar interests. To mine this relation the CONCEPT nodes associated with one AUTHOR node are compared against those associated with the other AUTHOR node using Wordnet dictionary and Freebase. Then we define the common interest similarity value ($CISV$) as per [14]. If $CISV$ is greater than 0.5 the two AUTHOR nodes are connected by a COMMON_INTEREST_SIMILARITY_EDGE (CISE) in which the $CISV$ is stored.

User-id Similarity Relationship. It represents whether two Authors are similar or not on the basis of their user-id. *Username*(**U**), *Gender*(**G**), *Description*(**D**), *Location*(**L**) and *Original name*(**O**) are compared. For *Username* Jaro-Winkler coefficient is used. For *Description* the $CISV$ is used. For rest of them Jaccard coefficient is used. We define the user-id similarity value (USV) as per [14].If USV is greater than 0.8 the two AUTHOR nodes are connected by a USERID_SIMILARITY_EDGE (USE) edge in which the USV is stored.

Relationship Between DOMAIN Nodes. For establishing relationships between DOMAIN nodes we create PAGERANK_SCORE_SIMILARITY_EDGE (PSSE) based on pagerank score. This relationship represents the domains which are similarly popular. So, *facebook.com* and *twitter.com* shall be connected by PSSE but *youtube.com* and *vimeo.com* will not have any such relationship.

Relationship Between LOCATION Nodes. Every document has location information associated with it. It represents from which location the document has been published. Here, we try to establish relationships between various locations. Every location has a co-ordinate (x°, y°) associated with it. We use it and apply Euclidean distance [15] based clustering. After the clustering is done we select an arbitrary LOCATON node from respective clusters and mark it as GEO-SPEAKER. All the other LOCATION nodes from respective clusters are connected with respective GEO-SPEAKER nodes via LOCATION_SIMILARITY_EDGE (LSE).

Relationship Between DOCUMENT and DOMAIN. We analyze the *Texthtml* of the DOCUMENT and find out different domain names and connect those corresponding DOMAIN nodes with the DOCUMENT node via PAGE_REFERENCE_EDGE (PRE).

Relationship Between DOCUMENT Node. There are two types of such relationships.

Document Concept Similarity Relationship. We connect two DOCUMENT nodes by a DOCUMENT_CONCEPT_SIMILARITY_EDGE (DCSE) if they have any similar concept. To mine this relation same procedure is followed as the process adopted while establishing CISE between AUTHOR nodes.

Document Reference Relationship. It represents if a DOCUMENT is referred by another DOCUMENT. To find this relationship we analyze *Texthtml* to find out if there exists any link to other document. If we find any such document then we connect those documents using DOCUMENT_REFERENCE_EDGE (DRE).

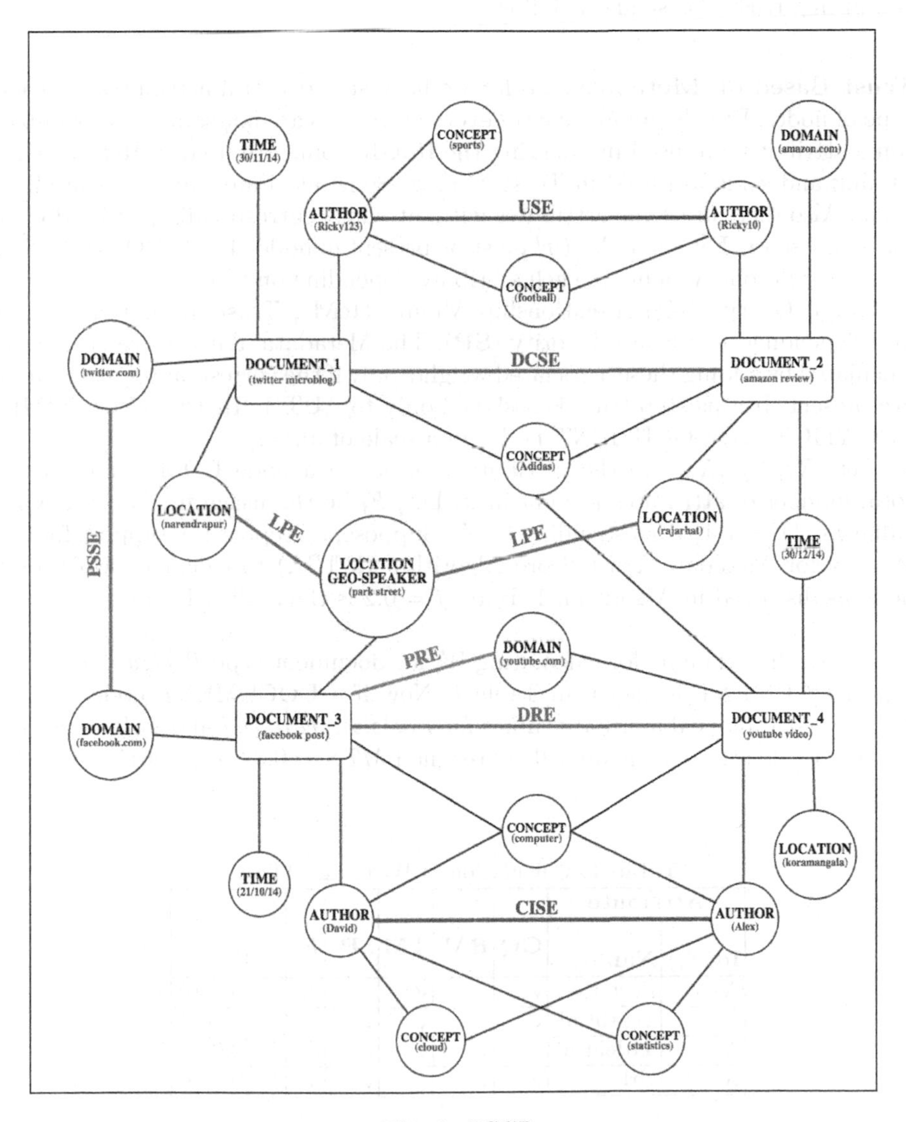

Fig. 3. RSGR

Finally we get Reduced node Social Graph with Relationships (RSGR) after the relationship mining as shown in Fig. 3. We shall use this RSGR graph for trust mining.

3.3 Trust Mining

We calculate the trust in 3 steps. At every step we consider different basis for computing trust. These are as follows:-

Trust Based on Metadata. So far we have stored useful attributes for each type of nodes. But the importance of every attributes are not same. For example, some attributes are used in querying the RSGR, some are used in Relationship Mining and some are used in Trust Mining. Some attributes are used in all of them. Also there are some attributes without which corresponding nodes do not make any sense. For example, *Url* must be present in node of type DOCUMENT. So we assign some weights to each attribute depending on whether they are used in Graph Query (**GQ**), Relationship Mining (**RM**), Trust Mining(**TM**) and also depending on a Static Priority (**SP**). The Metadata Trust Score (MTS) is calculated by adding these associated weights only if the corresponding attributes are present. In this step trust is assigned only to AUTHOR, DOMAIN, TIME, LOCATION and DOCUMENT nodes in a scale of $[0, 0.2]$.

Let, $X_1, X_2 ... X_m$ are the attributes present in a node DOMAIN m is the total number of attributes present in it. Let, F_j be the usage frequency having value $j \in \{1, 2, 3, 4\}$ corresponding to X_i. Suppose, S_j represents support for F_j. We develop Metadata Trust Score Algorithm (MTSA) to calculate MTS of a node as discussed in Algorithm 1. Here, $f = 0.2$ is the scaling factor.

Example. One example for calculating W_i for document type *Reviews* and node type DOCUMENT is shown in Table 1. Now if a DOCUMENT node of type *Reviews* contains real attribute values for *Postsize, Texthtml* and *null* values for the other fields then the node will be assigned $MTS = 0.0307 + 0.0615 = 0.0922$.

Table 1. Calculation of W_i using MTSA

<table>
<tr><th colspan="2">Attribute</th><th rowspan="2">GQ</th><th rowspan="2">RM</th><th rowspan="2">TM</th><th rowspan="2">SP</th><th rowspan="2">F_j</th><th rowspan="2">S_j</th><th rowspan="2">W_i</th></tr>
<tr><th>Id(X_i)</th><th>Name</th></tr>
<tr><td>X_1</td><td>Postsize</td><td>✓</td><td></td><td>✓</td><td></td><td>2</td><td>1</td><td>0.0307</td></tr>
<tr><td>X_2</td><td>Texthtml</td><td>✓</td><td>✓</td><td>✓</td><td>✓</td><td rowspan="2">4</td><td rowspan="2">2</td><td>0.0615</td></tr>
<tr><td>X_3</td><td>Titlehtml</td><td>✓</td><td>✓</td><td>✓</td><td>✓</td><td>0.0615</td></tr>
<tr><td>X_4</td><td>Url</td><td>✓</td><td>✓</td><td></td><td>✓</td><td>3</td><td>1</td><td>0.0461</td></tr>
</table>

Algorithm 1. Metadata Trust Score Algorithm (MTSA)

INPUT: Set of attributes $X_1, X_2 ... X_m$ of a node N
OUTPUT: MTS of N
  Initialize $MTS = 0$
  for all X_i **do**
    Calculate F_j by summing the number of times X_i is used in **GQ, RM, TM** and **SP**
  end for
  for all F_j **do**
    Calculate S_j by counting the number of attributes having usage frequency is F_j
  end for
  for all X_i **do**
    $W_i = \frac{F_j S_j}{\sum_{i=1}^{4} F_j S_j} \times \frac{1}{S_j} \times f = \frac{F_j}{\sum_{i=1}^{4} F_j S_j} \times f$
  end for
  for all X_i **do**
    if X_i is not null **then**
      $MTS = MTS + W_i$
    end if
  end for

Trust Based on Relationship. The relative trust represents the reliability of documents in the vicinity of other documents. We select deciding numeric attributes depending on the type of document and node type so that we can calculate relative trust score (RTS) based on them. Here we develop Relative Trust Score Algorithm (RTSA) to find RTS of AUTHOR, DOMAIN and DOCUMENT nodes. We use Z-score for this purpose. The z-score is a measure of how many units of standard deviation the raw attribute value is from the mean value. Thus, the z-score is a relative measure instead of an absolute measure. Then we define Dispersion Score (DS) for each numeric attribute values from respective attributes and finally find Relative Trust Score (RTS).

The RTSA for document type D and node type N is described in Algorithm 2. Here, $A^1, A^2 ... A^m$ are the m deciding attributes present in n number of nodes $N_1, N_2 ... N_n$ of type N. A_j^i represents the numeric value of A^i for N_j, $[1 \leq i \leq m, 1 \leq j \leq n]$. $Z(A_j^i)$ and $DS(A_j^i)$ represents Z-score and Dispersion score of A_j^i respectively. Clearly, $0 \leq DS(A_j^i), RTS(N_j) \leq 1$.

Now the deciding attributes are selected as follows,

Attributes for AUTHOR Nodes. For AUTHOR nodes we are assigning the trust in the range of $[0, 0.5]$. For docment type Videos we could not find any attributes. The attributes are,

Document type Boards and Microblogs: Twitter API [31] and Facebook API [30] are used to get some additional metadata of each Author from respective domains. The deciding attributes are as follows:

Number of Friends/Followers: It represents popularity of the author in the corresponding social network.

Algorithm 2. Relative Trust Score Agorithm (RTSA)

INPUT: $N_1, N_2...N_n$ and $A^1, A^2...A^m$ **OUTPUT:** $N_1, N_2...N_n$ with RTS

for all A^i **do**
  $A^i_{max} = max\{A^i_1, A^i_2...A^i_n\}$,
  $A^i_{min} = min\{A^i_1, A^i_2...A^i_n\}$,
  $\mu(A^i) = \frac{\sum_{j=1}^{n} A^i_j}{n}$,
  $\sigma(A^i) = \sqrt{\frac{\sum_{j=1}^{n}\left(A^i_j - \mu(A^i)\right)^2}{n}}$,
end for
for all A_j **do**
  for all N^i **do**
    $Z(A^i_j) = \frac{A^i_j - \mu(A^i)}{\sigma(A^i)}$;
    Calculate $median(Z(A^i))$
  end for
  for all N^i **do**
    if $Z(A^i_j) > median(Z(A^i))$ **then**
      $DS(A^i_j) = 0.6 + \frac{Z(A^i_j) - median(Z(A^i))}{Z\left(A^i_{max}\right) - median(Z(A^i))} \times 0.4$
    else if $Z(A^i_j) < median(Z(A^i))$ **then**
      $DS(A^i_j) = \frac{Z(A^i_j) - Z\left(A^i_{min}\right)}{median(Z(A^i) - Z\left(A^i_{min}\right)} \times 0.6$
    else if $Z(A^i_j) = median(Z(A^i))$ **then**
      $DS(A^i_j) = 0.6$
    else if $Z(A^i_j) = Z\left(A^i_{min}\right)$ **then**
      $DS(A^i_j) = 0.0$
    else if $Z(A^i_j) = Z\left(A^i_{max}\right)$ **then**
      $DS(A^i_j) = 1.0$
    end if
  end for
  for all N_j **do**
    $RTS(N_j) = \frac{\sum_{i=1}^{m} DS(A^i_j)}{m}$
  end for
end for

Follower to Following Ratio: Nobody relies on an author who is following 1000 accounts and has only 10 followers. So if the ratio of number of follower to the number of following is large then the author is more reliable.

Number of Distinct Tweets/Posts: We calculate how many of the last 5 tweets or posts of the author are distinct. More is the number of distinct posts more reliable the Author is.

Document Type Reviews: Generally review data come with *Author description* attribute which provides some information about the Author. For trip advisor data we get 5 deciding attributes:-

The *first field* represents the reputation of the reviewer i.e. whether he is a contributor of type **Top/Senior/Simple Contributor/Reviewer**. According to the presence of these 4 types of category 100, 70, 50 and 30 are considered as numeric values respectively.

The *second, third* and *fourth* field represent the number of total reviews, the number of hotel reviews and the number of helpful votes posted by the reviewer so far.

The *fifth* field represents for how many distinct cities the reviewer has voted so far.

For other documents that do not belong to trip-advisor a default static value 0.2 is assigned as RTS.

Attributes for Node Type DOCUMENT. In this part we assign the trust in the range of $[0, 0.5]$. The deciding numeric attributes are as follows:-

Document type Boards, Reviews, Microblogs
Review Rating/Blog Rating: More the review/blog rating more reliable the document is.

Number of Helpful Votes: It represents how much useful the document is.

Post-In-Thread: It represents how much discussion has been taken place regarding the document. More the number of post in the thread more reliable the document is.

Similarity Between Title and Actual Text: It represents with how much relevance with its title the document is written.

Length of the Content: Normally good documents are large as the author takes significant time to write them. Spams are more often small in size. So if the length of the content is large then probably the document is much reliable.

Number of Paragraphs in Content: Normally reliable documents are well structured and contain sections or paragraphs. But Spams usually contain many newline characters that is mistakenly interpreted as paragraph. So we also consider *Average length of paragraphs.*

Document Type Videos: The attributes are *Proximity between Titlehtml and Texthtml* and *Comments-in-post.*

Attribute for Node Type DOMAIN: The only deciding attribute is *Pagerank score.* In this part we assign the trust in the range of $[0, 0.8]$.

Now on these selected attributes found so far we apply RTSA to find out RTS of AUTHOR,DOCUMENT and DOMAIN node respectively. To make it in the required scale we multiply the calculated RTS value by 0.5, 0.5 and 0.8 respectively. Now for every node we add the RTS and the MTS to calculate Trust Before Propagation (TBP).

RTSA for LOCATION Nodes. For calculating the RTS of LOCATION nodes we use a different algorithm. Every location cluster has an unique GEO-SPEAKER node. In this step, we calculate the RTS as well as the trust before propagation (TBP).

Let, a particular location cluster has m number of AUTHOR nodes associated with it. Let, there are $L_1, L_2 ... L_x$ LOCATION nodes associated with that cluster. Let among those m nodes n number of AUTHOR nodes are trustable (those AUTHOR nodes having $(MTS + RTS) > 0.5$). The RTS of the GEO SPEAKER is calculated as per Eq. 1, where $w = 0.8$ is the scaling factor.

$$RTS(GEO - SPEAKER) = w \times \frac{n}{m} \tag{1}$$

The TBP of GEO-SPEAKER node is calculated as per Eq. 2,

$$\begin{aligned} TBP(GEO - SPEAKER) = & \frac{\sum_{j=1}^{x} MTS(L_j)}{x} \\ & + RTS(GEO - SPEAKER) \end{aligned} \tag{2}$$

Trust Propagation. TBP is propagated across different relationship edges depending on the relations and type of nodes associated. In this step, the trust is propagated in a range of $[0, 0.3]$. Let, N ={AUTHOR,DOMAIN,DOCUMENT}, Z ⊆{USE, CISE, CONTENT_LOCATION_EDGE, PRE, DCSE, DRE}. The Propagated Trust Score Algorithm (PTSA) to calculate propagated trust score (PTS) between nodes of type N via edges Z is shown in Algorithm 3. Here, w is the scaling factor and $sim\_score(E)$ is the similarity value stored in the edge E.

Algorithm 3. Propagated Trust Score Agorithm (PTSA)

INPUT: RSGR with TBP
OUTPUT: RSGR with PTS

for all $X \in N$ **do**
  if there exists any edge $E \in Z$ which connects node $Y \in N$ **then**
    Calculate number of such edges $N(E)$ associated with X and set the value of $sum = 0$
    for all E **do**
      Calculate $sum = sum + sim\_score(E) \times N(E)$
    end for
    Calculate $PTS = sum/N(E) \times w$
  else
    Set $PTS = 0$
  end if
end for

The steps for finding the final trust are:-

From AUTHOR Node to AUTHOR Node. If two Authors are similar either in terms of interest or in terms of user-id and we assign some trust value to one Author then some trust must be propagated to the other Author depending on the strength of their relation. We apply PTSA to calculate PTS. Here, X, Y are AUTHOR. Z is {USE, CISE}, $w = 0.3$ and $sim_score = CISVorUSV$.

From DOMAIN Node to DOCUMENT Node. The reliability of a document greatly depends on the domain where it gets published. It is pretty obvious that a *Wikipedia* document is much more reliable than a *chase.com* document even if they share the same content. We apply PTSA to calculate PTS. Here X is DOCUMENT and Y is DOMAIN, Z is {CONTENT_LOCATION_EDGE (CLE), PRE}, $w = 0.05$ and $sim\_score = 1$.

From LOCATION Node to DOCUMENT Node. Location of a document i.e. from where it has got published highly influences the quality of the document. If the documents of a particular location are proved to be reliable for some period of time then we shall have a tendency to trust the documents which are being published at that location. So, depending on the history of published documents associated with certain region some trust though being very small in magnitude, has to be propagated to the newly published documents those belong to the same location cluster. We apply PTSA to calculate PTS. Here X is DOCUMENT and Y is GEO-SPEAKER, Z is {LSE}, $w = 0.05$ and $sim\_score = 1$.

From TIME Node to DOCUMENT Node. The *date of creation* and *date of registration* plays an important role in determining the trust of a document. We rely a recently published document more than an old one. Yet if a newly registered Author creates a document we should trust the document less to some extent. Because we often come across some situations in which we are obligated to registered in some irrelavent social sites. For example, to view some *dailymotion* videos, sometimes we had to get registered in *lovelyfreindsvisit.com* and post some document. These type of documents are not written with much care because the prime objective uses to be watching the video. So, if such documents are encountered then these documents should not be trusted blindly. Taking into account both the cases an algorithm is developped to propagate trust from TIME node to the respective associated DOCUMENT nodes. For this purpose, we assign a *Grace Trust* ($GT = 0.8$) to every TIME node. The TBP of TIME node is calculated as per Eq. 3,

$$TBP(TIME) = MTS(TIME) + GT(TIME) \tag{3}$$

If the document is 1 year old then the TBP is halved. If it is 2 years old then TBP is further halved. After that, if the *date of registration* of the Author is more than 2 months only then we propagate the TBP. The algorithm for this purpose is described in Algorithm 4, where $w = 0.05$ is the scaling factor. The calculated PTS is propagated to the corresponding DOCUMENT nodes.

Algorithm 4. TRUST PROPAGATION FROM TIME TO DOCUMENT

INPUT: RSGR with TBP
OUTPUT: RSGR with PTS
for all TIME node **do**
  if $|date\ of\ registration\ \text{-}\ present\ date| \geq$ 60 days **then**
    Calculate $i = \frac{|date\ of\ creation - present\ date|}{365\ days}$, $[i \in \mathbb{Z}$ and $i \geq 0]$;
    Calculate $TBP = \frac{TBP}{2^i}$
    Calculate $PTS = TBP \times w$
  end if
end for
AUTHOR

From DOCUMENT Node to DOCUMENT Node. If two documents are similar either in terms of containing concepts and we assign some trust value to one document then some trust must be propagated to the other document depending on the strength of their relation.

We apply PTSA to calculate the propagated trust. Here X and Y is DOCUMENT, Z is DCSE, DRE, $w = 0.05$ and $sim\_score(DRE) = 1$.

From AUTHOR Node to DOCUMENT Node. There are two types of trust propagation. They are as follows,

Based on Common Concepts: If an Author is interested in a certain topic and he/she writes about it then we should rely the document. In other words some trust must be propagated from the author to the document. For each AUTHOR node we check if the AUTHOR node and the associated DOCUMENT node share any CONCEPT nodes. If they share then number of common CONCEPT nodes (say, *concept_intersection_count*) and the number of CONCEPT nodes associated with the DOCUMENT node (say *concept_count*) are calculated. Then we propagate the trust of the AUTHOR node multiplied by a static weight (0.05) and (*concept_intersection_count/concept_count*) to the associated DOCUMENT node. We then add the propagated trust to the existing trust of the DOCUMENT node.

Based on Popularity of Author: If the Author is highly reliable i.e. the trust value of the AUTHOR node is very high then whatever he/she writes, we trust the document to some extent. In other words some trust has to be propagated from that AUTHOR node to the associated DOCUMENT node. For each AUTHOR node we check if the trust is very high (say greater than 0.8). Then we propagate the trust of the AUTHOR node multiplied by a static weight (0.05) to the associated DOCUMENT node.

Finally we add the propagated trust with TBP of the DOCUMENT node and AUTHOR node to calculate Final Trust.

4 Experimental Results

In this section we show various experiments that support the effectiveness and the accuracy of our approach.

4.1 Experimental Setup

Data Sets. We use real life text documents especially from facebook.com, twitter.com, tripadvisor.com, amazon.com, youtube.com, dailymotion.com, hotel.com. It consists of 60,370 unsupervised documents. After the feature extraction step the resulting graph database consists of 8,01,350 nodes. For the purpose of experimental analysis we take a random sample of 600 documents consisting of 150 documents from each category i.e. Boards, Microblogs, Reviews and Videos. To test the accuracy of calculated trust, we adopt a user labeled dataset. In this dataset, those selected 600 documents are assigned 2 trust values one for the Author (i.e. AUTHOR node) and other for the document itself (i.e. DOCUMENT node) by manually looking into the actual document and looking into the Authors' profiles. Assignment of trust values is done by 10 unbiased persons. The trust value assigned has range [0, 1.0] and has precision up to one decimal place. These persons are not specialist in any domain and as the data used also do not belong to any specific category so we assume the labeling as the ground truth.

Method. At first we try to find out what amount of trust is assigned to corresponding AUTHOR and DOCUMENT node for individual documents at every steps of our trust mining approach. Then we show the trust distribution of AUTHOR and DOCUMENT node obtained at the end of each step of the discussed trust mining approach. Then we show the gradual improvement of the trust value assigned to the corresponding AUTHOR and DOCUMENT node through the steps of our algorithm. Finally we show the accuracy of our algorithm.

Let for a particular document i and for a particular node type j, A_{ij} represents the trust value assigned by user and B_{ij} represents the trust value assigned by our algorithm. Then the accuracy is calculated as per Eq. 4.

$$Accuracy_{ij} = 1 - \frac{|A_{ij} - B_{ij}|}{A_{ij}} \tag{4}$$

Let, N is the total number of nodes of type j. The overall accuracy for node type j is calculated as per Eq. 5.

$$Overall\ Accuracy_j = \frac{\sum_{i=1}^{N} Accuracy_{ij}}{N} \tag{5}$$

4.2 Trust Assigned at Every Step

The trust assigned only due to *MTSA, RTSA, PTSA* and the final trust are shown in Fig. 4(a) and Fig. 4(b) for AUTHOR and DOCUMENT nodes

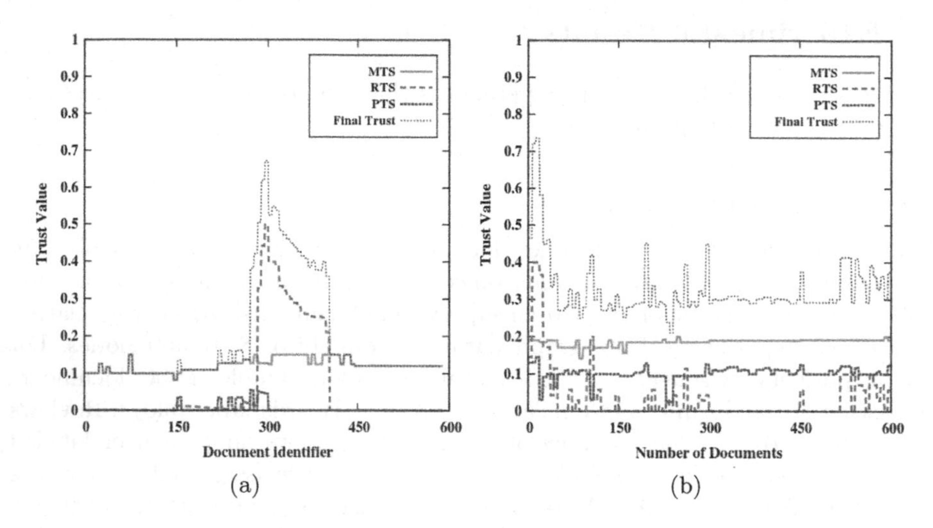

Fig. 4. Trust assigned at every step of our algorithm for (a) AUTHOR node; (b) DOCUMENT node

respectively. The documents are sorted according to their respective types. Indices from (1-150) represent Boards, (151-300) represent Reviews, (301-450) represent Microblogs and (451-600) represent Videos.

In Fig. 4(a) for AUTHOR node we get low MTS due to absence of some attributes.For Microblogs we have significant RTS which are mainly twitter documents because we have mined information about the Author using twitter api. For some Review documents we get significant RTS which are mainly trip-advisor documents. For other Review documents we get default RTS as 0.2. Since for document types Boards and Videos we have much less information about the Author, we do not get any RTS in most of the cases. Still for some Boards documents we get some RTS which are mainly facebook documents as we have used facebook api. We have significant PTS via CISE in case of Reviews because for other type of documents *Author description* field is generally empty. For AUTHOR node we have the most trustworthy Author (Trust value = 0.68) for document type Microblogs. In Fig. 4(b) for DOCUMENT node we get significant MTS. We see that except for Microblogs all documents are assigned significant RTS. Microblogs are small documents so the number of paragraphs, number of external links etc are negligible compared to other documents. On the other hand for some of the document of type Boards, we have got RTS upto 0.4 as for those documents they are well written with paragraphs, having large number of comments etc. For most of the documents we get significant PTS. For some Boards and Reviews documents we get very low PTS because for those documents the corresponding DOMAIN node has very low Final Trust and they do not have much common concepts with the other documents. Here we have the most reliable document (Final Trust = 0.72) for document type Boards.

4.3 Trust Distribution at the End of Every Step

The trust distribution of documents are shown in Fig. 5(a) and Fig. 5(b) for AUTHOR and DOCUMENT nodes respectively. In Fig. 5(a) for AUTHOR node we see that after applying MTSA most of them have trust less than 0.2. After applying RTSA some of the AUTHOR nodes are assigned more trust and few of them get trust value 0.5 and 0.6. After applying PTSA trust of some AUTHOR nodes are further increased.

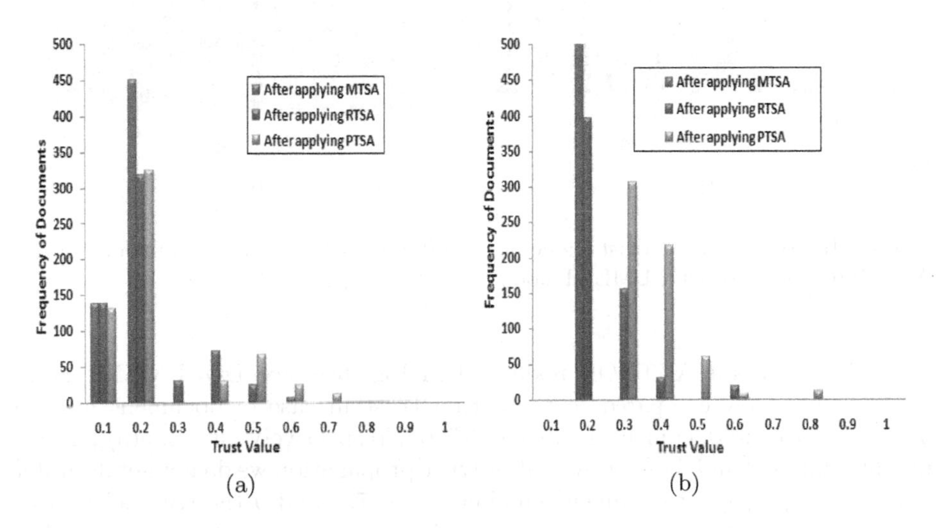

Fig. 5. Trust distribution for (a) AUTHOR node; (b) DOCUMENT node

In Fig. 5(b) for DOCUMENT node we see that after applying MTSA we have got trust almost equal to 0.2 for every document. After applying RTSA and PTSA we notice significant change in the trust distribution. If we are given individual documents in isolated manner we can find trust only based on the presence of metadata which would be very low. But if we are given all documents at the same time then we can calculate trust based on the richness of metadata as well as we can calculate trust of documents by comparing individual documents with the other documents. The low MTS for AUTHOR and DOCUMENT and relatively improved trust after applying RTSA and PTSA supports this fact.

4.4 Improvement of Trust Value

We take a random sample of 10 documents to show the improvement of the trust value assigned to AUTHOR and DOCUMENT node after end of each step of our algorithm as shown in Fig. 6(a) and Fig. 6(b) respectively. We compare these trust values with the trust based on user perception or actual trust.

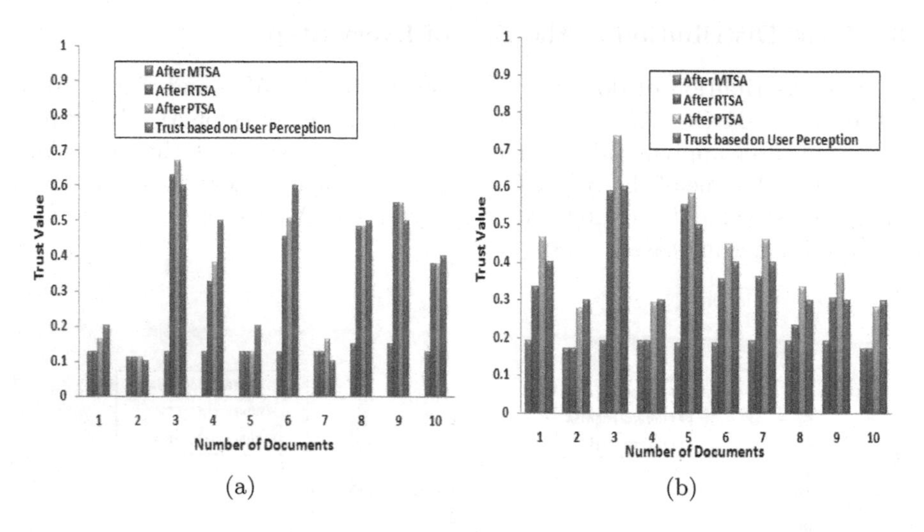

Fig. 6. Improvement of trust value for randomly selected 10 documents for (a) AUTHOR node; (b) DOCUMENT node

In Fig. 6(a) for AUTHOR node and in Fig. 6(b) for DOCUMENT node, MTS is very low compared to the actual trust in case of document number $3, 4, 6, 8, 9, 10$ and document number $1, 3, 5, 6, 7$ respectively. In Fig. 6(a) for document number 4 and 6 we see even after trust propagation we do not get desirable trust. In Fig. 6(b) for document number $1, 5, 6, 7, 8$ and 9 the trust after trust propagation exceeds the actual trust not with a large margin but in case of document number 3 we encounter some error. The possible reason for this is we are either increasing trust or leaving as it is. We are not decresing trust at any point because the social media documents are intrinsically not much reliable. The data we are dealing with are not much rich in terms of metadata. If we deal with some data which contain well structured attributes and highly rich metadata then we could implement the concept of decreasing trust. But overall we can clearly see that the difference between the user percepted trust and the calculated trust decreases after every step of our trust mining algorithm. This supports the effectiveness of our approach.

4.5 Accuracy

The comparison of the actual trust value with the trust value that we have calculated are shown in Fig. 7(a) and Fig. 7(b) for AUTHOR and DOCUMENT nodes respectively. We see that the calculated trust values more or less follow the actual trust values.

In Fig. 7(a) the *calculated mean trust value* is 0.19 and user defined or *actual mean trust value* is 0.16. In Fig. 7(b) the *calculated mean trust value* is 0.32 and user defined or *actual mean trust value* is 0.29.

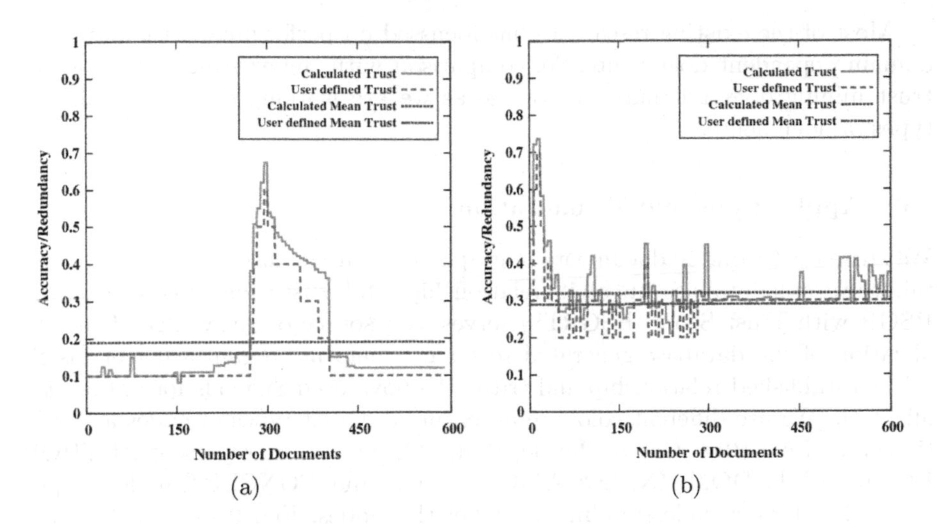

Fig. 7. Comparison of calculated trust with the actual trust for (a) AUTHOR node; (b) DOCUMENT node

In Fig. 8(a) and Fig. 8(b), we can see that the *overall accuracy* for AUTHOR node is calculated to be 0.82 and for DOCUMENT node the *overall accuracy* is calculated to be 0.88 which is quite good. The possible reason for the decrease in accuracy could be we calculate trust value having continuous value and the user percepted trust value has precision of only single decimal place.

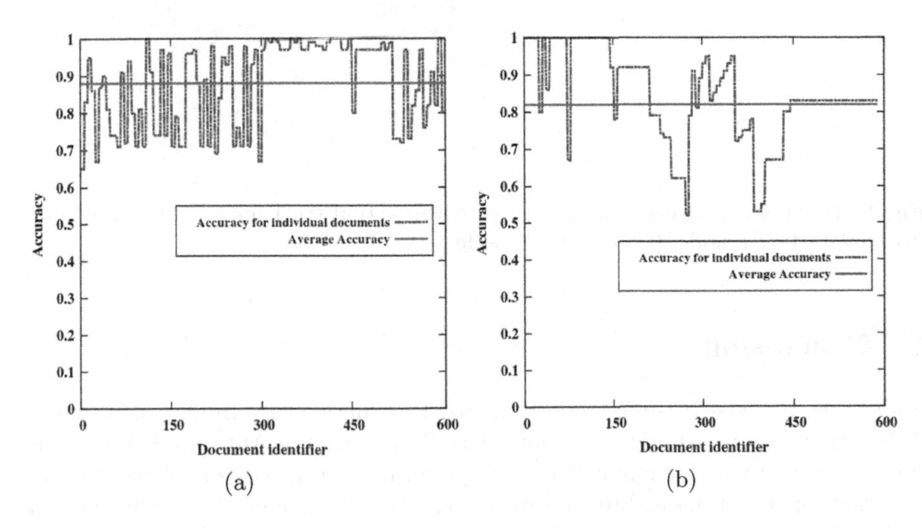

Fig. 8. Accuracy of calculated trust for individual documents for (a) AUTHOR node; (b) DOCUMENT node

Most of the existing researches has focussed on performing trust mining for domain dependent documents. So, comparison with the existing framework for trust mining does not make any sense as we are working on totally different types of documents.

4.6 Applications and Visualization

We store the incoming documents in graph structure as discussed and we maintain the graph database and all its relationships and trust value. So the generated RSGR with Trust Score (RSGRTS) serves as a source of knowledge. The visualization of the database generated so far can help us to gain a better insight of the established relationship and trust. We have used SigmaJs for graph visualization. We use different colors representing different types of nodes and relationships. Red, Blue, Green, Purple, Pink , Orange colors represent AUTHOR, DOCUMENT, DOMAIN, LOCATION, TIME and CONCEPT nodes respectively. For better understanding we label the nodes. Fig. 9(a), Fig. 9(b) and Fig. 9(c) represents a reliable Document, a trustworthy Author and a reputed Domain respectively. The trust value assigned at every step are also shown for those corresponding nodes.

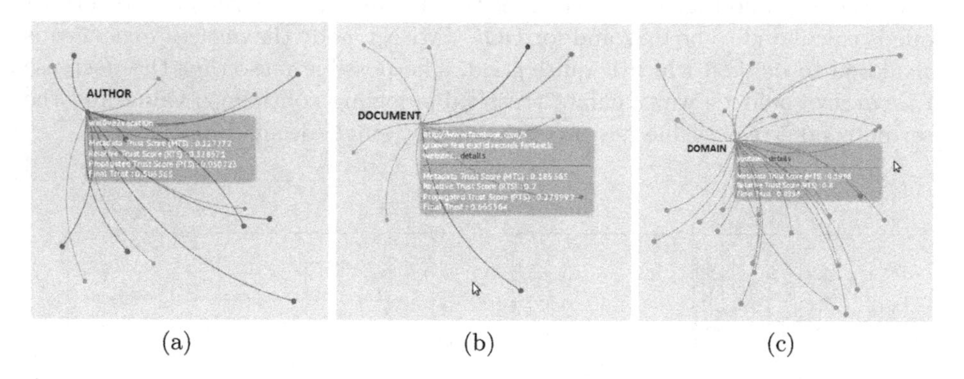

(a) (b) (c)

Fig. 9. Comparison of calculated trust with the actual trust for (a) AUTHOR node; (b) DOCUMENT node, (C) DOMAIN node

5 Conclusion

Finally we propose a three stage framework to mine trust from real life online documents step by step until we get trust distribution of Authors, Domains and the Content of the document itself. We generalized the concept of trust mining and developed an algorithm in which trust based on metadata, relative trust and propagated trust are considered to assign trust values. The experimental results show the effectiveness of our framework. We have used documents which are not very rich in terms of metadata overall. So we have used various api to mine additional information from respective domains. It is also the reason that we are not interested in penalizing trust. Our algorithm works fine even if there is large variation of the richness of metadata of the incoming documents.

Interesting problems related to the approach could be how to include the concept of different types of centrality measures in quantifying trust. Another problem could be how to incorporate the concept of event detection in trust mining. There is a huge amount of unstructured data available in the online social media. In future, we shall use segmentation analysis, opinion mining and sentimental analysis to conduct more accurate trust mining. Due to availability of large amount of social media data we have very good platforms to carry out our further work.

References

1. Golbeck, J., Parsia, B., Hendler, J.: Trust networks on the semantic web. In: Klusch, M., Omicini, A., Ossowski, S., Laamanen, H. (eds.) CIA 2003. LNCS (LNAI), vol. 2782, pp. 238–249. Springer, Heidelberg (2003)
2. Guha, R., Kumar, R., Raghavan, P., Tomkins, A.: Propagation of trust and distrust. In: Proceedings of the 13th International Conference on World Wide Web, NY, USA, 2004, pp. 403–412
3. Bizer, C., Oldakowski, R.: Using context- and content-based trust policies on the semantic web. In: Proceedings of the 13th International Conference on World Wide Web, NY, USA, pp. 228–229 (2004)
4. Agichtein, E., Castillo, C., Donato, D., Gionis, A., Mishne, G.: Finding high-quality content in social media. In: Proceedings of the International Conference on Web Search and Web Mining, pp. 183–194 (2008)
5. Blumenstock, J.: Size matters, word count as a measure of quality on Wikipedia. In: Proceedings of the 17th International Conference on World Wide Web, pp. 1095–1096. ACM, New York (2008)
6. Zolfaghar, K., Aghaie, A.: Mining trust and distrust relationships in social web applications. In: Proceedings of the 2010 IEEE International Conference on Intelligent Computer Communication and Processing, Cluj-Napoca, Romania, August 26-28, 2010, pp. 73–80 (2010)
7. McGuinness, D., Zeng, H., Da Silva, P., Ding, L., Narayanan, D., Bhaowal, M: Investigations into trust for collaborative information repositories: a Wikipedia case study. In: Proceedings of the Workshop on Models of Trust for the Web 2006, pp. 3–131 (2006)
8. Nguyen, V.A., Lim, E.P., Jiang, J., Sun, A.: To trust or not to trust? Predicting online trusts using trust antecedent framework. In: Proceedings of the 9th IEEE International Conference on Data Mining, Miami, Florida, USA, pp. 896–901 (2009)
9. Chua, F.C.T., Lim, E.P.: Trust network inference for online rating data using generative models. In: Proceedings of the 16th ACM SIGKDD Conference on Knowledge Discovery and Data Mining, Washington, DC, USA, July 25–28, 2010, pp. 889–898 (2010)
10. Matsuo, Y., Yamamoto, H.: Community gravity: measuring bidirectional effects by trust and rating on online social networks. In: Proceedings of the 18th International Conference on World Wide Web, Madrid, Spain, April 20–24, 2009, pp. 751–760 (2009)

11. Zhang, Y., Yu, T.: Mining trust relationships from online social networks. Journal of Computer Science and Technology **27**(3), 492–505 (2012)
12. Wu, Z., Palmer, M.: Verb semantics and lexical selection. In: Proceedings of the 32nd Annual Meeting of the Associations for Computational Linguistics, pp. 133–138 (1994)
13. Moturu, S.T., Liu, H.: Evaluating the trustworthiness of Wikipedia articles through quality and credibility. In: Proceedings of the 5th International Symposium on Wikis and Open Collaboration, NY (2009)
14. Sharma, T., Toshniwal, D.: A generalized relationship mining method for social media text data. In: Perner, P. (ed.) MLDM 2014. LNCS, vol. 8556, pp. 376–392. Springer, Heidelberg (2014)
15. Deza, M.M., Deza, E.: Encyclopedia of Distances. Springer, Heidelberg (2009)
16. Porter, M.F.: An algorithm for suffix stripping. Program: Electronic Library and Information Systems **14**(3), 130–137 (1980)
17. Zhang, J., Votava, P., Lee, T.J., Adhikarla, S., Kulkumjon, I.C., Schlau, M., Natesan, D., Nemani, R.: A technique of analyzing trust relationships to facilitate scientific service discovery and recommendation. In: 2013 IEEE International Conference on Services Computing (SCC), pp. 57–64. IEEE (2013)
18. Biryukov, M.: Co-author network analysis in DBLP: classifying personal names. In: Thi, H.A.L., Bouvry, P., Dinh, T.P. (eds.) Modelling, Computation and Optimization in Information Systems and Management Sciences. CCIS, vol. 14, pp. 399–408. Springer, Heidelberg (2008)
19. Huang, T.-H., Huang, M.L.: Analysis and visualization of co-authorship networks for understanding academic collaboration and knowledge domain of individual researchers. In: 2006 International Conference on Computer Graphics, Imaging and Visualisation, pp. 18–23. IEEE (2006)
20. Biryukov, M., Dong, C.: Analysis of computer science communities based on DBLP. In: Lalmas, M., Jose, J., Rauber, A., Sebastiani, F., Frommholz, I. (eds.) ECDL 2010. LNCS, vol. 6273, pp. 228–235. Springer, Heidelberg (2010)
21. Reitz, F., Hoffmann, O.: An analysis of the evolving coverage of computer science sub-fields in the DBLP digital library. In: Lalmas, M., Jose, J., Rauber, A., Sebastiani, F., Frommholz, I. (eds.) ECDL 2010. LNCS, vol. 6273, pp. 216–227. Springer, Heidelberg (2010)
22. Pang, B., Lee, L., Vaithyanathan, S.: Thumbs up?: sentiment classification using machine learning techniques. In: Proceedings of the ACL-02 Conference on Empirical Methods in Natural Language Processing, vol. 10, pp. 79–86. Association for Computational Linguistics (2002)
23. Zeng, Y., Yao, Y., Zhong, N.: DBLP-SSE: a DBLP search support engine. In: IEEE/WIC/ACM International Joint Conferences on Web Intelligence and Intelligent Agent Technologies, WI-IAT 2009, vol. 1, pp. 626–630. IET (2009)
24. Sun, Y., Han, J., Zhao, P., Yin, Z., Cheng, H., Wu, T.: Rankclus: integrating clustering with ranking for heterogeneous information network analysis. In: Proceedings of the 12th International Conference on Extending Database Technology: Advances in Database Technology, pp. 565–576. ACM (2009)
25. Taboada, M., Brooke, J., Tofiloski, M., Voll, K., Stede, M.: Lexicon-based methods for sentiment analysis. Computational Linguistics **37**(2), 267–307 (2011)
26. Lawrence, S.: Free online availability substantially increases a paper's impact. Nature **411**(6837), 521–521 (2001)
27. Zhang, X., Cui, L., Wang, Y.: Commtrust: computing multi-dimensional trust by mining e-commerce feedback comments. IEEE Transactions on Knowledge and Data Engineering **26**(7), 1631–1643 (2014)

28. Emayakumaari, T., Ananthi, G.: Mining E-commerce feedback comments for trust evaluation. In: 2015 IEEE International Conference on Engineering and Technology (ICETECH), pp. 1–5. IEEE (2015)
29. Yean, C.J., Yee, T.C., Tan, I.K.T.: Relative trust management model for Twitter: an analytic hierarchy process approach. In: International Conference on Frontiers of Communications, Networks and Applications (ICFCNA 2014-Malaysia), pp. 1–6. IET (2014)
30. https://developers.facebook.com/docs/reference/apis/
31. https://dev.twitter.com/docs/streaming-apis
32. https://www.quantcast.com/top-sites

A Comprehensive Analysis: Automated Ovarian Tissue Detection Using Type P63 Pathology Color Images

T.M. Shahriar Sazzad[1(✉)], L.J. Armstrong[1], and A.K. Tripathy[1,2]

[1] Edith Cowan University, Joondalup, WA, Australia
{t.sazzad,l.armstrong}@ecu.edu.au
[2] Don Bosco Institute of Technology, Mumbai, India
amiya@dbit.in

Abstract. Manual microscopic ovarian reproductive tissue analysis is a general routine examination process in the laboratory. This process requires longer processing time and prone to errors. Among all existing scanning devices ultrasound is commonly used but not optimal as it process grayscale images which do not provide satisfactory results. Computer based approaches could be a viable option as it can minimize the labor cost, effort and time. Additionally smaller tissues can be easily analyzed. In this paper a comprehensive analysis has been carried out and a new modified approach has been presented using type P63 histopathology ovarian tissues color images with different magnifications. Comparison of various existing automated approaches with manual identification results by experts indicates excellent performance of the proposed automated approach.

Keywords: Histopathology · Color digitized microscopic image · Image artifacts · Mean shift · Region fusion · Cluster · Ovarian reproductive tissues

1 Introduction

Modern digital scanners including ultrasound scanner, MRI (magnetic resonance imaging), CT (computerized tomography), PET (position emission tomography) are commonly used diagnosis modalities to analyze tissues in the pathology laboratory [1]. These modern digital scanners can quickly perform laboratory routine task and requires less human effort. Ultrasound is considered as commonly used option among all modalities in the pathology laboratory as it is cheaper, portable and less risky to patients [1]. However; it can only process grayscale images and suitable for large tissue analysis such as cancer tissues [2]. Smaller tissues are relatively hard to analyze using ultrasound scanners [2]. Ovarian tissues are smaller in size in compare to cancer tissues [1], [9]. Therefore; pathology experts prefer manual microscopic approach [9].

Research work of [3] mentioned that manual microscopic biopsy slide analysis is considered as “gold standard” for human ovarian tissue analysis. However; this method requires longer processing time and has observation variability issues [4]. Computer based approach could be more viable to overcome the issues associated with manual microscopic approach [4, 5].

Existing research study indicates that most computerized approaches are semi-automated rather automated and most analysis have been carried out on ovarian

P. Perner (Ed.): MLDM 2016, LNAI 9729, pp. 714–727, 2016.
DOI: 10.1007/978-3-319-41920-6_54

cancer cells or tumor cells rather than ovarian reproductive tissues [6]. Cancer cells or tumor cells analysis approaches are not suitable due to the fact that they have different shape, size and color in compare to ovarian reproductive tissue [6].

At present, only a few research works have been carried out on ovarian reproductive tissue analysis [2], [7]. Among all existing research works mostly are based on animal tissue analysis such as pigs and rats tissue analysis [7, 8]. Animal tissue structures and colors are different from human ovarian tissues for which existing animal ovarian tissue analysis approaches are not practically suitable approaches for human ovarian tissue analysis [7]. Therefore; it is essential to propose a suitable computerized approach for human ovarian tissue analysis.

In the pathology various types of biopsy slides with different magnifications are used [9]. Additionally, different methods and color substances are also used to prepare the biopsy slides [10]. This is due to the fact that typically all tissues inside the ovary are colorless and to distinguish different tissues color substances play an important role as color substances are mainly used to distinguish between tissues [10]. Different types of magnifications are used to enlarge the smaller tissues in bigger size for easy identification [9].

Among all existing available microscopic biopsy slides type Haematoxylin and Eosin (H&E) are commonly used as they are easy to prepare [8]. The limitation of H&E include intensity variation issues [4] which cause faulty identification [11, 12].

To minimize the issues associated with H&E [8] mentioned that type PCNA is more suitable as it has less intensity issues. Some research works [7, 8] already been carried out using type PCNA but are mainly animal tissues detection. Only one study to date on human ovarian tissues carried out by [3]. Type P63 could be a better choice in compare to type H&E and type PCNA suggested by [3].

Research study of [9], [13] worked with type P63 digitized color images where non-counter stained (tissue stained without eosin or haematoxylin) [9] and counter-stained (tissue stained with eosin or haematoxylin) images [13] were considered. Both approaches indicate an acceptable accuracy in comparison to laboratory manual microscopic identified result for 100x magnifications.

There are various types of magnifications used in the pathology laboratory which include (2x, 10x, 20x, 40x, 100x, 200x, 400x and 1000x). An example of non-counter and counter stained (100x and 200x magnification) digitized images acquired from P63 biopsy slides is shown in Figure 1. Figure 2 indicates the marked annotated images of Figure 1.

2 Related Work

Most related ovarian tissue analysis approaches [2], [7], [8] are based on cancer or tumor cell analysis and are mainly semi-automated. Only a limited number of automated research works have been carried out using human ovaries which include automated approach using type PCNA by [3], using type P63 non-counter stained by [9] and using type P63 counter stained by [13].

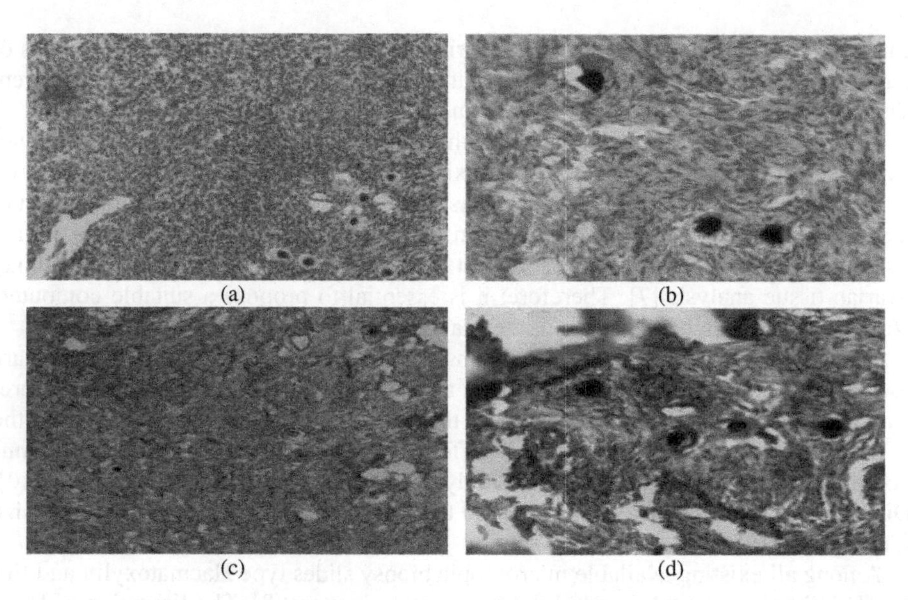

Fig. 1. (a) and (b) are research test type P63 (100x and 200x magnification) counter-stained color digitized ovarian images. (c) and (d) are research test type P63 (100x and 200x magnification) non-counter-stained color digitized ovarian images.

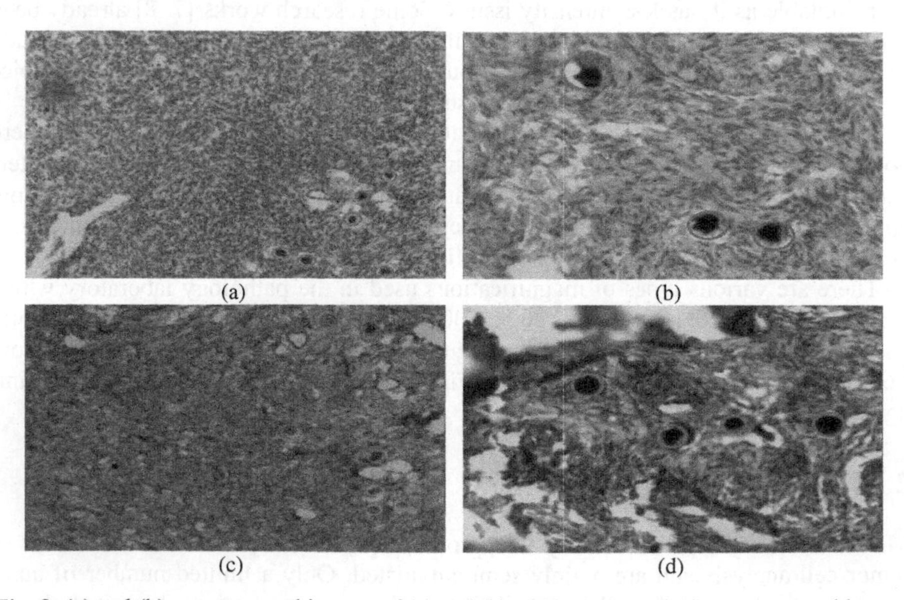

Fig. 2. (a) and (b) are annotated images of (a) and (b) of Fig 1. (c) and (d) are annotated images of (c) and (d) of Fig 1. Red circles identified ovarian nucleus marked by 2 experts and blue circles were marked by at-least 1 expert in the laboratory.

All existing semi-automated approaches related to animal ovarian cancer or tumor tissues analysis do not qualify for human ovarian tissue analysis as they differ from human tissues [9]. Automated research study by [3] incorporated conventional threshold based approach where for a new batch of images calibration of processing parameters are essential. An exception is the works of [9], [13] which do not require any human intervention.

Above mentioned all existing automated approaches related to human ovarian reproductive tissues [3], [9], [13] have considered 100x magnifications. Magnification is an important factor due to the fact that magnification is related to resolution and contrast [14, 15] and can cause intensity variations. Additionally, color chemicals used during slide preparation may also cause intensity variations [9]. Color chemicals are mainly used to assists experts to analyze tissues easily using microscopes but for computerized image processing techniques color variations creates processing difficulties [10] as it can lead to a faulty segmentation result and finally inaccurate identification [14]. Therefore; it is important to minimize intensity variations [15].

At present there are no such existing ovarian tissue detection approaches [2], [3], [8], [16], [17] except [9], [13] that have considered intensity variation issues. Most works have considered [2], [3], [8], [16], [17] grayscale image segmentation approaches for which image artifact issues were ignored because for a grayscale image processing correcting color variations is not necessary [10].

3 Proposed Method

This research study has considered the research work of [9], [13] and proposed a new modified approach to work with both types P63 digitized color images with 2 different magnifications (100x, 200x). Detailed working flowchart of this proposed research study is shown in Figure 3.

3.1 Correcting Image Artifacts

Previous research study [9], [13] mentioned that Gaussian low-pass filter is not a viable option for this process as the results are not satisfactory. Instead of Gaussian low-pass filter morphological operation is a more viable approach mentioned by [9], [13] where cell diameter was used as a morphological processing parameter to provide satisfactory results. Modification was made in this research study where calculated accurate cell diameter and image magnification were incorporated for morphological processing. Previous studies [9], [13] have considered approximate cell diameter collected from pathology experts but in this study region area was used to calculate the accurate cell diameter. As for example according to expert information for 100x magnification ovarian reproductive tissue region area ranges in between 700 – 1250 pixels. If we consider minimum region then the calculation can be done using the following:

i) A = 700 where A indicates regions area in pixel ;

ii) $(pi).r^2 = 700$;

iii) r = sqrt(700/pi) = 14.927 and

iv) D = 2 * r = 2 * 14.927 = 29.85

Although; this is a small difference between accurate values (i.e. 29.85) and expert given values (i.e. 30) but it changes the result shown in Figure 4. The modified approach provides satisfactory result which is almost similar to the original image shown in Figure 4. Figure 5 indicates the corrected images for this research study test images.

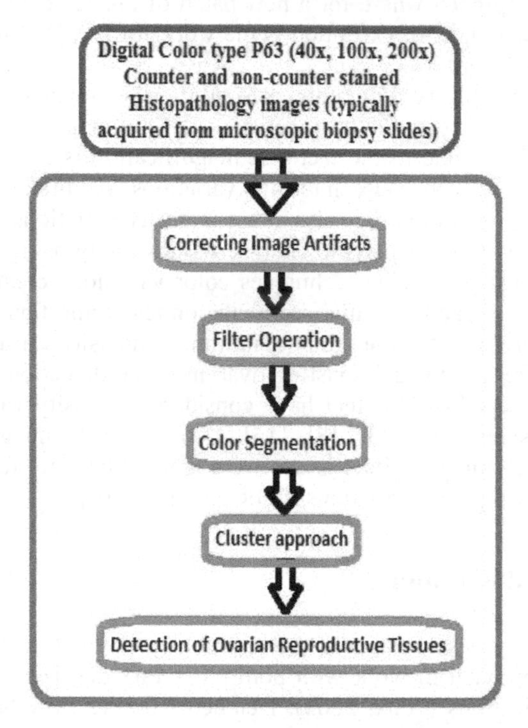

Fig. 3. Block diagram for automated identification process

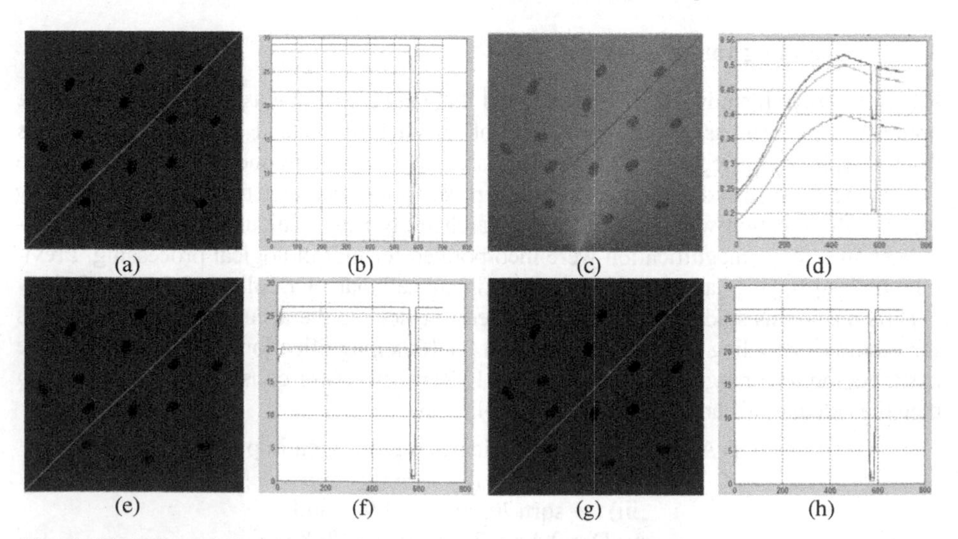

Fig. 4. (a) is the original image used from [9], (b) indicates intensity graph across the red line for image (a), (c) indicates artifact issues and (d) indicates intensity graph for image (c). (e) indicates corrected image proposed by [9] with intensity graph (f). (g) indicates corrected image using this research study proposed modified method and (h) indicates the intensity graph.

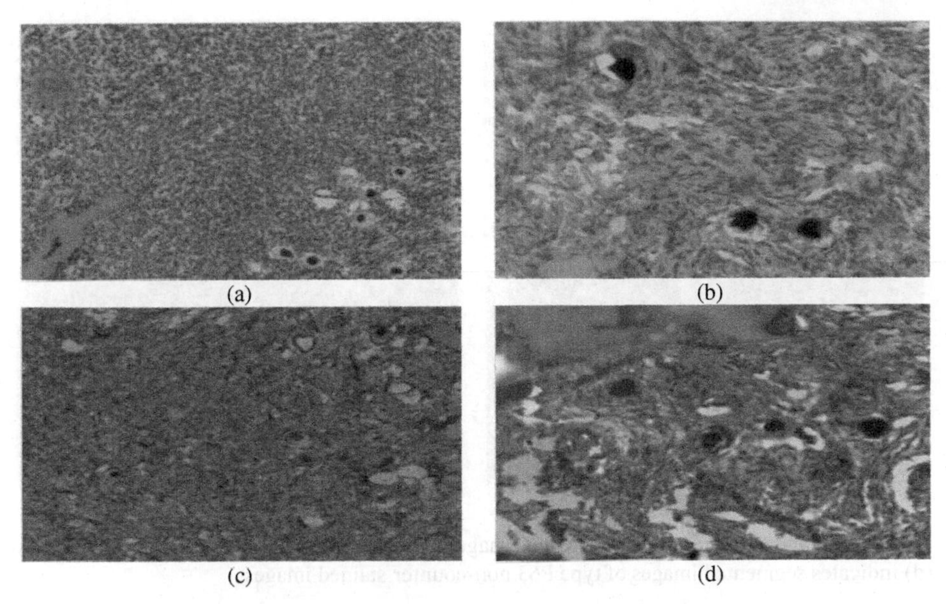

Fig. 5. (a), (b), (c) and (d) indicates corrected images using this study proposed modified morphological operation for correcting image artifacts.

3.2 Filter Operation

Modified pixel based mean shift filter was proposed by [9] has shown to provide suitable result in comparison to median or mean filter approaches. This study has considered the approach of [9] for this research study test images. Filtered results for this research study test images are shown in Figure 6.

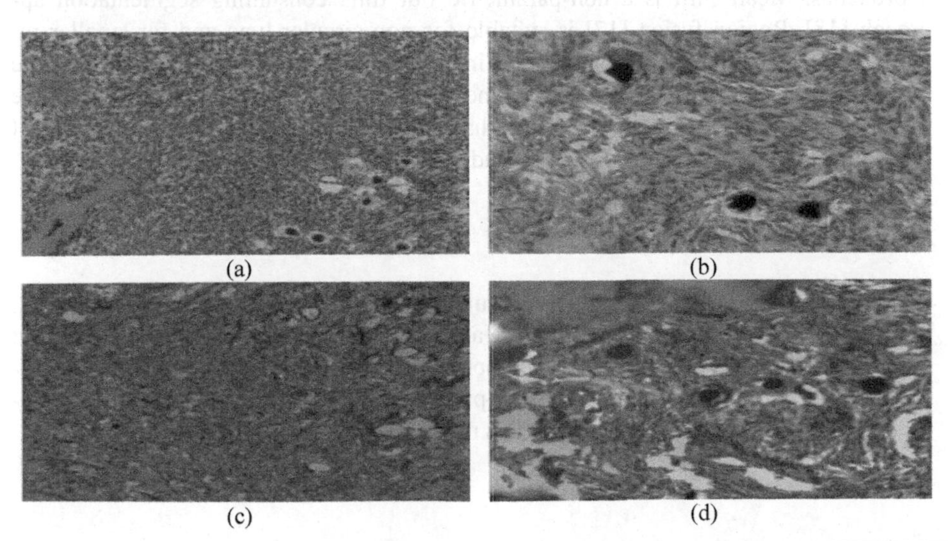

Fig. 6. (a), (b), (c) and (d) indicates filtered images using [9].

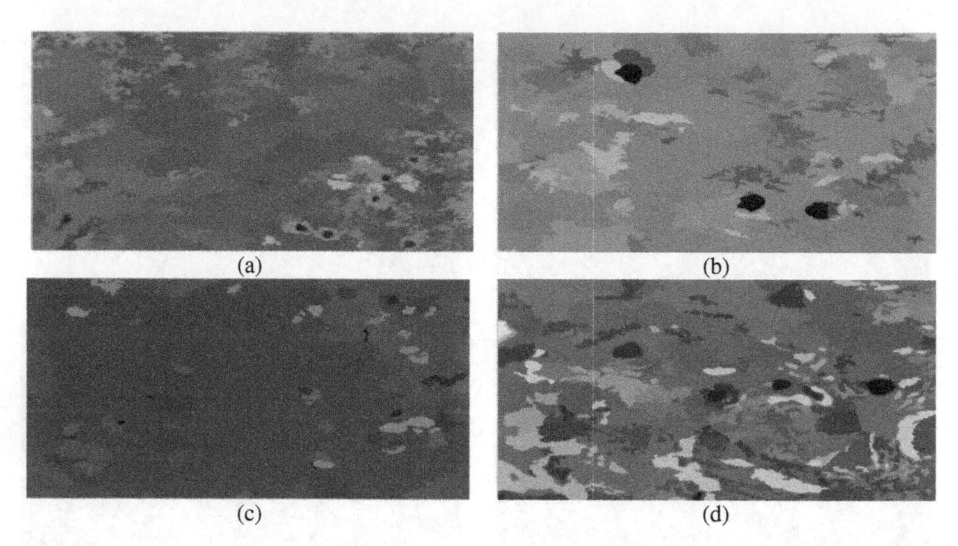

Fig. 7. (a) and (b) indicates segmented images of type P63 counter stained and (c) and (d) indicates segmented images of type P63 non-counter stained images.

3.3 Color Segmentation

Most related research works for ovarian tissue analysis [2], [3], [8], [16], [17], [18] have considered conventional threshold based and watershed based approaches as gray-scale images were processed during segmentation. Threshold based and watershed based approaches are not suitable for color image segmentation [10].

Existing work of [9] mentioned that there are two suitable color based segmentation approaches available which include mean shift [16] and region fusion [17] based approaches. Mean shift is a non-parametric but time consuming segmentation approach [18]. Region fusion [17] is suitable for non-circular type and for smaller regions as it needs a processing parameter in between 0-10. Study of [9] modified the regions fusion approach [17] to overcome the limitations which shown to provide satisfactory results. This research study has incorporated the modified approach of [9]. Sample test results of this research study test images are shown in Figure 7.

3.4 Cluster Approach

For color image processing perspective a suitable clustering approach is necessary [19]. Study of [9] has proposed a modified mean-shift cluster based approach of [20] and research work of [13] incorporated the approach of [9] as it provides satisfactory results. This research study incorporated the proposed approach of [9] for the study images. Figure 8 indicates cluster results for this research study test segmented images.

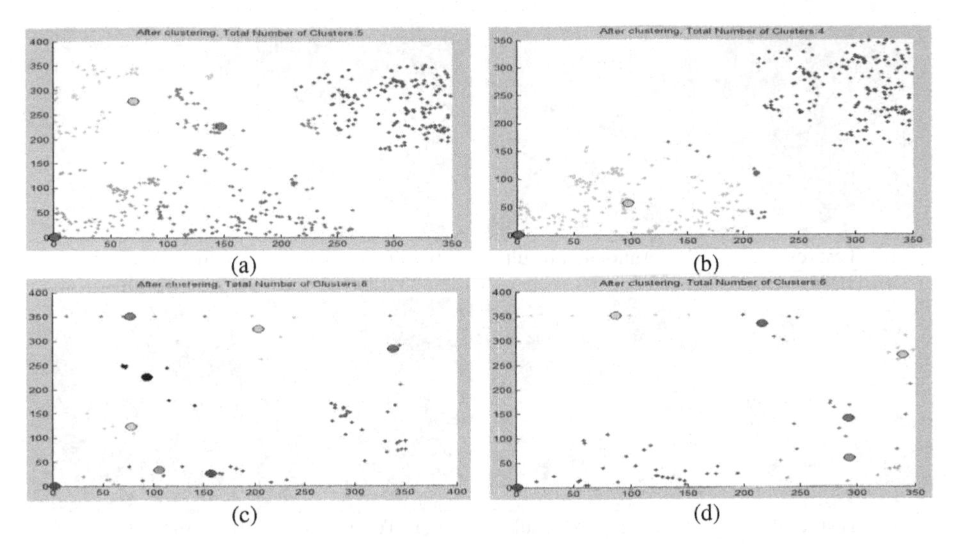

Fig. 8. (a) and (b) indicates the clustered number for (a) and (b) of Fig. 7. (c), (d) indicates clustered number for (c) and (d) of Fig. 7.

3.5 Identify Ovarian Nucleus

Use of important features is a vital part to identify the regions of interest [21]. According to [21] in digital image processing features can be categorized as object level features and spatial features. For ovarian reproductive tissue analysis shape, size and color are considered as most important features and almost all existing computerized approaches [2], [3], [8], [9], [13], [16], [17], [18] have performed ROI identification using shape, size and color features. For this research study most suitable features were incorporated which include intensity, area, major axis length, minor axis length, circularity, solidity and diameter to identify ovarian reproductive tissue for both type P63 and for 2 different magnifications (100x and 200x). In addition, border touching regions were eliminated in this research study identification process due to the fact that expert do not consider border touching regions as ROIs mentioned by [9]. Most viable algorithm to measure intensity is the work of [9] which was shown to provide satisfactory results has been incorporated in this research study. Figure 9 indicates the identified results of this research study proposed method.

3.6 Classify Identified Regions to Improve Accuracy

Some false regions were identified during identification process. To find out the accuracy rate it is necessary to use an appropriate classification technique. Unfortunately none of the existing approaches [2], [3], [8], [9], [16], [17], [18] have considered and incorporated any classification approach at all.

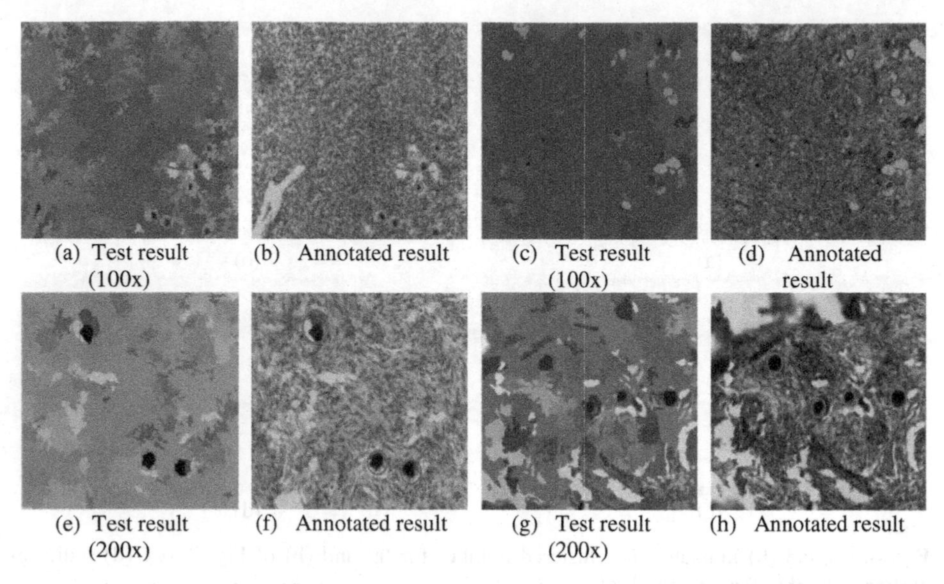

Fig. 9. (a) and (e) are identified test result for type P63 counter stained images. (b) and (f) are annotated images for (a) and (e). No regions missed for any magnifications. (c) and (g) are identified test result for type P63 non-counter stained images. No regions missed for any magnifications. No false regions identified for any magnifications and for any types of images

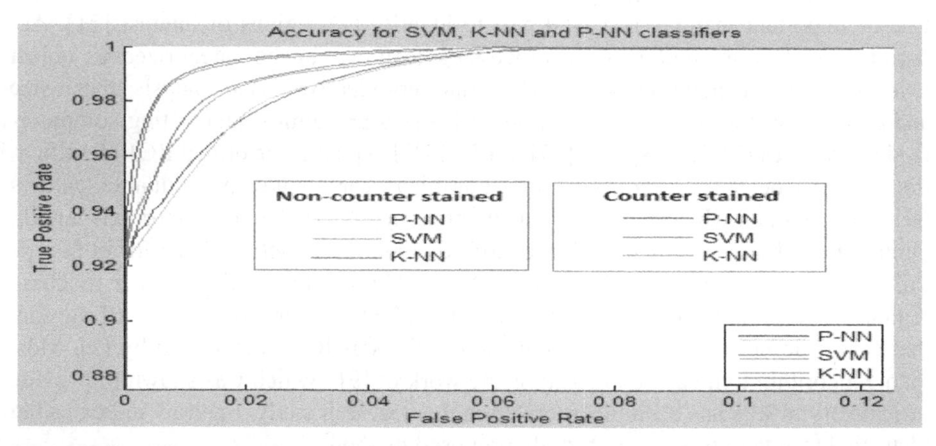

Fig. 10. Classification results of three different classifiers for both types P63 counter and non-counter stained images based on accuracy (100 x magnifications)

There are many classification approaches available at present but for medical image processing perspective SVM (support vector machine), K-NN and P-NN are most popular mentioned by [13]. This research study has considered all the above mentioned 3 classifiers (default value for each classifier) to classify the regions to increase the accuracy rate. For each magnification and for each type (counter and non-counter stained) several images were tested. Each test set were divided in 3 different groups

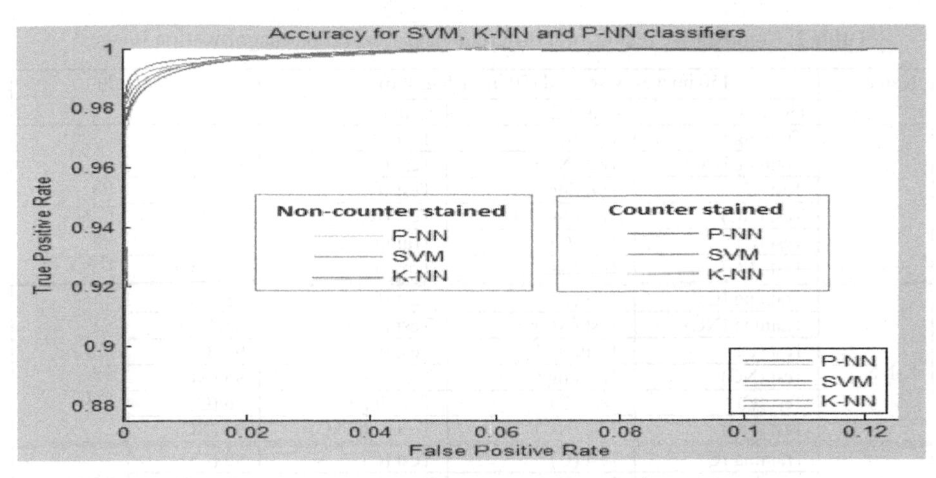

Fig. 11. Classification results of three different classifiers for both types P63 counter and non-counter stained images based on accuracy (200 x magnifications)

Table 1. Comparative results of 3 different classifiers (100x magnification images)

Name	150 images were used (50 in each group)			Accuracy %	
	Group 1	Group 2	Group 3		
SVM	Training (C)	Test (C)	Test (C)	96 (C)	96 (C), 95 (NC)
	Training (NC)	Test (NC)	Test (NC)	94 (NC)	
	Test (C)	Training (C)	Test (C)	97 (C)	
	Test (NC)	Training (NC)	Test (NC)	96 (NC)	
	Test (C)	Test (C)	Training (C)	95 (C)	
	Test (NC)	Test (NC)	Training (NC)	95 (NC)	
K-NN	Training (C)	Test (C)	Test (C)	94 (C)	94 (C), 92 (NC)
	Training (NC)	Test (NC)	Test (NC)	93(NC)	
	Test (C)	Training (C)	Test (C)	94 (C)	
	Test (NC)	Training (NC)	Test (NC)	91 (NC)	
	Test (C)	Test (C)	Training (C)	94 (C)	
	Test (NC)	Test (NC)	Training (NC)	92 (NC)	
P-NN	Training (C)	Test (C)	Test (C)	92 (C)	93 (C), 91 (NC)
	Training (NC)	Test (NC)	Test (NC)	91 (NC)	
	Test (C)	Training (C)	Test (C)	93 (C)	
	Test (NC)	Training (NC)	Test (NC)	92 (NC)	
	Test (C)	Test (C)	Training (C)	94 (C)	
	Test (NC)	Test (NC)	Training (NC)	90 (NC)	

and 3-fold cross validation was incorporated for each classifier to obtain the best possible result. Figure 10 and 11 provides the accuracy results for 100x and 200x magnification images. For all magnifications SVM classifier indicated the most appropriate result followed by K-NN and P-NN. Classification accuracy for type P63 counter stained and non-counter stained digitized color images for 2 different magnifications shown in Table 1 and Table 2.

Table 2. Comparative results of 3 different classifiers (200x magnification images)

Name	150 images were used (50 in each group)			Accuracy %	
	Group 1	Group 2	Group 3		
SVM	Training (C)	Test (C)	Test (C)	99 (C)	98 (C), 96 (NC)
	Training (NC)	Test (NC)	Test (NC)	98 (NC)	
	Test (C)	Training (C)	Test (C)	98 (C)	
	Test (NC)	Training (NC)	Test (NC)	98 (NC)	
	Test (C)	Test (C)	Training (C)	97 (C)	
	Test (NC)	Test (NC)	Training (NC)	94 (NC)	
K-NN	Training (C)	Test (C)	Test (C)	98 (C)	97 (C), 95 (NC)
	Training (NC)	Test (NC)	Test (NC)	95(NC)	
	Test (C)	Training (C)	Test (C)	97 (C)	
	Test (NC)	Training (NC)	Test (NC)	95 (NC)	
	Test (C)	Test (C)	Training (C)	96 (C)	
	Test (NC)	Test (NC)	Training (NC)	95 (NC)	
P-NN	Training (C)	Test (C)	Test (C)	95 (C)	95 (C), 95 (NC)
	Training (NC)	Test (NC)	Test (NC)	95 (NC)	
	Test (C)	Training (C)	Test (C)	95 (C)	
	Test (NC)	Training (NC)	Test (NC)	95 (NC)	
	Test (C)	Test (C)	Training (C)	95 (C)	
	Test (NC)	Test (NC)	Training (NC)	95 (NC)	

Table 3. Comparative results of different approaches for 2 magnifications

Number of test images 100x (412) 200x (205)	avg. processing time (sec)			Precision	Recall
Proposed method	100x	C	22.47	0.97	0.96
		NC	22.03	0.965	0.97
	200x	C	22.74	0.98	0.99
		NC	22.30	0.98	0.99
Automated approach [13]	100x	C	22.74	0.965	0.955
		NC	22.21	0.96	0.95
	200x	C	22.48	0.98	0.98
		NC	22.30	0.98	0.98
Automated approach [9]	100x	C	23.11	0.96	0.95
		NC	22.48	0.955	0.95
	200x	C	23.38	0.975	0.98
		NC	23.29	0.97	0.97
Automated approach [3]	100x	C	24.37	0.84	0.84
		NC	24.01	0.86	0.82
	200x	C	26.79	0.87	0.87
		NC	26.80	0.85	0.84

4 Experimental Results

This research study proposed modified approach was able to identify ovarian reproductive tissues with accuracy over 96% for 2 different magnifications. Research work of [9], [13] indicates accuracy over 95% and research work of [3] indicates 84% for the same test images. This study modified approach indicates improved accuracy in comparison to [9], [13]. Comparative results of all different existing approaches including this study proposed approach are shown in Table. 3 where approximate average processing time, precision and recall were compared and C refers to counter-stained and NC refers to non-counter stained type P63.

According to [3] ±10% error rate is acceptable. Table 3 indicates that this study proposed approach has accuracy over 96% for all magnification images and most accurate in compare to other existing approaches [3], [9], [13]. In comparison to 100x, 200x magnifications indicates higher accuracy rate.

5 Discussion and Conclusion

This paper has reviewed all existing [3], [9], [13] automated computerized approaches where human ovarian reproductive tissues were analyzed. This research study mainly focused on the work of [9], [13] due to the fact that these approaches incorporated type P63 images and are fully automated. Other automated human reproductive tissue analysis approach using type PCNA by [3] was also reviewed in this research study to compare the accuracy shown in Table 3.

In comparison to type PCNA type P63 requires less processing time. In comparison to type P63 counter stained digitized color images, P63 non-counter stained images requires less average processing time and are more cost effective. If processing time and cost is considered then type P63 non - counter stained digitized color images acquired from biopsy slides could be a more appropriate choice. Any type of P63 color digitized images with 100x and 200x magnifications can be analyzed using this research study proposed modified approach.

Processing time could be reduced if high performance computer can be used. This study is novel due to the fact that this is the first published study where one single approach was incorporated for both type P63 counter and non-counter stained images with different magnifications. A significant amount of human ovarian reproductive tissue images were tested in comparison with other approaches. In addition, 7 different image batches were used for analysis. It is not essential to be a laboratory expert to analyze ovarian reproductive tissues using this study proposed method but only requires knowing the image magnification.

Acknowledgement and Future Work. The authors would like to thank Assistant Professor and head of the department (Pathology) and domain expert Doctor S.I. Talukder (MBBS, M.Phil (Pathology), Shaheed Sayed Nazrul Islam Medical College, Kishoreganj, Bangladesh) for providing the test images, annotated images and necessary feature information. In future this research study will consider both type PCNA and type P63 images for a comparative study.

References

1. Kiruthika, V., Ramya, M.: Automatic segmentation of ovarian follicle using K-means clustering. In: 2014 Fifth International Conference on Signal and Image Processing (ICSIP), pp. 137–141 (2014)
2. Skodras, A., Giannarou, S., Fenwick, M., Franks, S., Stark, J., Hardy, K.: Object recognition in the ovary: quantification of oocytes from microscopic images. In: 2009 16th International Conference on Digital Signal Processing, pp. 1–6 (2009)
3. Kelsey, T.W., Caserta, B., Castillo, L., Wallace, W.H.B., Gonzálvez, F.C.: Proliferating Cell Nuclear Antigen (PCNA) allows the automatic identification of follicles in microscopic images of human ovarian tissue. arXiv preprint arXiv:1008.3798 (2010)
4. Muskhelishvili, L., Wingard, S.K., Latendresse, J.R.: Proliferating cell nuclear antigen—a marker for ovarian follicle counts. Toxicologic Pathology **33**, 365–368 (2005)
5. Lamprecht, M.R., Sabatini, D.M., Carpenter, A.E.: CellProfiler™: free, versatile software for automated biological image analysis. Biotechniques **42**, 71 (2007)
6. Chughtai, K., Heeren, R.M.: Mass spectrometric imaging for biomedical tissue analysis. Chemical Reviews **110**, 3237–3277 (2010)
7. Soucek, P., Gut, I.: Cytochromes P-450 in rats: structures, functions, properties and relevant human forms. Xenobiotica **22**, 83–103 (1992)
8. Picut, C.A., Swanson, C.L., Scully, K.L., Roseman, V.C., Parker, R.F., Remick, A.K.: Ovarian follicle counts using proliferating cell nuclear antigen (PCNA) and semi-automated image analysis in rats. Toxicologic Pathology **36**, 674–679 (2008)
9. Sazzad, T., Armstrong, L., Tripathy, A.: An automated detection process to detect ovarian tissues using type P63 digitized color images. In: 2015 IEEE 27th International Conference on Tools with Artificial Intelligence (ICTAI), pp. 278–285 (2015)
10. Magee, D., Treanor, D., Crellin, D., Shires, M., Smith, K., Mohee, K., et al.: Colour normalisation in digital histopathology images (2009)
11. Bolon, B., Bucci, T.J., Warbritton, A.R., Chen, J.J., Mattison, D.R., Heindel, J.J.: Differential follicle counts as a screen for chemically induced ovarian toxicity in mice: results from continuous breeding bioassays. Toxicological Sciences **39**, 1–10 (1997)
12. Bucci, T.J., Bolon, B., Warbritton, A.R., Chen, J.J., Heindel, J.J.: Influence of sampling on the reproducibility of ovarian follicle counts in mouse toxicity studies. Reproductive Toxicology **11**, 689–696 (1997)
13. Sazzad, T., Armstrong, L., Tripathy, A.: An automated approach to detect human ovarian tissues using type P63 counter stained histopathology digitized color images. In: IEEE-EMBS International Conference on Biomedical and Health Informatics (BHI), pp. 25–28 (2016)
14. Eramian, M.G., Adams, G.P., Pierson, R.A.: Enhancing ultrasound texture differences for developing an in vivo'virtual histology'approach to bovine ovarian imaging. Reproduction, Fertility and Development **19**, 910–924 (2007)
15. Li, Q., Nishikawa, R.M.: Computer-Aided Detection and Diagnosis in Medical Imaging. Taylor & Francis (2015)
16. Comaniciu, D., Meer, P.: Mean shift: A robust approach toward feature space analysis. IEEE Transactions on Pattern Analysis and Machine Intelligence **24**, 603–619 (2002)
17. Nock, R., Nielsen, F.: Statistical region merging. IEEE Transactions on Pattern Analysis and Machine Intelligence **26**, 1452–1458 (2004)

18. Wu, G., Zhao, X., Luo, S., Shi, H.: Histological image segmentation using fast mean shift clustering method. Biomedical Engineering Online **14**, 24 (2015)
19. Cheng, H.-D., Jiang, X., Sun, Y., Wang, J.: Color image segmentation: advances and prospects. Pattern Recognition **34**, 2259–2281 (2001)
20. Fukunaga, K., Hostetler, L.D.: The estimation of the gradient of a density function, with applications in pattern recognition. IEEE Transactions on Information Theory **21**, 32–40 (1975)
21. Gurcan, M.N., Boucheron, L.E., Can, A., Madabhushi, A., Rajpoot, N.M., Yener, B.: Histopathological image analysis: A review. IEEE Reviews in Biomedical Engineering **2**, 147–171 (2009)

Representation of 1-D Signals by a 0_1 Sequence and Similarity-Based Interpretation: A Case-Based Reasoning Approach

Petra Perner(✉)

Institute of Computer Vision and Applied Computer Sciences, Kohlenstrasse 2, 04107 Leipzig, Germany
pperner@ibai-institut.de

Abstract. Different spectrometer methods exist that have been developed over time to practical applicable systems. Researchers in different fields try to apply these methods to different applications especially in the chemical and biological area. One of these methods is RAMAN spectroscopy for protein crystallization or Mid-Infrared spectroscopy for biomass identification. For the applications are required robust and machine learnable automatic signal interpretation methods. These methods should take into account that not so much spectrometer data about the application are available from scratch and that these data need to be learnt while using the spectrometer system. We propose to represent the spectrometer signal by a sequence of 0/1 characters obtained from a specific Delta Modulator. This prevents us from a particular symbolic description of peaks and background. The interpretation of the spectrometer signal is done by searching for a similar signal in a constantly increasing data base. The comparison between the two sequences is done based on a syntactic similarity measure. We describe in this paper how the signal representation is obtained by Delta Modulation, the similarity measure for the comparison of the signals and give results for searching the data base.

Keywords: Computational methods · Delta Modulation · Feature extraction · Incremental knowledge acquisition · Spectrometer signal analysis · Similarity-based signal interpretation · Case-based reasoning

1 Introduction

Different spectrometer methods exist that have been developed over time to practical applicable systems. Researchers in different fields try to apply these methods to different applications especially in the chemical and biological area. One of these methods is RAMAN spectroscopy for protein crystallization [1], [2] or Mid-Infrared spectroscopy for biomass identification [3]. For the applications are required robust and machine learnable automatic signal interpretation methods. These methods should take into account the sparse available data for the application and that new data need to be acquired while using the spectrometer system. We propose a novel spectrometer analysis method based on Delta Modulation and similarity determination.

P. Perner (Ed.): MLDM 2016, LNAI 9729, pp. 728–739, 2016.
DOI: 10.1007/978-3-319-41920-6_55

We represent the spectrometer signal by a sequence of 0/1 characters obtained from a specific Delta Modulator. While doing this we preprocess the signal by smoothing at the same time. This prevents us from the extraction of a specific symbolic description of peaks and background from the basic spectrometer signal based on signal-theoretic methods [4]. The interpretation of the spectrometer signal is done by searching for a similar signal in a constantly increasing data base. The two 0/1 sequences of the spectrometer signal are compared based on a syntactic similarity measure.

The proposed new method has been tested on RAMAN spectrometer signals for screening of bio-molecular interactions but the method can be used for all kinds of spectrometer signals. With the aid of Raman spectroscopy, the vibrational spectrum of molecules can be examined. Functional groups like amino, carboxyl or hydroxyl groups can be identified through characteristic vibrational frequencies.

In this paper the architecture of the spectrometer-signal analysis system is described in Section 2. The calculation of the signal representation obtained by Delta Modulation is explained in Section 3 for three different kinds of delta modulation methods. Then we describe three different syntactical dissimilarity measures used for this study in Section 4. Finally, we give results in Section 5 for the three Delta Modulation methods and select the best one. This method is used for further studying the best dissimilarity measure. We show how good these measures can group similar spectra. At the end we use a prototype-based classifier to show how good we can classify the spectra based on the chosen representation and with the three different dissimilarity measures. In Section 6 we give conclusion.

2 Architecture of the Automated Spectrometer Signal Program

The architecture of the automatic spectrometer identification system is shown in Fig. 1.

After preprocessing the spectrometer-signal, the signal is coded into a 0/1 sequence by the delta modulator. While doing that the signal is step-wise smoothed by a linear function. The representation makes it unnecessary to develop special high-level features that describe all interesting properties of the spectra. The sequence itself can be interpreted in different ways. It can be ask for identity, similarity of the whole sequence or for partial identity or similarity. That allows identifying part-spectra, special single peaks or peak combinations within spectra.

This sequence is compared to sequences of reference spectra stored in a memory. The name of the spectrum where the coded sequence gives the highest similarity is given as output to the user. A side effect of the coding is also that the spectra is not stored with its real values but instead it is stored as 0/1 sequence. This saves memory capacity and makes it possible to implement the method into a special purpose processor.

When there is no similar sequence in the data base the input spectrum is stored into the data base after it has been coded by the delta modulator. The spectrum is labeled manually after it has been checked by other method what the spectrum is about.

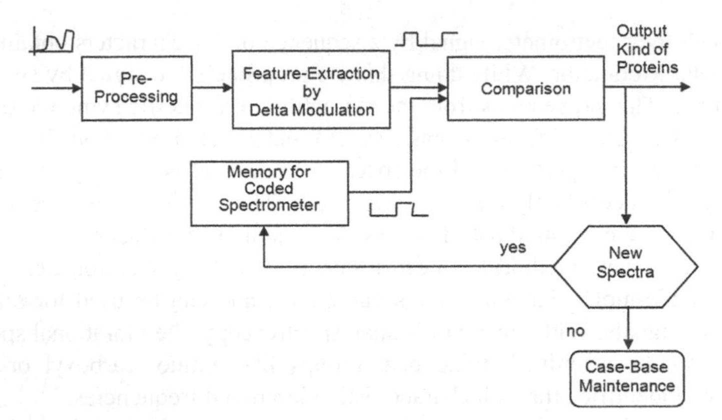

Fig. 1. Architecture of the Spectrum Interpretation System

This data collection is necessary since the appearance of the spectra for different proteins is not known yet.

The pre-processing of the RAMAN spectra is in this special case a baseline correction [5], a Fourier transformation to eliminate the influence of the special system device and its parts [6], and the calculation of the difference between the spectrum of the buffer and the liquid in the buffer.

3 Representation of the Spectra by Deltamodulation

The delta modulator compares the actual signal value $s(i)$ with an estimated signal value $r(i)$ of the coder. The difference $e(i)$ between these two signals is coded by only one bit. It mainly represents if the signal was increased or decreased by a certain constant. Three different methods exist to estimate actual signal value: Linear Delta Modulation (LDM) [7], Constant Factor Delta Modulation (CFDM) [8], and Continuously Variable Slope Delta Modulator (CVSD) [8], [9], [10].

3.1 Linear Delta Modulation

In case of the Linear Delta Modulation, the difference $e(i)$ between the actual signal value $s(i)$ and the estimated signal value $r(i)$ at sampling point *i* is calculated, see Fig. 2:

$$e(i) = s(i) - r(i) \tag{1}$$

If the difference is positive then the code *D* is equal *"1"* and *D* is equal *"0"* if the difference is negative. This binary signal D_n is stored in the memory. At the same time the magnitude of the signal to be expected at the next sampling point *i* is estimated from it. The corresponding rule is:

$$s(i) > r(i). D_n = 0. r = r(i-1) + \Delta u \tag{2}$$

$$s(i) \leq r(i). D_n = 1. r = r(i-1) - \Delta u \tag{3}$$

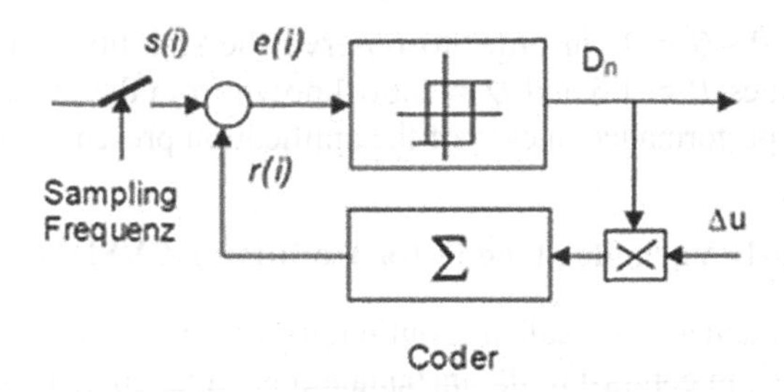

Fig. 2. Linear Delta Modulator

The incremental size Δu is a constant value which has to be selected in function of the standard-deviation δ_Δ of the first-order difference signal: $\Delta(i) = s(i) - s(i-1)$.

On the reproduction side (which is not necessary here since we do not want to reconstruct the signal) an inversely functioning decoder then generates the original curve by means of the binary signal stored in the memory. This approximated signal is $s'(i)$. The difference between the original signal $s(i)$and the approximated signal $s'(i)$is the approximation error$\varepsilon(i) = s(i) - s'(i)$, see Fig. 3

When process dynamics change, the linear delta modulator is not adjusted optimally anymore and the reconstruction error is increasing strongly. The adaptive delta modulators compensate this disadvantage. They dispose of a function block which takes over the control of the incremental size Δu in accordance with process dynamics. In the literature different adaptive delta modulators are known, two of which are presented in the following section.

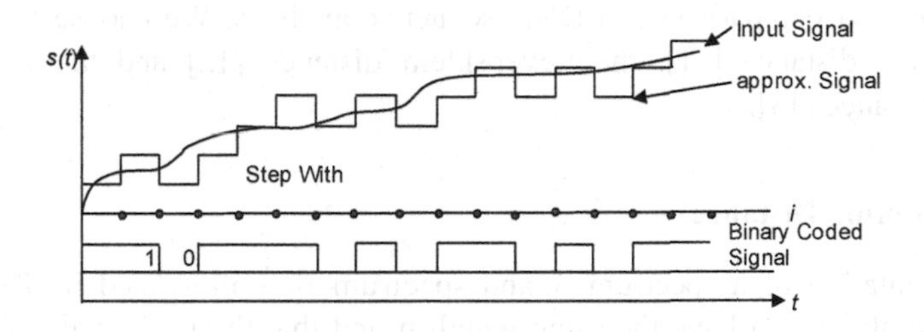

Fig. 3. Diagram with Input Signal, approximated Signal, and Binary Coded Signal

3.2 Constant Factor Delta Modulation

The instantaneous-value compander, also called "Constant Factor Delta Modulator" (CFDM), changes its increment size at each sampling point.

An adaptation-logic decides based on the input signal sequence $\langle D_n, D_{n-1}\rangle$ by which factor *k* the preceding increment size has to be multiplied:

$$\Delta u_i = \Delta u_{i-1} * k \tag{4}$$

with $D_n = D_{n-1}$ then $k = P$ and $D_n \neq D_{n-1}$ then $k = Q$.

It needs to be $P * Q = 1$, in order to observe the stability condition. For speech signals are the values $P = 1.5$ and $Q = 0.66$ known from the literature that have also been shown good performance in case of the application presented in this paper.

3.3 Continuously Variable Slope Delta Modulator CVSDM

The syllable compander, also called Continuously Variable Slope Delta Modulator (CVSDM), pursues, in contrast to the instantaneous-value compander, the tendency of the signal. Only when the same state has been recorded three times in a coincidence-register $\langle D_n = D_{n-1} = D_{n-2} \rangle$, the syllable compander increases its increment size. It is therefore more inert than the instantaneous-value compander. The rule for syllable companding is:

$$\text{3 bit coincidence} \quad k = 1;\ \Delta u_i = \Delta u_{i-1} + 1 \tag{5}$$

$$\text{no coincidence} \quad k = 0;\ \Delta u_i = \Delta u_{i-1} - 1 \text{ until } \Delta ui{=}0 \tag{6}$$

As Δu must not become zero, a minimum increment size u_{min} larger than 1 needs to be added. As standard value u_{min} can be assumed as $u_{min} = \sqrt{\delta_\Delta}$.

4 Similarity Determination Between Two SPECTRA

The spectra are represented by *0/1* sequences. To compare different spectra we need a distance measure that can work on such kind of representation. Different measures are known from text comparison and DNA sequence analysis. We choose for this work the Hamming distance [11], the Levenshtein distance [12] and the Levenshtein-Damerau distance [13].

4.1 Hamming Distance

The representation of a spectrum *A* and spectrum *B* is illustrated in Table 1. We assume that all spectra have the same length n and that the peaks are stable at their position (wavelength) in the spectra.

Then we have to compare two sequences *A* and *B*. The distance *d* between these two binary representations is the number of bits in which the two vectors are different.

Table 1. Representation of two Spectra, Sampling Points, and XOR connection

Spectrum A	1	0	0	0	1	1	0	0	0	...
Spectrum B	1	1	1	0	1	1	0	0	0	...
Sampling Points i	1	2	3	4	5	6	7	8	9	...
A **XOR** B	1	0	0	1	1	1	1	1	1	...

That is the well-known Hamming Distance:

$$d(A,B) = \|A - B\| = \sum_{i=1}^{n} |A_i - B_i| \quad (7)$$

4.2 Levenshtein Distance

Let $d_L(A,B) = D_{m,n}/n$ be the Levenshtein-Distance between the two 0/1 sequence A and B with $m = |A|$ and $n = |B|$. The Levenshtein distance is defined as the minimum number of modifications needed to transform the sequence A into B. The allowed operations are substitutions, insertions, and deletions. The dissimilarity in $D_{0,0}$should be $D_{0,0} = 0$. Then the dissimilarity is calculated as follows:

$$D_{i,0} = i, 1 \leq i \leq m$$
$$D_{o,j} = j, 1 \leq j \leq n$$

$$D_{i,j} = \min \begin{cases} D_{i-1,j-1} & +0 \quad if \quad A_i = B_j \\ D_{i-1,j-1} & +1 \quad Substitution \\ D_{i,j-1} & +1 \quad Insertion \\ D_{i-1,j} & +1 \quad Deletion \end{cases} \quad (8)$$

$$for\ 1 \leq i \leq m, 1 \leq j \leq n.$$

4.3 Damerau-Levenshtein-Distance

Let $D_{DL}(A,B) = D_{m,n}/n$ be the Damerau-Levenshtein-distance between the two 0/1 sequences A and B with $m = |A|$and $n = |B|$. The Damerau-Levenshtein distance is defined as the minimum number of modifications needed to transform the sequence A into B. Besides substitution, insertion, and deletion of a single character are allowed exchange of two adjacent single characters. The dissimilarity in $D_{0,0}$ should be $D_{0,0} = 0$. Then the dissimilarity is calculated as follow:

$$D_{i,0} = i, 1 \leq i \leq m$$
$$D_{0,j} = j, 1 \leq j \leq m$$

$$D_{i,j} = \min \begin{cases} D_{i-1,j-1} +0 \quad if \quad A_i = B_j \\ D_{i-1,j-1} +1 \quad Substitution \\ D_{i,j-1} \quad +1 \quad Insertion \\ D_{i-1,j} \quad +1 \quad Deletion \end{cases} \quad (9)$$

$$for\ (1 \leq i \leq 2, 1 \leq j \leq n)\ or\ (1 \leq i \leq m, 1 \leq j \leq 2)$$

$$D_{i,j} = \min \begin{cases} D_{i-1,j-1} + 0 & if \quad A_i = B_j \\ D_{i-1.j-1} + 1 & Substitution \\ D_{i,j-1} + 1 & Insertion \\ D_{i-1,j} + 1 & Deletion \\ D_{i-2.j-2} + c & Exchange \quad if \\ & A_i = B_{j-1} and \; A_{i-1} = B_j \end{cases} \tag{10}$$

$$for \;\; 3 \leq i \leq m, 3 \leq j \leq n.$$

5 Evaluation and Results

We have a data set of 30 different spectrometer signals. Each of the spectrometer signals have been preprocessed according to the methods described in Section 2, and afterwards processed and coded based on the delta modulation (see Sect. 3). The final outcome is a 0/1 sequence. The achieved results for the representation are presented in Section 4.A.

We calculated the pairwise distances between the thirty signals based on three distance measures: Hamming distance, Levenshtein distance, and the Damerau-Levenshtein distance.

We used the single-linkage clustering method to evaluate the goodness of the measures in Section 5.B.

5.1 Representation of the Spectrometer Signal by Delta-Modulation

The representation of the real signal by the approximated signal of the delta modulator is exemplary shown in Fig. 4 for Linear Delta Modulation and in Fig. 5 for Constant Factor Delta Modulation. The binary coded signal for both methods is shown in Table 2. It can be seen that the coded signal is different depending on the used delta modulation method. Table 3 shows the mean and maximum approximation error between the input signal and the approximated signal by the delta modulator. As expected the CFMD method shows the best result. The mean error is *1.677* increments and the maximum error is *16.04* increments. In the recent settings the CVSDM gave the worst results. It is left for further work to improve this method.

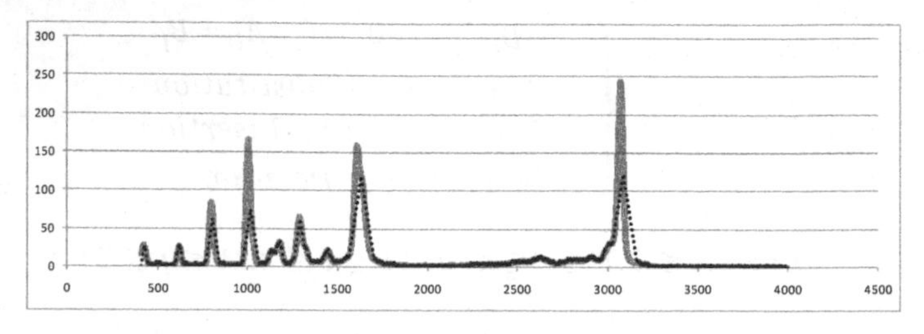

Fig. 4. Representation of Benzoic acid using LDM

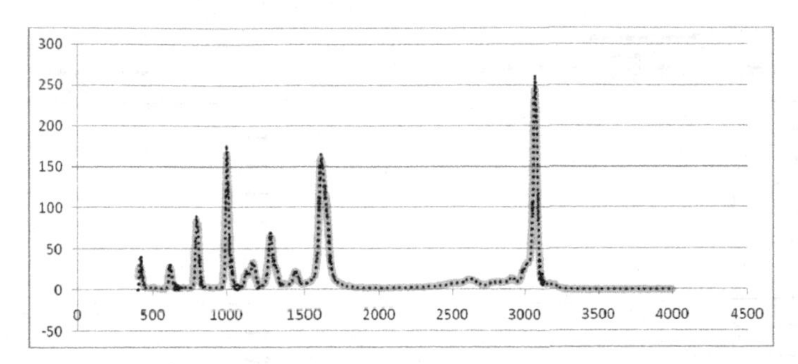

Fig. 5. Representation of Benzoic acid using CFDM

Table 2. Binary Representation of the Spectrum of Benzoic acid

Name of Compander	Sequence of Spectrum
LDM	... 1 0 0 1 0 1 0 0 1 0 1 0 1 0 1 0 1 0 1 0 1 0 1 0 1 1 0 1 0 1 0 1 0 1 0 1 0 0 1 ...
CFDM	... 0 1 0 0 1 0 0 1 0 0 1 0 0 1 0 1 0 0 1 0 1 0 1 1 0 0 1 1 1 1 0 1 1 1 0 1 1 1 0 0 ...

Table 3. Mean and Maximum Approximation Error between Input Signal and approx. Signal

Substance	Name of Delta Modulator			
	Linear Delta Modulator		CFDM	
	mean ε	max ε	mean ε	max ε
Acetone	1,74955958	4,642862	0,45178963	5,275991
Ascorbic acid	2,19114882	13,715031	1,06728145	10,356945
Benzamide	1,7514159	4,282954	0,40339027	2,977274
Benzoic acid	16,8602393	147,708368	4,78830105	45,56784
⋮	⋮	⋮	⋮	⋮
mean	5,6380909	42,5873038	1,6776906	16,045

In this study, we chose the CFMD method for the representation of the spectrometer signal.

5.2 Results for Similarity Between Two Spectra

It has been shown in Section 5. A that the **CFDM** delta modulator gives the best result for calculating the 0/1 sequence of the signal. The dendrogram for the different similarity measures between the thirty different spectra are shown in Figs. 6-8. The Hamming distance shows the highest differences in similarity but does not represent the similar groups well (see Fig. 6). The similarity measure will be sensitive to small changes in the spectra that might be caused by noise. Much better are represented similar groups in case of the Levenshtein (Fig. 7) and Damerau-Levenshtein similarity (Fig. 8). Both dendrograms show similarity in the group structure. They only slightly differ in the representation of the large group at the top of the dendrogram but in general the group structure is preserved.

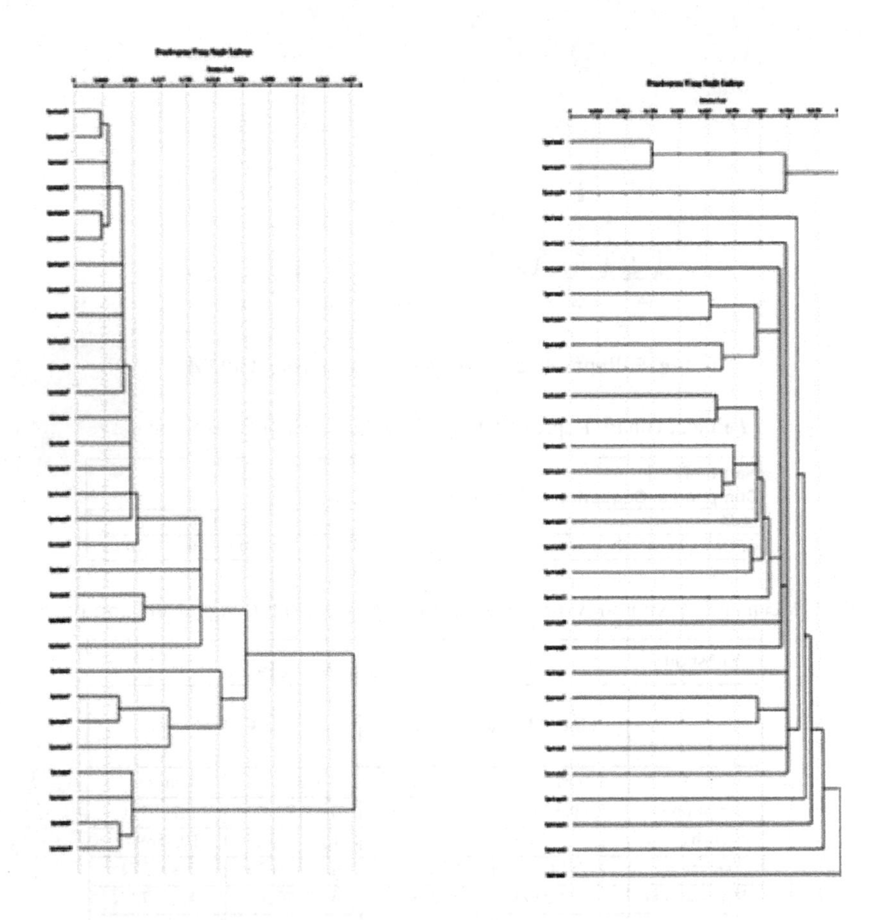

Fig. 6. Dendrogram using Levenshtein Distance using CFDM

Fig. 7. Dendrogram using Hamming Distance using CFDM

5.3 Accuracy of the Classification

We enlarged the data base by ten samples from the same spectrum. The final data base consists of three hundred samples. Our prototype-based classifier PROTOCLASS [14] was used for classification where 299 samples were the prototypes and one sample was classified against the 299 samples by searching for the three nearest neighbors. Crossvalidation was used for calculating the error rate. The results are show in Table 4.

The best results we have got for the Damerau-Levenshtein distance followed by the Levenshtein distance. The worst result we have got for the Hamming distance.

Table 4. Accuracy of prototype-based classifier for the different similarity measures

Distance Measure	Accuracy in %
Hamming	85,2
Levenshtein	90,5
Damerau-Levenshtein	91,2

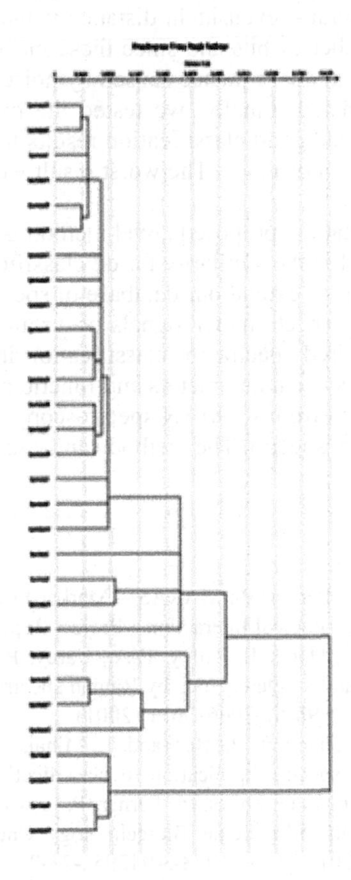

Fig. 8. Dendrogram using Damerau-Levenshtein Distance using CFDM

6 Conclusion

The representation of the spectra by a 0/1 sequence is a good representation for a spectrometer signal. While coding the signal in a 0/1 sequence it also smoothing the signal by a step-wise linear function. To keep the approximation error between the original signal and the coded signal small an adaptive delta modulator has to be selected. In the experiment above we used the CFDM delta modulation method instead of the linear delta modulator. A better method than this might be the continuously variable slope delta modulator. To construct such a modulator for this kind of signals is left for further work.

Three different similarity measures have been used: Hamming distance, Levenshtein distance, and the Damerau- Levenshtein distance. While the Hamming distance is very fast and simple to calculated, the latter two distances seem to represent the similar groups of spectra's very well. However, these two distance measures are more computationally expensive as the Hamming distance. The advantage of the Levenshtein and the Damerau-Levenshtein distance is that this distance can compare strings with different number of bits and since these measures can delete and substitute bits small differences in the sequence caused by noise or the behavior of the delta modulation can be eliminated. Finally, we tested the methods with our prototype-based classifier. We obtained good classification results for the Damerau-Levenshtein distance and the Levenshtein distance. The worst result we obtained for the Hamming distance.

In general we can say that the proposed novel method is a good method to represent spectrometer signals and that the similarity-based classification works very well. The proposed method allows us to extend our database of spectrometer signals very easily in the timely sequence the spectrometer signals occur and at the same time immediately to use the new acquired spectra for classification in daily work without going into a heavy update of the system parameters and functions.

We have tested it on data from RAMAN spectroscopy. However the method is not only applicable to RAMAN spectra. The method can be used for other spectra as well.

References

1. Janzen, C., Delbrück, H., Perner, P.: MARAS – Marker Free RAMAN Screening for Molecular Investigation of Biological Interactions, Project Report (2006)
2. Altose, M.D., Zheng, Y., Dong, J., Palfey, B.A., Carey, P.R.: Comparing protein–ligand interactions in solution and single crystals by Raman spectroscopy. Proceedings of the national Academy of Science **98**(6), 3006–3011 (2001)
3. Rammal, A., Perrin, E., Chabbert, B., Bertrand, I., Mihai, G., Vrabie, V.: Optimal preprocessing of mid infrared spectra. application to classification of lignocellulosic biomass: maize roots and miscanthus internodes. In: Perner, P. (eds.) Advances in Mass Data Analysis of Images and Signals in Medicine, Biotechnology, Chemistry and Food Industry, pp. 66–76. ibai-publishing (2013). ISBN: 978-3-942952-21-7
4. Bleghith, A., Collet, C.H., Armspach, J.-P.: A unified framework for peak detection and alignment: application to HR-MAS 2D NMR spectroscopy. In: Perner, P. (eds.) Advances in Mass Data Analysis of Images and Signals in Medicine, Biotechnology, Chemistry and Food Industry, pp. 106–118. ibai-publishing (2011). ISBN: 978-3-942952-02-6
5. Savitzky, A., Golay, M.J.E.: Smoothing and Differentiation of Data by Simplified Least Squares Procedures. Analytical Chemistry **36**(8), 1627–1639 (1964)
6. Zhao, J., Carrabba, M.M., Allen, F.S.: Automated Fluorescence Rejection Using Shifted Excitation Raman Difference Spectroscopy. Applied Spectroscopy **56**(7), 834–845 (2002)
7. Un, C.K., Lee, H.S.: A Study of Comparative Performance of Adaptive Delta Modulation Systems. IEEE Trans. on Communications **28**(1), 96–101 (1980)
8. Jayant, N.S.: Adaptive Delta Modulation with one-bit Memory. The Bell System Technical Journal **49**(3), 76–80 (1970)

9. Tazaki, S., Osawa, H., Shigematsy, Y.: A Useful Analytical Method for Discrete Adaptive Delta Modulation. IEEE Trans. on Communications **25**(2), 195–199 (1977)
10. Perner, P.: Datenreduktionsverfahren für technologische Industrierobotersteuerungen mit direkter Teach-in-Programmierung, 2nd unrevised edn. ibai-publishing, Leipzig. ISBN: 978-3-940501-16-5 (2010)
11. Hamming, R.W.: Error detecting and error correcting codes. Bell System Technical Journal **29**(2), 147–160 (1950)
12. Levenshtein, V.I.: Binary codes capable of correcting deletions, insertions, and reversals. Soviet Physics Doklady **10**(8), 707–710 (1966)
13. Damerau, F.: A technique for computer detection and correction of spelling errors. Communications of the ACM **7**(3), 171–176 (1964)
14. Perner, P.: Prototype-Based Classification. Applied Intelligence **28**(3), 238–246 (2008)

Rank Aggregation Algorithm Selection Meets Feature Selection

Alexey Zabashta, Ivan Smetannikov(✉), and Andrey Filchenkov

Computer Science Department, ITMO University, 49 Kronverksky Pr., 197101 St. Petersburg, Russia
{azabashta,ismetannikov,afilchenkov}@corp.ifmo.ru

Abstract. Rank aggregation is the important task in many areas, and different rank aggregation algorithms are created to find optimal rank. Nevertheless, none of these algorithms is the best for all cases. The main goal of this work is to develop a method, which for each rank list defines, which rank aggregation algorithm is the best for this rank list. Canberra distance is used as a metric for determining the optimal ranking. Three approaches are proposed in this paper and one of them has shown promising result. Also we discuss, how this approach can be applied to learn filtering feature selection algorithm ensemble.

Keywords: Meta-learning · Rank aggregation · Ensemble learning · Feature selection

1 Introduction

In many domains where multiple ranking algorithms are applied in practice, the task of rank aggregation arises. An example of such a field is computational biology, where several rank aggregation algorithms are used to detect how physiological characteristics depend on genes [7,10]. In the Web search problem, many different rank aggregation algorithms are also used. Each of them exploits some features for documents evaluation, such as document popularity, the quality of matching between a document and a query, information source authority and others [13,24].

For given set of algorithms, solving the rank aggregation problem is usually not that obvious, which one is the best. This situation is typical as well in other domains: different algorithms are better than others on different tasks. John Rice has formulated this problem in a general form in 1975 [22]: how to find the algorithm that minimizes an error function for given problem without executing each algorithm.

The solution of the algorithm selection problem is formulated under the meta-learning approach [6]. Meta-learning algorithms treat the algorithm selection problem as a prediction problem: they require a training set to learn a model for predicting the best algorithm for a new dataset.

P. Perner (Ed.): MLDM 2016, LNAI 9729, pp. 740–755, 2016.
DOI: 10.1007/978-3-319-41920-6_56

Since the problem of finding the best possible resulting rank is usually NP-hard, approximate algorithms are used to find it. Each of these algorithms is usually good for a certain type of rank optimization problem. However, the problem of rank aggregation algorithm selection was never solved in scientific literature.

In this work, we deal with selection of the best algorithm for aggregating given ranks of the same length. We extend the previous research on this problem [29] by suggesting new approaches involving meta-feature selection. Also we propose an application of the rank aggregation algorithm selection system to the problem of feature selection.

Some definitions and approaches for the rank aggregation problem (including description of all algorithms for permutation lists generating) are given in Section 2. Section 3 contains definitions and general scheme for meta-learning application to the algorithm selection problem. In Section 4, experiments are explained and thier results are presented and discussed. In Section 5, application of the proposed rank aggregation algorithm selection algorithm for designing novel ensemble-based feature selection algorithm is proposed. Section 6 concludes the paper with summarization of the obtained results and outlining future work.

2 Rank Aggregation

2.1 Metrics on Permutations

The mathematical formalization of rank involves permutations. Permutation π is an ordered set $\{\pi_1, \pi_2 \dots, \pi_n\}$ of distinct natural numbers between 1 and n, where n is the *length* of the permutation. The set of all permutations of length n will be denoted by Π^n. *Metric* $\mu(a, b)$ on the permutation space is a function from $\Pi^n \times \Pi^n$ to $\mathbb{R}$, which satisfies the following axioms [11]:

- $\mu(a, b) = \mu(b, a)$, symmetry;
- $\mu(a, b) = 0 \Leftrightarrow a = b$, coincidence axiom;
- $\mu(a, b) \geq 0$, non-negativity;
- $\mu(a, b) = \mu(a \cdot c, b \cdot c)$, right-invariance, where $a \cdot c$ is product of permutations a and c.

We use short term "metric" to refer to a metric on the permutation space. In this paper we use seven metrics on the permutation space described in [11] and [7].

The *Minkowski distance* between a and b is an adaptation of the classical Minkowski distance for vectors to the case of the permutation space [11]:

- $l_1(a, b) = \sum_i |a(i) - b(i)|$, the *Manhattan distance*;
- $l_2(a, b) = \sum_i (a(i) - b(i))^2$, the *Euclidean distance*;
- $l_\infty(a, b) = \max_i |a(i) - b(i)|$, the *Chebyshev distance*,

where $p(i)$ is the position of number i in the permutation p.

The *Canberra distance* between a and b is a modification of the Euclidean distance [7], which assigns higher priority to differences in the top positions in the permutation, and which is equal to

$$\sum_i \frac{|a(i) - b(i)|}{a(i) + b(i)}.$$

The *Cayley distance* between a and b is the minimum number of transpositions required to obtain b from a [11], which is equal to $n - \left|loops(a \cdot b^{-1})\right|$, where $loops(a \cdot b^{-1})$ is a set of loops in the permutation $a \cdot b^{-1}$ and b^{-1} is the inverse permutation to b.

The *Kendall tau rank distance* between a and b is the number of pairwise disagreements between a and b [11], which equals to

$$\sum_i \sum_{j<i} \mathrm{I}[(a(i) - a(j))(b(i) - b(j)) < 0],$$

where $\mathrm{I}[c]$ equals to 1 if condition c is true, and 0 otherwise.

The *Ulam distance* between a and b is the minimum number of edits required to obtain b from a.

All these metrics have $O(n \cdot \log(n))$ computational time complexity or less [1,19].

2.2 Rank Aggregation Problem and Algorithms

Formally, the task of rank aggregation comprises finding the permutation π for given permutation list Q, which minimizes an error function $E_\mu(\pi, \mathrm{Q})$. The error function depends on a metric μ. In this, paper we use

$$E_\mu(\pi, \mathrm{Q}) = \sum_{\theta \in \mathrm{Q}} \mu(\pi, \theta).$$

For most of the metrics described in previous subsection, the problem of rank aggregation is NP-hard [2]. However, there are many algorithms for approximate solution search. In this paper, we use four popular algorithms.

Let $m = |\mathrm{Q}|$ be the number of permutations in the given list Q. Each permutation from Q contains n elements. Define $v_{i,j}$ to be the number of permutations π in the list Q, which rank i over j and $l_{i,j} = m - v_{i,j}$:

$$v_{i,j} = \sum_{\pi \in \mathrm{Q}} \mathrm{I}[\pi(i) > \pi(j)].$$

Algorithm, which we will refer to as *weight-based algorithm* (WBA), assigns a weight w_i for each ith element and returns permutation $\pi = \{\pi_1, \pi_2 \ldots, \pi_n\}$. In this algorithm, $i \geq j \Leftrightarrow w(\pi_i) \leq w(\pi_j)$.

The *Borda Count* is a WBA, which assigns weight $w_i = \sum_{\pi \in \mathrm{Q}} f(n - i)$ to each ith element, where f is an increasing function [5]. We use the original increasing functions for Borda count, which is $f(x) = x$.

The *Copelands Score* is WBA, which assigns the following weight

$$w_i = \sum_{i \neq j} (\mathrm{I}[v_{i,j} > l_{i,j}] + 0.5 \cdot \mathrm{I}[v_{i,j} = l_{i,j}])$$

to each ith element [8].

The main idea of *Markov Chain* methods is to construct a Markov chain transition matrix T and find the stationary distribution for it [13], which we use as a weight vector in the same way as for WBA. We use two Markov chain methods. In the first method, $\mathrm{T}_{i,j} = 1/n$ if $v_{i,j} > 0$, and 0 otherwise. In the second method, $\mathrm{T}_{i,j} = 1/n$ if $v_{i,j} > 0.5 \cdot m$. In both methods, $\mathrm{T}_{i,i} = 1 - \sum_{i \neq j} \mathrm{T}_{i,j}$.

The *Local Kemenization* method sort elements in resulting permutation using following comparator: $\pi_i \leq \pi_j \Leftrightarrow v_{i,j} \geq v_{j,i}$ [13].

The *"Pick a perm"* method minimizes given error function using permutations from the input permutation list Q [7]. It returns

$$\arg\min_{\pi \in \mathrm{Q}} \mathrm{E}(\pi, \mathrm{Q}).$$

2.3 Algorithms for Generating Permutations

In this subsection, we introduce a scheme of an algorithm for generating permutations and describe three algorithms we use to generate a single permutation of fixed length. We follow [29].

An algorithm for generating permutations for given permutation length returns a permutation of this length. It is parameterized with variable $\sigma \in [0, 1]$, which can be understood as a degree of distinction between the resulting permutation and the identity permutation.

Algorithm A is a modification of the FisherYates shuffle [15]. In the original algorithm, on each ith step the transposition of the ith element with the jth element is performed, where j is a random number generated with the uniform distribution $\mathcal{U}(1, i)$. In the modified algorithm, we make a transposition on each step only if the following condition is satisfied: $(n - i + 1)b < n \cdot \sigma - s$, where b is a random number generated from the uniform distribution $\mathcal{U}(0, 1)$ and s is the number of transpositions that have been already made. It is the ordinary FisherYates shuffle, if $\sigma = 1$.

Algorithm B is another modification of the FisherYates shuffle. The difference is that on each step, we make the transposition of ith element with the element at position $i - |\sigma|$, where σ is a random number generated from the normal (Gaussian) distribution $\mathcal{N}(0, i \cdot \sigma)$ with zero mean and $i \cdot \sigma$ squared scale.

Algorithm C makes $n \cdot \sigma$ transpositions of elements at two random positions generated from $\mathcal{U}(1, n)$.

2.4 Algorithms for Generating Permutation Lists

In this subsection, we introduce a scheme of an algorithm for generating permutation lists and describe three algorithms we use to generate a list of permutation of a fixed length. We follow [29].

Algorithm for generating permutation list (AGPL) for given permutation length and number of permutations generates these permutations. It uses an algorithm for generating permutations and parameters $l, r \in [0, 1]$ to choose σ uniformly in the interval $[l, r]$ for the algorithm for generating permutations. Formally,

$$\sigma_i = l + \frac{(i-1) \cdot (r-l)}{m-1},$$

where m is the number of permutations in the resulting list and σ_i is the value for ith permutation from the resulting list.

We have described three algorithms for generating permutations in the previous subsection. Each of them can be used to generate permutation lists. We will refer to each algorithm for generation permutation list implying a certain algorithm for generating permutation as AGPL-A, AGPL-B and AGPL-C correspondingly.

3 Meta-Learning for Rank Aggregation Algorithm Selection

3.1 Meta-Learning

Meta-learning is the approach used to predict the best algorithms to solve given task without execution of these algorithms. The core idea of this approach is to use infomation about the given task, which hepls to make assumptions about the best algorithm. Usage of this information helps to avoid satisfying the No Free Lunch Theoems conditions [28].

Information about tasks is described with so-called meta-features, which are properties of the tasks that can be effectively measured (otherwise, it is faster to run each algorithm and compare their performance) [17]. A lot of different meta-feature types exists. Nevertheless, the meta-feature engineering step is the most complicated step in meta-learning application. After the set of meta-feature is chosen, each task is associated with a vector of the meta-features values. Thus, each task is a point in the meta-feature space.

Each meta-feature system has a set of algorithms it can choose of. To learn such a system, a training set of tasks is required. This training set contains tasks and information about the algorithms perfomance on these tasks it is used as labels for the tasks. The system is learnt on these tasks and can predict the best classifier for a new task. Thus, meta-learning reduces the algorithm selection problem to a supervised learning problem.

3.2 Baseline Approach

AGPL described in Section 2 uses two hidden variables l and r as parameters to generate a permutation list. We found that there is a certain dependency between optimal algorithm on dataset and l and r with which this dataset was generated. These dependencies can be seen Fig. 1, Fig. 2 and Fig. 3, in which

Fig. 1. Visualization of optimal rank aggregation algorithms distribution in HVA feature space for AGPL-A data. The domains of red points is the permutation lists, where the Borda Count is optimal, orange — the Markov Chain-1, green — the Markov Chain-2, cyan — the Copelands Score, blue — the Local Kemenization, magenta — "Pick a perm"

for each dataset described with l and r, the best algorithm for this dataset is presented; l and r are paratemeters with which this dataset was generated by means of AGPL-A, AGPL-B and AGPL-C correspondingly. These tree figures demonstate that a clear dependency exists between parameters l and r describing a dataset and the best algorithm for this dataset.

Hidden variable approach (HVA) provides description of permutation list with two features: l and r with which it was generated. We can apply this approach only to generic data, because we have no access to values of these hidden variables in real-world problems.

3.3 Approaches Without Meta-Feature Selection

In this subsection, we describe three approaches to generate meta-features. The first two are taken from [29], and the last one is novel.

Basic meta-features approach (BMFA) for each metric μ explores all possible pairs of permutations from the input list Q and then constructs sequence $X_\mu = \{\mu(a,b) : a, b \in Q\}$. After that, it extracts statistic characteristics

Fig. 2. Visualization of optimal rank aggregation algorithms distribution in HVA feature space for AGPL-B data.

(minimum, maximum, average, variance, skewness and kurtosis) from each sequence X_μ as meta-features. The meta-feature set thus consists of 7 (number of metrics) $\times$ 6 (number of features from each metric) that equals 42 meta-features.

BMFA can be accelerated. The slowest part in this approach is scanning of all possible pairs. It costs $O(m^2)$ time, and the algorithm works in $O(n \cdot \log(n) \cdot m^2)$ time. We are interested in decreasing its complexity, because some rank aggregation algorithms work faster. For example, the Borda count works in $O(n \cdot (\log(n) + m))$ time. That is why it cannot be applied on practice.

To solve this problem, we aggregate the input permutation list to a single permutation c by means of a faster method. In this paper, we use the Borda count. Then we construct a sequence $X_\mu = \{\mu(\pi, c) : \pi \in \mathrm{Q}\}$ containing distances between c and permutations from the input list, then we extract features in the same way as in BMFA. The new algorithm, which we will refer to as *accelerated meta-features approach* (AMFA), works in $O(n \cdot \log(n) \cdot m)$ time. AMFA creates 42 meta-features that do not intersect with the meta-features generated by BMFA.

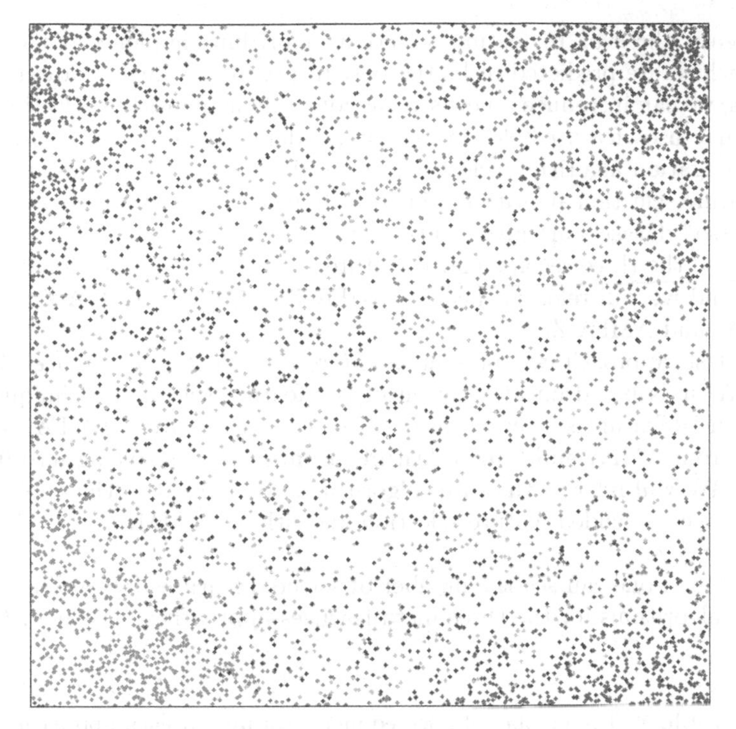

Fig. 3. Visualization of optimal rank aggregation algorithms distribution in HVA feature space for AGPL-C data.

Pairwise Meta-Features approach (PMFA) for each index i explores all possible pairs of permutations from input list Q sequence

$$X = \bigcup_i \{|a(i) - b(i)| : a, b \in \mathrm{Q}\}.$$

After that, it extracts as meta-features the same statistic characteristics from sequence X. This gives 6 meta-features.

3.4 Approaches with Meta-Feature Selection

The BMFA and AMFA meta-features approaches generate a lot of meta-features. Since the number of features is high and not all of them useful, it may reduce the accuracy and slow down the meta-classifiers. Therefore, we apply feature selection algorithms in order to reduce the number of features. The more detailed explanation of feature selection and its purpose can be found in Section 5.

Because this work is devoted to meta-learning, we used meta-learning for feature selection algorithm selection. To the best of our knowledge, only two papers are devoted to this topic: by Wang et al. [26] and by Filchenkov and Pendryak [14]. The latter improves the first one, this is why we use it. The system recommends three of 16 feature selection algorithms implemented in WEKA library, which are predicted to be the best to preprocess given dataset.

Thus, we can introduce a new scheme for ranking algorithm selection system design that uses feature selection recommender system, and an approach for rank aggregation algorithm selection. For a given dataset, the system suggests three feature selection algorithms. Then each algorithm is executed and then the preprocessed datasets are given as the inputs for the approach.

We used the three recommended feature selection algorithms for BMFA and AMFA, and we will refer to them as BMFA-X, BMFA-Y, BMFA-Z, AMFA-X, AMFA-Y, and AMFA-Z.

The last approach we present is *Merged Meta-Feature Space Approach* (MMFSA) it is based on application of the feature selection technique to the merged meta-feature space: we used 42 meta-features from BMFA, 42 meta-features from AMFA and 6 meta-features from PMFA, resulting into 90 meta-features. We will refer to the three algorithms obtained after the application of the three recommended feature selection algorithms as MMFSA-X, MMFSA-Y and MMFSA-Z.

In Table 1, one can see the number of selected features by the three feature selection algorithms for the three meta-features approaches on three AGPL.

Table 1. The number of selected meta-features for each approach.

	AGPL-A	AGPL-B	AGPL-C
BMFA-X	19	28	16
BMFA-Y	37	38	31
BMFA-Z	35	18	28
AMFA-X	13	22	12
AMFA-Y	34	36	29
AMFA-Z	28	17	30
MMFSA-X	35	54	20
MMFSA-Y	50	61	43
MMFSA-Z	82	36	80

4 Experiments

4.1 Experiment Setup

We conduct experiments for each approach described in Section 3. Each approach returns feature description of a permutation list. Each permutation list is labeled with an aggregation algorithm that minimizes the error function, which is based on the Canberra distance.

We use the following classifiers to solve the described classification problem: baggin (Bag), decision trees (DT), LogitBoost (LB), ClassificationViaRegression (Regres), support vector machine (SMV). All implementations were taken from WEKA library [16].

The result of the experiments is a confusion matrix C, where $C_{i,j}$ is the number of objects from ith class that classifier has classified to the jth class. We used F_1-measure as a quality measure to compare classification algorithms.

4.2 Experiment Details

We generated permutation lists of the permutation lists family: the length of permutations equals 80, the number of permutations in lists equals 20. For generating each permutation list, we randomly took l and r values for AGPL from the uniform distribution $\mathcal{U}(0,1)$.

We used six rank aggregation methods described in Section 2 as labels: the Borda count (BC), the Copelands score (CS), two Markov chain methods (MC-1 and MC-2), the Local Kemenization (LK), and "Pick a perm" (PP).

For each of the algorithms for generating permutation lists, we generated 1000 permutation lists for each class, therefore the sample contained of 6000 objects. We run each experiment 10 times and used cross-validation to measuring the efficacy of classification.

4.3 Results

Table 2, Table 3 and Table 4 contain resulting F1-score for each classifier used on meta-level for each approach and on rank lists generated with AGPL-A, AGPL-B and AGPL-C correspondingly.

Table 2. Results on rank lists generated with AGPL-A.

FAlg	Bag	DT	LB	Regres	SVM
HVA	0.623	0.625	0.587	0.636	0.506
PMFA	0.549	0.559	0.566	0.580	0.568
BMFA	0.648	**0.638**	0.649	0.645	0.645
BMFA-X	0.641	0.635	0.652	0.644	0.649
BMFA-Y	0.637	**0.638**	0.652	0.641	0.647
BMFA-Z	0.647	**0.638**	0.651	0.643	0.646
AMFA	0.631	0.623	0.638	0.626	0.632
AMFA-X	0.636	0.619	0.630	0.645	0.627
AMFA-Y	0.629	0.622	0.636	0.632	0.633
AMFA-Z	0.629	0.622	0.639	0.635	0.632
MMFSA-X	**0.651**	0.634	0.654	**0.658**	0.650
MMFSA-Y	0.645	**0.638**	0.652	0.655	0.649
MMFSA-Z	0.646	**0.638**	**0.655**	0.641	**0.659**

Table 3. Results on rank lists generated with AGPL-B.

Bag	DT	LB	Regres	SVM
0.434	0.442	0.422	0.445	0.335
0.391	0.379	0.386	0.415	0.359
0.452	0.440	0.456	0.449	0.460
0.449	0.441	0.455	0.451	0.465
0.452	0.440	0.456	0.449	0.460
0.448	0.441	**0.457**	**0.456**	0.461
0.455	0.447	0.450	0.443	0.453
0.457	**0.454**	0.463	0.452	0.455
0.455	0.447	0.456	0.446	0.456
0.454	0.442	0.452	0.450	0.448
0.456	0.447	0.455	0.439	0.456
0.465	0.447	0.455	0.449	**0.466**
0.459	0.446	0.453	0.442	0.462

As we see SVM meta-classifier under MMFSA-Z showed the highest result on datasets generated with AGPL-A. Confusion matrix for this approach is presented in Table 5.

Table 6, Table 7 and Table 8 constain top meta-features for each AGLP. We used Information Gain (IG) [21] as the feature importance measure.

Table 4. Results on rank lists generated with AGPL-C.

FAlg	Bag	DT	LB	Regres	SVM
HVA	0.570	0.579	0.556	0.574	0.484
PMFA	0.508	0.503	0.510	0.526	0.498
BMFA	0.590	0.576	0.585	0.577	0.582
BMFA-X	0.582	0.579	0.578	0.596	0.578
BMFA-Y	0.590	0.576	0.587	0.589	0.581
BMFA-Z	0.581	0.575	0.592	0.598	0.582
AMFA	0.582	0.563	0.587	0.585	0.592
AMFA-X	0.593	0.564	0.572	0.597	0.582
AMFA-Y	0.591	0.563	0.578	0.596	0.589
AMFA-Z	0.589	0.563	0.582	0.597	0.591
MMFSA-X	**0.632**	0.578	0.613	**0.649**	0.609
MMFSA-Y	**0.632**	**0.584**	0.611	0.633	0.616
MMFSA-Z	0.629	**0.584**	**0.617**	0.633	**0.631**

Table 5. Confusion matrix, for SVM under MMFSA-Z for datasets generated with AGPL-A.

	BC	MC1	MC2	CS	LK	PP
BC	956	44	0	0	0	0
MC1	108	825	0	12	36	19
MC2	145	53	615	23	152	12
CS	142	154	58	53	516	77
LK	21	123	49	47	638	122
PP	1	16	4	2	33	944

Table 6. Information gain values of meta-features for datasets geneterated with AGPL-A.

Meta-feature	IG
BMFA-E($CayleyDistance$)	1.0939
BMFA-E($CanberraDistance$)	1.0844
BMFA-E($UlamDistance$)	1.0802
PMFA-E	1.0695
BMFA-E($ManhattanDistance$)	1.0695
BMFA-E($KendallTau$)	1.0569
AMFA-E($EuclideanDistance$)	1.0567
BMFA-E($EuclideanDistance$)	1.0547
AMFA-E($KendallTau$)	1.0526
AMFA-E($ManhattanDistance$)	1.0418
AMFA-E($CanberraDistance$)	1.0382
PMFA-Skew	0.9889
AMFA-E($UlamDistance$)	0.9684
AMFA-Min($ManhattanDistance$)	0.9564
PMFA-Kurt	0.9512

Table 7. Information gain values of meta-features for datasets geneterated with AGPL-B.

Meta-feature	IG
BMFA-σ($CanberraDistance$)	0.5613
BMFA-σ($KendallTau$)	0.5600
BMFA-σ($ManhattanDistance$)	0.5571
BMFA-σ($EuclideanDistance$)	0.5472
BMFA-Kurt($CanberraDistance$)	0.5179
AMFA-E($CanberraDistance$)	0.5151
BMFA-Min($ManhattanDistance$)	0.5130
AMFA-Min($CanberraDistance$)	0.5058
BMFA-Min($CanberraDistance$)	0.5048
AMFA-E($ManhattanDistance$)	0.5013
BMFA-Min($EuclideanDistance$)	0.4999
BMFA-Min($KendallTau$)	0.4969
AMFA-E($EuclideanDistance$)	0.4950
AMFA-σ($ChebyshevDistance$)	0.4947
AMFA-Min($ManhattanDistance$)	0.4938

Table 8. Information gain values of meta-features for datasets geneterated with AGPL-C.

Meta-feature	IG
BMFA-E(*CayleyDistance*)	0.9610
BMFA-E(*UlamDistance*)	0.9563
BMFA-E(*CanberraDistance*)	0.9472
PMFA-E	0.9415
BMFA-E(*ManhattanDistance*)	0.9415
BMFA-E(*KendallTau*)	0.9195
BMFA-E(*EuclideanDistance*)	0.9108
AMFA-E(*KendallTau*)	0.9080
AMFA-E(*ManhattanDistance*)	0.8983
AMFA-E(*EuclideanDistance*)	0.8969
AMFA-E(*CanberraDistance*)	0.8855
PMFA-Skew	0.8540
AMFA-E(*UlamDistance*)	0.8265
AMFA-Min(*ManhattanDistance*)	0.8192
AMFA-Min(*KendallTau*)	0.8117

4.4 Discussion

As the results show, the rank lists generating approach is very impactful factor: the most complex task was predicting rank aggregation algorithms for data generated with AGPL-B. However, the approaches have lesser impact on the approach success. Nothing unexpected is in fact that in all the cases, the last approach involving all the meta-features outperformed all the other approaches in most of the cases. There are two notable exceptions:

- For decision tree, many algorithms showed the same performance. That can be simply explained by the fact that decision trees select features themselves.
- For data generated with AGPL-B, several classifiers showed better performance with other approaches. However, we may notice that these results were low, so we can claim that these classifiers work not well for this task (0.454 for DT, 0.457 for LB, and 0.456 for Regress versus 0.465 for Bag and 0.466 for SVM in Table 3).

In general, feature selection application has improved most of the initial approaches for most of the classifiers applied on the meta-level. As we can see from comparison of the most important meta-features, they are different for data generated with different AGPLs. However, the most important meta-features are generated under BMFA. Despite that PMFA generates only 6 meta-features, three of them are among the most important meta-features for AGPL-A and two — for AGPL-C.

5 Application to Feature Selection

5.1 Feature Selection

With development of social networks and other massive data tasks, the dimensionality reduction problem arises. In machine learning, two ways to reduce dimensionality are used: feature selection and feature extraction [18]. In this terminology, features are some attributes that suit to objects. These features are usually gathered or calculated directly and can belong to any type of data.

Feature extraction methods are commonly used for pattern recognition and image processing [20]. Some of these methods require deep data analysis and feature engineering, although if no such expert knowledge is available, they might still work [3]. Feature extraction techniques usually do not consider that some features may be redundant or irrelevant. In order to remove such features, feature selection techniques are used. This approach requires the resulting feature set to be a subset of the original feature set.

Feature selection algorithms can be divided into three groups: filters, wrappers and embedded algorithms [23]. Wrappers are usually considered as the most accurate methods, but having very high computational cost. This is result of gradual feature subspace search with a classifier training performed on each step. Another way to select features is to use embedded algorithms. Some classifiers, for example random forests, already have built-in feature selection. The main problem here is that this type of feature selection requires usage of this specific classifier. And the last ones are filtering methods. These methods evaluate single features or feature subsets using an inner measure of feature importance. Filters are quite fast and do not involve any classifier to select feature subset. They are not that accurate in terms of resulting classification. However, they are de-facto standard technique in many task, such as gene selection [9] or n-gram processing [27].

5.2 Feature Selection Algorithm Ensembling Involving Rank Algorithm Selection System

In order to improve filtering algorithms, filter ensembles are used. There are three ways to build a composition of them. The methods of the first group use ensemble of classifiers, which are learnt on datasets generated after filters application [12]. Another way to build a composition of filters is to aggregate their feature importance measures. After that, some cutting rule is applied [25]. And the third way to aggregate filters is to aggregate their ranking lists [4]. For this purpose, the approach described in this article can be used. We will explain the last one in details.

Ranking, or univariate filter is a feature selection algorithm, that uses a feature importance measure to estimate, how important different features are in given dataset for predicting true labels. All the features are ranked by the value of the used measure and the top of these features are selected.

There are a lot of different measures, which can be used in feature selection via feature filtering. And, as we mentioned above, the best way to use filters is to aggregate several of them together. But, the main problem is that there are a lot of different aggregation algorithms exist. The approach to create meta-learning-based filtering feature selection ensemble consists of three following steps. At first, being given a set of filters, apply these filters obraining different feature ranked lists. Then, for the obtained feature ranked lists select the most appropriate rank aggregation algorithm using meta-learning and apply this algorithm obtaining the final features ranks. Finally, the top features from the resulting rank are selected.

6 Conclusion

In this paper, we focused on rank aggregation algorithm selection problem. We described several rank aggregation algorithms and provided formalism to select the best one from the set. We also presented three algorithms to generate artificial ranking lists. These ranking lists were used to compare different rank aggregation algorithms. The baseline we used was the algorithm, which predicts the best ranking algorithm using hidden parameters of data generating process. The experiments showed that the best approach was based on the combination of the basic approaches for meta-feature description and application of feature selection algorithms to the resulting set. Also, we suggested an approach utilizing the presented meta-learning approach for feature selection algorithm ensemble learning.

In the further research, we are to improve meta-learning performance by applying more advanced machine learning techniques. This direction is related to developing a better rank aggregation algorithm selection system. Another direction of the future work is implementation and experimental comparison of the proposed feature selection algorithm ensemble learning.

This work was financially supported by the Government of Russian Federation, Grant 074-U01, and the Russian Foundation for Basic Research, Grant 16-37-60115 mol_a_dk.

References

1. Albert, M.H., Aldred, R.E., Atkinson, M.D., van Ditmarsch, H.P., Handley, B., Handley, C.C., Opatrny, J.: Longest subsequences in permutations. Australasian Journal of Combinatorics **28**, 225–238 (2003)
2. Bachmaier, C., Brandenburg, F.J., Gleißner, A., Hofmeier, A.: On maximum rank aggregation problems. In: Lecroq, T., Mouchard, L. (eds.) IWOCA 2013. LNCS, vol. 8288, pp. 14–27. Springer, Heidelberg (2013)
3. Benner, P., Mehrmann, V., Sorensen, D.C.: Dimension Reduction of Large-Scale Systems, vol. 45. Springer, Heidelberg (2005)
4. Bolón-Canedo, V., Sánchez-Maroño, N., Alonso-Betanzos, A., Benítez, J., Herrera, F.: A review of microarray datasets and applied feature selection methods. Information Sciences **282**, 111–135 (2014)

5. de Borda, J.C.: Mémoire sur les élections au scrutin (1781)
6. Brazdil, P., Carrier, C.G., Soares, C., Vilalta, R.: Metalearning: Applications to Data Mining. Springer Science & Business Media (2008)
7. Burkovski, A., Lausser, L., Kraus, J.M., Kestler, H.A.: Rank aggregation for candidate gene identification. In: Data Analysis, Machine Learning and Knowledge Discovery, pp. 285–293. Springer (2014)
8. Copeland, A.H.: A reasonable social welfare function. In: Seminar on Applications of Mathematics to Social Sciences. University of Michigan (1951)
9. Das, S., Das, A.K.: Sample classification based on gene subset selection. In: Behera, H.S., Mohapatra, D.P. (eds.) Computational Intelligence in Data Mining. AISC, vol. 410, pp. 227–236. Springer, India (2015)
10. DeConde, R.P., Hawley, S., Falcon, S., Clegg, N., Knudsen, B., Etzioni, R.: Combining results of microarray experiments: a rank aggregation approach. Statistical Applications in Genetics and Molecular Biology **5**(1) (2006)
11. Deza, M., Huang, T.: Metrics on permutations, a survey. Journal of Combinatorics, Information and System Sciences. Citeseer (1998)
12. Dietterich, T.G.: Ensemble methods in machine learning. In: Kittler, J., Roli, F. (eds.) MCS 2000. LNCS, vol. 1857, pp. 1–15. Springer, Heidelberg (2000)
13. Dwork, C., Kumar, R., Naor, M., Sivakumar, D.: Rank aggregation methods for the web. In: Proceedings of the 10th International Conference on World Wide Web, pp. 613–622. ACM (2001)
14. Filchenkov, A., Pendryak, A.: Datasets meta-feature description for recommending feature selection algorithm. In: AINL-ISMW FRUCT, pp. 11–18 (2015)
15. Fisher, R.A., Yates, F., et al.: Statistical tables for biological, agricultural and medical research. Statistical Tables for Biological, Agricultural and Medical Research **13**(Ed. 6.) (1963)
16. Garner, S.R., et al.: Weka: the waikato environment for knowledge analysis. In: Proceedings of the New Zealand Computer Science Research Students Conference, pp. 57–64. Citeseer (1995)
17. Giraud-Carrier, C.: Metalearning-a tutorial. In: Proceedings of the 7th International Conference on Machine Learning and Applications, pp. 1–45 (2008)
18. Guyon, I., Gunn, S., Nikravesh, M., Zadeh, L.A.: Feature Extraction: Foundations and Applications, vol. 207. Springer (2008)
19. Jones, N.C., Pevzner, P.: An Introduction to Bioinformatics Algorithms. MIT press (2004)
20. Kekre, H.B., Shah, K.: Performance Comparison of Kekre's Transform with PCA and Other Conventional Orthogonal Transforms for Face Recognition, pp. 873–879. ICETET (2009)
21. Kent, J.T.: Information gain and a general measure of correlation. Biometrika **70**(1), 163–173 (1983)
22. Rice, J.R.: The Algorithm Selection Problem (1975)
23. Saeys, Y., Inza, I., Larranaga, P.: A review of feature selection techniques in bioinformatics. Bioinformatics **23**(19), 2507–2517 (2007)
24. Schalekamp, F., van Zuylen, A.: Rank aggregation: together we're strong. In: Proceedings of the Meeting on Algorithm Engineering & Expermiments, pp. 38–51. Society for Industrial and Applied Mathematics (2009)
25. Smetannikov, I., Filchenkov, A.: Melif: filter ensemble learning algorithm forgene selection. In: Advanced Science Letters (2016, to appear)

26. Wang, G., Song, Q., Sun, H., Zhang, X., Xu, B., Zhou, Y.: A feature subset selection algorithm automatic recommendation method. Journal of Artificial Intelligence Research **47**(1), 1–34 (2013)
27. Wang, R., Utiyama, M., Goto, I., Sumita, E., Zhao, H., Lu, B.L.: Converting continuous-space language models into n-gram language models with efficient bilingual pruning for statistical machine translation. ACM Transactions on Asian and Low-Resource Language Information Processing **15**(3), 11 (2016)
28. Wolpert, D.H., Macready, W.G.: No free lunch theorems for optimization. IEEE Transactions on Evolutionary Computation **1**(1), 67–82 (1997)
29. Zabashta, A., Smetannikov, I., Filchenkov, A.: Study on meta-learning approach application in rank aggregation algorithm selection. In: MetaSel Workshop at ECML PKDD 2015, pp. 115–117 (2015)

Attention Region Based Approach for Tracking Individuals in a Small School of Fish for Water Quality Monitoring

Gang Xiao, Tengfei Shao, Tianqi Zhu, Yi Li, Jiafa Mao, and Zhenbo Cheng(✉)

Department of Computer Science and Technology,
Zhejiang University of Technology, Hangzhou, China
czb@zjut.edu.cn

Abstract. When fish schooling behavior is studied in a laboratory environment, video multi-tracking systems must automatically and correctly track individual fish. However, most multi-tracking systems do not perform well in this regard. To resolve this problem we develop a novel method for tracking fish in schools. The tracking process searches the state of the target fish by using the previous state of that fish, limiting the search to the attention region centered in the target fish. The attention region is then updated according to the new state of the target fish. We apply this method to track fish swimming in small schools, and demonstrate it to achieve up to 99% accuracy. Our method might find application in water quality monitoring.

Keywords: Attention · Fish · Shoal · School · Multi-object tracking

1 Introduction

Fish behaviors have been studied in a variety of applications from water quality monitoring to use in toxicity identification evaluation [1]. Water quality can be monitored by examining changes in fish behavior, such as their swimming speed, breathing frequency [2], and even tail-beating frequency [3]. Usually the behavior of individual fish is monitored [4], but this might reduce experimental robustness as differences among fish increase the uncertainty of an assay. Therefore, behavioral characteristics of a group of fish may be better for detecting the presence of toxins in water. As fish tend to swim in shoals or schools [5], their self-organized schooling behavior has attracted considerable interest from theoretical biologists [6], particularly those in the search of rules that explain the way a schools properties emerge from the behavioral mechanisms of its constituents [7].

Zhenbo Cheng—This work was supported by the National Natural Science Foundation of China (61272310), and Post-Doctor Natural Science Foundation of ZheJiang Province (BSH1502033) to Zhenbo Cheng.

P. Perner (Ed.): MLDM 2016, LNAI 9729, pp. 756–760, 2016.
DOI: 10.1007/978-3-319-41920-6_57

Laboratory studies on schooling behavior require video multi-tracking systems to automatically and accurately track individual fish. However, most tracking systems struggle in this regard [1]. Misidentification leads to error, including the loss of individual fish identity and the swapping of identity between individuals. We propose a tracking system based on attention map which automatically detects and tracks individuals within a school in the laboratory environment using video. Our method takes inspiration from the way humans perform visual tracking tasks by continually moving the fovea to follow the target in an attention region, and ignoring objects outside of that region [7] .

2 Methods

Red crucian carp (57 cm total length) were placed into an acrylic tank of 50 cm 30 cm filled with water to 5 cm depth. A CMOS camera (Imavision MER-200-20UC(-L); Daheng Science and Technology, Beijing, China) was placed at 35 cm above the tank to capture the entire arena (Fig. 1a). The camera took 15 frames per second, with images of 640 × 480 pixels. To test tracking performance, 900 consecutive frames of imagery of carp were taken at three densities (3, 4 and 5 fish) were examined. There were 10 trials for each of the densities. Before fish were added to the tank, we took 3 seconds video used for the background frame. Raw video files were processed using custom software written in C++ using the Open CV library.

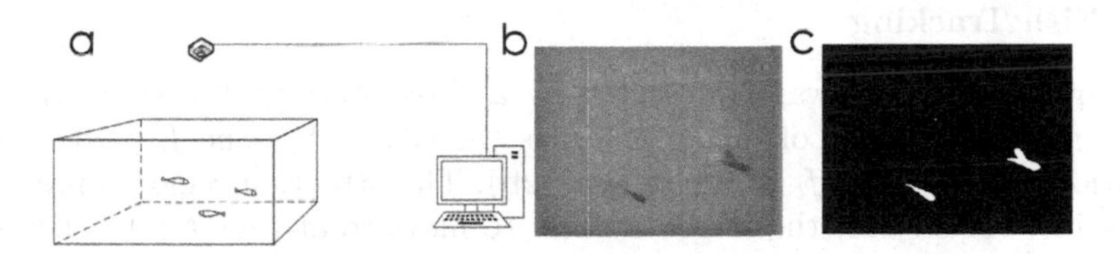

Fig. 1. a) Tank environment (50 × 30 × 5 cm). b) Typical (original) image. c) There are two blobs in the segmentation of the original image on background subtraction, of which the right comprises two overlapping fish.

Our method tracked individual fish within a school in two steps. After detecting the fish from the video frames by using the differential between each frame and the background image, our method first searched the state of the target fish by using the previous state of that fish. The state mainly included the contour, the position and the attention region of the target fish. The search was limited to the attention region centered in the target fish. Second, the attention region was then updated according to the new state of the target fish.

2.1 Fish Detection

The tank (Fig. 1a) background (Fig. 1b) was almost constant as the fish swam, and of a color that contrasted with the fish. Therefore, the background image was made by averaging the first 45 frames in the first 3 s, during which there are no fish in the test tank. Next, image segmentation was done by subtracting each frame from the background image. Each segmentation frame was then converted to grayscale and inverted to a blob (Fig. 1c). The blob included a set of contiguous pixels that belonged to one or several overlapping fish. In addition, the contour of each blob was identified by applying an edge-detection algorithm on the inverted frame. The centroid was then obtained from the average coordinates of the points on the contour.

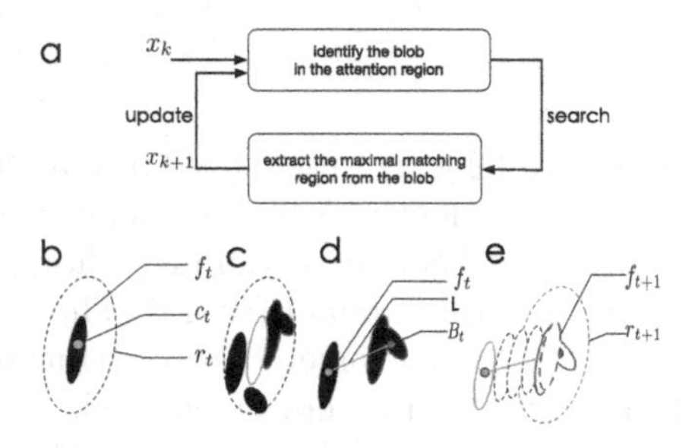

Fig. 2. a) Steps of our tracking method. b), c), d) for the first step and e) for the second step.

2.2 Fish Tracking

The target fishs state was represented by a three state vector $x_t = [f_t, c_t, r_t]$, where f_t is the contour of the target, c_t is the centroid of the f_t, and r_t is the attention region of the f_t at time t (Fig. 2b). The attention region represents a probability region that the target is likely to move to at time $t+1$. The r_t can be calculated from the maximum swimming distance of the fish in $1/15th$ of a second. Therefore, the targets next state x_{t+1} can be determined in the attention region r_t. After determining the region of the targets next state, we detected its contour at time $t+1$. Generally, the trajectories of the target fish and others may cross each other and form a blob including two or more fish. Therefore, we needed to extract the contour that belonged to the target individual from the blob that belonged to several individuals. In the event there were several blobs in the region r_t the blob with the maximum overlapping area with the target ft was chosen as the blob B_{t+1} that could include the target (Fig. 2c). We then calculated the centroid cb_{t+1} of the blob B_{t+1}. The contour of the target fish f_{t+1} at time $t+1$ was extracted from the blob B_{t+1} according to:

$$f_{t+1} = arg \max_{c \ along \ L} [area(B_{t+1}) - area(f_t(c))], \tag{1}$$

where the line L is drawn from the centroid (c_t) of the target fish at time t to the centroid cb_{t+1}. We moved the contour f_t along the line L to search the maximal matching region between the contour f_t and the blob B_{t+1} (Fig. 2d).

The maximal matching region was extracted from the blob B_{t+1} as the contour f_{t+1} of the target fish at time $t+1$ (Fig. 2e). Finally, the next state of the target fish $x_{t+1} = [f_{t+1}, c_{t+1}, r_{t+1}]$ was updated according to the contour f_{t+1} at time $t+1$.

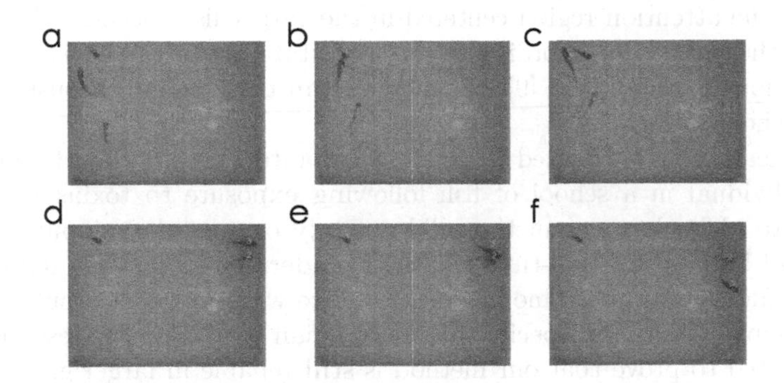

Fig. 3. Tracking result using our method in two kinds of fish cross. a) and d): three fish in the tank before crossing. b) and e): the trajectories of two fish cross each other. c) and f): the identities are correct after crossing. Fish are uniquely identified by a number.

3 Results

After manually validating identifications in portions of video without fish overlaps we found a mean performance of 99% correct trajectories. Individual fish identifications also remained accurate after their trajectories crossed (Fig. 3).

We compared fish trajectories using our method with those of idTracker [8](Fig. 4a, b) and found our method to maintain correct identities, even during fish overlaps. Furthermore, as fish density increased, the tracking performance of our method did not change significantly (Fig. 4c).

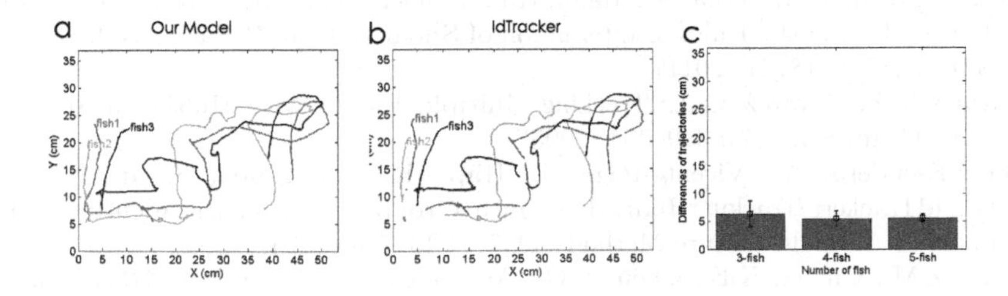

Fig. 4. Trajectories of fish obtained by a) our method and b) idTracker. The line discontinuity depicts where identities are lost. c) The differences of trajectories between two methods with different densities of fish (3, 4 and 5). Error bar indicates s.e.

4 Discussion

We propose an effective method for detecting and tracking individual fish within schools that integrates searching and updating steps. This method first searches the state of a target fish by using the previous state of that fish, limiting the search to the attention region centered in the target fish. Second, the attention region is then updated according to the new state of the target fish. Compared with previous methods [7] , [9], our method can track simultaneously all fish in a small school in real-time.

Our method has been used successfully for automatically detecting the death of an individual in a school of fish following exposure to toxins in water. In contrast to idTracker, our method did not rely on the whole input videos and thus could be used for real-time tracking, rendering our method more suitable for real-time water quality monitoring. To date we have verified our method on small groups of fish, and specifically, red crucian carp. Further research should be conducted to prove that our method is still reliable in larger groups and for other faster species.

References

1. Delcourt, J., Denoel, M., Ylieff, M., Poncin, P.: Video Multitracking of Fish Behaviour: a Synthesis and Future Perspectives. Fish Fish. **14**(2), 186–204 (2013)
2. Cairns Jr., J., Dickson, K.L., Sparks, R.E., Waller, W.T.: A Preliminary Report on Rapid Biological Information Systems for Water Pollution Control. Journal (Water Pollution Control Federation), 685–703 (1970)
3. Xiao, G., Feng, M., Cheng, Z.B., Zhao, M.R., Mao, J.F., Mirowski, L.: Water Quality Monitoring Using Abnormal Tail-Beat Frequency Of Crucian Carp. Ecotox. Environ. Safe. **111**, 185–191 (2015)
4. Kuklina, I., Kouba, A., Kozk, P.: Real-Time Monitoring of Water Quality Using Fish and Crayfish as Bio-Indicators: A Review. Environmental Monitoring and Assessment **185**(6), 5043–5053 (2012)
5. Lopez, U., Gautrais, J., Couzin, I.D., Theraulaz, G.: From Behavioural Analyses To Models Of Collective Motion In Fish Schools. Interface Focus. **2**(6), 693–707 (2012)
6. Herbert-Read, J.E., Perna, A., Mann, R.P., Schaerf, T.M., Sumpter, D.J.T., Ward, A.J.W.: Inferring the Rules of Interaction of Shoaling Fish. P Natl. Acad. Sci. USA **108**(46), 18726–18731 (2011)
7. Cavanagh, P., Alvarez, G.A.: Tracking Multiple Targets with Multifocal Attention. Trends. Cogn. Sci. **9**(7), 349–354 (2005)
8. Perez-Escudero, A., Vicente-Page, J., Hinz, R.C., Arganda, S., de Polavieja, G.G.: IdTracker: Tracking Individuals in a Group by Automatic Identification of Unmarked Animals. Nature Methods **11**(7), 743–748 (2014)
9. Qian, Z.M., Cheng, X.E., Chen, Y.Q.: Automatically Detect and Track Multiple Fish Swimming in Shallow Water with Frequent Occlusion. Plos One **9**(9), e106506 (2014)

FSMS: A Frequent Subgraph Mining Algorithm Using Mapping Sets

Armita Abedijaberi(✉) and Jennifer Leopold

Department of Computer Science, Missouri University of Science and Technology, Rolla, MO 65401, USA
{aan87,leopoldj}@mst.edu

Abstract. With the increasing prevalence of data that model relationships between various entities, the use of a graph-based representation for real-world problems offers a logical strategy for organizing information and making knowledge-based decisions. In particular, often it is useful to identify the most frequent patterns or relationships amongst the data in a graph, which requires finding frequent subgraphs. Algorithms for addressing that problem have been proposed for over 15 years. In the worst case, all subgraphs in the graph must be examined, which is exponential in complexity, and subgraph isomorphisms must be computed, which is an *NP-complete* problem. Frequent subgraph algorithms may attempt to improve the actual runtime performance by reducing the size of the search space, avoiding duplicate comparisons, and/or minimizing the amount of memory required for compiling intermediate results. Herein we present a frequent subgraph mining algorithm that leverages mapping sets in order to eliminate the isomorphism computation during the search for non-edge-disjoint frequent subgraphs. Experimental results show that absence of isomorphism computation leads to much faster frequent subgraph detection when there is a need to identify all occurrences of those subgraphs.

Keywords: Frequent subgraphs · Graph mining · Mapping sets

1 Introduction

Data are used every day to make knowledge-based decisions for the purposes of marketing, emergency management, law enforcement, and other applications. Large companies, such as Google, use data to provide predictive, quick searches based on related data [1]. Governments use a variety of inter-related data to help predict terrorist attacks [2]. Data mining also has been used in the health sciences to correlate multiple gene expression patterns for cancer to identify patients who are most at risk [3]. As more data become available, relationship mapping has become even more integral. The use of a graph-based representation for real-world problems offers a logical strategy for organizing information and making knowledge-based decisions. Currently, graph data are used for visualizing computer networks [4], biological networks [5], and the Internet [6].

P. Perner (Ed.): MLDM 2016, LNAI 9729, pp. 761–773, 2016.
DOI: 10.1007/978-3-319-41920-6_58

One of the most common applications of graph data mining (GDM) is to identify the most frequent relationships or patterns amongst the data in a graph, which requires finding frequent subgraphs. In a *graph-transaction setting*, the input will be a collection of relatively small graphs, which necessitates that the search for frequent subgraphs be performed over each individual graph in the collection before those results can be combined; this is in contrast to a *single-graph setting* where the input is a single (presumably large) graph. Formally, we define the Frequent Subgraph Mining (FSM) problem using Definitions 1-4 given below.

Definition 1. *A labelled graph $G = (V, E, L_V, L_E)$ consists of a set of vertices V, a set of undirected or directed edges E, and two labeling functions L_V and L_E that associate labels with vertices and edges, respectively.*

It should be noted that the labels of any two vertices (or any two edges) may not be unique. However, each vertex (and each edge) will have a unique *id*.

Definition 2. *A graph $S = (V_S, E_S, L_{V_S}, L_{E_S})$ is a subgraph of $G = (V, E, L_V, L_E)$ iff $V_S \subseteq V$, $E_S \subseteq E$, $L_{V_S}(v) = L_V(v)$ and $L_{E_S}(e) = L_E(e)$ for all $v \in V_S$ and $e \in E_S$.*

Definition 3. *A subgraph isomorphism of S to G is a one-to-one function f: $V_S \longrightarrow V$ where $L_{V_S}(v) = L_V(f(v))$ for all vertices $v \in V_S$, and for all edges $(u, v) \in E_S$, $(f(u), f(v)) \in E$ and $L_{E_S}(u, v) = L_E(f(u), f(v))$.*

Definition 4. *Let I_S be the set of isomorphisms of a subgraph S in graph G. Given a minimum support threshold τ, the frequent subgraph mining problem (FSM) is to find all non-edge-disjoint subgraphs S in G such that $|I_S| \geq \tau$.*

The advantage of limiting frequent subgraphs to only those with disjoint edges is computational tractability [7]. But this comes at the expense of disregarding potentially useful information. Consider the example shown in [Fig. 1]. If only edge-disjoint isomorphic subgraphs are considered, the vertex labelled g will not be included in any frequent subgraph when minimum support is ≥ 2. However, if we allow isomorphic subgraphs to share edges, then the subgraph containing the vertices labelled t, m, and g will have frequency 2.

The organization of this paper is as follows. Section 2 provides a brief overview of related work in graph data mining, focusing on graph theory based approaches. In Section 3, we discuss the FSMS algorithm. The results of running an implementation of our algorithm on a variety of graphs are provided in Section 4. Finally, we discuss our plans for future work in Section 5 and conclusions in Section 6.

2 Related Work

In [8], the authors divided existing GDM algorithms into three main categories: Graph Theory Based, Inductive Logic Programming, and Greedy Search.

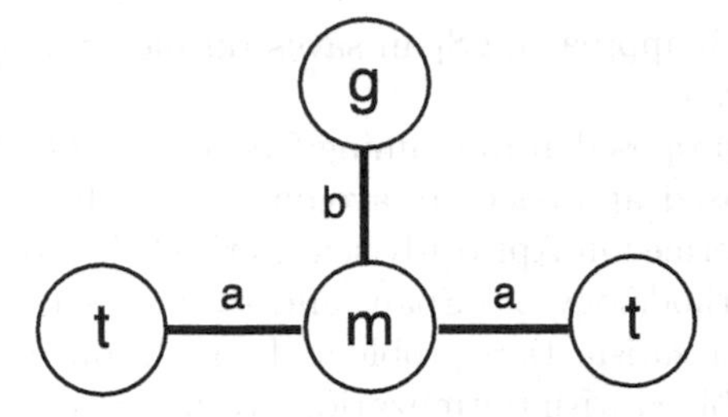

Fig. 1. Example of a graph where a particular vertex (i.e., g) will only be included in a frequent subgraph when minimum support is ≥ 2 if we allow isomorphic subgraphs to share edges (i.e., (g, m)).

Our research focuses on the Graph Theory Based category, which consists of two main groups: Apriori-based approaches and pattern growth-based approaches. Algorithms in the first group generate candidate subgraphs by joining two frequent subgraphs of size k (where k is the number of vertices or the number of edges, depending on the definition used in the particular algorithm). Two subgraphs must contain the same size $k-1$ connected subgraphs to form a new candidate subgraph of size $k+1$ in a level-wise manner. Pattern growth algorithms generate candidates by adding a new edge to a smaller frequent subgraph in all possible extensions.

AGM [9], an Apriori-based graph mining algorithm, uses a breadth-first search to discover frequent subgraphs. It starts with frequent subgraphs consisting of one edge, and iteratively combines frequent subgraphs of equal size to form larger frequent subgraphs. The algorithm is based on the fact that a $k+1$ subgraph cannot be frequent if its immediate parent subgraph of size k is not frequent [9]. FSG [10] is another Apriori-based approach that uses breadth-first search to find all frequent subgraphs by incrementally expanding size k frequent subgraphs by one edge to generate $k+1$ sized candidate subgraphs. Both algorithms do not scale well to large graphs because they tend to incur significant computational overhead due to large candidate pools and subgraph isomorphism comparisons either explicitly or implicitly. The breadth-first search strategy has an advantage in that it allows for the pruning of infrequent subgraphs at an early stage in the FSM process. However, that gain in efficiency comes at the cost of high I/O and memory usage.

Yan et al. proposed gSpan in [11], a pattern growth-based approach that uses depth-first search and a lexicographic ordering to map each subgraph pattern to construct a search tree of frequent subgraphs. The search code tree is built in a depth-first manner by one-edge expansion of the parent vertex. This tree only contains frequent subgraphs that have a smaller depth-first search code than any other isomorphic subgraph in the tree, thus reducing search cost. Additionally, gSpan uses a hash table to keep track of subgraphs already searched, thereby circumventing the processing of any subgraph more than once. Using an efficient coding of graphs, finding previously examined subgraphs is fast. Due to the use of a depth-first search approach, gSpan saves on memory usage in exchange for less efficient pruning.

A more recently proposed graph mining framework, GraMi [12], also employs a pattern growth-based approach. It avoids the costly enumeration of candidate generation performed in Apriori-based methods by adopting the depth-first search code tree methodology of gSpan, and evaluates the frequency of a subgraph as a constraint satisfaction problem. In addition, the authors propose a heuristic search combined with optimizations to improve the performance of the algorithm, employing strategies to prune the search space, postpone searches, and explore special graph types. However, GraMi will not necessarily find all frequent subgraphs because it will only extend a subgraph by frequent edges. GraMi can only guarantee that it will not return infrequent subgraphs (i.e., no false positives).

The above mentioned algorithms take different approaches to perform subgraph isomorphism comparison and to reduce the number of candidate subgraphs generated. Due to their computational complexity, these algorithms typically are used in graph-transaction settings where the individual graphs are relatively small in terms of number of vertices. In general, pattern growth methods tend to be faster than Apriori-based methods, with gSpan being one of the fastest of all FSM algorithms [13]. Like gSpan and GraMi, our algorithm, FSMS, is a pattern growth approach. FSMS iteratively extracts the frequent subgraphs using breadth-first search. However, it differs from other FSM algorithms in that it does not compute isomorphisms, which is the most expensive part of frequent subgraph mining. Instead, it keeps track of isomorphic subgraphs by using a $Mapping\text{--}Set$ system. In terms of runtime, FSMS cannot be compared fairly to gSpan, because gSpan cannot tell us how many occurrences of each frequent subgraph exist in a graph nor can the algorithm easily be modified to do so; in contrast, FSMS returns the actual instances of each frequent subgraph found in a graph. Hence, the preference for which algorithm to use (i.e., gSpan or FSMS) depends on the needs of the application. There is also another framework, MRFSE [14], bulit based on gSpan algorithm. MRFSE uses MapReduce to parallelise gSpan applicable on large datasets. MRFSE keeps the embedding for each frequent subgraph and instead of performing isomorphism, it compares their minimum DFS code [11] to check whether they are isomorphic. However FSMS keeps the bijection between vertices in the frequent subgraphs and group the equivalent vertices together using $MappingSets$. Hence, generated candidates in a group are already isomorphic. The current version of FSMS can be applied on relatively big datasets running on a centralized machine very efficiently.

3 FSMS

3.1 Preliminaries

FSMS can be applied to a directed or undirected graph $G = (V, E, L_V, L_E)$ that consists of a set of vertices V, a set of edges E, and two labeling functions L_V and L_E that associate labels with vertices and edges, respectively. The algorithm is based on an adjacency list representation of the graph. We assume that the label function $L_V(L_E)$ facilitates an efficient lookup of all vertices (edges) with a particular label (since those labels are not unique); similarly, we assume that, given a vertex (edge), we can find its label in $O(1)$ time. In discussions where distinction between vertex (edge) ids is not important, we will refer to vertices (edges) simply by their labels.

In the process of frequent subgraph mining, calculation of the frequency of a subgraph candidate involves subgraph isomorphism which is a *NP-complete* problem and is considered as a bottleneck for the algorithm. To tackle this issue, herein we present a novel idea for frequent subgraph mining without performing any subgraph isomorphism, which consequently leads to a faster process that is applicable on large graphs. The FSMS algorithm consists of k iterations in order to find all frequent subgraphs of different sizes from 1 to k. During the procedure of subgraph mining, by the end of iteration i, instances of frequent subgraphs of size i are found and candidate subgraphs of size $i+1$ will be built from them. The novelty of FSMS candidate generation process comes in replacing isomorphism computation by a *Mapping Sets* system to simplify subgraph frequency counting. This idea is based on the fact that when two or more isomorphic subgraphs expand from the equivalent vertices (in terms of *bijection* between isomorphic subgraphs) with same-labeled vertices and same-labeled edges, the extended graphs will be automatically isomorphic to each other.

In graph theory, an isomorphism of graphs G and H is a *bijection* between the vertex sets of G and H. A *bijection* is a one-to-one correspondence between the elements of two sets, where every element of one set is paired with exactly one element of the other set, and every element of the other set is paired with exactly one element of the first set. There are no unpaired elements. This kind of bijection is also known as *edge-preserving bijection*.

The FSMS algorithm starts with finding the frequent edges in the graph G as frequent subgraphs of size one. Each frequent subgraph S is represented by a triple of $(Key, Values, MappingSets)$. A Key is a set of *labels* of vertices in S, $Values$ is a list of instances of S, and a $MappingSet$ is a set with respect to each vertex in S that contains *ids* of corresponding vertices in instances of S in accordance with the *bijection* among them.

For example, in [Fig. 2] when $\tau = 2$, edge $\{R\text{–}a\text{–}B\}$ is a frequent subgraph S of size 1 with 3 instances in the graph G including $\{(1,3)\}$, $\{(7,8)\}$ and $\{(10,11)\}$ where each instance is sorted in lexicographic order. This frequent subgraph S can be represented by Key, $Values$ and $MappingSets$ as follows:

$$Key : \{(R\text{–}a\text{–}B)\}, \quad Values : \{\{(1,3)\}, \{(7,8)\}, \{(10,11)\}\}$$

$$MappingSets : \begin{Bmatrix} MS(R) : \{1,7,10\} \\ MS(B) : \{3,8,11\} \end{Bmatrix}$$

In one-edge subgraph S [Fig. 2], for each vertex, a $MappingSet$ is associated with it that includes those vertex *ids* in all instances of S that correspond to each other, $MS[l] = \{d_1, d_2, ..., d_n\}$ where d_i represents a unique vertex *id*. This means that all of these vertices are equivalent and play the same role in S, hence

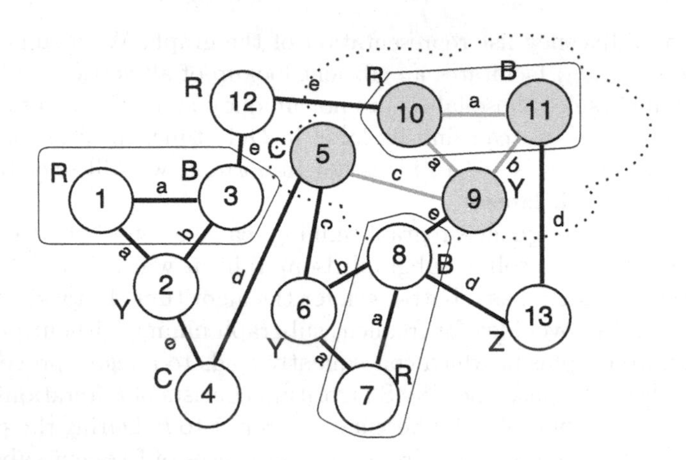

Fig. 2. Example of a graph G with 13 vertices and 18 edges, where each vertex (edge) has a *label* and a unique *id*. Solid lines outline all the instances of one-edge subgraph $\{R\text{–}a\text{–}B\}$ in G, where R and B are the vertex labels and a is the edge label. A dashed line outlines one of the instances of subgraph $\{\{R\text{–}a\text{–}B\}, \{B\text{–}b\text{–}Y\}, \{Y\text{–}a\text{–}R\}, \{Y\text{–}c\text{–}C\}\}$ in G.

they are grouped together in a $MappingSet$ such as $MS(R) : \{1, 7, 10\}$. In other words, a *bijection* function for 3 instances of S maps 3 corresponding vertices (i.e. one vertex per each instance) to each other [Fig. 3]. All $MappingSets$ in regard to a frequent subgraph have the same length. Consequently, the number of all $MappingSets$ and $Values$ for a frequent subgraph represents the size and the number of instances for that particular subgraph, respectively.

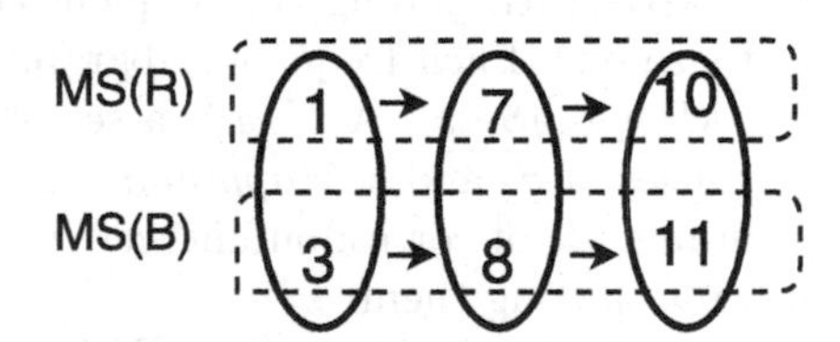

Fig. 3. Example of *bijection* mapping among the vertices of 3 isomorphic one-edge subgraph $\{R\text{–}a\text{–}B\}$ in graph G in [Fig. 2].

Candidate subgraphs of size i+1 are generated from subgraph S of size i by iterating through all $MappingSets$ of S. If the size of $Values$ for S (i.e. frequency of S in G) is greater than or equal to the frequency threshold τ, S is considered frequent and during the next iteration of FSMS it will be used to generate candidate subgraphs of larger size. For each $MappingSets$ MS of S, new candidate subgraphs are formed as follows: neighbors of vertices in each MS with the same label that are connected via same-labeled edges are grouped together. Since all vertices in one MS have the same role in S, if they extend with links that share the vertex label and the edge label, all the extended subgraphs will be automatically isomorphic to each other.

For example, suppose FSMS is in iteration 4, and a frequent subgraph of size 4 such as $\{\{R\text{–}a\text{–}B\}, \{B\text{–}b\text{–}Y\}, \{Y\text{–}a\text{–}R\}, \{Y\text{–}c\text{–}C\}\}$ is found in G, outlined by dashed lines in [Fig. 2]. This subgraph has 2 instances in the graph and is represented as:

$$Key : \{\{R\text{–}a\text{–}B\}, \{B\text{–}b\text{–}Y\}, \{Y\text{–}a\text{–}R\}, \{Y\text{–}c\text{–}C\}\}$$

$$Values : \left\{ \begin{matrix} \{(5,6),(6,7),(6,8),(7,8)\}, \\ \{(5,9),(9,10),(9,11),(10,11)\} \end{matrix} \right\}$$

$$MappingSets : \left\{ \begin{matrix} MS(R) : \{7,10\}, MS(B) : \{8,11\} \\ MS(Y) : \{6,9\}, MS(C) : \{5,5\} \end{matrix} \right\}$$

As depicted in [Fig. 2], in $MS[B]$ two of the vertices, 11 and 8 have a connection with vertex label Z and edge label d, and this connection is not already part of the subgraph $\{\{R\text{–}a\text{–}B\}, \{B\text{–}b\text{–}Y\}, \{Y\text{–}a\text{–}R\}, \{Y\text{–}c\text{–}C\}\}$. In those instances of a subgraph that 8 and 11 belongs to, we add an edge with label d from a vertex with label B to a vertex with label Z; hence 2 instances of a subgraph $\{\{R\text{–}a\text{–}B\}, \{B\text{–}b\text{–}Y\}, \{Y\text{–}a\text{–}R\}, \{Y\text{–}c\text{–}C\}, \{B\text{–}d\text{–}Z\}\}$ will be generated which are known to be isomorphic to each other. For the new candidate subgraph, Key, $Values$ and $MappingSets$ are shown below:

$$Key : \{\{R\text{–}a\text{–}B\}, \{B\text{–}b\text{–}Y\}, \{Y\text{–}a\text{–}R\}, \{Y\text{–}c\text{–}C\}, \{B\text{–}d\text{–}Z\}\}$$

$$Value : \left\{ \begin{matrix} \{(5,6),(6,7),(6,8),(7,8),(8,13)\}, \\ \{(5,9),(9,10),(9,11),(10,11),(11,13)\} \end{matrix} \right\}$$

$$MappingSets : \left\{ \begin{matrix} MS(R) : \{7,10\}, MS(B) : \{8,11\} \\ MS(Y) : \{6,9\}, MS(C) : \{5,5\} \\ MS(Z) : \{13,13\} \end{matrix} \right\}$$

As mentioned, all possible candidate subgraphs of size i+1 are generated after iteration of all available mapping sets of frequent subgraphs of size i. However, it is possible that among the candidate subgraphs, some of the $Values$ might become duplicated, due to the fact that instances of different subgraphs might become exactly the same after extension [Fig. 4]. To avoid listing the same instance of a candidate multiple times, a hash table is used to keep track of visited subgraphs. A key in this hash table is a sorted list of vertex ids, and values for this key are values of different subgraphs generated with those ids that have been encountered so far. In the hash table, each instance value is represented as a list of tuples sorted in a lexicographic order, where each tuple is representing an edge in the instance by vertex ids at both ends of that edge. Each time the $Values$ of a candidate subgraph are generated, FSMS will search for each value in the hash table using the sorted distinct vertex ids in that value. If a generated instance already exists in the hash table, it means that instance has already been encountered, so the value for that instance will be

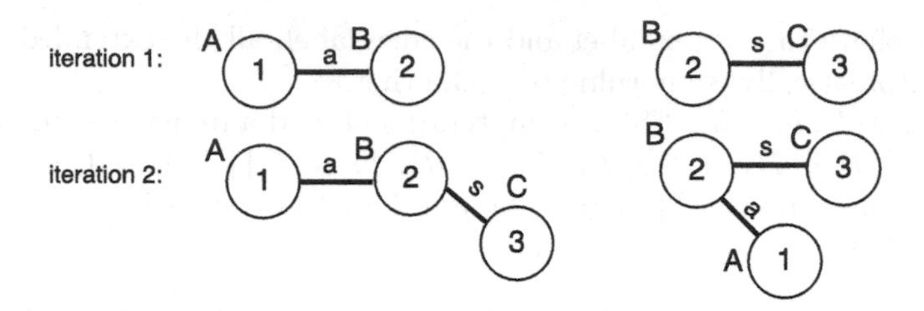

Fig. 4. In this example only a part of a bigger graph is shown. Assume that after iteration 1 in FSMS, two subgraphs $\{A\text{–}a\text{–}B\}$ and $\{B\text{–}s\text{–}C\}$ have been detected as frequent. Each of them have their own Key, $Values$ and $MappingSets$. During the iteration 2, using completely different $MappingSets$, two instances have become similar. To avoid counting an instance more than once, FSMS uses a hash table.

discarded. Otherwise, that value remains and the hash table will be updated. FSMS uses an in-memory pattern growth approach, and to perform frequent subgraph extraction on a centralized machine more efficiently by the end of each iteration the content of the hash table can be deleted.

3.2 Algorithm

Input to the FSMS algorithm is a graph $G = (V, E, L_V, L_E)$ and the value of minimum support τ for determining frequent subgraphs. Output from the algorithm is a set of frequent subgraphs, each of which contains at least one edge. In a graph-transaction setting, the FSMS algorithm would be executed for each transaction graph in an input collection, and then the resulting set of frequent subgraphs over the entire collection would be the intersection of the result sets obtained for each individual graph.

The FSMS algorithm initially builds a set containing all frequent edges in G based on the value of minimum support. Each frequent subgraph regardless of the size is represented as a triple of $(key, values, mappingSets)$. At this point all frequent subgraphs of size one have been found, and FSMS looks for candidate subgraphs that are one edge larger in size. Starting with a frequent subgraph S, each mappingSet MS contains a list of vertex ids where each belongs to one of the instances of S. By grouping neighbors of vertices in MS based on their label and their edge label, FSMS generates candidates of larger size. There also is $hashFreqSGs$ which is a hash table that keeps track of subgraphs that have been found. So any time a new instance of a subgraph is found, it is checked through the $hashfreqSubgraph$. If the length of values for each key is greater than or equal to the minimum support, that subgraph will be added to $freqSubgraphs$ and $hashFreqSGs$. Otherwise it will be discarded. This process continues until FSMS finds no more candidates.

Algorithm 1. FSMS

```
input  : Graph G(V, E, L_V, L_E) and minimum support τ ≥ 1
output: Collection of subgraphs with at least 1 edge that occurs in G a
        minimum of τ times
1  freqEdges: set of all frequent edges in G
2  freqSubgraphs: set of single-edge graphs built from freqEdges, each
   represented as a triple of (key, values, mappingSets)
3  mostRecentSubgraphs ← freqSubgraphs
4  hashFreqSGs ← ∅
5  while mostRecentSubgraphs ! = ∅ do
6      newSubgraphs ← ∅
7      for subgraph S ∈ mostRecentSubgraphs do
8          for mapSet MS ∈ S.mappingSets do
9              group neighbors of vertices ∈ MS with the same vertex label
               and same edge label
10             for each group g do
11                 candidKey ← extended S.key with the new vertex label and
                   edge label
12                 candidValues ← extended S.values with id ∈ g
13                 for value ∈ candidValues do
14                     index = a sorted list of distinct vertex ids in value
15                     if index ∉ hashFreqSGs then
16                         hashFreqSGs[index] ← ∅
17                     else if candidValues ∈ hashFreqSGs[index] then
18                         candidValues = candidValues \value
19                     else
20                         hashfreqSGs[index] ← hashFreqSGs[index] ∪ value
21                 candidMappingSets ← extended S.mappingSets based on
                   candidValues ∈ g
22                 if |candidValues| ≥ τ then
23                     S′ = (candidKey, candidValues, candidMappingSets)
24                     newSubgraphs ← newSubgraphs ∪ S′
25     hashFreqSGs ← ∅
26     mostRecentSubgraphs ← newSubgraphs
27     freqSubgraphs ← freqSubgraphs ∪ newSubgraphs
28 return freqSubgraphs
```

3.3 Algorithm Complexity

Let N and n be the number of vertices of graph G and subgraph S, respectively. In general, most algorithms that find frequent subgraphs (e.g., AGM, FSG) are composed of two parts. One part requires, in the worst case, the generation of all subgraphs of G, which takes $O(2^{N^2})$ time. The second part performs the calculation of the frequency of a subgraph S in G, which involves the computation of subgraph isomorphisms. Subgraph isomorphism is NP-complete with $O(N^n)$ time complexity. Therefore, the complexity of frequent subgraph mining is $O(2^{N^2}N^n)$, which is exponential in the problem size. However, one advantage of FSMS over other approaches is that it replaces isomorphism computation by creating and updating Mapping Sets for each frequent subgraph during each iteration. In the worst case, each MappingSets contains n MSs, one MS per vertex. For each MS, FSMS finds neighbors of n vertices and group them together. We are using adjacency list instead of adjacency matrix to store graph G, so for each vertex it takes $O(\frac{m}{n})$ to enumerate its neighbors and group them since the neighbor's labels are scanned and aggregated in one pass. Hence, the total complexity for each Mapping Sets in order to be updated for the next iteration is $O(mn)$. There are maximum number of $m-1$ iteration and the average number of discovered frequent subgraphs during each iteration is c. Therefore, the overall complexity of FSMS is cm^2n.

4 Experiments

4.1 Datasets

In this section, we experimentally evaluate FSMS, and for performance evaluation comparison, we have also implemented FSG [10]. Both FSMS and FSG follow a pattern growth approach and they return all the occurrences of frequent subgraphs of different sizes. All experiments are conducted using Python 3.3 on a Linux (Ubuntu 14.04) machine with 8 cores running at 2.67GHz with 64GB RAM. For the experiments we used real chemical datasets [15] with different numbers of vertices and edges. There are no multi-edges (two or more edges that are incident to the same two vertices) or self-loops (an edge from a vertex to itself) in the graphs. These datasets contain connected and disconnected graphs with labelled vertices and labelled edges.

4.2 Results

The difference between the amount of time required by FSG and FSMS to find all the frequent subgraphs of graphs with various sizes is shown in [Fig. 5]. In all of the experiments the minimum support for both FSG and FSMS was the same and they both discovered the same number of frequent subgraphs. However due to the isomorphism-free nature of FSMS, the performance of these two algorithms is drastically different from each other. In some cases, we aborted the computation for FSG because it was not able to finish the mining process in a reasonable amount of time.

In our experiments, several different values of minimum support were tested; [Table 1] shows FSMS s performance for a representative selection of the graphs. In these chemical datasets, the numbers of distinct vertex labels and distinct edge labels are very low. Hence there exist an extensive number of instances for different sizes of subgraphs in each graph. Because of that, we used relatively high

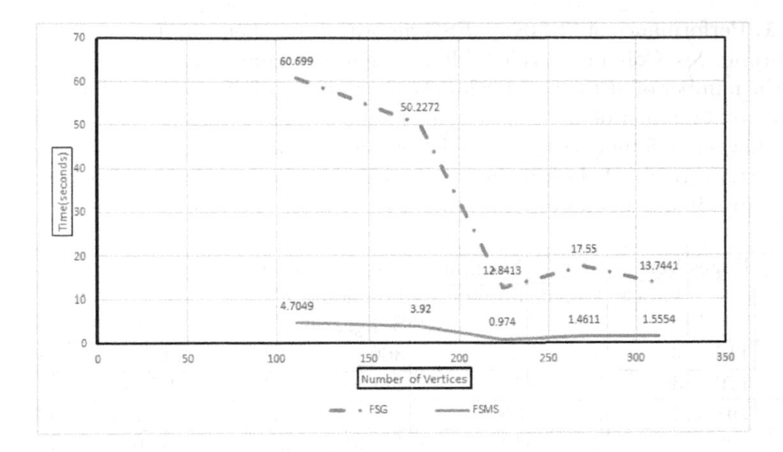

Fig. 5. Comparing the performance of FSMS vs. FSG for frequent subgraph mining in graphs with different numbers of labelled vertices and labelled edges. For both algorithms the same minimum support was selected. However we had to keep the minimum support threshold high to be able to compare the performance of FSG and FSMS; this was because FSG was not able to finish the processing for low minimum support.

minimum support for our experiments, and still the total number of discovered frequent subgraphs by FSMS and FSG are fairly large as shown in [Table 1]. Considering the large number of discovered frequent subgraphs and comparing the performance of FSG and FSMS, we can see that the FSMS algorithm requires a short amount of time to discover a large number of graphs without performing isomorphism.

5 Future Work

Many current graph mining algorithms assume that a graph can be read in to memory and processed on a single, commodity machine. However, that is not always the case; dataset sizes are growing exponentially. According to Cisco, global cloud data will reach approximately 715 *EB* per month by the end of 2018 [16]. Technologies to be considered for that purpose include Hadoop, as well as Pegasus [17], a big graph mining system built on top of MapReduce. This will require a redesign of the FSMS algorithm to remove linear dependencies. The distributed implementation of FSMS will be benchmarked against other existing distributed frequent subgraph mining programs such as MiRage [18] to compare its performance.

Additionally, we want to find more efficient ways of reading and writing graphs, which we believe will dramatically increase the speed, especially in a distributed implementation. Methods could include compression and/or SQL representation of the graph input and output. We also want to work on improving the memory usage of the algorithm. Depending upon the type of the graph, the

Table 1. Performance of FSMS vs. FSG algorithm on chemical datasets with different graph properties. Columns in this table represent the number of vertices, the number of edges, the number of distinct vertex labels, the number of distinct edge labels, minimum support, total number of discovered frequent subgraphs of different sizes, the maximum size of subgraphs found by FSMS and FSG, and running time of FSMS and FSG in seconds, respectively. A dash in the table means we had to abort the computation for FSG because it was taking too long time to complete execution.

Vertices	Edges	Vertex Labels	Edge Labels	Minimum Support	Total # of Frequent Subgraphs	Iterations	FSMS(sec)	FSG(sec)
111	119	3	2	30	360	21	4.70	839.69
111	119	3	2	13	11756	31	97.07	-
178	198	3	2	35	366	35	3.92	344.22
178	198	3	2	13	11840	30	84.64	-
224	252	3	2	52	79349	39	3646.12	-
270	306	3	2	60	6407	30	324.75	-
270	306	3	2	150	15	7	1.46	17.55
313	356	3	2	60	88381	38	4564.31	-
351	402	3	2	70	155973	31	8439.33	-
407	363	3	2	70	169879	31	9144.58	-
432	493	5	2	80	70981	30	5144.84	-
498	567	5	2	70	147645	31	9161.96	-

number of generated subgraphs could be exponentially large and since all of them are kept in the RAM, FSMS needs some heuristic improvements in order not to run out of memory.

Finally, we will investigate graph visualization and visual steering of the FSMS algorithm as discussed in [19]. This could provide a useful option for viewing the subgraphs as they are being found, and adding structural and semantic constraints during the process as suggested in [12].

6 Conclusion

Over the past 15 years, several methods have been developed to address the problem of identifying frequent subgraphs, particularly in graph-transaction settings. In this paper, we proposed a Frequent Subgraph mining algorithm, FSMS, that eliminates the process of isomorphism during frequent subgraph mining by using Mapping Sets. We demonstrated the efficiency of FSMS completely and accurately in finding (and listing) all existing frequent subgraphs in some real-world datasets. We are confident that, with additional work, the processing time and scalability of FSMS will be further improved, making it a viable FSM solution in the domains of graph query processing and frequent subgraph mining to discover recurrent patterns in scientific, spatial and relational datasets.

References

1. Google. Inside Search: Algorithms [Online] Written (2012) (accessed: 04–30-2015)
2. Clement, A.: NSA surveillance: exploring the geographies of internet interception. In: iConference 2014 Proceedings, pp. 412–425 (2014). doi:10.9776/14119
3. Rhodes, D.R., Yu, J., Shanker, K., Deshpande, N., Varambally, R., Ghosh, D., Barrette, T., Pander, A., Chinnaiyan, A.M.: ONCOMINE: A cancer microarray database and integrated datamining platform. Neoplasia **6**(1), 1–6 (2004). ISSN: 1476-5586 (accessed 04-30-2015). http://dx.doi.org/10.1016/S1476-5586(04)80047-2
4. Broder, A., Kumar, R., Maghoul, F., Raghavan, P., Rajagopalan, S., Stata, R., Tomkins, A., Wiener, J.: Graph structure in the web. Computer Networks **33** (2000)
5. Bader, D.A., Madduri, K.: A graph-theoretic analysis of the human proteininteraction network using multicore parallel algorithms. Parallel Comput. (2008)
6. Faloutsos, M., Faloutsos, P., Faloutsos, C.: On powerlaw relationships of the internet topology. In: SIGCOMM, pp. 251–262, August-September (1999)
7. Kuramochi, M., Karypis, G.: Finding frequent patterns in a large sparse graph*. Data Mining and Knowledge Discovery **11**(3), 243–271 (2005)
8. Gholami, M., Salajegheh, A.: A survey on algorithms of mining frequent subgraphs. International Journal of Engineering Inventions **1**(5), 60–63 (2012)
9. Inokuchi, A., Washio, T., Motoda, H.: An apriori-based algorithm for mining frequent substructures from graph data. In: Zighed, D.A., Komorowski, J., Żytkow, J.M. (eds.) PKDD 2000. LNCS (LNAI), vol. 1910, pp. 13–23. Springer, Heidelberg (2000)
10. Kuramochi, M., Karypis, G.: Frequent subgraph discovery. In: Proceedings of the 2001 IEEE International Conference on Data Mining. IEEE Computer Society (2001)
11. Yan, X., Han, J.W.: gSpan: graph-based substructure pattern mining. In: Proceedings of the 2002 IEEE International Conference on Data Mining. IEEE Computer Society (2002)
12. Elseidy, M., Abdelhamid, E., Skiadopoulos, S., Kalnis, P.: GRAMI: frequent subgraph and pattern mining in a single large graph. In: Proceedings of the VLDB Endowment, pp. 517–528 (2014)
13. Kuramochi, M., Karypis, G.: GREW - a scalable frequent subgraph discovery algorithm. In: Proceedings of ICDM, pp. 439–442 (2004)
14. Lu, W., et al.: Efficiently extracting frequent subgraphs using mapreduce. In: 2013 IEEE International Conference on Big Data. IEEE (2013)
15. National Center for Biotechnology Information. PubChem BioAssay Database; AID=2299, Source=Scripps Research Institute Molecular Screening Center (accessed February 22, 2011). http://pubchem.ncbi.nlm.nih.gov/assay/assay.cgi?aid=2299
16. Cisco. Cisco global cloud index: forecast and methodology 2013–2018 White Paper. [Online]. Written (2014) (accessed 04/27/2015). http://www.cisco.com/c/en/ussolutions/collateral/serviceprovider/global-cloud-index-gci/CloudIndexWhitePaper.html
17. Kang, U., Faloutsos, C.: Big graph mining: algorithms and discoveries. SIGKDDD Explorations **14**(2), 29–36 (2013)
18. Bhuiyan, M., Al Hasan, M.: MiRage: An iterative MapReduce based subgraph mining algorithm, July 22, 2013 (accessed 05/31/2015). arXiv:1307.5894
19. Puolamiki, K., Papapetrou, P., Lijffitj, J.: Visually controllable data mining methods. In: Proceedings of the 2010 IEEE International Conference on Data Mining Workshops, pp. 409–417, December 2010

Can Software Project Maturity Be Accurately Predicted Using Internal Source Code Metrics?

Mark Grechanik[1(✉)], Nitin Prabhu[1], Daniel Graham[2], Denys Poshyvanyk[2], and Mohak Shah[3]

[1] University of Illinois at Chicago, Chicago, IL 60607, USA
{drmark,nprabh3}@uic.edu
[2] College of William and Mary, Williamsburg, VA 23185, USA
{dggraham,denys}@cs.wm.edu
[3] Bosch Research, Palo Alto, CA 94568, USA
Mohak.Shah@us.bosch.com

Abstract. Predicting a *level of maturity (LoM)* of a software project is important for multiple reasons including planning resource allocation, evaluating the cost, and suggesting delivery dates for software applications. It is not clear how well LoM can be actually predicted – mixed results are reported that are based on studying small numbers of subject software applications and internal software metrics. Thus, a fundamental problem and question of software engineering is if LoM can be accurately predicted using internal software metrics alone?

We reformulated this problem as a supervised machine learning problem to verify if internal software metrics, collectively, are good predictors of software quality. To answer this question, we conducted a large-scale empirical study with 3,392 open-source projects using six different classifiers. Further, our contribution is that it is the first use of feature selection algorithms to determine a subset of these metrics from the exponential number of their combinations that are likely to indicate the LoM for software projects. Our results demonstrate that the accuracy of LoM prediction using the metrics is below 61% with Cohen's and Shah's $\kappa << 0.1$ leading us to suggest that comprehensive sets of internal software metrics alone cannot be used to predict LoM in general. In addition, using a backward elimination algorithm for feature location, we compute top ten most representative software metrics for predicting LoM from a total of 90 software metrics.

1 Introduction

Predicting various aspects of software quality and the *levels of maturity (LoM)* of software applications is important for multiple reasons including planning resource allocation for software development and maintenance, evaluating the cost, and suggesting delivery dates for software applications [8,12,13,18,27,33,45]. The essence of many machine learning approaches to predict certain characteristics of software is to obtain a model that maps software applications to classes that represent levels of software quality or maturity. In these approaches, software

P. Perner (Ed.): MLDM 2016, LNAI 9729, pp. 774–789, 2016.
DOI: 10.1007/978-3-319-41920-6_59

applications are represented as vectors of features, where features are measurements of some attributes of these applications. For example, LoM classes can be represented by alpha, beta, and other testing phases of software release life cycle and features may include the frequencies of language keyword usages for software applications that are assigned to these LoM classes. A classifier can be trained on a subset of feature vectors obtained from some software applications, where these applications are assigned to the LoM classes using information from a software repository. Once a model is learned as a result of training the classifier, it can be used to predict the LoM class for a software application.

Software metrics is a term that specifies a collection of measurements that describe software-related activities [15]. Many studies focused on predicting the level of software quality by using aggregated internal software metrics that represent different measurements of the source code of software applications [14,17,36,42,44]. However, to the best of our knowledge there is no research into a fundamental question of software engineering – *can LoM be accurately predicted using internal software metrics that are obtained from the source code of software applications?*

In this paper, we reformulated this problem as a as a supervised machine learning problem to verify if collectively 90 software metrics can predict software LoM level with a high degree of accuracy. Further, our goal is to determine if a subset of 90 software metrics can predict LoM more effectively, since it is likely that not all software metrics contribute equally to LoM. Unfortunately, to investigate all subsets of $\approx 2^{90}$ elements of the superset of all software metrics is infeasible. One of our main contributions is to use feature selection algorithms to determine if a subset of these metrics can do so to guard against noise and irrelevant attributes [20].

We make the following contributions. First, we conducted a large-scale empirical study with 3,392 open-source projects using six different classifiers. Next, we used different algorithms for selecting subsets of features to eliminate the noise that non-representative software metrics add and to determine what subset of these metrics best represent the LoM for software projects. We believe that this is the first empirical study that addressed a fundamental question of determining LoM for software projected using internal source code metrics alone. To the best of our knowledge there is no other work that addressed this question on a large scale and applied feature selection algorithms to determine a manageable subset of internal source code metrics that can summarily indicate a LoM. Our results show that the accuracy of software LoM level prediction reaches $\approx 61\%$ with Cohen's and Shah's $\kappa << 0.1$ leading us to suggest that internal, source code software metrics alone cannot serve as accurate predictors of LoM in general. All workflows and data files are available from the project's website http://www.cs.uic.edu/~drmark/prime.htm.

2 Problem Statement

Predicting different aspects of software quality and maturity with high precision using some attributes is a problem of great importance. Stakeholders can take timely corrective actions if an accurate prediction is issued that some quality aspects of a software application will worsen using a set of software metrics that are collected during development of this application. Taking corrective actions when software applications are still under development results in significantly lower costs and faster times to market, enabling stakeholders to reap huge economic benefits. Predicting LoM for software enables stakeholders to plan software release events, which is very important for the economic health of software projects. It is no wonder that predicting LoM draws significant attention of both academia and industry – companies that claim that they can predict different aspects of software quality can gain competitive market advantage, since they can systematically control production of software [25].

Selecting attributes from which the LoM of software applications can be predicted with a high degree of accuracy is a fundamental problem of software engineering. Various software metrics are used as attributes for predicting different aspects of the quality of software applications. As we show in our analysis of the related work in Section 5, a sheer majority of approaches follow the same methodology that consists of three steps: 1) some software or process metrics are selected ad-hoc as attributes, 2) a small set of software applications is selected with assigned values from some class attribute that describes the level of quality, and 3) a classifier is trained using these attributes to learn a prediction model. Different results are reported showing good accuracy of learned models where it reaches as high as over 90% for selected applications.

Unfortunately, many unanswered questions surround this methodology. Can internal, source code software metrics *alone* be used to build prediction models with a high degree of precision for the LoM of software applications? Do all software metrics contribute equally to building accurate predictors of software LoM? Do the learned models have good precision in predicting LoM of software applications that are developed by different programmers across many different domains? How sensitive these models are to the parameters of the learning algorithms? What is a minimal subset of attributes that can characterize LoM with a good level of precision? Answering these questions is the main goal of this paper.

In this paper we address the following question: can software LoM be accurately predicted using source code software metrics alone? Collecting various software metrics from the source code of applications is easy; determining subsets of these metrics that are useful to build a good predictor is very difficult. This problem is an instance of a bigger problem in ML, namely a supervised ML problem to verify if collectively these software metrics are predictors of software LoM. This problem focuses on constructing and selecting subsets of attributes that are important for creating high-quality prediction models [20]. The root of this problem is in difficulty of determining individual predictive powers for attributes using a specific ML algorithm. For a small number of attributes it is possible to train classifiers with all the subsets from the powerset of these attributes, however, this brute-force solution quickly becomes computationally prohibitive even when the number of attributes increases to couple of dozens.

In this paper, we deal with over 90 different software metrics, and trying all their subsets is not feasible. Further, we perform feature selection to determine if a subset of these metrics can do so to guard against noise and irrelevant attributes.

The problem is not only in training predictors using all subsets of attributes – different ML algorithms may perform differently, so these ML algorithms should be used to build predictors using subsets of attributes. Parameter spaces of these ML algorithms should be explored as part of sensitivity analysis to check to see if varying values of these parameters significantly changes the precision of the predictor. These and many other variables will be explored as part of addressing our problem.

3 Experiments

In this section, we describe variables, explain our methodology, provide background on ML algorithms for learning models and on algorithms for computing an optimal set of software metrics for learning predictors, describe our experimental design, and discuss threats to validity.

3.1 Measuring LoM of Software Projects

The quality of a software application is a set of measures that describe how this application is expected to fulfill some need and meet some standards [19, pages 15-37]. There are many different measures including but not limited to correctness, reliability, performance, robustness, maintainability, and usability. *Phases of testing* (e.g., alpha, beta, release candidate) are widely used in software release life cycle to indicate levels of maturity of software. The concept of phase of testing was introduced in 1950s and widely used in software development to indicate certain aspects of the level of maturity and quality of software applications [6, pages 515-516] [19, page 400][22]. Since it is difficult to obtain other indicators of the LoM for thousands of open-source software project, we use the phase of testing that is specified for each project as an approximation for its LoM.

There are three main phases of testing. When a software application is built and tested within a development organization, this application is usually assigned the *alpha phase*, where testing is performed by dedicated teams within this organization. Alpha phase indicates the lowest level of maturity for software applications. During the work on the alpha version of the application, software engineers fix bugs and add and change features (i.e., units or functionality) among many other things. When the collective quality of the application improves to a certain level, stakeholders assign this application to the *beta phase*, where the application is shipped to selected customers who will use it and give detailed feedback to the development organization. At some point, software engineers further improve the quality of the application, and stakeholders assign it the *production phase*. Depending on the organization, LoM may include other intermediate phases besides these three major ones. Of course, no single external software metric can serve as a strong measure of software quality, but different metrics can reflect on certain integral aspects of software quality. A testing phase is the quality aspect of software applications that is reflected in their LoM.

3.2 Methodology

A main purpose of our methodology is to increase the confidence in the results of experiments with which we attempt to answer the main research question in this paper – *can the LoM level of a software application be accurately predicted using internal software metrics alone?* We answer this question through experimentation by aligning our methodology with the guidelines for statistical tests in software engineering and machine learning [3,24]. Our goal is to collect large and highly representative samples of data when applying different approaches, perform statistical tests on these samples, and draw conclusions from these tests. Our main objective is to reduce a bias in our experimental design to minimum.

In most existing studies [21], experimental bias comes from different sources: a small number of subject applications, stakeholders who develop and maintain applications in some peculiar ways, ML algorithms that use a specific representation of the model that benefits from certain metrics, limited software metrics that reflect a certain aspect of software, and skewed representations of software LoM levels. We design our methodology to address these biases.

Selection of ML algorithms is critical for generalization of the results of this experiment. Different ML algorithms employ different function spaces for mapping inputs to the (expected) outputs. We explore various learning paradigms, and hence, a variety of function or classifier spaces so as to not introduce dependency of a choice of space on the results. We choose a representative set of ML algorithms ranging from simple (naive) bayesian formulation to more complex kernel spaces for the majority of different representations of learners [11].

Since one of our goal is to determine best performances of learned models for subsets of data, we experiment using ML algorithms with multiclass data (i.e., all applications in a single dataset with Alpha, Beta, Production LoM phases) and with binary classification (i.e., application from any two classes). Often, binary classification yields better classifier performance results when compared with multiclass performance [1,43]. Therefore, we experiment with four subsets of the data: Alpha/Beta, Alpha/Production, Beta/Production, Alpha/Beta/ Production.

With four subsets of application's software metrics, we carry out two main experiments: the first with six ML algorithms using three subsets of software metrics based on their weights and the second one with six ML algorithms using backward elimination. In total, we perform 96 experiments $= 4 \times 6 \times 3 + 4 \times 6$. We carry out experiments using RapidMiner, an open-source tool that implements various ML and data mining techniques such as data transformation, feature selection, classification and model evaluation [32]. We discarded applications that were categorized ambiguously (e.g., an application that is assigned both Alpha and Beta phases). All experiments were carried out using five-fold cross validation and stratified sampling, a standard methodology for ML algorithms.

We do not report execution times for learning models, since performance measures of RapidMiner do not address research questions that we formulate here.

3.3 Why We Choose ML Algorithms

Since our goal is to provide evidence that may show that the LoM level can be accurately predicted, a natural question is why we did not use the entire universe of all statistical methods instead of concentrating on selected few. More specifically, are the ML methods that we use in some way superior to some other statistical methods to answer our research question?

An answer is that the techniques that we chose are superset of a lot of statistical approaches and being data-driven scale better. For instance, decision based classifiers are supersets of the classes of conjunctions and disjunctions and decision trees that approximate disjunctions of conjunctions, and hence can approximate quite a few linear combinations. *Support Vector Machines (SVM)* are of course generalizations of linear combinations to reproducing kernel Hilbert space (i.e., generalizations of Euclidean spaces). Many a data-driven approaches are generalizations of univariate/multi-variate statistical approaches and have shown both to scale and give state-of-the-art performances. By selecting a set of ML algorithms, we intend to cover a variety of these classifier spaces.

3.4 Background

In this section, we provide background on a feature selection method called backward elimination that we use in our experiments to determine a minimal subset of software metrics with which a maximum performance can be achieved, and on Cohen's and Shah's κ-measures that we use in this paper as a dependent variable to measure the performance of ML algorithms.

Backward Elimination. Backward elimination is a popular approach as part of the wrapper methodology that considers an ML algorithm as a black-box and uses it to assess the usefulness of subsets of features [29,30]. The essence of backward elimination, which is considered the best-performing approach for automatically selecting subsets of features with a high predictive power, is in a search algorithm that performs a direct objective optimization by maximizing the goodness-of-fit (i.e., the performance) and minimzing the number of features that are used as the input to this ML algorithm. As the name of this approach suggests, it starts with the full set of features (i.e., software metrics in our case) and uses a measure of the goodness-of-fit to eliminate some features from the input that make the lowest contribution to this goodness-of-fit. The process repeats until a minimum subset of features is found. Backward elimination approach is computationally intensive; for example, one of our experiments using this approach took 43 hours to complete using Intel Xeon Dual Core CPU X5680 at 3.33GHz and RAM 24Gb. However, using backward elimination gives us a high level of confidence that we did not overlook certain subsets of software metrics that may lead to high accuracy of prediction of LoM.

Cohen's κ-measure. Often, precision is not enough to determine if a predictor is good enough. While 50% precision is essentially a random choice, does 60% or 70% precision makes it a good predictive model? To answer this question, Cohen's κ measures the agreement obtained that two raters classifying items in two mutually exclusive classes would obtain above and beyond chance (coincidental concordance). Raters in complete agreement would yield $\kappa = 1$ while raters achieving the same agreement as can be obtained by chance (owing to rater biases) would yield $\kappa = 0$. Raters in complete disagreement (worse than chance) can results in κ values as low as -1 [7,40]. In our case, we compare the predictive model as a rater with a random rater. This measure is computed as $\kappa = \frac{P_r - P_c}{1 - P_c}$, where P_r is the relative observed agreement among raters, and P_c is the hypothetical probability of chance agreement, using the observed data to calculate the probabilities of observers randomly selecting categories. Cohen's κ in its classical version applies to two-rater two-class scenario. While various generalizations have been proposed to extend it to multi-class multi-rater case, these rely on the marginalization argument. Since our experiments are an instance of fixed rater case, we employ the Shah's κ_S [40].

3.5 Variables

Main independent variables are the subject applications, their software metrics, the values of LoM levels, ML algorithms with which we experiment, and approaches for selecting subsets of software metrics. A dependent variable is the learned model (and its performance measures) and the subset of software metrics with which this model was learned. We measure the performance of the learned model both in terms of precision, P, as the Cohen's κ-measure that specifies the agreement between two raters who each classify items into mutually exclusive categories, and the number of software metrics used as features to build the classifier. The effects of other variables (the structure of software applications and the types and semantics of data they process) are minimized by the design of this experiment.

Subject Applications. A small number of subject applications is a threat to external validity, since it is unclear how results can be generalized. Our goal is to experiment with a large representative sample of software applications that are written and maintained by different stakeholders who assigned LoM levels to these applications independently. Even though some stakeholders may have personal biases when assigning LoM levels to some applications, we assume that the general trend is dictated by definitions of the LoM levels that we described in Section 3.1. We use 3,392 open-source projects, comprised of 958 alpha-phase projects, 1,292 beta-phase projects, and 1,142 production-phase projects. The source code for these open source projects was downloaded from Metrobase, an online database that contains the source code for open source Java projects collected from ten repositories: Sourceforge, Tigris, dev.java.net, Netbeans, nongnu.org, gnu.org, gna.org, apache.org, Javaforge.com, and googlecode.com. For each application, its creators and maintainers assigned one of the three LoM levels: alpha, beta, or production. These labels were extracted from the project repository FLOSSMole (http://flossmole.org).

We chose this dataset because it is the largest dataset that we found that contains LoM information. We accomplished the following steps to arrive to this dataset.

- We took all the Java projects from Merobase repository. We computed the metrics for all the classes from these projects. We cleaned data e.g.,, we excluded projects that did not contain source code or contained only a few classes (toy projects).
- We took all the metadata (including LoM info) from FLOSSMole repository, however, this dataset does not include the source code.
- We intersected these two sets (using the name of the projects) and arrived to the set of projects with metrics and LoM information.

Sourceforge has seven LoM levels for the project status: planning, pre-alpha, alpha, beta, production/stable, mature, and inactive. Some of these levels, however, have nothing to do with LoM. For example, the level *inactive* is used to mark the projects that are essentially dead meat (which could be as well mature or beta). Also, projects in the *planning* phase often do not even contain the source code, which means that we can not compute the metrics for those. Moreover, we did not find that many projects (less than ten) in certain LoM levels, so we discarded those. The only aggregation we did was to merge *pre-alpha* with *alpha* and we merged *production/stable* with *mature*.

ML Algorithms. Since selection of a machine learning algorithm affects the accuracy of a learned model, we experiment with six different ML classification algorithms: Decision Tree, Naïve Bayes, Random Forest, NBTree, JRIP, and Support Vector Machine (SVM). The rationale for choosing this set of ML algorithms is that they are representative of the various classifier spaces explored and have shown to be very effective on a variety of domains [11].

Software Metrics. For software metrics, we chose a large number of fundamentally different software metrics that are computed from the source code of these applications, thus effectively diversifying different criteria that may be used in the decision-making process when stakeholders assigned LoM levels to applications. The Software metrics were extracted using a tool, named `Columbus` that analyzes software systems and extracts 90 software metrics, errors, and warnings from project, package, class, attribute and method-level artifacts [16]. Software metrics were computed using the class granularity and aggregated for the application using their averages.

3.6 Hypotheses

We introduce the following null and alternative hypotheses to evaluate relations between LoM levels and internal software metrics.

H_0 The primary null hypothesis is that there is no difference between performances of different ML algorithms that are used to learn prediction models of LoM levels from software metrics.

H_1 An alternative hypothesis to H_0 is that there is significant difference between performances of different ML algorithms that are used to learn prediction models of LoM levels from software metrics.

To test the null hypothesis H_0, we are interested in the comparing precisions, recalls, and Cohen's k-measures to understand how ML algorithms perform better than random assignments. In particular, our experiments are designed to examine the following null and alternative hypotheses:

H1: Alpha and Beta. The effective null hypothesis is $\kappa_\alpha \approx k_\beta \approx 0$, while the true null hypothesis is that $p_\alpha \approx 50\%$ and $p_\beta \approx 50\%$. Conversely, the alternative hypothesis is $\kappa_\alpha \approx k_\beta >> 0.5$ and $p_\alpha >> 50\%$ and $p_\beta >> 50\%$.

H2: Production and Beta. The effective null hypothesis is $\kappa_\pi \approx k_\beta \approx 0$, while the true null hypothesis is that $p_\pi \approx 50\%$ and $p_\beta \approx 50\%$. Conversely, the alternative hypothesis is $\kappa_\pi \approx k_\beta >> 0.5$ and $p_\pi >> 50\%$ and $p_\beta >> 50\%$.

H3: Alpha and Production. The effective null hypothesis is $\kappa_\alpha \approx k_\pi \approx 0$, while the true null hypothesis is that $p_\alpha \approx 50\%$ and $p_\pi \approx 50\%$. Conversely, the alternative hypothesis is $\kappa_\alpha \approx k_\pi >> 0.5$ and $p_\alpha >> 50\%$ and $p_\pi >> 50\%$.

H4: Alpha and Beta and Production. The effective null hypothesis is $\kappa_\alpha \approx k_\beta \approx k_\pi \approx 0$, while the true null hypothesis is that $p_\alpha \approx 50\%$ and $p_\beta \approx 50\%$ and $p_\pi \approx 50\%$. Conversely, the alternative hypothesis is $\kappa_\alpha \approx k_\beta \approx k_\pi >> 0.5$ and $p_\alpha >> 50\%$ and $p_\beta >> 50\%$ and $p_\pi >> 50\%$.

The rationale behind the alternative hypotheses to `H1`, `H2`, `H3` and `H4` is that it is possible to create a model with some ML algorithm with which LoM levels of software applications can be predicted with a high precision using internal software metrics.

3.7 Threats to Validity

In this section, we discuss four kinds of threats: to statistical conclusion validity, internal validity, construct validity, and external validity [39, pages 61-65]. Our goal is to show that our experimental design gives an answer to the research question about the absence or relation between LoM levels (i.e., phases of testing) and internal software metrics with a high degree of confidence.

In the context of this paper, statistical conclusion validity refers to the question of presumed causal relationship between software metrics and LoM levels or absence thereof. Essentially, a threat to statistical conclusion validity is that we missed covarying between software metrics and LoM levels. To address this threat, we show that our experiments are sufficiently sensitive and able to detect small differences and we have evidence that software metrics and LoM levels do not covary. Specifically, this threat is addressed by using six ML algorithms instead of one with varying parameters and a feature selection approach to establish that no relation exists between software metrics and LoM levels. In addition, we experimented on thousands of different software projects to ensure that we significantly reduced the bias. Most importantly, the Cohen's κ-measure evaluates if a classifier performs better than a random choice, thus indicating that the possibility of covarying between independent and dependent variables. With a comprehensive set of measures and a thorough experimental design we address a threat to statistical conclusion validity.

Threats to internal validity refer to as confounding variables that may introduce a possible rival hypothesis to our hypotheses. In the context of our paper, confounding variables could be a threat if we uncover a strong relation between LoM levels and internal software metrics. In this case, a confounding variable could be the bias of the stakeholders who assign LoM levels to software projects based on some software metrics. In our case, the situation could be reversed, that is, a threat to internal validity could come from the bias of the stakeholders who assign LoM levels to software projects based on some software metrics which we did not take into consideration. However, it is highly unlikely, since we use many software applications whose LoM levels were assigned independently by many independent stakeholders, and the possibility that they all agreed on some obscure software metric that we did not cover in this paper is minuscule.

A threat to external validity refers to generalization of relation between LoM levels and internal software metrics beyond the confines of our experimental design. This is a main threat for our experimental design and it lies in the selection of subject applications and their software metrics. Since we operated on a very large set of diverse subject applications, we address a threat to external validity.

4 Empirical Results

To evaluate the null hypothesis H_0, we carried out experiments with six different ML algorithms using binary and multiclass classification to learn predictive models. We computed confusion matrices for six ML algorithms for hypotheses H1, H2, H3, and H4. A confusion matrix is a table that illustrates the performance of an ML algorithm. Each column of the matrix represents the instances in a predicted or modeled class, while each row represents the instances in an actual or true class, thereby illustrating how an algorithm confuses two classes (i.e. mislabels one as another) [41]. Symbols α, β, and π stand for Alpha, Beta, and Production LoM levels respectively. Results from these confusion matrices confirm that there are no significant differences between random choices and the results of classification, all Cohen's κ-measures are less than 0.2. In addition, precision is often lower than 50%, the highest is for testing H3 with the algorithm NBTree when it gets to 60.6%.

For the multiclass classification, the results are even worse, with the precision mostly in 30%, getting even to 28% for the Naïve Bayes classifier. That is, the results of prediction of LoM levels using software metrics is effectively the random choice. Based on the values of precision, P and Cohen's κ measures, we accept hypotheses H1–H4. In addition, results from experiments that use the feature selection backward elimination approach are consistent from both experiments.

Hence, we accept the null hypothesis H_0 and reject the alternative hypothesis H_1, meaning that **there is no difference between performances of different ML algorithms that are used to learn prediction models of LoM levels from software metrics**. We can summarize these results as following.

- Classified Alpha-phase applications have the poorest precision and the worst recall values in general, and the multiclass classification gives the worst performance. We explain it as a result of large variances between software metrics for Alpha-phase applications, since these are the least stable versions of applications when compared to software metrics for applications that are assigned other LoM phases. Some stakeholders do a better job when releasing Alpha-phase applications, while others know that the applications should go through many changes before assigning these applications to more advanced phases (i.e., Beta and Production). Because of disparities in many software metrics, Alpha-phase applications are the most difficult ones to classify with precision.
- Performance differences are the most significant when considering binary classification between Alpha and Production-phase applications, while the differences between Alpha and Beta-phase applications are the least significant. We explain it as a result of maintenance and evolution work on applications between LoM phases. Since the amount of work on a software application is significant between Alpha and Production phases, its software metrics differ a lot, and subsequently, it makes classifiers to learn better predictive models.
- Backward elimination rarely resulted in the improved performance of the predictive model, and when it did (i.e., for Random Forest for H1 and H2, for Decision Tree for H1 and H3, for JRIP for H1 and H3) it was not significant, on the order of three to five percent or less. The best and most consistent improvement using backward elimination was for the SVM algorithm for multiclass classification for all LoM phases, even though it was small to affect our conclusions (less than five percent).

We interpret the results of our experiments as follows.

1. *We strongly suggest that software metrics* ***cannot*** *be used alone, but only with some other indicators for estimation of LoM levels.* We think that the overall estimation of quality of software is based on some other factors, which we suggest depend on certain characteristics of the applications, finding which is a subject of future work. While this sounds like an unexpected result, the recent work in the similar direction showed that process metrics might be better predictors of software quality as compared to product metrics, which corroborates our result from a different point of view [37].
2. *We observe that the variance in precision and recall values is the largest for application in the Alpha phase.* For the algorithm NBTree, the precision and recall for projects in the Alpha-phase for H1 was zero percent with recall values for the algorithm Decision Tree as low as 5.3%. At the same time, other ML algorithms performed much better using the same data set, with Random Forest showing the precision of 42.1% and recall 23.5% for the same experiment H1. Such large variations in the learned predictive models show

that it is difficult to rely on results of a single ML algorithm, since they are very sensitive to different dependencies in the input data.

3. *Our experimentation shows that the precision for multiclass experiments are in general lower than the precision for binary classification.* We explain it as a difficulty for classifiers to find discriminative attributes with more predictive power. Indeed, given that multiple software metrics have overlapping ranges of values for applications that are assigned to different LoMs, it is difficult to find a model that can discriminate among different LoMs. Adding more classes will likely exacerbate this problem, since finer granularity of LoM levels mean that there will be more overlapping among ranges of values for different software metrics, and therefore, worse precision.
4. *For the multiclass experiment, JRIP and SVM gave the worst values of recall for the alpha phase, and JRIP returned the worst value of recall, which is 2.5%.* This abnormality correlates with our binary experiments where alpha gives much lower recall when classified against beta LoM. However, in binary experiments where applications in alpha and production LoMs are classified, recall values are much closer to 50/50%. We explain it as a result of the difficulty to find strict discriminative software metrics for applications in alpha in beta phases, since they are much closer to one another. Indeed, different indirect evidence point out toward this explanation: the time intervals between releasing alpha and beta LoMs are shorter, fewer changes are made, since the software applications are not released in production yet, and only critical and major bugs are fixed. The differences between production/beta and especially between production/alpha are much bigger, and it is reflected in the precision and recall values.
5. *Our experiments show it is possible to obtain good models by selecting a small subset of applications (less than 50) in a way that intervals are large between average values of software metrics that are collected from applications that belong to different LoMs.* In addition, when only a small subset of software metrics is chosen, it is easy to learn a model with a very high degree of accuracy. **Thus, we demonstrate the danger of making non-generalizable conclusions from a small set of data – increasing dimensions in classification problem illustrates the fallibility of small controlled experiments**. This is known as the curse of dimensionality [11], when a growing number of examples reduces the precision of classification algorithms drastically.

As part of using backward elimination algorithm for feature location, we computed top ten most representative software metrics for predicting LoM from a total of 90 software metrics. Even though using these metrics only leads to better prediction results, they still fall short of giving accurate prediction of software quality. These top ten metrics which are the following.

1. Long function is a bad smell metric that indicates that the sizes of methods grow too large.
2. Long parameter list is a bad smell metric that indicates that the number of parameters in methods is large.

3. Shotgun surgery is a bad smell metric that indicates that programmers change the source code in many locations when implementing a small feature.
4. The number of nonempty non-commented lines of code.
5. Number of incoming invocation summarizes the cardinality of the set of methods which invoke other methods.
6. Number of methods is a metric where locally defined and inherited methods and declarations and definitions are counted but if there is implementation for a declaration, the declaration is not counted.
7. Number of attributes is a metric that counts the local and inherited attributes of the class in an application.
8. Number of foreign methods accessed is a metric that summarizes the cardinality of the set of method invocations of the method where the invoked methods belong to other classes than the method itself.
9. Raw exception avoidance metric that describes the use of specialized exception classes rather than more general exception classes like Exception.
10. A metric that indicates a frequent use of method-level synchronization versus a block-level synchronization.

5 Background and Related Work

Software quality and LoM respectively are important aspects of modern software systems. Estimating and tracking quality and LoM of software systems is essential for various development and maintenance decisions. The ISO/IEC 9126 standard [23] defines six high-level product quality characteristics which are widely accepted both by industrial experts and academic researchers: functionality, reliability, usability, efficiency, maintainability and portability. The characteristics are impacted by the low-level quality properties, that can be *internal* (measured by looking inside the product, e.g. by analyzing the source code or deriving product metrics) or *external* (measured by execution of the product, e.g. by performing testing or deriving process metrics) [4]. Many research papers have been published proposing software quality models ranging from purely theoretical to more applicable approaches [2,4,5,9,10,26,31,34,35]. While all these models are built from different combinations of product metrics, only a few of those were actually deployed and tested in industrial settings [4][28][38]. Unfortunately, they did not answer the question that we posed in this paper.

Finally, inter- and intra- system prediction of fault-prone classes or simply bug prediction is an active research areas with a number of research publications in the last decade. While we are not classifying or discussing all these models in-depth in this paper, we refer the reader to the recent systematic literature review on fault prediction of 208 studies [21]. The results of that review motivate our study. Also, feature selection appears to be one of the major success factors for building good models in our work. *However, none of the reported 208 studies is attempted on such a scale and uses such an exhaustive combination of internal product metrics.*

6 Conclusion

We answered the question negatively whether software quality can be accurately predicted using internal source code software metrics alone by conducting a large-scale empirical study with 3,392 open-source projects using six different classifiers. Further, we performed feature selection to determine if a subset of these metrics can do so to guard against noise and irrelevant attributes. Our results show that the accuracy of software quality prediction reaches $\approx 61\%$ leading us to suggest that comprehensive sets of internal software metrics alone cannot accurately predict software quality in general. Our result affects the research and deployment of hundreds of different approaches in academia and industry, since it shows that the claim is wrong that it is possible to effectively control and predict software development activities by using internal source code metrics alone to predict the quality of software projects in general.

Acknowledgments. We would like to thank to Rudolf Ferenc and Tibor Gyimothy for providing academic license for Columbus. This material is based upon work supported by the National Science Foundation under Grants No. 0916139, 1017633, 1217928, 1017305, and 1547597.

References

1. Allwein, E.L., Schapire, R.E., Singer, Y.: Reducing multiclass to binary: a unifying approach for margin classifiers. J. Mach. Learn. Res. **1**, 113–141 (2001)
2. Alves, T., Ypma, C., Visser, J.: Deriving metric thresholds from benchmark data. In: 26th ICSM 2010, pp. 1–10. IEEE, Timisoara, September 12–18, 2010
3. Arcuri, A., Briand, L.C.: A practical guide for using statistical tests to assess randomized algorithms in software engineering. In: ICSE, pp. 1–10 (2011)
4. Bakota, T., Hegedus, P., Krtvlyesi, P., Ferenc, R., Gyimthy, T.: A probabilistic software quality model. In: 27th ICSM 2011, Williamsburg, Virginia, USA, pp. 243–252, September 25–30, 2011
5. Bansiya, J., Davis, C.: A hierarchical model for object-oriented design quality assessment. TSE **28**(1), 4–17 (2002)
6. Beizer, B.: Software Testing Techniques, 2nd edn. Van Nostrand Reinhold Co., New York (1990)
7. Carletta, J.: Assessing agreement on classification tasks: the kappa statistic. Comput. Linguist. **22**(2), 249–254 (1996)
8. Cataldo, M., Nambiar, S.: On the relationship between process maturity and geographic distribution: an empirical analysis of their impact on software quality. In: Proceedings of the 7th ESEC/FSE 2009, pp. 101–110. ACM, New York (2009)
9. Correia, J., Kanellopoulos, Y., Visser, J.: A survey-based study of the mapping of system properties to ISO/IEC 9126 maintainability characteristics. In: 25th ICSM 2009, pp. 61–70. IEEE, Edmonton, September 20–26, 2009
10. D'Ambros, M., Lanza, M., Robbes, R.: Evaluating defect prediction approaches: a benchmark and an extensive comparison. Empirical Softw. Engg. **17**(4–5), 531–577 (2012)

11. Domingos, P.: A few useful things to know about machine learning. Commun. ACM **55**(10), 78–87 (2012)
12. Dubey, S.K., Rana, A., Dash, Y.: Maintainability prediction of object-oriented software system by multilayer perceptron model. SIGSOFT Softw. Eng. Notes **37**(5), 1–4 (2012)
13. Fenton, N., Krause, P., Neil, M.: Probabilistic modelling for software quality control. In: Benferhat, S., Besnard, P. (eds.) ECSQARU 2001. LNCS (LNAI), vol. 2143, p. 444. Springer, Heidelberg (2001)
14. Fenton, N.E., Neil, M.: A critique of software defect prediction models. IEEE Trans. Softw. Eng. **25**(5), 675–689 (1999)
15. Fenton, N.E., Pfleeger, S.L.: Software Metrics: A Rigorous and Practical Approach, 2nd edn. PWS Publishing Co., Boston (1998)
16. Ferenc, R., Beszédes, A., Tarkiainen, M., Gyimthy, T.: Columbus - reverse engineering tool and schema for c++. In: Proceedings of the ICSM 2002, pp. 172–181. IEEE Computer Society, Washington, DC (2002)
17. Ferzund, J., Ahsan, S.N., Wotawa, F.: Empirical evaluation of hunk metrics as bug predictors. In: Abran, A., Braungarten, R., Dumke, R.R., Cuadrado-Gallego, J.J., Brunekreef, J. (eds.) IWSM 2009. LNCS, vol. 5891, pp. 242–254. Springer, Heidelberg (2009)
18. Genero, M., Olivas, J.A., Piattini, M., Romero, F.: Using metrics to predict OO information systems maintainability. In: Dittrich, K.R., Geppert, A., Norrie, M. (eds.) CAiSE 2001. LNCS, vol. 2068, p. 388. Springer, Heidelberg (2001)
19. Ghezzi, C., Jazayeri, M., Mandrioli, D.: Fundamentals of Software Engineering: Prentice Hall PTR, Upper Saddle River, NJ, USA (2002)
20. Guyon, I., Elisseeff, A.: An introduction to variable and feature selection. J. Mach. Learn. Res. **3**, 1157–1182 (2003)
21. Hall, T., Beecham, S., Bowes, D., Gray, D., Counsell, S.: A systematic literature review on fault prediction performance in software engineering. IEEE TSE **38**(6) (2012)
22. Humble, J., Farley, D.: Continuous Delivery: Reliable Software Releases through Build, Test, and Deployment Automation, 1st edn. Addison-Wesley Professional (2010)
23. ISO/IEC. Iso/iec 9126. software engineering - product quality (2001)
24. Japkowicz, N., Shah, M.: Evaluating Learning Algorithms: A Classification Perspective. Cambridge University Press, New York (2011)
25. Jones, C.: Applied Software Measurement (2nd edn.): Assuring Productivity and Quality, 3rd edn. McGraw-Hill Inc., Hightstown (2008)
26. Jung, H.-W., Kim, S.-G., Chung, C.-S.: Measuring software product quality: a survey of ISO/IEC 9126. IEEE Software **21**(5), 88–92 (2004)
27. Khoshgoftaar, T.M., Allen, E.B., Xu, Z.: Predicting testability of program modules using a neural network. In: Proceedings of the 3rd Symposium on ASSET 2000, pp. 57–62. IEEE Computer Society, Washington, DC (2000)
28. Kim, S., Zimmermann, T., Whitehead, J.E., Zeller, A.: Predicting faults from cached history. In: 29th ICSE 2007, Minneapolis, MN, USA, pp. 489–498, May 20–26, 2007
29. Kohavi, R., John, G.H.: Wrappers for feature subset selection. Artif. Intell. **97**(1–2), 273–324 (1997)
30. Koller, D., Sahami, M.: Toward optimal feature selection. In: In 13th International Conference on Machine Learning, pp. 284–292 (1995)
31. Menzies, T., Greenwald, J., Frank, A.: Data mining static code attributes to learn defect predictors. IEEE TSE **33**(1), 2–13 (2007)

32. Mierswa, I., Scholz, M., Klinkenberg, R., Wurst, M., Euler, T.: Yale: rapid prototyping for complex data mining tasks. In. In Proceedings of the 12th ACM SIGKDD, pp. 935–940. ACM Press (2006)
33. Moses, J.: Learning how to improve effort estimation in small software development companies. In: 24th COMPSAC 2000, pp. 522–527. IEEE Computer Society, Washington, DC (2000)
34. Oman, P., Hagemeister, J.: Metrics for assessing a software system's maintainability. In: IEEE ICSM 1992, pp. 337–344. IEEE, Orlando, November 1992
35. Ozkaya, I., Bass, L., Nord, R., Sangwan, R.: Making practical use of quality attribute information. IEEE Software **25**(2), 25–33 (2008)
36. Raaschou, K., Rainer, A.W.: Exposure model for prediction of number of customer reported defects. In: Proceedings of the ESEM 2008, pp. 306–308. ACM, New York (2008)
37. Rahman, F., Devanbu, P.: How, and why process metrics are better. In: 35th IEEE/ACM ICSE 2013, San Francisco, CA (2013)
38. Rahman, F., Posnett, D., Hindle, A., Barr, E., Devanbu, P.: Bugcache for inspections: hit or miss? In: 8th ESEC/FSE 2011, pp. 322–331. ACM, Szeged, September 5–9, 2011
39. Rosenthal, R., Rosnow, R.L.: Essentials of Behavioral Research: Methods and Data Analysis, 2nd edn. McGraw-Hill (1991)
40. Shah, M.: Generalized agreement statistics over fixed group of experts. In: Gunopulos, D., Hofmann, T., Malerba, D., Vazirgiannis, M. (eds.) ECML PKDD 2011, Part III. LNCS, vol. 6913, pp. 191–206. Springer, Heidelberg (2011)
41. Stehman, S.V.: Selecting and interpreting measures of thematic classification accuracy. Remote Sensing of Environment **61**(1), 77–89 (1997)
42. Tosun, A., Bener, A., Turhan, B., Menzies, T.: Practical considerations in deploying statistical methods for defect prediction: A case study within the turkish telecommunications industry. Inf. Softw. Technol. **52**(11), 1242–1257 (2010)
43. Vapnik, V.N.: The Nature of Statistical Learning Theory. Springer-Verlag New York Inc., New York (1995)
44. Vogelsang, A., Fehnker, A., Huuck, R., Reif, W.: Software metrics in static program analysis. In: Dong, J.S., Zhu, H. (eds.) ICFEM 2010. LNCS, vol. 6447, pp. 485–500. Springer, Heidelberg (2010)
45. Wake, S.A., Henry, S.M.: A model based on software quality factors which predictsmaintainability. Technical report, Virginia Polytechnic Institute & State University, Blacksburg, VA, USA (1988)

Multiple Consensuses Clustering by Iterative Merging/Splitting of Clustering Patterns

Atheer Al-najdi(✉), Nicolas Pasquier, and Frédéric Precioso

Univ. Nice Sophia Antipolis, CNRS, I3S, UMR 7271, 06900 Sophia Antipolis, France
{alnajdi,pasquier,precioso}@i3s.unice.fr

Abstract. The existence of many clustering algorithms with variable performance on each dataset made the clustering task difficult. Consensus clustering tries to solve this problem by combining the partitions generated by different algorithms to build a new solution that is more stable and achieves better results. In this work, we propose a new consensus method that, unlike others, give more insight on the relations between the different partitions in the clusterings ensemble, by using the frequent closed itemsets technique, usually used for association rules discovery. Instead of generating one consensus, our method generates multiple consensuses based on varying the number of base clusterings, and links these solutions in a hierarchical representation that eases the selection of the best clustering. This hierarchical view also provides an analysis tool, for example to discover strong clusters or outlier instances.

Keywords: Unsupervised learning · Clustering · Consensus clustering · Ensemble clustering · Frequent closed itemsets

1 Introduction

Although the last decades witnessed the development of many clustering algorithms, getting a "good" quality partitioning remains a difficult task. This problem has many dimensions, one of them is the fact that the results of clustering algorithms are data-dependent. An algorithm can achieve good results in some datasets while in others it does not. This is because each is designed to discover a specific clustering structure in the dataset. Another aspect of the problem is the effect of algorithm parameters on the results since changing the settings may produce different partitioning in terms of the number and size of clusters. Defining what should be a "correct" (or a "good") clustering also contributes to the problem, despite the existence of many validation measures whether internal or external.[1] External validation measures are not always applicable, because usually class labels are not provided, especially for large datasets. Moreover, Färber *et al.* [6] states that using such measures, usually applied to synthetic datasets, may not be sound for real datasets because the classes may contain

[1] More details about validation measures can be found in [4], [12] and [19].

P. Perner (Ed.): MLDM 2016, LNAI 9729, pp. 790–804, 2016.
DOI: 10.1007/978-3-319-41920-6_60

internal structures that the present attributes may not allow to retrieve, or also the classes may contain anomalies. On the other hand, internal validation measures may overrate the results of a clustering algorithm which targets the same underlying structure model as the one targeted by the measure.

From many available clustering algorithms with variable outcome, researchers focused recently on the possibility of combining multiple clusterings, called *base clusterings*, to build a new consensus solution that can be better than what each single base clustering could achieve. Such process is called *consensus clustering* or *aggregation of clusterings*. It involves 2 steps: first, building an ensemble of partitions (i.e. the combination of all the partitions provided by the base clustering algorithms), then applying a consensus function. Some consensus clustering approaches impose limitations on the ensemble, such as all the base clusterings must produce the same number of clusters. Other approaches require consensus function with high storage or time complexities. In this work, we present a new category of consensus clustering methods, that is, a pattern-based consensus generation using the frequent closed itemset technique from the frequent pattern mining domain. Since clustering quality depends more on their meaningfulness to the analyst, our method involves generating multiple consensuses by varying the number of base clusterings. The results are presented in a tree of consensuses that not just facilitates the selection of the preferred partitioning, but also depicts how the clusters are generated, and what clusters are more stable than others. We also present a recommended solution, which is the consensus that is the most similar to the ensemble.

In the next section, we summarize some of the consensus clustering methods. Section 3 details the proposed approach. Experimental results are shown in Sect. 4. We conclude in Sect. 5.

2 Related Work

Consensus clustering methods can be organized into several categories according to the underlying approach [10],[24]:

- **Graph partitioning methods:** The problem of consensus clustering is formulated as a graph partitioning problem where the instances and clusters of the base clusterings are used to build the vertices and edges of the graph respectively. Examples of such consensus methods: Cluster-based Similarity Partitioning Algorithm (CSPA), HyperGraph Partitioning Algorithm (HGPA), Meta CLustering Algorithm (MCLA) (Strehl and Ghosh [21]), and Hybrid Bipartite Graph Formulation (HBGF) (Fern and Brodley [7]).
- **Voting methods:** The objective is to match the cluster labels in all base partitions, then perform a voting procedure to find the final grouping of the instances. One limitation in voting methods is that they require the clusterings in the ensemble to produce the same number of clusters as the targeted consensus. Example: the Plurality Voting method (Dudoit and Fridlyand [5], Fischer and Buhmann [8]).

- **Co-association based methods:** A co-association matrix can be used to record how many times 2 instances belong to the same cluster in the ensemble of all partitions. Thus, the matrix defines a new measure of similarity between the instances. As an example, Fred and Jain [9] applied a minimum spanning tree algorithm on the matrix to generate a consensus clustering.
- **Information based methods:** The consensus function here tries to find a clustering that shares the maximum information with the ensemble. Topchy et. al. [22] proposed to use the Category Utility function in order to define the similarity between partitions. The resulting consensus represents the "median" partition which is the most similar to the ensemble.

More details can be found in the surveys by Ghaemi *et al.* [10], Sarumathi *et al.* [20], and Vega-Pons & Ruiz-Shulcloper [24].

3 Pattern-Based Consensus Generation

Our approach is based on discovering *clustering patterns* among the ensemble. Pattern mining and association rule discovery aims at identifying relationships between items in very large datasets [17]. If we consider each cluster in the ensemble as an item, then using pattern mining enables us to discover relationships between the clusters, by identifying the sets of instances that are clustered together by sets of base clusterings.

One of the powerful techniques is the *Frequent Closed Itemset* (FCI) [17]. Its objective is to find the maximal sets of items (clusters) that are common to sets of objects (instances). FCI generates patterns of fewer items only if they become more frequent than their maximal sets, thus eliminating many redundant clustering patterns for our approach, and reducing memory consumption and execution times compared to the approaches based on generating all frequent itemsets.[2]

Our consensus generation process starts by creating a clustering ensemble, from which a binary membership matrix is built. FCI technique is applied on the binary matrix to find clustering patterns, and finally we apply our proposed algorithm to generate the consensus partition, as explained in the following subsections.

3.1 Clustering Ensemble

We do not impose any restriction on the selection of the base clustering algorithms or their settings, as long as they produce hard partitions, that is, each instance belongs to only one cluster. However, changing the base algorithms (partition-based, hierarchical, density-based, etc) and/or theirs settings is preferable to ensure the diversity in the ensemble, which makes the consensus more powerful as it can benefit from the different shapes and sizes of base clusters.

[2] See [2] for an extensive survey on association rule mining.

3.2 Cluster Membership Matrix

After generating a clustering ensemble by partitioning the dataset considering different clustering algorithms, a cluster membership matrix $\mathcal{M}$ is built. $\mathcal{M}$ is used in several consensus clustering methods, as in [1] and [21]. It consists of n rows and m columns, where n is the number of instances, and m is the total number of clusters of all base clusterings. The membership matrix records the binary relation between instances and clusters as given in definition 1.

Definition 1. *A cluster membership matrix $\mathcal{M}$ is a triplet $(\mathcal{I}, \mathcal{C}, \mathcal{R})$ where $\mathcal{I}$ is a finite set of instances represented as rows, $\mathcal{C}$ is a finite set of clusters represented as columns, and $\mathcal{R}$ is a binary relation defining relationships between rows and columns: $\mathcal{R} \subseteq \mathcal{I} \times \mathcal{C}$. Every couple $(i, c) \in \mathcal{R}$, where $i \in \mathcal{I}$ and $c \in \mathcal{C}$, means that instance i belongs to cluster c. This binary relation is represented in the matrix by 1 at $\mathcal{M}_{ic}$, and 0 if there is no relation.*

Let us take as an illustrative example a dataset of nine instances $\mathcal{D} = \{1, 2, 3, 4, 5, 6, 7, 8, 9\}$. Suppose that we used 5 base clustering algorithms to produce 5 partitions of $\mathcal{D}$ as follows: $P1 = \{\{1, 2, 3, 4\}, \{5, 6, 7, 8, 9\}\}$, $P2 = \{\{1, 2, 3\}, \{4, 5, 6, 7, 8, 9\}\}$, $P3 = \{\{1, 2, 3, 4, 5\}, \{6, 7, 8, 9\}\}$, $P4 = \{\{6, 7\}, \{1, 2, 3, 4, 5, 8, 9\}\}$, and $P5 = \{\{4, 5, 6, 7, 8, 9\}, \{1, 2\}\}$. Table 1 presents the membership matrix. Each column P_j^i represents cluster j in partition i as a binary vector where values '1' identify the instances that belong to the cluster. In pattern mining domain, each column in $\mathcal{M}$ represents an item, as defined below.

Table 1. Example cluster membership matrix.

Instance ID	P_1^1	P_2^1	P_1^2	P_2^2	P_1^3	P_2^3	P_1^4	P_2^4	P_1^5	P_2^5
1	1	0	1	0	1	0	0	1	0	1
2	1	0	1	0	1	0	0	1	0	1
3	1	0	1	0	1	0	0	1	1	0
4	1	0	0	1	1	0	0	1	1	0
5	0	1	0	1	1	0	0	1	1	0
6	0	1	0	1	0	1	1	0	1	0
7	0	1	0	1	0	1	1	0	1	0
8	0	1	0	1	0	1	0	1	1	0
9	0	1	0	1	0	1	0	1	1	0

Definition 2. *An item of a cluster membership matrix $\mathcal{M} = (\mathcal{I}, \mathcal{C}, \mathcal{R})$ is a cluster identifier $c \in \mathcal{C}$.*
An itemset is a non-empty finite set of items $C = \{c_1, ..., c_n\} \subseteq \mathcal{C}$ in $\mathcal{M}$.
An itemset $C \subseteq \mathcal{C}$ is frequent in $\mathcal{M}$ if and only if its frequency, called support, in $\mathcal{M}$ defined as $support(C) = |\{I \in \mathcal{I} \mid \forall i \in I, \forall c \in C$, we have $(i, c) \in \mathcal{R}\}|$ is greater than or equal to the user-defined minsupport threshold.

3.3 Clustering Patterns

The rows in $\mathcal{M}$ present binary patterns (frequent itemsets) that tell the clusters to which each instance belongs. Using FCI methodology, we can find all the sets of instances that share the same clustering pattern. FCI discovers patterns from not only the full set of base clusterings, but also from their subsets, as long as they satisfy the FCI properties defined in [17]. We call the combination of the set of instance identifiers with its corresponding set of cluster identifiers a *Frequent Closed Pattern* (FCP). Table 2 shows the set of the FCPs extracted from Table 1.[3]

Definition 3. *A Frequent Closed Pattern* $P = (C, I)$ *in the cluster membership matrix* $\mathcal{M} = (\mathcal{I}, \mathcal{C}, \mathcal{R})$ *is a pair of sets* $C \subset \mathcal{C}$ *and* $I \subset \mathcal{I}$ *such that:*
i) $\forall i \in I$ *and* $\forall c \in C$, *we have* $(i, c) \in \mathcal{R}$.
ii) $|I| \geq minsupport$, *i.e.,* C *is a frequent itemset.*
iii) $\nexists i' \in \mathcal{I}$ *such that* $\forall c \in C$, *we have* $(i', c) \in \mathcal{R}$.
iv) $\nexists c' \in \mathcal{C}$ *such that* $\forall i \in I$, *we have* $(i, c') \in \mathcal{R}$.

3.4 Generating Multiple Consensuses

This process involves filtering the FCPs based on the number of clusterings in the itemset, and build a consensus from each group. The idea is that, since the base partitions represent clustering decisions and we do not know which of these decisions is/are better than the others, then we search for the item membership similarities between different combinations of these decisions to build a final consensus decision for each possible valid combination. A *Decision Threshold* (**DT**) is used for the filtering process, as it defines the number of base clusters involved in the clustering pattern. Thus, to generate multiple consensuses, we start by building the first consensus from the instance sets of FCPs whose cluster identifier set (itemset) defines patterns shared by all base clusterings (DT= number of base clusterings[4]). Then, we sequentially decrement DT towards 1, and at each DT value, we generate a consensus from the instance sets of FCPs built from DT clusterings, plus the clusters of the previous consensus (the clusters of the consensus at DT+1).

Definition 4. *Given the first consensus* $L^{MaxDT} = \{P_1, P_2, ..., P_m\}$, *and the definition* $B^{DT} = I^{DT} \cup L^{DT+1}$, *where* I^{DT} *is the instance sets of the FCPs built from DT base clusterings, and* L^{DT+1} *is the instance sets (clusters) of the previous consensus, a new consensus* L^{DT} *is the result of applying a consensus function* $\mathcal{Y}$ *on* B^{DT}, *that is,* $L^{DT} = \mathcal{Y}(B^{DT}) = \{L_1, L_2, ..., L_k\}$ *such that* $L_i \cap L_j = \emptyset$, $\forall (i, j) \in \{1, ..., k\}$, $i \neq j$, *and* $\bigcup_{i=1}^{i=k} L_i = \mathcal{I}$.

[3] We can see that the number of generated clustering patterns is larger than dataset size. This happens only for a small dataset, while for a large dataset, most of its instances will share the same clustering pattern, resulting in a much lower number of patterns compared to dataset size.

[4] The clusters in the first consensus are known in [25] as "data fragments".

Table 2. Frequent Closed Patterns extracted from Table 1.

FCP ID	Itemset (FCIs)	Instance ID set
1	$\{P_1^1, P_1^2, P_1^3, P_2^4, P_1^5\}$	{3}
2	$\{P_1^1, P_2^2, P_1^3, P_2^4, P_1^5\}$	{4}
3	$\{P_2^1, P_2^2, P_1^3, P_2^4, P_1^5\}$	{5}
4	$\{P_1^1, P_1^3, P_2^4, P_1^5\}$	{3,4}
5	$\{P_2^2, P_1^3, P_2^4, P_1^5\}$	{4, 5}
6	$\{P_1^1, P_1^2, P_1^3, P_2^4, P_2^5\}$	{1,2}
7	$\{P_2^1, P_2^2, P_2^3, P_1^4, P_1^5\}$	{6,7}
8	$\{P_2^1, P_2^2, P_2^3, P_2^4, P_1^5\}$	{8,9}
9	$\{P_1^3, P_2^4, P_1^5\}$	{3,4,5}
10	$\{P_1^1, P_1^2, P_1^3, P_2^4\}$	{1,2,3}
11	$\{P_2^1, P_2^2, P_2^4, P_1^5\}$	{5,8,9}
12	$\{P_1^1, P_1^3, P_2^4\}$	{1,2,3,4}
13	$\{P_2^2, P_2^4, P_1^5\}$	{4,5,8,9}
14	$\{P_2^1, P_2^2, P_2^3, P_1^5\}$	{6,7,8,9}
15	$\{P_1^3, P_2^4\}$	{1,2,3,4,5}
16	$\{P_2^4, P_1^5\}$	{3,4,5,8,9}
17	$\{P_2^1, P_2^2, P_1^5\}$	{5,6,7,8,9}
18	$\{P_2^2, P_1^5\}$	{4,5,6,7,8,9}
19	$\{P_2^4\}$	{1,2,3,4,5,8,9}
20	$\{P_1^5\}$	{3,4,5,6,7,8,9}

At each DT, an instance set $I \subseteq \mathcal{I}$ has one of the following three properties:

i) Uniqueness: it does not intersect with any other set $I' \subseteq \mathcal{I}$, that is, $I \cap I' = \emptyset$.
ii) Inclusion: it is a subset of another set $I' \subseteq \mathcal{I}$, that is, $I \subseteq I'$.
iii) Intersection: it intersects with another set $I' \subseteq \mathcal{I}$, that is, $I \cap I' \neq \emptyset$, $I \setminus I'$ $\neq \emptyset$ and $I' \setminus I \neq \emptyset$.

Note that at the first consensus, all the instance sets are unique, because they are generated from clustering patterns shared by all base clusterings. However, for the next consensuses, the sets of instance identifiers can have any of the above properties, because when we consider fewer base clusterings, instances can belong to several patterns.

The objective of the consensus function is to build unique clusters from sets of instance identifiers. Thus, to decide how to deal with intersecting sets, we use the size of intersection as a measure for merging or splitting them. The idea of measuring the similarity between sets based on their intersection is not new. Jaccard index [14] is based on calculating the ratio between intersection size over the size of the union of 2 sets:

$$Jaccard(X,Y) = \frac{|X \cap Y|}{|X \cup Y|} = \frac{|X \cap Y|}{|X| + |Y| - |X \cap Y|}$$

Let us consider the three cases of sets intersection in Fig. 1. By calculating Jaccard for each case we get:

$$Jaccard(A, B) = \frac{3}{27}\ ,\ Jaccard(B, C) = \frac{7}{23}\ ,\ Jaccard(D, E) = \frac{7}{23}$$

Jaccard measure provides the same score for cases 2 and 3, while they are actually different: Indeed, the set B is mostly part of set C while the sets in case 3 share only about the half of their instances. Thus, instead of using Jaccard, we take a merge/split decision based on comparing intersection size over the size of each set. Let us then define $I(X|Y)$ as the ratio of intersection between sets X and Y over the size of set X, that is:

$$I(X|Y) = \frac{|X \cap Y|}{|X|}$$

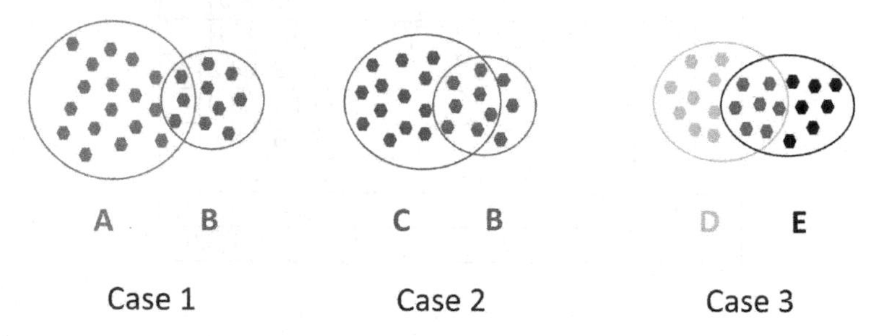

Fig. 1. Examples of sets intersection

$$\text{Case 1: } I(B|A) = \frac{|A \cap B|}{|B|} = \frac{3}{10}\ ,\ I(A|B) = \frac{|A \cap B|}{|A|} = \frac{3}{20}$$

$$\text{Case 2: } I(B|C) = \frac{|B \cap C|}{|B|} = \frac{7}{10}\ ,\ I(C|B) = \frac{|B \cap C|}{|C|} = \frac{7}{20}$$

$$\text{Case 3: } I(E|D) = \frac{|E \cap D|}{|E|} = \frac{7}{14}\ ,\ I(D|E) = \frac{|D \cap E|}{|D|} = \frac{7}{16}$$

From the above scores, we can see that $I(B|C)$ is the highest, and that the scores provided for cases 2 and 3 are different, compared to Jaccard which assigned them the same score. Thus the question is: how will we use all these information to decide to merge or split intersecting sets? We propose the following method:

To decide to either merge or split intersecting sets, a *Merging Threshold* (**MT**) can be used. MT is the minimum intersection ratio of a set (X) to decide to merge it with another set (Y). That is, sets X and Y are merged only if the intersection ratio of any of them, $I(X|Y)$ or $I(Y|X)$, is bigger than MT, otherwise they are split. While the merge operation is simply the union of two sets, the split operation involves removing the common instances from one of the sets. Since the sets represent clusters, the winner is the smaller set, as it is fundamentally more coherent. Thus, the shared instances are kept in the smaller cluster and removed from the bigger one. Any set that is subset of another set is removed. The process checks all the sets and the newly generated sets of the merge/split operation until obtaining all the remaining sets as unique sets. Algorithm 1 explains the full process of generating multiple consensuses.

Going back to the FCPs in Table 2, the first consensus (DT=5) consist of 6 clusters (data fragments) which are the instance sets of FCPs 1, 2, 3, 6, 7, and 8. For the next consensus (DT=4), we add the instance sets of FCPs 4, 5, 10, 11, and 14. However, the clusters of the previous consensus will be removed because they are just subsets of the new sets. Therefor, we will have the following sets: {3,4}, {4,5}, {1,2,3}, {5,8,9}, and {6,7,8,9}. With MT=0.5, we start with set {3,4} that intersects with set {4,5} by 0.5 of their instances, thus they are merged to form the set {3,4,5}. Next, set {3,4,5} intersects with set {1,2,3} but by 0.3 which is less than MT, thus they are split into sets {3,4,5} and {1,2}. The same split process is performed for sets {3,4,5} and {5,8,9} to form sets {3,4,5} and {8,9}. For the remaining sets, we find that set {8,9} is a subset of {6,7,8,9}, thus it is removed. The final clusters at DT=4 are: {3,4,5}, {1,2}, and {6,7,8,9}. The same process is performed for the remaining DTs.

By varying DT, it is possible that a consensus at a given value is identical to the previous one. Therefore, redundant consensuses are removed, and a *Stability counter* (**ST**) is used to record how many times a consensus is generated. The ST value is assigned to the consensus with the highest DT, suggesting that there is no better solution found for ST consecutive consensuses. Although the recommended consensus is the one with the highest similarity to the ensemble, a stable consensus can also be considered as another good solution.

3.5 ConsTree

The ConsTree is a Hasse diagram of consensuses, where each level depicts the clusters of a consensus as nodes, with node's size and label reflecting the cluster size. The bottom of the tree is the first consensus, then the tree goes up to the root having, at maximum, number of levels equals the number of base clusterings. Each cluster in a consensus can be linked to several clusters at the higher level, because the merge/split operations can result in regrouping some instances in a different manner at a higher level. Figure 2 shows a ConsTree of applying 9 base clusterings to partition a dataset of 400 instances, with the recommended consensus circled by a red line.

Definition 5. *A tree of consensuses is an ordered set* $(\mathcal{L}, \preceq)$ *of consensuses* $\mathcal{L} = \bigcup_{DT=MaxDT}^{DT=MinDT} L^{DT}$ *ordered in descending order of DT values. Let us denote* $L^{\alpha} = \{P_1^{\alpha}, ..., P_m^{\alpha}\}$ *and* $L^{\beta} = \{P_1^{\beta}, ..., P_n^{\beta}\}$ *the consensuses generated for* α *and* β *DT values respectively. Let us denote* P_q^{α} *the* q^{th} *cluster in* L^{α} *and* P_r^{β} *the* r^{th} *cluster in* L^{β}*, with* $1 \leq q \leq m$ *and* $1 \leq r \leq n$*. For* $\alpha > \beta$ *we have* $L^{\alpha} \preceq L^{\beta}$*, that is* $\forall P_q^{\alpha} \in L^{\alpha}$, $\exists P_r^{\beta} \in L^{\beta}$ *such that* $P_q^{\alpha} \cap P_r^{\beta} \neq \emptyset$. L^{α} *is a predecessor of* L^{β} *in the tree of consensuses.*

In the tree of Fig. 2, few shifting of instances occur, while the majority is just merging clusters from low level into 1 cluster at a higher level. However,

```
   Input  : Dataset to cluster, merging threshold MT
   Output : ConsTree tree of consensuses, list of consensus clustering vectors
 1 Generate clusterings ensemble of the dataset;
 2 Build the cluster membership matrix M;
 3 Generate FCPs from M for minsupport = 0;
 4 Sort the FCPs in ascending order of the size of their instance set;
 5 MaxDT ← Number of base clusterings;
 6 BiClust ← {instance sets of FCPs built from MaxDT base clusters};
 7 Assign a label to each set in BiClust to build the first consensus vector and store it in a
   list of vectors ConsVctrs;
 8 for DT = (MaxDT - 1) to 1 do
 9     BiClust ← BiClust ∪ {instances sets of FCPs built from DT base clusters};
10     N ← |BiClust| ;
11     repeat
12         for i = 1 to N do
13             B_i ← i^th set in BiClust;
14             for j = 1 to N, j ≠ i do
15                 B_j ← j^th set in BiClust;
16                 IntrscLeng ← |B_i ∩ B_j|;
17                 if IntrscLeng = 0 then
18                     Next j ;
19                 else if IntrscLeng = |B_i| then
20                     Remove B_i from BiClust;
21                     Next i;
22                 else if IntrscLeng = |B_j| then
23                     Remove B_j from BiClust;
24                     Next j;
25                 else if (IntrscLeng ≥ |B_i| × MT) or (IntrscLeng ≥ |B_j| × MT)
                   then
26                     B_j ← B_i ∪ B_j;
27                     Remove B_i from BiClust;
28                     Next i;
29                 else
30                     if |B_i| ≤ |B_j| then
31                         B_j ← B_j \ B_i ;
32                     else
33                         B_i ← B_i \ B_j ;
34
35
36             end
37         end
38     until All sets in BiClust are unique;
39     Assign a label to each set in BiClust to build a consensus vector and add it to
       ConsVctrs;
40 end
41 Find stable consensuses in ConsVctrs and remove extra duplicates;
42 For each remaining consensus, calculate its average similarity to the ensemble using
   Jaccard index;
43 Build a tree from the consensuses in ConsVctrs, with a recommended solution as the
   one that has the highest average similarity to the ensemble;
```

Algorithm 1. Generate Multiple Consensuses

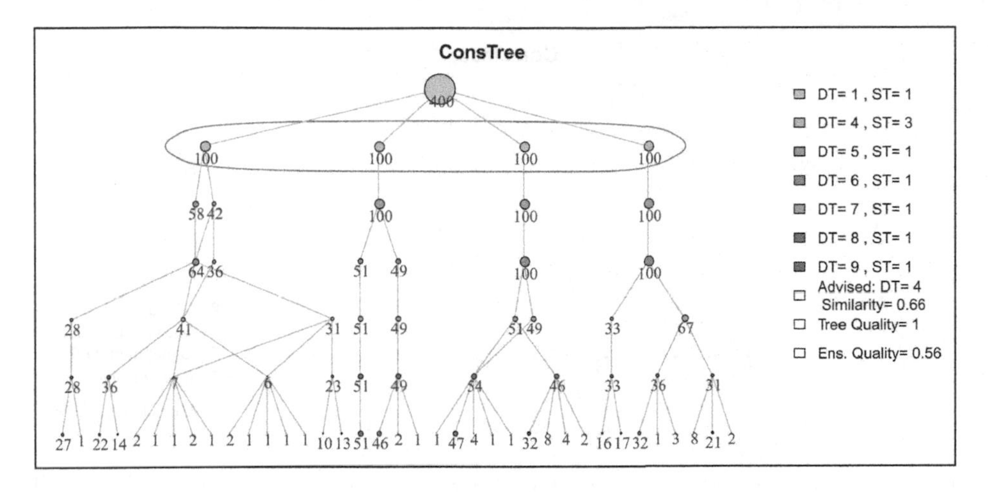

Fig. 2. Example of a ConsTree.

to enhance the visualization of the tree and to make it easier for the analyst to recognize the stable clusters or their cores (those that do not change over several tree levels), we developed a *tree refinement process* on which the few instances that shift are removed. Figure 3 is the result of performing the refinement process on the tree in Fig. 2. The refinement process does not alter the original consensuses, it just simplify the visualization. The *Removed Instances* (RI) at the bottom of the figure tells how many instances are removed. It is a set of instance identifiers that the analyst can retrieve to investigate why these instances are not stable and cause conflicts between the base clusterings on where they should belong. If she/he prefers the clusters generated by the refinement process, she/he can simply remove the RI from the selected consensus.

The tree represent an important tool to analyze the dataset and discover the hidden cluster structure. For example, we can recognize in Fig. 2 4 columns of node-structures (the heads of these columns are the children nodes of the root). The ST value of DT=4 (which happened to be also the recommended solution) tells that this consensus of 4 clusters is the most stable one, which also adds to our discovery of a hidden structure of 4 clusters. We can also recognize other strong clusters, as the clusters of sizes 51 and 49 in the second column from the left, telling a strong intra-cluster similarity between their instances compared to other clusters. The fact that these 2 clusters are then merged into 1 cluster tells that their instances are close in the data space. The refinement process allowed us also to discover in Fig. 3 that 36 instances have strong similarity between them so they did not change over 4 consecutive tree levels. Based on such analysis, the analyst is not restricted to choose the recommended consensus, since the meaningfulness of the clusters depends more on the analyst preference. This is why we generate multiple consensuses from different combinations of base clusters (different views), rather than presenting just one solution.

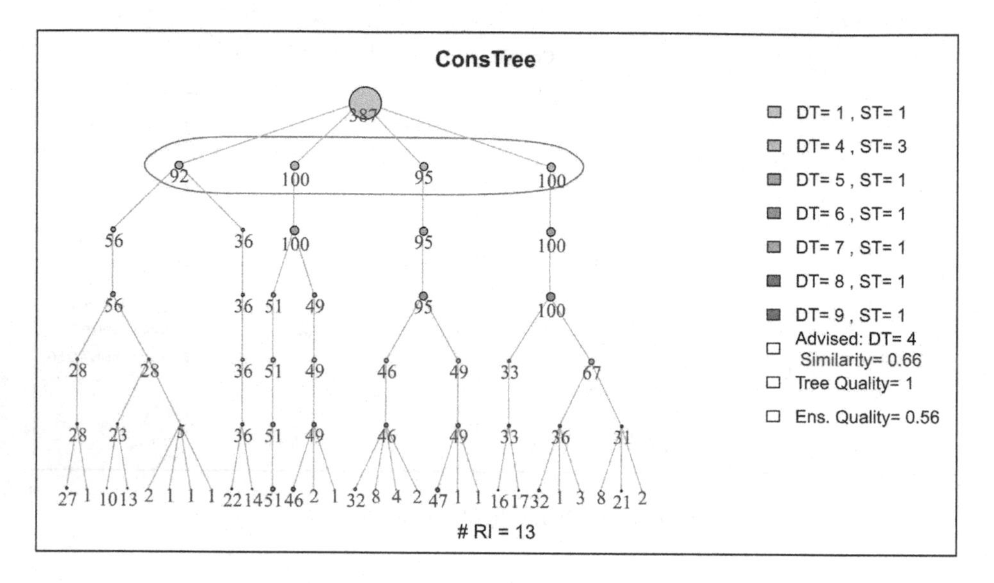

Fig. 3. The result of refining the tree in Fig. 2.

4 Experiments

Tests were run on a laptop with Intel® Core™ i7-4710MQ @ 2.50GHz, and 32 GB RAM. The proposed method was implemented using R language [18]. To find the clustering patterns, function *apriori* in *arules* R package [11] was used, by setting the *target* parameter to "closed frequent itemsets".[5] To draw the ConsTree, each cluster in a consensus was represented as a node in a graph. Nodes of consecutive consensuses were linked by edges based on the shared instances between them. A data frame that defines these edges was used to build the graph, then the ConsTree was plotted by the *plot* function in the *igraph* R package [3]. The refinement process keeps only the edges that has the maximum number of shared instances between 2 nodes.

Table 3 summarizes the performed tests. In each test, the ensemble was generated by random selection of the following clustering algorithms with random settings: K-means, PAM, agglomerative hierarchical clustering, AGNES, DIANA, MCLUST (Gaussian Model-Based Clustering), C-Means, FANNY, Bagged Clustering, and SOM. All these clustering algorithms are available in R. If the random generation of the ensemble results in creating identical clusterings, 1 cluster , or 1 dominant cluster that involves 90% of the instances, then these partitions are removed. "Ensemble size" specifies the number of base clusterings, that each partitioned the dataset into a number of clusters within the range "K range". We measured the quality of the ensemble as the average Jaccard similarity between

[5] A faster algorithm for generating closed itemsets called FIST is proposed by [16], and an implementation of it in java is available on the website of the authors.

Table 3. Tests validation.

Dataset	Seeds	Mushroom	Zoo	E.Coli	Iris	Breast Cancer	Wine	EngyTime	Wingnut
Dataset Size	210	8124	101	336	150	699	178	4096	1016
# of attributes	7	22	16	7	4	9	13	2	2
# of true classes	3	2	7	8	3	2	3	2	2
Ensemble size	9	7	10	12	8	9	9	8	12
K range	[2,8]	[2,5]	[2,12]	[4,10]	[2,8]	[2,6]	[2,8]	[2,8]	[2,7]
In-ensemble similariy	0.56	0.58	0.62	0.59	0.53	0.67	0.58	0.49	0.58
Ensemble Min.	0.30	0.34	0.45	0.28	0.29	0.42	0.45	0.27	0.37
Ensemble Max.	0.74	0.69	0.90	0.69	0.73	0.82	0.89	0.88	1.00
Our method	0.74	0.67	0.84	**0.67**	0.69	**0.87**	0.78	**0.86**	**0.98**
# of clusters in our method	3	3	5	5	3	2	3	2	2
SE	0.64	**0.70**	0.75	0.42	0.62	**0.87**	0.82	**0.86**	0.88
GV1	0.67	**0.70**	0.82	0.38	**0.71**	0.86	0.80	0.85	0.88
DWH	0.67	**0.70**	0.81	0.51	0.62	0.85	0.73	0.75	0.89
HE	0.65	**0.70**	0.81	0.50	0.62	0.85	0.78	**0.86**	0.88
GV3	0.72	**0.70**	0.82	0.44	0.59	**0.87**	**0.84**	**0.86**	0.91
SM	0.64	**0.70**	0.79	0.50	0.62	0.85	0.78	**0.86**	0.88
soft/symdiff	**0.75**	**0.70**	0.69	0.42	0.59	0.86	0.47	**0.86**	0.92
Medoids	0.74	0.69	**0.90**	0.64	0.55	0.81	0.75	0.64	0.52

each pair of clusterings in the ensemble. We call this the "in-ensemble similarity", while "ensemble min" and "ensemble max" denotes the minimum and maximum similarity to the true class. These information are to ensure that we did not use very similar high quality clusterings in the ensemble to generate a high quality consensus. In all tests, we compared the "recommended consensus" of our method against voting-based consensus methods available in R package CLUE [13], which include the following: SE, GV1, DWH, HE, SM, GV3, soft/symdiff, and consensus medoid. To justify the quality of the results, the consensus solution is compared using Jaccard index against true class labels available for each tested dataset. Note that CLUE methods require specifying the number of required clusters in the consensus, thus we used the true number of classes, while our method do not require this. For the MT parameter, we used MT = 0.5 as the default value.

Seeds, Mushroom, Zoo, E.Coli, Iris, Breast Cancer and Wine are real datasets available on the UCI repository [15]. EngyTime and Wingnut are synthetic datasets from [23]. Table 4 shows the execution time, in seconds, of the consensus methods used. For our method, we separated between the time required to generate the patterns, and the time used by the method to generate all the consensuses and calculate the recommended one.[6] We can see that the total time of the proposed method is acceptable, and it does not depend on the dataset

[6] The time required to display the ConsTree is not considered, as it depend on the I/O device.

Table 4. Execution time of the consensus methods (in seconds).

Dataset	Seeds	Mushroom	Zoo	E.Coli	Iris	Breast Cancer	Wine	EngyTime	Wingnut
Patterns	0.116	0.353	0.085	0.342	0.053	0.147	0.147	0.210	0.568
Our method	0.417	0.491	0.262	1.492	0.229	0.488	0.688	3.567	1.892
SE	0.011	0.064	0.010	0.037	0.013	0.017	0.009	0.066	0.023
GV1	0.080	0.047	0.048	0.076	0.019	0.018	0.026	0.051	0.031
DWH	0.006	0.051	0.006	0.010	0.016	0.009	0.006	0.030	0.015
HE	0.011	0.062	0.017	0.018	0.001	0.016	0.008	0.087	0.020
GV3	1.483	3532.199	0.664	7.061	0.768	13.031	0.848	673.044	30.901
SM	0.722	22.917	0.817	9.305	0.675	2.172	0.833	11.455	4.720
soft/symdiff	10.918	21925.92	4.433	55.181	5.301	175.815	8.518	5332.099	414.718
Medoids	0.028	0.222	0.031	0.067	0.016	0.047	0.030	0.154	0.101

size, but on the number of the base clusterings used, and the similarity between them. For example, in the test of the Mushroom dataset, the total number of generated patterns is 106, while the dataset size is 8124 instances. This is a huge pruning of the search space.

5 Conclusions

We presented a new method that can generate multiple consensus clustering solutions, and recommend to the user the solution the most similar to the ensemble. Frequent closed itemsets technique is used to detect similarities between the base partitions, and define clustering patterns common to sets of instances. The similarity between clustering patterns is calculated based on their shared instances. The tests showed that the proposed method achieved generally good results in terms of quality and the number of discovered clusters. In addition, it does not require the user to specify K (the number of clusters in the generated consensus), as this parameter is difficult to predict in the absence of domain knowledge about the number of hidden clusters in the dataset.

An additional advantage of the proposed method is the ConsTree. As an analysis tool, it enables the user to discover strong clusters, that is, those that do not change over several tree levels, pointing out to strong intra-cluster similarity among the instances. A stable consensus (identified on the tree by ST>1) suggests usually the existence of a well separated clusters structure, thus the user is advised to investigate this consensus in addition to the recommended solution.

Execution time of our method is not related directly to the dataset size as in other consensus methods. Instead, it depends on the size of the ensemble and whether there are many similarities or conflicts among the base clusterings, as this will determine the generated clustering patterns. Thus, for large datasets, we can get smaller number of patterns compared to dataset size if there are many agreements between the base clusterings. In addition, to generate a consensus, the proposed method requires only accessing a small subset of the available clustering patterns. Tests showed that the CLUE methods "GV3" and "soft/symdiff" cost longer execution time compared to all other methods, and are not applicable on large datasets because of their high storage complexity.

References

1. Asur, S., Ucar, D., Parthasarathy, S.: An ensemble framework for clustering protein-protein interaction networks. Bioinformatics **23**(13), i29–i40 (2007)
2. Ceglar, A., Roddick, J.F.: Association mining. ACM Computing Surveys **38**(2) (2006)
3. Csardi, G., Nepusz, T.: The igraph software package for complex network research. InterJournal Complex Systems **1695** (2006). http://igraph.org
4. Dalton, L., Ballarin, V., Brun, M.: Clustering algorithms: on learning, validation, performance, and applications to genomics. Current Genomics **10**(6), 430 (2009)
5. Dudoit, S., Fridlyand, J.: Bagging to improve the accuracy of a clustering procedure. Bioinformatics **19**(9), 1090–1099 (2003)
6. Färber, I., Günnemann, S., Kriegel, H.P., Kröger, P., Müller, E., Schubert, E., Seidl, T., Zimek, A.: On using class-labels in evaluation of clusterings. In: KDD MultiClust International Workshop on Discovering, Summarizing and Using Multiple Clusterings, p. 1 (2010)
7. Fern, X.Z., Brodley, C.E.: Solving cluster ensemble problems by bipartite graph partitioning. In: Proceedings of the Twenty-First International Conference on Machine Learning, p. 36. ACM (2004)
8. Fischer, B., Buhmann, J.M.: Bagging for path-based clustering. IEEE Transactions on Pattern Analysis and Machine Intelligence **25**(11), 1411–1415 (2003)
9. Fred, A.L., Jain, A.K.: Combining multiple clusterings using evidence accumulation. IEEE Transactions on Pattern Analysis and Machine Intelligence **27**(6), 835–850 (2005)
10. Ghaemi, R., Sulaiman, M.N., Ibrahim, H., Mustapha, N.: A survey: Clustering ensembles techniques. WASET **50**, 636–645 (2009)
11. Hahsler, M., Gruen, B., Hornik, K.: arules – A computational environment for mining association rules and frequent item sets. Journal of Statistical Software **14**(15), 1–25 (2005)
12. Halkidi, M., Batistakis, Y., Vazirgiannis, M.: On clustering validation techniques. Journal of Intelligent Information Systems **17**(2), 107–145 (2001)
13. Hornik, K.: A CLUE for CLUster Ensembles. Journal of Statistical Software **14**(12) (2005)
14. Jaccard, P.: The distribution of the flora in the alpine zone.1. New Phytologist **11**(2), 37–50 (1912). http://dx.doi.org/10.1111/j.1469-8137.1912.tb05611.x
15. Lichman, M.: UCI machine learning repository (2013). http://archive.ics.uci.edu/ml
16. Mondal, K.C., Pasquier, N., Mukhopadhyay, A., Maulik, U., Bandhopadyay, S.: A new approach for association rule mining and bi-clustering using formal concept analysis. In: Perner, P. (ed.) MLDM 2012. LNCS, vol. 7376, pp. 86–101. Springer, Heidelberg (2012)
17. Pasquier, N., Bastide, Y., Taouil, R., Lakhal, L.: Efficient mining of association rules using closed itemset lattices. Inf. Systems **24**(1), 25–46 (1999)
18. R Core Team: R: A Language and Environment for Statistical Computing. R Foundation for Statistical Computing, Vienna, Austria (2015). https://www.R-project.org/

19. Rendón, E., Abundez, I., Arizmendi, A., Quiroz, E.: Internal versus external cluster validation indexes. International Journal of Computers and Communications **5**(1), 27–34 (2011)
20. Sarumathi, S., Shanthi, N., Sharmila, M.: A comparative analysis of different categorical data clustering ensemble methods in data mining. IJCA **81**(4), 46–55 (2013)
21. Strehl, A., Ghosh, J.: Cluster ensembles – a knowledge reuse framework for combining multiple partitions. JMLR **3**, 583–617 (2003)
22. Topchy, A., Jain, A.K., Punch, W.: Clustering ensembles: Models of consensus and weak partitions. IEEE Transactions on Pattern Analysis and Machine Intelligence **27**(12), 1866–1881 (2005)
23. Ultsch, A.: Clustering with SOM: U*C. In: Proc. WSOM Workshop, pp. 75–82 (2005)
24. Vega-Pons, S., Ruiz-Shulcloper, J.: A survey of clustering ensemble algorithms. IJPRAI **25**(03), 337–372 (2011)
25. Wu, O., Hu, W., Maybank, S.J., Zhu, M., Li, B.: Efficient clustering aggregation based on data fragments. IEEE Trans. Syst. Man Cybern. B Cybern. **42**(3), 913–926 (2012)

Author Index

Printed in the United States
By Bookmasters